Vertebrate Paleontology in the Neotropics

The Miocene Fauna of La Venta, Colombia

Edited by Richard F. Kay, Richard H. Madden, Richard L. Cifelli, and John J. Flynn

SMITHSONIAN INSTITUTION PRESS
Washington and London

Production Editor: Deborah L. Sanders

Library of Congress Cataloging-in-Publication Data

Vertebrate paleontology in the Neotropics : the Miocene fauna of La
    Venta, Colombia / Richard F. Kay . . . [et al.], editors.
       p. cm.
    Includes bibliographical references and index.
    ISBN 1-56098-418-X
    1. Vertebrates, Fossil—Colombia. 2. Paleontology—Miocene.
    3. Paleontology—Colombia. I. Kay, Richard F.
    QE841.V415   1995
    566′.0986—dc20   94-12118

British Library Cataloguing-in-Publication Data are available.

Manufactured in the United States of America
03 02 01 00 99 98 97 96 5 4 3 2 1

For permission to reproduce illustrations appearing in this book, please
correspond directly with the owners of the works, as listed in the
individual captions. The Smithsonian Institution Press does not retain
reproduction rights for these illustrations individually or maintain a file
of addresses for photo sources.

The epigraph to chapter 1 (found on p. 3) is as quoted by M.
Boussingault in *Viajes científicos a los Andes ecuatoriales ó colección de
memorias sobre física, química é historia natural de la Neuva Granada, Ecuador
y Venezuela* (Paris: Libreria Castellana, 1849).

The unnumbered drawings by Nicholas J. Czaplewski interspersed
throughout the text are life reconstructions of some of the animals of La
Venta. These drawings show much more than is actually known from
skeletal parts; they are included to convey a collective opinion of what
these animals looked like in their habitat—the lost world of the Miocene
Neotropics. Mark Twain once described how he had helped Henry
Fairfield Osborn construct an exhibit in a similar situation: "Professor
Osborn and I built the colossal skeleton brontosaur that stands fifty-seven
feet long and sixteen feet high in the Natural History Museum, the awe
and admiration of all the world, the stateliest skeleton that exists on the
planet. We had nine bones, and we built the rest of him out of plaster of
paris" (Twain, *Is Shakespeare Dead? From My Autobiography,* New York:
Harper & Bros., 1909, 40−41). The reconstruction drawings herein
contain far more bones and much less plaster.

—*Eds.*

Page i: *Pseudoprepotherium confusum* (sloth). Page xii: *Megadolodus
molariformis* (proterothere) and *Anachlysictis gracilis* (saber-toothed
marsupial). Page xvi: *Top,* the El Cardón Red Beds, now Duke Locality
32, at the head of Quebrada el Cardón in 1944−45; *bottom,* the Servicio
Geológico/University of California camp in 1944−45, at the Monkey
Beds, now Duke Locality 22, looking eastward toward San Nicolás.
Cerro Gordo is visible on the horizon. (Both photos by José Royo y
Gomez, courtesy of INGEOMINAS Geological Museum, Bogotá,
Colombia.) Page 12: Two *Miocochilius anamopodus* (interatheres, in
foreground) and a giant caiman (in background, partially concealed).
Page 60: Doradidae (armored catfish), *Podocnemis medemi* (turtle),
Arapaima sp. (pirarucu), *Mourasuchus atopus* (duck-billed crocodilian),
Hydrolycus sp. (fish), cf. *Myletes* sp., and *Noctilio* cf. *N. albiventris* (bat).
Page 92: *Pericotoxodon platignathus* (toxodont, in death), *Sebecus huilensis*
(sebecid crocodilian), and *Jabiru* cf. *J. mycteria* (stork). Page 184:
Pericotoxodon platignathus (toxodont). Page 432: *Stirtonia tatacoensis*
(howler-like monkey, above left), *Neotamandua borealis* (anteater, below
left), *Micoureus laventicus* (mouse opossum, above right), and *Lagonimico
conclucatus* (marmoset-like monkey, below right). Page 496:
Granastrapotherium snorki (astrapothere) and cf. *Hapalops* (sloth).

Contents

Editors' Note   v

Acknowledgments   vii

List of Contributors   ix

List of Abbreviations   xiii

Part One. Introduction   1

1. **A History of Vertebrate Paleontology in the Magdalena Valley**   3
   Richard H. Madden, Donald E. Savage, and Robert W. Fields

Part Two. Geology and Geochronology   13

2. **Stratigraphy, Sedimentary Environments, and the Miocene Uplift of the Colombian Andes**   15
   Javier Guerrero

3. **Geochronology of the Honda Group**   44
   John J. Flynn, Javier Guerrero, and Carl C. Swisher III

Part Three. Invertebrates and Fishes   61

4. **Trichodactylid Crabs**   63
   Gilberto Rodríguez

5. **Freshwater Fishes and Their Paleobiotic Implications**   67
   John G. Lundberg

Part Four. Reptiles, Amphibians, and Birds   93

6. **Limbless Tetrapods**   95
   Max K. Hecht and Thomas C. LaDuke

7. **A Reassessment of the Fossil Tupinambinae**   100
   Robert M. Sullivan and Richard Estes

8. **Crocodilians, *Gryposuchus,* and the South American Gavials**   113
   Wann Langston, Jr., and Zulma Gasparini

9. **Turtles**   155
   Roger Conant Wood

10. **Birds**   171
    D. Tab Rasmussen

Part Five. Nonprimate Mammals   185

11. **New Clues for Understanding Neogene Marsupial Radiations**   187
    Francisco J. Goin

12. **New Caenolestoid Marsupials**   207
    Elizabeth R. Dumont and Thomas M. Bown

13. **Armored Xenarthrans: A Unique Taxonomic and Ecologic Assemblage**   213
Alfredo A. Carlini, Sergio F. Vizcaíno, and Gustavo J. Scillato-Yané

14. **A New Giant Pampatheriid Armadillo**   227
Gordon Edmund and Jessica Theodor

15. **Xenarthrans: Pilosans**   233
H. Gregory McDonald

16. **Locomotor Adaptations in Miocene Xenarthrans**   246
Jennifer L. White

17. **Paleobiology and Affinities of *Megadolodus***   265
Richard L. Cifelli and Carlos Villarroel

18. **Litopterns**   289
Richard L. Cifelli and Javier Guerrero

19. **A New Leontiniid Notoungulate**   303
Carlos Villarroel and Jane Colwell Danis

20. **Enamel Microstructure in Notoungulates**   319
Mary C. Maas

21. **A New Toxodontid Notoungulate**   335
Richard H. Madden

22. **Uruguaytheriine Astrapotheres of Tropical South America**   355
Steven C. Johnson and Richard H. Madden

23. **Sirenia**   383
Daryl P. Domning

24. **Rodents**   392
Anne H. Walton

25. **Chiroptera**   410
Nicholas J. Czaplewski

Part Six. Primates   433

26. **A New Small Platyrrhine and the Phyletic Position of Callitrichinae**   435
Richard F. Kay and D. Jeffrey Meldrum

27. **Postcranial Skeleton of Laventan Platyrrhines**   459
D. Jeffrey Meldrum and Richard F. Kay

28. **Fossil New World Monkeys**   473
John G. Fleagle, Richard F. Kay, and Mark R. L. Anthony

Part Seven. Summary   497

29. **The Laventan Stage and Age**   499
Richard H. Madden, Javier Guerrero, Richard F. Kay, John J. Flynn, Carl C. Swisher III, and Anne H. Walton

30. **Paleogeography and Paleoecology**   520
Richard F. Kay and Richard H. Madden

Literature Cited   551

Taxonomic Index   585

Editors' Note

It has long been recognized that the La Venta badlands in central Colombia hold special significance for understanding the evolution of vertebrate faunas in South America. First, although the continent has been considered a model of completeness in terms of the evolutionary record of mammalian evolution (e.g., Patterson and Pascual 1972), a glance at that record reveals more gaps than data. Even where the record is most complete—in Argentina—major hiatuses exist, some of which may be up to 12 million years in length (MacFadden 1990). Second, even the better-known faunas of Argentina are often poorly constrained temporally—at present there are no dated horizons to constrain the ages of any of the Eocene land mammal ages, and dates available for Oligocene and early Miocene faunas are in flux. Further, only a handful of studies exist that assess the paleobiology of the known mammalian paleofaunas, although a few notable exceptions are now emerging (e.g., Pascual and Ortiz Jaureguizar 1990). In this book we fill one important gap in the middle Miocene fossil record of South America by describing the vertebrate fauna from the La Venta area (including many new species), interpreting their phylogeny and paleobiology, and placing them into a secure temporal and geologic framework.

For two reasons the La Venta area and its paleofaunas are especially important for understanding South American faunal evolution. The first has to do with its geographic position. A major impediment to understanding the continent-wide evolution of vertebrate and mammalian faunas has been the paucity of good fossil sites within the tropical zone. Even though 70% of the land mass of South America is situated within the tropics (that is, between approximately 23° north and south of the equator), La Venta is practically the only place where a Tertiary tropical lowland paleofauna can be studied in geochronologic context. The only other vertebrate fauna comparable in number of species and geochronologic control is the late Oligocene/early Miocene fauna from Salla in Bolivia, at approximately 17° south; there, however, the fish, reptile, bird, and small mammal faunas are comparatively poorly known.

It is generally thought that lowland humid tropical environments do not favor the preservation of animal remains, and consequently it is unusual to find such faunas preserved in such abundance and richness. The explanation for the exceptional diversity and excellent preservation of the fossil vertebrates at La Venta resides, we believe, in the extraordinary rates of sediment accumulation that occurred on the piedmont area at the foot of the volcanic central Andes.

Second, the La Venta fauna holds special significance because if its temporal position. It is well known that the paleofaunas of South America underwent a massive read-justment called the "Great Faunal Interchange" (Stehli and Webb 1985), beginning in the late Miocene, when the Isthmus of Panama was formed. The fossil record of this readjustment is best known from southern South America, and our knowledge about the consequences of this biotic interchange on tropical lowland faunas in particular is limited. La Venta is one of the few places in South America where the paleoecology of a lowland tropical forest fauna can be studied from a time prior to the Great Interamerican Faunal Interchange.

Building upon an already extensive literature about the Honda Group and its faunas, the contributors to this book present detailed information and interpretation about the paleobiology of La Venta animals and the environment in which they lived, thereby painting a picture of what vertebrate life was like in the South American tropics before the Great Faunal Interchange. An introductory chapter summarizes the long and curious history of the study of this important vertebrate fauna. The immediately following chapters present evidence for the geologic age and sedimentary environments in which Miocene animals lived and were preserved. In each succeeding chapter, the contributors have used geological, phylogenetic, or anatomical evidence to infer the paleobiology of all the animals known from La Venta. These animals include a fossil crab, fish, amphibians, reptiles, birds, and mammals. Two summary chapters at the end of the book propose a new chronostratigraphic unit, the Laventan Land Mammal Stage,

for the interval of time represented by the La Venta fauna, and a paleoenvironmental reconstruction based upon the data from prior chapters and comparisons with selected mammal faunas from the South American tropics today.

The careful reader will note an occasional lack of correspondence among chapters in the way the taxonomic headings are presented. This was intentional and reflects the contributors' differing stances on taxonomic matters. For example, some of the contributors use taxonomic names only for monophyletic groups, whereas others accept paraphyletic taxa. Likewise, some use names in the Linnaean hierarchy, whereas others eschew them. Despite the taxonomic arguments and borborygmi, two of us (R.F.K. and R.H.M.) attempted to bring some order to the contributors' diverse paleontological findings by compiling the interpretive summary given in appendix 30.1.

References

MacFadden, B. J. 1990. Chronology of Cenozoic primate localities in South America. Journal of Human Evolution 19: 7–21.

Pascual, R., and E. Ortiz Jaureguizar. 1990. Evolving climates and mammal faunas in Cenozoic South America. Journal of Human Evolution 19: 23–60.

Patterson, B., and R. Pascual. 1972. The fossil mammal fauna of South America. In A. Keast, F. C. Erk, and B. Glass, eds., Evolution, mammals, and southern continents. Albany: State University of New York Press. Pp. 247–309.

Acknowledgments

Our research program for the collection of fossil vertebrates and study of the stratigraphy and geochronology of the La Venta area was undertaken after a preliminary visit to Colombia in February 1984. Richard Kay and Richard Madden, with Colombian geologist Carlos Ulloa of INGEOMINAS and Javier Guerrero Diaz, then a student at the Colombian National University, visited the La Venta area on February 23, 1984. The hospitality, as well as logistic and technical assistance, of many researchers at INGEOMINAS stimulated us to begin the project on a larger scale in 1985.

Since then the following individuals have participated in Duke University–INGEOMINAS vertebrate paleontology expeditions: Richard F. Kay (1985–91), Richard H. Madden (1985–92), Richard L. Cifelli (1985–87), Javier Guerrero (1985–92), Jaime Alberto Fúquen (1985), José Franklin Lugo Buendia (1985, 1987, 1989, 1990), Rodrigo Marín (1985, 1990), César Agosto Carvajál (1985), Leonardo Sepúlveda (1985), Carlos Ulloa (1985), D. Tab Rasmussen (1986), Manuel González (1986), Anne H. Walton (1987–89), Larry G. Marshall (1987), John J. Flynn (1987), Jorge Espitía (1987), J. Michael Plavcan (1988), Callum Ross (1988), Nick Czaplewski (1988), Francisco Ballén (1988), Hans Thewissen (1989), Jennifer L. White (1989), Mario Cozzuol (1989), Arley de Jesús Gómez Cruz (1989, 1990), D. Jeffrey Meldrum (1990), Carlos Villarroel (1991), Elvira Cristina Ruíz (1990), Germán Marquinez (1991), Victor Ramirez (1991), and Marcelo Sánchez Villagra (1992).

Many individuals have been helpful over the years. At INGEOMINAS we wish especially to thank Hermann Duque-Caro of Stratigraphy; former directors Alfonso Lopez Reina, Luis Jaramillo Cortes, and Alberto Lobo-Guerrero Uscategui; Francisco Zambrano Ortiz, Carlos Jairo Vesga Ordoñez, and Dario Mosquera Torres, former subdirectors of Geologic Mapping; Victor Laverde Eastman and Luis Felipe Rincón of the Museum of Geology; and Alberto Núñez, formerly director at the INGEOMINAS regional office in Ibague.

For special assistance in the Magdalena valley, we wish to thank Darío Valencia-Caro, formerly head of exploration at HOCOL S.A. (Tenneco), and Kim Butler and William Frank (Tenneco, Houston). We wish to thank Steven Schamel and Michael Waddell at the University of South Carolina Earth Sciences and Resources Institute, and Takeshi Setoguchi at Kyōto University. In Bogotá, we wish to thank Gustavo Guerrero Gomez (formerly of INGEOMINAS), Marta Espitía Avilez of the Asociación Colombiana para el Avance de la Ciencias, and Jorge Hernández-Camacho of INDERENA. Also in Bogotá, we owe thanks to our friends Tom Wells, David Wells, Ruth Aschmann, Don Butler, and Ricardo de la Espriella for their generous hospitality and numerous courtesies. In Neiva, Aurelio Pastrana of INGEOMINAS, Arnol Tobar, and Doli Andrade of the Instituto Huilense de Cultura have been helpful throughout our project. In Chaparral, we thank Severo Hernández and "La Gorda" Ayala. Special thanks are due to Don Savage and J. Howard Hutchison of the Museum of Paleontology, University of California, who have supported our work in Colombia from the beginning.

We thank Peter Cannell, science acquisitions editor at Smithsonian Institution Press, for his help and understanding. At Duke University, we especially wish to thank Mrs. Rachel Hougom.

Contributors

Mark R. L. Anthony
Department of Biological Anthropology and Anatomy
Duke University Medical Center
Durham, NC 27710

Thomas M. Bown
U.S. Department of the Interior
U.S. Geological Survey
Box 25046
Denver Federal Center
Denver, CO 80225

Alfredo A. Carlini
Departamento Científico de Paleontología de Vertebrados
Museo de Ciencias Naturales de La Plata
Paseo del Bosque s/n
1900 La Plata
Argentina

Richard L. Cifelli
Oklahoma Museum of Natural History
University of Oklahoma
1335 Asp Avenue
Norman, OK 73019

Nicholas J. Czaplewski
Oklahoma Museum of Natural History
University of Oklahoma
1335 Asp Avenue
Norman, OK 73019

Jane Colwell Danis
Tyrrell Museum of Paleontology
P.O. Box 7500
Drumheller, Alberta T0J 0Y0
Canada

Daryl P. Domning
Department of Anatomy
Howard University
520 W Street, NW
Washington, DC 20059

Elizabeth R. Dumont
Department of Neurobiology, Anatomy, and Cell Science
School of Medicine
Scaife Hall
University of Pittsburgh
Pittsburgh, PA 15261

Gordon Edmund
Department of Vertebrate Paleontology
Royal Ontario Museum
100 Queen's Park
Toronto, Ontario M5S 2C6
Canada

Richard Estes (deceased)
formerly,
Department of Biology
San Diego State University
San Diego, CA 92115

Robert W. Fields (deceased)
formerly,
Department of Geology
University of Montana
Missoula, MT 59812

John G. Fleagle
Department of Anatomical Sciences
State University of New York
Stony Brook, NY 11794

John J. Flynn
Field Museum of Natural History
Roosevelt Road at Lake Shore Drive
Chicago, IL 60605

Zulma Gasparini
Departamento Científico de Paleontología de Vertebrados
Museo de Ciencias Naturales de La Plata
Paseo del Bosque s/n
1900 La Plata
Argentina

Francisco J. Goin
Departamento Científico de Paleontología de Vertebrados
Museo de Ciencias Naturales de La Plata
Paseo del Bosque s/n
1900 La Plata
Argentina

Javier Guerrero
INGEOMINAS
Apartado Aereo 4865
Bogotá, D.E.
Colombia

Max Hecht
Department of Biology
Queens College
City University of New York
Flushing, NY 11367-0904

Steven C. Johnson
Department of Anthropology
Diablo Valley College
Pleasant Hill, CA 94523

Richard F. Kay
Department of Biological Anthropology and Anatomy
Duke University Medical Center
Durham, NC 27710

Thomas C. LaDuke
Department of Biological Sciences
East Stroudsburg University
East Stroudsburg, PA 18301-2999

Wann Langston, Jr.
University of Texas at Austin
Vertebrate Paleontology Laboratory
Texas Memorial Museum
Austin, TX 78712

John G. Lundberg
Department of Ecology and Evolutionary Biology
University of Arizona
Tucson, AZ 85721

Mary C. Maas
Department of Biological Anthropology and Anatomy
Duke University Medical Center
Durham, NC 27710

Richard H. Madden
Department of Biological Anthropology and Anatomy
Duke University Medical Center
Durham, NC 27710

H. Gregory McDonald
Hagerman Fossil Beds National Monument
Hagerman, ID 83332

D. Jeffrey Meldrum
Department of Biological Sciences
Idaho State University
Campus Box 8007
Pocatello, ID 83209-8007

D. Tab Rasmussen
Department of Anthropology
Washington University
St. Louis, MO 63130

Gilberto Rodríguez
Centro de Ecologia
Instituto Venezolano de Investigaciones Cientificas
Apartado 21827
Caracas 1020A
Venezuela

Donald E. Savage
Oklahoma Museum of Natural History
University of Oklahoma
1335 Asp Avenue
Norman, OK 73019

Gustavo J. Scillato-Yané
Departamento Científico de Paleontología de Vertebrados
Museo de Ciencias Naturales de La Plata
Paseo del Bosque s/n
1900 La Plata
Argentina

Robert M. Sullivan
State Museum of Pennsylvania
Third and North Streets
Harrisburg, PA 17108-1026

Carl C. Swisher III
Berkeley Geochronology Center
Institute of Human Origins
2455 Ridge Road
Berkeley, CA 94709

Jessica Theodor
Department of Integrative Biology
University of California
Berkeley, CA 94720

Carlos Villarroel
Facultad de Geociencias
Universidad Nacional de Colombia
Apartado Aereo 56833
Bogotá, D.E.
Colombia

Sergio F. Vizcaíno
Departamento Científico de Paleontología de Vertebrados
Museo de Ciencias Naturales de La Plata
Paseo del Bosque s/n
1900 La Plata
Argentina

Anne H. Walton
University of Texas at Austin
Vertebrate Paleontology Laboratory
Texas Memorial Museum
Austin, TX 78712

Jennifer L. White
Department of Biology and Environmental Science
Simpson College
Indianola, IA 50125

Roger Conant Wood
Faculty of Natural Sciences and Mathematics
Stockton State College
Pomona, NJ 08240

Abbreviations

Standard abbreviations for chemical elements, taxonomic terms, and common SI (Système International) units of weight and measure are used without definition. Other abbreviations are defined at first use in the text, and, for convenience, most are also listed here. Because the contributors subscribed to various systems of dental nomenclature, such terms are not included here but instead are defined in the individual chapters.

AF	alternating field
AMNH	American Museum of Natural History, New York
ASA	average ameloblast secretory area
BIOM	software program for biostatistics
biot	biotite
BM	British Museum (Natural History), London
BODYMASS	software program
CAS	California Academy of Sciences
CD	distance between adjacent prism centers (of teeth)
CI	(1) consistency index; (2) confidence interval
CNRS	Centre National de Recherche Scientifique
Congl.	conglomerate beds
CV	coefficient of variation
DGM	Divisão de Geologia e Mineralogia, Departamento Nacional da Produção Mineral, Rio de Janeiro
DGP	Departamento de Geociencias, Universidad Nacional de Colombia, Bogotà
DU	Duke University, Durham, North Carolina
EDJ	enamel dentine junction
EPN(Q)	Escuela Politécnica Nacional, Quito, Ecuador
FC	Fish Canyon Tuff sanidine
FHSS	femoral head shape score
Fm.	formation
FMNH	Field Museum of Natural History, Chicago
FRI	faunal resemblance index
GMPTS	geomagnetic polarity time scale
HM-GI	Hamburger Mineralogische-Geologischen Institut
hnbl	hornblende
IGM	*see* INGEOMINAS

IGMM	Instituto Geología y Mineralogía, Mexico City	NACSN	North American Commission on Stratigraphic Nomenclature
INGEOMINAS	Instituto Nacional de Investigaciones en Geociencias, Mineria y Química, Museo Geológico, Bogotá; *also abbreviated* IGM	NRM	natural remanent magnetism
		OTU	operational taxonomic unit
		p *or* P	probability of type I error; significance level
ING-KU	Kyōto University expeditions in cooperation with INGEOMINAS; specimens deposited in the Museo Geológico, Bogotá	*PA*	prism area (of tooth)
		PA/ASA	measure of the relative amounts of prismatic and interprismatic enamel
		PAUP	Phylogenetic Analysis Using Parsimony [computer program]
IRF	Instituto Roberto Franco		
ISSC	International Subcommission on Stratigraphic Classification	PCHP	P. C. H. Pritchard [paleontological collection]
JEOL	trade name for a microscope	pers. com.	personal communication
K	apicocervical compression or distention of prisms (of teeth)	pers. obs.	personal observation
		PGR	patellar groove ratio
KI	Kälin index	PIMUZ	Paläontologisches Institut und Museum der Universität Zürich
KU	Kyōto University		
LACM	Los Angeles County Museum of Natural History, Los Angeles, California	plag	plagioclase
		PSR	proximal shaft ratio
LCA	last common ancestor	Qda.	quebrada (=ravine)
LMA	land mammal age	ROM	Royal Ontario Museum, Toronto
Ma	megannum; millions of years before present	SALMA	South American land mammal age
		SD	standard deviation
MACN	Museo Argentino de Ciencias Naturales, "Bernardino Rivadavia," Buenos Aires	SDSNH	San Diego Natural History Museum, San Diego, California
		SE	standard error
Mbr.	member	SEM	scanning electron microscope
MBUCV	Museo de Biología, Universidad Central de Venezuela	SGNC	Servicio Geológico Nacional de Colombia
MCZ	Museum of Comparative Zoology, Harvard University	Ss.	sandstone beds
		SYSTAT	statistical software program
MLP	Museo de Ciencias Naturales de La Plata, La Plata, Argentina	TATAC	specimens collected from the Tatacoa Desert and housed at the Departamento de Geociencias, Universidad Nacional de Colombia, Bogotá
MNHN	Muséum National d'Histoire Naturelle, Paris		
MPLR	medial patellar lip ratio	TMM	Texas Memorial Museum, Austin
MPTS	magnetic polarity time scale	TSC	tubercle for the M. scapulohumeralis caudalis
MPV	Museo Paleontológico de Villavieja, Instituto Huilense de Cultura, Villavieja, Colombia		
		UCLA	University of California, Los Angeles
mT	millitesla	UCMP	University of California Museum of Paleontology, Berkeley
n *or* N	number of subjects in sample		

UFAC	Laboratório de Pesquisas Paleontológicas, Universidade Federal do Acre, Brazil	USNM	U.S. National Museum of Natural History, Smithsonian Institution, Washington, D.C.
UMMZ	University of Michigan Museum of Zoology	VGP	virtual geomagnetic pole
UNC	Facultdad de Geociencias, Universidad Nacional de Colombia, Bogotá	YPM-PU	collections of Princeton University housed at the Yale Peabody Museum, New Haven, Connecticut
UPGMA	unweighted pair-groups using mathematical averages	ZMUZ	Zoologisches Museum der Universität Zürich

Part One
Introduction

1
A History of Vertebrate Paleontology in the Magdalena Valley

Richard H. Madden, Donald E. Savage, and Robert W. Fields

> En efecto, la historia de la tierra ésta escrita con esqueletos, los restos
> mortales de la organización son los únicos indicios, que han sobrevivido á
> todas las catástrofes, de lo que fué la vida en cada época, y este vasto
> cementerio que llamamos corteza mineral, encierra todos los elementos
> para enumerar las vicisitudes de nuestro globo, profunda materia de
> estudio para el filósofo.
>
> CORONEL JOAQUÍN ACOSTA

The Republic of Colombia is well known for its record of Cretaceous plesiosaurs (*Alzadasaurus colombiensis* Welles 1962) and for its Tertiary and Pleistocene vertebrates (Stirton 1947a, 1947b, 1953a). Above all, however, Colombia is renowned for its Miocene fauna of land vertebrates, especially the collection that is known as the La Venta paleofauna, obtained from the badland exposures in the Tatacoa Desert northeast of the village of Villavieja in the upper valley of the Magdalena River.

The upper valley of the Magdalena was originally called Gran Tolima, from the Carib word for "land of snows," referring to the high stratovolcanoes of the Colombian Central Cordillera. The slopes of Volcán Puracé (4,700 m) at the union of the Colombian Eastern and Central Cordilleras are the source of the Magdalena River. From these lofty beginnings the Magdalena flows 400 km through the wide, semiarid Valle de la Tristeza. Frederic Edwin Church painted the valley landscape in *Scene on the Magdalena* from drawings made during his travels through Honda in 1853 (Kelly 1989).

By interrupting moisture-laden northeast trade winds, the Eastern Cordillera casts a rain shadow over the upper Magdalena valley. The shadow explains the occurrence of semiarid desert scrubland at low elevation (440 m) near the terrestrial equator (3°20′ N latitude). The Eastern Cor-

dillera was uplifted, beginning in the middle Miocene, to elevations greater than 1,500 m in the Pliocene (chap 2; also Hammen et al. 1973; Hammen 1979).

Early Collecting Efforts

Vertebrate paleontology in Miocene strata of the Magdelena valley began with the collecting efforts of Maurice Antoine Rollot, better known as Brother Ariste Joseph, of the Christian Schools of La Salle. In 1923 Brothers Ariste Joseph and Nicéforo María undertook a scientific expedition into the Magdalena valley, collecting zoological specimens and fossil invertebrates, plants, and vertebrates. According to their mapped itinerary, they ascended the valley southward as far as Girardot where the Christian Schools of La Salle maintained a colegio. Fossil vertebrates collected by local residents from various localities in the Honda Group in the upper valley were brought to the colegio by students. Brother Ariste Joseph transported this material back to the museum of the Instituto de La Salle in Bogotá (Ariste Joseph and Nicéforo María 1923).

By means of diplomatic correspondence with J. B. Reeside, Jr., of the United States Geological Survey in Washington, Brother Ariste Joseph shipped fossil vertebrate, mammal, and plant specimens from the upper Magdalena valley to the United States National Museum in exchange for scientific literature relating to mineral and petroleum geology. Among these fossils received in Washington between September 1922 and October 1924 were USNM 10889, the type of *Caiman neivensis* (Mook), collected somewhere between Neiva and the Río Bache (Mook 1941; Langston 1965) and USNM 10870, the type of *Metaxytherium ortegense,* collected in August 1920 from the municipality of Ortega in Tolima Department. Other specimens in this collection include vertebrate and mammal material from Coyaima, Aipe, Neiva, La Union (possibly between Neiva and Villavieja), and near the village of Villavieja.

The fossil and living vertebrates and plants from the Magdalena valley that were gathered during this expedition comprised a very small part of the extensive natural history collections at the Museo de Ciencias Naturales of the Instituto de La Salle in Bogotá (López 1989). By the late 1940s the Museo de Ciencias Naturales housed one of the largest natural history collections anywhere in South America. Of a total 73,000 natural history specimens collected by the clerics of La Salle before 1929, more than 8,000 were fossils (invertebrates and vertebrates), 350 fossil wood, and 9,480 mineral and rocks (López 1989). Regrettably, these extensive collections, the museum archives, furnishings, and library were utterly destroyed by fire on April 10, 1948, during the Bogotazo, by an angry mob seeking revenge for the assassination of Jorge Eliecer Gaitán. In the Bogotazo, the central commercial district of Bogotá was ransacked or destroyed and burned. The Instituto de La Salle filed a claim against the Colombian government for damages arising from the Bogotazo. Estimated property losses, including the collections, library, and archives, were 2 million pesos, roughly 1.14 million in 1948 U.S. dollars (Braun 1985).

Fossil vertebrate collecting efforts in the Magdalena valley were renewed in the 1930s as the Colombian government and private petroleum companies began an assessment of the mineral and oil reserves of the valley. Enrique Hubach discovered fossil mammals in the heights southeast of the village of Coyaima while studying the geology of the lowland plains of Tolima for the Colombian Geological Service (Hubach 1930). Gerardo Botero (1937) published an early map indicating the occurrence of fossil vertebrates farther south, east of the river, near Villavieja. More intensive exploration for petroleum reserves in the upper valley before the Second World War led to further fossil discoveries, including the type of *Scleromys schurmanni* Stehlin 1940, collected from the Miocene Honda "Formation" on the Finca Llano Redondo near Carmen de Apicalá, Tolima Department, by H. M. E. Schürmann of the Shell Oil Company (Stehlin 1940).

George R. Hyle of the Richmond Petroleum Company mapped much of the surface geology of the upper valley. During his fieldwork in 1938, Hyle discovered fossil vertebrates at several localities. Hyle's unpublished studies include several maps of the surface geology, including (1) the West Neiva anticline (1:25,000) showing two vertebrate fossil localities and one plant macrofossil site; (2) the Villavieja-Baraya area (1:50,000) in July 1940; (3) the Aipe area (1:25,000) in August 1940; and (4) the Gigante-Yaguará area (1:50,000), the vicinity 5 km west of Hobo, where vertebrate fossils were found in the Honda Group.

José Royo y Gómez

José Royo y Gómez, professor of mineralogy and geology at the Museo Nacional de Ciencias Naturales and director of the Museo de Anthropología, Ethnografía, y Prehistoria in Madrid, fled Spain at the end of the Spanish civil war and came to reside in Colombia where he worked for the Servicio Geológico Nacional de Colombia (SGNC) between 1939 and 1951. While with the service, Royo y Gómez filed some ninety-six scientific reports and published twenty papers on the economic geology, stratigraphy, and historical geology of Colombia. His most enduring contribution to vertebrate paleontology in Colombia was the creation and organization of the Museo Geológico Nacional at INGEOMINAS (Instituto de Investigaciones en Geociencias, Minería y Química). In 1951 Royo y Gómez emigrated from Colombia to Venezuela where he served as professor in the Departamento de Geología at the Universidad Central and the Instituto Pedagógico Nacional in Caracas (Alvarado 1959; Linares 1988). While in Venezuela, Royo y Gómez made the first important collections of fossil vertebrates from the Urumaco Formation of Falcón State (Royo y Gómez 1960).

On March 7, 1940, during a field expedition to survey the mineral resources of Huila Department, Royo y Gómez discovered fossil turtles, crocodilians, notoungulates, and rodents weathering from the Honda beds at several points along the trail between San Alfonso and Villavieja (Alvarado 1940, 1941; Royo y Gómez 1941, 1942a, 1942b). Through that part of the Tatacoa Desert, the trail follows the crest of thick channel sandstones through an area of sparsely vegetated ranchland. As documented in his field notes archived at the Museo Geológico in Bogotá, Royo y Gómez discovered fossil vertebrates at four localities along the San Alfonso–Villavieja trail: (1) north of Quebrada (Qda. = ravine) Totumo, (2) south of Qda. Totumo and west of Cerro Gordo, (3) at "an oasis in the desert" just north of Qda. Pechoyo, and (4) just north of the San Nicolás house. These fossil occurrences were described and some of the fossil mammals and other vertebrate remains were illustrated in publications (Royo y Gómez 1942a, 1942b, 1946).

University of California Museum of Paleontology

Fossil vertebrates collected by George Hyle were brought to the attention of the Museum of Paleontology (UCMP) at the University of California, Berkeley, through a shipment received on November 9, 1939, from William Effinger of the Richmond Oil Company. These fossil mammals led to the first collaborative American/Colombian vertebrate paleontology expedition in 1944–45, led by Rueben Arthur Stirton of the UCMP and José Royo y Gómez of the SGNC.

R. A. Stirton's interest in fossil vertebrates from South America may have originated during his undergraduate training at the University of Kansas. At that time Handel T. Martin was curator of fossil vertebrates at Kansas and Elmer S. Riggs and George Sternberg were both collecting in Patagonia. In addition, Stirton had had previous experience with fieldwork in El Salvador between 1925 and 1927 and again in 1941–42 (Burt and Stirton 1961).

Stirton recognized an opportunity in the Colombian fossils collected by Hyle. Reasoning that the history of faunal migration between North and South America could

best be studied in Colombia, Stirton applied for and received a fellowship from the John Simon Guggenheim Foundation for this work. Stirton reached Colombia later than expected in 1944 (Stirton 1951, 1953b, and field notes in the archives of the UCMP) because the ship that he and his family were taking to Colombia caught fire and sank in Acapulco. As all their clothing and equipment was lost with the ship, Stirton's wife and son were forced to return to California. Stirton's perseverance paid off, however, and some financial assistance was obtained in Bogotá, primarily from the Tropical Oil Company through the efforts of J. Wyatt Durham. Additional support was also obtained from the Gulf Oil Company, Richmond Petroleum Company, Socony Vacuum Oil Company, and Texas Petroleum Company. Information about fossil localities was obtained from the geologists and paleontologists of these and other companies.

Vertebrate paleontology in Colombia was strongly supported in the late 1940s and early 1950s by Roberto Sarmiento Soto, director of the SGNC. Benjamín Alvarado of the SGNC took a special interest in the proposed work and recommended the addition of José Royo y Gómez and a mapmaker, Manuel I. Varón, to the effort. The basic outline of Stirton's arrangement with the SGNC is recorded in official government documents (Del Rio 1945). B. Alvarado assisted in obtaining an adequate budget for the field expenses of the Colombian personnel and for the use of a field vehicle.

Stirton, Royo y Gómez, and their colleagues began fieldwork in the Tertiary strata of the Department of Tolima on October 13, 1944 (fig. 1.1). Following up on Stehlin's description of *Scleromys schurmanni*, they established their first base in the town of Carmen de Apicalá and collected in the "Honda Formation" north of town. The first fossils collected from this area included turtle skeletal fragments, crocodile teeth, and "ungulate" teeth and bone fragments, which were given field numbers in the system employed by Royo y Gómez. During the ensuing days of October, pieces of fish, ground sloth, notoungulate, astrapothere, and additional teeth of the rodent *Scleromys* were found in strata around Carmen de Apicalá.

On November 8, 1944, the five expedition members moved from Carmen de Apicalá to the village of Coyaima on the Río Saldaña. Following Hubach's original discovery,

Figure 1.1. Scientific staff of the first University of California/Colombian Geological Service paleontology expedition to the Magdalena River valley (1944–45). *Left,* José Royo y Gómez (Servicio Geológico Nacional de Colombia); *right,* Reuben A. Stirton (University of California).

they prospected several small hills of exposed sandstone, conglomerate, and red mudstone beds east and south of town. A variety of fragmentary fossils, including an interatheriid notoungulate, rodent, ground sloth, armadillo, toxodont, glyptodont, astrapothere, crocodilian, turtle, fishes, and freshwater gastropods were found weathering from these beds. As is often the case in the Miocene of the Magdalena valley, turtle and crocodile were represented by the most abundant specimens. Based upon his study of the interathere remains, Stirton (1953b) concluded that the "Coyaima fauna" was distinct from and older than that of the Tatacoa Desert farther south.

Following a tip obtained through W. C. Hatfield of the Texas Petroleum Company, the party prospected an exposure of the Gualanday Group near Chaparral, a village west of Coyaima in the foothills of the Central Cordillera (Stirton 1947a, 1953a). On the farm of Severo Hernández, on the low pastures just above Qda. Tuné, Stirton and party found fossil fishes, amphibians, turtles, crocodilians, a sloth, a possible condylarth, a proterotheriid litoptern, a nesodontine toxodont, and isolated and broken teeth of an astrapothere. On the basis of this assemblage Stirton (1953b) inferred an Oligocene (Deseadan) age for this fauna. The vertebrate fossils were recovered by quarrying a discontinuous gravel bed within a minor channel sandstone. The fossil bed must have been small because it was quarried out

in six days. Repeated visits by Duke University parties between 1985 and 1990 added only a few crocodile teeth to the known material.

Hepatitis put Stirton under a doctor's care in Bogotá through the Christmas holidays of 1944. In January Stirton returned to the field with Royo y Gómez and a party from the SGNC. At this time they established themselves in the village of Villavieja. On January 13, 1945, Stirton, Royo y Gómez, son José Royo González, Manuel I. Varón, the invaluable José E. S. Périco (employee of the SGNC and field assistant par excellence), and two local young men hiked northeastward from Villavieja on the San Alfonso trail, following the path of Royo's earlier exploration. "Eighty minutes out of Villavieja" in the first patch of Miocene badlands the party prospected, Manuel Varón and José Royo González discovered part of the skeleton and an incomplete skull of a fossil monkey, described and named *Cebupithecia sarmientoi* by Stirton and Savage (1951; Stirton's field notes for 1944–45).

The discovery of *Cebupithecia* may be the most significant fossil vertebrate discovery in the Tatacoa Desert badlands. The discovery helped Stirton obtain the financial support of the Associates in Tropical Biogeography at the University of California, Berkeley. Moreover, the discovery of *Cebupithecia* and other monkey fossils motivated many later expeditions to the Tatacoa Desert, including those of

Kyōto University, the Field Museum of Natural History, and Duke University.

Between 1944 and 1949 Stirton and Royo y Gómez developed an informal stratigraphy in the Tatacoa Desert, recognizing such subdivisions of the Honda as Monkey unit, Ferruginous sandstone, Lower red beds, and so on. This nomenclature was later supplemented and refined by Fields (1959). It was employed by Savage and Henao-Londoño as they collected in the district in 1950, and it has been followed by Langston (1965) and all later workers.

At the close of the 1944–45 field season, Stirton remarked in his field notes that six months of additional work in the area would prove profitable. This was an understatement. Although the first full field season of vertebrate paleontology in the upper Magdalena valley had been a tremendous success, every later expedition also proved fruitful. Most of the fossil vertebrates collected during this and subsequent UCMP/SGNC expeditions are housed at the Museum of Paleontology of the University of California although important specimens are also to be found in the Museo Geológico Nacional at INGEOMINAS in Bogotá. Published results by José Royo y Gómez and R. A. Stirton opened the eyes of the international scientific community to the significance of the fossil vertebrates of this area.

Continuing the original arrangement with the Tropical Oil Company to correlate the nonmarine Cenozoic of Colombia, Stirton returned to Bogotá in May 1946 for a second field season devoted almost entirely to prospecting the extensive mapped Tertiary in the vicinity of Villavicencio in the eastern llanos region of Colombia. Between May and August Stirton worked with seismograph crews of the Tropical Oil Company in the llanos. On May 18 Stirton savored the opportunity to fly from Bogotá to Villavicencio over the westernmost llanos between the Tertiary foothills and mesas and the flat plains. Between May and July Stirton's efforts suffered considerably because of the rainy season. Torrential rains slowed prospecting, and swollen rivers hindered travel. Although Stirton collected many living vertebrates, he found no fossils east of the Cordillera.

At the request of the Tropical Oil Company Stirton then briefly prospected the Oligocene-Eocene beds of the middle Magdalena valley for fossil vertebrates between July 24 and August 10. Along with Waldo W. Waring, geologist, Stirton prospected both the La Cira and Mugrosa fossil invertebrate horizons. On August 3 they found a badly eroded astrapothere mandible with one broken lower second molar from a level below the Mugrosa freshwater invertebrate horizon, the only fossil mammal recovered that year. At the end of the 1946 season, Stirton observed, "This closes the most expensive and yet most unproductive field season I have ever experienced." Regrettably, in the only extensive published account of Stirton's efforts in 1946 (Simpson 1984, 187–88), Simpson mistook the term *Villavo* in Stirton's notes to mean Villavieja instead of Villavicencio, thereby shifting the field of Stirton's operations hundreds of kilometers westward.

In 1949 Stirton returned to the Magdalena valley in company with Alden Miller (vertebrate zoologist from the University of California), Herbert Mason (botanist from the University of California), and graduate student Robert W. Fields. These men, along with Diego Henao-Londoño and José Périco and other employees of the SGNC, supported in part by funds from the Associates in Tropical Biogeography, conducted field studies of the stratigraphy, paleontology, botany, and zoology of the upper Magdalena valley, especially in the environs of the Tatacoa Desert (Fields field notes for 1949). Stirton, Fields, Henao-Londoño, and Périco collected many important Miocene vertebrates over a six-month period between January and June 1949 and added greatly to the taxonomic list comprising the La Venta paleofauna. On February 7 Diego Henao-Londoño discovered the type mandible of *Stirtonia* ("*Homunculus*") *tatacoensis* east of Qda. Tatacoa.

On April 5, Robert Fields, prospecting in the vicinity of the *Cebupithecia* monkey locality, discovered the right and left mandibles of a new small primate, which Stirton (1951) named *Neosaimiri fieldsi*. "In the same spot I recovered elements belonging to a small armadillo and the right mandible of a rodent. The elements of all three of these animals may be represented but tentatively I have separated what I believe to be the elements of a primate. All the isolated fragments found on the spot I have put with the small armadillo. There may even be some reptile bones mixed with these but as they were found, they were all together so all fragments were picked up." A few moments later on the same day, Fields "found an isolated Primate molar and incisor" about 36 m south of the San Alfonso trail. Three days later, on April 8, Fields received two letters from Stirton expressing general pleasure with the progress of his work. When he wrote these letters, Stirton did not yet know about the discovery of the new primates. Fields noted in his journal with obvious delight, "I can imagine how he will react to the two new monkeys" (117).

On the return to Berkeley, Fields stopped in Chicago to visit Bryan Patterson at the Field Museum. Fields showed Patterson the new primates and an odd "condylarth" or "pyrothere" tooth found near Cerro Gordo (see McKenna 1956). While in Chicago, Fields learned from Patterson that the new primates were unrelated to Santacrucian *Homunculus* and that other elements of the La Venta fauna were morphologically distinct from those of the Santacrucian. Among the more significant work accomplished by Fields in 1949 was a thorough description of the geology of the La Venta badlands (Henao-Londoño and Fields 1949; Fields 1959).

With the continued support of the Associates in Tropical Biogeography, Donald E. Savage joined with O. P. Pearson (vertebrate zoologist), S. Smith (botanist), Diego Henao-Londoño, José Périco, and other employees of the SGNC during the summer of 1950 for additional work in the Tatacoa Desert (Savage 1951b; 1952). Among the objectives of the 1950 field season was the exploration of the periphery of the desert. Despite the severe shortage of gasoline for truck transportation at that time owing to La Violencia, the field party managed to work between May and July and was fortunate enough to find and collect splendid skulls of crocodilians previously unknown from the Honda Group and numerous ground sloths and excellent leontiniid and astrapothere material (Savage, unpublished field notes).

Most of the collection of fossil vertebrates from the La Venta badlands at the University of California Museum of Paleontology was prepared by W. Langston and D. Savage. The scientific productivity that this collection inspired is prodigious. In addition to those papers already cited, the University of California collection inspired many postgraduate theses and other scientific studies (e.g., Reinhart 1951; Savage 1951a; Miller 1953; Fields 1957; Estes 1961, 1983; Estes and Wassersug 1963; Colwell 1965; Langston 1965; Hershkovitz 1970; Auffenberg 1971; Hirschfeld 1971, 1976, 1985; Marshall 1976a, 1977; Wood 1976a; Bondesio and Pascual 1977; Krohn 1978; and Johnson 1984).

Later Collecting Efforts

Between 1950 and 1985 local collectors in Colombia accumulated many excellent specimens. The most active local collector has been José Antonio Calderón Avendaño of Villavieja, who has made numerous important discoveries. José Antonio is thought to have made his first excursion in search of fossils in company with Hoffstetter, de Porta, and Etayo in 1966. From that time on José Antonio has supported himself by selling colmillos (fossil teeth) to the tourists and other interested visitors to Villavieja. Most of the valuable pieces of his collection were purchased by Miguel Rubiano, a prominent citizen of Villavieja. Heightened local public interest eventually gave rise to provincial concern over the destiny of the fossil vertebrates collected by outsiders, tourists, amateurs, and, especially, professional collectors. Through the admirable efforts of Arnol Tovar of the Instituto Huilense de Cultura, with the collaboration of Carlos Villarroel of the Departamento de Geociencias at the Universidad Nacional de Colombia in Bogotá and Javier Guerrero, the largest local private collections have been brought together into the Museo Paleontológico of Villavieja. The museum was set up between 1985 and 1986 in the chapel of Santa Barbara on the northwest corner of the town plaza. This building, built in 1748, is different from the more recently constructed chapel and bell tower on the south side of the plaza where in 1950 O. P. Pearson made an informative collection of large phyllostomid bats. The twin bell towers of the chapels of Villavieja and Aipe, and their adjacent discotecas (dance halls), are important cultural landmarks in this part of the Magdalena valley. Some fossil vertebrates from the Tatacoa Desert are also housed in the offices of Arnol Tovar at the Instituto Huilense de Cultura on the Plaza Santander in Neiva.

Robert Hoffstetter of the Muséum National d'Histoire Naturelle in Paris has had a long professional interest in the vertebrate paleontology of the South American tropics, especially the central Andean republics. His professional labors began in 1946 during six years residence in Quito as professor at the Escuela Politécnica Nacional. Between 1962 and 1981 Hoffstetter visited Colombia on nine separate occasions, traveling throughout the country visiting fossil-bearing exposures to collect vertebrate fossils and examining local collections.

Jaime de Porta, of the Departament de Geología Dinámica, Geofisica i Paleontología of the Universitat de Barcelona, arrived in Colombia in 1958 where he worked as professor in the Facultad de Ingeniería de Petróleos at the Universidad Industrial de Santander in Bucaramanga, the principal training center for the petroleum industry centered in the middle Magdalena valley. Between 1959 and 1963, de Porta served as professor of paleontology at the Instituto de Ciencias Naturales of the National University in Bogotá, and between 1961 and 1963 worked as chief of the Section of Stratigraphy and Paleontology at the SGNC. During this time de Porta devoted much of his professional activities to studies of the geology and paleontology of the Tertiary of the Magdalena valley (Porta 1962, 1965, 1966), describing the pampathere and glyptodonts (Porta 1962) and publishing on general aspects of the fossil vertebrates of the Honda Group (Porta 1961, 1969). He returned to the Universidad Industrial de Santander between 1964 and 1967 and from there eventually returned to Spain. In 1974 de Porta's comprehensive lexicon of the Tertiary and Quaternary geology of Colombia was published.

In 1966 de Porta, Hoffstetter, Fernando Etayo (National University), and José Périco of the SGNC, with funding from the CNRS and the National Museum of Natural History in Paris, visited the area northeast of Villavieja where they made a modest but interesting collection of fossil vertebrates, including snakes, rodents, xenarthrans, notoungulates, and litopterns from the Monkey Beds and Fish Bed (for correct stratigraphic nomenclature, see chap. 2) outcropping in Qda. La Venta near Los Mangos (Hoffstetter 1967a, 1967b, 1970a, 1970b, 1971, 1976; Hoffstetter and Rage 1977; Hoffstetter and Soria 1986). The collection they made was divided: a portion is now in Paris, part is at the Instituto de Ciencias Naturales in Bogotá, and part is at the Universidad Industrial de Santander in Bucaramanga. In 1980 Hoffstetter again visited the area of Villavieja with F. Etayo and D. Garcia of the Universidad Nacional de Colombia and the collection made that year is housed at the Instituto de Ciencias Naturales, Ciudad Universitaria, Bogotá.

Franklyn B. Van Houten of the Geology Department of Princeton University directed doctoral dissertation research for several students on the Tertiary geology of the upper Magdalena valley, including the Honda Group (Van Houten and Travis 1968; Wellman 1970; Anderson 1972; Howe 1974; Van Houten 1976). Some large fossil mammals collected during this fieldwork are now in the Peabody Museum of Natural History, Yale University, New Haven, Connecticut. Ernest L. Lundelius, Jr., of the University of Texas at Austin, briefly visited the area in 1972 with Van Houten and made a small collection, which is now in the museum of the University of Texas.

Fernando Etayo, professor of geoscience and member of the Instituto de Ciencias Naturales, Universidad Nacional de Colombia in Bogotá, has maintained an interest in the vertebrate paleontology of the area. His interest in the La Venta fauna, in particular, was increased during his graduate studies in geology at the University of California at Berkeley where he obtained a Ph.D. The Instituto de Ciencias Naturales houses some fossil vertebrates from the Miocene Honda Group that he collected in 1966 and 1980. Alberto Cadena, professor of mammalogy and member of the Instituto de Ciencias Naturales, has been involved in the study of fossil vertebrates, especially the small mammals, in collaboration with the Kyōto University expeditions (Setoguchi et al. 1986).

Kubet Luchterhand of the Field Museum of Natural History, Chicago, with support from the National Geographic Society, made a small collection of fossil vertebrates from the Villavieja Formation near the San Nicolás house in 1980. This small collection is now at the Field Museum and includes the type specimen of the monkey *Mohanamico hershkovitzi* (Luchterhand et al. 1986).

Carlos Villarroel of the Departamento de Geociencias, Universidad Nacional de Colombia, Bogotá, has organized a number of geology and paleontology field trips to the Tatacoa Desert since about 1985 (fig. 1.2) and has assembled an important collection of fossil vertebrates from the Honda Group at the department (Villarroel and Guerrero 1985). Late Pleistocene fossil mammals from surface soils in the Tatacoa Desert have recently been described (Villarroel et al. 1989).

Kyōto University

Beginning in 1977 and continuing throughout the 1980s expeditions from Kyōto University in Japan (originally directed by Kondo Shiro and, most recently, by Nogami

Figure 1.2. Paleontologists from the Universidad Nacional de Colombia (Bogotá) in the Tatacoa Desert in 1991. *Left,* Carlos Villarroel (National University of Colombia); *right,* Javier Guerrero (INGEOMINAS).

Yasuo and funded through the Japanese Ministry of Education) have collected fossil vertebrates in the Tatacoa Desert. The principal objective of these field efforts is directed toward recovering additional fossil monkeys and placing them in a chronological framework. This research has been reported in the series *Kyōto University Overseas Research Reports of New World Monkeys.*

The first collecting trip was undertaken by Tsuyoshi in January 1977 and then between December 1977 and February 1978 by Watanabe, Takeshi Setoguchi, and Nobuo Shigehara. Initially, the emphasis was on surface collecting throughout the Qda. La Venta area with special attention on the vicinity of San Nicolás (Watanabe et al. 1979). The following year, in 1979, Shiro Kondo, Tsuyoshi Watanabe, Takeshi Setoguchi, and Toshiro Mouri discovered the up-

per dentition of *Stirtonia tatacoensis* (Setoguchi 1980; Setoguchi et al. 1981) near Qda. Tatacoa northeast of Villavieja.

As was the case following the discovery of *Cebupithecia,* the discovery of a fossil monkey inspired Kyōto University to continue work in the Tatacoa Desert. Their screenwashing operation, the first to be undertaken at Tatacoa, produced additional isolated teeth of monkeys and other small mammals and set the pattern for most of Kyōto University's later collecting efforts, screening fine-grained sediments at or near the site of the discovery of a fossil monkey. These intensive collecting efforts have been directed by Takeshi Setoguchi. Subsequent work by Kyōto University researchers in the Tatacoa Desert involved some geology and geochronology. For example, during the 1981 field season, an effort was made to study the geology of the Kyōto site and along the east side of the Magdalena River between Neiva and Villavieja (Takemura 1983). In addition, pumice samples from the Gigante Formation were collected and zircon fission-track determinations for the Gigante Formation were later reported (Takemura and Danhara 1983).

During the 1983 field season a "monkey graveyard" was discovered (Setoguchi 1983), and since this discovery, Takeshi Setoguchi and Alfred L. Rosenberger of the Department of Anthropology, University of Illinois at Chicago, have collaborated fruitfully on the description of the fossil monkeys collected by Japanese field crews at the Kyōto site (Setoguchi 1985; Setoguchi and Rosenberger 1985a, 1985b). The fossil primates discovered during 1983 were summarized by Setoguchi and others (1986). After this season a short paleomagnetic column was published for the upper Honda Group in the vicinity of the Kyōto Site (Hayashida 1984). Bats and small marsupials collected by screening at the Kyōto Site have also been reported (Takai et al. 1988).

The 1985 field season concentrated again on the Kyōto Site, and fission-track dates were reported for the Villavieja Formation (Takemura and Danhara 1986). After the description of *Mohanamico hershkovitzi* in 1986 (Luchterhand et al. 1986), the 1987 field season endeavored to find new primate-bearing localities near the San Nicolás house (El Dinde). This work led to the discovery of the type of *"Aotus dindensis"* by José Antonio Calderón Avendaño, who had been employed by the Japanese group.

Figure 1.3. First Duke University/ INGEOMINAS expedition to the Tatacoa Desert in 1985, sponsored by the National Geographic Society. *Left to right:* Leonardo Sepúlveda, Carlos Ulloa (INGEOMINAS), Richard H. Madden (Duke University), Richard L. Cifelli (University of Oklahoma), Richard F. Kay (Director, Duke University), and Javier Guerrero (INGEOMINAS).

Duke University

Duke University and INGEOMINAS began annual prospecting in the Miocene Honda Group in 1985 under the direction of Richard F. Kay. This work was originally supported by the National Geographic Society (1985 and 1986) and later by the U.S. National Science Foundation (1987–89, 1990–92). Multinational and multidisciplinary research teams principally from Duke University and INGEOMINAS worked in the Honda Group yearly between 1985 and 1992 (fig. 1.3). In addition, associates from the University of Oklahoma, the Field Museum of Natural History (Chicago), the Museo de Ciencias Naturales de La Plata (Argentina), Southern Methodist University (Texas), the Institute of Human Origins (Berkeley), and the Universidad Nacional de Colombia (Bogotá) collaborated with both the fieldwork and later scientific research.

Between 1985 and 1992 Duke University/INGEOMINAS collections were made, numbering 3,272 cata-

loged specimens from more than 140 localities at 52 different stratigraphic levels within the exposed Honda Group. These new collections are permanently housed at the Museo Geológico at INGEOMINAS and include important accessions of nearly all vertebrates and, especially, small mammal fossils. Richard L. Cifelli (University of Oklahoma) and Anne H. Walton (Southern Methodist University) established an ambitious screenwashing operation for small vertebrates, using the geology laboratory facilities of Hocol at Campo Tello. The screenwashing has been maintained in parallel with extensive surface prospecting covering the entire Tatacoa Desert, from the Río Cabrera and Cerro Chacarón in the north, to the Polonia Red Beds to the south. The fossil vertebrates have been and continue to be made available to interested scientists in Argentina, Colombia, Canada, and the United States. These new collections provide the principal inspiration for the descriptions and progress reports published in this volume.

Part Two
Geology and Geochronology

2
Stratigraphy, Sedimentary Environments, and the Miocene Uplift of the Colombian Andes

Javier Guerrero

RESUMEN

El Grupo Honda del Valle Superior del Magdalena está limitado, en su base y tope, por discordancias (13,5 Ma en la base del Grupo Honda y 10,1 Ma en la base del Grupo Huila) que reflejan importantes episodios de volcanismo y levantamiento tectónico en los Andes Colombianos.

El depósito de la unidad inferior del Grupo Honda, la Formación La Victoria (13,5 a 12,9 Ma), se inició durante una fase de intensa actividad volcánica y tectónica de la Cordillera Central. La mayoría de la Formación La Victoria fue depositada con una tasa de subsidencia de aproximadamente 950 mm/1.000 años y está constituída por depósitos de ríos meandriformes de hasta 8–10 m de profundidad, que fluían hacia el oriente-suroriente. Los ríos que depositaron la Formación La Victoria eran aparentemente continuos hacia el este, con los depósitos costeros de la Formación Pebas en el Amazonas. Las Capas del Conglomerado Cerbatana en el tope de la Formación La Victoria fueron depositadas por ríos trenzados, como resultado de un pulso de levantamiento rápido de la Cordillera Central.

El depósito de la Formación Villavieja (12,9 a 11,5 Ma) se inició durante el ya activo levantamiento y volcanismo de la Cordillera Central, y también durante el primer levantamiento de la Cordillera Oriental. La Formación Villavieja se depositó con una tasa de subsidencia de aproximadamente 413 mm/1.000 años, y está constituida en su mayoría por ríos meandriformes de hasta 2–3 m de profundidad, que fluían al comienzo hacia el oriente y luego se invirtieron hacia el occidente y adquirieron un patrón anastomosado.

El levantamiento de la Cordillera Oriental se inició hace aproximadamente 12,9 Ma, cuando la tasa de sedimentación y la energía del depósito decrecían considerablemente y se presentaban algunas direcciones de corriente hacia el norte y sur, además de las todavía presentes hacia el oriente y suroriente. La continuidad regional de las capas rojas en la parte superior del Grupo Honda, y las direcciones de paleocorriente hacia el occidente en las Capas Rojas de Polonia, indican que existía una cadena montañosa continua hace aproximadamente 11,8 Ma. Este evento tectónico de levantamiento de la Cordillera Oriental de 12,9 a 11,8 Ma coincidío con una extensa actividad tectónica global y con el hiato oceánico NH3.

Aunque los paleosuelos son muy comunes, éstos son más maduros hacia el tope de la secuencia, en el Miembro Cerro Colorado de la Formación Villavieja. En general, los paleosuelos del Grupo Honda se formaron bajo una fuerte influencia de las tasas de sedimentación, controladas por levantamiento relativo de las áreas de aporte y subsidencia en una cuenca muy activa tectónicamente, con un clima tropical sub-húmedo a húmedo y marcada estacionalidad.

In this chapter the stratigraphy and sedimentary environments of the Honda Group in the La Venta area are described and compared with those in other localities. The primary goal is to provide a stratigraphic framework for a

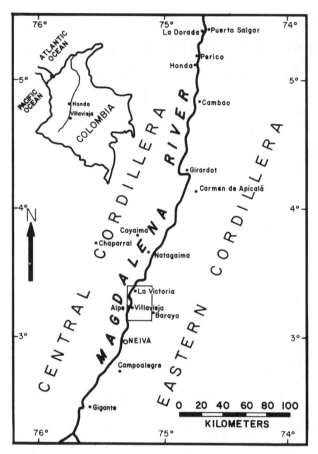

Figure 2.1. Outcrop localities of the Honda Group along the upper Magdalena River valley, including the mapped La Venta area.

large collection of fossil vertebrates made between 1985 and 1992. The sedimentary environments of the approximately 1,250-m-thick Honda Group are considered in light of the evolution of the Colombian Andes during the Miocene.

The Honda Group crops out in several localities along the Magdalena River valley between the Central and Eastern Cordilleras of the Colombian Andes (fig. 2.1). Although the type locality of the Honda Group (Hettner 1892) is located near Honda and La Dorada in the middle Magdalena River valley, the best exposures and most fossiliferous sections are located near Villavieja and La Victoria in an area known as La Venta in the upper Magdalena valley.

In the La Venta area the Honda Group deposits, mainly mudstone and sandstone, are poorly cemented and erode easily, producing a semidesertic area of badlands with vegetation composed predominantly of thorn-scrub acacia and

woody cactus. Outcrops are excellent, and large numbers of fossil vertebrates have been recovered, including fish, reptiles, birds, and mammals. The Honda Group is preserved in a structural depression (fig. 2.2) bounded on the west by the west-dipping Chusma Thrust System and on the east by the north prolongation (Baraya Thrust) of the east-dipping Suaza Thrust System (INGEOMINAS 1988). The Honda Group overlies with angular unconformity the Jurassic andesites of the Saldaña Formation, and Lower Cretaceous to Oligocene deposits; it postdates the predominantly west-dipping thrust faults of early to middle Miocene age associated with an uplift event in the Central Cordillera.

In the La Venta area and further south near Gigante (fig. 2.1), the Honda Group is overlain by the Neiva and Gigante formations (Howe 1969, 1974). The boundary between the Honda Group and the Neiva Formation is a disconformity; the available chronostratigraphic information implies a hiatus between the two from approximately 11.6 to 10.1 Ma (Guerrero 1993; also chap. 3).

The application of the name Mesa Group proposed by Howe (1969, 1974) to include the Neiva and Gigante formations and the Ceibas Conglomerate has been abandoned because the lithology and tectonic setting of the Mesa Formation (Pliocene) northwest of Honda in the middle Magdalena valley cannot be correlated with the sequence of the upper Magdalena valley. The late Miocene Neiva and Gigante formations were included by Guerrero (1993) in the Huila Group. The Honda and Huila Groups are unconformably covered by Quaternary alluvial fans in the study area.

The main source area of the Honda Group was located west of the present Magdalena valley. Several volcanic bodies are exposed more than 150 km to the northwest of the La Venta area along what is today the western border of the Central Cordillera of the Colombian Andes. The isotopic ages of these volcanic bodies range between 17 and 6.5 Ma (Restrepo 1983).

The first stratigraphic and geologic work dealing with the Honda sediments in the La Venta area was by Royo y Gómez (1942a, 1946). Much of the stratigraphic nomenclature proposed by Royo y Gómez (1946) was retained by Stirton (1951) and followed by Fields (1959) in a geologic map and detailed description of the stratigraphy. More recently, Wellman (1970) proposed to extend the type lo-

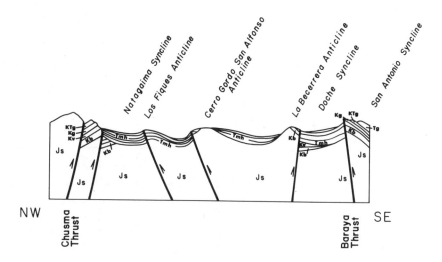

Figure 2.2. Cross section illustrating the structure of the upper Magdalena River valley. Abbreviations: *Js,* Jurassic Saldaña Formation; *Kb,* Cretaceous Caballos Formation; *Kv,* Cretaceous Villeta Formation; *Kg,* Cretaceous Guadalupe Formation; *KTg,* Cretaceous-Tertiary Guaduas Formation; *Tg,* Tertiary Gualanday Group; *Tmh,* Tertiary Honda Group. (Section from figure 2.8 and from INGEOMINAS Mapa Geológica de Colombia [1988].)

calities of the Honda Group along selected localities of the Magdalena valley and divided it into two formations, each with two members. The type locality of the Villavieja Formation, the upper unit of the Honda Group, is in the La Venta area.

Based on paleontological information, Fields (1957) stated that the La Venta area was a broad flood-plain through which the ancestral Magdalena River and many tributaries meandered; he proposed that fish, crocodiles, snails, frogs, turtles, and sirenians lived in broad rivers and swamps. Wellman (1970) interpreted the lower part of the Honda Group outcropping in the La Venta area as deposited by probably braided, high-gradient, low-sinuosity streams. The red beds of the upper part of the Honda Group were interpreted as deposited by low-gradient, high-sinuosity (meandering) streams. In addition, from cross-bedding measurements, he concluded that the Honda Group rivers flowed to the east at a time when the Eastern Cordillera of the Colombian Andes had not yet been uplifted.

Methods

Field work was undertaken in one-month seasons each year from 1987 to 1991. The information collected was plotted on air photos and topographic maps from the Instituto Geográfico "Agustín Codazzi." The photos of the study area include flight paths C-1564 images 22 to 40, C-2277 244 to 254, C-2099 14 to 19, and C-2165 81 to 85. The 1:25,000 scale topographic maps covering the study area are

302-II-D, 302-III-C, 302-IV-B, 302-IV-D, 303-i-C, and 303-III-A of the Carta Nacional of Colombia published by the Instituto Geográfico "Agustín Codazzi" in Bogotá.

The sections were measured in meters by using a Jacob's staff, Brunton transit, and measuring tape. Sediment size and composition were obtained in the field by using a hand lens, a grain-size scale, and standard chemical tests.

Grain size of the sediments is reported according to the Wentworth (1922) scale. Pebble sizes are additionally divided into very fine (4–8 mm), fine (8–16 mm), medium (16–32 mm), and coarse (32–64 mm). Cobbles are divided into fine (6.4–12.8 cm) and coarse (12.8–25.6 cm). Mineralogical classification of the sandstones follows Folk (1980).

Stratigraphy

The Honda Group is a sequence of light gray "salt and pepper" volcanic litharenite, reddish-brown mudstone, greenish-gray mudstone, and minor conglomerate. These sediments commonly are interlayered, forming simple fining-upward sequences or couplets a few meters thick. Spherical, postdepositional concretions of calcite-cemented sandstone are fairly common.

The heterogeneity of the sediments is very impressive in the field, and many facies occur adjacent to one another. Lateral and vertical changes occur at outcrop scale. Sandstone bodies change laterally to mudstone in distances that range from tens of meters to kilometers.

The units are difficult to correlate, and stratigraphic divi-

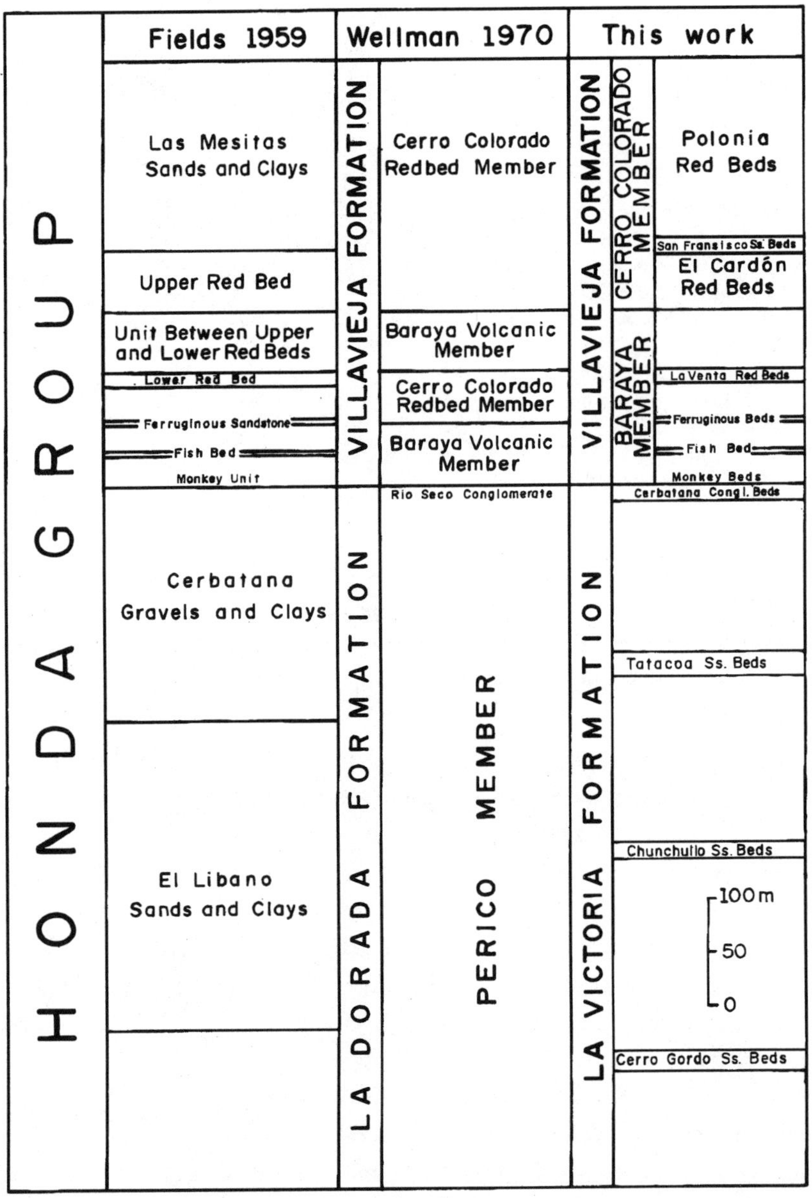

HONDA GROUP	Fields 1959	Wellman 1970		This work		
		VILLAVIEJA FORMATION		VILLAVIEJA FORMATION	CERRO COLORADO MEMBER	
	Las Mesitas Sands and Clays	Cerro Colorado Redbed Member				Polonia Red Beds
						San Fransisco Ss. Beds
	Upper Red Bed					El Cardón Red Beds
	Unit Between Upper and Lower Red Beds	Baraya Volcanic Member			BARAYA MEMBER	
	Lower Red Bed	Cerro Colorado Redbed Member				La Venta Red Beds
	Ferruginous Sandstone					Ferruginous Beds
	Fish Bed	Baraya Volcanic Member				Fish Bed
	Monkey Unit					Monkey Beds
		Rio Seco Conglomerate				Cerbatana Congl. Beds
	Cerbatana Gravels and Clays	LA DORADA FORMATION	PERICO MEMBER	LA VICTORIA FORMATION		
						Tatacoa Ss. Beds
	El Libano Sands and Clays					Chunchullo Ss. Beds
						100 m — 50 — 0
						Cerro Gordo Ss. Beds

Figure 2.3. Stratigraphic nomenclature and correlation of the Honda Group in the upper and middle Magdalena River valley. The lower boundary of the Honda Group as proposed in this work is a regional unconformity. Abbreviations: *Ss.,* sandstone; *Congl.,* conglomerate.

sion must be based on major changes. Some features are, however, easy to recognize and can be followed basinwide in the Magdalena valley. The most clear division of the Honda Group is marked by a clast-supported coarse-pebble conglomerate, the top of which constitutes the boundary between its two component formations. The top of the upper formation is composed predominantly of red and reddish-brown mudstone, making possible a division of that formation into two members.

Divisions of the Honda Group sediments for biostratigraphic purposes were made by Royo y Gómez (1946a)

and Fields (1959). Most of their names are retained here. Units with good lateral continuity in the upper part are easier to recognize in the field than units in the lower part of the Honda Group. In fact, the approximately 570 m of the lower part of the Honda Group outcropping in La Venta were undivided by Wellman (1970) and were recognized as two informal units by Fields (1959).

Herein, the Honda Group that crops out in the vicinity of Polonia, Villavieja, and La Victoria is subdivided into nineteen units contained in two formations (fig. 2.3). A modification of the stratigraphic nomenclature presented

HONDA AREA, MIDDLE MAGDALENA VALLEY				UPPER MAGDALENA VALLEY	
Hettner 1892	Buttler 1942	Porta 1966	Wellman 1970	Wellman 1970	This Work
Honda	Upper Honda	Los Limones Fm.	Villavieja Fm.	Villavieja Fm.	**HONDA GROUP** — Villavieja Fm.
		San Antonio Fm.	La Dorada Fm. / Perico Mbr.	La Dorada Fm. / Perico Mbr.	La Victoria Fm.
≈≈≈≈≈≈≈≈≈≈	≈≈≈≈≈≈≈≈≈≈	≈≈≈≈≈≈≈≈≈≈	≈≈≈≈≈≈≈≈≈≈	≈≈≈≈≈≈≈≈≈≈	≈≈≈≈≈≈≈≈≈≈
	Lower Honda	Cambrás Fm.	La Dorada Fm. / Puerto Salgar Mbr.	La Dorada Fm. / Puerto Salgar Mbr.	Tuné Fm.
	La Cira Fm.	Colorado Fm.	La Cira Fm.	La Cira Fm.	La Cira Fm.

Figure 2.4. Correlation table of Honda Group units in the La Venta area. Abbreviations: *Fm.,* formation; *Mbr.,* member.

by Royo y Gómez (1946a), Fields (1959), and Wellman (1970) is formally proposed here (figs. 2.3, 2.4), following the recommendations of the International Subcommission on Stratigraphic Classification (Hedberg 1976) and the North American Stratigraphic Code (NACSN 1983).

LA VICTORIA FORMATION

Name and Definition

The new name, La Victoria Formation, is formally proposed for the lower unit of the Honda Group in the upper Magdalena valley. The La Victoria Formation is composed of alternations in fining-upward sequences of gray "salt and pepper" pebbly volcanic litharenite and variegated mudstone. The base is a regional angular unconformity of middle Miocene age, and the top includes a clast-supported coarse-pebble conglomerate. The La Victoria Formation is conformably covered by the fine-grained deposits of the Villavieja Formation (Wellman 1970). The new name establishes more clearly the lower boundary of the Honda Group and thereby clarifies some stratigraphic confusion introduced in the past.

The La Victoria Formation may be subdivided by several marker beds useful for correlation and reference in the La Venta area; from bottom to top, these are (new names) the Cerro Gordo Sandstone Beds, the Chunchullo Sandstone Beds, and the Tatacoa Sandstone Beds. These thick (up to 31.5 m) associations of multistory sandstones (figs. 2.5, 2.6A) occur in belts 2–4 km wide and tens of kilometers in length and are used to separate finer units of alternating mudstone and sandstone. The latter are organized similarly to the simple fining-upward sequences of the Baraya Member of the Villavieja Formation (figs. 2.6B, 2.7D) that do not have overlapping sandstones.

The sheetlike, clast-supported Cerbatana Conglomerate Beds (new name) constitute the uppermost unit of the La Victoria Formation. The unit can be recognized throughout the upper Magdalena valley and is equivalent to the unit that Wellman (1970) called the Río Seco conglomerate east of Honda. It also conforms to the uppermost conglomeratic part of the unit that INGEOMINAS (1959) mapped as T6a south of Neiva.

The type locality and sections of the La Victoria Formation and its constituent beds, here formally proposed, are in the La Venta area (secs. A–F in fig. 2.8). The formation is named after the village of La Victoria where it is best exposed.

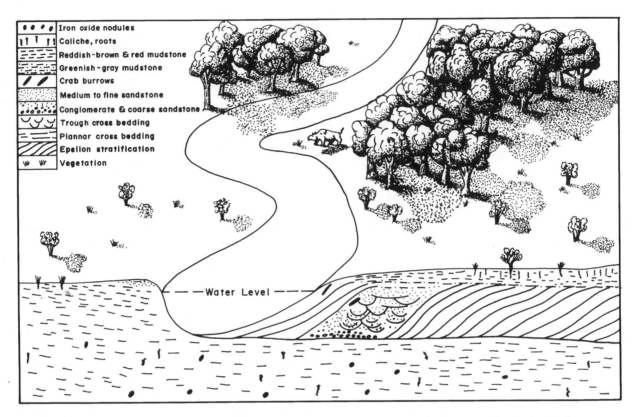

Iron oxide nodules
Caliche, roots
Reddish-brown & red mudstone
Greenish-gray mudstone
Crab burrows
Medium to fine sandstone
Conglomerate & coarse sandstone
Trough cross bedding
Plannar cross bedding
Epsilon stratification
Vegetation

Water Level

Figure 2.5. Sedimentary structures and grain size associations in the meandering channel and flood-plain deposits of the Honda Group.

Historical Background

The name Honda was first used by Hettner (1892) to refer to greenish-gray tuffaceous sandstones outcropping in proximity to the city of Honda (fig. 2.1). Later, in response to some stratigraphic confusion, the type locality was redefined (Butler 1942) as the east-dipping volcanic-sandstone ridges of the Cordillera San Antonio on the east side of the Magdalena River north of Honda. Butler included at the base of the Honda some red beds that were not in the original concept of the unit and divided it into a nonandesitic "Lower Honda" and an andesitic "Upper Honda" (fig. 2.4).

The lower boundary and the stratigraphic nomenclature of the red beds at the base of the Honda Group (nonandesitic "Lower Honda") has been very confusing because Butler (1942) applied the name "La Cira Formation" to the lower part of the red bed sequence and considered the Honda Group as conformably overlying the "La Cira Formation." He based his correlation on the occurrence of an

invertebrate fossil horizon that contained some of the forms present in the La Cira fossil horizon of the Colorado Formation (Wheeler 1935) 230 km north-northeast of Honda. Since then several authors (INGEOMINAS 1957a, 1957b, 1959; Porta 1966; Van Houten and Travis 1968; Wellman 1970) have included some red beds in the lower part of the Honda Group and some in the "La Cira Formation" in several localities along the Magdalena River valley.

Porta (1966) divided the Honda Group in the Honda area into three formations. He included the nonandesitic "Lower Honda" of Butler (1942) in the "Cambrás Formation" and the andesitic "Upper Honda" in the San Antonio and Los Limones formations. Unaware of the work of Porta (1966), Wellman (1970) divided the Honda Group into the "La Dorada" and Villavieja formations and further subdivided the "La Dorada Formation" into the "Puerto Salgar" and "Perico" Members. The "Puerto Salgar Member" included the nonandesitic red beds of the "Lower Honda," and the "Perico Member" included the lower part of the andesitic "Upper Honda." Neither Porta (1966) nor

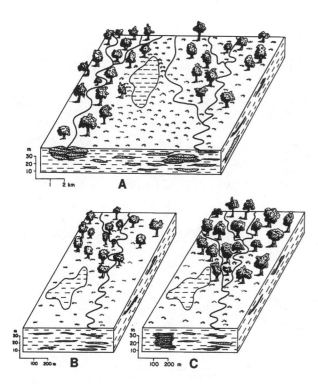

Figure 2.6. Sedimentary models for the Honda Group in the La Venta area. *A.* La Victoria Formation (except the Cerbatana Conglomerate Beds), with large meander belts forming multistory sandstone bodies (e.g., Tatacoa Sandstone Beds) and isolated meandering channels that leave single-story sandstones alternating with mudstone. *B.* Baraya Member and lower part of the Cerro Colorado Member with simple fining-upward sequences deposited by small meandering rivers with sandstone bodies one channel thick. *C.* The upper part of the Cerro Colorado Member with simple fining-upward sequences deposited by meandering rivers and very narrow multistory sandstones deposited by anastomosed rivers.

Wellman (1970) provided a stratigraphic section of the approximately 1,100-m-thick "Cambrás Formation" or "Puerto Salgar Member" in its type locality.

Because the proposed type locality of the "La Dorada Formation" of Wellman (1970) is near the town of Honda, where the earlier nomenclature proposed by Porta (1966) has priority, the name is considered invalid. The nomenclature proposed by Porta should, however, be applied only to the Honda area and should be used with caution because most, if not all, of the red beds included in the "Cambrás Formation" are related to the Gualanday Group sequence.

Wellman (1970) extended the name "La Dorada Formation" from its proposed type locality north of Honda to the entire Magdalena valley and applied the names "La Cira Formation" and "Puerto Salgar Member" without estab-

lishing clearly the lithologic boundary between them or to the reddish-brown and red mudstone that is on top of the uppermost clast-supported pebble conglomerate of the Gualanday Group. The correlation with the "La Cira Formation" should be avoided, however, because in the type locality of the Colorado Formation, the upper 115 m that contain the La Cira fossil horizon are made up of carbonaceous gray and black claystone with minor layers of pale green medium-grained sandstone. According to Morales et al. (1958), Porta (1966), and Porta et al. (1974), the name La Cira should be used only as "La Cira fossil horizon" in the type locality of the Colorado Formation.

The thick horizons of red beds that Wellman (1970) included in the "Puerto Salgar Member" of the "La Dorada Formation" were not recognized by him as outcropping in the La Venta area. He indicated, however, that the "Puerto Salgar Member" and "La Cira Formation" were apparently conformable in the subsurface of the Dina oil field 15 km southwest of the La Venta area. The "Puerto Salgar Member" was only included in a stratigraphic section of the South Neiva area (Quebrada [Qda.] Guandinosita, northwest of Gigante) on top of an interval of 10 m of red-brown mudstone named "La Cira Formation" (Wellman 1970, 2363).

In the localities where the Honda Group overlies older rocks with a clear angular unconformity (e.g., at La Venta, Coyaima, and Chaparral), the red mudstone beds of the "Puerto Salgar Member" are absent; these red beds were identified only in areas where the Honda Group paraconformably overlies the Gualanday Group. In the localities where the Gualanday Group is incompletely exposed, or not exposed at all, the Honda Group begins with volcanic litharenites interlayered with variegated mudstone, indicating that the nonvolcanic red beds included in the "Puerto Salgar Member" are part of the Gualanday Group sequence.

To avoid further stratigraphic confusion, I have applied the new name, La Victoria Formation, to the lower part of the Honda Group exposed in the upper Magdalena valley south of the Honda area. The middle Miocene angular unconformity at the base of the La Victoria Formation coincides with a change in composition and grain size where gray "salt and pepper" pebbly volcanic litharenites first appear and alternate with variegated mudstone. The

Figure 2.7. *A.* Channel sandstone from the lower part of the La Victoria Formation northwest of Duke Locality 21, with well-exposed epsilon surfaces and scoured contact at the base. *B.* Scalloped border of a channel sandstone body from the La Victoria Formation at Duke Locality 49. Channel flow is toward the viewer. The escarpment in the background (west) is the Gualanday Group. *C.* Rhizoliths and caliche nodules on a paleosol (inceptisol) from the fossiliferous Monkey Beds (Duke Locality 22 screenwash site, where many small mammals and bats have been recovered) at the base of the Villavieja Formation. The escarpment in the background (west) is the Gualanday Group. *D.* Fining-upward sequences of point bars from the Baraya Member north of Duke Locality 32. The channel sandstone at the top is 2 m thick. *E.* Lower part of the Polonia Red Beds close to Duke Locality 140. Most of the deposit is flood-plain mudstone; channel sandstones constitute a minor part. *F.* Cross-bedding toward the west in a multistory channel sandstone body above Duke Locality 131, in the upper part of the Polonia Red Beds.

concept of a volcaniclastic unit follows the original concept of the "Honda" of Hettner (1892). The Honda Group is unconformable or paraconformable over different portions of the Gualanday Group or other older units. The names "La Cira Formation," "La Dorada Formation," "Puerto Salgar Member," and "Perico Member" should be abandoned.

The rocks of the upper Magdalena valley (Gigante, Campoalegre, Dina oil field near La Venta, Chaparral, Carmen de Apicalá, and Girardot) previously included in the "La Cira Formation" or in the "Puerto Salgar Member" of the "La Dorada Formation" should be referred to the Tuné Formation (figs. 2.1, 2.4), which conformably overlies the uppermost conglomerate of the Gualanday Group and is unconformably overlain by the La Victoria Formation of the Honda Group. The Tuné Formation (Stirton 1953b) is a predominantly red bed sequence that contains vertebrate and invertebrate remains of late Oligocene to early Miocene age; it has been mapped as T5 by INGEOMINAS (1957b) in the Chaparral area and 20 km west of the La Venta area. The type locality of the Tuné Formation is located in the Qda. Tuné, 3 km northeast of Chaparral.

Distribution and Thickness

The La Victoria Formation is exposed in several locations (Gigante, Campoalegre, La Venta, Coyaima, Chaparral, Carmen de Apicalá, and Girardot) along the upper Magdalena valley (fig. 2.1) between the Central and Eastern Cordilleras of the Andes. Its northern boundary is the Honda area, middle Magdalena valley, where the stratigraphic nomenclature of the Honda Group proposed by Porta (1965, 1966) is used. Some of the formally proposed component units (Cerro Gordo, Chunchullo, and Tatacoa sandstone beds) of the La Victoria Formation can be recognized only in the La Venta area, where they are useful for biostratigraphic purposes. The Cerbatana Conglomerate Beds in the uppermost part of the formation can be recognized in the upper Magdalena valley in all the localities where the Honda Group outcrops.

Because the Honda Group begins with an angular unconformity and progressively covers an irregular topography, the thickness of the La Victoria Formation varies from 462 to 570 m in its type area (fig. 2.1). To the south, in the Gigante area, its thickness is approximately 1,000 m (after taking out the lower 270 m of the "Puerto Salgar Member" of Wellman [1970] or "Sequence A" of Wiel [1991] that should be referred to the Tuné Formation). In the Honda area, it can be inferred from the sections presented by Porta (1966) that the thickness of the deposits equivalent to the La Victoria Formation (the lower part of the San Antonio Formation) is approximately 960 m.

Lithology

The lithology and internal geometry of the La Victoria Formation are described next. The most important characteristic is the appearance of gray, "salt and pepper," coarse- to fine-grained pebbly lithic arenites with important amounts of volcanic fragments and plagioclase alternating vertically and laterally with reddish-brown and greenish-gray mudstone. The mudstones are very rich in fossil vertebrates. The sandstones have erosive bottoms and include pebble clasts and pockets of conglomerate. The only clast-supported coarse-pebble conglomerate with lateral extent is a thin interval of 9 m that is included in the uppermost part of the formation. Rounded, calcite-cemented concretions produced during incipient diagenesis are present.

UNIT BELOW THE CERRO GORDO SANDSTONE BEDS   The unit below the Cerro Gordo Sandstone Beds is the lowermost unit of the La Victoria Formation and of the Honda Group, outcropping on top of the Jurassic andesites of the Saldaña Formation on the Cerro Chacarón and Cerro Gordo hills. It is exposed on the north side of Cerro Gordo in the axis of an anticline that dips to the south-southwest so that progressively older beds are in contact with the Jurassic rocks to the north-northeast. The unit is composed mainly of reddish-brown mudstone with interbedded "salt and pepper" volcanic litharenites that occasionally are conglomeratic with fine- to medium-sized pebbles of mixed intraformational and extraformational nature. The best outcrops of this unit are south of Cerro Chacarón where the measured deposits are approximately 100 m thick. Although the unit is composed predominantly of reddish-brown mudstone, multistory sandstone bodies as thick as 16 m also occur.

The sandstones have erosive bottoms, fine upward, and

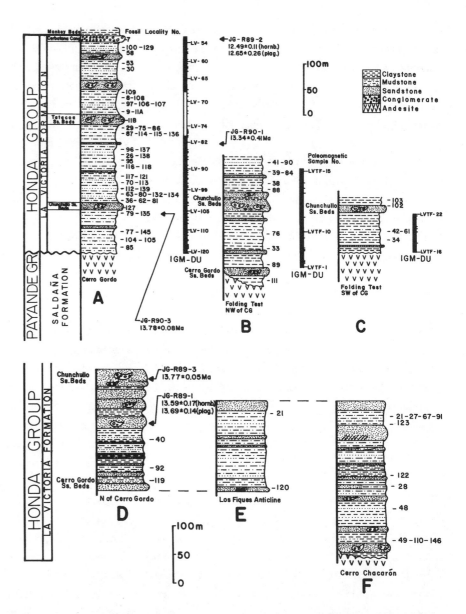

Figure 2.8. Stratigraphic sections of the Honda Group in the La Venta area. *JG-R numbers are dated rock samples. Fossil locality numbers refer to Duke localities. LV numbers are paleomagnetic sampling horizons. Large boldface capital letters refer to locations on the map* (fig. 2.9). *This page: Top (secs. A–C). Stratigraphic sections of the La Victoria Formation south and west of Cerro Gordo. Bottom (secs. D–F). Stratigraphic sections of the lower part of the La Victoria Formation north of Cerro Gordo. Continued on facing page.*

include trough and planar cross-bedding. Some display evidence of lateral deposition including epsilon stratification (epsilon-cross-stratification of Allen 1963, 1970) perpendicular to the direction of flow indicated by the cross-bedding.

CERRO GORDO SANDSTONE BEDS   The new name, Cerro Gordo Sandstone Beds, is formally proposed to designate a

15-m-thick sandstone that forms the flat broad surface on which the Cerro Gordo house is built on the northwest side of the hill named Cerro Gordo. The unit is composed of gray volcanic litharenite horizons interlayered with minor conglomeratic sandstone lenses. Multiple scouring surfaces and separate individual sandstone bodies are common.

The unit constitutes a useful stratigraphic marker bed that can be followed in the field and on aerial photos on the

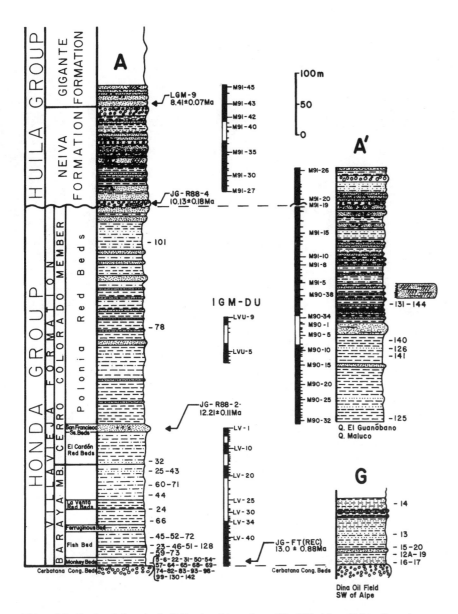

Figure 2.8. *Continued. Secs. A, A', G.* Stratigraphic sections of the Villavieja and Neiva formations.

northwest side of the asymmetric Cerro Gordo–San Alfonso anticline. The fold is gently dipping on the southeast side and steeply dipping and accompanied by a reverse fault on the northwest side (fig. 2.9).

The Cerro Gordo Sandstone Beds are the product of meandering streams that kept their positions within a meander belt for some time, as is the case for the Chunchullo and Tatacoa sandstone beds. The resulting multistory sandstone belt is elongated in the direction of the main drainage pattern to the east-southeast and changes

into flood-plain mudstone both on the north and south sides of the belt. The width of the meander belt is about 4 km, and it has scalloped borders that represent the transition from meandering channels to overbank mudstone.

UNIT BETWEEN THE CERRO GORDO AND CHUNCHULLO SANDSTONE BEDS This unit presents the same general characteristics of the unit below the Cerro Gordo Sandstone Beds and differs from that unit only in its stratigraphic position. The thickness varies from 83.3 m in section A to

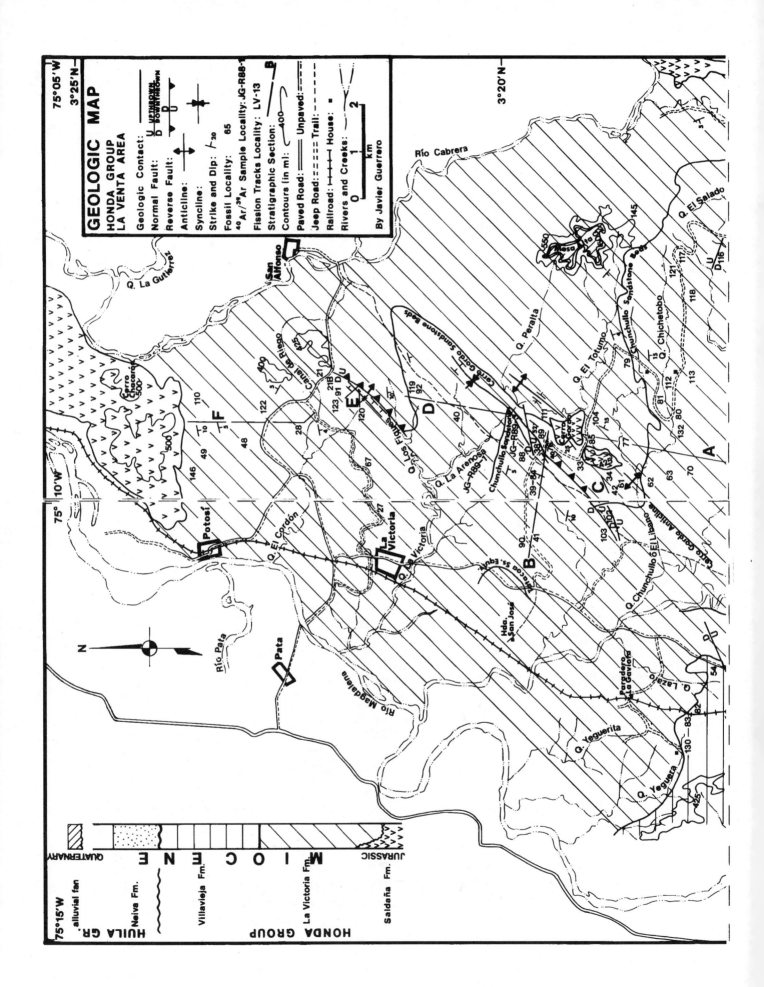

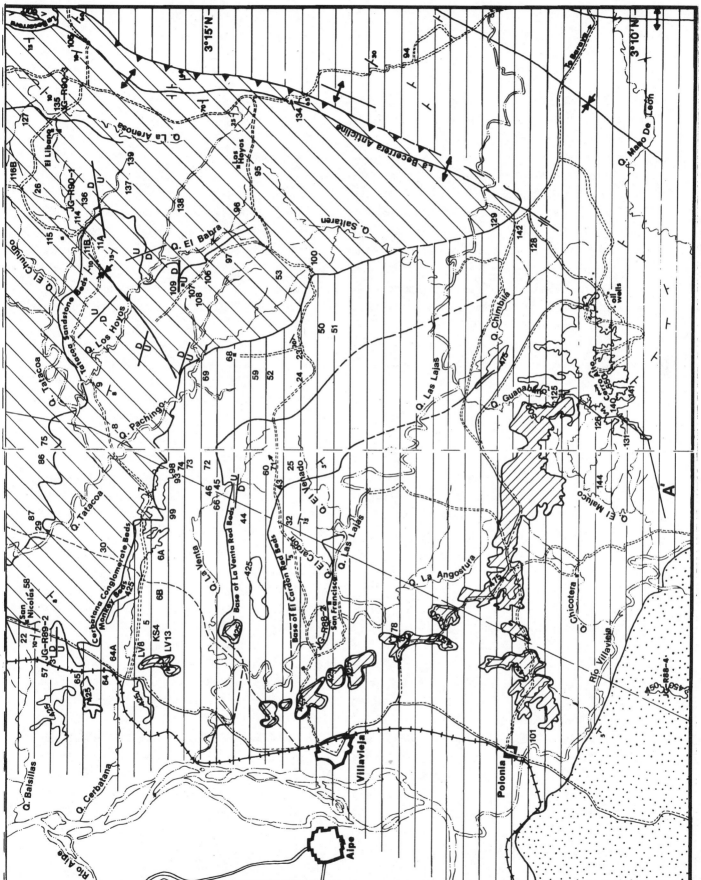

Figure 2.9. Geologic map of the Honda Group in the La Venta area, divided into four parts that join along the dashed lines. Abbreviations: *Fm.*, formation; *Hda.*, hacienda; *Q.*, quebrada; others are defined in the key.

157 m in section D. Although the deposits are composed predominantly of mudstone, the interlayered sandstones are much thicker than those higher in the unit between the Tatacoa and Chunchullo sandstone beds.

During deposition of this unit the sediments lapped onto and progressively covered the Jurassic basement now outcropping at Cerro Gordo; the unit is thicker toward the north end of Cerro Gordo. The deposits that overlap this basement contain pebble- to cobble-sized debris derived locally from the purple porphyritic andesite. Most of the grains, however, reflect a source area farther to the west in the Central Cordillera; clasts are mostly metamorphic quartzite, igneous rocks, and chert; minor white pumice and porphyry fragments of intermediate composition, from middle Miocene volcanoes, are also present.

Along section A (fig. 2.8) most sandstones of this unit are ± 1.5 m thick; there is, however, one 6.0-m-thick sandstone with conspicuous cross-bedding and internal erosive surfaces. Shown in section B are two sandstones, 6.7 m and 6.0 m thick, respectively. Along section C the sandstones are also 1.5 m thick. In section D the sandstone bodies are as thick as 21.3 m; two stacked sandstone bodies, 16.8 m and 16.3 m thick, respectively, combine to make a body 33.1 m thick. In section E is a 6.1-m-thick sandstone body, and two sandstones, each 4.0 m, make a body 8.0 m thick. In section F the thickest sandstone body is 26.8 m thick. The amount of sandstone varies from 5% (sec. C) to 37% (sec. D).

CHUNCHULLO SANDSTONE BEDS A new name, Chunchullo Sandstone Beds, is formally proposed for medium to coarse sandstone beds with horizons of conglomeratic sandstone and coarse pebble conglomerate. The unit is named where it is best exposed, at Qda. Chunchullo, south of Cerro Gordo. Individual sandstone bodies are separated by multiple scouring surfaces.

The Chunchullo Sandstone Beds are 6.0 m thick in section A (fig. 2.8). Along section B the unit measures 31.5 m thick. Toward the top, their color is gray with lenses of medium to coarse pebble conglomerate. The bottom third is yellowish conglomeratic sandstone. Thin lenses of pale brown and gray claystone are present. This "shoestring" sandstone trends to the east-southeast similarly to the Tatacoa sandstone.

Along section D (fig. 2.8) the unit is coarser than in any other section. The 30.0-m-thick Chunchullo Sandstone

Beds can be divided here into an upper 20.0 m of gray medium to coarse sandstone, conglomeratic sandstone, and coarse pebble conglomerate. At the bottom are 10.0 m of yellowish-gray and pale-brown conglomeratic sandstone with lenses of coarse pebble conglomerate up to 0.8 m thick.

Sandstones are characterized by conspicuous trough and planar cross-bedding. Epsilon stratification also can be observed in many sandstones. The conglomeratic portions are commonly on the bottom of scoured surfaces and on the cross-bedding that follows. Large tree trunks from forested stream borders are abundant; very well-preserved logs up to 0.5 m in diameter and as long as 2.0 m can be observed.

UNIT BETWEEN THE CHUNCHULLO AND TATACOA SANDSTONE BEDS The sediments of this unit are mostly fine grained and composed of very fine sandstone and mudstone alternating with minor 1- to 2-m medium-grained sandstones. Along section A the unit measures 149.3 m thick. Only 19.6 m (about 13%) of the total thickness of the unit is medium-grained sandstone. The unit is well exposed in the area north of the Qda. Tatacoa (fig. 2.9). The easiest access is via the jeep track from Qda. Cerbatana to El Libano School. The unit can also be reached in the upper course of Qda. Los Hoyos on the road from Villavieja to the Los Hoyos house. A good portion of the unit is exposed along an eastward deviation from the San Alfonso Road east of Hacienda San José.

The mudstone to sandstone ratio varies from 4:1 toward the top of the unit to more than 10:1 toward the middle and lower parts. Reddish-brown mudstone and interspersed gray, very fine- to fine-grained sandstone are the dominant lithologies. Fine- to medium-grained sandstone lenses less than 1 m thick are massive or display only trough and planar cross-bedding. Epsilon stratification is not observed.

A typical fining-upward sequence within the unit begins with gray, lithic-rich medium-grained sandstone 1.5 m thick with epsilon stratification, followed by 2.5 m of gray, muddy, very fine-grained sandstone and brownish-gray claystone. Then follows 1.5 m of reddish-brown claystone and, finally, on top of the sequence occurs a 1.5 m brownish-gray claystone.

Individual sandstones have very little lateral extent and change rapidly to mudstone; there are no marker beds with enough lateral extension to establish correlations. The base of the sandstones contains some granule to medium-sized mudstone intraclasts. Sandstones are volcanic litharenites.

TATACOA SANDSTONE BEDS The new name, Tatacoa Sandstone Beds, is formally proposed to designate a 19.0-m-thick unit composed of multistory medium- to coarse-grained sandstone and conglomeratic sandstone beds. Individual sandstone bodies are separated by erosive surfaces. Grains in the conglomerate and the conglomeratic sandstone reach medium pebble size; very minor pockets of conglomerate may contain coarse pebbles. Grains are mostly of metaquartzite, chert, and white porphyritic dacite and andesite.

The unit is named after the Qda. Tatacoa where the best exposures are found. It is easily followed on aerial photos and constitutes an excellent marker bed within the La Venta area.

Where the sediments can be followed to the north or south, the body thins laterally. It exhibits scalloped borders that represent the lateral limit of migration of meandering streams in a belt composed of many channels that kept its position for some time. This situation is best observed at Localities 11 A and B along the road to El Libano School on the axis of a gentle syncline dipping to the southwest. The thinning of the sandstone can also be observed toward the south (east of Duke Locality 109) and along Qda. El Babrá.

From the elongated shape of the unit and cross-bedding directions, the trend of the streams that formed the Tatacoa Sandstone was toward the east-southeast. The width of the meander-belt sandstone perpendicular to paleoflow direction is 1.5–2.0 km. Mudstones of overbank deposits are in lateral contact with the scalloped sandstones of the border of the channel belt. Duke Localities 114 and 115 are in an area of fine-grained sedimentation adjoining the area of sedimentation of the Tatacoa Sandstone Beds.

UNIT BETWEEN TATACOA SANDSTONE AND CERBATANA CONGLOMERATE BEDS In section A, this is a unit composed of 98.6 m of fine-grained deposits (mudstone and very fine- to fine-grained sandstone) and 40.1 m of coarse-grained deposits (coarse- to medium-grained sandstones) for a total of 138.7 m.

The easiest access to this unit is along the jeep road from Villavieja–Qda. Cerbatana to El Libano School (Duke Lo-

calities 8, 9, and 11). Excellent exposures can also be seen along the jeep road from Villavieja to Los Hoyos, especially on the detour around the heads of Qda. Pachingo (Duke Localities 100, 53, 97, 106, 107, 108, and 109). The foot-trail from Villavieja to San Alfonso also has good exposures (Duke Locality 30).

The sandstones at the base of the simple fining-upward sequences are up to 3 m thick. Locally, the sandstones are deposited on top of one another and have eroded the mudstones of previous flood-plains, producing multistory sandstone bodies up to 15 m thick. Maximum grain size (medium pebbles 16–32 mm) is attained in scarce pockets of conglomeratic sandstone in these composite sandstone bodies.

In a typical fining-upward sequence, 2.8 m of conglomeratic gray sandstone to gray medium-grained sandstone with epsilon stratification is followed by 5.5 m of horizontal layers of brownish-gray, fine-grained sandstone and mudstone. This sequence is followed by 2.0 m of gray fine-grained sandstone and bioturbated brownish-gray mudstone that are, in turn, covered by 3.5 m of gray sandy mudstone and fine-grained sandstone. Finally, 1.0 m of medium-grained limonite-cemented sandstone constitutes the base of the next fining-upward sequence.

Mudstones are less colored than those of the Villavieja Formation. Of the 98.6 m of flood-plain deposits, less than 25.0 m are reddish-brown mudstone. The remaining portion is mostly greenish-gray mudstone.

CERBATANA CONGLOMERATE BEDS The new name, Cerbatana Conglomerate Beds, is proposed for a 9-m-thick unit (sec. A) composed of coarse-pebble (3.2–6.4 cm), clast-supported conglomerate, best exposed at the headwaters of Qda. Cerbatana. Characteristic multiple scouring surfaces are the result of high-energy channel migration. Clast size increases to small cobble (6.4–12.8 cm) in minor sedimentary pockets. Conglomeratic sandstone is also present. Trough cross-bedding and planar bedding alternate.

The pebble fragments are mostly metaquartzite, igneous rocks, and chert. Weakly consolidated (middle Miocene) white pumice fragments and porphyries of andesitic and dacitic composition with biotite and hornblende are present in minor amounts.

This is the only extensive sheet of clast-supported pebble

conglomerate within the Honda Group that can be correlated along the Magdalena valley. The Cerbatana Conglomerate Beds represent only the topmost conglomerate of the unit that Fields (1959) named "Cerbatana gravels and clays." This distinction is made because the conglomeratic sandstones with interbedded mudstones that Fields included differ little from the underlying units and, thus, the boundary between the "Cerbatana gravels and clays" and "El Libano sands and clays" cannot be established throughout the La Venta area. Another problem is that in his section Fields (1959, 418) duplicated part of the deposits in the area west of the San Nicolás house where the Monkey Beds ("San Nicolás clays") are displaced below the conglomerate beds on the north side of a normal fault. This error added 68 m to the total thickness of the "Cerbatana gravels and clays."

The name Cerbatana Conglomerate Beds is now restricted to the uppermost clast-supported conglomerate of coarse pebbles that caps the La Victoria Formation.

VILLAVIEJA FORMATION

The type locality of the Villavieja Formation (Wellman 1970) is the La Venta area, where the unit is composed mainly of red and gray mudstone with minor layers of medium-grained volcanic litharenites. The lower part of the Villavieja Formation is very fossiliferous and is composed mainly of gray mudstone and sandstone with minor layers of red mudstone. The upper part is less fossiliferous and is composed of thick horizons of red mudstone with a very small amount of volcanic litharenite and chert litharenite. The fine-grained deposits of the Villavieja Formation contrast sharply with the coarser-grained deposits of the La Victoria Formation.

Wellman (1970, fig. 5) recognized two alternations of the red bed Cerro Colorado Member and the volcanic Baraya Member in the La Venta area. According to Wellman, the succession is from bottom to top, Baraya Member, Cerro Colorado Member, Baraya Member, and Cerro Colorado Member. Unfortunately, using two names to refer to four units introduces confusion because the members are not bounded by a single lower and a single upper surface and the name then recurs in a normal stratigraphic succession. It is very confusing to include a thin and arbitrary interval of red beds in the (lower?) Cerro Colorado Member and to apply the same name to the uppermost thick horizons of red beds.

Red horizons of mudstone are a conspicuous feature of the entire Villavieja Formation, but only in the upper part are they of considerable thickness. In the lower part of the formation these red beds are only 1–3 m thick, and the maximum thickness attained is 12.5 m. To avoid further stratigraphic confusion it is formally proposed to restrict the Baraya Member to the lower 164 m of the Villavieja Formation composed predominantly of gray mudstone and sandstone with minor layers of red beds. It is also formally proposed to restrict the Cerro Colorado Member to the upper 414 m of the Villavieja Formation composed predominantly of red beds.

The Baraya Member in the La Venta area is composed of all the layers between the Cerbatana Conglomerate Beds and the El Cardón Red Beds (new name), including the Monkey Beds, the Fish Bed, the Ferruginous Beds, the La Venta Red Beds (new name), and the four unnamed units between the Monkey Beds and El Cardón Red Beds. The Cerro Colorado Member is restricted in the La Venta area to the El Cardón Red Beds, the San Francisco Sandstone Beds (new name), and the Polonia Red Beds (new name).

The Villavieja Formation conformably overlies the Cerbatana Conglomerate Beds of the La Victoria Formation. The uppermost red beds of the Villavieja Formation are overlain disconformably by the first clast-supported coarse-pebble to fine-cobble polygenetic conglomerate of the Neiva Formation. Above this basal conglomerate, no red beds appear. The red beds that Howe (1974) included in the lower part of the Neiva Formation in the La Venta area in the Manueleón (Qda. Mano de León) section are part of the Villavieja Formation.

The type locality and sections of the Villavieja Formation, its members, and component beds are in the La Venta area (secs. A, A′ in fig. 2.8). The type locality includes the sections designated by Wellman (1970).

Distribution and Thickness

The Villavieja Formation is exposed in several localities (Gigante, Campoalegre, La Venta, and elsewhere) along the upper Magdalena valley (fig. 2.1). For the Honda area in the middle Magdalena valley the stratigraphic nomenclature proposed by Porta (1966) has priority over the one proposed by Wellman (1970). The named beds of the Villavieja Formation can be recognized only in the La Venta area where they are useful for biostratigraphic purposes. Outside this area, only the component members of the formation can be identified.

The thickness of the Villavieja Formation in the La Venta area is 578 m. South of the La Venta area near Gigante, the thickness of the formation, to the top of the last-occurring red mudstone, is 390 m as reported by Wellman (1970). In the Honda area the deposits equivalent to the Villavieja Formation (the uppermost 254 m of the San Antonio Formation and the whole Los Limones Formation of Porta 1966) are truncated by the Cambrás Thrust and measure approximately 450 m according to the sections of Porta (1966) and Wellman (1970).

Lithology

The Villavieja Formation consists mainly of red mudstone alternating vertically and laterally with medium- to fine-grained, reddish-yellow sandstone; fining-upward sequences are a common feature. In minor proportion there are gray-greenish mudstones interlayered with gray, "salt and pepper," pebbly, medium- to fine-grained volcanic litharenites in fining-upward sequences. Epsilon stratification and cross-bedding can be observed in the sandstones. Fossil vertebrates are very abundant, especially in the gray-green facies of the lower part of the formation. Calcite-cemented concretions similar to those of the La Victoria Formation are also present. The component members and beds of the Villavieja Formation are described next.

Baraya Member

MONKEY BEDS A change in rank of the "Monkey unit" of Fields (1959) is introduced formally. The name Monkey Beds is proposed to designate 14.8 m of extremely fossiliferous greenish-gray mudstone and sandstone at the base of the Baraya Member. The base of the Monkey Beds forms a sharp but conformable contact between the Villavieja and La Victoria formations.

Although the Monkey Beds are composed predominantly of mudstone, the interlayered sandstones are slightly thicker than those in the overlying units of the Baraya Member. A typical fining-upward sequence of the Monkey Beds (figs. 2.6B, 2.7D) is a 1-m-thick medium-grained sandstone followed by 4.5 m of reddish-brown sandy mudstone, 1.2 m of pale-green mudstone, and, on top, 0.8 m of reddish-brown mudstone.

The richly fossiliferous sediments that outcrop west of the San Nicolás house are part of the Monkey Beds, not a lens of the underlying unit as had been stated by Fields (1959). The Monkey Beds outcrop to the west of the San Nicolás house because a normal fault repeats the sequence (Luchterhand et al. 1986, 1757). The name "San Nicolás Clays" should be avoided.

The paleomagnetic section presented by Hayashida (1984) is fault repeated and miscorrelated. It has been restricted to a short interval (Guerrero 1990) with only one normal and one reverse polarity horizon between the Monkey Beds and the Ferruginous Beds and cannot be used to establish the age of the Honda Group. The fission-track dates (Localities LV8, KS4, LV13) reported by Takemura and Danhara (1986) come from this same interval and have been recalculated as 13.00 ± 0.88 Ma by Guerrero (1990).

UNIT ABOVE THE MONKEY BEDS This unit is composed of 18.1 m of greenish-gray and reddish-brown mudstone with minor thin layers of medium-grained sandstone of which the thickest measures 1.5 m.

In section A (fig. 2.8) the unit is composed of a sequence of alternating thin sandstones and siltstones very similar to the ones of the unit above the Fish Bed and can be differentiated only by its stratigraphic position. One of the sandstones toward the base of the unit is characterized by its high content of biotite and hornblende. Above this 1.2 m sandstone are 5 m of greenish-gray sandy mudstone.

FISH BED This green claystone and mudstone is characterized by enormous quantities of disarticulated fish remains (spines, vertebrae, and teeth). The unit is between 1.2 m (sec. A) and 6 m (Fields 1959) thick. The best exposures are located north of Qda. La Venta on the Villavieja–San Alfonso foot-trail and on the jeep track to Los Hoyos near the headwaters of Qda. Pachingo.

The Fish Bed appears to be a lake or swamp deposit in

which the fragmentary nature of the fossils and abundance of coprolites suggest that many of the fragments are predator residues. Besides the fish remains, isolated mammal teeth and snake vertebrae are abundant. The lateral continuity of the bed (at least in an east-west direction) is also indicative of a depositional environment such as a lake or swamp in an area at least 10 km in diameter.

UNIT ABOVE THE FISH BED   This is a 22.5-m-thick predominantly mudstone unit identified by its position between the Fish Bed and the Ferruginous Beds. The mudstones are greenish-gray and reddish-brown. Toward the top the claystone is very rich in biotite. The mudstone component is thicker than the sandstone component by about 8:1; the thickest sandstone is 1.5 m. A typical sequence has about 1 m of gray medium-grained sandstone, then 5.0 m of alternating gray mudstone and fine sandstone, and, on top, 3.5 m of reddish-brown mudstone.

FERRUGINOUS BEDS   The Ferruginous Sandstone of Fields (1959) is here designated formally as the Ferruginous Beds. These pale-brown, medium-grained sandstone beds are cemented with calcite and limonite and form a 10.3-m-thick prominent ledge, which laterally changes to red and reddish-brown mudstone. Fields (1959, 421) stated that "to the east the color changes to deep rust-red and the beds are composed of fine- to medium-grained sand" in contrast to the more coarse and conglomeratic sandstone to the west. In section A the sandstone is medium-grained and rich in volcanic fragments, hornblende, and biotite. Trough and planar cross-bedding were also observed.

UNIT BETWEEN THE FERRUGINOUS BEDS AND THE LA VENTA RED BEDS   This is a 26.8-m-thick greenish-gray mudstone unit that crops out between two darker red units, the La Venta Red Beds and the Ferruginous Beds. There are very few sandstones, the thickest one being about 1 m thick; in some of these, hornblende crystals, biotite, and plagioclase are abundant. Minor horizons of reddish-brown and red mudstone are also present.

LA VENTA RED BEDS   The new name, La Venta Red Beds, is formally proposed to designate the red beds that Fields (1959) included in the "Lower Red Beds," excluding the

gray units that he included in the bottom and top of the unit. This 12.5-m-thick unit of red beds consists mainly of fine-grained deposits. On the line of section A the unit is entirely mudstone, but when followed laterally, it also includes minor sandstones. Iron oxide nodules 0.5–3.0 cm in diameter are commonly present in the mudstones.

The best exposures of the La Venta Red Beds are on the north side of the head of Qda. La Venta. The unit is found also along the jeep road from Villavieja to Los Hoyos and on the south drainage of Qda. La Venta.

UNIT BETWEEN THE LA VENTA RED BEDS AND THE EL CARDÓN RED BEDS   The unit is exposed in the south drainage and heads of the Qda. La Venta where it is 57.7 m thick. It includes the unit that Fields (1959) named "Unit between Upper and Lower Red Bed" and also the gray units that he included in the base of the "Upper Red Bed" and on the top of his "Lower Red Bed." The sandstones are 1–3 m thick and alternate with greenish-gray and very minor reddish-brown mudstones. The alternation of fining-upward sequences is well exposed. The sandstones are usually rich in hornblende, biotite, and pumice fragments.

The unit displays good examples of the lateral and vertical variation characteristic of small meandering streams. The bases of the channel sandstones contain mud intraclasts. Cross-bedding is difficult to detect, but epsilon stratification is beautifully exposed in the sandstone bodies. Mudstone colors are between reddish-brown and brown. The sandstones can be traced laterally up-dip of the epsilon stratification (in the line of lateral movement of the stream) within about 200 m, where the change to mudstone is gradual in the inner side of the meanders.

Cerro Colorado Member

EL CARDÓN RED BEDS   The new name, El Cardón Red Beds, is proposed to designate part of the unit that Fields (1959) called "Upper Red Bed" and in which he included gray units in its base and top. I do not include those gray beds in the El Cardón Red Beds. The unit is named El Cardón after a highly variable stream course where the unit outcrops on the road from Villavieja to Los Hoyos. It is composed almost entirely of fine-grained deposits, including red claystone and mudstone with very minor sandstone.

The unit is 52.2 m thick on section A. The name "Upper Red Bed" is abandoned because it introduces confusion regarding the stratigraphic position of the overlying Polonia Red Beds. The sediments of the El Cardón Red Beds are more mature than the ones of the underlying units and contain higher amounts of rounded quartz. The base of the El Cardón Red Beds is the base of the Cerro Colorado Member, where red and reddish-brown mudstone dominate the section.

SAN FRANCISCO SANDSTONE BEDS The new name, San Francisco Sandstone Beds, designates a gray lithic sandstone with some lenses of conglomeratic sandstone and minor siltstone. Its total thickness is 12 m on section A. The unit outcrops at Hacienda San Francisco north of Qda. Las Lajas between the Polonia Red Beds and El Cardón Red Beds. Its gray color and coarse-grained nature contrasts with the red claystone units on top and below. The pebble fraction contains fresh pumice fragments of dacitic and andesitic composition. Because it is a "shoestring" sandstone deposited in a belt of meandering channels with an easterly trend, this unit does not outcrop in section A' (fig. 2.8). In the Qda. Guanábano section the El Cardón and Polonia Red Beds constitute a continuous section of red beds.

POLONIA RED BEDS The new name, Polonia Red Beds, is used to designate a unit composed predominantly of red claystone and siltstone. A few sandstone horizons 4–6 m thick (exceptionally, 25 m thick), composed of several stacked sandstone bodies, are present toward the top of the Polonia Red Beds. They represent a minor amount of the total thickness of the unit, which is 350 m (fig. 2.6C).

The Polonia Red Beds are the uppermost part of the Cerro Colorado Member and of the Honda Group. It is the upper part of the unit that Fields (1959) called "Las Mesitas." These red beds are exposed east of Polonia and Villavieja and can be observed in the jeep roads connecting those towns to Baraya. There are also excellent and continuous exposures south of this road in Qdas. Guanábano and Maluco, west of the Andalucía oil field. These exposures are located in the area that Wellman (1970, fig. 4) referred to as Cerro Colorado, a name that unfortunately cannot be found on any published topographic map. Instead, the hill is named Cerro Alto on maps of the Instituto

Geográfico "Agustín Codazzi." In these Guanábano-Maluco areas the unit can be followed to its upper contact with the Neiva Formation.

The Polonia Red Beds rest conformably on the San Francisco Sandstone Beds; the lower section is composed mostly of red claystone that forms spectacular badlands (fig. 2.7E, F) in contrast with more abundant sandstone bodies toward the top. The uppermost red beds of the unit are overlain disconformably by the Neiva Formation, a conglomeratic unit 160 m thick in which gray to white rock fragments of andesitic and dacitic natures are abundant.

A noteworthy change in sedimentation pattern and current direction of the river system that produced the deposits occurs in the upper part of the Polonia Red Beds. Above Duke Localities 126 and 131 (sec. A' in fig. 2.8) are two multistory sandstone bodies 20 and 25 m thick that show very restricted lateral extent between 150 and 350 m wide. These sandstone bodies are elongated approximately 1 km to the west and display cross-bedding sets, indicating that the direction of the water flow was toward the west with the source area located to the east.

This change in flow direction provides the earliest clear evidence that the Eastern Cordillera was already uplifted and had enough elevation to be the source area for the upper part of the Cerro Colorado Member. In contrast, all the lower units within the Villavieja Formation, and the underlying La Victoria Formation, appear to have a sediment source to the west in the Central Cordillera (fig. 2.10).

Sedimentary Environments

Most of the Honda Group is dominated by fining-upward sequences that include channel sandstone and overbank mudstone deposited in a meandering river setting (fig. 2.6). The only portions of the section that have a different sedimentary pattern are the uppermost parts of the La Victoria and Villavieja formations. The top of the La Victoria Formation includes a basinwide sheet of clast-supported pebble conglomerate deposited by braided streams. The top of the Villavieja Formation, in addition to a few fining-upward sequences, includes very narrow and elongate multistory sandstone bodies encased in red mudstone. These show little variation in current direction and have

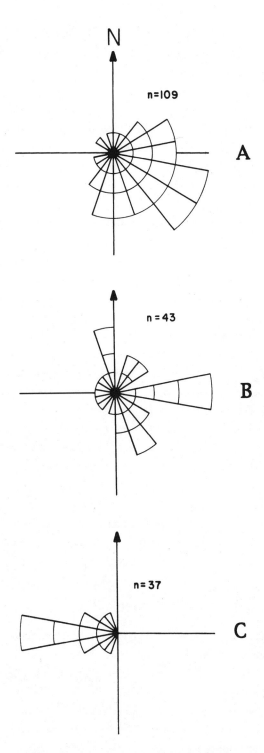

Figure 2.10. Paleocurrent directions in the Honda Group of the La Venta area. *A*. Paleocurrent directions from cross-bedding sets in the La Victoria Formation (with the exception of the Cerbatana Conglomerate Beds). *B*. Paleocurrent directions from cross-bedding sets in the Baraya Member and the lower part of the Cerro Colorado Member.
C. Paleocurrent directions from cross-bedding sets in the upper part of the Cerro Colorado Member.

width : thickness ratios of between 6 and 14 perpendicular to flow direction; they were apparently deposited by anastomosing streams.

In the lower part of the La Victoria Formation coarse channel sandstone and conglomerate predominate where meander belts kept their position for some time, producing multistory sandstone bodies as thick as 31.5 m. The continuous lateral migration of large channels erodes previous deposits and produces "shoestring" sandstone bodies extending tens of kilometers in the direction of stream flow and encased in mudstone. Perpendicular to paleoflow, the sandstone bodies are 2–4 km wide and have a width : thickness ratio ranging from 63 to 125. These sandstone bodies are scalloped (fig. 2.7A, B) on the sides of the meander belt where overbank deposits are preserved. In exposures oriented perpendicularly to paleoflow, epsilon stratification (inclined and inclined heterolitic stratification of Thomas et al. 1987) is present. Similar multistory channel sandstones deposited in point bars were described by Kraus (1980, 1985) from the Eocene of Wyoming.

The most common type of succession through the section is composed of medium-grained sandstones (2–6 m thick) overlain by siltstone and claystone arranged in a simple fining-upward trend. The fine member is usually as thick as the coarse member in these sandstone-mudstone couplets. The bases of individual sandbodies are coarser than their tops. A conglomeratic lag composed of mudstone intraclasts and minor reworked calcite nodules, similar to those described by Edwards et al. (1983) from the Permian of Texas, is usually present at the basal erosive surfaces of these sandstones. Mixed conglomeratic lags containing extraclasts and intraclasts are also common. Sandstones are tabular (width : thickness ratios from 100 to 130) and present limited lateral extent of usually 200–800 m oblique to paleoflow. The upper boundary of each set is transitional into siltstone and claystone. Epsilon stratification is present in many of these tabular sandstones and can be followed up dip into the mudstones where it becomes horizontal. Epsilon surfaces within the medium-grained sandstone bodies are indicated by 1- to 2-cm discontinuities of either mudstone or fine-grained sandstone.

Trough and planar cross-bedding of various scales is present in the sandstone; their size is related to the size of the sand body. The largest trough sets occur in the large chan-

Table 2.1. Measured and calculated dimensions (m) of the Honda Group meandering rivers

D	W	W_m	W_{mr}
2–3	20–37	188–351	150–350
6	107	1,020	—
8–10	167–235	1,588–2,240	2,000–4,000

Notes: D = depth, or thickness of epsilon units. W = width, $6.8D^{1.54}$. W_m = meander belt width, $64.6D^{1.54}$. W_{mr} = width of sandstone bodies in the field.

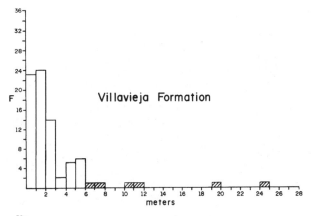

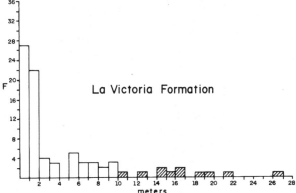

Figure 2.11. Frequency distribution of thicknesses of Honda sandstones. The hatched bars correspond to multistory sandstone bodies in meander belts. Values are from table 2.1.

nels of the lower portion of the La Victoria Formation. The slip faces of sand waves up to 4 m high can be distinguished by granule or pebble layers one clast thick. The sandstone fines upward to planar, cross-bedded, medium-grained sandstone that constitutes the bulk of the sand body. Then, fine sandstone and silt are followed by bioturbated reddish-brown mudstone with roots (rhizoliths) that have been replaced by calcite. Occasionally, there are very minor amounts of dark gray and black shales that probably represent swamp sediments or filled abandoned channels.

Fossil wood and vertebrate remains are common throughout the entire section. Mammal remains are extremely abundant and well preserved in the fine-grained deposits of overbank origin, especially in the La Victoria Formation and lower part of the Villavieja Formation. Transported tree fragments as large as 2 m long and 50 cm in diameter are preserved within the thickest sand bodies of the lower part of the section in small pockets of muddy sand or in the conglomeratic sandstone. Very large crocodiles, sirenians, fish, and complete turtle remains are fairly common, indicating the existence of permanent water bodies. Among the invertebrates, a few small fresh-water bivalve and gastropod casts and common crab claws are preserved.

CHANNEL SANDSTONE FACIES

Epsilon units, as those described elsewhere (Allen 1963, 1970; Stewart 1981; Flach and Mossop 1985; Smith 1987), are common in the fine- and medium-grained sandstones of the Honda Group. The epsilon surfaces have either sigmoidal or planar profiles and, commonly, inclinations of 5–15°. Foreset dip directions of trough and planar crossbeds are approximately perpendicular to the dip of the ep-

silon surfaces. The fine sediments on top are either horizontally bedded or bioturbated. The measured thickness of the tabular sandstone bodies with epsilon stratification is taken as the channel depth (ignoring compaction) of the meandering rivers that produced them (table 2.1; fig. 2.11). The horizontally bedded siltstone and claystone with interspersed very fine- and fine-grained sandstone represent overbank deposits.

The bankfull depth of the channels can be estimated confidently from the simple fining-upward sequences where it is clear that the sandstone body, encased in floodplain deposits, is produced by the lateral migration of a channel, leaving epsilon surfaces of point bars during the meandering process. Erosion does not affect this simple sandstone body, which fines upward to mudstone. When, some time later, another channel develops on top, erosion affects only the intervening flood-plain fines. These individual channels had maximum depths of 8–10 m during the deposition of the La Victoria Formation and 2–6 m during deposition of the Villavieja Formation.

The width of meandering rivers (assuming sinuosity >1.7) and the width of the meander belts can be estimated from channel depth using formulas derived from measurements of modern river systems. The width of the channel (W) is related to its depth (D) by the relationship $W = 6.8 \times D^{1.54}$ (Leeder 1973). The width of the meander belt (W_m) is related to channel depth by the relationship $W_m = 64.6 \times D^{1.54}$ (Collinson 1978). The values obtained for the Honda Group rivers are shown in table 2.1.

A channel 2.0–3.0 m deep is 20.0–37.0 m wide and has a meander belt width of 188.0–351.0 m. A channel 8.0–10.0 m deep is 167.0–235.0 m wide and has a meander belt width (W_m) of 1,588.0–2,240.0 m. This meander belt width represents the width of the active meander at any given time, not the width of the sandstone belt produced in the meander belt ridge through time. The meander belt ridge sandstones (W_{mr}) of the La Victoria Formation are 2.0–4.0 km wide in the La Venta area, about twice the calculated width (W_m) of the meander belt.

The Tatacoa Sandstone Beds consist of an elongate multistory-sandstone body 19.0 m thick, about 2 km wide and several kilometers long with an east-southeast trend. The Chunchullo Sandstone Beds are 31.5 m thick in the axis of the belt, which is about 3.5 km wide. The Cerro Gordo Sandstone is 15.0 m thick on one side of its belt, which is about 4.0 km wide. The large channels and meander belts of the La Victoria Formation are comparable to the ones described by Smith (1987) from the Permian of South Africa (with meander belt widths [W_m] of 3.0 km and calculated channel depths of 9.2–12.8 m and widths of 209.0–346.0 m) and by Gardner (1983) from the Carboniferous of Kentucky (meandering channels 9.5–10.5 m deep that produced meander belt ridge sandstones [W_{mr}] 3.3 km wide).

The large channels of the La Victoria Formation were previously identified by Wellman (1970) as deposits of high-gradient braided streams. When lateral changes are followed (perpendicular to flow direction), however, it is observed that most of the deposit is made up of flood-plain mudstone and that large channels constitute a minor fraction. The abundance of laterally equivalent mudstone, including only minor amounts of smaller (2–6 m) sandstones

in fining-upward sequences with epsilon stratification, indicates that a meandering environment is a better interpretation for the lower part of the La Victoria Formation. As indicated by Walker and Cant (1984), the sand-body geometry of meandering streams will be essentially an elongated "shoestring" bounded below and on both sides by flood-basin fines. The marked differences in paleoflow directions (fig. 2.10) within sandstone bodies of the La Victoria Formation in the La Venta area also indicate deposition by meandering rather than by braided streams. The interpretation of braided streams by Wellman (1970) is valid only for the Cerbatana Conglomerate Beds in the uppermost part of the La Victoria Formation.

After the episode of large meandering channels and associated flood-plains that deposited most of the La Victoria Formation, an episode of braided stream sedimentation occurred, represented by the Cerbatana Conglomerate Beds. Trough and planar cross-bedding alternates with horizontal bedding in this clast-supported conglomerate. Massive or crudely stratified conglomerate is also common. Only a few lenses of trough and planar cross-bedded and massive conglomeratic sandstone are present. Cross-bedding directions indicate a paleoflow to the east and southeast. The uppermost clast-supported pebble and, occasionally, fine cobble-sized conglomerate beds of the La Victoria Formation can be followed north and south in the Magdalena valley in a direction perpendicular to the trend of the rivers that deposited them. The episode of braided streams that produced this basinwide conglomerate sheet was most probably the result of increased uplift and volcanism of the source area. Subsequent to this event, the sedimentary regime changed drastically, and sediments became finer during deposition of the Villavieja Formation as a new barrier, the Eastern Cordillera, slowly began to uplift and closed the sedimentary basin on its eastern border.

The lower unit of the Villavieja Formation, the Baraya Member, was deposited by small (2–3 m deep) meandering channels that left classic point bar sequences. The Baraya Member channels were approximately of the same size as those reported by Edwards et al. (1983) from the Permian of Texas. Sandstones constitute a minor portion of the member. They have sheetlike geometry, are up to 3 m thick and can be followed laterally 200–800 m in a lithology dominated by gray mudstone. Sedimentary structures, geometry, and lateral relationships are the same as those present in the thinnest (4–6 m) simple fining-upward sequences of the La Victoria Formation. Current directions (fig. 2.10) have a wide range of variability, indicating the beginning of a major change in sedimentary pattern and flow direction.

The upper unit of the Villavieja Formation, the Cerro Colorado Member, is composed of large amounts of overbank red mudstone and minor amounts of channel sandstone that also have well-developed point bar sequences with epsilon stratification. Very narrow (150–350 m) and elongate (up to 1 km exposed in the flow direction) multistory sandstone bodies 20 and 25 m thick, however, encased in mudstone from the Polonia Red Beds in the upper part of the Cerro Colorado Member (above Duke Localities 126 and 131) suggest the presence of anastomosing rivers. These multistory sandstones are composed of several fining-upward sandstone bodies 2–6 m thick containing planar cross-bedding sets up to 1 m thick. The most interesting feature is that paleoflow directions have a trend to the west, opposite to the one recorded in the La Victoria Formation to the east-southeast (fig. 2.10). In addition, current directions were much less variable than those from the meandering channels of the La Victoria Formation and Baraya Member of the Villavieja Formation.

These multistory sandstone bodies are coarse- to fine-grained and have sharp bottoms with intraformational lags composed of pebble- to boulder-sized fragments of red mudstone. The upper boundary with the overlying red mudstone is commonly gradational but can be sharp. Epsilon stratification indicating lateral accretion is very rare but could be observed. The lateral boundaries (perpendicular to paleoflow) are sharp and change abruptly to red mudstone. The Polonia Red Beds overbank deposits (very fine sandstone, silt, and clay) predominate over channel deposits. Soil development is most conspicuous, indicating highly stable banks and longer periods of soil-forming processes. The deposits are notably different from those lower in the section, and in no other part of the Honda Group is red mudstone so abundant.

An anastomosed fluvial network was also interpreted from sandstone bodies that have approximately the same dimensions, sedimentary structures, and lateral relations by Smith and Putnam (1980), Putnam and Oliver (1980), and Putnam (1983) from the Cretaceous Mannville Group in Canada. They reported sandstone bodies thick (up to 35 m), narrow (300 m perpendicular to paleoflow), and elongated several kilometers in the paleoflow direction with a general "shoestring" shape. In the Polonia Red Beds, however, the interchannel, horizontally bedded red mudstone and very fine sandstone facies do not contain coal beds.

OVERBANK FACIES AND PALEOSOLS

In the overbank deposits that comprise most of the Honda Group section paleosols are very common and display a continuous color range from almost unaltered greenish-gray to deeply altered red and purple mudstones. Reddish-brown mudstone paleosols with calcite-replaced roots and bioturbation cap the fining-upward sequences throughout the entire section. Evidence for paleosols in the mudstones also include small (0.5–3 cm) dispersed calcite nodules, desiccation cracks, slickensides, drab-haloed root traces, small (0.5–1.5 cm) dispersed iron oxide nodules, and soil horizons.

Abundant wood and vertebrate remains are also present. Complete skeletons with no evidence of transport are common. Most of the remains appear to have been either broken and dispersed by predators or just buried in situ with little disturbance. Duke Localities 28 and 115 include several rodent skeletons buried in burrows. The preservation of vertebrate remains in the Honda Group was especially favored by the early precipitation of calcite on the bones.

Horizontally laminated sequences, 1–6 m thick, of gray, fine- to very fine-grained sandstone (1–20 cm and up to 1.5 m thick) with thinner interspersed gray siltstone are distributed throughout the section and are thought to represent levee and crevasse splay deposits. The sandstones that have channel-like or lenticular shapes and coarsen upward or display cross-bedding sets are interpreted as crevasse-channels following Bridge (1984). Although the sequences are commonly fining-upward, some of them are coarsening-upward or alternate from fining to coarsening in the centimeter to decimeter thick layers. These sequences are both, on top of the fining-upward sandstone sheets that display epsilon stratification and laterally adjacent to the large channel belts that deposited the Cerro Gordo, Chunchullo, and Tatacoa sandstone beds.

Dispersed calcite nodules (1–3 cm) and dispersed to abundant rhizoliths (fig. 2.7C) are present.

A large portion of the Honda Group is composed of sheetlike deposits (up to 50 m thick and several hundred meters in lateral extent) of variegated very fine-grained sandstone, siltstone, and claystone thought to have been deposited by vertical aggradation in flood-plains. Colors range from reddish-brown, orange, red, and purple. These mudstone deposits can be laminated, but they are more commonly bioturbated.

In the Villavieja Formation channels are smaller than those of the La Victoria Formation, and overbank deposits are far more abundant. The mudstones are increasingly deep red with nodules of hematite, and the thickness of red beds increases. Abundant kaolinite and smectite were reported by Wellman (1970). The presence of hematite and abundant kaolinite indicates that there were prolonged periods of alteration (Weaver 1989; Retallack 1990) of iron-bearing magnetite, ilmenite, hornblende, and biotite along with the alteration of feldspars and other aluminum silicates during soil genesis. This explains, in part, the fact that the sediments in the Cerro Colorado Member are comparatively enriched in quartz.

Johnsson and Meade (1990) illustrated the importance of time in chemical weathering of fluvial sediments in a point bar in the Upper Amazonas River in northwestern Brazil approximately 500 km east of Colombia. They showed that under the same climatic conditions (humid-tropical with 2,600 mm annual rainfall) the fraction of smectite increased (up to 68%) with sediment age within the point bar and active channel. The oldest studied sediments in a terrace contained more kaolinite (up to 40%) than any of the younger point bar sediments. The terrace also contained a large amount of a clay that was either Mg-rich vermiculite or Fe-rich aluminum-chlorite. They also reported abundant (up to 80% in the very fine sand fraction) sand-size ferricrete fragments thought to represent incipient laterites. Among the silts, increasing sediment age was marked by an enrichment in quartz relative to plagioclase. For the sand-size fraction a steady reduction in lithic fragments and enrichment in quartz with age was also reported.

For the Honda Group, Wellman (1970) found that red mudstone is always richer in kaolinite than the associated green mudstones; the latter have more abundant smectite plus mixed-layer illite-smectite and illite. The highest percentage (57%) of kaolinite found was in a dark red-brown mudstone of the Cerro Colorado Member that contrasts with 87% smectite and 2% kaolinite in a pale green mudstone of the Baraya Member. Wellman also noted that the mudstones are enriched in ilmenite or ilmenite-hematite if green, and in hematite or hematite-ilmenite if red. He related this to apparent weathering of magnetite and iron-bearing silicates that occurred both in the source area and in the site of deposition, possibly the result either of better drainage or more rainfall. He did not identify any paleosols.

The sedimentologic evidence for the most mature soils of the Honda Group in the La Venta area agrees with the conclusion that the kaolinite is the product of advanced alteration of feldspars and the hematite is the final product of the oxidation of magnetite, ilmenite, and iron-bearing silicates. All of those are abundant constituents of Honda Group sandstones. The flood-plain deposits are fine-grained so that the decomposition and transformation of the constituent minerals easily occur through weathering and biological activity during soil genesis. In the less mature greenish-gray soils ilmenite enrichment relative to hematite would attest to a shorter time for soil formation. Smectite and illite have some Mg, Fe, K, and Ca in their structures and would point in the same direction as chemically intermediate stages in the alteration of plagioclase, hornblende, and biotite. The release of Ca from the alteration of plagioclase and hornblende probably contributed to the formation of small calcite nodules during soil genesis.

The best line of evidence for recognition of fossil soils in the La Venta area rests in the sedimentologic information available. The identification of meandering channels with their overbank deposits (and associated paleosols) preserved in fining-upward sequences, as those characterized elsewhere (Allen 1970; Bown and Kraus 1987; Smith 1990), explain and predict the relative vertical and lateral position of facies. In one point bar sequence of the Honda Group the flood-plain reddish-brown to red, either uniform or bioturbated mudstone with desiccation cracks, calcite nodules, and rhizoliths (vertisols) overlap burrowed, pale brown mudstone containing rhizoliths and dispersed, small calcareous nodules (inceptisols). These overlap greenish-gray mudstone and very fine sandstone with rhizoliths (entisols) of the levee and crevasse splays, which, in turn, overlap the fine- to medium-grained sandstones deposited by the channels and displaying the epsilon stratification that results from lateral migration. The sequence is not always the same, and very frequently only one type of soil is present between two channel sandstones. Lateral changes of soils can also be observed and have the same sequence of vertical changes. The range of colors from pale brown, reddish-brown to deep red or purple, related to the amount of hematite and other iron compounds, is similar to that presented by Bown and Kraus (1987) for progressive stages of soil maturity with distance from the main channels and time available for pedogenesis. A few paleosols of the Honda Group appear to be the product of continuous sedimentation that did not interrupt the soil genesis; they were not developed in static parent material, and soil horizons are not evident.

The Honda Group paleosols range from calcic entisols, calcic inceptisols, and calcic vertisols to oxisols thought to have been produced by chemical weathering during increasing lengths of time as a result of different depositional rates. The available chronostratigraphic information indicates that average depositional rates in the La Victoria Formation (~950 mm/1,000 yr) were much faster than in the Villavieja Formation (~413 mm/1,000 yr) (Guerrero 1993; also chap. 3). Calcic entisols and calcic inceptisols are common in the La Victoria Formation and lower part of the Villavieja Formation. Calcic vertisols are present throughout the entire section but are more abundant in the upper part of the Villavieja Formation in the Cerro Colorado Member. Oxisols represented by kaolinite-rich, mottled (drab-haloed) red to purple mudstone with small and dispersed hematite nodules are restricted to the Cerro Colorado Member.

There is neither faunal nor any other evidence between the La Victoria and Villavieja Formations that could indicate that the different paleosols were affected by a climatic change. Apparently, the major cooling event during the establishment of a large Antarctic ice cap (Shackleton and Kennett 1975; Woodruff et al. 1981), which took place at 12.9–11.8 Ma (Keller and Barron 1987) during deep sea Neogene Hiatus NH3, had little influence on continental climate near the equator. If there was any climatic change

affecting the paleosols of the Villavieja Formation, it was very minor compared with the effect resulting from dramatic changes in sedimentary rates.

Paleosols in the entire section include small and dispersed calcite nodules, indicating that carbonate was not effectively leached by available soil water (Retallack 1990, 107) during dry periods; desiccation cracks are also indicative of seasonal dry periods. In contrast, drab-haloed root mottles would indicate periodically waterlogged soils (Retallack 1990, 29). Fossil wood indicating a subhumid-tropical to humid-tropical climate (*Goupioxylon stutzeri* of the family Celastraceae) was reported from the Honda Group in the Honda and Villavieja areas by Pons (1969). The family is represented today by tall trees (25–40 m high) along river valleys with annual rainfalls of at least 2,000 mm and temperatures between 23 and 30°C.

The climate of this part of the continent during the late Miocene, after the deposit of the Honda Group, continued unchanged. A humid-tropical lowland assemblage of woodland plants was reported by Howe (1974) for the Neiva Formation in the La Venta area. Plant remains that also indicate a humid-tropical climate were reported by Berry (1936) from the middle Magdalena valley; these remains apparently come from the Bagre Formation, which correlates with the Neiva and Gigante formations of the Huila Group in the upper Magdalena valley.

The Honda Group paleosols were strongly influenced by changes in sedimentary rates controlled by relative uplift and subsidence in a tectonically active basin where the climate was subhumid-tropical to humid-tropical with marked seasonality. Evans (1991) reported calcic entisols and inceptisols from a Paleogene alluvial fan in Washington, indicating that rapid uplift, high erosion rates, and rapid subsidence partly concealed the distinctive signature of intense humid-tropical climate.

Perhaps the most comparable environment of deposition of the Honda Group (except for the Cerro Colorado Member when the basin became enclosed between the Central and Eastern Cordilleras) is eastern Colombia, the area of the Orinoco Llanos. The area is characterized by large meandering tributary systems of the larger Guaviare, Vichada, and Meta rivers of the Orinoco River Basin. The vegetation is that of herbaceous savannah with small (100 m) patches of dryland wooded forest and larger gallery forests along streams. The climate (Etter and Botero 1990) is characterized by a three- to five-month dry season, annual rainfall of 1,800–2,200 mm and mean annual temperatures of 25–30°C.

PALEOCURRENT DIRECTIONS AND UPLIFT OF THE COLOMBIAN ANDES

Paleocurrent directions from cross-beds of the La Victoria Formation indicate an east-southeast paleoflow (fig. 2.10A). The "shoestring" sandstone trends of the Cerro Gordo, Chunchullo, and Tatacoa sandstones also indicate a direction to the east-southeast during a time in which the Eastern Cordillera did not exist.

The Baraya Member in the lower part of the Villavieja Formation has indications of changes in paleoflow directions to the north and northeast (fig. 2.10B) besides the still predominant direction to the east and southeast. The Polonia Red Beds in the upper part of the Cerro Colorado Member show a complete shift in paleocurrent directions toward the west, indicating a source area to the east in the newly uplifted Eastern Cordillera (fig. 2.10C).

Wellman (1970) reported paleoflows to the east and southeast from measurements that come from the lower part (La Victoria Formation and lower part of the Villavieja Formation) of the Honda Group. Wiel (1991) concluded that paleocurrent directions of the Honda Group were toward the east and that the uplift of the Garzón Massif on the east side of the basin had little influence on the Honda Group deposits.

The inversion of the drainage, toward the west, after the uplift of the Eastern Cordillera, was undetected by Wellman (1970) and Wiel (1991) because no paleocurrent measurements of the upper part of the Cerro Colorado Member were available. Paleocurrent indicators are rare in this part of the section because most of the Polonia Red Beds are composed of mudstone, and only very small amounts of sandstone are present.

The Central Cordillera was still the source area of the San Francisco Sandstone Beds in the lower part of the Cerro Colorado Member as indicated by its content of pumice clasts from middle Miocene volcanism. Above this, the lower part of the Polonia Red Beds in the La Venta area are composed of approximately 150 m of red mudstone of

overbank origin where there is no record of paleocurrent directions. Immediately above, the upper part of the Polonia Red Beds contain the sandstone bodies with paleocurrent directions to the west, opposite to the directions from the underlying units in the Baraya Member and the La Victoria Formation. This change in paleocurrent directions provides the first direct evidence of substantial uplift in the Eastern Cordillera at about 11.8 Ma (using the time scale of Berggren et al. 1985) and further refines the results of 13.9 ± 2.3 Ma to 9.2 ± 2.0 Ma reported by Wiel (1991), using apparent fission-track ages on apatite-bearing samples from the Eastern Cordillera. The wide (north-south) regional distribution of thick red beds in the upper part of the Honda Group suggests that at 11.8 Ma the Eastern Cordillera was already a continuous range that completely closed the basin on its eastern side. The changes in paleoflow in the Baraya Member indicate that the uplift of the Eastern Cordillera began as early as 12.9 Ma.

Clast composition and paleocurrent directions (to the north and northeast) of the Neiva Formation (10.1–8.4 Ma) in the La Venta area indicate both a source area to the west in the Central Cordillera and a source area to the east in the newly uplifted Eastern Cordillera. The braided-stream conglomerates of the Neiva Formation are the oldest known ancestral deposits of the Magdalena River.

Howe (1974) concluded that the regional paleocurrent system of the Neiva Formation showed an eastward-directed drainage network. He unfortunately included in his sections (fig. 4 in Howe 1974) deposits of what is now known to be the Villavieja Formation (Qda. Guandinosita and Qda. Mano de León) or deposits of what is now known to be the Gigante Formation (Campoalegre). True sections of the Neiva Formation north of Neiva (Desengaño, Qda. Michio, and Palogrande), however, show paleocurrent directions to the northeast (fig. 10 in Howe 1974). More recently, Wiel (1991) reported north-northeast paleocurrent directions for the Neiva Formation in the Gigante Area, which includes the section at Qda. Guandinosita.

Stratigraphic and Structural Relations

The Honda Group overlies with angular unconformity older units that range from Jurassic to early Miocene in age. Prior to Honda Group deposition, the area was subjected to tectonism and subsequent erosion so that the predominantly west-dipping thrust faults of early to middle Miocene age predate the Honda Group; this tectonic event was related to the uplift of the eastern side of the Central Cordillera. The east-dipping thrust faults related to the uplift of a new mountain range, the Eastern Cordillera, postdate the Honda Group. These east-dipping thrust faults of the western side of the Eastern Cordillera postdate not only the middle Miocene Honda Group but also the late Miocene Huila Group.

The available evidence indicating that uplift of the Eastern Cordillera began at about 12.9 Ma, during deposition of the Villavieja Formation, suggests that this uplift began on the east and progressed to the west. The earliest uplifted area of the Eastern Cordillera was located far enough to the east of the present day Magdalena valley that the effect on Honda Group deposition was only to tilt and close the basin. The exposed Honda Group sequence in the La Venta and Gigante areas does not include synsedimentary faults or folds.

The structural trend in the upper Magdalena valley is to the northeast. At La Venta three anticlines affect the Honda Group; they are from northwest to southeast the Los Fiques, Cerro Gordo, and La Becerrera anticlines (new names). The Los Fiques and Cerro Gordo anticlines (figs. 2.2, 2.9) are accompanied by thrust faults that dip to the east-southeast and have minor displacements. In contrast, accompanying the La Becerrera Anticline, a major thrust fault dips to the west-northwest. This thrust puts the lower part of the La Victoria Formation in contact with the upper part of the Villavieja Formation, the Cerro Colorado Member. The amount of displacement is larger to the north where the base of the Honda Group is exposed. At Cerro La Becerrera the La Victoria Formation rests unconformably on the volcanic rocks of the Jurassic Saldaña Formation and also on the marine deposits of the Cretaceous Caballos and Villeta formations.

The Doche Syncline, east of the La Becerrera Thrust, exposes the upper part of the Villavieja Formation; then, the contact, further east, between the Cretaceous units and the Honda Group is a thrust fault, not an unconformity. This thrust fault can be seen in Baraya and is named after it (fig. 2.2). The Baraya Thrust can be followed toward the south, up to the village of Rivera where it is continuous

with the major regional Suaza Thrust System. Because of uplift and subsequent erosion, only minor exposures of the lower part of the Honda Group are present east of the Baraya Thrust.

Another dominant feature in the La Venta area is the presence of normal faults of very high angle; the trend of these faults (N70°W) is approximately normal to the trend of the main structures. In addition, two sets of calcite-filled vertical joints with approximately east-west and north-south directions can be observed. Most of the rounded calcareous concretions (torpedo-shape concretions of Fields 1959) have their axes oriented with these vertical joints. These structural elements were discussed by Guerrero (1985) who concluded that the evidence indicated a local compression in a northwest-southeast direction.

An interesting fact to be noticed is that despite the northeast trend of the structures, the sandstone beds of the La Victoria Formation can only be followed in a direction normal to this trend. Sandstone belts encased in mudstone were deposited by rivers moving to the east-southeast, leaving narrow and elongate bodies in these directions. The perpendicular trends of the structure and channel belts, however, provide additional exploration targets for oil in the Honda Group. Oil is being recovered from the Andalucía and Tello oil fields west of the La Becerrera and Baraya Thrusts, respectively. The La Becerrera is a west-dipping fault that thrust the Saldaña Formation against marine Cretaceous deposits prior to the Honda Group deposition. The thrust was reactivated after the deposition of the Honda-Huila sequence, creating the anticline structure of the Andalucía oil field.

Age and Correlation

Following the time scale of Berggren et al. (1985), the available chronostratigraphic evidence (Guerrero 1990, 1993; also chap. 3) indicates that the Honda Group was deposited in the interval between about 13.5 and 11.5 Ma. The La Victoria and Villavieja formations were deposited from about 13.5 to 12.9 Ma, and 12.9 to 11.5 Ma, respectively. Honda Group deposition began with a wide regional unconformity (Guerrero 1990).

Another set of important data for biostratigraphic and geochronologic correlation comes from calibrations between zones of planktonic foraminifera, palynological zones, and molluscan faunas. In lithology and stratigraphic position the La Victoria and Villavieja formations of the Honda Group in the upper Magdalena valley are equivalent to the San Antonio and Los Limones formations of the Honda Group in the Honda area and to the lower part (Lluvia, Chontorales, Hiel, and Enrejado formations) of the Real Group in the middle Magdalena valley. North of Honda, the Honda Group and the Real Group overlie (paraconformably and unconformably, respectively) strata that contain the La Cira fossil horizon of the upper part of the Colorado Formation.

The Colorado Formation and the Real Group of the middle Magdalena valley have palynological ages. Hopping (1967) indicated that the base of the Zone of *Verrutricolporites rotundiporis* contained in the upper part of the Colorado Formation coincided more or less with the top of the foraminiferal Zone of *Globorotalia ciperoensis ciperoensis* (Oligocene-Miocene boundary). Thus, the age of the Colorado Formation would be broadly late Oligocene to early Miocene. Hopping also correlated the base of the Zone of *Crassoretitriletes vanraadshooveni* that is contained in the lower part of the Real Group with the lower part of the foraminiferal Zone of *Globorotalia fohsi fohsi*. This would mean that the base of the Real Group is younger than about 14.5 Ma and, therefore, of broadly middle Miocene age.

The Pebas Formation in eastern Peru (Costa 1981) extends into the Colombian Amazonas Basin where it was dated palynologically by Horn (1990). She reported a middle Miocene age based on the presence of an association that contains *Crassoretitriletes vanraadshooveni* and *Grimsdalea magnaclavata*. Thus, the Pebas Formation would correlate in time of deposition to the lower part of the Real Group and to the Honda Group. The Pebas Formation contains benthic foraminifera in some sections, and the depositional environment was interpreted by Horn (1990) as coastal with brackish lagoons. The affinities of the molluscan fauna studied by Nutall (1990) from the La Tagua Locality in the Pebas Formation in Colombia led him to conclude that there was a brackish-water connection with the Caribbean at that time.

A very important aspect of the Honda Group relative to the uplift of the Colombian Andes places the unit in a global context. It is very interesting to note that an uplift event in the Andes beginning at about 12.9 Ma coincides with widespread global tectonism and with the deep sea hiatus NH3 dated as 12.9–11.8 Ma by Keller and Barron (1987). The uplift of the Eastern Cordillera of the Colombian Andes coincides with the final closure of the eastern end of the Mediterranean Sea (Steininger 1983), with the collision of the Australian subcontinent with Indonesia (Kennett et al. 1985), and with a major cooling event during the establishment of an Antarctic ice cap (Shackleton and Kennett 1975; Woodruff et al. 1981).

Conclusions

The middle Miocene Honda Group of the upper Magdalena valley is a distinct lithostratigraphic unit bounded above and below by unconformities (the base of the Honda Group at about 13.5 Ma and the base of the Huila Group at 10.1 Ma) that reflect important episodes of volcanism and uplift of the Colombian Andes.

The deposition of the La Victoria Formation (about 13.5–12.9 Ma) was initiated during active uplift and volcanism of the Central Cordillera, which was the source area on the west side of the basin. Paleocurrent directions from the La Victoria Formation to the east and southeast indicate that the Eastern Cordillera did not exist at that time. The river deposits of the La Victoria Formation on the modern Magdalena River valley were apparently continuous to the east with the coastal and brackish deposits of the Pebas Formation in the present Amazon basin.

The deposition of the Villavieja Formation (about 12.9–11.5 Ma) began during active volcanism and the uplift of the Central Cordillera and also during the uplift of a new mountain range, the Eastern Cordillera. Paleocurrent directions of the Baraya Member in the lower part of the Villavieja Formation are very variable, and although most of them point to the east and southeast, some have north and northeast trends, indicating that the uplift of the Eastern Cordillera was initiated as early as 12.9 Ma. In the upper part of the Cerro Colorado Member (Polonia Red Beds), the paleocurrent directions are to the west, indicating that the Eastern Cordillera was already uplifted at about 11.8 Ma and had closed the eastern side of the basin. Thick red beds in the upper part of the Honda Group are regionally continuous and can be followed along the Magdalena River valley, suggesting that the Eastern Cordillera was a continuous range at about 11.8 Ma. The age (beginning at 12.9 Ma) of the tectonic event of the Colombian Andes that led to the uplift of the Eastern Cordillera coincides with widespread global tectonism during deep sea hiatus NH3 (12.9–11.8 Ma).

Most of the La Victoria Formation was deposited by large meandering rivers that were up to 8–10 m deep and flowed to the east-southeast. Meander-belt ridges were produced by various episodes of channel sedimentation in places where the river position persisted for some time. Medium- and coarse-grained conglomeratic multistory-sandstone bodies with "shoestring" shape (and encased in mudstone) are up to 31.5 m thick and 2–4 km wide. The Cerbatana Conglomerate Beds in the upper part of the La Victoria Formation were deposited by braided rivers that also flowed to the east and left a basinwide pebble to cobble conglomerate derived from the Central Cordillera. The change in sedimentary regime was probably the result of a rapid pulse of uplift in the Central Cordillera just before the uplift of the Eastern Cordillera.

The Villavieja Formation was deposited by meandering rivers smaller than those of the La Victoria Formation. The rivers of the Baraya Member and lower part of the Cerro Colorado Member were up to 2–3 m deep and flowed mostly to the east. The upper part of the Cerro Colorado Member was deposited by meandering and anastomosing rivers up to 6 m deep that flowed to the west.

Paleosols form a very important portion of the sequence. They are present with different stages of maturity at the top of the fining-upward sequences except where the following channel eroded the upper portion of the preceding cycle. Although paleosols are common in the entire sequence, they are most mature (oxisols and calcic vertisols) and abundant in the Cerro Colorado Member of the Villavieja Formation where fossils are scarce. In contrast, fossils are most common in the soils with less maturity (calcic entisols and calcic inceptisols) in the La Victoria Formation and Baraya Member of the Villavieja Formation. Calcic vertisols are found through the entire section but are more abundant in the Cerro Colorado Member.

The La Victoria Formation was deposited much faster (~950 mm/1,000 yr) than the Villavieja Formation (~413 mm/1,000 yr). The finer sediments and more mature soils of the Villavieja Formation indicate less energetic environments and prolonged periods of subaerial alteration during which soils reached almost complete mineral degradation. The Honda Group paleosols were strongly influenced by sedimentary rates controlled by relative uplift and subsidence in a tectonically active basin where the climate was subhumid- to humid-tropical with marked seasonality.

Acknowledgments

The project of which this chapter is part has been made possible by a cooperative agreement between Duke University and the Colombian Geological Survey (INGEOMINAS). Very special thanks go to R. F. Kay and R. H. Madden from Duke University and to D. Mosquera, A. Loboguerrero, and L. Jaramillo from INGEOMINAS.

This work was mainly supported by grants from the U.S. National Science Foundation NSF BSR 8614533 and NSF BSR 8614133 to R. F. Kay, to whom I am indebted. Additional funding came from grants of the U.S. National Science Foundation to J. Flynn (NSF BSR 8614133/8896178) and from grants of INGEOMINAS, Colciencias, and Icetex in Colombia.

The magnetostratigraphic analyses were made by J. J. Flynn and the author in the Laboratory of Paleomagnetism of the Field Museum. The $^{40}Ar/^{39}Ar$ analyses were made by C. C. Swisher III in the Geochronology Center of the Institute of Human Origins.

Some of the results were reported earlier in a master's thesis (Guerrero 1990) presented to the Geology Department at Duke University under the supervision of D. Heron. The paleomagnetic results from the upper part of the Honda Group and from the overlying Neiva and Gigante formations (samples JG-M90 and JG-M91) constitute part of my Ph.D. dissertation (Guerrero 1993).

I had valuable discussions with R. F. Kay, R. H. Madden, J. J. Flynn, L. G. Marshall, R. L. Cifelli, C. Villarroel, A. Gómez, C. Ruíz, V. Ramírez, and G. Marquinez, who were my partners during the field seasons. Two anonymous reviewers helped improve the manuscript. The final versions of the drawings were by J. Smith.

3
Geochronology of the Honda Group

John J. Flynn, Javier Guerrero, and
Carl C. Swisher III

RESUMEN

Se reporta un estudio magnetoestratigráfico de detalle del Grupo Honda en el Valle Superior del Magdalena, basado en 210 horizontes paleomagnéticos, y seis dataciones ^{40}Ar/^{39}Ar asociadas. También se reportan dataciones adicionales de las suprayacentes Formaciones Neiva y Gigante.

Las edades ^{40}Ar/^{39}Ar indican que el Grupo Honda y la fauna de La Venta son más jóvenes que 13,8 Ma y más antiguos que 10,0 Ma. La más precisa correlación paleomagnética indica que el Grupo Honda abarca el intervalo de tiempo entre 13,5 y 11,6 Ma durante los intervalos C5ABn–C5An.1n (de la geocronología de Cande y Kent 1992). Esta edad para la fauna de La Venta indica que ésta sería temporalmente correlativa con algunas faunas que han sido asignadas a la edad mamífero "Friasense," y que es 4–6 Ma más jóven que la Edad Santacrucense.

Las tasas de acumulación de la Formación La Victoria fueron un poco más que el doble de las de la Formación Villavieja. Así, la Formación La Victoria representa aproximadamente 0,6 millones de años, mientras que la Formación Villavieja representa aproximadamente 1,4 millones de años. El límite entre las Formaciones La Victoria y Villavieja no representa un hiato erosivo o deposicional importante. Las diferencias litológicas entre las dos formaciones, aunque resultan principalmente de los cambios en las tasas de sedimentacion, podrían también indicar cambios paleoambientales y tectonicos significativos.

Precise geochronologic information is essential to any study of the temporal patterns of faunal evolution and geologic history. Early studies of the La Venta fauna (Stirton 1951, 1953b; Fields 1959) lacked any independent geochronologic constraint. Biochronology, in the form of "stage-of-evolution" of the mammalian fauna, suggested that the La Venta fauna was broadly comparable to Friasian, or possibly Santacrucian, Land Mammal Age faunas from Patagonian South America. The Santacrucian and Friasian, in turn, were considered Miocene in age, although until only recently these mammal ages also lacked reliable, direct correlation to standard global chronostratigraphic series (such as the Miocene) or stages (such as the Serravalian, middle Miocene). Even though the La Venta fauna is one of the most important South American mammal assemblages by virtue of its rarely sampled tropical paleogeographic position, high diversity (including one of the most significant fossil assemblages of New World primates), and unusual paleoecology, the first numerical dating of the La Venta sequence did not occur until the 1980s. Takemura and Danhara (1986) reported fission-track dates for two horizons within the Honda Group (lower Villavieja Formation), which were complemented by preliminary paleomagnetic studies of a short stratigraphic interval by Hayashida (1984).

Beginning in 1986, the U.S. National Science Foundation supported an extensive collaborative study of the paleontology (coordinated by Richard Kay) and geochronology (coordinated by John Flynn) of the Honda Group in the La Venta area (locally known as the Desierto de la Tatacoa) of Colombia. This chapter summarizes some of the geochronologic results of that study, especially paleomagnetic and $^{40}Ar/^{39}Ar$ radioisotopic analyses, constraining the numerical age and temporal correlation of the approximately 1,250-m-thick Honda Group (La Victoria and Villavieja formations; chap. 2) and the contained La Venta fauna. Current understanding of the stratigraphy of this sequence is presented in detail in Guerrero (1990; also chap. 2), building upon earlier stratigraphic studies (Hettner 1892; Butler 1942; Royo y Gómez 1942a, 1946; Stirton 1951, 1953b; Fields 1959; Wellman 1970). We focus on developing a geochronology for this sequence and using that geochronologic information to (1) evaluate and test patterns and rates of sedimentation within the Honda Group and (2) evaluate the age of the tropical La Venta fauna relative to higher latitude faunas exemplifying South American Land Mammal Ages, thereby providing a test of the applicability of previously recognized Land Mammal Ages over wide geographic areas.

Summary of Paleomagnetic Analyses

Hayashida (1984) presented preliminary paleomagnetic analyses from a short stratigraphic interval within the lower Villavieja Formation, which were affected adversely by unrecognized structural geologic complexity. Because this volume is intended as a complete summary of the current geological and paleontological research in the La Venta area, we provide here a brief summary of those paleomagnetic results. This summary also will facilitate evaluation of the data used to construct the magnetic polarity stratigraphy analyzed in this chapter.

We collected more than 210 sites from more than 1,000 m of stratigraphic section, yielding an average sampling interval of 1 site per 5 m. We collected 21 additional sites near the base of the section (stratigraphically overlapping the main section), from the limbs of a fold (axis trends NE–SW) near Cerro Gordo, to perform a fold test for magnetic

stability. At least three to four individually oriented samples were collected from each site.

All sample preparation and paleomagnetic measurements were performed in the Field Museum Paleomagnetism Laboratory. Detailed thermal (16 steps, to 700°C) and alternating field (AF) (14 steps, to 100 mT) demagnetization studies were performed on pilot samples from about 50% of the total sites. Detailed thermal and AF studies were performed on complementary subsamples from the same sample block, for each pilot sample site, to provide direct comparison of magnetic vector behavior under differing demagnetization treatments of the same sample. Based on vector analysis of the magnetization behavior of pilot samples, we chose to demagnetize all remaining samples in up to 9 steps (10 mT, 30 mT, 200°C, 400°C, 450°C, 500°C, 530°C, 560°C, 600°C) to be certain to identify all low to moderate coercivity and blocking temperature magnetization components. All magnetic remanence measurements were made with a Molspin Minispin MS-1 spinner magnetometer. Magnetic susceptibility measurements were made following the natural remnant magnetism (NRM) measurement and every thermal demagnetization step, to monitor physicochemical changes in the magnetic mineralogy of samples.

Demagnetization studies indicate that it is possible to isolate a stable, characteristic remanence in most samples. This characteristic remanence very likely represents a primary remanence (acquired at, or shortly following, deposition) accurately reflecting the orientation of the earth's magnetic field at the time of sediment accumulation. Both high blocking temperature (probably representing hematite or titanohematite) and moderate blocking temperature (presumably representing magnetite or titanomagnetite) magnetizations are present in various samples, but the moderate blocking temperature component (which almost certainly carries a primary, depositional remanence) dominates most samples.

Following isolation of the characteristic remanence for each sample, we used principal components analysis (least-squares fit program "Manualft" of Gillette; based on Kirschvink 1980) to determine the direction and intensity of the remanence vector (using at least three individual data points [demagnetization steps] to define a remanence vector component). The best-fit directions for all samples from each site were composited (Fisher 1953) to determine a site mean direction of magnetization and associated calculation of site mean virtual geomagnetic pole (VGP). The site mean directions were plotted versus stratigraphic position to construct a magnetic polarity stratigraphy. Only Class I reliability sites, having statistically significant magnetization directions, *after* thermal demagnetization (using statistical tests of Watson 1956a, 1956b; modified from reliability classification of Opdyke et al. 1977), were used to define polarity intervals. We follow the convention of not recognizing polarity intervals defined only by a single site.

Summary of Radioisotopic Dating

Eight volcanic units found interbedded in the terrestrial sediments of the La Victoria and Villavieja formations of the Honda Group and the overlying Neiva and Gigante formations were collected for radioisotopic age determination (tables 3.1, 3.2). Samples of the volcanic units were collected by Javier Guerrero (J.G.) and were subsequently submitted at various times during the study to Carl Swisher (C.C.S.) at the Geochronology Center, Institute of Human Origins, for ^{40}Ar/^{39}Ar analysis. The volcanic samples consist primarily of pumice pieces found in volcaniclastic sandstones. Generally, the pumice from the Neiva and Gigante formations were much larger and fresher than those recovered from the underlying Honda Group, possibly reflecting more primary deposition. In this regard, the ages because some reworking may have occurred since the time of eruption. Given the agreement of the ages derived from these units with the general stratigraphy, however, these volcanics are considered, for the most part, penecontemporaneous with sediment deposition.

The sample preparation and ^{40}Ar/^{39}Ar analytical procedures follow those of Deino and Potts (1990), Deino et al. (1990), Swisher and Prothero (1990), Renne and Basu (1991), and Swisher et al. (1992). The handpicked minerals, together with a centrally located monitor mineral (Fish Canyon Tuff sanidine [FC]), were irradiated in the hydraulic rabbit core facility of the Omega West research reactor, Los Alamos National Laboratory. After irradiation, single crystals of the monitor mineral and single crystals of hornblende, plagioclase, and biotite of the unknowns were

Table 3.1. Summary of ^{40}Ar/^{39}Ar analyses of volcanic units from the Honda Group and overlying formations

Volcanic Sample	Mineral Dated	N	Age (Ma), Weighted Mean ± SE
Gigante Formation LGM-LV9	Hornblende	8	8.412 ± 0.065
Neiva Formation JG-R88-4	Plagioclase	5	10.133 ± 0.184
Villavieja Formation JG-R88-2	Hornblende	6	12.230 ± 0.117
	Plagioclase	6	12.111 ± 0.259
	Overall	12	12.210 ± 0.107
La Victoria Formation JG-R89-2	Hornblende	6	12.486 ± 0.111
	Plagioclase	6	12.649 ± 0.258
	Overall	12	12.512 ± 0.102
JG-R90-1	Hornblende	4	13.342 ± 0.408
JG-R90-3	Biotite	2	—
	Hornblende	2	—
	Overall	4	13.778 ± 0.081
JG-R89-3	Hornblende	5	13.608 ± 0.210
	Biotite	5	13.801 ± 0.056
	Plagioclase	6	13.565 ± 0.170
	Overall	16	13.767 ± 0.052
JG-R89-1	Hornblende	5	13.590 ± 0.171
	Plagioclase	6	13.690 ± 0.138
	Overall	11	13.651 ± 0.107
JG-R90-3, 89-3, and 89-1	Overall	31	13.754 ± 0.040

Note: N = number of analyses.

analyzed at the Geochronology Center of the Institute of Human Origins. Laser heating, gas purification, and mass spectrometry were completely automated following computer programmed schedules.

The ^{40}Ar/^{39}Ar ages of the unknowns are calculated using a J-value determined from at least six replicate analyses of individual grains of the co-irradiated monitor mineral Fish Canyon Tuff sanidine. The age of the Fish Canyon sanidine adopted in this study (27.84 Ma) is similar to that recommended by Cebula et al. (1986) but slightly modified as a result of in-house intercalibration at the Geochronology Center with Minnesota hornblende MMhb-I (published age of 520.4 ± 1.7 Ma, Samson and Alexander

Table 3.2. Radioisotopic analyses of volcanic units from several locations in the Honda Group and from two overlying formations

Mineral Dated[a] and Lab Sample	$^{40}Ar/^{39}Ar$	$^{37}Ar/^{39}Ar$	$^{36}Ar/^{39}Ar$	$^{40}Ar\star/^{39}Ar$	%$^{40}Ar\star$	Age (Ma) Mean (or Weighted Mean, in Bold)	SD, 1σ (or SE, in Bold)
GIGANTE FORMATION SAMPLE LGM–LV9							
Hornblende (S9D):							
1131-01	2.52455	5.243834	0.005989	1.18667	46.83	8.969	0.162
1131-02	2.73660	4.401004	0.006778	1.09554	39.93	8.282	0.167
1131-03	1.34363	5.102880	0.002224	1.10641	82.03	8.364	0.171
1131-04	1.51858	4.265698	0.002615	1.09652	72.03	8.289	0.147
1131-05	1.65573	4.790506	0.003167	1.11410	67.03	8.422	0.334
1131-06	1.96001	6.492087	0.004180	1.26061	64.03	9.527	0.198
1131-07	4.33244	5.931083	0.012430	1.14795	26.43	8.677	0.279
1131-08	2.73896	3.101322	0.006932	0.94480	34.43	7.145	0.175
All (N = 8)	—	—	—	—	—	**8.412**	**0.065**
NEIVA FORMATION SAMPLE JG–R88–4							
Plagioclase (15B):							
1595-01[b]	1.54204	5.308253	0.004246	0.72282	46.73	6.306	0.373
1595-02	1.56483	9.902993	0.004007	1.19850	76.03	10.443	0.496
1595-03	1.38568	7.069369	0.002728	1.16244	83.43	10.130	0.362
1595-04	1.59700	5.805780	0.003135	1.14886	71.63	10.012	0.285
1595-05	1.62962	7.902843	0.003759	1.17086	71.43	10.203	0.464
1595-06	1.41514	9.360951	0.003508	1.15105	80.83	10.031	1.159
All (N = 5)	—	—	—	—	—	**10.133**	**0.184**
VILLAVIEJA FORMATION SAMPLE JG–R88–2							
Hornblende (28A2):							
2444-01	30.81295	6.771322	0.012370	27.79802	89.83	12.486	0.407
2444-02	34.19724	6.317772	0.026020	27.10261	78.93	12.174	0.527
2444-03	35.59240	6.478473	0.030071	27.31686	76.43	12.270	0.280
2444-04	40.01713	6.504565	0.046919	26.76315	66.63	12.022	0.219
2444-05	34.33003	6.731263	0.025514	27.42556	79.53	12.319	0.493
2444-06	34.41988	6.833303	0.025871	27.41956	79.33	12.316	0.200
All (N = 6)	—	—	—	—	—	**12.230**	**0.117**

Continued on next page

Notes: Constants and corrections used in the calculations are as follows:

Plagioclase	Hornblende	90-s
D = 1.005	D = 1.005000	D = 1.005
±D = 0.00015	±D = 0.000150	±D = 0.00015
Ca 39/37 = 0.00067	Ca 39/37 = 0.000670	Ca 39/37 = 0.00067
Ca 36/37 = 0.0002582	Ca 36/37 = 0.000258	Ca 36/37 = 0.0002582
K 38/39 = 0.01077	K 38/39 = 0.010770	K 38/39 = 0.01077
K 40/39 = 0.0024	K 40/39 = 0.002400	K 40/39 = 0.0024

"90-s" denotes constants and corrections for analyses JG-R90-1, -1B, -2, and -3.

Note: $^{40}Ar\star$ = radiogenic ^{40}Ar.

[a]In parentheses after the mineral is the irradiation number.

[b]Analysis not included in mean calculation.

Table 3.2. Continued

Mineral Dated[a] and Lab Sample	$^{40}Ar/^{39}Ar$	$^{37}Ar/^{39}Ar$	$^{36}Ar/^{39}Ar$	$^{40}Ar\star/^{39}Ar$	$\%^{40}Ar\star$	Age (Ma)	
						Mean (or Weighted Mean, in Bold)	SD, 1σ (or SE, in Bold)
Plagioclase (28A):							
2437-10	32.90914	18.919290	0.026347	26.90575	80.73	12.028	0.588
2437-11	48.71430	20.064740	0.079816	27.02040	54.73	12.079	0.810
2437-12	39.59354	19.104730	0.047379	27.39889	68.33	12.247	1.039
2437-13	33.16259	18.955410	0.026102	27.23938	81.13	12.177	0.452
2437-14	32.02505	18.031190	0.022408	27.10427	83.63	12.116	0.506
2437-15	82.22982	22.647780	0.195407	26.61665	31.93	11.899	1.099
All (N = 6)	—	—	—	—	—	**12.111**	**0.259**
Hnbl, plag:							
All (N = 12)	—	—	—	—	—	**12.210**	**0.107**

LA VICTORIA FORMATION SAMPLE JG-R89-2

Mineral Dated[a] and Lab Sample	$^{40}Ar/^{39}Ar$	$^{37}Ar/^{39}Ar$	$^{36}Ar/^{39}Ar$	$^{40}Ar\star/^{39}Ar$	$\%^{40}Ar\star$	Mean	SD/SE
Hornblende (28A2):							
2443-01	52.99926	6.878727	0.085972	28.24708	53.13	12.687	0.411
2443-02	40.13709	5.458227	0.044242	27.57837	68.53	12.387	0.169
2443-03	40.62971	5.691529	0.044857	27.91274	68.43	12.537	0.192
2443-04	55.90072	7.235060	0.098784	27.39244	48.83	12.304	0.584
2443-05	59.12641	5.978217	0.107709	27.86362	46.93	12.515	0.386
2443-06	41.05070	5.780835	0.044189	28.54193	69.33	12.819	0.514
All (N = 6)	—	—	—	—	—	**12.486**	**0.111**
Plagioclase (28A):							
2432-10	61.89320	9.611100	0.082157	28.53035	54.63	12.751	0.519
2432-11	84.51453	12.525970	0.195897	27.81358	32.63	12.432	0.809
2432-12	180.22200	22.451000	0.520604	28.52321	15.63	12.748	6.681
2432-13	95.72075	9.415843	0.230453	28.51774	29.63	12.746	0.461
2432-14	128.78440	14.712710	0.345751	28.01119	21.53	12.520	2.022
2432-15	39.54900	9.197359	0.041779	28.07561	70.63	12.549	0.459
All (N = 6)	—	—	—	—	—	**12.649**	**0.258**
Hnbl, plag:							
All (N = 6)	—	—	—	—	—	**12.512**	**0.102**

LA VICTORIA FORMATION SAMPLES JG-R90-1 AND -1B

Mineral Dated[a] and Lab Sample	$^{40}Ar/^{39}Ar$	$^{37}Ar/^{39}Ar$	$^{36}Ar/^{39}Ar$	$^{40}Ar\star/^{39}Ar$	$\%^{40}Ar\star$	Mean	SD/SE
Hornblende (57B):							
4837-01	58.52090	7.272837	0.121444	23.30018	39.63	13.660	0.759
4837-02	44.05065	9.211583	0.074259	22.94917	51.83	13.455	1.198
4837-03	49.36831	7.339304	0.092868	22.59448	45.53	13.248	2.669
4842-01[b]	74.38980	8.540560	0.188483	19.45353	26.03	11.412	0.527
4842-02	89.78168	7.823687	0.230275	22.44760	24.93	13.162	0.540
4842-03[b]	52.44941	8.098479	0.114732	19.26613	36.53	11.302	0.524
All (N = 4)	—	—	—	—	—	**13.342**	**0.408**

Table 3.2. Continued

Mineral Dated[a] and Lab Sample	$^{40}Ar/^{39}Ar$	$^{37}Ar/^{39}Ar$	$^{36}Ar/^{39}Ar$	$^{40}Ar\star/^{39}Ar$	$\%^{40}Ar\star$	Age (Ma) Mean (or Weighted Mean, in Bold)	SD, 1σ (or SE, in Bold)
			LA VICTORIA FORMATION SAMPLE JG-R90-3				
Biotite (57B):							
4835-01	32.86034	0.009728	0.031742	23.47921	71.53	13.765	0.197
4835-02[b]	37.46959	0.045275	0.038145	26.19967	69.93	15.353	0.357
4835-03	29.68090	0.011339	0.020701	23.56246	79.43	13.813	0.095
Hornblende (57B):							
4844-01[b]	48.27403	5.859094	0.076857	26.10987	53.93	15.300	0.655
4844-02	31.06536	5.525230	0.028384	23.18283	74.33	13.591	0.478
4844-03	45.58593	5.919340	0.078160	23.03036	50.33	13.502	0.317
Biot, hnbl:							
All (N = 4)	—	—	—	—	—	**13.778**	**0.081**
			LA VICTORIA FORMATION SAMPLE JG-R89-3				
Hornblende (28A2):							
2445-01	46.91004	7.906872	0.057934	30.55326	64.83	13.718	0.564
2445-02	51.05243	8.684949	0.072402	30.49538	59.43	13.693	0.333
2445-03[b]	40.69625	9.578953	0.044930	28.32978	69.23	12.724	0.291
2445-04	69.44096	9.153598	0.136333	30.03466	43.03	13.486	0.484
2445-05	45.16644	7.982059	0.053736	30.05493	66.23	13.496	0.480
2445-06	45.96754	9.851885	0.056666	30.17128	65.23	13.547	0.738
All (N = 5)	—	—	—	—	—	**13.608**	**0.210**
Biotite (28A2):							
2448-01	58.48811	0.015683	0.093586	30.83265	52.73	13.843	0.096
2448-02	50.81936	0.025323	0.068287	30.64068	60.33	13.757	0.304
2448-03	59.55398	0.055583	0.100106	29.97571	50.33	13.460	0.250
2448-05	50.15366	0.024553	0.065759	30.72171	61.33	13.794	0.087
2448-06	56.90808	0.027161	0.088166	30.85523	54.23	13.853	0.148
All (N = 5)	—	—	—	—	—	**13.801**	**0.056**
Plagioclase (28A):							
2435-10	56.80755	18.904580	0.096015	30.25842	52.63	13.521	0.281
2435-11	54.51643	17.775010	0.089703	29.71684	53.93	13.280	0.425
2435-12	37.70918	13.506360	0.031929	29.56994	77.73	13.214	0.869
2435-13	50.15741	18.521660	0.072038	30.66150	60.43	13.700	0.325
2435-14	61.97900	15.564120	0.108728	31.36205	50.13	14.012	0.455
2435-15	148.54510	14.431860	0.411191	28.41154	18.93	12.699	1.009
All (N = 6)	—	—	—	—	—	**13.565**	**0.170**
Biot, hnbl, plag:							
All (N = 16)	—	—	—	—	—	**13.767**	**0.052**

Continued on next page

Table 3.2. Continued

Mineral Dated[a] and Lab Sample	$^{40}Ar/^{39}Ar$	$^{37}Ar/^{39}Ar$	$^{36}Ar/^{39}Ar$	$^{40}Ar\star/^{39}Ar$	$\%^{40}Ar\star$	Age (Ma) Mean (or Weighted Mean, in Bold)	SD, 1σ (or SE, in Bold)
LA VICTORIA FORMATION SAMPLE JG-R89-1							
Hornblende (28A2):							
2442-02	40.59898	6.629370	0.039017	29.70490	72.83	13.339	0.452
2442-03	33.12465	5.636874	0.011453	30.28242	91.13	13.597	0.265
2442-04	46.39849	8.427279	0.057309	30.27516	64.93	13.594	0.876
2442-05	39.51211	6.993673	0.031554	30.86382	77.73	13.857	0.488
2442-06	32.57550	6.838804	0.010040	30.26673	92.53	13.590	0.325
All (N = 5)	—	—	—	—	—	**13.590**	**0.171**
Plagioclase (28A):							
2341-10	36.88348	11.332210	0.026238	30.22182	81.33	13.505	0.693
2341-11	33.33941	12.877710	0.013884	30.47992	90.63	13.620	0.303
2341-12	52.94499	12.189390	0.079501	30.63020	57.43	13.687	0.301
2341-13	36.23683	11.720120	0.022726	30.65374	83.93	13.697	0.282
2341-14	33.86688	13.420450	0.015077	30.70931	89.93	13.722	0.277
2341-15	112.34890	10.770920	0.278683	31.04147	27.43	13.870	0.512
All (N = 6)	—	—	—	—	—	**13.690**	**0.138**
Hnbl, plag:							
All (N = 11)	—	—	—	—	—	**13.651**	**0.107**
LA VICTORIA FORMATION SAMPLES, OVERALL (JG-R90-3, -R89-3, -R89-1)							
Biot, hnbl, plag:							
All (N = 31)	—	—	—	—	—	**13.754**	**0.040**
LA VENTA (LOCATION UNCERTAIN) SAMPLE JG-R88-3							
Plagioclase (28A):							
2436-10	3.35276	4.157873	0.009249	0.93726	27.93	0.420	0.097
2436-11	3.90747	4.414541	0.008436	1.75432	44.83	0.787	0.095
2436-12	10.18914	4.451125	0.031717	1.15749	11.33	0.519	0.101
2436-13	5.49634	4.180739	0.014447	1.54809	28.13	0.694	0.093
2436-14	4.14591	4.799000	0.011265	1.18465	28.53	0.531	0.089
2436-15	9.35127	4.866961	0.030199	0.79910	8.53	0.358	0.114
LA VENTA (LOCATION UNCERTAIN) SAMPLE JG-R90-2							
Hornblende (57B):							
4839-01	66.99966	7.692637	0.185715	12.77131	19.03	7.500	1.572
4839-02	18.20245	7.204002	0.020908	12.63238	69.13	7.419	0.271
4839-03	17.86219	0.123554	0.318446	−76.23782	−426.8	−45.433	20.976

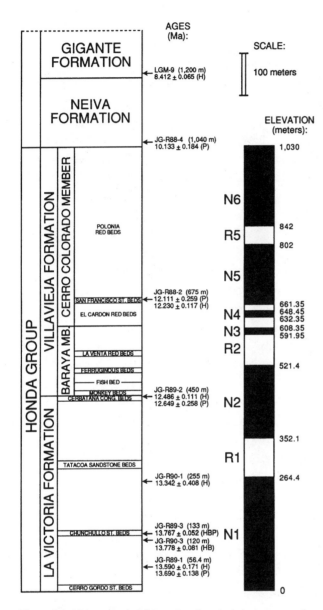

AGES (Ma):

- LGM-9 (1,200 m) — 8.412 ± 0.065 (H)
- JG-R88-4 (1,040 m) — 10.133 ± 0.184 (P)
- JG-R88-2 (675 m) — 12.111 ± 0.259 (P) / 12.230 ± 0.117 (H)
- JG-R89-2 (450 m) — 12.486 ± 0.111 (H) / 12.649 ± 0.258 (P)
- JG-R90-1 (255 m) — 13.342 ± 0.408 (H)
- JG-R89-3 (133 m) — 13.767 ± 0.052 (HBP)
- JG-R90-3 (120 m) — 13.778 ± 0.081 (HB)
- JG-R89-1 (56.4 m) — 13.590 ± 0.171 (H) / 13.690 ± 0.138 (P)

SCALE: 100 meters

ELEVATION (meters): 1,030 — 842 — 802 — 661.35 — 648.45 — 632.35 — 608.35 — 591.95 — 521.4 — 352.1 — 264.4 — 0

Figure 3.1. Magnetostratigraphic section plotted relative to measured stratigraphic section of Guerrero (chap. 2). Abbreviations: *Mb.*, member; *cong.*, conglomerate; *st.*, sandstone; *H*, hornblende; *P,* plagioclase; *B*, biotite; *N1–N6*, normal polarity intervals; *R1–R5*, reversed polarity intervals.

1987). Samples were analyzed from four different irradiations over the course of the four-year study. The subsequent J-values are as follows: for Irr. no. 9 (8 hr) J = 0.004200 ± 0.00001; for Irr. no. 15 (7 hours) J= 0.0048441 ± 0.000002; for Irr. no. 28A-28A2 (20 min) J = 0.0002486 ± 0.0000005 and 0.0002498 ± 0.0000005, respectively; and for Irr. no. 57 (30 min) J = 0.0003262 ± 0.0000005. Ca and

K corrections were determined from laboratory salts: $(^{36}\text{Ar}/^{37}\text{Ar})_{\text{Ca}} = 2.557 \times 10^{-4} \pm 4.6 \times 10^{-6}$, $(^{39}\text{Ar}/^{37}\text{Ar})_{\text{Ca}} = 6.608 \times 10^{-4} \pm 2.53 \times 10^{-5}$ and $(^{40}\text{Ar}/^{39}\text{Ar})_{\text{K}} = 2.4 \times 10^{-3} \pm 7.0 \times 10^{-4}$. Mass discrimination during this study, as determined by replicate air aliquots delivered from an on-line pipette system, was 1.005 ± 0.002. Decay constants are those recommended by Steiger and Jager (1977) and Dalrymple (1979). The uncertainties associated with the individual ages are 1 sigma errors, whereas those that accompany the calculated weighted mean ages of the replicate analyses are standard errors (SE) following Taylor (1982).

Three pumice samples from the lower part of the La Victoria Formation were dated (fig. 3.1; table 3.2). JG-R90-33 is from the main stratigraphic section A, and JG-R89-1 and JG-R89-3 are from section D. Biotite and hornblende from JG-R90-3 yielded four concordant é^{40}Ar/^{39}Ar ages, with a weighted mean age of 13.78 ± 0.08 Ma. At nearly the same stratigraphic interval, pumice from two horizons in section D yielded similar ^{40}Ar/^{39}Ar ages on hornblende, biotite, and plagioclase. A weighted mean age of 13.75 ± 0.04 Ma is calculated for JG-R89-1, whereas JG-R89-3 yielded a weighted mean age of 13.77 ± 0.05 Ma. Although separated by approximately 75 m of section, JG-R89-1 and JG-R89-3 possibly represent pumice from the same eruption; consequently, these ages must be regarded as minimum ages. Alternatively, these two horizons may represent two distinct volcanic events whose ages could not be distinguished.

Analysis of one biotite grain and one hornblende from JG-R90-3 yielded consistent but significantly older ages of 15.3 Ma (table 3.2). These analyses probably reflect contaminant grains from a volcanic event that took place 1.5 million years earlier. If so, these analyses would indicate that volcanic activity in this region may extend back to at least 15 Ma. Similarly, some of the older, anomalous fission-track ages reported by Takemura and Danhara (1983) from the Honda Group may be a result of reworked zircon from these older volcanics.

A third pumice sample (JG-R90-1), collected from between the Chunchullo and Tatacoa sandstone beds in the middle part of the La Victoria Formation, yielded mixed ages. Four hornblende analyses resulted in a weighted mean ä^{40}Ar/^{39}Ar age of 13.34 ± 0.41 Ma (table 3.2), an age that appears consistent with its stratigraphic level above

JG-R90-3, JG-R89-3, and JG-R89-1. Two hornblendes yielded consistent, albeit younger, ages of 11.3 and 11.4 Ma, however (table 3.2). These two ages appear anomalous with those obtained from overlying units and are perhaps young as a result of alteration, although this cannot be demonstrated analytically. It is unlikely, however, that all samples from the upper part of the La Victoria and from the overlying Villavieja Formation are reworked.

Hornblende and plagioclase extracted from a fourth sample, JG-R89-2, collected from the Cerbatana Conglomerate Beds at the top of the La Victoria Formation, yielded dates of 12.49 ± 0.11 Ma (H) and 12.65 ± 0.26 Ma (P) with a mean concordant $^{40}Ar/^{39}Ar$ age for the horizon of 12.51 ± 0.10 Ma (table 3.2).

Stratigraphically higher in section A, from the San Francisco Sandstone Beds in the overlying Villavieja Formation, a fifth horizon yielded a weighted mean $^{40}Ar/^{39}Ar$ age of 12.21 ± 0.11 Ma on analyses of hornblende and plagioclase (table 3.2).

Although no paleomagnetic data are yet published for the overlying Neiva and Gigante formations, two volcanic horizons yielded ages which are stratigraphically consistent with those from the underlying Honda Group and are reported here to substantiate further the minimum possible age for the La Venta fauna. Plagioclase from a volcanic unit within the Neiva Formation, JG-R88-4, resulted in a weighted mean age of 10.13 ± 0.18 Ma, and hornblende separated from a unit in the overlying Gigante Formation (LGM-LV9) was dated at 8.41 ± 0.07 Ma (table 3.2). Attempts to date other mineral phases in these two units resulted in widely disparate ages that ranged from 400,000 years to more than 12 Ma, most likely owing to the low radiogenic yields resulting from alteration. Consequently, these results are not discussed further in this study.

Interpretation of Radioisotopic Date Constraints on Correlation of the Magnetic Polarity Stratigraphy

There are at least three ways in which the available radioisotopic dates can be used to constrain the correlation of the Honda Group magnetic polarity stratigraphy to the standard magnetic polarity time scale (MPTS). All three methods assume that the dates associated with each horizon represent a *maximum* age for that horizon because the dated material must be derived either from contemporaneous volcanism or from material reworked or transported from older events. In all cases the younger limit on the age of the upper part of the Honda Group is provided by $^{40}Ar/^{39}Ar$ ages from overlying units: 10.13 ± 0.18 Ma (JG-R88-4) from the lower Neiva Formation and 8.41 ± 0.07 Ma (LGM-LV9) from the lower Gigante Formation. Similarly, the cluster of dates near the base of the Honda Group averaging about 13.75 Ma indicate that the Honda Group must be younger than about 14 Ma.

The first method is to assume that the series of dates from horizons within the Honda Group only broadly constrain the age of the sequence. A variation of this approach ("Method 1, variation A") assumes that each date represents an independent volcanic event and an independent estimate of the age of its associated horizon, each date is a maximum age for that horizon, and each date has some chance of being incorrect or inaccurate because of analytical error or problems associated with the dated mineral (diagenesis, nonclosed system behavior, etc.). Using this assumption, the Honda Group is constrained in age to be approximately 1.5 million years long, ranging in age between 12.2 and 13.7 Ma (although it could range slightly younger but is minimally constrained by the 10.1 and 8.4 Ma dates from overlying strata).

A second method is to assume that (1) the radioisotopic dates all are extremely accurate (accepting that they truly lie somewhere within the interval defined by the mean age determination and the associated error range), and (2) each individual date must be contemporaneous with, or older than the horizon sampled. This assumption and the available dates provide maximum age constraints for the upper (Villavieja Formation) Honda Group of 12.2–12.5 Ma (representing at least two separate episodes of volcanism) and for the lower (La Victoria Formation) Honda Group of 13.3–13.8 Ma (representing at least two different volcanic episodes).

A third method is to assume that most, or all, of the horizons analyzed by $^{40}Ar/^{39}Ar$ techniques sampled volcanic material representing the same volcanic event. This would require the entire section to be younger than 12.2 Ma, based on the age of the youngest Honda Group horizon. The relatively consistent decrease in age of dated horizons moving upward in the section indicate that this method probably cannot be applied validly to the Honda Group sequences.

BIVARIATE MAGNETIC POLARITY / GEOMAGNETIC POLARITY TIME SCALE PLOTS

We used "Method 1, variation A" (in which the dates are interpreted to represent maximum ages for the associated horizons, recognizing that error/inaccuracy associated with the dates preclude rigorously requiring age estimates based on magnetic polarity correlations to be younger than all of the associated dates) in interpreting the radioisotopic age constraints on correlation of the magnetic polarity stratigraphy. In analyzing magnetostratigraphic and radioisotopic data from the Miocene of the East African Rift, Tauxe et al. (1985; see also Deino et al. 1990) used an approach similar to the "Method 1, variation A" discussed previously. We used the available dates and comparison to time scales of Berggren et al. (1985), Deino et al. (1990), McDougall et al. (1984), and Cande and Kent (1992), to constrain the possible correlation of the Honda Group magnetostratigraphy (figs. 3.2–3.4). Within the limitations imposed by the radioisotopic dates, there are only three reasonable alternative correlations to the magnetic polarity time scale, as illustrated in the bivariate plots of figures 3.2–3.4. All other possible correlations would violate constraints imposed by the radioisotopic ages because correlation to older magnetic polarity chrons would make all available dates younger than associated magnetic polarity intervals (even though the dates must be contemporaneous with, or older than, the horizons they date), and correlation to younger magnetic polarity chrons would make polarity events in the Honda Group younger than the two dated horizons within the overlying Neiva and Gigante formations. Figures 3.2–3.4 each show correlations of the Honda Group magnetostratigraphy to two "end-member" magnetic polarity time scales (McDougall et al. 1984; Berggren et al. 1985). These two time scales differ by about 7% in the numerical ages assigned to the same polarity chrons (Berggren et al. [1985] estimate about 7% younger than McDougall et al. [1984] estimate), and the two other comparative time scales provide intermediate estimates. These

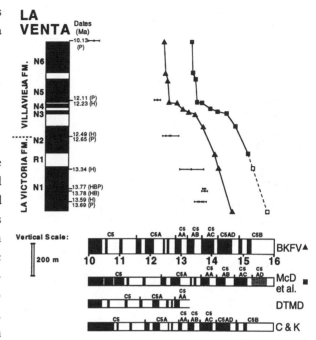

Figure 3.2. Correlation 1 for Honda Group magnetostratigraphy: C5Ar.1n–C5ADn (terminology of Cande and Kent 1992). Correlation of Honda Group magnetostratigraphy *(polarity intervals N1–N6, vertical axis)* and radioisotopic dates *(filled circles with error bars)* to a younger (Berggren et al. 1985; *BKFV, triangles*) and older (McDougall et al. 1984; *McD et al., squares*) "end-member" time scale *(horizontal axis).* Correlation to the Deino et al. (1990; *DTMD)* or Cande and Kent (1992; *C & K*) time scales would be similar to, but temporally between, the two "end-member" scales.

correlations assume there are no major hiatuses, and associated nonpreservation of a polarity chron, in the section. The radioisotopic dates from the Honda Group and overlying Neiva and Gigante formations constrain the entire Honda Group to be deposited in less than 4 million years (approximately 10–14 Ma) and, most likely, in less than 1.5–2.0 million years (approximately 12.2–13.8 Ma), based on the Honda Group dates alone. Given the extremely high sedimentation rates implied by deposition of about 1,000 m of section in less than 1.5–2.0 million years, it is very likely that all Miocene polarity events would be represented in the section. This is supported by evidence that major hiatuses probably are not present within the Honda Group, even at the boundaries between major lithologic units. For instance, the La Victoria Formation / Villavieja Formation boundary lies within a single normal polarity interval (N2), suggesting no major depositional hiatus. These correlations also assume that all polarity events preserved in the Honda

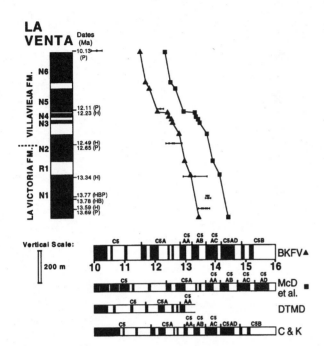

Figure 3.3. Correlation 2, our preferred correlation, for Honda Group magnetostratigraphy: C5An.1n–C5ABn (terminology of Cande and Kent 1992). See caption to figure 3.2 for definitions and clarifications.

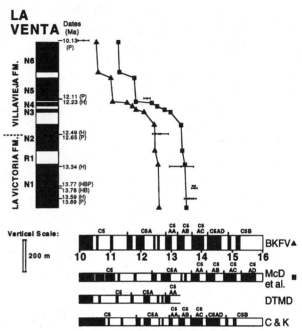

Figure 3.4. Correlation 3 for Honda Group magnetostratigraphy: C5r.1n–C5Ar.2n (terminology of Cande and Kent 1992). See caption to figure 3.2 for definitions and clarifications.

Group section represent true polarity chrons rather than possible short-term polarity reversals (less than 30,000 years duration) or intensity fluctuations hypothesized from some sea floor magnetic anomaly profiles ("tiny wiggles" of La-Brecque et al. 1977). This assumption is supported by the lack of any hypothesized "tiny wiggles" in the middle Miocene (approximately 10.8–17.8 Ma; Cande and Kent 1992).

The three possible Honda Group correlations each span approximately 1–2 million years. Using the polarity chron terminology of Cande and Kent (1992), the magneto-stratigraphic correlations are

Correlation 1: Chrons C5Ar.1n–C5ADn
Correlation 2: Chrons C5An.1n–C5ABn
Correlation 3: Chrons C5r.1n–C5Ar.2n

Of these three, Correlation 2 (fig. 3.3) clearly is the most reasonable correlation of the Honda Group to the magnetic polarity time scale of Berggren et al. (1985), as well as the "intermediate" age time scales of Cande and Kent (1992) and Deino et al. (1990). This correlation implies ages for the Honda Group strata that are younger than all the avail-

able radioisotopic date constraints, except that the boundary between the La Victoria and Villavieja formations is slightly too old for the 12.51 Ma horizon mean date (weighted mean of hornblende: 12.49 ± 0.11 Ma; of plagioclase: 12.65 ± 0.26 Ma). In Correlation 2 there is a slowing in the sedimentation rate in the Villavieja Formation as we would predict from sedimentologic differences between the Villavieja and La Victoria formations (see also chap. 2). If we estimate "instantaneous" sedimentation rates for individual polarity units, the rate during polarity interval N2 was two to three times higher than during the immediately preceding and following polarity intervals, suggesting a speed-up in sedimentation rate across the Villavieja/La Victoria formation boundary. Such an increase might be predicted by the presence of the Cerbatana Conglomerate Beds, the only braided stream deposit within the otherwise meandering stream sedimentary regime of the Honda Group. Within the Villavieja Formation, there also appears to be an exceptionally low relative rate of sedimentation during R4 (although this does fall within a red bed interval).

Reasonable tests of the validity of Correlation 2 would be provided by additional dates from within the upper Villavieja Formation (difficult because of the paucity of large, fresh pumices within this dominantly fine-clastic interval). If Correlation 2 is correct, there is probably a short (about 1-million-year) hiatus between the Honda Group and the Neiva Formation. A long normal polarity interval (Chron C5n.2n) should begin at the base, or within the lower part, of the Neiva Formation.

Each of the other two alternative correlations has serious deficiencies. Correlation 1 (fig. 3.2) seems highly unlikely. The ages calculated for horizons throughout the Honda Group are all significantly older than the available radioisotopic date constraints, contrary to expectations. This is true for comparisons to all of the four standard magnetic polarity time scales although, of course, it is most serious for the oldest "end-member" (McDougall et al. 1984) scale in which the correlation suggests horizon ages about 1.3–2.0 million years older than the associated dates. Even the youngest "end-member" scale (Berggren et al. 1985) is seriously contradicted because the predicted horizon ages are about 0.4–1.0 million years older than associated dates.

Correlation 3 (fig. 3.4) is more reasonable than Correlation 1 (fig. 3.2), in that the horizon ages all (except for the 12.51 Ma and 13.34 Ma dates relative to the McDougall et al. 1984 scale) are younger than the associated radioisotopic dates. These horizon ages, however, are up to 1.7 million years younger than the associated dates (using Berggren et al. 1985), implying reworking from much older sources, in a manner in which all of the horizons consistently are 1.0–1.7 million years younger than the associated dates. This seems unlikely, given the relatively large and fresh pumice clasts that were dated from these horizons and the difficulty in envisioning a mechanism for recording uniformly offset dates throughout the section. In addition, "instantaneous" sedimentation rates for individual polarity intervals vary dramatically and unpredictably (not correlated with lithology; see extreme slope changes between adjacent polarity intervals, fig. 3.4). For example, rates range over more than 2 orders of magnitude within the entire Honda Group, between the La Victoria and Villavieja formation averages, within the generally "low-energy" fine clastic deposits of the Villavieja Formation (vary up to 1–2 orders of magnitude), and even between adjacent polarity intervals. Sedimentation rates appear unusually high for polarity intervals

N5 and N6 within the red bed sequence of the upper Villavieja Formation.

DISCUSSION

Chronostratigraphy and Geochronology

One of the most exciting results of our geochronologic study is the first detailed and precise chronostratigraphic correlation for the entire Honda Group in the study area. This correlation permits us to establish that the La Venta fauna spans the interval of at least 12.2–13.8 Ma (based on dates alone) or 11.6–13.5 Ma (based on magnetostratigraphic correlations); the biochronologic duration of this fauna may have been longer, but there are no older strata exposed, the uppermost 200 m of the Honda Group is poorly fossiliferous in this area (a single edentate skull, not age diagnostic, is known from the top of the section), and the overlying Neiva Formation is unfossiliferous. This age for the tropical La Venta fauna indicates that it is temporally correlative with at least some faunas that have been assigned to the "Friasian" (this and other ages sometimes considered coeval or as subdivisions, referred to below in quotes, are currently under revision; also see chap. 30) and Chasicoan Land Mammal Ages in temperate latitude Patagonia, although there is limited faunal similarity between these contemporaneous faunas. Contrary to some previous suggestions, it is clear that the La Venta fauna is not coeval with Santacrucian Land Mammal Age faunas from higher latitudes but instead is up to 4–6 million years younger than the Santacrucian.

Before our work on the Honda Group at La Venta, the only available geochronologic information consisted of fission-track dates on three horizons near the middle of the Honda Group (near the base of the Villavieja Formation; Takemura and Danhara 1986) and paleomagnetic information (Hayashida 1984) on a very short stratigraphic interval (85–90 m; less than 10% of the thickness of the Honda Group section we analyzed) bracketing the fission-track horizons. The magnetic polarity stratigraphy consisted mainly of a normal polarity interval overlain by a shorter reversed polarity interval. This pattern originally (Hayashida 1984) was correlated to either part of Chrons C5AC–C5AD ("Epoch 15"), based on assumed correlation between the La Venta fauna and "Friasian" faunas dated at about 14–15 Ma (Marshall et al. 1977), or Chron C5N ("Epoch 9," the longest Miocene normal polarity chron), assuming their "thick" normal represented a long time and correlation to Chron C5N (Lowrie and Alvarez 1981) was not contradicted by the approximately 8 Ma dates from the upper part (Gigante Formation) of the overlying Mesa Group (no dates were then available from the Neiva Formation).

Takemura and Danhara (1986) reported three fission-track dates (15.7 ± 1.1 Ma, believed to represent the "lower red bed" of Fields [1959], but now [see Guerrero 1990] known to be correlative with the stratigraphically lower Ferruginous Sands with a recalculated age of 13.00 ± 0.88 Ma; 14.6 ± 1.1 Ma from the "Unit below Fish Bed"; and 16.1 ± 0.9 Ma from near the "Monkey Unit") from the upper Honda Group. They concluded that the magnetostratigraphy of Hayashida (1984) was best correlated to part of Chrons C5AC–C5AD ("Epoch 15") and "Friasian" faunas from the Argentine Collón Curá Formation (14.0–15.4 Ma; Marshall et al. 1977). Interestingly, Hayashida (1984, 82) previously noted that in samples from the red beds of the upper Honda Group "unfortunately, zircon crystals extracted from the bentonite samples had scuffed surfaces, which are characteristics of reworked material. Consequently, fission-track dating of these samples was abandoned."

Taken at face value, those results (Hayashida 1984; Takemura and Danhara 1986) indicated an approximate temporal correlation of the "La Venta fauna" with classic Santacrucian or earliest "Friasian" ("Colloncurense") faunas from Argentina, rather than middle or late "Friasian" ("Friasense," "Mayoense") faunas as suggested by many previous workers (Hirschfeld and Marshall 1976; and time scales in Marshall, Drake, et al. 1986, and in Marshall, Cifelli, et al. 1986).

After our first field season of geochronologic work, we collected and ^{40}Ar/^{39}Ar dated pumice samples from the upper Honda Group and the overlying Neiva and Gigante formations. Those results (Flynn et al. 1989) constrained the age of the Honda Group to be older than about 10 Ma, based on preliminary ^{40}Ar/^{39}Ar dates of 12.7 Ma from the San Francisco Sandstone (upper Villavieja Formation), 10.1 Ma from low in the Neiva Formation, and 8.4 Ma from the

base of the overlying Gigante Formation. The dates from the Neiva and Gigante formations are derived from large and very fresh pumice clasts that are unlikely to have been reworked or transported very far from the source, suggesting contemporaneity between volcanic source eruption/pumice age and the horizons in which the pumices are preserved. The dates for the Neiva and Gigante formations are likely to date accurately the horizons in which the pumice is preserved.

Consideration of those preliminary ^{40}Ar/^{39}Ar dates and the fission-track dates from the middle of the Honda Group suggested that sedimentation rates might have been relatively constant across the entire Honda and Mesa Groups, and that the fossiliferous Honda Group (containing the "La Venta fauna") might span a temporal interval as long as 10 million years. Such a lengthy time span for the Honda Group, equivalent to the time represented by at least three to four land mammal ages in southern South America, with little change in the La Venta fauna (assignable to a single land mammal age) would have profound implications for concepts of rates of evolution in the tropics, biogeography and faunal interchange, and biochronologic correlation. There were, however, no ^{40}Ar/^{39}Ar dates from the lower two-thirds of the Honda Group, and it was impossible to force a preliminary magnetostratigraphy (Flynn et al. 1989) to fit such a lengthy time span because there was no reasonable correlation of the relatively small number of magnetic polarity intervals present in the Honda Group to any sequence of polarity intervals spanning 10 million years of the Miocene. These points emphasize the dangers inherent in "overinterpreting" preliminary geochronologic data, short magnetostratigraphies, studies with only one or a small number of dates not broadly spaced over the stratigraphic interval of interest, or dates in which it is difficult to determine contamination by older grains or nonclosed system behavior of the dated minerals.

We now have three additional ^{40}Ar/^{39}Ar dates, one from the middle Honda Group and two from near the base of the Honda Group. Reanalysis (Guerrero 1990, 1993) of the fission-track data of Takemura and Danhara (1986) showed that the grains analyzed in that study were derived from at least two different source populations, indicating detrital contamination of the samples (as suggested by Hayashida 1984 for samples from the section ["LV13"] that produced one of the three samples we dated) and a resultant anomalously old age for both of the fission-track dates. The section sampled by Hayashida (1984) and Takemura and Danhara (1986) also is fault repeated, miscorrelated, and actually is quite thin, probably representing only 50–70 m of the almost 1,250-m-thick Honda Group. Graphic presentation (figs. 3.1–3.4) of all the available $^{40}Ar/^{39}Ar$ ages versus stratigraphic position show clearly that the Honda Group was deposited over a very short temporal interval of probably less than 2 million years. Additionally, it is clear (figs. 3.2–3.4) that sedimentation rates were not constant throughout the entire Honda Group–Neiva Formation– Gigante Formation sequence. Rather, there were significant changes in sedimentation rate within the Honda Group. Near the Honda Group/Neiva Formation contact is either another sediment accumulation rate change (fig. 3.4) or, more likely, an unconformity between the upper Honda Group and the Mesa Group (figs. 3.2, 3.3).

These conclusions, based on the radioisotopic data alone, are supported and enhanced by the paleomagnetic data (cited previously) and sedimentology (chap. 2). Within the Honda Group, sedimentologic evidence (chap. 2) from paleosols (abundance, maturity, mineral alteration, etc.); lithology and grain size; sedimentary structures; and channel geometry, lithology, and abundance indicate that the La Victoria Formation was deposited much faster than the Villavieja Formation. Further, the maturity and abundance of paleosols (and concurrent reduction in size and abundance of meandering channels) suggest especially slow sedimentation rates in the red beds of the Cerro Colorado Member of the Villavieja Formation. Only the coarse-pebble Cerbatana Conglomerate Beds, which cap the La Victoria Formation, represented braided stream deposition. Lateral continuity of this unit throughout the entire basin suggests a short phase of rapid tectonic change during uplift and volcanism of the source area in the Central Cordillera (chap. 2).

Sediment Accumulation Rate

Using our preferred Correlation 2 (fig. 3.3), we estimate the temporal durations of the La Victoria Formation and Villavieja Formation at about 0.6 million years and 1.3 million years, respectively. Our preferred correlation (in fact any of the three possibilities) indicates that the sediment accumulation rates were about three times faster in the La Victoria Formation than in the Villavieja Formation, implying a major change in the depositional regimes in the La Venta area somewhere near the time of the La Victoria/Villavieja Formation boundary. The relationship between regional and global climatic, tectonic, and sedimentary regime influences and the observed changes in sediment accumulation rates are being explored by us. The major change in sediment accumulation rates is consistent with the sedimentological evidence and depositional system interpretations presented by Guerrero (chap. 2).

When the available magnetic polarity data presented here are evaluated in comparison to the sedimentologic data, we draw the following conclusions about sedimentation rates and patterns. First, sedimentologic data indicate that the Cerbatana Conglomerate Beds, at the top of the La Victoria Formation, represent the only episode of high-gradient, extremely rapid, braided-channel deposition within the Honda Group. Sedimentation in the remainder of the underlying La Victoria Formation represents slower, meandering-channel deposition, whereas the overlying Villavieja Formation represents the slowest deposition within the sequence. The paleomagnetic data show a long normal polarity interval (N2) spanning the upper La Victoria Formation (including its capping Cerbatana Conglomerate Beds) through lower Villavieja Formation, making it impossible to use paleomagnetic information to constrain variations in sedimentation rate across this hypothesized short pulse of very rapid sedimentation. Comparison of inferred sedimentation rates indicates, however, that sedimentation was significantly more rapid during the normal polarity interval (N2) than during the intervals immediately preceding (R1) and postdating (R2) it, in agreement with the sedimentologic data from the Cerbatana Conglomerate Beds. In fact, using any of the possible reasonable magnetic polarity correlations discussed previously (figs. 3.2–3.4), deposition rate during this interval (N2) was as fast as, or faster than, the rate in any other part of the La Victoria Formation and was much faster than during any part of the Villavieja Formation, again in agreement with the sedimentologic evidence. Second, the sedimentologic hypothesis of faster sedimentation rate for the La Victoria Formation than for any part of the Villavieja Formation is corroborated by the paleomagnetic information (figs. 3.2–3.4).

Time Scales

The La Venta magnetostratigraphic correlations have some implications for calibration of standard time scales, which differ by up to 7% in ages assigned to Miocene polarity chrons. It is noteworthy that our preferred Correlation 2 and Correlation 1 (and, possibly, Correlation 3), and associated radioisotopic date information, are incompatible with the polarity ages assigned in the McDougall et al. (1984) time scale. There has been debate over the validity of that time scale and its calibration dates (e.g., Tauxe et al. 1985; Deino et al. 1990), and our data suggest that the numerical ages assigned to Miocene polarity chrons by McDougall et al. (1984) indeed are too old. Our data currently are not sufficiently resolved, however, to decisively resolve which of the remaining three time scales (Berggren et al. 1985; Deino et al. 1990; Cande and Kent 1992) provides the best numerical age estimates for Miocene polarity chrons.

Biostratigraphy and Biochronology

Our paleomagnetic and geochronologic results have significant implications for interpreting the biostratigraphy of the La Venta sequence.

The La Victoria Formation clearly spans very little time and represents a consistent depositional paleoenvironment. We would not expect temporally significant or environmentally driven changes in the fauna nor major evolutionary changes within individual lineages. Lineages with marked transformations within the La Victoria Formation must have been evolving extremely rapidly in situ or had an influx of representatives with significantly different morphologies from other areas. The La Victoria/Villavieja Formation boundary almost certainly does not represent a major depositional or erosional hiatus, based on the close concordance of radioisotopic ages above and below the boundary, and indications of continuous, rapid rates of sediment accumulation across the boundary. The Cerbatana Conglomerate Beds mark a short pulse of very high-energy, braided stream deposition within a continuous series of somewhat lower energy deposits dominated by meandering stream systems, without any associated marked temporal gap. Similarly, the overlying Monkey Beds–Fish Bed interval (at the base of the Villavieja Formation) falls within the same interval of rapid sediment accumulation and normal magnetic polarity as the upper La Victoria Formation, implying continuity of sedimentation across this entire interval. The lack of evidence for any significant depositional hiatus stands in contrast to the biostratigraphic evidence from the Monkey Beds–Fish Bed interval, which suggests some faunal change in the Villavieja Formation relative to the older units (see composite biostratigraphic chart; figs. 29.1, 29.5). Although absolute sediment accumulation rates remained very high throughout the entire La Victoria and Villavieja formations, there are dramatic absolute and relative sedimentation rate and lithologic changes beginning in the lower Villavieja Formation (Monkey Beds–Fish Bed interval). These lithologic differences indicate significant paleoenvironmental changes, probably to more humid, higher rainfall, less seasonal, riparian and flood-plain environments. Therefore, the faunal changes beginning across the La Victoria/Villavieja formation boundary may be the result of regional tectonic climate/environmental changes, and possibly biostratigraphic sampling of different facies/paleoenvironments (creating taxonomic biases because of different taxic environmental distributions), rather than to any dramatic faunal turnover resulting from extinction/immigration/temporal differences.

CONCLUSIONS

This study presents a detailed paleomagnetic stratigraphy (more than 210 sites) and associated single-crystal laser fusion ^{40}Ar/^{39}Ar dates (6 horizons) for more than 1,000 m of the Honda Group in the La Venta area, Magdalena River valley, Colombia. Additional dates were determined for the overlying Neiva and Gigante formations. Dates alone indicate that the La Venta fauna and sampled Honda Group strata certainly are younger than 13.8 Ma and older than 10 Ma. Our dating and the most reasonable correlation of the magnetostratigraphy indicates that the Honda Group spans approximately 1.8–2.0 million years (about 11.6–13.5 Ma) during Magnetochrons C5An.1n–C5ABn (terminology of Cande and Kent 1992).

This approximately 11.6–13.5 (or 13.8) Ma age for the tropical La Venta fauna indicates that it is temporally correlative with at least some faunas that have been assigned to the "Friasian" and Chasicoan Land Mammal Ages in temperate latitude Patagonia, although there is little faunal similarity between these contemporaneous faunas. Contrary to some previous suggestions, it is clear that the La Venta fauna is not coeval with Santacrucian Land Mammal Age faunas from higher latitudes but instead is up to 4–6 million years younger than the Santacrucian (Flynn and Swisher in press).

Our preferred correlation (fig. 3.3; in fact, any of the three possibilities) indicates that the sediment accumulation rates were about three times faster in the La Victoria Formation than in the Villavieja Formation, implying a major change in the depositional regimes in the La Venta area somewhere near the time of the La Victoria/Villavieja formational boundary. Sedimentologic data indicate that the Cerbatana Conglomerate Beds, at the top of the La Victoria Formation, represent the only episode of high-gradient, braided-channel deposition within the Honda Group. Sedimentation in the remainder of the underlying La Victoria Formation represents slower, meandering-channel deposition, whereas the overlying Villavieja Formation represents the slowest deposition within the sequence.

The La Venta magnetostratigraphic correlation has implications for calibration of standard time scales, which differ by up to 7% in ages assigned to Miocene polarity chrons. Our preferred correlation (fig. 3.3, as well as the next most reasonable alternative, fig. 3.4) and associated radioisotopic date information are incompatible with the polarity ages assigned in the McDougall et al. (1984) time scale. Our data are not yet sufficiently resolved to permit decisive resolution of which of three other time scales (Berggren et al. 1985; Deino et al. 1990; Cande and Kent 1992) provides the best numerical age estimates for Miocene polarity chrons.

The La Victoria Formation clearly spans very little time (about 0.6 million years) and represents a consistent depositional paleoenvironment.

The La Victoria/Villavieja formational boundary almost certainly does not represent a major depositional or erosional hiatus, based on the close concordance of radioisotopic ages above and below the boundary and indications of continuous, rapid rates of sediment accumulation across the boundary.

Lithologic differences between the La Victoria and Villavieja formations are the result of changes in sedimentation rates and may also indicate significant paleoenvironmental changes. Therefore, the faunal changes across the La Victoria/Villavieja formation boundary may be the result of regional tectonic and climate/environmental changes and, possibly, biostratigraphic sampling of different facies/paleoenvironments (creating taxonomic biases because of different taxic environmental distributions) rather than to any dramatic faunal turnover resulting from extinction/immigration/temporal differences.

Acknowledgments

The research presented in this paper was supported by United States National Science Foundation grants (NSF BSR-8896178/8614133 to J. Flynn and NSF BSR-8614533 to R. Kay). We thank the Field Museum (U.S.), INGEOMINAS (Colòmbia), Duke University (U.S.), and the Institute for Human Origins (U.S.) for use of facilities, personnel, and equipment and for logistical and financial support. From the Field Museum, we thank C. Nuñez and D. Eatough for extensive and thorough assistance in paleomagnetic laboratory analyses, E. Gyllenhaal for figure drafting and artistic advice, E. Zeiger for typing, and R. Masek for assistance in paleomagnetic sample preparation. We benefited greatly from discussion and sampling assistance during field work in Colombia with R. Kay, L. Marshall, R. Madden, R. Cifelli, and A. Walton. R. Cifelli and B. MacFadden provided helpful reviews which significantly improved this paper.

Part Three
Invertebrates and Fishes

4
Trichodactylid Crabs

Gilberto Rodríguez

RESUMEN

La presencia en el Grupo Honda del cangrejo *Sylviocarcinus piriformis* (Trichodactylidae) se establece en base a la morfología de los quelípedos, el número y la posición de los dientes grandes, y la posición del tubérculo de gran tamaño en la superficie externa del dedo móvil. El ápice bífido del dedo inmóvil es una característica diagnóstica para la familia Trichodactylidae. Actualmente *Sylviocarcinus piriformis* se encuentra desde el nivel de mar hasta 350 m de altura en tres áreas disyuntas: la cuenca de Maracaibo, el valle del Río Cesar (y los riachuelos que bajan desde la Sierra Nevada de Santa Marta), y el sector del valle del Río Magdalena entre Puerto Boyacá y el Río Gualanday. El registro de la especie en el valle alto del Magdalena durante el Mioceno medio cuando los ríos desembocaban al este de la Cordillera Oriental sugiere que la distribución geográfica desconectada actual corresponde a una distribución relictual interrumpida por cambios paleogeográficos.

During the 1987 INGEOMINAS–Duke University Paleontology Expedition to the La Venta area in the upper Magdalena River valley, numerous specimens of the right cheliped of a trichodactylid crab were obtained from middle Miocene fossiliferous outcrops of the Honda Group. The Trichodactylidae are at present a conspicuous element of the fauna of rivers and lakes of the lowland basins of South America, east of the Andes, and of an isolated area in southern Mexico (Rodríguez 1986, 1992), but heretofore no fossil remains have been described. Furthermore, the only fossil freshwater crabs reported until now are species of the genus *Potamon,* Family Potamonidae, from Pliocene to recent sites in central Europe, Turkey, and India (Rathbun 1904; Glaessner 1969; Pretzmann 1972), and a Pseudothelphusidae from the Quaternary of Venezuela (Rodríguez and Díaz 1977).

The material studied is deposited in the fossil collections of INGEOMINAS, Bogotá, Colombia. In the text, the abbreviation cel. stands for "cutting edge length," that is, the length of the toothed portion of the fingers measured from the first proximal tooth to the tip, and IGM refers to the numbers used by INGEOMINAS in their collections.

Systematic Paleontology

Family Trichodactylidae H. Milne Edwards 1853
Subfamily Dilocarcininae Pretzmann 1978
Tribe Holthuisiini Pretzmann 1978
Sylviocarcinus H. Milne Edwards 1853
Sylviocarcinus piriformis (Pretzmann 1968)
(fig. 4.1)

SYNONYMY

Valdivia (*Valdivia*) *piriformis* Pretzmann 1968b, 73. For a detailed synonymy and references, see Rodríguez (1992).

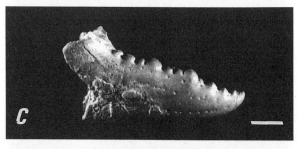

Figure 4.1. *A.* External view of the hand and fingers of the left cheliped of *Sylviocarcinus piriformis* (Pretzmann 1968) from a Recent specimen collected in the Tocuyo River basin, Venezuela, carapace length 46 mm. *B.* Mobile finger from a Miocene specimen of *S. piriformis,* IGM 250583, external view. *C.* Same, immobile finger, external view. *D.* Same, immobile finger, externo-dorsal view to show the bifid tip characteristic of the family Trichodactylidae. Scale bars 4 mm.

MATERIAL

All specimens were collected by the INGEOMINAS–Duke University Paleontology Expedition from fossiliferous outcrops of the Honda Group within the 100-km² in the La Venta area on the east side of the Magdalena River valley, Huila Department, Colombia, at the following localities. Duke Locality 32, El Cardón Red Beds: one immovable finger cel. = 19.0 mm, one movable finger cel. approximately 23 mm (tip broken) (IGM 250583). Duke Locality 25: 1 immovable finger cel. = 14.2 mm, 2 movable fingers cel. = 14.6 and 13.5 mm (IGM 184482). Duke Locality 31, Monkey Beds: one movable finger cel. approximately 14 mm (tip broken), fingers and part of palm, movable finger cel. = 15 mm (IGM 183285). Duke Locality 54, Monkey Beds: one movable finger cel. approximately 20 mm (tip broken) (IGM 183531). Duke Locality 74: six finger fragments (IGM 183097). Duke Locality 98, Monkey Beds: palm with fragment of immovable finger (IGM 184559). Duke Locality 8: immovable finger cel. approximately 19 mm (tip broken), two movable fingers cel. approximately 17 and 18 mm (both tips broken), eighteen finger fragments (IGM 183699). Same locality: two immovable fingers cel. = 15.7 and 13.8 mm, one movable finger cel. = 12.8 mm, three fragments of fingers (IGM 184260). Same locality: three finger fragments, one fragment of left finger (IGM 183707). Duke Locality 86: one immovable finger cel. = 15.5 mm, one movable finger cel. approximately 17 mm (tip broken), one fragment immovable finger (IGM 184278). Duke Locality 25: one immovable finger cel. = 17.7 mm, one movable finger cel. = 20.0 mm, both fingers probably from the same cheliped (IGM 183110). Duke Locality 63: two immovable fingers cel. approximately 14.6 mm (tip broken) and 10.2 mm, three movable fingers cel. = 19.7, 14.9, and 12.0 mm (IGM 183758). Duke Locality 103: two movable fingers, cel. = 23.6 and 16.9 mm, immovable and movable fingers embedded in rock, seven fragments of fingers (IGM 184642). Duke Locality 67: one immovable finger cel. = 13.0 mm, two movable fingers cel. = 14.8 and 12.4 mm, fingers with fragment of palm, cel. of movable = 12.3 mm (IGM 184748). Duke Locality 91: two immovable fingers cel. approximately 16.0 mm (tip broken) and 11.5 mm, three movable fingers cel. approximately 16.8, 17.0, and 22 mm (all tips broken), palm and bases of fingers partially embedded in rock, fragment of palm, fragment of immovable finger (IGM 184420). Duke Locality 40: fragment of palm and fingers (IGM 184391).

DESCRIPTION (FIG. 4.1)

The fingers of the left cheliped nearly meet; both fingers as seen in dorsal and ventral views are conspicuously and regularly bent inward. The immovable finger is almost one-fifth as high as long proximally (length/height = 0.18, IGM 250583); the lower margin is almost straight for two-thirds of its extent, rounded in its distal third; the adjacent part of the palm is also nearly straight; the series of teeth of the cutting edge (IGM 250583) begin proximally with three small cusps, followed by a large tooth, two small cusps, another large tooth, and three intermediate-sized cusps; the tip is bifid, with the inner point placed at an angle with the axis of the segment and less developed than the outer one; the interdental spaces between the last two cusps and the outer point are well defined by grooves; the more proximal of these grooves extends for a short length over the outer surface of the finger, the more distal one is an extension of an external groove (IGM 183265); both grooves are well defined. The movable finger is almost one-third as high as it is long (length/height = 0.27, IGM 250583), the upper margin is regularly and gently rounded; the series of teeth of the cutting edge (IGM 250583) arise proximally with two intermediate-sized teeth that are continued outwardly by a strong tubercle followed by a small cusp, a large tooth, two small cusps, a large tooth, a small cusp, and two middle-sized teeth. The tip is directed more internally in relation to the last tooth of the series and, consequently, the last dental interspace faces distally; this interdental space extends, in some specimens (IGM 1844681, IGM 184642, IGM 184642), as a short groove over the outer surface of the finger. In some specimens the upper surface of the movable finger is covered with papillae (IGM 184260, IGM 184642).

DIAGNOSTIC CHARACTERS

The bifid tip of the immovable finger is present in all Trichodactylidae, and it has diagnostic value for the family. In some marine crabs there is, occasionally, an internal papilla, but it is arranged differently than that of the Trichodactylidae. This is the case in all four species of Grapsidae, *Sesarma curacaoense* de Man 1892, *Geograpsus lividus* (Milne Edwards 1837), *Hemigrapsus nudus* (Dana 1851), and *Goniopsis cruentata* (Latreille 1803), where an internal papilla results from the division of the spoon-shaped tip found in other Grapsidae.

The chelipeds display the same general shape in almost all

members of the family Trichodactylidae except in some aberrant cases such as old males of *Trichodactylus fluviatilis* Latreille 1828, *Valdivia gila* Pretzmann 1978, and *Forsteria venezuelensis* (Rathbun 1904), but the large tubercle on the outer surface of the movable finger is characteristic of the species. The number and position of the large teeth also tends to be constant for the species, but the number of small teeth may vary. In the fossil material the number of small cusps between the large teeth could be one or two, but in recent material examined by me (Rodríguez 1992), the number of small cusps vary from zero to three. Some small denticles may be present on the external surface at the base of the teeth. In recent specimens the chelipeds are strongly unequal. In old males the larger chela becomes enormously developed (hand >1.5 times as long as the carapace), and the fingers have a large gap between them. This last type of cheliped was not observed among the fossil material.

BIOGEOGRAPHY

At present, *Sylviocarcinus piriformis* is found up to 350 m above sea level in three disjunct areas (fig. 4.2): the Maracaibo Depression, where it inhabits almost all river basins. This depression is separated from the Magdalena Basin by the Cordillera Oriental and the Sierra de Perija. On the eastern side this depression is closed by the Sierra de Merida and the high mountains that end on the Falcón coastal plain. The records on the Falcón coastal plain and the Río Tocuyo basin are probably extensions of this general area. It is also found in the Río Cesar valley and the small river basins draining the Sierra de Santa Marta and the upper Magdalena valley at several localities between Puerto Boyacá (Boyacá Department) and the Río Gualanday (Tolima Department) (Von Prahl 1982; Rodríguez 1992). The fossil records described here come from about 200 km upriver from the Río Gualanday. The species has not been recorded from the middle and lower Magdalena valley except for a problematic record at La Regla, an undetermined locality, presumably in Bolívar Department (type locality of *Valdivia torresi* Pretzmann 1968, a synonym of *V. piriformis*).

The characteristic disjunct pattern of distribution in three areas suggests a relict distribution interrupted as a

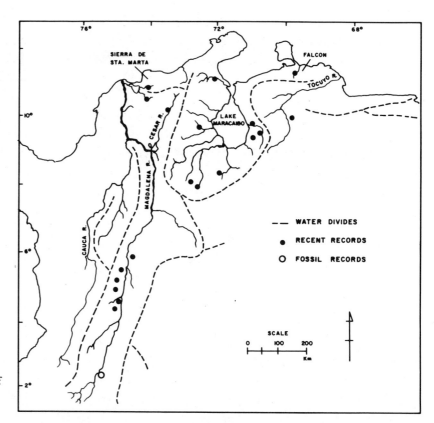

Figure 4.2. Fossil and Recent distribution of *Sylviocarcinus piriformis*. Abbreviations: *R.*, river; *Sta.*, Santa.

result of paleogeographic changes. Regarding the Cesar–Sierra de Santa Marta area, the similarity of the fish fauna of the Maracaibo and Cesar basins (Dahl 1971) indicates a close relationship between the two areas. Occupancy of the Cesar–Sierra de Santa Marta area by freshwater organisms, however, must be a relatively recent event because marine conditions prevailed in the lower Magdalena from mid-Miocene to Pliocene times (Harrington 1962), and, consequently, the crab populations from the Cesar–Santa Marta

basins probably are extensions of the populations of the Maracaibo Depression. Regarding the Magdalena valley, Wellmann (1970) and Guerrero (chap. 2) have shown, by the study of paleocurrents during deposition of Miocene sediments, that the rivers flowed to the east in the lower and middle reaches of this valley. The presence of the species in the Magdalena valley in the middle Miocene reinforces the possibility of a continuous distribution of the species in the upper Magdalena and the Maracaibo basin at this time.

5
Freshwater Fishes and Their Paleobiotic Implications

John G. Lundberg

RESUMEN

Las extensas aguas dulces de América del Sur sostienen en la actualidad la más rica fauna de peces sobre la Tierra. Una comprensión cabal de la evolución de esta fauna ha sido imposible debido a la naturaleza del archivo fósil. La abundancia de especímenes de peces del Grupo Honda, la mayoría huesos y dientes desarticulados pero también algunos cráneos y mandíbulas, sirven para aumentar el número de taxa reconocidos para el Neógeno. La fauna de La Venta abarca el conjunto más diverso de peces de agua dulce que se conoce en América del Sur. Hasta ahora, solamente cinco especies habían sido descritas para la fauna. En este estudio, dieciocho peces distintos son identificados y descritos y las implicaciones de éstos para la historia biótica y ambiental de esta región son discutidos.

Morfológicamente estos peces se asemejan mucho a géneros, y en algunos casos hasta especies, actuales. Las formas presentes incluyen los Potamotrygonidae (Elasmobranchii); *Lepidosiren paradoxa* (Dipnoi); el osteoglosiforme *Arapaima* (Arapaimidae); characiformes *Hoplias* (Erythrinidae), *Hydrolycus* (Cynodontidae), y *Leporinus* (Anostomidae); charácidos Tetragonopterinae no determinado y dos Serrasalminae (*Colossoma macropomum*, y *Serrasalmus, Pygocentrus,* o *Pristobrycon*); los siluriformes Pimelodidae *Brachyplatystoma* cf. *B. vaillanti*, una forma semejante a los del grupo "*Malacobagrus,*" cf. *Pimelodus*, y *Phractocephalus hemiliopterus*, un Ariidae no determinado, algunos Doradidae, *Hoplosternum* (Callichthyidae), y Loricariidae *Acanthicus* (Ancistrinae), cf. *Hypostomus* (Hypostominae) y dos otros taxa; y material perciforme Cichlidae.

Los peces representan una gran diversidad de tipos ecológicos, abarcando tanto especies que habitan el bentos, el semibentos, y las zonas pelágicas de grandes y amplios ríos, como especies de la orilla y zonas marginales de aguas poco profundas y calmadas. Cuatro peces tienen mecanismos especializados para la respiración aérea (*Lepidosiren, Hoplias, Hoplosternum,* y *Arapaima*), los primeros tres de los cuales se encuentran frecuentemente en aguas anóxias hasta efímeras. Los peces de agua dulce de La Venta representan un amplio espectro de categorías tróficas, abarcando carnívoros generalizados hasta especializados, piscívoros, herbívoros, algívoros, detritívoros, y formas especializadas para frugivoría y granivoría. En conjunto, la piscifauna se asemeja a aquella de los extensos panoramas acuáticos de las cuencas bajas y centrales de los ríos Amazonas y Orinoco, de elevaciones menores a 250 m sobre el nivel del mar.

La mayor parte de los taxa guardan relaciones filogenéticas con peces actuales que habitan la Amazonia. Gran parte de las semejanzas morfológicas se interpretan como sinapomorfías, indicando relaciones filéticas estrechas, y así proveen puntos claves para la historia de estos linajes. Los fósiles de *Colossoma macropomum* y *Phractocephalus hemiliopterus*, taxa terminales en marcos filogenéticos disponibles, implican diversificación a nivel genérico dentro de Serrasalminae y Pimelodinae ya para el Mioceno medio. La descripción de una piraña en la fauna implica que los géneros de Serrasalminae *Pygopristis*, *Catoprion*, y *Metynnis* se remontan al menos hasta el Mioceno medio. Para los bagres Pimelodidae, la presencia de una especie cercana a *Pimelodus* y dos especies de *Brachyplatystoma* indica una diversificación ya marcada para el Mioceno medio en los grupos basales o troncales de muchos géneros de este gran grupo monofilético. Estos y los demás elementos de la ictiofauna de La Venta demuestran que la piscifauna del Mioceno medio en América del Sur ya había alcanzada una riqueza taxonómica esencialmente moderna.

Aunque rica en taxa, la piscifauna del Grupo Honda representa una pequeña fracción de la ictiofauna moderna de las aguas dulces de América del Sur. Gran parte de los taxa de La Venta son taxa endémicos en las cuencas de la Amazonia y Orinoquia y de estrecha relación filética con peces que habitan la Amazonia hoy en día. Ninguno de los peces de La Venta pertenece a taxa ahora endémicos al sistema hidrográfico del Río Magdalena. Al contrario, los peces indican una relación zoogeográfica directa con el Río Amazonas y el Río Orinoco y así corroboran evidencias neontológicas para una distribución más amplia durante el Mioceno medio.

With perhaps as many as five thousand living species, South America has the richest freshwater fish fauna on Earth (Böhlke et al. 1978). An understanding of the patterns and processes that underlie the evolutionary generation and ecological maintenance of this diverse fauna has been frustrated by inadequate systematic, distributional, and paleontological information. Although paleontological localities are widespread from Venezuela to Argentina and Chile, knowledge of Cenozoic freshwater fishes from South America has rested on few specimens, and most of these have not been treated phylogenetically. Fortunately, the abundance of fossil fish specimens recently collected from the Miocene Honda Group of Colombia promises to expand considerably the range of taxa known from the Neogene. These fossils belong to an assemblage known as the La Venta fauna. Previous papers on La Venta fishes reported a total of just five species (Savage 1951b; Bondesio and Pascual 1977; Lundberg et al. 1986; Lundberg and Chernoff 1992). In this study an additional eighteen kinds of fishes are identified to at least family level (app. 5.1).

In general, the Miocene La Venta fishes are morphologically very similar to extant genera and, in some cases, even living species (Lundberg et al. 1986; Lundberg

and Chernoff 1992). Much of this similarity is interpretable as synapomorphy, thus indicating close phylogenetic relationship between the fossil and modern fishes. Because all of the La Venta fishes belong to extant Neotropical groups, they provide important benchmarks in the latest 13-million-year history of their lineages. This fish fauna also has considerable biogeographic significance (Lundberg et al. 1986; Lundberg and Chernoff 1992). It is located in the present Magdalena River valley in south-central Colombia, but it contains distinctive taxa that are presently endemic to the Amazon and Orinoco faunas to the east of the high drainage divide formed by the Cordillera Oriental (eastern Andes). Eight or nine of fourteen taxa identified to generic level or below are most closely related to fishes that occur today in the Amazon and, in some cases, the Orinoco basin, but not in the present Magdalena region. Five fish genera present in the La Venta fauna contain species living now in the Magdalena basin as well as the Amazon and Orinoco. None of the La Venta fishes, however, belong to taxa currently endemic to the Magdalena.

This chapter gives a systematic summary of the fish taxa known in the La Venta fauna. Descriptive information and illustrations bearing on the identification of fossils is given. The implications of these fishes for the late Tertiary environmental and biotic histories of the upper Magdalena River Valley region are discussed.

Materials and Methods

This report is based on specimens that were collected during the Duke University–INGEOMINAS expeditions to the upper Magdalena valley between 1985 and 1991 (chap. 1). Well over one thousand skeletal elements of fishes were collected. The vast majority of the fossils are three-dimensional, disarticulated, and usually broken bones and teeth. Details of surface features are generally well preserved. Rare articulated specimens include a few crania and mandibles. Most of the larger articulated specimens have been identified to at least family. Many of the small isolated teeth and spines obtained from screen washing remain to be studied in detail.

Specimens are identified by museum catalog numbers at INGEOMINAS (IGM). Comparative materials include dry and cleared and alizarin-stained skeletons and preserved specimens of modern fishes deposited at the American Museum of Natural History (AMNH), California Academy of Sciences (CAS), Duke University (DU), Field Museum of Natural History (FMNH), Museo de Biología, Universidad Central de Venezuela (MBUCV), University of Michigan Museum of Zoology (UMMZ), and the National Museum of Natural History, Smithsonian Institution (USNM). These specimens are listed in appendix 5.2.

Despite their fragmentary condition, many Cenozoic fish fossils from South America can be identified confidently to family, genus, and, in a few cases, species where adequate modern reference specimens are available for comparison (Roberts 1975; Cione 1986; Lundberg et al. 1986, 1988). Skull and postcranial elements of fishes offer complex and detailed characteristics of form and surface texture that are diagnostic over a wide range of the taxonomic hierarchy. It is rare, however, to find such diagnostic characteristics described in the literature, so identification requires ample comparative collections of skeletal specimens of modern species. Many of the identifications of South American fossil fishes will be better resolved as more complete samples of modern skeletal materials are obtained and analyzed comparatively.

The identification and systematic placement of fossils is a problem in phylogenetic systematics that is especially important where paleontological data are used to infer taxic and biogeographic patterns. Levels of confidence in the identification of Cenozoic fossil fishes vary widely. The soundest identifications, at all levels of the taxonomic hierarchy, are based on uniquely derived features (synapomorphies) plus strong overall similarity (i.e., lack of autapomorphic and plesiomorphic differences) shared by a fossil and its closest extant relatives. At the other extreme, identifications are most tentative where fossils exhibit similarities and differences of uncertain diagnostic value and where not all pertinent taxonomic comparisons to modern (and other fossil) taxa have been made. In this chapter the bases for identifications are made as explicit as possible. Four conventions of taxonomic syntax are used to indicate the general nature of evidence for identification. (1) Fossils assigned at the species level without equivocation are so identified on the basis of strong overall similarity, including diagnostic species-level apomorphic character states, determined by comparison to samples of all phylogenetically

related species. (2) Specimens assigned to a genus followed by *sp.* share diagnostic synapomorphies of the genus but cannot be confidently identified either as a known species or recognized as an undescribed species because of lack of species-level apomorphic diagnostic characters and / or because of incomplete sampling of congeneric species. (3) Fossils identified to family (or other suprageneric taxon) followed by *incertae sedis* share diagnostic synapomorphies of the taxon but cannot be placed at finer levels because they lack lower-level apomorphic diagnostic characters and because of incomplete comparison to relevant genera and species. (4) Specimens with an identification preceded by *cf.* share strong overall similarity with their taxon but no clear synapomorphies based on known material (which is often incomplete).

Systematic Paleontology

Class Chondrichthyes
Subclass Elasmobranchii
Order Myliobatiformes
Family cf. Potamotrygonidae
Gen. et sp. indet.
(figs. 5.1, 5.2B)

MATERIAL

IGM 251283, five isolated teeth (fig. 5.1) from the Fish Bed; and IGM 250484, three partial tail stings, two proximal pieces lacking serrated edges (10–15 mm overall length), one distal serrated piece (15 mm overall length; fig. 5.2B) from Duke Locality 32.

DESCRIPTION

Two teeth, likely from the upper jaws of dimorphic males, with elevated central cusps (fig. 5.1). Labial and lingual sides of crown demarcated by sharp-cornered transverse crest; labial surface of crown convex, coarsely ornamented with labio-lingually oriented ridges and a deep median sulcus; lingual surface smooth; transversely convex and dorso-ventrally concave. Root bilobed; each lobe with triangular, slightly convex basal face. Three teeth, likely from females or juvenile males, with low central cusp crossed by trans-

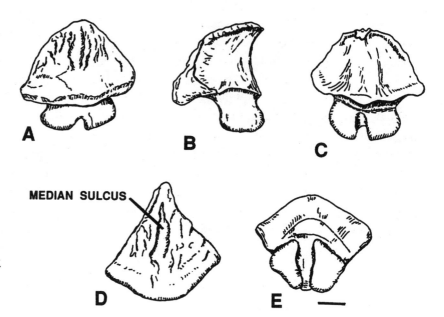

Figure 5.1. Fossil tooth of ?male sting ray, cf. Potamotrygonidae, IGM 251283. Views of five surfaces: *A*, labial; *B*, lateral; *C*, lingual; *D*, occlusal; *E*, root. Scale bar 0.4 mm.

MEDIAN SULCUS

verse crest; crown surfaces scarcely ornamented, mostly smooth.

Dorsal surface of stings (fig. 5.2B) flattened, marked by shallow parallel ridges and grooves, without median groove. Ventral surface of serrated piece elevated and rounded in section. Serrae low, sharp, retrorse.

DISCUSSION

Identification of these fossils as Potamotrygonidae is thus far based only on overall similarity. The teeth resemble specimens and illustrations representing all potamotrygonid genera (*Plesiotrygon, Potamotrygon, Paratrygon*) provided by Ricardo Rosa. The serrae of the tail sting are relatively weakly developed as in *Plesiotrygon* (fig. 5.2A) and unlike the much more strongly ornamented stings of dasyatid sting rays. Because a complete sample of modern ray teeth and stings is lacking, however, it is not certain if any of the characters of the features preserved in the fossils always distinguish freshwater potamotrygonids from dasyatid rays that may enter the lower reaches of South American rivers. Frailey (1986) referred to Potamotrygonidae fossil tail stings from the Miocene Río Acre deposits of western Amazonia.

Potamotrygonid rays are endemic to South American freshwaters and now range in the north from the Magdalena and Maracaibo basins east and south to the Paraguay–

Paraná–Río de la Plata. These fishes commonly occur in large flowing rivers and still backwaters and lagoons.

Class Osteichthyes
Subclass Dipnoi
Order Lepidosireniformes
Family Lepidosirenidae
Lepidosiren paradoxa
(fig. 5.3)

MATERIAL

IGM 184734, partial skull, jaws, and tooth plates from Duke Locality 81. IGM 183026 from Duke Locality 6, IGM 183685 from Duke Locality 60, IGM 184531 from Duke Locality 53, IGM 184613 from Duke Locality 102, lower jaws and tooth plates. IGM 182832 from Duke Locality 6, IGM 182866 from Duke Locality 23, IGM 182888 from Duke Locality 24, IGM 183102 from Duke Locality 32, IGM 183267 from Duke Locality 39, IGM 183406 from Duke Locality 46, IGM 183441 from Duke Locality 50, IGM 183482 from Duke Locality 53, IGM 183509 from Duke Locality 54, IGM 183540, IGM 183543, IGM 183673, IGM 183677 and IGM 183568 from the Fish Bed, IGM 183632 from Duke Locality 6, IGM 183755 from Duke Locality 62, IGM 183933 and IGM 183927 from

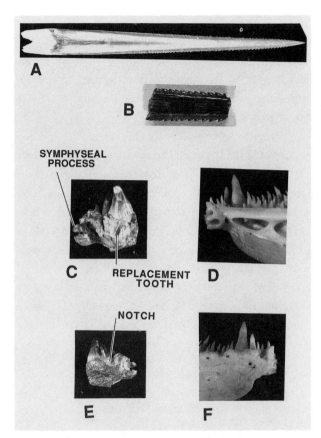

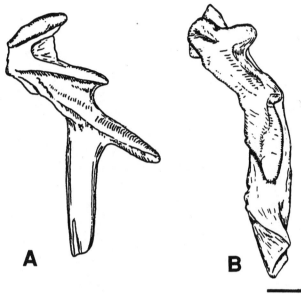

Figure 5.3. Fossil lungfish, *Lepidosiren paradoxa*, IGM uncat., jaws and tooth plates. *A*. Palatal tooth plate, left side, ventral view. *B*. Mandible and its tooth plate, right side, dorsal view. Not drawn to the same scale. Scale bar 3 mm for *A*, 10 mm for *B*.

Figure 5.2. *A*. Dorsal view of tail spine of modern potamotrygonid sting ray *Plesiotrygon*, Río Negro, Brazil, spine length 67 mm. *B*. Dorsal view of partial tail spine of fossil potamotrygonid sting ray, specimen length 15 mm. *C*. Fossil *Hoplias* sp., IGM 182935, fragmentary symphyseal region of mandible, posterior view, overall length 13.1 mm. *D*. Modern *Hoplias* sp., DU F2006, Venezuela, symphyseal region of mandible, posterior view. *E*. *Hoplias* sp., IGM 182935, fragmentary symphyseal region of mandible, anterior view. *F.* Modern *Hoplias* sp., DU F2006, Venezuela, symphyseal region of mandible, anterior view.

Duke Locality 46, IGM 184078 from the Fish Bed, IGM 184097 and IGM 184382 from Duke Locality 40, IGM 184180 from Duke Locality 84, IGM 184316 from Duke Locality 88, IGM 184355 from Duke Locality 90, IGM 184505 and IGM 184547 from Duke Locality 96, and IGM 184648 from Duke Locality 104, isolated upper and lower jaw tooth plates.

DESCRIPTION

Figure 5.3 illustrates palatal and mandibular tooth plates. Bondesio and Pascual (1977) described and illustrated dipnoan specimens from the Honda beds. The partial skull is under study by Hans Peter Schultze.

DISCUSSION

Lepidosiren paradoxa is the sole extant species of South American lungfish. Savage (1951b) and Bondesio and Pascual (1977) reported lungfish remains from the La Venta fauna as *Lepidosiren* cf. *L. paradoxa*. The La Venta lungfish fossils show a detailed resemblance to and no apparent differences from the distinctive modern tooth plates and jaws of *L. paradoxa*. Thus, there is presently no basis for excluding the fossil elements from the species taxon. Fernández et al. (1973) assigned remains of lungfishes from the lower Tertiary of Argentina to *L. paradoxa*.

A nominal fossil species of lungfish, *L. megalos*, was described from the Miocene Río Acre region of Brazil (Silva Santos 1987). Recognition of this fossil species was justified only on the basis of its much larger size in comparison with the modern specimens. Size difference alone is a questionable criterion for taxonomic differentiation among fishes, particularly in comparisons of middle Tertiary with later or Recent specimens. I recommend that *L. megalos* be treated as a synonym of *L. paradoxa* until nonsize related diagnostic characters are located that separate it from samples of modern lungfish.

Modern *L. paradoxa* occur in the lowlands of central Amazonia and the Paraguay system. Lungfishes are absent from the Orinoco and modern Magdalena region. Because lungfishes are among the most abundant fish fossils in the La Venta fauna, their ecology is of special interest as a paleoenvironmental indicator. *Lepidosiren* typically inhabit quiet, shallow, often anoxic waters of marginal lagoons and flooded marshes. These fishes are obligate air breathers and estivate in mud burrows during periods of temporary desiccation, although they are not restricted to regions of periodic drought. *Lepidosiren* reach a maximum size of somewhat more than 1 m in length, and the fossil described by Silva Santos (1987) was estimated to have come from an individual of about 2 m. Some of the specimens from the La Venta fauna may have approximated that larger size.

Subclass Actinopterygii
Order Osteoglossiformes
Family Arapaimidae
Arapaima sp.

MATERIAL

IGM 183402, basioccipital complex with anterior vertebrae attached, from Duke Locality 46. This specimen was illustrated and described by Lundberg and Chernoff (1992).

DISCUSSION

Lundberg and Chernoff (1992) compared this fossil to other osteoglossiforms and provided evidence for its identification as *Arapaima* sp. The fossil differs in a minor way from modern *Arapaima gigas* in being slightly wider relative to length. The taxonomic significance of this shape difference is uncertain but it led Lundberg and Chernoff (1992) away from assigning the specimen to the modern species *A. gigas*.

Among living fishes, the African *Heterotis niloticus* is hypothesized to be the sister taxon of *Arapaima*. In addition, Taverne (1979) considered two extinct Cretaceous species to be arapaimids and most closely related to *Heterotis: Laeliichthys* (Aptian, Brazil) and *Paradercetis* (Cenomanian,

Africa). If correctly placed, these older fossil species provide indirect evidence for the minimum dates of late early Cretaceous origin for both the *Heterotis* and *Arapaima* lineages (Lundberg 1993). In addition to the present record, scales of *Arapaima* sp. were reported from Miocene rocks of the Río Acre region, Brazil (Taverne 1979).

Modern *Arapaima gigas* occur throughout the Amazon lowlands and the Guianas. The species is now absent from the Orinoco basin and Magdalena region. *Arapaima gigas* is a species that lives in large, lowland waters. Individuals are found in rivers but are most common in marginal lagoons and backwaters. Like the lungfish this species is capable of aerial respiration and, thus, can withstand anoxic conditions. *Arapaima gigas* is said to be the largest species of strictly freshwater fish; individuals can attain an immense size, about 3.7 m and 130 kg. The La Venta *Arapaima* fossil is estimated to have come from a modest-sized individual of about 1.1 m (Lundberg and Chernoff 1992).

Order Characiformes
Family Erythrinidae
Hoplias sp.
(fig. 5.2 C–F)

MATERIAL

IGM 182932, fragmentary right dentary 13.1 mm in overall length, including mandibular symphysis, three large erupted caniniform teeth with broken crowns and a few small, partly concealed replacement teeth from an unknown locality.

DESCRIPTION

Mandibular ramus coarsely rugose externally; narrowed abruptly near symphysis, which bears three peg-like processes separated by three conversely-shaped sockets (fig. 5.2C); lower two symphyseal processes dorsoventrally flattened and posteriorly offset from tip of jaw. Two outer plus one inner, nondepressible canine teeth laterally offset from symphysis; second canine the largest. Outer face of the dentary (fig. 5.2E) with a wide rounded notch medial to first canine. In modern specimens this notch receives the medial enlarged canine tooth of the premaxilla. Replacement teeth with labiolingually flattened crowns and sharp cutting edges running mediolaterally over the crowns.

DISCUSSION

The fossil jaw is attributed to *Hoplias* based on detailed overall similarity, including apparently unique detailed forms of the symphyseal processes, tooth crowns, and tooth arrangement. Further, the fossil exhibits no apparent differences from modern specimens of *Hoplias*.

The genus *Hoplias* contains three nominal species (Gery 1977), but these are in need of review. The overall range of *Hoplias* is very broad in South and lower Central America and includes the present Magdalena region. Roberts (1975) reported fossil teeth of *Hoplias* sp. from Miocene rocks of the Cuenca basin, Ecuador, and the genus is represented by a fragmentary premaxilla in the Miocene Río Acre deposits along the Peruvian–Brazilian border (pers. obs.).

Hoplias occur in diverse habitats, but they are especially common in quiet waters of lakes, backwaters, sluggish streams, and swamps. These fishes are capable of some degree of aerial respiration during stagnant, anoxic situations and drying pools.

Family Cynodontidae
Hydrolycus sp.
(fig. 5.4A, B)

MATERIAL

IGM 183259, two fragments of large canine tooth 7.6 mm and 16.0 mm in height, from Duke Locality 38; IGM 183692, one distal fragment of large canine tooth 17.5 mm in height, from the Fish Bed.

DESCRIPTION

Canine teeth large, bladelike, labiolingually compressed; sharp, raised cutting edge along midline of symphyseal face and apex, cutting edge placed more labially on articular

face. Cutting edge finely and regularly serrate on IGM 183259 (fig. 5.4A); cutting edge sharp and nonserrate on IGM 183692 (fig. 5.4B).

DISCUSSION

The two fragmentary specimens are identified as the greatly enlarged mandibular canines of *Hydrolycus* based on their distinctive form, large size, and sharp cutting edges. The serrated cutting edge of IGM 183259 is matched exactly by the unerupted and recently erupted mandibular canines of modern specimens. The unserrated cutting edge of IGM 183259 is like those of the erupted and worn teeth of the same modern specimens. In addition to two nominal species of *Hydrolycus* (*H. scomberoides* and *H. pectoralis*) there are at least two undescribed species (N. Menezes pers. comm.). Skeletal material of only *H. scomberoides* was available for comparison with the fossils. Because it is not known if there are interspecific differences in the large canine teeth, the fossils are not assigned at the species level.

Smaller canine teeth (fig. 5.4C) with flattened labial sides, rounded lingual sides, and nonserrate cutting edges displaced laterally are common fossils in the Honda Group beds. These teeth are identical to the medium and small canines of *Hydrolycus* and also to those of *Cynodon gibbus* and *Rhaphiodon vulpinus*, the other two taxa of the Cynodontidae. Identification of these smaller canines below family level is not possible.

Hydrolycus is widely distributed in the Orinoco and Amazon drainages and the Guianas. These fishes are absent from the present Magdalena region. *Hydrolycus* are large pelagic piscivores found in big rivers and open-water lakes. The immense mandibular canines are presumably used to injure prey.

Family Anostomidae
cf. *Leporinus*
(fig. 5.4D, E)

MATERIAL

IGM 253295, symphyseal premaxillary tooth, 2.3 mm in overall height, from Duke Locality 26.

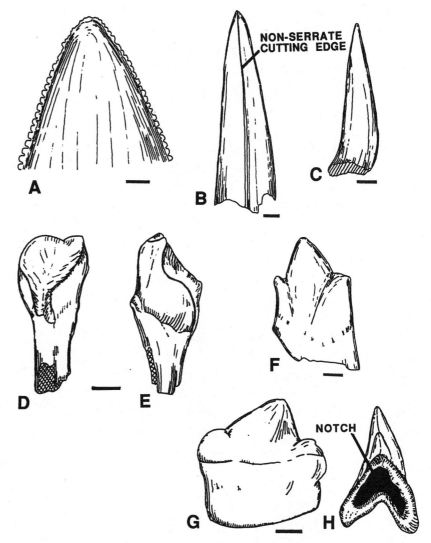

Figure 5.4. *A.* Fossil *Hydrolycus* sp., IGM 183259, tip of large replacement mandibular canine tooth in labial (or lingual) view showing unworn serrated cutting edge; scale bar 0.5 mm. *B.* Fossil *Hydrolycus* sp., IGM 183692, tip of large mandibular canine tooth in symphyseal-end (or articular-end) view; scale bar 0.65 mm. *C.* Fossil cynodontid canine tooth, IGM 183692, in lingual view; scale bar 0.75 mm. *D.* Fossil *Leporinus* sp., IGM 253295, premaxillary tooth, lingual side view. *E.* Same tooth in symphyseal-end view. Scale bar for *D, E,* 0.45 mm. *F.* Fossil cf. Tetragonopterinae, gen. et sp. incertae sedis, IGM 184787, tooth in labial side view; scale bar 0.17 mm. *G.* Fossil piranha (*Serrasalmus, Pygocentrus,* or *Pristobrycon*), IGM 251277, tooth in labial side view. *H.* Same tooth in symphyseal-end view. Scale bar for *G, H,* 0.17 mm.

DESCRIPTION

Crown spatulate (fig. 5.4D, E), convex labially, with a deep concave trough lingually; cutting edge bilobed on crown apex.

DISCUSSION

The symphyseal premaxillary teeth of *Leporinus* have a distinctive form that permits tentative identification based on overall close similarity. The fossil is esentially identical to the left-side, symphyseal premaxillary teeth of specimens in the *Leporinus friderici* group. Lacking a more complete tax-onomic range of anostomid skeletal materials, it is uncertain if any of the features of similarity are uniquely apomorphic. The other anostomids examined (*Leporinus* cf. *L. fasciatus, Abramites, Schizodon*) differ in various features from the fossil.

Roberts (1975) reported fossil *Leporinus* sp. teeth of a form similar to that illustrated here from the Miocene rocks of the Cuenca basin in Ecuador. *Leporinus* are widespread from northwestern South America, including the Magdalena basin through the Orinoco, Guianas, and Amazon to the Río de la Plata. These are lowland fishes that inhabit a wide range of habitats, but they tend to avoid extremes, such as very fast currents or stagnant pools.

Family Characidae
Subfamily cf. Tetragonopterinae
Gen. et sp. indet.
(fig. 5.4F)

MATERIAL

IGM 184787, isolated tooth crown, 1.2 mm in height, from the Fish Bed.

DESCRIPTION

A small tricuspid tooth with large central cusp flanked by smaller lateral cusps; cutting edge nonserrate. Crown biconvex in frontal section, vertical profile of labial side convex, vertical profile of lingual side slightly concave.

DISCUSSION

The fossil tooth is generally similar to the tricuspid teeth possessed by many tetragonopterine taxa. Robert's (1975) reported indeterminate tetragonopterine teeth from the Miocene of the Cuenca basin, Ecuador. The large taxon Tetragonopterinae is considered nonmonophyletic (Weitzman and Fink 1983). Without more anatomical information and detailed character surveys among characids, a finer placement of the fossil is impossible. Lacking a finer identification, no biotic or paleoenvironmental inferences can be drawn.

Subfamily Serrasalminae
Colossoma macropomum
(fig. 5.5)

MATERIAL

IGM 183434, partial left premaxilla preserving four outer row teeth, the first (symphyseal) and fourth teeth and two inner row teeth represented only by their attachment scars, 38.2 mm in overall length, from Duke Locality 50; IGM 253265, partial premaxilla, 10.5 mm in overall length, from the Fish Bed.

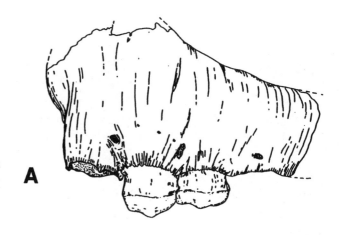

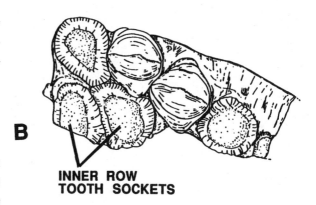

INNER ROW TOOTH SOCKETS

Figure 5.5. Fossil *Colossoma macropomum*, IGM 183434, left premaxilla. *A*, anterior view; *B*, ventral view. Overall length of specimen 38.2 mm.

DESCRIPTION

Lundberg et al. (1986) described and illustrated premaxillae, dentaries, and dentitions of *C. macropomum* from the La Venta fauna. Figure 5.5 illustrates the distinctive form of the premaxillary teeth and their diagnostic arrangement.

DISCUSSION

Lundberg et al. (1986) compared the fossil upper and lower jaw elements to other serrasalmines and provided evidence for their identification as *C. macropomum*. Several additional isolated teeth and infraorbital bones (IGM 182876, left second infraorbital bone; IGM 183543, two complete right fourth infraorbitals) from the Honda Group beds are very similar to *C. macropomum* but cannot be assigned to this species with certainty from among related serrasalmines.

As discussed by Lundberg et al. (1986), the discovery of

13-million-year-old fossils attributable to the extant species *C. macropomum* was of significance in indicating (1) conservative phyletic evolution of an ecologically specialized fish, (2) a high degree of diversification by the middle Miocene of related serrasalmine genera, (3) former extension of lowland Amazonian-Orinocan fishes into what is now the Magdalena region, and (4) subsequent extinction of much of that fish fauna in the later history of the Magdalena. These points are corroborated by several of the taxa recently found to be members of the La Venta biota. Further, Cione (1986, and pers. comm.) reported *Colossoma* from the Miocene of the Paraná basin, Argentina, a record that indicates a significant southern extension of the genus.

Serrasalmus, Pygocentrus, or *Pristobrycon* sp.
(fig. 5.4G, H)

MATERIAL

IGM 251277, isolated tooth, 0.8 mm in height, from the Fish Bed.

DESCRIPTION

Crown extremely compressed labiolingually, covered with smooth enamel; central cusp triangular, half as high as long, apex rounded; symphyseal-end cusplet low, scarcely set off from central cusp by shallow notch (fig. 5.4G); articular-end cusplet rounded, offset labially from axis of central cusp, apex rounded; cutting edge very sharp, nonserrate, complete over all cusps. Frontal-plane profile of crown biconvex; vertical profile of labial face scarcely convex; vertical profile of lingual face scarcely concave (fig. 5.4H); ventral profile of base flattened except at rounded corners; symphyseal end of base with prominent inverted V-shaped notch for articular-end cusplet of preceding tooth (fig. 5.4H); articular-end of base with small notch below cusplet.

DISCUSSION

The single tooth is identified as a piranha belonging to one of the genera *Serrasalmus, Pygocentrus,* or *Pristobrycon* based on its extreme compression, shape of cusps with very sharp cutting edges, and distinctive notch on its symphyseal end that would receive the articular-end of the preceding tooth. These genera were identified as a monophyletic group by Machado-Allison (1983) based in part on the apomorphic tooth form that is related to cutting flesh.

Although the identification of this fossil tooth is not as finely resolved as the material of *Colossoma macropomum,* its significance is similar. This fossil provides direct evidence that the relatively distinctive tooth form of piranhas evolved by 13 Ma. In addition, within the phylogenetic framework of the Serrasalminae developed by Machado-Allison (1983), the middle Miocene age of the piranhas indicates indirectly that at least the stem groups of the following related genera were coeval: *Metynnis, Catoprion,* and *Pygopristis.*

Piranhas, like *Colossoma* and some other fishes in the La Venta fauna, do not occur in the Magdalena basin today, but they are widely distributed in Atlantic drainages from Venezuela to Argentina. Ecologically, *Serrasalmus, Pygocentrus,* and *Pristobrycon* are generally piscivorous fishes found in habitats ranging from standing water to moderately fast-flowing, small and medium-sized rivers.

Order Siluriformes
Family Pimelodidae
Subfamily Pimelodinae
Brachyplatystoma cf. *B. vaillanti*
(fig. 5.6)

MATERIAL

IGM 253274, IGM 183605, IGM 183674, IGM 253266, IGM 184129, partial mesethmoids, 23.0–37.3 mm range of greatest widths; IGM 183675, IGM 184093, two partial left sphenotics, 17.3 and 20.1 mm greatest lengths; IGM 253267, two right pterotics, 13.6 and 16.1 mm greatest lengths; IGM 183543, two supraoccipital processes, 11.1 and 17.5 mm greatest longitudinal dimensions; IGM 183543, IGM 183605, IGM 184079, IGM 184790, four postcleithral processes, 20.5–23.1 mm range of greatest lengths; all from the Fish Bed.

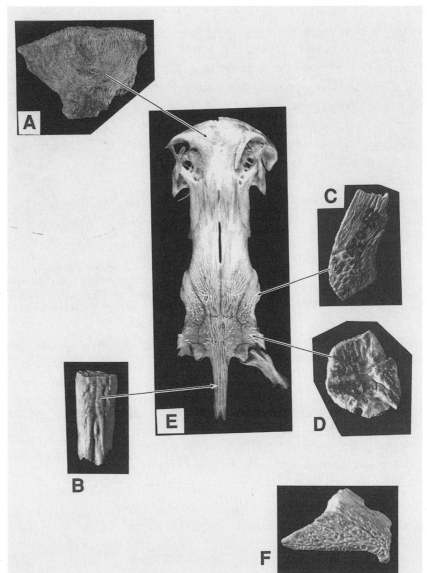

Figure 5.6. *A–D, F.* Fossil *Brachyplatystoma* cf. *B. vaillanti. A.* Partial mesethmoid, IGM 183543, dorsal view, 3.1 mm greatest width. *B.* Supraoccipital process, IGM 183543, dorsal view, 17.5 mm length. *C.* Right sphenotic, IGM 184093, dorsal view, 20.1 mm length. *D.* Right pterotic, IGM 253267, dorsal view, 13.6 mm length. *E.* Modern *Brachyplatystoma vaillanti* skull, DU F1034, dorsal view, 115 mm overall length. *F.* Fossil *B.* cf. *B. vaillanti,* right side posterior cleithral process, IGM 184790, lateral view, 21.1 mm length.

DESCRIPTION

Mesethmoids strongly depressed, without anterior median notch but margin broadly convex and rugose; dorsum marked with concentric, fine ridges centered around an anteromedian focus. Supraoccipitals, sphenotics, pterotics, and postcleithral processes ornamented with short ridges arranged in subparallel, roughly longitudinal rows, ridges occasionally expanded as scarcely broader and higher rounded tubercles. Sphenotics with concave muscle scar above hyomandibular facet; anterior end of facet truncate and elevated above side wall of bone; lacking a pit for an anterior hyomandibular process. Pterotics with a sharp-angled posterolateral process. Supraoccipital processes with scarcely tapering sides, weak middorsal sulcus, and low midventral keel. Posterior cleithral process relatively deep and blunt-tipped.

DISCUSSION

These fossils are listed as *Brachyplatystoma* cf. *B. vaillanti* based on overall similarity plus correspondence in some possibly unique features. The type of dermal bone ornamentation is more like that found in *Brachyplatystoma vaillanti* than in any other pimelodid examined. The ornamentation, however, is more extensive on the fossil mesethmoids than in modern skeletons of *B. vaillanti* from the Orinoco. The systematic significance of these differences will be uncertain until larger geographic samples of modern *B. vaillanti* and related species are examined.

Several fragmentary basioccipitals and anterior (Weberian complex) centra are in the collection of fossils that are similar to *B. vaillanti* but cannot be distinguished from those elements in some species of *Brachyplatystoma, Goslinia, Pseudoplatystoma, Platysilurus, Paulicea,* and *Sorubim.* These include IGM 253269, IGM 184475, two basioccipitals; IGM 184095, partial basioccipital and Weberian complex; IGM 253270, two basioccipitals, one anterior part of Weberian complex centrum; and IGM 253271, two partial Weberian complexes.

Brachyplatystoma sp. (*"Malacobagrus"* subgroup)
(fig. 5.7A)

MATERIAL

IGM 182652, articulated first centrum, compound Weberian centrum (fused centra 2–4) with the base of the left fourth transverse process, fifth and sixth centra, 89.9 mm in overall length, from Duke Locality 6.

DISCUSSION

This is a massive uncrushed fossil of the Weberian complex centra that exhibits apomorphically diagnostic features of the *"Malacobagrus"* subgroup within the genus *Brachyplatystoma:* coarsely spongelike bone of the first centrum and a hypertrophied bony attachment site for the swim bladder on the ventral side of the fifth centrum. Figure 5.7 illustrates its distinctive form in ventral view. The *"Malacobagrus"* subgroup of *Brachyplatystoma* contains *B. filamentosum* and *B. rousseauxi* (Nass 1991). The placement of the fossil within this group is uncertain because of incongruent similarities shared among the species and the fossil and the more extensive swim bladder attachment site of the fossil.

Brachyplatystoma filamentosum and *B. rousseauxi* are widely distributed in the lowland large rivers in the Amazon and

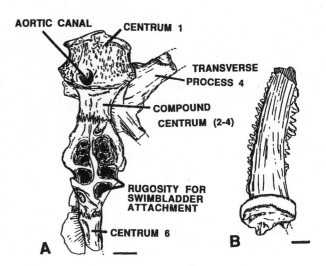

Figure 5.7. *A.* Fossil *Brachyplatystoma (“Malacobagrus”)* sp., Weberian complex centra (first through sixth), IGM 182652, ventral view; scale bar 10 mm. *B.* Fossil cf. *Pimelodus,* IGM 183541, right side pectoral spine, dorsal view; scale bar 1 mm.

Orinoco basins. These species do not occur in the Magdalena region today. They are fishes of open river channels, and they make extensive seasonal longitudinal migrations that sometimes involve ascending swift rapids (Goulding 1980). The large size of the fossil vertebrae (~90 mm in overall length and first centrum width ~25 mm) indicates a fish about 1.5 m in standard length.

cf. *Pimelodus*
(figs. 5.7B; 5.8A, B)

MATERIAL

IGM 253272, supraoccipital fragment, 10.0 mm greatest length; IGM 183541, IGM 184099, IGM 250426, seventeen pectoral-fin spines, 14.1–22.4 mm greatest lengths; IGM 183606, IGM 253273, two dorsal-fin spines, 14.4 and 15.7 mm greatest lengths; all from the Fish Bed.

DESCRIPTION

Supraoccipital process (fig. 5.8A, B) broad based, with tapering sides, strongly and evenly convex in cross section, with a well-developed ventral keel; surface ornamentation evenly and finely tuberculate.

Pectoral spine (fig. 5.7B) ovoid in section; anterior dentations antrorse, strong and regularly spaced except a small

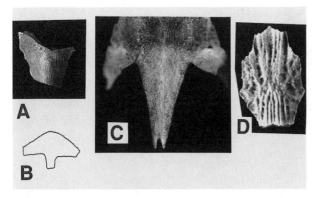

Figure 5.8. *A.* Fossil cf. *Pimelodus,* supraoccipital fragment, IGM 253272, dorsal view, 10 mm greatest overall length. *B.* Same specimen, outline of cross-sectional shape. *C.* Modern *Pimelodus blochi,* DU F1037, dorsal view, 210 mm standard length. *D.* Fossil Doradidae, gen. et sp. incertae sedis 2, complete supraoccipital, IGM 253279, dorsal view, 18.7 mm length.

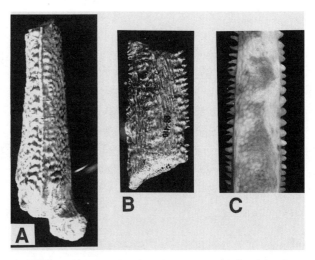

Figure 5.9. *A.* Fossil Ariidae, gen. et sp. incertae sedis, dorsal-fin spine, IGM 182899, anterior view, 58.5 mm length. *B.* Fossil *Phractocephalus hemiliopterus,* fragment of large pectoral-fin spine, IGM 184467, anterior edge to right, 54.7 mm length. *C.* Modern *P. hemiliopterus,* pectoral-fin spine, middle section of shaft, anterior to right, 45.7 mm length.

cluster of crowded dentations proximally; posterior dentations retrorse, strong, not exceptionally flattened transversely; posterior dentations attached to dorsal half of spine shaft proximally. Dorsal spine slender, its anterior ridge strong, adenticulate proximally and with a prominent knob for ligamentous attachment to dorsal spinelet.

DISCUSSION

This material is tentatively identified as *Pimelodus* based on close overall similarity and lack of differences. No uniquely apomorphic features, can be specified, however, and *Pimelodus* is a paraphyletic taxon containing a diverse array of nominal species. Pectoral spine morphology is variable among species of *Pimelodus,* but the fossil spines are closest to *P. blochi.* Comparative material in this study included nine species. There are some, small "segregate" genera allied to *Pimelodus* (e.g., *Cheirocerus, Parapimelodus,* and *Iheringichthys*) that are similar in some of the features exhibited by the fossil. *Pimelodus* sp. has been listed by Cione (1986) among the fossil fishes of Miocene through Pleistocene age from Argentina.

Pimelodus has a wide distribution from lower Central America to Argentina. The fossil is similar to the two nominal species of *Pimelodus* in the present Magdalena basin: *P. grosskopfii* and *P. clarias.* These catfishes commonly occur across a wide range of habitats from deep, fast-flowing rivers to small lagoons and streams.

The following fossil cleithra and postcleithral processes are noted here under cf. *Pimelodus* although their placement is uncertain: 253275, 253276, 253277, 253278. In overall shape and ornamentation of the postcleithral process these pectoral girdle elements are most similar to several pimelodids, doradids, and auchenipterids. Their assignment to either Doradidae or Auchenipteridae is precluded by absence of the diagnostic deep accessory spine-articulating socket anterior to the main socket on the inner face of the cleithrum (J. Friel pers. comm.). Among pimelodids the raised and elongated form and extensive, dense tuberculation of the fossil postcleithral processes are most like those of *Pimelodus* species such as *P. blochi, P. grosskopfii,* and *P. clarias,* but the fossils differ in being less triangular.

Phractocephalus hemiliopterus
(fig. 5.9B)

MATERIAL

IGM 253267, partial mesthmoid, 23.4 mm greatest longitudinal dimension, from the Fish Bed. IGM 184467, fragment of pectoral spine shaft, 54.7 mm greatest longitudinal dimension, from Duke Locality 93.

DESCRIPTION

Posterior dorsal surface of mesethmoid coarsely ornamented with circular to ovoid pits surrounded by raised rims; slightly concave anteromedial margin; cornua laterally truncated. Pectoral spine shaft (fig. 5.9B) dorsoventrally flattened, coarsely ornamented with striae and shallow, irregular pits; anterior and posterior dentations strong, erect, irregularly spaced.

DISCUSSION

These fragmentary fossils are identified as *Phractocephalus hemiliopterus* based on overall similarity plus diagnostic ornamentation characters of the bones. No pimelodids or other catfishes are known to have an osseous ornamentation pattern like that of *Phractocephalus* (Lundberg et al. 1988).

Phractocephalus hemiliopterus is a member of the subfamily Pimelodinae and most closely related to *Leiarius* (Amazon, Orinoco) and *Perrunichthys* (Maracaibo) (Lundberg et al. 1986). Modern *P. hemiliopterus* occur in the Orinoco and Amazon systems and the large rivers of the Guianas. The species is absent from the present fish fauna of the Magdalena region. Lundberg et al. (1988) described fossil skulls indistinguishable from *P. hemiliopterus* from the late Miocene Urumaco Formation of Falcon State, Venezuela, where the species also does not occur today.

Phractocephalus hemiliopterus is a species of large, lowland rivers. Individuals are found in fast-flowing channels or slack-water habitats such as marginal lagoons and backwaters. The species attains a large size, about 1.3 m and 80 kg. The La Venta *Phractocephalus* fossil spine represents a very large fish, well over 1 m in length based on crude extrapolation from available much smaller skeletal materials.

Family Ariidae
Gen. et sp. incertae sedis
(fig. 5.9A)

MATERIAL

IGM 182899, proximal half of dorsal-fin spine with left side of basal articulating processes intact, 58.5 mm long, from Duke Locality 24.

DESCRIPTION

Anterior median ridge bearing a few tubercles but lacking large, evenly spaced dentations; anterior and lateral surfaces of shaft covered with many irregular rows of rugose tuberculations and ridges; posterior side with a prominent median groove, lacking dentations. Shaft cordate in cross section.

DISCUSSION

This fossil is tentatively identified as an ariid based on the multiple tuberculated ridges on the front and sides of the shaft. Similar ornamentation is known to occur in some modern species of South American Ariidae such as *Arius proops* (USNM 214860). Other Neotropical catfishes that possess a robust dorsal-fin spine have either no anterior ornamentation or a single midline row of dentations.

Ariid catfishes generally occur in coastal marine waters. On all tropical coasts of South America, however, ariid species are known that enter at least the lower reaches of large rivers. Unfortunately, the unresolved identification of this fossil spine and the current poor state of knowledge of ariid ecology in South American fresh waters inhibit any conclusion on the paleoenvironmental significance of this specimen. Fossil Ariidae have been reported but not described in detail from several estuarine and fluvial Tertiary deposits from Venezuela (Lundberg et al. 1988) to Argentina (Cione 1978, 1986).

Family Doradidae
Gen. et sp. incertae sedis 1
(fig. 5.10A)

MATERIAL

IGM 183545, occipital region of skull, plates surrounding dorsal-fin base (nuchal plates), first (spinelet) and base of second dorsal spines, and, on the ventral surface, badly crushed pectoral girdle, 61.7 mm in overall length, 39.3 mm in greatest width, from an unknown locality.

DESCRIPTION

Skull roof elements and nuchal plates ornamented with coarse, tuberculate ridges, those near dorsal midline trend-

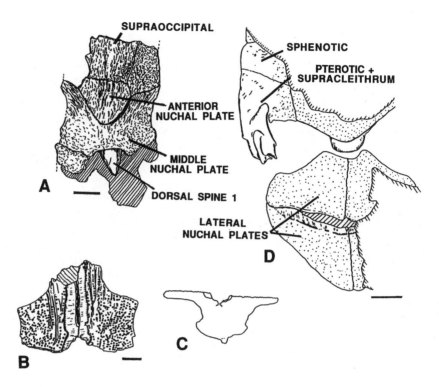

Figure 5.10. *A.* Fossil Doradidae, gen. et sp. incertae sedis 1, occipital region and dorsal-fin nuchal plates, IGM 183545, dorsal view; scale bar 10 mm. *B.* Fossil Doradidae, gen. et sp. incertae sedis 3, frontal bones, IGM 183886, dorsal view; scale bar 3 mm. *C.* Same specimen, outline of cross-sectional shape showing also orbitosphenoid and parasphenoid in section. *D.* Fossil cf. *Hoplosternum,* occipital region and dorsal-fin plates, IGM 253268, dorsal view; scale bar 5 mm.

ing longitudinally, those near lateral margins irregular or trending obliquely. Supraoccipital posterior process truncate, its joint with anterior nuchal plate (supraneural) broadly tranverse and serrate. Anterior nuchal plate convex in cross section but with a weak midline sulcus, its form approximating equilateral triangle with rounded corners and apex directed posteriorly, short sutural contact with epioccipital lateral to supraoccipital process, laterally and posteriorly bounded by limbs of middle nuchal plate. Middle nuchal plate very broad, with long and robust anterolateral limbs reaching epioccipitals, posterior edge notched for first dorsal spine (spinelet). Posterior nuchal plate expanded behind middle nuchal plate and lateral to bases of dorsal spines.

DISCUSSION

Among the doradids examined this fossil is most similar to *Orinocodoras* and *Doraops,* but these differ in several details. There are greater differences between the fossil and *Leptodoras, Hemidoras, Pterodoras, Megalodoras, Amblydoras, Platydoras,* and *Pseudodoras.* Many genera of this diverse group have not been available for study. Lacking a more

precise identification and because the family is so widespread and ecologically diverse, nothing can be said about the biotic or ecological significance of the fossil. Fossils from the Miocene of Argentina have been identified as indeterminate doradids by Cione (1986).

Doradidae, gen. et sp. incertae sedis 2
(fig. 5.8D)

MATERIAL

IGM 253279, complete dorsal side of supraoccipital, 18.7 mm long, from the Fish Bed.

DESCRIPTION

This well-preserved bone exhibits weakly serrate joints for frontals, sphenotics, pterotics, epioccipitals, and anterior nuchal plate (supraneural); posterior process truncated; surface ornamentation includes anterior median sulcus flanked on each side by three ridges, these flanked by sulci for parietal sensory canal branch, and these flanked by several deep pits; posterior process ornamented with about nine longitudinal ridges; all ridges bear fine bony tubercles.

DISCUSSION

See discussion under Doradidae, gen. et sp. incertae sedis 1, for comparative materials, none of which exhibit close similarity to the fossil.

Doradidae, gen. et sp. incertae sedis 3
(fig. 5.10B, C)

MATERIAL

IGM 183886, articulated frontals plus underlying orbitosphenoid and parasphenoid, 12.5 mm in greatest width, from Duke Locality 50.

DESCRIPTION

Frontals notably flat and evenly covered with small bony tubercles, no ridges or pits; orbital (lateral) margins deeply concave; anterior cranial fontanelle narrowly open; orbitosphenoid with prominent horizontal shelves; parasphenoid projecting as a low keel below orbitosphenoid.

DISCUSSION

See discussion under Doradidae, gen. et sp. incertae sedis 1, for comparative materials, none of which exhibit close similarity to the fossil.

Family Callichthyidae
cf. *Hoplosternum*
(fig. 5.10D)

MATERIAL

IGM 253268, partial occipital and temporal region of skull roof and anterior trunk plates and vertebrae, one-half skull width (left side) 16.1 mm, width across anterior pair of lateral nuchal plates 20.7 mm, from the Fish Bed; IGM 184375, partial crushed skeleton including parts of crania and anterior vertebrae, lateral nuchal plates and dorsal-fin base with median nuchal plates, and pectoral spines, from Duke Locality 40; IGM 184793 and IGM 184802, two pectoral spines, 10.5 and 11.7 mm greatest lengths, from the Fish Bed.

DESCRIPTION

From IGM 253268 (fig. 5.10D) dermal skull roofing bones and anterior lateral trunk plates with irregularly spaced tiny pits. Skull roof slightly arched. Supraoccipital posteriorly with a convex surface margin subtended by a short posterior process (not truncate contra Gosline 1940) extending ventral to symphyseal notch in first pair of anterior lateral nuchal plates (plates not fused contra Gosline 1940 but meeting in a long suture). Pterotic fused with supracleithrum; bearing large pores and open groove of temporal laterosensory canal.

From IGM 184375 (not illustrated) middle nuchal plate on first dorsal-fin radial moderately expanded, its anterior margin rounded. Posterior nuchal plate on second dorsal-fin radial greatly elongated, curved lateral processes from below dorsal-fin spine facet to adjacent lateral trunk plates. Trunk plates subrectangular and very deep, anterior edges in region of plate overlap slightly thickened, posterior edges with single file of odontode bases.

Pectoral-fin spines (not illustrated) heavily ossified; shaft deeply ovoid in section; shaft ornamentation weakly striate plus small, circular odontode bases anteriorly and anterodorsally. No anterior groove, dentations, or serrae. Posterior dentations strong, numerous, and crowded, many with bases coalesced to neighbors; dentation cusps erect to retrorse except at proximal end where antrorse. Posterior groove obsolete, proximal posterior dentations attached to dorsal half of spine shaft. Anterior articulating process much larger than ventral process, bilobed and marked by fine grooves.

DISCUSSION

This material is identified as cf. *Hoplosternum* based on close overall similarity and lack of differences. Although the close resemblance is compelling, at present none of these points of similarity can be certainly identified as uniquely apomorphic. Other generally similar callichthyids are *Callichthys* that stongly differ from the fossil in having flatter occiput and nape and pectoral-fin spines that lack posterior dentations and *Dianema* that have more arched nape and slender pectoral-fin spines. Fossil pectoral spines resembling those from La Venta have been collected from the Miocene Río Acre deposits of western Amazonia (pers. obs.).

The genus *Hoplosternum* contains about ten nominal taxa

variously treated as species and subspecies. Its overall range includes most of lowland South America east of the Andes from the Paraná-Paraguay system north through the Amazon and Orinoco and river systems of the Guianas. The nominal *H. magdalenae* ranges from the Magdalena basin into eastern Panama. These catfishes typically are found in quiet, shallow lagoons and flooded marshes; they are capable of aerial respiration and, thus, can survive in anoxic and temporary waters.

Family Loricariidae
Subfamily Ancistrinae
cf. *Acanthicus*
(figs. 5.11, 5.12A)

MATERIAL

IGM 183605, seventh and eighth vertebrae, 21.6 mm length; IGM 253280, IGM 184190, three pectoral-fin spine distal tips, 16.7–21.6 mm in overall length. All material from the Fish Bed.

DESCRIPTION

Flattened and widened seventh and eighth centra suturally united (fig. 5.11) with a continuous, pitted midventral (aortic?) groove; eighth centrum elongated. Anteriorly, seventh centrum with expanded accessory lateral condyles for articulation with corresponding posterolateral expansions of the sixth centrum; lateral condyles with small dorsally directed processes. Intervertebral joint between centra six and seven nonsutural across ventral midline.

Pectoral-fin spine tip (fig. 5.12A) circular in cross section, covered with large, circular crowded odontode (external teeth) pedestals except for narrow naked strip along posterior edge.

DISCUSSION

The tentative identification is based on overall similarity, lack of difference, and of the loricariids examined only *Acanthicus* exhibits ventral midline sutures between the sev-

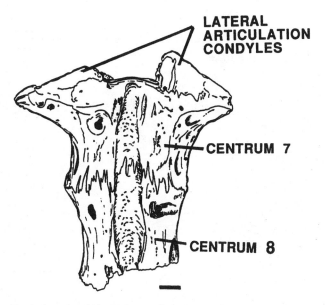

Figure 5.11. Fossil cf. *Acanthicus,* united seventh and eighth centra, IGM 183605, ventral view; scale bar 2 mm.

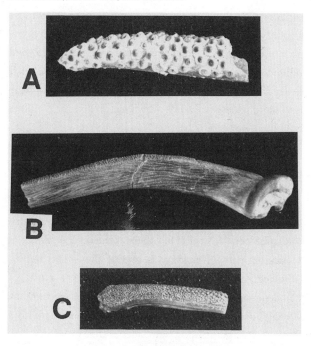

Figure 5.12. *A.* Fossil cf. *Acanthicus,* left pectoral-fin spine distal tip, IGM 253280, dorsal view, 21.6 mm length. *B.* Fossil cf. *Hypostomus,* left pectoral-fin spine distal tip, IGM 183562, dorsal view, 32.2 mm length. *C.* Fossil Loricariidae, gen. et sp. incertae sedis 2, proximal section of left pectoral-fin spine, IGM 253281, ventral view, 12.3 mm length.

enth and eighth vertebrae and large odontode pedestals nearly encircling the spine tip.

Acanthicus is a monotypic genus in the large subfamily Ancistrinae. *A. hystrix* occurs in large river channels of the Amazon and Orinoco. Owing to the tentative nature of this identification pending comparison to several other loricariids, the biogeographic and ecological significance of these fossils is uncertain.

Subfamily Hypostominae
cf. *Hypostomus*
(fig. 5.12B)

MATERIAL

IGM 183562, pectoral-fin spine with base and proximal section of shaft, 32.2 mm long, from Duke Locality 6.

DESCRIPTION

Dorsal flange of pectoral spine base low, broad, with angular edges; articulating surface cross-striate. Dorsal and ventral surfaces of shaft weakly striate with tiny odontode pedestals aligned as subparallel rows; odontodes about twice as dense on venter. Anterior surface of shaft with densely crowded odontode pedestals, but not markedly raised above shaft surface; odontode pedestals of moderate diameter. No large odontode pedestals.

DISCUSSION

The tentative identification is based on overall similarity and lack of difference compared to *Hypostomus* among the loricariids examined. The fossil, however, may well turn out to be equally similar to others of the several genera of Hypostominae. *Hypostomus* is one of the largest genera of loricariids with at least 116 nominal species. Fossils of Miocene to Pleistocene age are referred to *Hypostomus* by Cione (1986). The overall range of the genus extends from Panama to Argentina, on both sides of the Andes.

Family Loricariidae
Gen. et sp. incertae sedis 1
(fig. 5.13)

MATERIAL

IGM 183400, large pectoral spine with base and proximal section of shaft, 42.7 mm long, from Duke Locality 46.

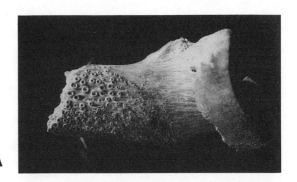

A

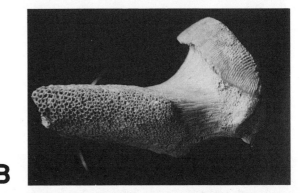

B

C

Figure 5.13. Fossil Loricariidae, gen. et sp. incertae sedis 1, proximal section of left pectoral-fin spine, IGM 183400, 42.7 mm length. *A*, dorsal view; *B*, anterior view; *C*, ventral view.

DESCRIPTION

Dorsal flange of pectoral spine base low, broad, with a sharp crest projecting outward along its dorsolateral margin; articulating surface cross-striate. Dorsal surface of shaft (fig. 5.13A) with moderate to large, sparsely distributed odontode pedestals not aligned in files. Anterior and anteroventral surface (fig. 5.13B, C) with densely crowded odontode pedestals of moderate diameter, a proximal group of odontode pedestals distinctly raised above shaft surface. Posteroventral surface with small, sparsely distributed odontode bases.

DISCUSSION

Among the loricariids examined this fossil is most similar to *Panaque*. *Panaque*, however, differs sharply from the fossil in having a distinctive enlarged row of odontodes along the posterodorsal shaft margin. Many genera of this diverse family are not available for study. Lacking a more precise identification and because the family is so widespread and ecologically diverse, nothing can be said about the biotic or ecological signficance of the fossil.

Loricariidae, gen. et sp. incertae sedis 2
(fig. 5.12C)

MATERIAL

IGM 253281 and 184787, spine fragment lacking distal tip and proximal articulation, 12.3 mm long.

DESCRIPTION

Dorsal surface of shaft with moderate to sparsely distributed odontode pedestals of variable diameter; not aligned in rows. Anterior and ventral surfaces with densely crowded odontode pedestals of moderate diameter and a proximal group of odontode pedestals sharply and prominently raised above shaft surface. Posteroventral surface with small, sparsely distributed odontode bases.

DISCUSSION

The proximal ventral odontode pedestals are more extensive and denser in this loricariid fossil than in others from the Honda Group beds (e.g., compare figs. 5.12C and 5.13). None of the modern comparative materials available matches the fossil in this feature.

Order Perciformes
Family Cichlidae
Gen. et sp. incertae sedis
(fig. 5.14)

MATERIAL

IGM 253282, IGM 253283, IGM 253284, three partial premaxillae, 12.3–16.6 mm in maximum dimensions.

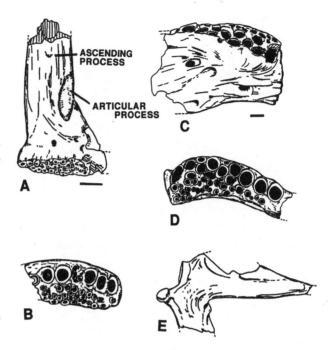

Figure 5.14. Fossil Cichlidae, gen. et sp. incertae sedis. *A–B*. Partial right premaxilla, IGM 253283. *A*, posterior view; *B*, ventral view. *C–D*. Partial right dentary, IGM 184798. *C*, anterior view; *D*, ventral view. *E*. Anguloarticular, IGM 253285, lateral view. *A–B*, scale bar 1 mm; *C–E*, scale bar 1 mm.

IGM 253285, IGM 253286, IGM 184798, three partial dentaries, 11.4–18.0 mm long. IGM 253287, IGM 253288, two anguloarticulars, 13.2 and 19.7 mm long. IGM 253289, IGM 253290, IGM 253291, IGM 253292, and IGM 253293, several median fin spines. All material from the Fish Bed.

DESCRIPTION

Premaxillae (fig. 5.14A, B) preserving only proximal parts of ascending and articular processes. Symphyseal end of premaxillary dentigerous arm with circular pitlike tooth bases arranged in two to four rows; outer tooth row without diastemae; outer teeth enlarged, basal diameters about three times those of inner rows.

Dentaries (fig. 5.14C, D) with two to five rows of circular tooth bases; diastemae lacking; a single stoutly conic tooth preserved on one jaw; outer two to three with enlarged bases, about three times diameter of inner rows; outer row tooth bases near symphyses turned anteriorly outward and subtended by thickened shelf. Mandibular laterosensory pores in ventrolateral row. Anguloarticular (fig.

5.14E) with relatively slender ventral process below articular condyle; ventral process directed anteriorly and lacking posterior extension; postarticular process short and blunt.

Median fin spines stout and slightly to strongly asymmetrical; surfaces of many specimens coarsely ridged.

DISCUSSION

Identification of the jaw elements as Cichlidae is based on their close similarity to several modern genera and species, but a more resolved determination has not been possible. Skeletal material of several genera is not available, and polarities of the differentiating features are uncertain. Of the modern taxa examined the fossils are phenetically most similar to many or all available species of *Cichlasoma* (five species), *Acarichthys heckeli*, *Heros severus,* and *Geophagus surinamensis.* The fossil jaw elements differ in dentition and/or bone shape features from *Petenia splendida, P. motaguense, Crenicichla vittata, Acaronia nassa, Aequidens awani, Astronotus ocellatus, Chaetobranchus flavescens, Mesonauta festivum, Geophagus jurupari,* and *Symphysodon discus.*

The median fin spines are tentatively assigned to Cichlidae. Several spines are coarsely striate like the anal-fin spines of *Astronotus ocellatus.* In comparison to other perciform groups found today in South American inland waters these spines are too large to belong to nandid leaf fishes, and of the Sciaenidae at least *Plagioscion* have nonstriate median fins spines.

Without a finer taxonomic identification the cichlid fossils of the La Venta fauna have little phylogenetic, biogeographic, or ecologic signficance. Nothing in the morphology of the fossils suggests a remote relationship to modern cichlids. Living Neotropical cichlids are found in a variety of habitats, but most occur in the relatively shallow marginal parts of lakes, streams, and rivers, generally away from strong currents.

General Discussion

The La Venta fauna contains the most diverse Neogene assemblage of tropical freshwater fishes from South America. One chondrichthyan, one dipnoan, and twenty-one actinopterygian species have thus far been identified to at least family level. Additional species will likely be found in ongoing studies of hundreds of small fish fossils obtained by screen washing. Other described assemblages of Neogene freshwater fishes from South America each have fewer than ten species (Travassos and Silva Santos 1955; Bardack 1961; Roberts 1975; Cione 1978, 1986; Frailey 1986). The La Venta fish fauna is also a little larger than the largest Miocene and Pliocene fish faunas from North America and Africa, which contain about nineteen species (Greenwood 1974; Kimmel 1975; Smith 1975). Nevertheless, the La Venta fish assemblage may represent only 1–2% of the total regional ichthyofauna if it is a subsample of the lowland Miocene Amazonian-Orinocoan fauna, as suggested by Lundberg et al. (1986), Lundberg and Chernoff (1992), and herein. Despite this apparent small sample size, the La Venta fishes represent a broad range of ecological types and taxonomic groups. The overall picture suggested by the fauna is that fish diversity in northern South America by the middle Miocene was much like that of today.

As far as they are known, the most general ecological characterization of the La Venta fishes is that they belong to a lowland assemblage. None of the closest living relatives of La Venta fishes occur in steep-gradient, mountain or piedmont streams above just a few hundred meters above sea level. Rather, these are fishes that today live in the vast aquatic expanses of the central and lower Amazon and Orinoco basins, and, in some cases, the Magdalena, below about 200 m above sea level. Beyond this commonality the La Venta fishes present ample ecological diversity. The assemblage includes species of both large river and inshore habitats. In the large, open-river group, the potamotrygonid ray and loricariid catfish cf. *Acanthicus,* and the large pimelodid catfishes *Brachyplatystoma* and *Phractocephalus* and ariid catfish are benthic to semibenthic fishes, whereas the tarponlike characiform *Hydrolycus* is pelagic. The cichlid and *Leporinus* are fishes of marginal, shallow, relatively still waters. In the shallow water group also are four fishes with specialized mechanisms for aerial respiration: *Lepidosiren, Hoplias, Hoplosternum,* and *Arapaima*. The first three of these are often found in anoxic, sometimes even temporary, waters. The La Venta fauna includes a wide spectrum of trophic types from generalists to specialized carnivores and herbivores. The sting ray, cichlid, *Pimelodus,* and *Leporinus* are generalized carnivores. The large pimelodid catfishes *Brachyplatystoma* and *Phractocephalus* and the piranhas and *Hydrolycus* are primarily piscivores. The last two of these have remarkably specialized dentitions for cutting and piercing flesh. The armor-plated loricariid catfishes are scraping algivores, or wood and detritus eaters, and the plated callichthyid catfish *Hoplosternum* is a detritivore. *Colossoma* is a herbivore with a bizarre dentition specialized for crushing fleshy fruits and seeds.

As far as they are known, the fishes of the La Venta fauna, like other Cenozoic freshwater fishes from South America, very closely resemble and are closely related to living taxa. Fossils of at least twelve Neotropical fish taxa do not exhibit any major morphological differences from living genera and, in some cases even species. These fishes are: lepidosirenids (Fernández et al. 1973; Silva Santos 1987; also this chap.), arapaimids (Lundberg and Chernoff 1992), parodontids, erythrinids, anostomids (Roberts 1975; also this chap.), characids (Eigenmann and Myers 1929; Schaeffer 1947), "herbivorous" serrasalmines (Lundberg et al. 1986; also this chap.), pimelodids (Lundberg et al. 1988; also this chap.), callichthyids (Bardack 1961; also this chap.), loricariids (Malabarba 1988; also this chap.), cichlids (Bardack 1961; also this chap.) and percichthyids (Schaeffer 1947; Chang et al. 1978; Rubilar and Abad 1990). The fishes added here to the La Venta fauna extend the list of apparently conservative groups to include the Cynodontidae, Ariidae, Doradidae, and possibly Potamotrygonidae. Usually, where adequate modern comparative material is available, the fossils can be identified to below the family level. Cases for which identifications have not been resolved at intrafamilial levels appear to be the result of poor preservation and uncertainties about the diagnostic values of the similarities shared by fossils and related modern species.

Cladistic analyses yield hypotheses on the relative ages of nested monophyletic groups. Additionally, accurately dated fossils placed in a cladistic framework provide estimates of minimum absolute ages of taxic origins (i.e., ages of "origin" of Hennig [1966, p. 161]). The Miocene fossils of *Colossoma macropomum* (Characidae, Serrasalminae) and *Phractocephalus hemiliopterus* (Pimelodidae, Pimelodinae) offer clear examples of this (Lundberg et al. 1986, 1988). From their relatively "distal" positions in available phylogenetic frameworks (figs. 5.15, 5.16), *Colossoma* and *Phractocephalus* imply minimum degrees of "generic"-level diversification within serrasalmine characids and pimelodine catfishes, re-

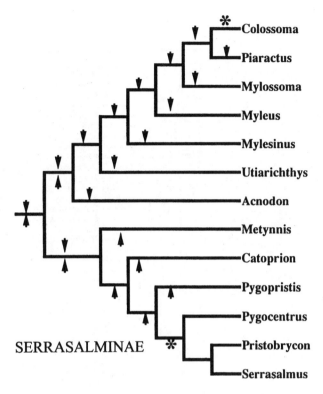

Figure 5.15. Phylogenetic relationships among the genera of the characiform clade Serrasalminae (based on Machado-Allison 1983). *Asterisks* mark the lineages of lowest rank to which middle Miocene fossils have been identified, that is, *Colossoma macropomum* and the stem group of the piranhas. *Arrows pointing down* mark the lineages that are implied by these relationships to be at least as old as the *Colossoma* lineage; *arrows pointing up* mark the lineages that are implied to be at least as old as the piranha stem group.

spectively, by the middle Miocene. These two cases are corroborated and extended by the additional fossil serrasalmines and pimelodines in the La Venta fauna. For the serrasalmines this study reports a piranha (*Serrasalmus, Pygocentrus,* or *Pristobrycon*). Within the phylogenetic framework of the Serrasalminae (fig. 5.15; also Machado-Allison 1983) the middle Miocene age of the piranhas indicates that three sequential outgroup genera are at least that old: *Pygopristis, Catoprion, Metynnis*. For the Pimelodinae, this work adds to the Miocene fauna a species of (or close to) *Pimelodus* and two species of *Brachyplatystoma*. Placed in the phylogenetic hypothesis of pimelodine catfishes illustrated in figure 5.16, these fossils point to considerable diversification by about 13 million years ago of at least the stem groups of many genera of this large clade.

For many other clades represented by fossils in the La

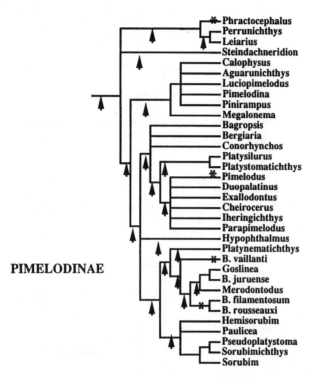

PIMELODINAE

Figure 5.16. Phylogenetic relationships among the nominal genera of the catfish subfamily Pimelodinae (based on Lundberg et al. 1991 and Nass 1991). *Asterisks* mark the lineages of lowest rank to which middle Miocene fossils have been identified, that is, *Phractocephalus,* and cf. *Pimelodus, Brachyplatystoma* cf. *B. vaillanti,* and the stem group of *Brachyplatystoma filamentosum–B. rousseauxi* (=the *"Malacobagrus"* subgroup). *Arrows* mark the lineages that are implied by these relationships to be at least as old as the lineages represented by fossils.

Venta fauna, lack of phylogenetic resolution prevents detailed statements about levels of their diversification. Exact higher-level cladograms aside, the fossils of *Hoplias, Hydrolycus, Arapaima,* and *Lepidosiren;* of at least close relatives of *Leporinus, Hoplosternum, Acanthicus,* and *Hypostomus* and other loricariids; and of some doradids and cichlids show that the middle Miocene fish fauna of South America was taxically rich and, apparently, essentially modern in composition.

Information on ages of taxic origin and diversification is especially important for constraining correlative and/or causal hypotheses relating Neotropical fish diversity to biogeographic and earth history events. For example, the great diversity of the Neotropical biota is often hypothesized to be a product of allopatric divergence among populations inhabiting fragmented habitat refugia that presumably

waxed and waned during Pleistocene climatic fluctuations (Haffer 1982). Further, Roberts (1972) suggested that most of the taxonomic richness of South American fishes has resulted from a few million years of evolution. Weitzman and Weitzman (1982), however, found no support for the Pleistocene refuge "species pump" among forest-stream fishes for which cladistic frameworks were available. And, these authors questioned such youthfulness for the Neotropical ichthyofauna, arguing instead for early to middle Tertiary diversification of modern genera. Indeed, the conclusion to be drawn here from a broader base of phylogenetic and paleontological information is that considerable diversification to the level of modern genera and even species of characins, catfishes, and other South American groups predated the Pleistocene by 11 million years at the very least.

The location and geographic relationships of the La Venta fauna are critically important for understanding the history of fishes in the Magdalena region. Although the fossil assemblage is located in the present Magdalena River valley in south-central Colombia, eight or nine of its members (roughly one-third of the total) are now absent from there. Instead, these fishes live in the Amazon basin (and some are also in the Orinoco). There are lungfish *Lepidosiren paradoxa;* pirarucu *Arapaima* sp.; three characiforms, *Hydrolycus* sp., *Colossoma macropomum,* and piranha (*Serrasalmus* sp., *Pygocentrus* sp., or *Pristobrycon* sp.); three pimelodid catfishes, *Phractocephalus hemiliopterus, Brachyplatystoma* (*"Malacobagrus"*) sp., and *Brachyplatystoma* cf. *B. vaillanti;* and possibly the loricariid catfish cf. *Acanthicus* sp. Five genera (four tentatively identified) in the La Venta fauna contain species living now in the Magdalena basin as well as the Amazon and Orinoco: two characins, *Hoplias* sp. and cf. *Leporinus* sp.; and three catfishes, cf. *Pimelodus* sp., cf. *Hoplosternum* sp., and cf. *Hypostomus* sp. None of the La Venta fishes yet known belong to taxa now endemic to the Magdalena. Therefore, the fossil fishes of the La Venta fauna strongly indicate a direct relationship to the Amazon and Orinoco faunas.

This paleoichthyological relationship of the La Venta fauna strongly corroborates geological and neontological evidence for former connections between the Magdalena region and the Amazon and/or Orinoco. Julivert (1970) and Megard (1989) report that the eastern Andes were not greatly uplifted until after Honda time. During Honda time the La Venta region was a piedmont fluvial plain flanking the Cordillera Central and sloping to the east to southeast (Wellman 1970). The fossil fishes suggest that there was no divide between the Magdalena and the Amazon and Orinoco systems. Local massifs in the eastern Andes began to form in the late Oligocene but the high drainage divide and concomitant formation of the isolated Magdalena watershed did not develop until the late Miocene or early Pliocene when Mesozoic and early Tertiary rocks were uplifted to 3,500 m or more (Megard 1989).

More than seventy years ago, Carl Eigenmann (1920) proposed that the modern Magdalena fish fauna has a closer relationship with the large riverine faunas of the Atlantic slope than with trans-Andean (Pacific slope) faunas. His argument was based on shared species and presumed closely related species pairs, at least some of which have been confirmed by recent cladistic analyses (e.g., Vari 1989; Walsh 1990). Although Eigenmann (1920, 26) mentioned the possibility of westward dispersal of eastern fishes into the Magdalena system "perhaps via Lake Maracaibo," he clearly favored a straightforward vicariance explanation. That is, before the rise of eastern Andes, large, low-gradient Amazonian and/or Orinocoan tributaries extended farther westward with their biota into the present Magdalena region. The formation of the high Andean drainage divide caused a vicariance event that fragmented the widespread predivide fauna. Today, undifferentiated species or sister taxa shared by the Magdalena and Amazon or Orinoco mark that earlier fauna. The La Venta fossil fishes fully support Eigenmann's vicariance explanation. Specifically, the eight La Venta taxa that are presently restricted to Atlantic slope rivers provide additional direct evidence, beyond the shared modern fishes, for a more widespread Amazonian/Orinocoan fauna into the Magdalena region in the middle Miocene. The alternative, dispersalist hypothesis for the present-day shared taxa between the Amazon/Orinoco and Magdalena would predict paleofaunas showing less similarity, just the opposite of that observed.

An interesting question in Neotropical fish biogeography is why the Magdalena fauna, with about 150 species, is relatively depauperate by an order of magnitude compared to the Amazon and Orinoco faunas. At once it can be said

that this is expected because the Magdalena basin has a much smaller drainage area than the immense Atlantic slope watersheds. In general, drainage basin area is positively correlated with fish species richness (e.g., G. R. Smith 1981). Area-diversity relationships are static, however, and do not of themselves suggest particular patterns or causes of historical development of differential interbasin diversity. The absolute number of fish taxa recorded in the La Venta fauna is most likely only a small fraction of the modern Magdalena and Amazon faunas, and surely this sample grossly underestimates the real size of the Miocene fish fauna. Nevertheless, the high relative abundance (~30%) of taxa present in the La Venta fauna yet absent today from the Magdalena region demonstrates that extinction has occurred there during the last 13 million years or so.

It is suggested that late Cenozoic local extinction of fishes in the Magdalena watershed was caused by the habitat instability accompanying volcanism in the central Andes, tectonism in the eastern Andes, and local loss of habitat as a result of increasing aridity in parts of the upper section of the basin (Lundberg et al. 1986; Lundberg and Chernoff 1992). Furthermore, after the basin was isolated by the uplifted Cordillera Oriental in the late Miocene to Pliocene, it was cut off from recolonization by range expansion of Amazonian and Orinocoan populations. The proximate causal factors underlying particular extinctions remain uncertain, in part, because there are no obvious correlations between extirpation and ecological characteristics. The taxa that suffered extinction from the Magdalena are lowland fishes of both large open-water habitat (i.e., *Hydrolycus, Phractocephalus, Brachyplatystoma* ["*Malacobagrus*"], *Brachyplatystoma* cf. *B. vaillanti,* and cf. *Acanthicus*) and shallower, marginal habitats (i.e., *Lepidosiren, Arapaima, Colossoma,* and piranha). Among these all major trophic types are represented. The present Magdalena basin contains a wide range of aquatic habitats that support an ecologically diverse fish fauna, including species that are roughly comparable in their habitat requirements or preferences to those that disappeared. Suitable conditions now exist for many of the large lowland Amazonian species, particularly in the lower reach of the Magdalena in the wet *cienagas* region of Magdalena, Bolivar, and Cesar departments. In fact, *Colossoma* has been successfully "re"-introduced there by fish culturists. Although nothing is known directly of the long-term, late-Cenozoic stability of the wet lowland habitats in the region, either the cienagas area or other refuges obviously had to be available continuously to the fish groups that have survived the 13-million-year isolation of the Madgalena system.

The middle Miocene La Venta fauna contains an ecologically and taxonomically varied assemblage of fishes that provide evidence of four important patterns in the evolutionary history of the rich South American fish fauna. First, many of the morphologies that characterize and diagnose several living genera and even some species evolved by the middle Miocene. Second, also by the middle Miocene, there was extensive taxic diversification of modern genera within large lineages, such as serrasalmine characins and pimelodid catfishes. Third, at least in northern South America, the lowland Amazonian/Orinoco fauna was more widespread in the Miocene than it is today. Fourth, local extinction has been a significant factor in molding the composition of the modern Magdalena fauna. A large sample of fish remains from the Miocene rocks of Colombia remains to be sorted, identified, and phylogenetically placed among the myriad clades of Neotropical fishes. Without question these fossils, and others from scattered localities from the north coast of Venezuela, along the Andean chain and into northern Patagonia, can contribute uniquely valuable information in our attempt to understand the phylogenetic and biogeographic history of the Earth's richest inland fish fauna.

Appendix 5.1. Fish from the La Venta fauna, Honda Group, middle Miocene, Colombia

Fossils identified to at least the family level are listed.

Family cf. Potamotrygonidae
    incertae sedis
Family Lepidosirenidae
    Lepidosiren paradoxa
Family Arapaimidae
    Arapaima sp.
Family Erythrinidae
    Hoplias sp.
Family Cynodontidae
    Hydrolycus sp.
Family Anostomidae
    cf. *Leporinus*

Family Characidae
    Subfamily cf. Tetragonopterinae
        incertae sedis
    Subfamily Serrasalminae
        Colossoma macropomum
        Serrasalmus, Pygocentrus, or *Pristobrycon* sp.
Family Pimelodidae
    Subfamily Pimelodinae
        Brachyplatystoma cf. *B. vaillanti*
        Brachyplatystoma sp. (*"Malacobagrus"* subgroup)
        cf. *Pimelodus*
        Phractocephalus hemiliopterus
Family Ariidae
    incertae sedis
Family Doradidae
    incertae sedis 1
    incertae sedis 2
    incertae sedis 3
Family Callichthyidae
    cf. *Hoplosternum*
Family Loricariidae
    cf. *Acanthicus*
    cf. *Hypostomus*
    incertae sedis 1
    incertae sedis 2
Family Cichlidae
    incertae sedis

Appendix 5.2. Comparative materials

Dry and cleared/stained skeletal specimens of the following taxa were examined. Material is listed alphabetically by family. Uncat. = uncataloged.

Anostomidae
    Abramites sp. DU F1163
    Leporinus "fasciatus" DU F1162
    Leporinus "friderici" DU F1167, DU F1195
    Schizodon sp. DU F1170

Arapaimidae
    See materials list in Lundberg and Chernoff (1992)

Ariidae
    Ariopsis seemani DU F1017, DU F1019
    Arius felis DU F995
    Arius jordani DU F1020
    Arius kessleri DU F1015; DU F1016
    Arius proops USNM 214860
    Bagre panamensis DU F1018
    Potamarius nelsoni UMMZ 143496
    Sciadops troscheli USNM 214864

Auchenipteridae

Entomocorus benjamini DU F1146

Trachycorystes striatulus SU (at CAS) 2151

Callichthyidae

Callichthys sp. USNM 144506

Corydoras sp. DU F1147

Dianema USNM 179006

Hoplosternum littorale MBUCV-V-ES003

Hoplosternum thoracatum DU F1061, DU F1062, FMNH 84055

Characidae

Acestrorhynchus sp. DU F1164

Colossoma macropomum DU F1072

Markiana sp. DU F1196

Metynus sp. DU F1178

Mylossoma duriventris DU F1054

Piaractus brachypomum DU F1050, DU F1073

Pygocentrus caribe DU F1173, DU F1174, DU F1176, DU F1177

Serrasalmus sp. DU F uncat.

Cichlidae

Acarichthys heckeli UMMZ 215932-S-4

Acaronia nassa UMMZ 215930-S

Aequidens awani UMMZ 204640-S

Astronotus ocellatus UMMZ 204579-S, DU F uncat.

Chaetobranchus flavescens UMMZ 204678-S

Cichla ocellaris UMMZ 204679-S

Cichla temensis UMMZ 215931-S

Cichlasoma spp. UMMZ 196595-S-1, UMMZ 198800-S-1, UMMZ 188030-S-2, UMMZ 189574-S, UMMZ 196648-S-1

Crenicichla vittata UMMZ 207540-S-1

Geophagus jurupari UMMZ 204527-S

Geophagus surinamensis UMMZ 204939-S-3

Heros severus UMMZ 204390-S

Mesonauta festivum UMMZ 204530-S-2

Petenia motaguense UMMZ 199603-S-1

Petenia splendida UMMZ 189987-S-2

Symphysodon discus UMMZ 182551-S-2

Cynodontidae

Cynodon gibbus DU F uncat.

Hydrolycus scomberoides DU F uncat.

Raphiodon vulpinus DU F1175

Doradidae

Leptodoras sp. DU F847

Megalodoras irwini DU F925, DU F996, DU F1041, DU F1045, DU F1069, DU F1126

Orinocodoras eigenmanni DU F926

Platydoras costatus DU F1105, DU F1106

Pseudodoras niger DU F1092, MBUCV-V-ES uncat.

Pterodoras angeli DU F1128, MBUCV-V-ES uncat.

Erythrinidae

Hoplias malabaricus DU F uncat.

Erythrinus DU F uncat.

Lepidosirenidae

Lepidosiren paradoxa FMNH 51331

Loricariidae

Acanthicus histrix DU F1093

Loricaria sp. DU F1066, DU F1127

Panaque sp. DU F1067

Sturisoma sp. DU F1064, DU F1065

Pimelodidae

Brachyplatystoma filamentosum DU F1052, DU F1079, DU F1080

Brachyplatystoma juruense DU F983, DU F1071, MBUCV-V-ES13

Brachyplatystoma rousseauxi DU F981, DU F1051, DU F1057, DU F1078, MBUCV-V-uncat., MBUCV-V-uncat.

Brachyplatystoma vaillanti DU F994, DU F1034, MBUCV-V-CT235, MBUCV-V-uncat.

Phractocephalus hemiliopterus DU F925, MBUCV-V-CT328, MBUCV-V-uncat.

Pimelodus altissimus AMNH 40138

Pimelodus blochi DU F988; DU F1037; DU F1077; DU,F1094; DU F1096; DU F1101; MBUCV-V-CT242, -CT243, -CT244, -CT284

Pimelodus clarias DU F922

Pimelodus coprophagus MBUCV-V-CT150

Pimelodus grosskopfi DU F919

Pimelodus ornatus MBUCV-V-CT145

Pimelodus pictus DU F866; MBUCV-V-CT70; DU F1012

Pimelodus sp. 1 ("altipinnis" group) SU (at CAS) 54122

Pimelodus sp. 2 ("altipinnis" group) DU F1006, DU F1010

Potamotrygonidae

Plesiotrygon, Potamotrygon spp., and *Paratrygon* (teeth and photographs provided by Ricardo Rosa)

Part Four
Reptiles, Amphibians, and Birds

6
Limbless Tetrapods

Max K. Hecht and Thomas C. LaDuke

RESUMEN

Los tetrápodos sin extremidades son escasos en el archivo fósil de la región Neotropical, en contraste con la alta diversidad de estos animales en la actualidad. El conjunto de la fauna de La Venta es uno de los más diversos hasta ahora conocido en el continente. El nuevo material consiste mayormente de vértebras aisladas y articuladas y algunos elementos craniales. La fauna incluye los serpientes *Colombophis portai* (Aniliidae), otro Aniliidae no determinado, el Boidae *Eunectes* (la anaconda), otro Boidae no determinado, nuevo material de Colubroidea, y una Scolecophidia. Gymnophiona, el grupo de los anfíbios sin miembros, esta representada también. Esta faunula de La Venta parece incluir taxa de hábitos acuáticos y también unos que indican ambientes subterráneos húmedos.

The limbless tetrapods are poorly represented in the Neotropical fossil record, a condition that contrasts with the modern rich diversity of this region. The La Venta fauna is distinguished from other fossil faunas of the region by its higher diversity, which is described in the following annotated list.

Before this preliminary report, there has been only one study of snakes from the La Venta fauna (Hoffstetter and Rage 1977). Their study included descriptions of a new aniliid snake, *Colombophis portai,* and the boid *Eunectes stir-*

toni. They also recorded the presence of a vertebra of an unidentified colubrid snake.

The new material available to us consists primarily of vertebrae of snakes and apodans, and some snake skull elements, representing at least two boid snake taxa, two aniliids, one scolecophidian, one colubroid, and one apodan. The preservation of this material ranges from badly abraded to well preserved.

The following section lists taxa present, justification for and discussion of identification, and localities where found. The morphological terminology for snake vertebrae is based on LaDuke (1991); that for the apodan vertebrae, on Wake (1980) and Taylor (1977).

Systematic Paleontology

Serpentes
Aniliidae
Colombophis portai Hoffstetter and Rage 1977

REFERRED SPECIMENS (VERTEBRAE)

IGM 184285 (Duke Locality 6), IGM 183642 (Duke Locality 31), IGM 183249 (Duke Locality 37), IGM 183464 (Duke Locality 46), IGM 184579 (Duke Locality 50), IGM

183988 (Duke Locality 75), IGM 184126 (Duke Locality 84), IGM 184786 (Duke Screenwash Locality CVP-5), IGM 184037 (Duke Locality 39), and IGM 184797 (Duke Screenwash Locality CVP-10).

Hoffstetter and Rage (1977) diagnose this species by its large size and the following vertebral characteristics: the notched posterior border of the neural arch, the slight inclination of the zygapophyseal facets, reduction of the neural spine to a tubercle, the broad, indistinct, posteriorly placed hemal keel, and the undifferentiated synapophyses (sensu Hoffstetter and Rage 1977; =paradiapophyses of Hecht and LaDuke 1988). To this, we add the unusual placement of the subcentral foramina, which occur on the ventral face of the centrum, close to the midsagittal plane and just posterior to the level of the synapophyses (fig. 4 in Hoffstetter and Rage 1977). These foramina are anterior to the abbreviated hemal keel and are usually subequal in size, but in some cases one member of the pair may be extremely enlarged or reduced, or absent altogether.

cf. Aniliidae
Unnamed gen. and sp.

REFERRED SPECIMENS

IGM 184184, associated skull and vertebral elements (Duke Locality 21); and the following vertebrae: IGM 184176 (Duke Locality 84), IGM 183274 (Duke Locality 27), IGM 183533 from the Fish Bed, and IGM 182930 from the Villavieja Formation.

An incomplete skeleton with skull material (IGM 184184) has vertebrae that bear some similarities to those of *Colombophis portai*, such as the neural arch notch weakly developed and the hemal keel obsolete. The vertebral column is composed of a series of segments of articulated or clustered vertebrae with ribs and rib fragments. Many of the elements are largely obscured by crystalline matrix and are badly abraded. This taxon differs from *Colombophis* in having vertebrae that are shorter and broader, with strong, cylindrical neural spines and robust zygantra. The ventral face of the centrum is broadly rounded and lacks a distinct hemal keel. Subcentral foramina face ventrolaterally but are located more laterally than in *Colombophis*. The following elements of the skull are preserved and in need of further

preparation: a compound bone of the lower jaw, ?pterygoid (exposed ventrally) with tooth sockets; anterior fused cranial elements, and a posterior piece of dentary bearing three teeth.

Boidae
Eunectes sp.

REFERRED SPECIMENS

IGM 182840 (Duke Locality 22), and IGM 184636 (Duke Locality 103).

Hoffstetter and Rage (1977) described *E. stirtoni* on the basis of the basisphenoid and prootic. They also mentioned the presence of associated badly preserved vertebrae. We have examined the vertebral material associated with the type, and we question its assignment to *Eunectes* on the basis of size and shape and on the basis of other morphological features (Kluge 1991).

Available to us are eight fragmented vertebrae (IGM 184636) that compare favorably with large vertebrae of the living genus, *Eunectes*. Three vertebrae are complete enough to measure. Figure 6.1 is a composite drawing based on these specimens. These vertebrae have an approximate central length of 15–16 mm and appear to be of the posterior trunk region. These measurements are well within the range of medium-sized specimens of the living species. As in our comparative material, the paradiapophyses are very protuberant.

Boidae
Genus and species undetermined

REFERRED SPECIMEN

IGM 184344 (Duke Locality 90).

Twenty-one associated, abraded vertebrae document the presence of a small boid in the fauna. In this sample are eight isolated vertebrae, five pairs of articulated vertebrae, and one set of three articulated vertebrae. They are small and, in dorsal outline, square to slightly longer than wide. The most complete specimen of the series bears the following features: an elongate centrum 4.58 mm long; no paracotylar foramina; interzygophyseal ridge sharp in lateral view; he-

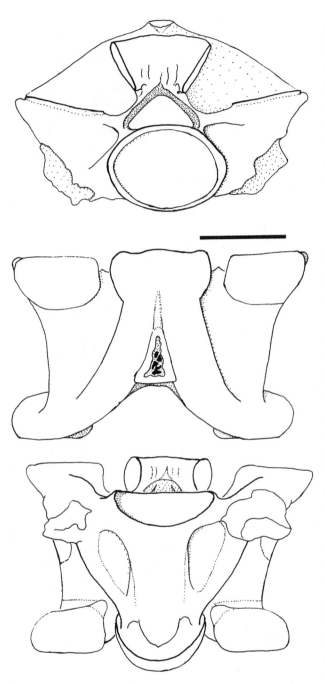

Figure 6.1. Diagrammatic reconstruction of posterior trunk vertebra of *Eunectes* sp., IGM 184636. *From top:* anterior view; dorsal view; ventral view. Scale bar 10 mm.

mal keel present but flattened; deep neural arch notch; neural spine running from base of zygosphene to neural notch with neural spine not high; zygosphene thin with sinuous lip; prezygapophyseal process obsolete and not protruding beyond prezygapophyseal facet; diapophyseal facet

higher than broad and with two distinct mounds. Considering the above description, the general appearance of the specimens available allows us to conclude that the vertebrae do not represent a close relative of xenopeltine, erycine, tropidophine, or other peripheral boids (Underwood 1967; McDowell 1987). We therefore consider it a "henophidian" in the gradal sense (Underwood 1967) but neither related to the aniliids and their relatives nor to generalized larger boids.

Colubroidea

REFERRED SPECIMENS

IGM 184185 (Fish Bed), IGM 250517 (Duke Locality CVP-14), IGM 183296 (Duke Locality 40).

Hoffstetter and Rage (1977) and Rage (1984) record a single, incomplete, unidentified midtrunk vertebra of a colubroid from the Fish Bed. The fossil, as recorded by Hoffstetter and Rage, is shorter in outline than many colubroids, bears a high, strong neural spine, large prezygapophyses, and a hemal keel that occupies the full length of the centrum and is protruding and inflated ventroposteriorly. The features that identify the vertebrae as colubroid are a high, lateral, flattened neural spine, a deep neural arch notch, a distinct hemal keel, and indications of a well-developed prezygapophyseal process.

On morphological grounds it is difficult to assign colubroid vertebrae to taxa lower than family. This difficulty is compounded in this case by a lack of sufficient osteological material from the Neotropics. Of the seventy-one colubroid genera listed by Dixon (1979) for the lowland tropical forests, only 5% of the characteristic genera are available for comparison. Among the North American genera, these specimens are comparable to taxa like *Hypsiglena*. We do not, however, imply a close relationship to this genus.

Scolecophidia

REFERRED SPECIMEN

IGM 253294, vertebra, from Duke Screenwash Locality CVP-5.

The presence of this group in the fauna is documented by a single vertebra with a central length of 1.9 mm (fig. 6.2).

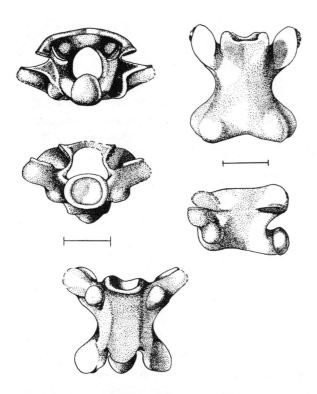

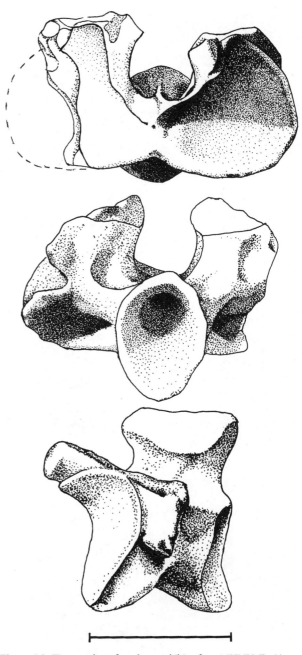

Figure 6.2. Diagrammatic reconstruction of trunk vertebra of scolecophidian IGM 253294. *Top left,* posterior view; *top right,* dorsal view; *middle left,* anterior view; *middle right,* lateral view; *bottom,* ventral view. Scale bars 2 mm.

Scolecophidian vertebra are extremely difficult to identify even at the familial level because there are few characters of the trunk vertebrae that distinguish the families (List 1966; Rage 1984) within the infraorder. This taxon is distinguished by a highly depressed neural arch, paradiapophyses with single facet, neural spine absent, postneural arch notch absent, hypapophyses absent in most vertebrae, and hemal keel absent.

Gymnophiona

REFERRED SPECIMENS

IGM 183404 (Duke Locality 46), IGM 184791 (Duke Screenwash Locality CVP-10), and IGM 182186 (Duke Locality 6b).

Three isolated vertebrae are identified as those of apodan amphibia on the basis of their amphicoelous condition, the presence of a neural arch ridge (=nuchal crest of Wake

Figure 6.3. First vertebra of apodan amphibian from MHNH (Paris). *From top:* anterior view; posterior view; lateral view. Scale bar 5 mm.

1980), the presence of an anterior, protruding pa-papophysial process, and a prominent ventral keel on the centrum (Wake 1980; Estes 1981). Furthermore, the vertebrae have a protruding ventral lip on the posterior cotyle, as is characteristic of this group.

All specimens appear to be from the anterior part of the

vertebral column (Taylor 1977; Wake 1980; Azpelicueta et al. 1987), but all are incomplete. All the characteristic papapophyseal processes are broken and are represented by only their bases. The three vertebrae with complete neural arches bear neural arch ridges which indicate that they are from the anterior half of the column. The most complete vertebra, which is probably from between the third and tenth vertebrae, has a central length of 9.1 mm and a maximum height of 10.0 mm. Our incomplete osteological collections have required us to compare this material with radiographs of apodans made available by S. Renous. It should be noted that the fossils are three to four times the size of the available comparative material. On the basis of available comparative material, we cannot definitively relate these fossils to any living taxon. The best evidence for the eventual identification of these vertebrae may be an undescribed atlas (no number) that was collected by Hoffstetter in 1966 from Honda Fish Beds at Villevieja (fig. 6.3) and is held in the collections of the Institut de Paléontologie (Paris). Based on size and similar appearance, we might speculate that this atlas belongs to the same taxon as the specimens described above.

Discussion and Conclusions

The limbless tetrapod fossils of the Honda Group indicate an assemblage bearing only partial resemblance to the modern Neotropical fauna. The modern Neotropical fauna of South America has a great diversity of caenophidian snake genera, whereas the La Venta fauna appears to have only one genus. For the lowland tropical forests of South America, Dixon (1979) lists seventy-one genera of colubroids, eight genera of boids, three genera of viperids, five genera of scolecophidians, one genus of elapid, and one aniliid

genus. Roze (1966) lists thirty-six colubroid genera, four boid genera, three genera of viperids, two elapid genera, four scolecophidian genera, and one genus of aniliid. If the frequency of colubrid elements in the La Venta snake fauna indicates relative abundance rather than a taphonomic bias, then this group is underrepresented. The aniliids are more diverse and larger in size than would be predicted on the basis of comparison with the living genus. Estes and Baez (1985) attempted to analyze the interchange of the herpetofaunas of North and South America and provided some weak evidence for an interchange of ophidian elements during the Miocene. If the La Venta ophidian diversity is not an artifact, then the low colubroid diversity and the lack of viperid and elapid fossils indicate that, as far as the snakes are concerned, the sample is probably preinterchange or early interchange.

The apodans or caecilians are represented at present by six families (Nussbaum and Wilkinson 1989), of which three are in South America; the La Venta fauna appears to have at least one or possibly two taxa that indicate minimally humid subterranean environments and, possibly, aquatic forms. In addition to the above, the genus *Eunectes* (the anaconda) is largely aquatic.

Of the total diversity represented, three taxa are fossorial or leaf-litter forms, at least one is aquatic, and the remainder are indeterminable as to habitat.

Acknowledgments

We thank S. Renous of the Muséum National d'Histoire Naturelle (Paris), who made comparative material available to us, and the Research Foundation of the City University of New York for Research Grant no. 6-69180, which made this study possible.

7
A Reassessment of the Fossil Tupinambinae

Robert M. Sullivan and Richard Estes

RESUMEN

Paradracaena colombiana, **un nuevo género grande de lagarto Teiidae, está basado en un cráneo casi completo articulado a un esqueleto parcial.** *Paradracaena colombiana* **es el grupo hermano primitivo del actual** *Dracaena guianensis.* **La morfología de** *Paradracaena colombiana* **es distinta y más primitiva que aquella de** *Dracaena* **en base a un número mayor de dientes, la retención de los huesos postorbitales y postfrontales separados, proyección anterior de la apófisis coronoide en alineamiento casi vertical con el surangular y la apófisis angular, y la retención tanto de un plano ancho del isquión como de un contacto neto entre el lacrimal y el prefrontal.** *Paradracaena colombiana* **comparte un número de rasgos derivados con** *Dracaena,* **entre ellos dientes molariformes grandes con una curvatura pronunciada del parapeto del dentario, un cuadrado robusto, un surco esplenial posterior subdental internamente, y el jugal en contacto con el postfrontal. El taxon** *Tupinambis huilensis* **es un sinónimo más reciente subjetivo de** *Paradracaena colombiana.* **Un solo espécimen de lagarto fósil del Grupo Honda es asignada con certeza al género** *Tupinambis.*

The Teiidae is a relatively heterogenous group of lizards that is now largely confined to the New World tropics (Central America, South America, and West Indies), and to the middle and southern portions of North America. Estes (1983b) recognized two teiid subgroups, the Polyglyphanodontinae (extinct) and the Teiinae (which includes the Teiini Presch 1974 and the Tupinambini Presch

1974); Estes et al. (1988) later elevated the latter to subfamily rank (Tupinambinae), which is the focus here. The Teiidae is recognized by modern herpetologists as monophyletic. The relationships among taxa (extinct and extant) within this group, however, have not been well established, and a thorough revision and rigorous character assessment is lacking. Thus, it is generally assumed, the species diversity of the teiids is not well understood. A number of teiid species have been named and many modern taxa are diagnosed only by external features (i.e., scale morphology, toe length, etc.), making comparison to fossil material impossible. Moreover, detailed osteological characterization of extant teiids has yet to be made. Although such a study is beyond the scope of this chapter, we hope that the following contribution concerning tupinambines will be a starting point for understanding the taxonomic diversity of these Neotropical lizards. Until a major revision of extant teiids, using both external and osteological features, is published, we rely on the characters used by Presch (1974) and supplemental characters listed in appendix 7.1.

Fossil Record of the Teiidae

Teiid lizards have a long geological history that begins in the Mesozoic. Both teiines and tupinambines have been found in Late Cretaceous sediments of Campanian age in New Mexico (Armstrong-Ziegler 1980; Sullivan 1981; Estes 1983a) and in Maastrichtian age deposits of Wyoming and Montana (Estes 1964, 1969; Estes et al. 1969). Gao and Fox (1991) reported on "new" teiids from the Campanian (Judithian) of Canada, but some of these are most certainly *Leptochamops,* whereas others are not diagnosable owing to a lack of viable characters. Polyglyphanodontine teiids are confined to Late Cretaceous strata of North America and the People's Republic of Mongolia (Estes 1983a, 1983b).

In the Cenozoic, there have been few records of teiine teiids until the Miocene; earlier Cenozoic occurrences are inaccurately identified or doubtful (Estes 1970). Donadio (1985) reported on a new teiid, *Lumbrerasaurus scaglia,* from the Eocene of Argentina, but this taxon, established on the basis of cross-section tooth shape and subtle differences in vertebral morphology, is inadequately diagnosed and is, therefore, a nomen dubium. Other putative teiids have been named (i.e., *Diasemosaurus* and *Diblosodon* from the lower Miocene of Argentina). Báez and Gasparini (1977) considered the former taxon to be near *Tupinambis* but designated it a nomen dubium. Estes (1983b) considered the latter taxon a lizard of unknown affinities.

Tupinambines do not occur in North America and were apparently restricted to South America at the end of the Maastrichtian. Although no Cretaceous teiids have been found in South America, biogeographical considerations suggest that they were present (Estes 1983a); no evidence suggests that they immigrated to South America at or near the end of the Cretaceous although this is possible.

Previous Work

Fossil vertebrates from the upper Magdalena valley, Colombia, were collected by the University of California-Berkeley during the summer of 1950 and reviewed by Savage (1951b), who also commented on previous collecting efforts in this region. Among the lower vertebrates listed were occurrences of an unidentified iguanid lizard (UCMP 39644) from the Villavieja Formation (Monkey Beds), and teiid lizards, also from the Villavieja Formation (El Cardón Red Beds and La Venta Red Beds and San Nicolás clays

[=Monkey Beds]) Honda Group, upper Magdalena valley, Colombia, both of late Miocene age. One decade later, Estes (1961) described these fossil lizards in more detail and assigned the teiids to the extant taxa *Tupinambis* and *Dracaena* represented by the species cf. *T. teguixin* and *D. colombiana* (UCMP 39643), respectively. Additional *Tupinambis* material (UCMP 56303) was also recorded from putative late Oligocene (now considered early Miocene; see Estes and Báez 1985) Coyaima fauna (Estes 1961). Estes (1983b) later referred the late Miocene *Tupinambis* material (UCMP 38856) to a new species, *T. huilensis.* All these specimens are here placed in *Paradracaena colombiana* (gen. nov., comb. nov.). Estes (1983b) placed the iguanid specimen (UCMP 39644) in Iguanidae incertae sedis.

Additional collections of fossil teiids, including a complete skull from the same strata, were recently amassed by the paleontological expeditions of INGEOMINAS and Duke University (1986–90) and are the subject of this chapter. We describe this new material and reassess the identity of the earlier teiid fossils referred to *Dracaena* and *Tupinambis.*

We note here, in passing, that among the INGEOMINAS-Duke fossil lizards from the La Victoria Formation is a single medial part of a right dentary, bearing four tricuspid teeth (IGM 250986), possibly belonging to a tropidurid iguanian (G. Pregill, pers. obs.). Because the focus of this chapter is on the teiids, this specimen is not considered further.

Biostratigraphy

All the lower vertebrate material described in this chapter was collected from the La Victoria and Villavieja formations (Honda Group). Fission track and ^{40}Ar/^{39}Ar dates for the Honda Group, the section from which the lizard specimens came, range from 13.5 to 11.8 Ma (Guerrero 1990; also chap. 3).

Systematic Paleontology

The taxonomy of fossil and extant Teiidae follows that of Estes (1983b), Estes et al. (1988), and Presch (1974).

Squamata Merrem 1820
Scincomorpha Camp 1923
Teiidae Gray 1827
Tupinambinae Presch 1974
Paradracaena gen. nov.

SYNONYMY

Dracaena Daudin 1802 (in part). *Tupinambis* Daudin 1803 (in part).

TYPE SPECIES

Dracaena colombiana Estes 1961, p. 4.

HOLOTYPE

UCMP 39643, right dentary (holotype of *Dracaena colombiana* Estes 1961, p. 4.).

KNOWN DISTRIBUTION AND GEOLOGIC HORIZON

Honda Group (Miocene), La Venta area and Coyairra, Colombia, South America.

ETYMOLOGY

From the Greek *para,* meaning near or beside, referring here to its affinities to extant *Dracaena.*

DIAGNOSIS

A large tupinambine teiid most closely related to *Dracaena guianensis* from which it differs in having a higher tooth count (15); lesser expansion of the molariform teeth (with round to suboval molariform tooth outline); a nonexpanded quadrate process; a nonbulbous quadrate; anterior extensions of the coronoid, angular, and surangular in near vertical alignment; the jugal not in contact with the squamosal; a rodlike squamosal; pterygoid teeth; the ischial blade not reduced; the lacrimal not reduced, with a strong lacrimal-prefrontal contact; separate postfrontal-postorbital. Differs from *Dracaena, Tupinambis,* and *Crocodilurus* in having pterygoid teeth and from *Crocodilurus* in that the

ascending process of the premaxilla does not contact the frontal. Differs from all non-*Dracaena* tupinambines in tooth replacement morphology; striated molariform teeth coupled with pronounced curvature of the parapet of the dentary; and a splenial (posteromedial) groove.

Paradracaena colombiana (Estes 1961) comb. nov.
(figs. 7.1, 7.2)

SYNONYMY

Dracaena colombiana Estes 1961, p. 4. *Tupinambis huilensis* Estes 1983b, p. 92.

REVISED DIAGNOSIS

Same as for genus.

REFERRED MATERIAL

UCMP 37899, sacrum and associated first caudal vertebra; UCMP 37945, numerous postcranial and limb fragments; UCMP 38856 (holotype of *Tupinambis huilensis* Estes 1983b) left dentary and splenial; UCMP 38927, posterior half of right maxilla; UCMP 40277, distal and proximal ends of a left femur; UCMP 56303, maxillary fragment; IGM Specimens (La Victoria Formation) Duke Locality 33, IGM 183211, nearly complete left dentary with the remains of seven (visible) teeth and miscellaneous bone fragments encased in matrix; Duke Locality 40, IGM 183297, left mandible (broken at the symphysis anteriorly and just behind the coronoid posteriorly) with fragments of splenial, angular, and surangular preserved, anterior half of right maxilla with six (visible) teeth and anterior tip slightly broken, relatively complete frontal, parietal lacking the distal end of the right supratemporal process, centrum of axis vertebra, two fragmentary anterior trunk vertebrae, two fragmentary posterior trunk vertebrae, sacral vertebrae, two caudal vertebrae, distal ends of both humeri, left ilium with acetabular portions of ischium and pubis, distal left tibia, left astragalocalcaneum, an unidentifiable fragment of limb bone; Duke Locality 41, IGM 183308, anterior part of right maxilla (concreted) with four (visible) teeth and space for five; Duke Locality 42, IGM 183368, partial left dentary

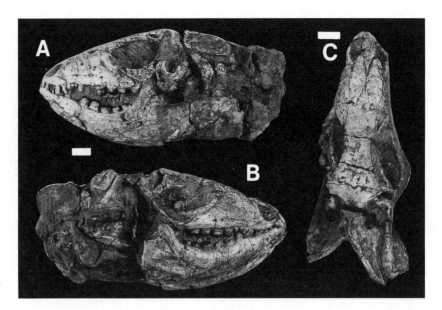

Figure 7.1. Skull of *Paradracaena colombiana* (Estes 1961) gen. nov., IGM 251105, from the Miocene Honda Group of Colombia, South America. *A,* left lateral view; *B,* right lateral view; *C,* dorsal view. *A–C.* Scale bars 1 cm. See figure 7.2 for osteological key.

with four teeth, adherent ectopterygoid, jugal, and other bone mass; Duke Locality 85, IGM 184228, fragment of ?right dentary with one (very large) molariform tooth; Duke Locality 87, IGM 184309, medial part of right dentary bearing two teeth and ?maxilla with five (visible) teeth (both elements heavily concreted); Duke Locality 90, IGM 184343, anterior tip of left maxilla bearing two teeth and incomplete left dentary bearing eight teeth; Duke Locality 97, IGM 184513, posterior part of frontal, posterior parts of both dentaries, fragment of left ilium and acetabular region, proximal end of right femur, distal end of left femur, proximal end of left tibia, incomplete proximal end of right humerus, posterior fragment of right maxilla, medial fragment of right dentary, incomplete angular region of both mandibles, six incomplete vertebrae and unidentified bone fragments; Duke Locality 97, IGM 184518, frontal, maxillae fragments bearing teeth, incomplete ilium and ischium, left coronoid, metacarpal and phalangeal fragments, fragments of vertebrae and miscellaneous bone fragments; Duke Locality 102, IGM 184622, right dentary fragment with one tooth, ?left dentary fragment with one tooth and approximately six incomplete trunk vertebrae and unidentified bone fragments (all concreted); Duke Locality 79, IGM 184671, posterior part of left maxilla bearing two teeth; Duke Locality 121E, IGM 250717, anterior part of left dentary with two teeth; Duke Locality 21, IGM 250787, anterior part of left dentary with four teeth; Duke

Locality 113, IGM 250929, anterior part of left mandible (include parts of the dentary, coronoid, splenial, and angular) with eight (visible) teeth (heavily concreted); and Duke Locality 113, IGM 250959, incomplete right dentary with six (visible) teeth and single trunk vertebra (both concreted); and Duke Locality 132, IGM 251105, a skull and partial skeleton of a large individual.

REFERRED MATERIAL (VILLAVIEJA FORMATION)

Duke Locality 31, IGM 183286, medial fragment of left dentary and splenial with two complete teeth; Duke Locality 54, IGM 183526, three tooth-bearing fragments and limb fragments, including two metatarsal or metacarpal fragments and two other limb bone fragments, all lacking articular ends; Duke Locality 68, IGM 184124, posterior part of ?left dentary bearing two teeth; Duke Locality 83, IGM 184138, incomplete right maxilla bearing five teeth and incomplete trunk vertebra; Duke Locality 22, IGM 184146, right angular, posterior part of right dentary with five teeth, incomplete left jugal, three fragmentary vertebrae and unidentified bone fragments; Duke Locality 6, IGM 184286, posterior part of right maxilla bearing two teeth; Duke Locality 98, IGM 184550, medial part of left dentary bearing two teeth; and Duke Locality 98, IGM

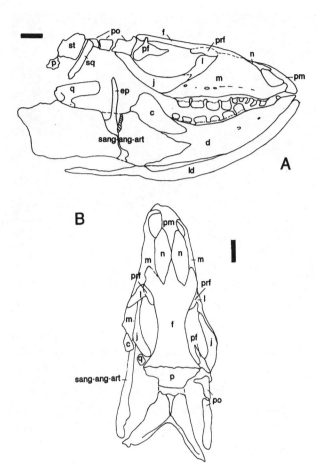

Figure 7.2. Line drawings of skull of *Paradracaena colombiana*, IGM 251105. *A*, right lateral view; *B*, dorsal view. Scale bars 1 cm. See appendix 7.3 for osteological abbreviations.

184558, posterior part of very large ?right dentary with one tooth.

Mixed in with IGM 251105 is the much less complete remains of at least one smaller individual, which includes the following elements: retroarticular region of left mandible, two dentary fragments each bearing one tooth, proximal end of right humerus, distal end of ?left humerus, proximal end of right femur and distal end of right femur (with unfused epiphyses), distal end of left femur and proximal end of left tibia (in articulation), distal end of ?left femur, portion of metacarpals or phalanges, acetabular region and proximal part of left ilium, acetabular portion of left ischium, articulated posterior trunk vertebra and anterior sacral vertebra, portions of five isolated trunk vertebrae, portions of six caudal vertebrae, proximal ends of ribs and unidentified skeletal elements. This material is distinguishable from the rest of IGM 251105 on the basis of smaller size and bone coloration. Because of the uncertainty of the affinities of this material, it is not considered in the description or diagnosis of *Paradracaena colombiana*.

ADDITIONAL REFERRED MATERIAL (UNKNOWN LOCALITY AND HORIZON)

Kyōto University 8-VI-87-13, incomplete parietal, incomplete proximal end of left femur, nearly complete left tibia, five trunk vertebrae (one incomplete), and two caudal vertebrae (associated with paired avian frontals, four boiid vertebrae and unidentified mammal bones).

COMPOSITE DESCRIPTION

The following description of *Paradracaena colombiana* is based on IGM 251105 (excluding the smaller individual) except where noted. Only the major osteological elements that are visible are described. IGM 251105 (figs. 7.1, 7.2) is significant because it is the largest and most complete individual yet found of *P. colombiana*.

Skull

FRONTAL   The frontal is flat with less-pronounced concave orbital borders compared to *Dracaena*. The prefrontal-frontal contact is broad while the postfrontal-frontal contact is slightly developed. Medially, a small flange of the frontal projects posteriorly along the midline to the parietal. The frontal has a lightly rugose surface, a feature also seen in the frontals of IGM-183297 and IGM-184513.

PARIETAL   There is no parietal foramen. Parietal tabs underlie the frontoparietal suture, and the supratemporal processes are deeply excavated for the epaxial musculature. The parietal table is lightly rugose as in the frontals, particularly evident in Kyōto University specimen no. 8-VI-87-13.

LACRIMAL   The lacrimal is preserved only in IGM 251105. In lateral view, both lacrimals lie adjacent to the maxilla and have a broad contact with the prefrontal. Posteriorly, the contact with the anterior projection of the jugal is equally broad.

POSTFRONTAL   Preserved only on the right side of IGM 251105, the postfrontal is distinct and disarticulated, being displaced down and anterior to the frontal-parietal juncture. Posteriorly, there is wide contact with the postorbital, which is excluded from participation with the orbit.

POSTORBITAL   The postorbital is preserved only in IGM 251105 and forms a distinct, stout bar. Anteriorly, the bar flares medially to contact the postfrontal. The anterolateral surface is sculpted with a cluster of small rugosities. Anteroventrally the postorbital contacts the ascending process of the jugal.

SQUAMOSAL   The right squamosal, preserved in IGM 251105, is rodlike and rotated to a vertical position so that it lies nearly perpendicular to the postorbital.

SUPRATEMPORAL   Both supratemporals are preserved in IGM 251105 and are disarticulated from their respective supratemporal processes of the parietal.

MAXILLAE AND TEETH   There are ten teeth on the right maxilla of IGM 251105 and eleven on the left. The anteromost tooth is missing from the right maxilla. The second and third teeth are elongate, caniniform, and recurved. The fourth, fifth, and sixth maxillary teeth are smaller, pointed, and conical. The seventh maxillary tooth is much enlarged and cylindrical on the right side, whereas on the left it is much less enlarged. The right eighth to tenth teeth are enlarged and molariform; the occlusal surfaces are covered with fine, vertical striations that produce a wrinkled crown surface. On the left maxilla, the eighth tooth is missing. The ninth, tenth, and eleventh teeth are enlarged and molariform, although the eleventh is much smaller than the preceding teeth.

PREMAXILLA   The premaxilla of IGM 251105 is crushed on the right side. The proximal end of the nasal process of the premaxilla has a moderately broad contact between the nasals. The premaxilla preserves only one complete tooth and the bases of two others. The premaxillary tooth is blunt with a slight curvature posteriorly.

MANDIBLE   Here, discussion of the mandible covers the dentary, splenial, coronoid, surangular, angular, and retroarticular process.

IGM 251105 has an estimated fourteen teeth on the right dentary and fifteen on the left. The tooth count is difficult to estimate on the right dentary owing to overlap by the maxillary teeth; the left dentary, however, is relatively unobscured. There are seven small, conical, and more or less subequal anterior dentary teeth of which the fourth is missing. The eighth dentary tooth is also conical but about 75% larger than those preceding it. On the right dentary, the ninth tooth is enlarged and cylindrical, and projects into the concavity in the upper dentition. On the left, the ninth tooth is missing. On both dentaries the tenth to fourteenth teeth are enlarged, molariform, and striated on their occlusal surfaces as in the posterior maxillary teeth; the fourteenth is relatively smaller than the preceding teeth. On the left dentary is a tiny, molariform, fifteenth tooth.

Most of the posterior mandibular elements are well preserved in IGM 251105. The coronoids are robust, each bearing a strongly developed anterolateral border and a blunt apex, sculpted with crenulated rugosities. In labial view, the anterolateral processes of the coronoid extends anteriorly on the dentaries to below the posterior part of the penultimate molariform tooth and is in nearly vertical alignment with the anterior extension of the angular and surangular. The inferior border of the coronoid overlaps the surangular. Lingually, neither coronoid is visible on IGM 251105. The left surangular is complete, but the right surangular is missing the posterior portion of the right coronoid process. Along with the surangulars, the articulars, angulars, and, presumably, the retroarticular processes are preserved and are strongly fused together forming a single element (surangular-angular-articular; see fig. 7.2A). Traces of their respective suture articulations, however, are vaguely discernible.

JUGAL   Both jugals are preserved in IGM 251105, but the right is more discernible. The anterior process of the jugal and the inferior border of the jugal blade form a strong articulation with the maxilla and the lacrimal. The posterior ascending process of the jugal is well developed and contacts the ventral surface of the postorbital.

PTERYGOID   Both pterygoids are partially preserved. The posterior aspect of the right pterygoid is partially visible, being obscured by the right quadrate. A single row of large pterygoid teeth is present on the tooth-bearing region of the right pterygoid. The tooth-bearing region of the left pterygoid is not visible. The epipterygoid contacts the pterygoid.

EPIPTERYGOID   The inferior part of the right epipterygoid is visible where it ascends from the pterygoid. In lateral view the epipterygoid is tapered, forming a distinct edge on this otherwise rodlike bone.

QUADRATE   Both quadrates are preserved. The right quadrate has been displaced medially, exposing the ventral border in lateral view. The quadrate is massive with a moderately developed tympanic crest. The left quadrate is preserved but lies next to the jugal owing to postmortem disarticulation.

Postcrania

GENERAL CONSIDERATIONS   Owing to the fragmentary remains of most of the postcranial material and the fact that many are concreted, detailed postcranial descriptions are, for the most part, severely hindered. Therefore, no attempt was made to characterize the postcrania.

SCAPULOCORACOID   An element, possibly the scapulocoracoid, is associated with IGM 251105.

HUMERUS   Only the distal end of the left humerus, embedded in matrix, is preserved.

VERTEBRAE   The vertebral centra are bluntly triangular in ventral view, their apices pointed posteriorly. Each condyle is separated from its respective centrum by a slight constric-tion and is slightly inclined dorsally. The corresponding cotyle faces slightly ventrad. The zygosphenes and zygantrae are robust.

SACRUM   The right ilial blade, acetabular region, and ischium are visible in a block of matrix. The ilial protuberance is broken, and part of the ischial blade is preserved in a counterblock. Impressions of sacral vertebrae in surrounding matrix and the ?left pubis are preserved on the counterblock.

FEMUR   The distal end of a right femur (in articulation with the proximal end of the right tibia) and distal end of left femur show the presence of a deep patellar groove, cited by Estes (1961) for a left femur (UCMP 40277) and typical of teiids in general.

TIBIA   The proximal end of the right tibia in IGM 251105 (in articulation with the distal end of the right femur) is visible but too incomplete and obscured by matrix to allow a description.

cf. *Paradracaena colombiana*

REFERRED MATERIAL (LA VICTORIA FORMATION)

IGM 184518, frontal, left coronoid process, anterior tip of left maxilla bearing one tooth, anterior part of right maxilla with two teeth, posterior fragment of right maxilla, maxilla fragment with one tooth, nearly complete left ilium, acetabular region of left pubis, incomplete proximal end of left humerus, and miscellaneous unidentified postcranial elements; IGM 184522, two nearly complete and two partially complete trunk vertebrae, one nearly complete caudal vertebra and two vertebrae fragments; and IGM 250936, proximal end of left femur, distal end of left humerus, proximal end of left tibia, distal end of right femur articulated with the proximal end of right tibia, medial part of left mandible with attached coronoid process, an isolated tooth, incomplete trunk and caudal vertebrae, and miscellaneous bone fragments.

REMARKS

In the absence of diagnostic characters these specimens are tentatively referred to *Paradracaena colombiana* primarily on the basis of size.

Outgroup Comparisons and Phylogenetic Systematic Position of *Paradracaena*

Presch (1974) presented a phylogeny of the Tupinambinae based on osteological characters. He recognized two distinct groups: the Teiinae (=Teiini of Presch 1974), consisting of *Ameiva, Cnemidophorus, Kentropyx, Teius,* and *Dicrodon,* and the Tupinambinae (=Tupinambini of Presch 1974), composed of *Callopistes, Tupinambis, Crocodilurus,* and *Dracaena.* Estes (1983b) included the fossil taxon *Chamops* in this latter group. From his analysis of primitive and derived character states, Presch (1974) presented a phenogram illustrating the "degree of relatedness" among these taxa.

Our character analysis uses thirty-one osteological characters. We have taken eighteen characters listed in Presch (1974) and combined them with an additional eleven characters from this study (table 7.1; app. 7.1) and two from Estes (1983b). We used PAUP version 2.4.1 (Swofford 1985) to analyze these data. Some of the character states were reassessed (e.g., Presch's character 15 [anterior] supraangular foramen [window] is present in *Dracaena* and absent in *Callopistes*). One tree was generated with a length of 37 and a rescaled consistency index of 0.892. Figure 7.3 and appendix 7.1 present the resulting cladogram and characters, respectively.

Our results, which corroborate those of Presch (1974), include *Paradracaena* as the sister taxon of *Dracaena* and *Crocodilurus* as the sister taxon to *Paradracaena-Dracaena.* Numerous autapomorphies separate *Dracaena* from *Paradracaena,* suggesting unequivocal generic distinction between fossil and Recent taxa.

Comments and Comparisons

Several skull characters distinguish *Dracaena* from *Paradracaena* and thereby support recognition of two distinct genera. In *Dracaena,* the postfrontal and postorbital are fused into a single element, whereas both bones remain unfused in *Paradracaena* and all other tupinambines. Although slightly displaced (in IGM 251105), it is clear that the postfrontal in *Paradracaena* was in contact with the jugal as in *Dracaena.*

Crocodilurus, Paradracaena, and *Dracaena* all have a robust quadrate. The anterolateral expansion and lunate curvature of the tympanic crest in *Paradracaena* is not as developed as in *Dracaena,* and is more like that of *Crocodilurus.* The massive nature of the quadrate (as seen in IGM 251105) may be (in part) attributed to its large size.

In *Paradracaena* and *Crocodilurus,* the anterolabial process of the coronoid, angular, and surangular are aligned vertically as in some *Tupinambis* (such as IGM 184037, described later) although in some other *Tupinambis* specimens the anterolabial coronoid process does not reach the anterior extent of the surangular and angular. In *Dracaena* the anteriolabial process of the coronoid lies well in advance of the angular and surangular projections. The former condition is reminiscent of that seen in the large glyptosaurine lizards (Anguidae) and may be correlated with massive mandibles and teeth (Sullivan 1979, 1986). Moreover, the anteriolabial process of the coronoid in *Dracaena* may be a derived feature that arises as the result of developing large molariform teeth.

Paradracaena, as with all other tupinambines, departs from *Dracaena* in that the ascending process of the jugal and anterior projecting process of the squamosal remain separate.

Dracaena is the only tupinambine lizard to modify the squamosal into a tapered blade. *Paradracaena* retains the more primitive barlike construction, a condition seen in all other tupinambines. In addition, *Paradracaena* retains a broad prefrontal-lacrimal contact, another primitive feature based on outgroup comparison. In *Dracaena* the lacrimal is reduced and the contact with the prefrontal is weak.

Only one postcranial feature of *Paradracaena* is sufficiently preserved to compare with modern tupinambines: *Paradracaena* retains an expanded ischial blade, which, based on our analysis, is a primitive feature. The ischial blade in *Dracaena* is reduced.

The only known unique (autapomorphic) character for *Paradracaena* is the presence of pterygoid teeth. Pterygoid teeth in *Paradracaena* is a character reversal because of their absence in *Tupinambis, Crocodilurus,* and *Dracaena.*

Table 7.1. Character state matrix for the Tupinambinae

Taxon	Character																														
	1	2	3	4	5	6	7	8	9	10	11	12	13	14	15	16	17	18	19	20	21	22	23	24	25	26	27	28	29	30	31
Ancestor	0	0	0	0	0	0	0	0	0	0	0	0	0	0	0	0	0	0	0	0	0	0	0	0	0	0	0	0	0	0	0
Chamops	?	?	?	?	?	?	?	?	?	?	?	?	?	?	?	?	?	?	?	?	?	?	?	?	?	?	?	0	0	?	0
Callopistes	0	0	1	0	0	1	1	0	1	0	0	1	0	0	1	1	1	0	0	0	0	0	0	1	0	0	0	1	1	0	0
Tupinambis	0	0	1	1	2	1	1	1	1	0	0	1	1	0	0	1	0	0	0	0	0	0	0	1	0	0	1	1	1	0	0
Crocodilurus	0	0	1	1	1	1	1	1	1	0	1	1	1	0	0	1	0	0	0	0	0	?	0	1	1	0	1	1	1	1	0
Paradacaena	0	0	1	0	3	?	?	?	?	0	?	?	?	1	0	1	0	0	0	1	0	0	0	1	1	0	1	1	1	0	0
Dracaena	1	1	1	1	3	1	1	1	1	0	1	1	1	1	0	1	0	0	1	1	1	1	1	1	1	1	1	1	1	0	1

Sources: Characters 1–18 from Presch (1974); 19–27, 30, 31, this chapter; 28, 29, Estes (1983b).

Note: Consult appendix 7.1 for the key to the thirty-one characters.

Paradracaena and *Dracaena* share two characters that set them apart from other tupinambines. First, Estes (1961) noted that the splenial of the type specimen of *"Tupinambis huilensis"* (*Paradracaena*) was concave, unlike the flat or con-vex surface seen in Recent individuals of *Tupinambis* and other tupinambines. This feature, referred to as the splenial groove, is situated posteromedially on the lingual surface of the splenial. Based on our analysis, this is a synapomorphy that unites *Dracaena* and *Paradracaena*.

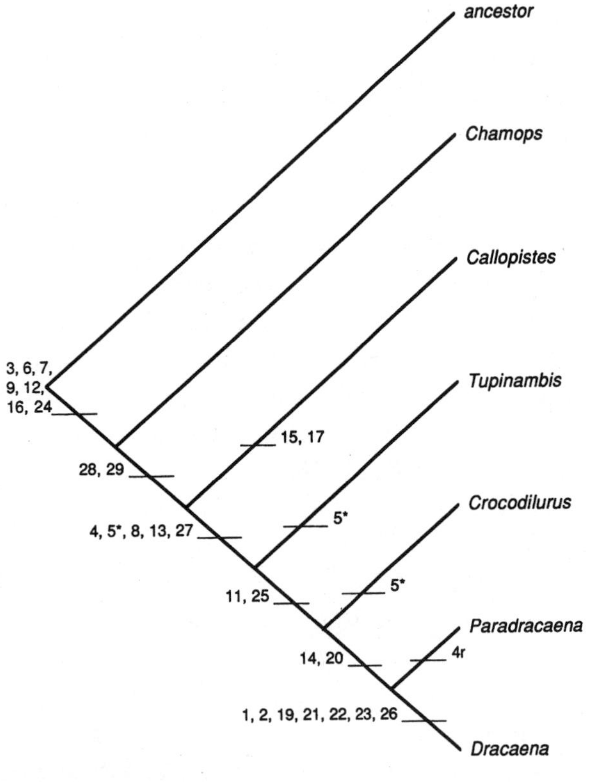

Another feature shared by both *Paradracaena* and *Dracaena* is the presence of striated, mushroom-shaped (=molariform) teeth. In occlusal view, the molariform teeth of *Paradracaena* are round to suboval in outline, whereas those of *Dracaena* tend to be more square. In *Paradracaena* these teeth are distinct from the large, blunt teeth seen in some large specimens of extant *Tupinambis* in being more molariform. Presumably correlated with their molariform dentition is the deep, curved parapet of the dentary.

Other features of *Paradracaena* are worth noting but their phylogenetic significance is uncertain. The premaxillary teeth are blunt and concave labially, with a slight convexity lingually, and slightly recurved posteriorly. This shape departs from the pronounced tricuspid premaxillary teeth of *Tupinambis*, the sharply pointed, recurved premaxillary teeth of *Callopistes*, and the less-pronounced heterodont (bicuspid and tricuspid) premaxillary teeth of *Crocodilurus*. They are similar to those of *Dracaena* but differ in that they are not finely striated. Estes (1961) referred postcranial material (UCMP 40277 and 37899) to the fossil species in question and, based on comparisons with specimens collected by INGEOMINAS and Duke University, there is little doubt that this material was properly referred. The overall size and form of the femora is consistent with the left femur (UCMP 40277) described by Estes (1961), and the

Figure 7.3. Phylogenetic hypothesis for fossil and Recent Tupinambinae. Characters are listed in appendix 7.1. *Asterisk* denotes unordered multicharacter state; *r* denotes reversal. Characters 10 and 18 are constant characters and have no phylogenetic value.

sacral, vertebral, and other postcranial material (UCMP 37899 and 37945) conform to the newly discovered specimens described herein.

It is worth noting that, in addition to the features mentioned, the large, mature specimen of *Paradracaena colombiana* (IGM 251105) has a skull length of 110 mm; by comparison, a large extant individual of *Tupinambis* (BM 53.3.7.27) has a skull length of 105 mm, comparable in size with the fossil specimen. In contrast, extant *Dracaena* has smaller skull lengths, on the order of 75–85 mm, based on adult comparative material of *Dracaena guianensis* (see app. 7.2) and Presch (pers. comm. 1991 to Sullivan), as well as the material referred by Estes (1961). Given this size difference in the skull, and other evidence, such as size differences between sacral vertebrae (UCMP 37899) and extant *Dracaena*, *Paradracaena colombiana* was gigantic, nearly one-third larger. The small size of modern *Dracaena* may be a derived feature, but differences in size alone cannot necessarily be assumed to be of any phylogenetic value. We, therefore, take a more conservative approach in not accepting size as a viable derived feature. It is interesting to note that *P. colombiana* exceeds all known tupinambines in robustness. The massive nature (relative robustness) of this lizard is already reflected in a number of characters used in the diagnosis.

Systematic Paleontology

Tupinambis Daudin 1802
Tupinambis sp.
(fig. 7.4)

REFERRED MATERIAL (LA VICTORIA FORMATION)

IGM 184037, dorsally crushed skull of a young specimen, mandibles with teeth, lower parts of maxillae with teeth, partial palate, basicranial area, quadrate, upper temporal fenestra, and part of dorsal skull roof.

DESCRIPTION

IGM 184037, an immature specimen, consists of a skull crushed dorsoventrally. The skull roof is not preserved ex-

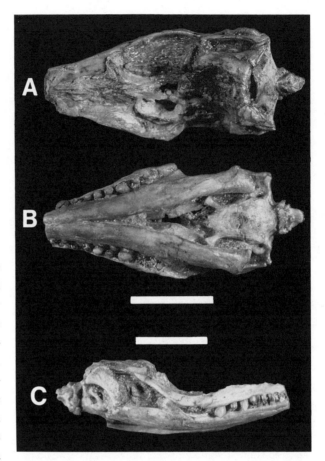

Figure 7.4. Incomplete skull of *Tupinambis* sp., IGM 184037, from the Miocene Honda Group of Colombia, South America. *A.* Dorsal view, showing palatal region and basicranium. *B.* Ventral view, showing mandibles, maxillae with teeth, and basicranium. *C.* Right lateral view, showing quadrate, basicranium, mandible, jugal, postorbital, and tooth-bearing region of the right maxilla. Scale bars 5 mm.

cept posteriorly; the posterior region of the parietal and the bones surrounding the left temporal emargination are preserved.

The basisphenoid is about the same width from the suture with the basioccipital to the base of the basipterygoid process. Anterior to the latter, the bone narrows, projecting anteriorly as the parasphenoid rostrum, of which only the base is preserved. There is a slight, rounded elevation midway between the basipterygoid processes, and behind this, the basisphenoid is slightly concave.

An unossified basicranial fontanelle occurs on the basioccipital-basisphenoid suture. The basipterygoid processes extend laterally and are about two-thirds the length of the basisphenoid bone as preserved.

The basioccipital is gently convex and is quadratic. The basioccipital condyle is poorly developed and is covered by elements of the atlas-axis complex. The lateral surface of the basioccipital forms the floor of the fenestra rotundum. Lateral to the condyle occur the large vagus foramen and the small, more ventral hypoglossal foramen. Above the large crista interfenestralis occurs the fenestra ovalis.

The large supraoccipital is a rhomboid bone with dorsal and ventral excavations for parietal attachment and spinal cord passage. The prootic is only partially exposed.

The palate is obscured by the mandibles. Only the anterior parts of the quadrate processes are visible in ventral view. In dorsal view there is partial exposure of the palatine. Its maxilloectopterygoid process is sutured to the maxilla over two-thirds of the length of the latter bone. The anterior portion of the pyriform space is also visible. The ectopterygoid is elongated anteroposteriorly with its medial surface bounding the suborbital fenestra.

Only the tooth-bearing area of the maxillae is preserved. The maxilla articulates posteriorly with the long, narrow jugal. The jugal contacts the postorbital region, which is partially preserved. The posterior process of the jugal contacts the slender squamosal. The articulation region of quadrate, squamosal, supratemporal, and parietal shows no unusual features.

The azygous parietal is preserved only posteriorly, its supratemporal processes reaching the quadrate area. The processes are short and blunt.

The mandibles are complete except for their retroarticular processes, but articular, prearticular, angular, surangular, coronoid, and dentary are visible only in ventral view. The juxtaposition of the anterior processes of the coronoid, surangular, and angular are in a vertical plane.

The dentition is well preserved. There are nine maxillary teeth on the right, ten on the left. The first maxillary tooth is slender and conical, recurved and pointed. The second tooth is much larger but has a similar shape to the first. The third tooth is also conical but is tiny relative to the anterior two, thus forming a diastema. It is followed by a fourth tooth that is relatively large. The fifth through ninth teeth are tricuspid and enlarged, with the eighth being swollen and globose. The ninth and tenth teeth are tiny cones, minute in comparison with the enlarged eighth tooth. The mandibular dentition is equally well preserved but is covered by the maxillary dentition. The mandibular dentition is similar to that of the maxillary except that the fifth mandibular tooth is relatively much larger than the fifth maxillary tooth; nonetheless, both are tricuspid and similarly rounded in outline. In addition, the ninth and tenth mandibular teeth are large and globose in contrast to their maxillary homologues.

REMARKS

Only a single fossil specimen (IGM 184037, fig. 7.4) from the Honda Group of Colombia is referable to the taxon *Tupinambis*. Unfortunately, no immature extant *Tupinambis* of comparable size to the immature fossil specimen was available for comparison. A specimen about twice the size of the present example (SDSNH 66269, skull length 52 mm), however, shows much the same distribution of tooth shape and form as does the fossil, although it has thirteen teeth rather than ten. From this admittedly incomplete comparison it can be suggested that Miocene *Tupinambis* went through much the same ontogenetic tooth changes as in extant individuals. Further comments are premature.

Conclusions and Summary

Based on the study of this new material and the review of previously collected teiid specimens the following conclusions were reached:

1. *Paradracaena colombiana* (gen. nov., comb. nov.) is a large extinct teiid distinct from extant *Dracaena*.
2. *Tupinambis huilensis* is a subjective junior synonym of *Paradracaena colombiana* based on the presence of a distinct splenial groove (posteromedially), large bulbous molariform teeth in association with pronounced curvature of the parapet of the dentary, and the overall massive form of the dentary.
3. Only one specimen (IGM 184037) is referable to the genus *Tupinambis* with any degree of certainty. Because of its juvenile nature, affinities to modern *Tupinambis* are not clear.

4. *Paradracaena colombiana* is more primitive than modern *Dracaena* based on a higher tooth count; retention of separate postorbital and postfrontal; anterior projection of coronoid process in near-vertical alignment with surangular and angular processes; and retention of both a wide ischial blade and strong lacrimal-prefrontal contact. Because of the absence of these features in modern *Dracaena*, coupled with the magnitude of the differences between the fossil and Recent species, we believe that the fossil material represents a genus distinct from *Dracaena*.

5. *Paradracaena colombiana* shares a number of derived features with extant *Dracaena*. These include large molariform teeth with associated pronounced parapet curvature of the dentary; robust quadrate (a feature also present in *Callopistes*); posterior subdental splenial groove (medially); and jugal in contact with postfrontal.

In summary, *Paradracaena colombiana* displays a mosaic of both primitive and derived characters based on comparison between the fossil and extant material. In this way it appears that *Paradracaena colombiana* is morphologically intermediate between *Crocodilurus* and *Dracaena* with many features clearly shared by the latter taxon.

Acknowledgments

We gratefully acknowledge R. Kay and R. Madden (Duke University) for allowing us the privilege of studying the fossil lizards collected by the INGEOMINAS-Duke paleontology expeditions.

R. M. Sullivan thanks the following individuals: R. Etheridge (San Diego State University) and W. Presch (California State University, Fullerton) for the loan of extant specimens. R. M. Sullivan thanks W. Presch for discussions concerning teiid characters and identification; E. N. Arnold (British Museum [Natural History]) and G. K. Pregill (University of San Diego) for the loan of extant comparative material; J. H. Hutchison (University of California, Museum of Paleontology) for the loan of the type and referred specimens of *Paradracaena colombiana* and the holotype of *Tupinambis huilensis;* M. Burgers (San Diego State University) for allowing R. M. Sullivan to examine AMNH 46290 (*Crocodilurus*) on loan to her and for help translating some of the Spanish literature; and T. Deméré for taking the photographs.

R. M. Sullivan also thanks W. Presch, G. Pregill, and S. G. Lucas (New Mexico Museum of Natural History) for reading an earlier version of this chapter; J. A. Holman (Michigan State University) and J. H. Hutchison for formally reviewing the manuscript. Their respective critiques are very much appreciated.

Appendix 7.1. Derived characters

1. Postfrontal and postorbital fused
2. Expanded quadrate process of the pterygoid
3. Dorsal squamosal process lost
4. Pterygoid teeth lost
5. Biconodont or triconodont premaxillary teeth present
6. Rodlike clavicle
7. Clavicular hooks present
8. Short interclavicular median process
9. Scapular fenestra lost
10. Divergent process series of caudal vertebrae lost, or lack of caudal process
11. Less than sixty caudal vertebrae
12. Twenty-five presacral vertebrae; eleven postxiphisternal ribs
13. Second ceratobranchial absent
14. Molariform teeth (associated with pronounced curvature of the dentary)
15. Supraangular fenestra absent
16. Reduced ectopterygoid to inferior orbital foramen
17. Frontal parietal shape flat (non-concave)
18. Fifth toe reduced
19. Coronoid process in advance to anterior projections of the angular/surangular processes
20. Posteromedial splenial groove
21. Jugal contacts squamosal
22. Ischial blade reduced
23. Squamosal modified into tapered blade
24. Enlarged epaxial muscle surface attachment and reduction of parietal table
25. Quadrate robust
26. Lacrimal reduced; contact with prefrontal weak
27. Postfrontal not in contact with jugal
28. Strong heterodonty
29. Parietal foramen absent
30. Ascending process of the premaxilla contacts the frontal
31. Robust quadrate bulbous

Appendix 7.2. Specimens examined for this study

Specimens were complete, extant skeletons, except for fossil specimens, unless otherwise noted.

American Museum of Natural History, New York
    AMNH 46290 *Crocodilurus lacertinus* (skull and mandibles)

British Museum (Natural History), London
    1964.1825 *Tupinambis teguixin nigropunctatus* (skull and mandibles)
    53.3.7.27 *Tupinambis teguixin nigropunctatus* (skull and mandibles)

San Diego Natural History Museum, San Diego
    SDSNH 64802 *Ameiva chrysolaema*
    SDSNH 64932 *Tupinambis teguixin nigropunctatus*
    SDSNH 65165 *Amieva ameiva*
    SDSNH 65496 *Tupinambis teguixin nigropunctatus*
    SDSNH 66269 *Tupinambis teguixin nigropunctatus*
    SDSNH 66582 *Ameiva festiva*

University of California Museum of Paleontology, Berkeley
    UCMP 37899 *Paradracaena colombiana*
    UCMP 37945 *Paradracaena* cf. *P. colombiana*
    UCMP 38856 *Paradracaena colombiana* (holotype of *Tupinambis huilensis*)
    UCMP 38927 *Paradracaena colombiana*
    UCMP 39643 *Paradracaena colombiana* (holotype)
    UCMP 40277 *Paradracaena colombiana*
    UCMP 56303 *Paradracaena colombiana*

Richard Etheridge (Private Collection)
    No. 656 *Dracaena guianensis*
    No. 1297 *Callopistes maculatus* (skull and mandibles)

William Presch (Private Collection)
    No. 185 *Tupinambis teguixin nigropunctatus* (skull and mandibles)
    No. 534 *Dracaena guianensis*
    No. 539 *Callopistes flavipuntatus*

Appendix 7.3. Osteological abbreviations

c	coronoid
d	dentary
ep	epipterygoid
f	frontal
j	jugal
l	lacrimal
ld	left dentary
m	maxilla
n	nasal
p	parietal
pf	postfrontal
po	postorbital
prf	prefrontal
q	quadrate
sang-ang-art	fused surangular, angular, and articular
sq	squamosal
st	supertemporal

Contributor's note added in proof

While this volume was in press, Denton and O'Neill (1995) published a paper naming a new teiid taxon, *Prototeius stageri*, based on incomplete, and isolated, skull elements from the Late Cretaceous of New Jersey. This lizard is a composite taxon whose diagnosis was based on associated elements rather than on just the holotype (a right dentary with teeth). This taxon must be viewed with skepticism, as association of isolated skull elements (recovered by screenwashing) is equivocal. Denton and O'Neill's (1995) erection of the subfamily Chamopsiinae (which includes the taxa *Chamops*, *Leptochamops*, *Meniscognathus*, and *Prototeius*) is not defensible owing to the fact that *Chamops* (type material: right maxilla with five teeth [Gilmore 1928:24; Marsh 1892]), *Leptochamops* (holotype left dentary with twenty-three teeth [Estes 1964; Gilmore 1928:26]), and *Meniscognathus* (left dentary [Estes 1964:113]) are all composite taxa, whose diagnoses rely heavily on referred material which may, or may not, belong to the taxon represented by the type material. Only two characters diagnose the Chamopsinae (Denton and O'Neill 1995): (1) teeth striated longitudinally and (2) parietal fossa anteriorly displaced. "Striated

teeth" is a dubious character because this condition is known in *Paradracaena* and other lizards, and therefore is probably a plesiomorphic character. The position of the parietal fossa is not known in these lizards with any certainty. None of the holotype specimens are known by parietals (only questionably referred parietal material), and so this character is equivocal. The rest of the characters presented were cited as plesiomorphic with respect to the remaining teiids; thus, these characters have no phylogenetic value. The taxon Chamopsinae is not justifiable.

R. M. Sullivan
November 1995

Denton, R. K., Jr., and R. C. O'Neill. 1995. *Prototeius stageri,* gen. et sp. nov., a new teiid lizard from the Upper Cretaceous Marshalltown Formation of New Jersey, with a preliminary phylogenetic revision of the Teiidae. Journal of Vertebrate Paleontology 15: 235–253.

Estes, R. 1964. *See* References Cited.

Gilmore, C. W. 1928. Fossil lizards of North America. Memoirs of the National Academy of Sciences 22: 1–297.

Marsh, O. C. 1892. Notice of new reptiles from the Laramie Formation. American Journal of Science 43: 449–453.

8
Crocodilians, *Gryposuchus,* and the South American Gavials

Wann Langston, Jr., and Zulma Gasparini

RESUMEN

Nuevos restos de cocodrilos fósiles del Grupo Honda proveen importantes datos sobre varios taxa. Comparado con *Sebecus icaeorhinus* de Argentina, *Sebecus huilensis* tiene el sector rostral del premaxilar más alto y angosto, el cuarto diente premaxilar relativamente más grande, un gran diastema premaxilar-maxilar no indentado ni perforado, narinas externas no rodeadas por una suave depresión, un puente internarial completo formado por nasales y premaxilares, y vértebras caudales anficélicas. Las coronas bulbosas características de *Balanerodus logimus* han sido registradas por primera vez en el Grupo Honda. *Gryposuchus colombianus* está representado por excelente material craneano, indicando que comparte sinapomorfías con los Gavialidae, especialmente *Eogavialis* del Eoceno tardío-Oligoceno temprano de Africa. El cráneo es deprimido, no verticalizado, y de contorno menos anguloso. Al menos uno de los nasales contacta con el premaxilar. Posee cuatro dientes en cada premaxilar, veintiúno a veintidós en el maxilar, y veintitrés en el dental. Las narinas son más anchas que largas y carecen de excrecencia ósea y zócalo internarial. Falta la espina postorbital, y grandes y elongadas bullas pterigoideas están presentes. Los arcos óseos que limitan las fenestras supratemporales son angostos, y el proceso retroarticular termina en una característica expansión en forma de gancho. En *Gryposuchus, Ikanogavialis,* y *Hesperogavialis,* el surangular termina anterodorsalmente en o por detrás del último diente mandibular. En este carácter estos taxa difieren de otros gaviales y tomistomas. Las evidencias disponibles sugieren que los gaviales arribaron a América del Sur desde Africa, cruzando el Atlántico Sur en el Eoceno medio o aun antes. Si *Caiman* cf. *C. lutescens* es verdaderamente *Caiman latirostris,* su presencia en el Grupo Honda es el registro fósil más antiguo de una especie viviente de cocodrilo.

The middle Miocene rocks exposed in the Tatacoa Desert northeast of the village of Villavieja, Huila Department, Colombia, are fertile ground for fossil vertebrates. A major collection was obtained here in the 1940s and 1950s by parties from the University of California (see Savage 1951b and Fields 1957 for general summaries of this work). The area, including many of the University of California localities, was revisited from 1985 to 1992 by joint field parties from the geological survey of Colombia, INGEOMINAS (IGM), and Duke University. Among the fossils recovered by these parties are a number of crocodilian specimens, some of which are the subject of this chapter.

Most of the crocodilian fossils in the IGM-Duke collection are fragmentary and pertain to taxa already recognized in the La Venta fauna (Langston 1965). They add, however, some information on the stratigraphic and local geographic ranges of *Sebecus, Mourasuchus,* and *Charactosuchus* in Colombia. The chief interest of the collection, however, attaches to the gavials, which are represented by some excellent cranial and mandibular material. Problems attending delivery of the specimens have made it impossible to prepare them fully in time for this study, but the best material has been sufficiently developed to permit some description and observations bearing upon the taxonomy of South American Tertiary gavials.

Alligatorid remains are sparingly represented in the collection, and with the exception of the enigmatic *Balanerodus* they are not dealt with here.

Our knowledge of South American Tertiary crocodilians has been greatly expanded in recent decades by, among others, Gasparini (1968, 1972, 1981, 1984), Sill (1970), Buffetaut and Hoffstetter (1977), Bocquentin Villanueva and Buffetaut (1981), Buffetaut (1982), Bocquentin Villanueva (1984), Bocquentin Villanueva et al. (1989), Souza Filho and Bocquentin Villanueva (1989), Bocquentin Villanueva and Souza Filho (1990), and Gasparini et al. (1993). Previous systematic assignments of the South American Tertiary gavials have been called into question by Buffetaut and Hoffstetter (1977), and we discuss these issues herein.

Our treatment of the IGM-Duke material is largely morphological, and no attempt is made to interpret the geological context of the specimens. Localities and stratigraphy are mentioned only briefly in the text, and the reader is referred to chapter 2 for a fuller description and analysis of the provenance of the specimens.

Systematic Paleontology

Crocodylia
Sebecosuchia
Sebecidae
Sebecus Simpson 1937
Sebecus huilensis Langston 1965

MATERIAL AND PROVENANCE

IGM 250816, right premaxilla and fragmentary maxilla, anterior part of a left dentary, part of the dorsal edge of a rostrum, and two vertebrae, Duke Locality 57, Villavieja Formation, Baraya Member (Monkey Beds), Honda Group, middle Miocene.

AMENDED DIAGNOSIS

To the original diagnosis by Langston (1965) may be added: premaxilla higher and narrower than in *Sebecus icaeorhinus* Simpson; the fourth premaxillary tooth hypertrophied; a large premaxilla-maxilla diastema without indentation or perforation; external naris lateral, lacking the smooth marginal depression seen in *S. icaeorhinus;* complete internarial bridge formed by nasals and premaxillae.

DESCRIPTION

The premaxilla of IGM 250816 (fig. 8.1A, B) is very deep; it surrounds most of the external nares and takes part in the internarial bridge. The external surface of the premaxilla is roughened by numerous irregular furrows. Anteriorly the surface of the narial bridge is marked by fine, mostly vertical striations (paralleling the longitudinal axis of the premaxillary component of the bridge). The slightly roughened surface of the rostrum, particularly the maxillary part, is a character shared with *Sebecus icaeorhinus* Simpson 1937, and *Baurusuchus pachecoi* Price 1945, but in some other sebecosuchians (*Ayllusuchus* Gasparini 1984; *Bretesuchus bonapartei* Gasparini et al. 1993), the external walls are much more rugose. Like all sebecosuchians, *S. huilensis* had four teeth in the premaxilla. The premaxilla (IGM 250816) is preserved from the posterior edge of the second alveolus to the fourth. The fourth alveolus is twice the size of the third, and judging by the space between the posterior tip of the premaxilla and the posterior border of the second alveolus, the first two alveoli were small. The interalveolar spaces 2–3 and 3–4 are constricted laterally, indicating that strong mandibular teeth sheared past the premaxilla. *Sebecus icaeorhinus* lacks this lateral constriction; instead, the lateral wall of the premaxilla is expanded outward between the second and third teeth. Moreover, in *S. icaeorhinus* the largest tooth in the premaxilla is the third, whereas in *S. huilensis* it is the fourth tooth that is hypertrophied.

Behind the fourth alveolus a large diastema extends onto the maxilla. Outwardly, the diastema is not constricted or perforated, but it contains a deep excavation that extends beyond the midheight of the rostrum. This indicates the great length of the fourth mandibular tooth, which may have extended upward for 3.5 cm when the jaws were closed.

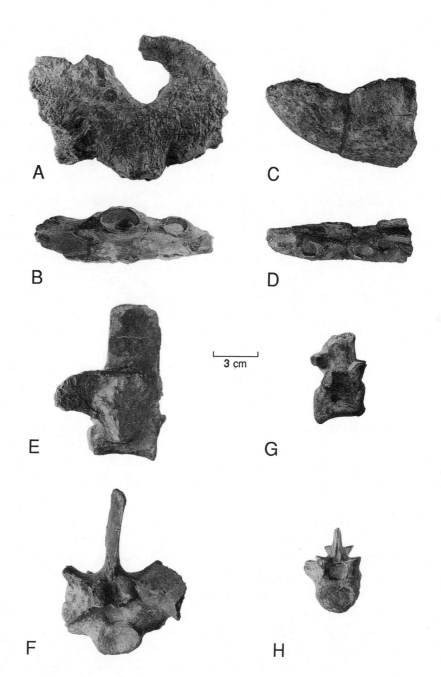

Figure 8.1. *Sebecus huilensis* Langston, IGM 250816. *A–B.* Right premaxilla. *A,* lateral view; *B,* palatal view. *C–D.* Anterior end of left dentary. *C,* lateral view; *D,* lingual view. *E–F.* Second sacral vertebra and rib. *E,* right lateral aspect; *F,* posterior aspect. *G–H.* Anterior caudal vertebra. *G,* right lateral aspect; *H,* anterior aspect.

3 cm

The palate of *S. huilensis* is very narrow and is little vaulted in relation to the level of the tooth row, a character shared with *S. icaeorhinus.* The maxillary fragment, which retains three alveoli, displays the strong compression and elevated, almost smooth lateral wall characteristic of *Sebecus.* The three alveoli, although incomplete, are 1.5 times as long as they are wide. The most complete alveolus contains a strongly compressed tooth with serrated edges.

The external naris is lateral in position as in *S. icaeorhinus.* The internarial bridge is formed by the premaxilla and nasal, but unlike other sebecosuchians and some other ziphodonts (e.g., *Peirosaurus* Price 1955), a smooth depressed area bordering the naris does not exist.

The anterior part of the dentary is preserved back to the fourth alveolus (fig. 8.1C, D). It is very high and compressed laterally. The first alveolus is low and projects forward in the holotype of *S. huilensis*. The second alveolus is subcircular and lies close behind the first as in the holotype of *S. huilensis;* the interalveolar space 2–3 is 3.5 times longer than the space between the first two alveoli. This space allowed for lateral passage of the hypertrophied fourth premaxillary tooth. This wide diastema is followed by an abrupt elevation of the dorsal edge of the dentary lateral to the third and fourth alveoli. The third alveolus is similar in size to the second and projects forward. The fourth is hypertrophied and possesses a noticeable collar.

Measurements (cm)

Height of premaxilla from the internarial bridge to the edge of the third alveolus	8.5
Width of premaxilla at the third alveolus (ventral aspect)	2.4
Anteroposterior diameter of the fourth alveolus	2.6
Height of dentary at the level of the second alveolus	3.7
Width of dentary at the level of the second alveolus	2.4
Height of dentary at the level of the fourth alveolus	6.2
Anteroposterior diameter of first alveolus	~1.5
Anteroposterior diameter of second alveolus	1.1
Anteroposterior diameter of third alveolus	1.3
Anteroposterior diameter of fourth alveolus	~2.6

Bearing the same field number as the cranial parts are two amphyplatyan or slightly amphicoelous vertebrae. One is a posterior sacral with the right rib attached (fig. 8.1E, F); the other is an anterior caudal, possibly the first (fig. 8.1G, H). The neural spine of the sacral is broad and roughly rectangular in lateral aspect. The articular faces of the posterior zygapophyses are slightly inclined downward toward the midline. The sacral rib is firmly fused to the centrum and is broadly flared posteriorly, suggesting that the ilium was expanded in this direction. The anterior face of the centrum is higher than the posterior face, which is wider than high (principal diameters of 3.3 and 2.0 cm, respectively). This vertebra is 10.7 cm high; the centrum is 4.5 and 5.5 cm long at the bottom and top, respectively. The estimated width at the posterior end of the sacral ribs is about 10 cm.

The second vertebra lacks the distal part of the neural spine. The neural spine is slightly shorter at its base than that of the sacral. The anterior zygapophyseal facets are widely separated and face dorsomedially at an angle corresponding to the posterior facets of the sacral. The posterior zygapophyseal facets are more strongly inclined than the anterior facets, confirming that this is a caudal vertebra. The neural canal is very large for the size of the vertebra, having, posteriorly, a cross-sectional area of more than one-half that of the centrum. The centrum is slightly higher posteriorly than anteriorly. Its longitudinal axis is inclined caudoventrally. The anterior face is 2.8 cm wide and 2.1 cm high. Corresponding dimensions of the posterior face are 2.6 cm and 2.2 cm. The dorsal and ventral edges of the anterior face of the centrum extend farther forward than the lateral rim, imparting a somewhat cupped appearance to the lateral profile. The posterior face is straight in the same aspect. Chevron facets are absent, but the posterolateral ventral corners of the centrum are thickened and strongly beveled. At its base the caudal rib is thick and rounded in cross section. It is inclined backward at an angle of about 35° to the sagittal plane.

A few isolated ziphodont teeth from various localities and stratigraphic levels in the La Venta area are present in the collection (table 8.1). IGM 184165, presumably a right anterior maxillary tooth, which has a root 2.5 times as long as the crown, is narrowly suboval in section and displays some gentle fluting on the lingual side at the base of the crown. Fluting has not been noted previously in *Sebecus huilensis* teeth. Teeth whose crowns are almost twice as high as they are wide predominate (e.g., IGM 184427). They resemble UCMP 40186 illustrated by Langston (1965, pl. 1B). Other teeth have proportions of 4:1 (e.g., IGM 250505) and are very similar to UCMP 37877, which was found with the holotype of *S. huilensis* (pl. 1A in Langston 1965). The number of marginal serrations on these teeth, varying from three to six per millimeter, are within the range of *S. huilensis* teeth measured by Langston (1956, table 1). There are four and five serrations per millimeter on the anterior and posterior edges, respectively, in *S. icaeorhinus*. The larger teeth have fewer serrations, a pattern not as apparent in previously measured teeth of *S. huilensis*. Two weathered and incomplete teeth, IGM 250457 and 250541, resemble the large tooth UCMP 40186 figured by

Table 8.1. *Sebecus huilensis* teeth in the IGM collection

IGM Specimen	Duke Locality	Crown Height (mm)	Diameter of Crown at Base (mm)		Basal Length of One Serration (mm)	Serrations per 5 mm (Anterior Edge)
			Transverse	Anteroposterior		
184427	91	36	9	17	0.17	29.0
250541	196	≥39	12	27	0.31	32.0
250457	106	≥52	15	30 (estimated)	—	—
184378	40	30	8	17	—	—
184165	41	≥20	9	12	—	—

Langston (1965, pl. 1B) in profile, but both are relatively more compressed transversely and IGM 250457 is larger than any previously reported sebecosuchian tooth from Colombia. The variability in ziphodont crocodilian teeth, particularly in sebecosuchians, has not been studied in detail. Serrated teeth with differing degrees of lateral compression occur not only in different sectors of the maxilla and mandible but also in unrelated taxa, for example, Sebecidae and Pristichampsinae, and similar variability is noted among such South American "mesosuchian" crocodilians as *Cariridsuchus* (Kellner 1987), *Itasuchus* Price (1950), and the peirosaurids (Price 1955; Gasparini et al. 1991).

DISCUSSION

Langston (1965) described part of a mandible found at UCMP Locality V4517, 4.1 km northeast of Villavieja, Huila, and assigned it to the new species *Sebecus huilensis*. Busbey (1986a) referred additional material (TMM 41658-8) from the same locality to *S.* cf. *S. huilensis*. IGM 250816 is referred to *S. huilensis* because the morphology of the dentary corresponds to that of the holotype. The anterior sector of the palate, which is deep and very narrow, agrees with TMM 41658-8, which is likewise referred to *S. huilensis*. Buffetaut and Hoffstetter (1977) assigned to *S.* cf. *S. huilensis* a rostrum from the late Miocene of Peru. Busbey (1986a) rejects this assignment because in the Peruvian specimen the rostrum is wider, the alveoli are closer together and less compressed, and there is a pronounced osteodermal sculpture not present in *S. huilensis*. These characters are, however, found in the Bretsuchidae, a new family of sebecosuchians (Gasparini et al. 1993).

Many sebecosuchian specimens, mostly teeth, have been collected in the La Venta badlands. The bulk of the material is from the Monkey Beds of the Villavieja Formation. University of California specimens are listed by Langston (1956, 1965) and Texas Memorial Museum material by Busbey (1986a). All IGM specimens are from previously known localities or are within or near the stratigraphic bounds encompassing these localities.

Material encountered to date is so fragmentary that no conclusions can be drawn concerning relationships to the varied sebecosuchians from elsewhere in South America. Except for the questionable occurrence in the upper Ucayali basin of eastern Peru, *S. huilensis* appears to have been restricted to a narrow span of middle Miocene time in the Magdalena basin.

Sebecosuchian crocodiles seem to hold special fascination for paleobiologists, perhaps because of the remarkable convergence in tooth morphology with carnivorous dinosaurs and certain eusuchian crocodiles and because of their somewhat uncrocodilian appearance and, possibly, behavior. Whatever the reason, they have received a great deal of attention over the years since they were first recognized by G. G. Simpson in 1937 (for summaries of current knowledge, see Buffetaut 1980 and Gasparini et al. 1993). Despite an extensive literature, however, the group is still inadequately known, and almost nothing has been published concerning the postcranial skeleton of any taxon. The IGM–Duke University material appears to reinforce existing but ambiguous evidence that sebecosuchian vertebrae are amphyplatyan or amphicoelous, at least in the caudal series.

Eusuchia
Crocodylidae
cf. Tomistominae
Charactosuchus Langston 1965
Charactosuchus fieldsi Langston 1965

MATERIAL AND PROVENANCE

IGM 183870, section of mandible comprising united dentaries from just in front of mandibular tooth 4 posteriorly to tooth 10, including a small part of the splenial symphysis, Duke Locality 69, Monkey Beds ("Monkey Unit" of Fields 1959), at base of the Villavieja Formation (Baraya Member), Honda Group, middle Miocene.

DESCRIPTION

This badly weathered specimen, although slightly larger, differs little from the holotype of *C. fieldsi* UCMP 39646 (see fig. 14 in Langston 1965). The splenials extend slightly farther forward in the symphysis, to the level of the ninth tooth, compared to the tenth tooth in the holotype. A feature present in both the holotype and IGM 183870 is a low, V-shaped, ventral tumescence that crosses the rostrum diagonally on either side, subjacent to the long root of the fourth tooth. This character is present in differing degrees and locations in some other longirostral crocodiles, including *Tomistoma;* however, the character appears to be undeveloped in *Gavialis.* Only basal sections and unerupted tips of teeth are preserved, and coronal fluting is scarcely developed.

Measurements (mm)

Length of splenial symphysis	22
Width of jaw at posterior end of symphysis	38(e)
Height of jaw at posterior end of symphysis	17
Least transverse diameter of 3–4 dental interspace	18
Least transverse diameter of 5–6 dental interspace	18.5

DISCUSSION

IGM 183870 furnishes no new morphological information about *C. fieldsi.* The specimen is from the same stratal unit as the holotype and a referred specimen, UCMP 37093 (see Langston 1965).

A second species of *Charactosuchus, C. sansoai,* has been described from Neogene deposits in Acre, Brazil (Souza Filho 1991). An incomplete mandible with alveoli is said to have more laterally everted alveoli opposite the posterior end of the symphysis and stronger alveolar collars than *C. fieldsi.* A purported third species of *Charactosuchus, C. kugleri,* from the middle Eocene Chapelton Formation on the island of Jamaica, is distinguished by a slightly shorter symphysis, extending almost to the tenth tooth, and the absence of lateral undulations on the labial edge of the jaw posterior to the level of the symphysis. The ridges and grooves on the lingual surface of the symphysis of *C. kugleri* are absent in *C. fieldsi.* Although *C. kugleri* appears less derived than *C. fieldsi,* the evolutionary conservatism displayed by the mandible of *Charactosuchus* over some thirty million years would be quite remarkable. Clark and Domning, however, believe that *C. kugleri* may belong to the Paleogene genus *Dollosuchus* (J. M. Clark, pers. comm. 1992).

Webb and Tessman (1967) tentatively refer a slender curved tooth with a multiribbed (fluted) crown and fore and aft carinae from Pliocene estuarine deposits of Florida to *Charactosuchus.* They also cite a fragmentary maxilla from South Carolina that contains a comparable tooth. The holotype of *C. fieldsi* contains partial crowns of only four teeth. The large first tooth is not fluted, but the tenth, eleventh, and fourteenth display varying degrees of fluting. The eleventh tooth, like the tooth from Florida, has weak carinae on the anterior and posterior edges and smooth but generally distinct flutes extending the full length of the crown. There are about nine such flutes on the lingual surface in UCMP 39646 compared to seven in the Florida specimen. A report by Steel (1973) citing Auffenberg (1954), of *Charactosuchus* in the Miocene/Pliocene of South Carolina evidently refers to the maxilla mentioned above, but Auffenberg compared it to *Gavialosuchus,* not *Charactosuchus.* The taxonomic usefulness of dental characters in crocodilians is, of course, debatable and surely varies at different taxonomic levels. Owing to their fragmentary and equivocal nature, we do not regard either of these last two assignments as evincing *Charactosuchus* in North America.

Nettosuchidae
Mourasuchus Price 1964

SYNONYMY

Nettosuchus Langston 1965. *Carandaisuchus* Gasparini 1985.

Mourasuchus atopus (Langston 1965)

MATERIAL AND PROVENANCE

IGM 184040, several pieces of a skull and jaw, Duke Locality 49, La Victoria Formation, basal part; IGM 250860, right maxilla and part of a left mandible, Duke Locality 22, Villavieja Formation, Monkey Beds. IGM 250385, an incomplete maxilla, three fragments of a left dentary, and an angular, Duke Locality 26, La Victoria Formation, between the Chunchullo Sandstone and the Tatacoa Sandstone units; IGM 250481, fragmentary left maxilla containing seven alveoli, Duke Locality 32, Villavieja Formation, El Cardón Beds.

DESCRIPTION

This material differs only in details from the holotype specimen of *M. atopus*, UCMP 38012. IGM 184040 comprises a piece of the left pterygoid, right and left prefrontals, and a fragmentary right exoccipital, a symphyseal segment of the left dentary containing one partial and six complete alveoli, the posterior half of the left surangular, and other fragments. This individual was about 30% larger than the holotype.

The prefrontal bones of IGM 184040 are considerably more rugose than those of the type of *M. atopus*, there being a pronounced, rounded tumescence at the edge of the orbit where the prefrontal met the lacrimal. The elevation was obviously carried over onto the lacrimal, no doubt producing something like the large boss seen in *M. amazonensis* (Price 1964). Although this area is thickened and roughened in the type specimen of *M. atopus*, it does not display the hypertrophy seen in IGM 184040, which appears intermediate between conditions in the types of *M. atopus* and *M. amazonensis*. The same may be true of the hypertrophied transverse "spectacle bridge" that crosses the prefrontals.

The mandible is damaged, but it accords fairly closely with the type dentary of *M. atopus*. Although the first four alveoli are larger than the fifth and sixth (see "Measurements"), the disparity in size among the first four teeth is less marked than in the type where the first tooth is the largest in the dentary. The first alveolus in IGM 184040 is, however, no more than half the size of the fourth. Whether this condition is abnormal is unclear, but the other IGM specimens agree with the holotype. The first, second, fifth, and sixth alveoli are round, whereas the third and fourth are oval, the fourth being twice as long as it is wide. Price (1964) states that the first four mandibular alveoli in *M. amazonensis* are the largest in the jaw and are subequal in diameter.

Anteroposterior diameters (mm) of the six anterior dentary alveoli in *M. atopus* (IGM 184040) are 10.00, 11.00, 13.00, 19.00, 8.00, and 8.75.

The maxillae, IGM 250481 and IGM 250860, confirm the remarkable flatness of the rostrum of *M. atopus*. The maxilla of IGM 250860 contains the posterior edge of the perforation for the passage of the fourth dentary tooth, present in the type of *M. atopus*. The posterior part of the mandible of IGM 250860 resembles the type specimen, but its retroarticular process seems relatively a little longer. In both specimens, however, the process is remarkably short and weak in view of the great length of the head.

DISCUSSION

The Nettosuchidae were evidently endemic to South America, occurring in middle to late Miocene and possibly early Pliocene deposits in the Venezuelan and Brazilian pericratonic areas. The oldest species is *M. atopus* (Langston 1965) from the Friasian of Colombia. Other species, from the upper Miocene/lower Pliocene, are *M. arendsi* Bocquentin Villanueva 1984, Urumaco, Venezuela; *M. amazonensis* Price 1964, Acre, Brazil; *M. nativus* (Gasparini 1985), Paraná, Argentina, and Acre, Brazil.

Gasparini (1985) named *Carandaisuchus nativus* on the basis of two skull tables with huge tumescent bony excrescences on the squamosals that produce a massive elevated transverse occipital crest. Better material from Acre (Bocquentin Villanueva and Souza Filho 1990) shows that such masses are sometimes present in *Mourasuchus* to which *C.*

nativus is assigned. A large nettosuschid occiput (AMNH 14441) identified by its massive squamosal crest is purported to have came from the well-known Pleistocene field at Tarija, Bolivia. If the source can be confirmed, this will be the most recent example of a nettosuchid, but its origin is very uncertain because after its discovery it was used as a canoe anchor on the Río Acre and then passed successively through the hands of several owners (E. Manning, pers. comm. 1992).

Of the four *Mourasuchus* species, *M. atopus* and *M. amazonensis* are best known. According to Price (1964), the type specimen of *M. amazonensis* was adult and probably an old male individual. It is about twice the size of the type of *M. atopus* (UCMP 38012). Excrescent bone is less developed on the prefrontal of UCMP 38012 than in the type of *M. amazonensis* and is also less hypertrophied than in IGM 184040, suggesting that differences in sculpture are owing to ontogenetic or sexual dimorphism. In the absence of more complete material the limited differences noted previously do not seem sufficient justification to recognize a new species of *Mourasuchus* inasmuch as IGM 184040 is derived from the same general provenance as the type of *M. atopus.*

The IGM specimens extend the geological range of *M. atopus* in the Tatacoa Desert downward from the lower part of the Villavieja Formation (Fish Bed) to near the base of the La Victoria Formation and upward to the San Francisco Sandstone Beds of the Villavieja Formation. Owing to the fragmentary condition of the material and a lack of knowledge about variability within species of *Mourasuchus,* it is not possible to comment on evolutionary changes that may have occurred in *M. atopus* during that span of time.

In the absence of information about the squamosal bone in *M. atopus,* it is impossible to say if the tumescence present in *M. nativus* and incipiently developed in *M. arendsi* also occurred in the species from Colombia, but Price's failure to mention it in the large skull of *M. amazonensis* indicates that it was not uniformly developed in the genus *Mourasuchus.* It may represent a secondary sexual character and, thus, is less developed or absent from half of the population.

Individuals of *Mourasuchus* attained substantial size despite the apparent fragility of the head. Although known specimens of *M. atopus* are not conspicuously large (skull length 735 mm as reconstructed in UCMP 38012), Price's

specimen of *M. amazonensis* exceeded 1.18 m in length, and an undescribed skull from Urumaco, Venezuela, in the Museum of Comparative Zoology (Harvard University), is 1.25 m long (C. R. Schaff, pers. comm. 1991).

Alligatoridae incertae sedis
Balanerodus Langston 1965
Balanerodus logimus Langston 1965

MATERIAL AND PROVENANCE

IGM 250668, fragment of a right maxilla containing two teeth. Duke Locality 118, La Victoria Formation, between the Chunchullo and Tatacoa Sandstone Beds, Honda Group, middle Miocene.

DESCRIPTION

Langston described this probable alligatorid on the basis of 118 scattered teeth from late Oligocene/early Miocene rocks northeast of Chaparral (UCMP Locality V4406, Chaparral Fauna). The present specimen (fig. 8.2) is probably from the middle part of the maxilla. The external wall is high and nearly smooth with various surficial pits ranging from isolated medium-sized excavations to larger foramina arranged parallel to the alveolar border. Tooth morphology accords with the holotype tooth UCMP 45787 (pl. 1e, f in Langston 1965). The bulb-shaped crown displays many minute radiating crenulations interspersed among somewhat coarser ridges. There is a tiny point at the top of the crown. An irregular anteroposterior crest crosses the top of the crown (fig. 8.2C). The teeth have a marked annular constriction ("neck") at the base of the crown, and the root is cylindrical and three times as high as the maximum diameter of the crown.

DISCUSSION

Bulbous tooth crowns have long been regarded as indications of durophagy. This interpretation has recently been questioned by Aoki (1977, 1989), who correlated heterodont dentitions in some extant crocodilians with a certain

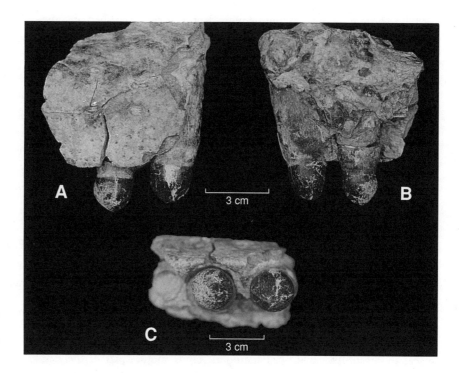

Figure 8.2. *Balanerodus logimus* Langston, IGM 250668, fragment of a right maxilla containing two teeth. *A*, lateral aspect; *B*, medial aspect; *C*, teeth, occlusal aspect.

degree of masticatory capability. Bartels (1984) reached similar conclusions about such "durophagous" forms as *Allognathosuchus* and *Ceratosuchus,* viewing their broad and often worn coronal surfaces as a form of molarization. He pointed out a number of anatomical features that seemed inconsistent with an inertial feeding strategy and suggested that these animals fed on "nonrigid" tetrapods and fishes besides invertebrates. In all the so-called durophagous taxa, the bulbous teeth display worn occlusal surfaces, now presumably produced by grinding against opposing teeth or horny pads in the jaws. Although many *Balanerodus* teeth have pristine crowns, examples of grinding exist; the two teeth in IGM 250668 show no sign of occlusal wear, however. Currently known only from the Deseadan and Friasian of Colombia, *Balanerodus* remains an enigma.

Gavialidae
Gryposuchus Gürich 1912

SYNONOMY

Rhamphostoma Ambrosetti 1887 (preoccupied). *Rhamphostoma* Wagler 1830. *Rhamphostomopsis* Rusconi 1933.

AMENDED DIAGNOSIS

Skull and jaw resemble *Gavialis* most closely among existing eusuchians but differ as follows: skull more depressed with more gradual craniofacial transition horizontally and vertically. Squamosal more drawn out behind otic recess, plan of cranial roof more strongly trapezoidal, widest behind, with supratemporal arcade more strongly expanded laterally above otic recess; posttemporal arcade narrower. Interorbital skull roof less concave, orbital rims less elevated, orbital flange of jugal less flared. Paroccipital processes broader, more drawn out posteriorly, more broadly exposed from above. Rostrum shorter, broader, more depressed, edges with stronger dental salients; Greifapparat less expanded transversely, lacks octagonal outline when seen from above. At least one nasal bone makes contact with its respective premaxilla, jugal and lacrimal longer in relation to rostral length, prefrontal relatively shorter. Postorbital bar lacks spina. Parietals overhang postoccipital processes and invest superior and posterior surfaces of the nuchal prominences of the supraoccipital bone. Trochlear notch between quadrate condyles wider and deeper. Palatal fenestrae relatively narrower, palatine bridge wider anteriorly; contact between palatine and pterygoid bones broadly arcuate posteriorly. Braincase more depressed; laterosphenoid lacks a lateral bridge anteriorly; basioccipital tubera less expanded; foramen magnum transversely oval; median auditory canal opens into slitlike transverse excavation instead of a round foramen. Mandibular and splenial symphyses shorter, no median pit at base of symphysis; surangular not inserted anteriorly between dentary and splenial, excluded from lateral mandibular fenestra by posterodorsal process of dentary; retroarticular process narrower transversely, medial flange of articular wider and more regularly triangular, dorsal crest ends in narrow hooklike process. Radiating grooves and deep pits lacking on frontals and jugals. Siphonial foramina in quadrate and articular greatly reduced or absent. Fewer, more widely spaced teeth (4 + 21[22]/23).

Differs from *Ikanogavialis:* rostrum much shorter, alveolar salients much less developed, much greater transverse expansion of the premaxillae, much wider interorbital space, more abrupt craniofacial transition, orbits more rounded with more pronounced anterior and posterior rims, powerful eavelike process of postorbital present posteriorly above orbit, palatine bridge shorter and wider; fewer, more closely spaced teeth. Both nasal bones make contact with premaxillae.

Differs from *Hesperogavialis:* One or both nasal bones contact premaxillae, palatine bridge narrower, one more tooth in ?upper and lower jaws, tooth row more lateral in position.

Gryposuchus colombianus (Langston 1965)

SYNONOMY

Gavialis colombianus Langston 1965.

MATERIAL AND PROVENANCE

IGM 184696, an almost complete skull and articulated mandible, the atlas-atlas complex, an incomplete thoracic vertebra, proximal ends of the left scapula and coracoid, and rib fragments of one individual, Duke Locality 79, just below the Chunchullo Sandstone Beds, La Victoria Formation. IGM 184057, a badly weathered, dorsoventrally crushed, incomplete skull and articulated mandible, Duke Locality 79, just below the Chunchullo Sandstone in the lower part of the La Victoria Formation; IGM 250712, an incomplete mandible comprising the rostrum and a segment of the right posterior ramus and fragmentary articular bone, Duke Locality 121W, between the Chunchullo and Tatacoa sandstones, La Victoria Formation; IGM 250480, a mandible lacking the midsection of the rostrum and the posterior half of the left posterior ramus, Duke Locality 106, above the Tatacoa Sandstone, La Victoria Formation; IGM 183569, several segments of an incomplete mandibular rostrum, from the Ferrugenous Sandstone Beds, Baraya Member, Villavieja Formation; IGM 183477, a short segment of a left dentary containing five truncated teeth and a piece of an attached splenial, Duke Locality 23, lower (Fish Bed) part of the Baraya Member, Villavieja Formation; IGM 183933, a supraoccipital bone and posterior section of the united parietals, Duke Locality 46, lower (Fish Bed) part of the Baraya Member, Villavieja Formation; IGM 183547, a short section of a presplenial mandibular rostrum containing three truncated teeth on each side, locality unknown (specimen acquired from a local collector), Villavieja Formation.

AMENDED DIAGNOSIS

Differs from *Gryposuchus jessei:* nares much wider and shorter, lacks pronounced "shouldered" appearance of the premaxillary-maxillary notch seen in holotype of *G. jessei;* excepting first premaxillary alveolus, alveoli smaller and teeth more widely spaced.

Differs from *G. neogaeus:* nares wider and shorter, supratemporal fenestrae more strongly trapezoidal in plan. Uniquely "hooked" ending of retroarticular process probably better developed.

Although the skulls IGM 184696 and 184057 are from one locality (Duke Locality 79), IGM 184696 was enclosed in a fine-grained, poorly sorted, gray, clayey sandstone, whereas IGM 184057 occurred in a gray, iron-stained, blocky mudstone, the relative incompetence of which has allowed greater damage to the specimen. IGM 184057 has been only partly prepared for study, and IGM 184696, which is well preserved, serves as the principal basis of the following description.

DESCRIPTION

We believe that the most useful comparisons between crocodilians of similar habitus employ examples of similar size, so we have compared the fossils with the following skulls (basal length given in centimeters): *Gavialis gangeticus:* ZUMZ 100–197, 78.5; PIMUZ A/III 752, 71.5; ZSM 62/1959 [male], 67.5; TMM M5485, 48.2; TMM M5486, 56.9; MLP unnumbered [male], 63.8. *Tomistoma schlegelii:* ZSM 370/1907, 84.0; TMM 6342, 35.6. In this description, any reference to "large *G. gangeticus*" and "large *T. schlegelii*" arbitrarily means a basal cranial length exceeding 70 cm. No specimen of either taxon available to us had a basal length greater than 84 cm; hence, all are somewhat shorter than *Gryposuchus colombianus* IGM 184696, whose basal length is 103 cm.

Kälin (1933) provided data for fourteen species of extant crocodilians. He presented these data in the form of indices (Kälin indices = KI), but except for the basal length of each skull he did not give the measurements from which his indexes were derived. Thus, except where basal length is a component of an index, the original data are unavailable from Kälin (Telles Antunes 1961). Müller (1927) measured many of the same specimens (in the Zoologische Staatssammulung München) that Kälin used, and derivation of KIs from Müller's measurements yields sufficiently close agreement to assure us that his mensuration was very similar to Kälin's. Because simple ratios are slightly less cumbersome than indices and because Müller's data are readily available, we use ratios (table 8.2; fig. 8.3) here instead of Kälin indices. Comparative measurements of *Gryposuchus* are presented in table 8.3.

The skull of IGM 184696 (fig. 8.4) lacks the lateral walls of the braincase and the posterior part of the palate, and it is slightly crushed dorsoventrally. It has not been possible to separate the cranial and mandibular rostra, so their lingual surfaces are not visible. Despite its generally excellent preservation, sutures are often obscure, especially in heavily sculptured areas where they are extremely irregular. Other sutures that are parallel to the "grain" of the bone are easily confused with cracks, so some of the sutures indicated on the accompanying maps of the skull and jaw are uncertain (fig. 8.5). IGM 184057, which is slightly larger than IGM 184696, has suffered significant dorsoventral crushing,

faulting, and disarrangement of parts. The anterior ends of both cranial and mandibular rostra and most of the superior surface of the skull have been lost to weathering. Still obscured by matrix or absent are the braincase and paroccipital processes; however, certain features not visible in IGM 184696 are present here.

Among the mandibles, IGM 250712 represents the largest gavial in the collection. It is roughly the same size as a mandible in the University of California collections (UCMP 40062), referred to cf. *Rhamphostomopsis* by Langston (1965). The other mandibular specimens are useful mainly to augment the morphological data obtained from the mandibles attached to the two skulls. As is usual with gavial-like jaws, all the mandibles have been crushed dorsoventrally posterior to the symphysis.

Skull

FORM AND PROPORTIONS The skull and jaw of *Gryposuchus colombianus* (figs. 8.4, 8.5) present an interesting combination of gavial- and *Tomistoma*-like features. The width of the skull at the jaw hinge relative to the basal length of the skull is closer to *T. schlegelii* than to *Gavialis gangeticus* (table 8.2: ratio 1; for graphic representation, in dorsal view, of this and following proportions, refer to fig. 8.3). The cranial rostrum is *Tomistoma*-like in both length and width (table 8.2: ratios 2, 3). Major gavial-like features include the great width of the interorbital space, which in IGM 184696 is even broader than in *G. gangeticus* (table 8.2: ratio 4), the circular orbits (table 8.2: ratios 5, 6) with their flared anterolateral rims, the huge supratemporal fenestrae (table 8.2: ratio 7), and the width of the face at the front of the orbits that is the same in relation to the length of the rostrum as in *G. gangeticus* (table 8.2: ratio 8). While approximating the shape in *T. schlegelii*, the strongly trapezoidal plan of the postrostral cranium contrasts sharply with the more regular outline of most large *G. gangeticus* skulls (table 8.2: ratio 9). Similarly, the cranial table is geometrically closer to *T. schlegelii* (but is relatively larger; table 8.2: ratio 11). Its trapezoidal plan contrasts with the almost rectangular outline of the table in *G. gangeticus* (table 8.2: ratio 10). Reflecting its relatively great expanse, the skull roof is wider anteriorly in relation to the width of the skull at this point than in all but one of the *G. gangeticus* skulls

Table 8.2. Cranial and mandibular ratios for selected longirostral crocodilians

Ratio Number[a]	Measurement	Key to Fig. 8.3[b]	Gryposuchus colombianus			Gryposuchus neogaeus, MLP 26-413	Ikanogavialis gameroi[d]	Tomistoma schlegelii			Gavialis gangeticus[e]		
			IGM 184696[c]	UCMP 41136	UCMP 38358			Range	Mean	N	Range	Mean	N
1	Skull width at quadrates / Basal length of skull	2/1	0.42	—	—	—	<0.35	0.39–0.46	0.44	8	0.37–0.39	0.38	4
2	Snout length from orbit / Skull length from occipital crest	4/5	0.76	—	—	—	~0.80	0.72–0.77	0.75	8	0.75–0.87	0.79	4
3	Snout width at 6th maxillary tooth / Snout width at anterior end of orbit	28/3	0.51	—	—	—	(0.38)	0.51–0.59	0.52	8	0.42–0.47	0.45	2
4	Least interorbital distance / Length of orbit	6/7	1.37; 1.56	1.14	—	~1.25	(0.80)	0.32–0.50	0.47	8	1.10–1.54	1.34	5
5	Length of orbit / Skull length from occipital crest	7/5	0.08; 0.07	—	—	—	(~0.06)	0.11–0.13	0.12	8	0.08–0.11	0.09	5
6	Width of orbit / Length of orbit	14/7	0.93; 1.11	1.17	—	—	(1.22)	0.65–0.79	0.72	8	0.86–1.14	0.97	5
7	Width, supratemporal fenestra / Length, supratemporal fenestra	22/23	1.50; 1.63	—	1.33	1.20	(1.35)	0.82–1.09	0.94	8	1.03–1.20	1.14	4
8	Snout width, anterior end of orbit / Snout length from orbit	3/4	0.29	—	—	—	0.27	0.30–0.39	0.35	8	0.25–0.34	0.29	4
9	Skull width, anterior end of orbit / Skull width at quadrates	3/2	0.50	0.49	—	—	0.61	0.49–0.61	0.56	8	0.46–0.68	0.58	5
10	Skull roof width anteriorly / Skull roof width posteriorly	8/10	0.81	0.85[f]	0.89[f]	0.85	(0.82)	0.67–0.74	0.70	8	0.88–0.94	0.92	4
11	Skull roof length / Skull roof width posteriorly	19/10	0.44	—	0.46	0.42	(0.61)	0.56–0.66	0.61	8	0.47–0.58	0.53	5
12	Skull roof width anteriorly / Skull width at postorbital bar	8/11	0.91	—	—	—	(0.80)	0.47–0.68	0.54	8	0.77–0.93	0.83	5
13	Height of mandible posteriorly / Mandible length	41/34	0.15	—	—	—	—	0.12–0.15	0.13	3	0.09–0.16	0.12	3
14	Symphysis length / Mandible length	35/34	0.52	—	—	—	(~0.60)	0.43–0.48	0.46	6	0.54–0.56	0.55	3
15	Retroarticular length / Mandible length	40/34	0.10	—	—	—	—	0.14–0.17	0.15	6	0.11–0.13	0.12	3
16	Nares width / Nares length	15/16	1.43	—	—	1.20	—	0.75–1.12	0.91	8	0.89–1.20	1.04	5
17	Palatal fenestra width	25/—	0.31	0.43	—	—	(0.52)	0.38–0.42	0.39	8	0.53–0.58	0.57	3

#	Measurement	ratio										
18	Mandibular width posteriorly / Mandible length	44/34	0.39	—	—	—	0.33–0.40	0.36	6	0.29–0.33	0.32	3
19	Mandible height at 6th tooth / Mandible width at 6th tooth	42/43	0.45	—	—	—	0.66	0.66	1	0.54	0.54	1
20	Mandibular fenestra length / Mandibular fenestra height	37/38	<4.43	—	—	4.24	2.20–2.80	2.60	6	1.62–2.39	1.90	4

[a]Sequential numbers identify the ratios cited in the text. They are arranged approximately in the sequence mentioned in the descriptions.

[b]The numbers in this column refer to the measurements depicted in the accompanying mensuration diagram (fig. 8.3).

[c]Two ratio values denote left and right of the same specimen.

[d]Ratios for *Ikanogavialis gameroi* that are based on measurements by Sill (1970) are in parentheses and may not be strictly comparable to similar measurements of other specimens made by us. Differences, however, are probably insignificant at the confidence levels prevailing in the case of these large fossils.

[e]Data for *Tomistoma schlegelii* and *Gavialis gangeticus* are from skulls exceeding 70 cm in basal length.

[f]Estimated.

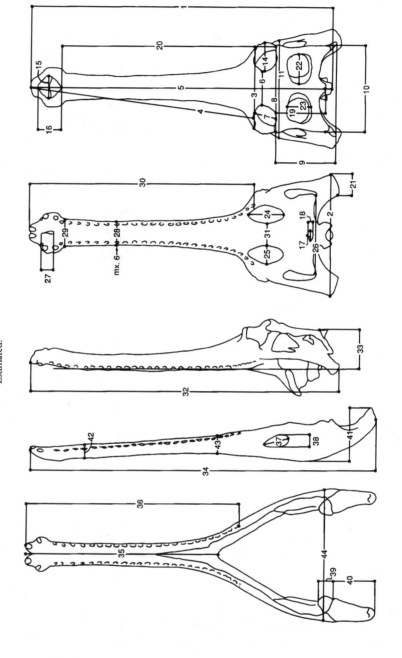

Figure 8.3. Key to measurements in table 8.2.

Table 8.3. Comparative measurements (cm) for various *Gryposuchus* specimens

Variable Measured	G. colombianus			G. neogaeus, MLP 26-413	G. jessei, HM-GI
	IGM 184696	UCMP 41136	UCMP 38358		
SKULL MEASUREMENTS					
1. Basal length	103.0	—	—	—	—
2. Greatest width	43.0	26.2e	—	32.4	—
3. Width of rostrum, posterior	21.5	12.7	—	—	—
4. Length of rostrum	73.6	—	—	—	—
5. Length of skull, posterior edge of table to tip of rostrum	96.6	—	—	—	—
6. Interorbital distance	10.3	5.4	5.5	7.0	—
7. Orbit length	7.5(L) 6.6(R)	4.7	—	~5.6	—
8. Skull table width, anterior	27.2	17.6	15.0e	23.0	—
9. Skull table length	14.8	—	7.7	13.0	—
10. Skull table width, posterior	33.6	20.8e	16.8e	27.0	—
11. Skull width across postorbital bars	30.0	—	—	—	—
12. Occipital condyle width	5.3	—	3.0	5.4	—
13. Occipital condyle height	4.8	—	2.2	4.3	—
14. Orbit width	7.0(L) 7.3(R)	5.5	~6.4	—	—
15. Nares width	7.0	—	—	5.9	6.2
16. Nares length	4.9	—	—	4.9	4.9
21. Quadrate condyle width	9.3	—	—	6.9	—
22. Supratemporal fenestra width	12.0(L) 13.0(R)	—	6.0	10.0	—
23. Supratemporal fenestra length	8.0	—	4.5	8.3	—
24. Palatal fenestra length	14.5	9.9	—	—	—
25. Palatal fenestra width	4.5	4.2	—	—	—

compared. It is significantly wider than in any *T. schlegelii* (table 8.2, no. 12).

The transition between rostral and cranial segments and the flaring of the lateral rims of the orbits are less marked than in *G. gangeticus* but are more so than in *Tomistoma* where orbital rims are little expanded. The medial rims of the orbits are lower and the interorbital space is, thus, less concave than in *G. gangeticus*. Partly owing to lower rims above the orbits, the cranial profile appears depressed, as noted by Langston (1965) in the type of *"Gavialis" colombianus*. The left nasal bone is in narrow contact with the premaxilla of the same side in IGM 185696, but the right nasal and premaxilla do not meet. The jugal extends slightly farther anteriorly than in *G. gangeticus*, whereas the prefrontal and especially the lacrimal are relatively shorter than in the gavial, and, all the aforementioned bones are much shorter than in *T. schlegelii*. The lateral margins of the rostrum are but slightly serrated. The Greifapparat is more highly developed than in *T. schlegelii*. Strong alveolar collars are present around the first and second teeth, but the premaxillae are less expanded transversely at the second tooth position than is usual in large *G. gangeticus* (the expansion contains the second and third teeth in the gavial), and the emargination of the upper jaw edge above the fourth mandibular tooth thus appears shallower in IGM 184696. Noteworthy is the transverse expansion of the dorsolateral edge of the squamosal above the external otic recess at the level of the transverse occipital crest. Here the skull table of IGM 184696 is 35 cm wide compared with a greatest width of 27.2 cm at the posterolateral edge of the orbits. In large *G.*

Table 8.3. Continued

Variable Measured	G. colombianus			G. neogaeus, MLP 26-413	G. jessei, HM-GI
	IGM 184696	UCMP 41136	UCMP 38358		
SKULL MEASUREMENTS *(continued)*					
26. Pterygoid flanges width	14.5	—	—	—	—
27. Incisive foramen length	1.6	—	—	3.0	—
28. Rostrum width at secondary dental peak	11.0	—	—	—	11.8e[a]
29. Rostrum width at notch for 4th mandibular tooth	4.0	—	—	6.7	8.0
30. Tooth row length	77.5	—	—	—	—
31. Palatine bar width	10.8	—	—	—	—
32. Skull length	102.5	—	—	—	—
33. Skull height	9.5	—	—	~8.0	—
MANDIBLE MEASUREMENTS					
34. Mandible length	117.0	—	—	—	—
35. Symphysis length	61.0	—	—	—	—
36. Tooth row length	69.7	—	—	—	—
37. Lateral fenestra length	6.2	—	—	8.9	—
38. Lateral fenestra height	≥1.4	—	—	2.1	—
39. Glenoid fossa length	5.2	—	—	~4.0	—
40. Retroarticular process length	12.1	—	—	—	—
41. Mandible height	17.3	—	—	—	—
42. Mandible width at 6th tooth	77.0	—	—	—	—
43. Mandible width at end of symphysis	165.0	—	—	—	—

Notes: See figure 8.3 for key to measurements: sequential numbering of measurements corresponds to standard mensuration of Langston (1965); inapplicable measurements are omitted. (Measurements 17–20 cannot be made in any of these specimens.) e = estimated; (R) = right; (L) = left.

[a] At eighth maxillary tooth.

gangeticus and *T. schlegelii,* the corresponding area is but slightly, if at all, expanded. Behind this expansion the cranial table narrows to a width of 33.6 cm at the posterolateral corners of the squamosals. Posteriorly, the edge of the cranial table is broadly emarginated on either side of an hypertrophied nuchal crest formed, apparently, of parietals and the supraoccipital. The bone in the cranial table is greatly reduced by the large supratemporal fenestrae. Thus, the parietal roof has a least width of 2.4 cm and the flat-topped post-temporal arcade, composed of parietal and squamosal, has a least anteroposterior diameter of 1.2 cm, both measurements being approximately the same as in large *G. gangeticus* skulls ZSM 28/1912 and ZSM 100–197, which are only about three-quarters as large as IGM 184696. In parasagittal section the post-temporal arcade is T-shaped, with the lower bar of the "T" greatly constricted anteroposteriorly by the presence of the transverse sulcus that contains the anterior opening of the posttemporal fenestra. The lateral supratemporal arches have a least width of 2.2 cm. The occipital face of the skull is more strongly inclined posteriorly than in *Gavialis gangeticus,* and, hence, more of the occipital structures are exposed in dorsal aspect in *Gryposuchus.*

SCULPTURE   Except near the free edges of some circumorbital bones, for example, the lacrimal, jugal, and postorbital, the dermal bones are less deeply sculptured than usual in large *Gavialis gangeticus.* The elongate radiating pits and

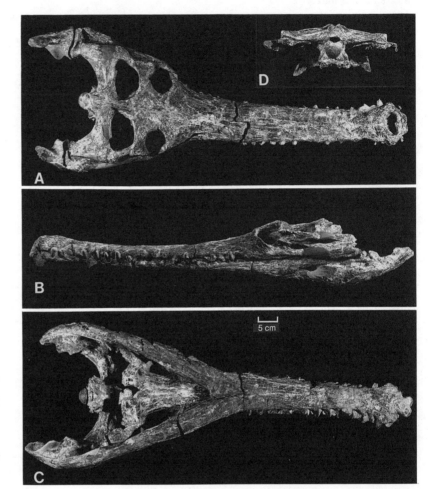

Figure 8.4. *Gryposuchus colombianus* (Langston), IGM 184696, skull and mandible. *A,* dorsal aspect; *B,* left lateral aspect; *C,* ventral aspect; *D,* occipital aspect. Most of the palatal expanse of the pterygoids is missing and the roof of the choanal passage is exposed posteriorly in *C.* The mandible is attached to the skull in *A, B,* and *C;* in *D,* the mandible and quadrate condyles have been removed. For interpretation see figure 8.5.

ridges characteristic of the interorbital skull roof of *Gavialis gangeticus* are absent as are the deep more or less vertical sulci on the anterolateral flange of the jugal, whose external surface appears pitted and "wrinkled." The pitting on the squamosal, as seen at the posterolateral corner of the cranial table in *G. gangeticus,* is little developed. Elsewhere on the skull osteodermal irregularities comprise more or less longitudinal anastomosing ridges and grooves of low relief, but some coarsening of the sculpture occurs on the premaxillae anteriorly around the nares and the Greifapparat. The "frothy" texture sometimes seen on the skull of large male gavials is absent. The lingual surfaces of the cranial rostrum is marked by irregular, shallow, more or less longitudinal grooves.

EXTERNAL NARES The narial opening is broadly cardiform (figs. 8.4A, 8.6B). Possibly correlated with the large size of IGM 184696, the nares are relatively wider than in *Gavialis gangeticus* (table 8.2: ratio 16); this is as expected because the nares become relatively wider in the gavial with increasing size (Kälin 1933). The rim of the opening is not much elevated above the level of the rostrum and is interrupted anteromedially by a broad and shallow, V-shaped notch. The unique bony structures associated with the narial "excrescence" (i.e., a plateaulike shelf within the narial aperture anteriorly and nearly vertical anterior wall of the opening) in adult male *G. gangeticus* (fig. 8.6A) are absent (see Martin and Bellairs 1977 for an account and interpretation of the narial excrescence).

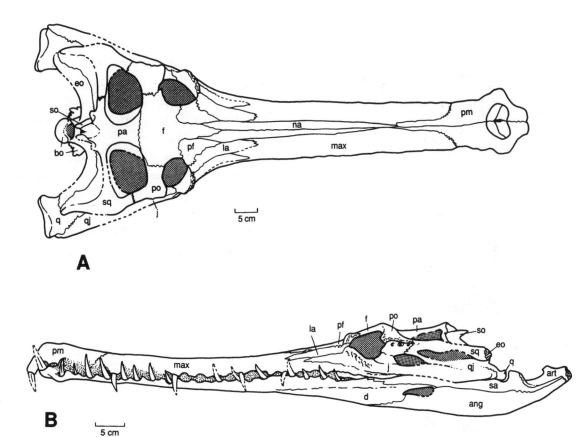

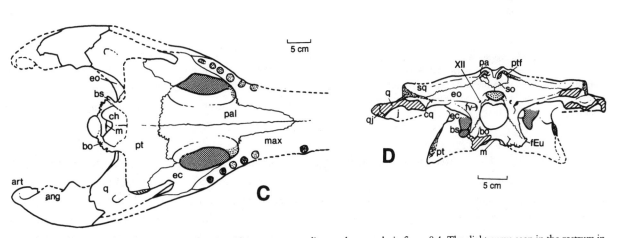

Figure 8.5. *Gryposuchus colombianus*, IGM 184696. Cranial maps corresponding to photographs in figure 8.4. The slight curve seen in the rostrum in figure 8.4 *A* and *C* has been corrected in the drawings. Abbreviations: *ang*, angular; *art*, articular; *bo*, basioccipital; *bs*, basisphenoid; *ch*, choanae; *cq*, cranio-quadrate passage; *d*, dentary; *ec*, ectopterygoid; *eo*, exoccipital; *f*, frontal; *fEu*, lateral Eustachian foramen; *fv*, foramen vagi; *j*, jugal; *la*, lacrimal; *m*, median Eustachian foramen; *max*, maxilla; *na*, nasal; *pa*, parietal; *pal*, palatine; *pf*, prefrontal; *pm*, premaxilla; *po*, postorbital; *pt*, pterygoid; *ptf*, posttemporal fenestra; *q*, quadrate; *qj*, quadratojugal; *sa*, surangular; *so*, supraoccipital; *sq*, squamosal; *stf*, supratemporal fenestra; *XII*, foramen for twelfth cranial nerve.

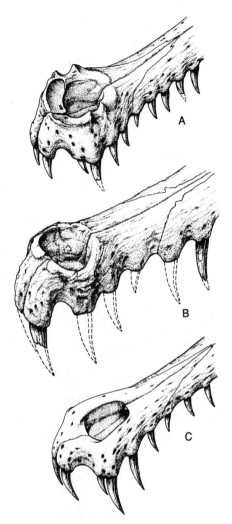

Figure 8.6. Narial architecture in gavials. *A. Gavialis gangeticus,* large male. *B. Gryposuchus colombianus,* IGM 184696. *C. Gavialis gangeticus,* immature. Female and immature male *Gavialis gangeticus* lack the excrescences and the excavations on the dorsal side of the premaxilla and the ledgelike plateau within the nares anteriorly.

INCISIVE FORAMEN   The incisive foramen, seen through the nares in IGM 184696, is a narrowly oval slit 16 mm long and 13 mm wide. Its edges on the floor of the narial opening are elevated into low parasagittal ridges.

ORBITS   The left orbit is 9 mm longer than the right and 3 mm wider (figs. 8.4A, B; 8.5A, B). The anterolateral rim, which comprises a semicircular crest formed by the thickened posterior edge of the lacrimal and anterior ramus of the jugal, is less sharply elevated than in *Gavialis gangeticus,* but it is slightly everted and is more pronounced than in

Tomistoma. The rim of the orbit displays a notch between the lacrimal and prefrontal, but the conspicuous wide notch that carries the lacrimal-jugal suture across the rim in *G. gangeticus* is scarcely discernible. A flattened, anterolaterally directed and strongly sculptured, eavelike lappet of the postorbital overhangs the orbit posterodorsally.

SUPRATEMPORAL FENESTRAE   The supratemporal fenestrae are relatively wider and shorter than in *Gavialis gangeticus* and somewhat more irregular in outline, widening considerably posteriorly (figs. 8.4A, 8.5A). The rims of the opening are not elevated. The frontal forms part of the anterior border of the fenestra but does not descend into the temporal fossa. The fossa is not greatly constricted by the bones that surround it at depth. The quadrate, squamosal, and parietal, best seen in the referred cranium UCMP 38358, are considerably more attenuated transversely in the posterior wall of the temporal fossa than in *G. gangeticus.* These bones and the laterosphenoid are also narrower dorsoventrally, reflecting the flatter lateral profile of the *Gryposuchus colombianus* cranium.

LATERAL TEMPORAL FENESTRAE   The lateral temporal fenestrae (figs. 8.4B, 8.5B) are damaged in both IGM skulls and few details can be made out. As preserved, the openings in IGM 184696 appear much longer than high. The quadratojugal spine is not preserved, but its broken base is barely visible in UCMP 41136 (see Langston 1965). Anterodorsally, the edge of the fenestra is formed by the squamosal and postorbital. It is not clear whether the posterodorsal edge is formed exclusively by the quadratojugal as in *Gavialis gangeticus.*

POSTTEMPORAL FENESTRAE   The posttemporal fenestrae are slitlike openings passing through the occipital plate between the supraoccipital, parietal, squamosal, and exoccipital (fig. 8.5D). The fenestra opens into the posteromedial corner of the temporal fossa deep within a transverse sulcus that extends across the posterior wall of the fossa from the posterolateral end of the parietal to the foramen that connects the meatal cavity with the temporal fossa (fig. 8.7). The structure of this groove and its associated foramina is complex, and we are unaware of any detailed account of its features in the literature. Lacking soft tissue preparations,

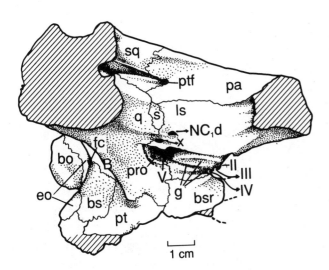

Figure 8.7. *Gryposuchus colombianus*, referred cranium (UCMP 38358), viewed diagonally from below and anteriorly. The superior temporal arcade has been removed. Abbreviations: *B*, mandibular adductor crest B of Iordansky; *bo*, basioccipital; *bs*, basisphenoid; *bsr*, basisphenoidal rostrum; *eo*, exoccipital; *fc*, posterior carotid foramen; *g*, grooves in the laterosphenoid for N. ophthalmicus (V_1) and a recurrent branch of the internal carotid artery; *ls*, laterosphenoid; *pa*, parietal; *pro*, prootic; *pt*, pterygoid; *ptf*, posttemporal fenestra and foramen for head vein; *q*, quadrate; *s*, supernumerary—possibly prootic—bone; *II–V*, formina for cranial nerves; NC_1d, foramen for motor branch of trigeminal nerve to the levator bulbi muscle (after Hopson 1979); *x*, foramen for an unidentified branch of N. V?—not Hopson's Nx, which apparently does not emerge onto the surface of the braincase. Scale bar 1 cm.

we were unable to determine what structures occupy the groove in living gavials. Its bony expression appears to be unique to the Gavialidae among existing eusuchians. It is not present in *Tomistoma*.

PALATAL FENESTRAE   The palatal fenestrae have the shape of elongate, irregular ovals (figs. 8.4C, 8.5C). Owing to the broadly arcuate transverse section of the palatine bones and lack of well-defined margins, measurements of width and, hence, length-width relationships of the palatal fenestrae are of limited value for comparisons. Measured somewhat arbitrarily, the width of the opening is less than one-third of its length in IGM 184696 and may be even less in IGM 184057. The palatines are less expanded in the type of *Gryposuchus colombianus* than in the much larger IGM specimens, and the fenestra is considerably wider in relation to its length, falling just outside the range of eight large *T. schlegelii* (table 8.2: ratio 17). It is relatively narrower than the fenestra in large *G. gangeticus*. In existing crocodilians,

excepting *G. gangeticus*, the fenestra increases in length more rapidly than in width with increasing body size (Kälin 1933). *G. colombianus* agrees with nongavials in this regard. The fenestra ends anteriorly, even with the anterior edge of the nineteenth maxillary alveolus. The pterygoid enters the opening only narrowly behind.

CHOANAE   The choanal opening has been largely destroyed in all specimens. The roof of the nasal passage is, however, preserved posteriorly in IGM 184696. It rises steeply against the floor of the basicranium. Judging from what remains, the opening was broad (apparent width ~8.6 cm) as in *Gavialis* rather than rounded as in *Tomistoma*. Its posterior boundary lay close to the median Eustachian foramen (= basisphenoidal + basioccipital foramina of Tarsitano et al. 1989). As in *Gavialis* and *Tomistoma*, the choanae were undivided by a median septum.

FORAMEN MAGNUM   The foramen magnum is twice as wide as it is long in IGM 184696, but it has been affected by vertical compression (figs. 8.4.D, 8.5D). As preserved, it is 4 cm wide and 1.8 cm high. As in other gavials, it is visible from above owing to the posteroventral inclination of the occipital surface. The foramen magnum is bounded largely by exoccipitals with a narrow ventral component supplied by the basioccipital. Other cranial foramina are described later.

PREMAXILLAE   The premaxillae form a powerful Greifapparat (Kälin 1933). This structure, when seen from above, lacks the distinctive octagonal outline peculiar to large male *Gavialis gangeticus* (fig. 8.6A, B). The minutely interdigitating suture between premaxilla and maxilla crosses over the side of the rostrum in the posterodorsal direction just in front of the first maxillary tooth. Posteriorly, the facial processes of the right and left premaxillae end, respectively, at the level of the interspace between the fourth and fifth maxillary teeth and the fifth maxillary tooth. The relationship between the premaxillae and the nasals is described later. The palatal side of the premaxillae is obscured by the mandibular rostrum in IGM 184696 and is not preserved in IGM 184057 or UCMP 41136. Absence of secondary sexual features present within the nares in large male *G. gangeticus* is noted previously. The two osteodermal elevations and associated excavations that appear dorsally on the premaxilla of old male *G. gangeticus* are also lacking (fig. 8.5B). Interestingly, however, a photograph of a gavial skull from Brazil, which we believe belongs to *Gryposuchus,* displays irregularities on the premaxillae that may, in fact, correspond to these structures (W. D. Sill, pers. comm. 1990).

MAXILLA   Opposing maxillae do not meet in the midline dorsally, being separated narrowly by the junction of the left nasal and premaxilla (figs. 8.4A, B; 8.5A, B). The contact between the maxilla and the lacrimal ends obtusely at the level of the fourteenth maxillary tooth. In the palate, the maxillae are separated posterior to the thirteenth to fourteenth maxillary teeth by a long isosceles-shaped wedge of united palatines (fig. 8.5C). The narrow posterior ramus of the maxilla, which contained three to four teeth, forms most of the lateral and anterior edges of the palatal fenestra. Posteroventrally, the maxilla is inserted posteriorly between the maxillary processes of the ectopterygoid and the jugal and ends acutely below the center of the postorbital bar.

A notable difference between *Gryposuchus colombianus* and both *Gavialis gangeticus* and *T. schlegelii* is the dorsoventral flattening of the cranial rostrum (figs. 8.4B, 8.5B). Thus, the united maxillae have a narrowly ovate cross section with transverse and vertical diameters of 12.0 and 2.8 cm, respectively, at the level of the eleventh maxillary tooth in IGM 184696. The relative positions of the maxillary tooth rows also differ from *G. gangeticus,* lying closer to the labial edge of the rostrum in IGM 184696. In *G. gangeticus* there is a tendency for those edges of the maxilla (and premaxilla) to expand laterally during ontogeny so that in large males, at least, a considerable amount of bone extends laterally above the tooth row, obscuring the alveolar salients from above. Maxillary alveoli are separated by interalveolar distances of 1.4–2.3 times the anteroposterior diameters of adjacent alveoli, compared to interalveolar distances of only about 0.66 of alveolar diameters in *G. gangeticus.* Alveolar salients appear stronger in *G. colombianus* owing to the relatively lateral position of the tooth row in the rostrum.

NASAL   The left nasal meets the premaxilla narrowly on the left side of the rostrum in IGM 189696, level with the interspace between the fourth and fifth maxillary teeth. The right nasal, which is narrower than the left, ends even with the posterior edge of the fifth tooth, failing to reach the premaxilla superficially by a distance of about 18 mm (fig. 8.5A).

LACRIMAL   The lacrimal forms a smaller proportion of the orbital rim than in *Gavialis gangeticus* and is approximately one-fifth as long as the rostrum (about one-sixth in *G. gangeticus*). Anteriorly, the lacrimal is separated from the nasal by insertion of a short and narrow posterior wedge of maxilla in IGM 184696 (fig. 8.5A, B). The lacrimal-nasal contact is variable anteriorly in *G. gangeticus,* appearing occasionally as in IGM 184696 but usually lacking the wedge of maxilla.

PREFRONTAL   The prefrontal invades the interorbital space to a greater extent than in *Gavialis gangeticus.* Anteriorly, the bone wedges out acutely between the nasal and lacrimal (fig. 8.5A, B). It ends about opposite the seventeenth maxillary tooth. The prefrontal pillar is not exposed in the IGM skulls but is partly visible in the type of *Gryposuchus colombianus.* It appears lower than in either the gavial or tomistoma and is somewhat inclined dorsoventrally in the anterior direction.

FRONTAL   The anterior process of the frontal plate wedges out between the nasals at the level of the fifteenth or sixteenth maxillary tooth. Posteriorly, the transverse frontalparietal suture is even with the edge of the supratemporal fenestra; the frontal does not enter the opening (fig. 8.5A).

POSTORBITAL   Below the massive anterolateral eavelike process of the postorbital bone is a longitudinally triangular, lateral excavation in which lie three broadly oval, presumably pneumatic foramina, the middle one being the largest (figs. 8.4B, 8.5B). One or more such foramina are common in eusuchians, those of crocodylids usually being smaller than those of alligatorids, but the openings are best developed in gavials. The thick descending ramus of the postorbital appears to form more of the massive, pillarlike postorbital bar than in *Gavialis gangeticus.* The characteristic

G. gangeticus postorbital spina is not present. A thick anteroposterior crest occurs laterally at about midheight of the postorbital bar. For further comments about these two structures, see appendix 8.1.

PARIETAL   The parietal roof (figs. 8.4A, 8.5A) is slightly arched transversely in IGM 184696 but is flat in UCMP 38358. Its lateral edges (cristae cranii parietales of Iordansky 1973) are sharp but unelevated. Just beneath the crista the wall of the temporal fossa contains a well-marked, peripheral channel that extends from the anterolateral end of the parietal posteriorly along the side of the braincase. The channel widens posteriorly, disappearing at the posteromedial corner of the temporal fossa. Posteriorly, dorsolateral to the supraoccipital, the parietal gives rise to a triangular, eavelike flange or lappet of bone (figs. 8.4A, D; 8.5A, D). This structure overhangs and partly occludes the fossa containing the occipital opening of the posttemporal fenestra and also the postoccipital processes of the supraoccipital. Flanges are incipient in the much smaller referred cranium of *Gryposuchus colombianus* (UCMP 38358). This architecture is so unlike that of most examples of *Gavialis gangeticus* (fig. 8.8A) that it seems possible that the parietals have "overgrown" the top of the supraoccipital in IGM 184696 (a similar overgrowth of the parietals may have been developing in *G. gangeticus* skulls ZUMZ 100–197, but none is evident in ZSM 28/1912).

SQUAMOSAL   The occipital plate of the squamosal (figs. 8.4A, D; 8.5A, D) is broader and more widely exposed in dorsal view than in *Gavialis gangeticus*. Seen from the side, the squamosal (and the subjacent quadrate) (figs. 8.4B, 8.5B) appears depressed in comparison to the relatively elevated profile of the suspensorial part of the skull in most *Gavialis* species. The type cranium of *Gryposuchus colombianus*, which is little if any distorted dorsoventrally, is similarly depressed (fig. 7 in Langston 1965). The posterior wall of the squamosal that descends laterally to meet the quadrate behind the otic recess is much more attenuated caudad and presents a flatter expanse when seen from the side, contrasting sharply with *G. gangeticus* where, although lengthening somewhat with age, the squamosal-quadrate region is relatively abbreviated posterior to the otic recess.

JUGAL   The jugal (figs. 8.4A, B; 8.5A, B) appears less derived in most respects than that of *Gavialis gangeticus*. The anterolateral rim of the orbit—whose flangelike elevation is so striking in *Gavialis*—is much less developed in *Gryposuchus* but is, nevertheless, more reminiscent of the gavial state than that of any other eusuchian (fig. 8.8A). As in *G. gangeticus* and in contrast to other eusuchians, the subdermal surface of the infratemporal ramus of the jugal is devoid of sculpture. It is slender and oval in cross section, with vertical and horizontal diameters of 22 and 15 mm, respectively, beneath the middle of the lateral temporal fenestra in IGM 184696. Noted also in the type of *Gryposuchus colombianus* (see Langston 1956), this geometry differs from the frequent descriptions of the bar in *G. gangeticus* as being dorsoventrally compressed and somewhat wider than high (e.g., see Kälin 1933; Iordansky 1973). This characterization may be more applicable to very large examples of *G. gangeticus* because in all skulls up to at least 48 cm long available to us, the infratemporal bar is higher than wide. The massive postorbital bar, best seen in the holotype UCMP 41136, arises from the dorsal side of the jugal as in *G. gangeticus* and displays characteristic gavial architecture. The usual large gavial posterolateral pneumatic foramen is present at the base of the bar. The base of the bar has an anteroposterior diameter of 6.5 cm in IGM 184696.

QUADRATOJUGAL   The quadratojugals are crushed or are incomplete in all specimens. It appears from UCMP 41136 and 38358 that the quadratojugal bounds the lateral temporal fenestra behind, but it is not clear whether, as in *Gavialis gangeticus*, it sends a small anterior process forward along the dorsomedial edge of the jugal in the posteroventral corner of the opening. It appears possible that the quadratojugal fails to meet the postorbital anterodorsally, thus allowing the quadrate to enter the edge of the lateral temporal fenestra. Posterolaterally, the quadratojugal extends almost to the end of the quadrate.

PALATINES   The palatine bones (figs. 8.4C, 8.5C) are greatly expanded transversely in IGM 184696, having a least width across the palatine bridge of 19 cm. The path of the palatine-pterygoid suture is the reverse of that in

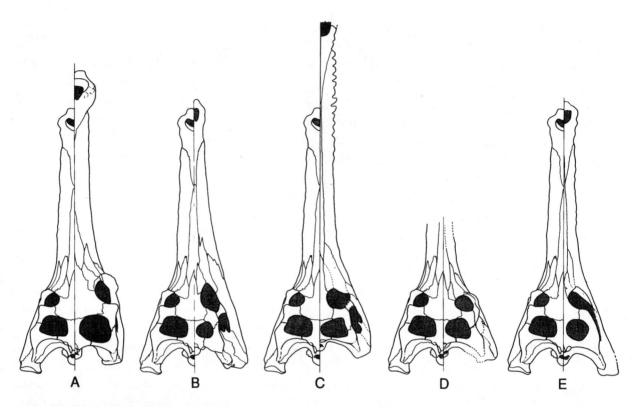

Figure 8.8. Longirostral eusuchian skulls compared. Drawings reduced to same scale (greatest transverse diameter posteriorly) and somewhat simplified. *A–E, left of the midline:* Skull of *Gryposuchus colombianus,* IGM 184696 (right side reversed). *Right of the midline: A, Gavialis gangeticus* (composite drawing modified to eliminate distortion present in specimens); *B, Tomistoma schlegelii; C, Ikanogavialis gameroi* (after Sill 1970); *D, Gryposuchus colombianus,* composite drawing of UCMP 41136 and UCMP 38358 (after Langston 1965); *E, Eogavialis gavialoides* (after Andrews 1906; left side reversed).

Gavialis gangeticus, the palatines extending posteriorly into the pterygoid plate, then turning sharply toward the midline instead of describing a broad forwardly directed "V" projecting into the posterior end of the palatine bar. In UCMP 41136 the suture is more regularly V-shaped than in IGM 184696, but it is still the reverse of the gavial pattern. Anteriorly, the combined palatines have a hastate form with very small lateral lobes extending about halfway across the anterior end of the palatal fenestrae. Anteriorly, they end more obtusely than in *G. gangeticus.* In IGM 184696 they extend forward to the level of the thirteenth or fourteenth maxillary tooth as in the type of *Gryposuchus colombianus.* The bones are shorter in *G. gangeticus.* In *T. schlegelii* the palatines end bluntly against the maxillae.

VOMER   Unlike in *Tomistoma,* the vomer is not exposed in the palate.

PTERYGOID   The posterior part of the transverse pterygoid plate is damaged and partly missing in all specimens. One torus transiliens is preserved in each IGM skull. In IGM 184696 the right torus is 10 cm long, and it is clear that the pterygoid wings were relatively short and of gavial proportions. What remains of the choanal opening in IGM 184696 indicates that its posterior boundary lay almost at the same level in relation to the medium Eustachian foramen as it does in *Gavialis gangeticus,* attesting to the relatively small degree of verticalization in the braincase (for exposition of the verticalization concept, see Tarsitano 1985; Tarsitano et al. 1989).

Ovoid pterygoid bullae are present in IGM 184696, but they do not expand into the palatal vacuities as they do in adult male *Gavialis* skulls. The left bulla is crushed dorsoventrally, the bony walls having responded like egg shells (which they resemble in thickness and texture) to pressures during fossilization. The bulla was ovoid, at least 10 cm long, and had a greatest width, posteriorly, of 6.5 cm. It rests in a shallow spoon-shaped depression on the superior and lateral surfaces of the palatine bone. We see no evidence that it was divided into two communicating chambers as is reported in *G. gangeticus* (Martin and Bellairs 1977). Bullae are not present in the type of *Gryposuchus colombianus,* owing probably to immaturity or sex.

It is often stated that, like the aberrations of the narial structures, large pterygoid bullae are a secondary sexual character of adult male gavials. If this is true, the onset of these hypertrophic effects is evidently heterochronic in ontogeny. A skull in the Munich collection (ZSM 62/1959) with a basal length of 67.5 cm, which is probably from a young adult male, displays both "male" narial features and bullae, whereas a slightly larger but more gracile skull at Zurich (PIMUZ A/III752), with a basal length of 71.5 cm, lacks the narial features but has pterygoid bullae of hen-egg proportions. If A/III752 is not paedomorphic with respect to the narial features, it may represent a female, in which event it would be clear that large female *G. gangeticus* may occasionally develop expanded bullae. Indeed, Martin and Bellairs (1977) cite a skull of *G. gangeticus* only 49.5 cm long, which has bullae but lacks the male narial features. They report six other skulls, ranging from 57 to 61 cm in length, that display the same conditions, and they note that the bullae may, therefore, not be a secondary sexual character as supposed by Kälin (1933) and widely stated by authors. IGM 184696, which lacks the narial aberrations, may, thus, have been female.

ECTOPTERYGOID   The ectopterygoid (figs. 8.4C, 8.5C) is relatively short and broad as in gavials generally. The maxillary ramus extends forward to the level of the twentieth maxillary tooth. Posteriorly, the infratemporal ramus, best seen in UCMP 41136, is powerfully developed as in *Gavialis*. The contribution of the ectopterygoid to the postorbital bar cannot be determined.

LATEROSPHENOID   The laterosphenoids are crushed and incomplete in IGM 184696 and are not exposed in IGM 184057. In the referred cranium of *Gryposuchus colombianus* (UCMP 38358) the laterosphenoids are low and greatly

expanded in front in correlation with the broad frontal plate above. The anterior edge of the bone slopes upward more gradually than in existing gavials, resulting in a considerable lengthening of the braincase. The wide optic foramen (fig. 8.7) faces downward to a greater degree than in eusuchians generally. This condition appears natural, however, because there is no evidence of dorsoventral crushing in the specimen. The exits of cranial nerves III and IV are indicated by small notches in the anterolateral edge of the laterosphenoid. (For a useful account of cranial nerve foramina and related matters in crocodilians, particularly *Caiman,* see Hopson 1979.) The laterosphenoid bounds the trigeminal foramen anteriorly and dorsally. The foramen is broadly oval anteroposteriorly with principal diameters of 7.0 and 13.2 mm (in *Gavialis gangeticus* the longer diameter is dorsoventral). The large vestibular excavation occupied in life by the Gasserian ganglion is sharply defined laterally by the laterosphenoid anteriorly and the quadrate posteriorly. In most extant eusuchians there is a so-called lateral bridge (Iordansky 1973) on the side of the laterosphenoid, just anterior to the trigeminal foramen, beneath which the ophthalmic branch (V_1) of the trigeminal nerve passes forward and a recurrent branch of the internal carotid artery passes caudad (Miall 1878). No bridge is present in UCMP 38358 (fig. 8.7) and its existence in IGM 184696 is doubtful. The bridge is variably expressed in *G. gangeticus* (Iordansky 1973), and we have never seen a complete bridge in *T. schlegelii.* Three distinct grooves pass forward out of the excavation for the Gasserian ganglion, crossing the side of the laterosphenoid and eventually leaving its surface near the anterior edge of the bone. These grooves no doubt mark the paths of the nerve and blood vessel that normally pass beneath the downgrowth of the laterosphenoid. Hopson (1979), citing Schumacher (1973), identifies two narrow canals that pass upward out of the excavation for the Gasserian ganglion into the side wall of the braincase in *Caiman crocodilus* (figs. 1A, 2C in Hopson 1979). The more anterior canal transmits a small motor nerve from the dorsal side of the ganglion to the levator bulbi muscle (Hopson 1979). A similar canal is present in UCMP 38358 (fig. 8.7, NC_1d), evinced by its foramen in the side of the laterosphenoid, and in *G. gangeticus,* but we have not recognized it in *T. schlegelii.* A second oval foramen of about the same size occurs in the dorsolateral wall of the excavation for the Gasserian ganglion a few millimeters below and behind the NC₁d foramen. Hopson (1979, figs. 1A, B; 2C) shows a small "unidentified branch of the trigeminal nerve" labeled Nx in his figures of an endocranial cast of *Caiman crocodilus*. He does not say whether this branch also emerges onto the side of the braincase in *Caiman,* but we have been unable to find any such opening in skulls of other crocodilians available to us, except in an immature *G. gangeticus* (TMM M5485). The purpose of foramen X in figure 8.7 is unknown.

PROOTIC   The prootic cannot be defined in either IGM skull, nor does it appear on the lateral surface of the braincase in UCMP 38358. It is narrowly visible, however, in the UCMP specimen (fig. 8.7), deep to the sharp ventral edge of the trigeminal foramen, more as in *Alligator* or *Caiman* than in either *Gavialis gangeticus* or *T. schlegelii* in which the bone is more broadly exposed on the lateral wall of the braincase.

A small, diamond-shaped, supernumerary bone with irregular sutural boundaries with the quadrate, parietal, and laterosphenoid occurs some 8 mm posterodorsal to the trigeminal foramen in UCMP 38358 (fig. 8.7). It may be the prootic, whose position here would, however, be anomalous if not unique among eusuchians.

EXOCCIPITAL   The exoccipitals (figs. 8.4A, D; 8.5A, D) are broadly exposed in dorsal view owing to the natural inclination of the occipital plate. The lateral wing of the exoccipital is narrower dorsoventrally than that of *Gavialis gangeticus* and is overlapped laterally to a greater extent by the squamosal. The sharp posteroventral edge of the bone that projects over the cranioquadrate canal is broadly arcuate when seen from above, reflecting the relatively wide occiput. The various foramina that issue from the exoccipital (for cranial nerve XII, the foramen vagi, and the posterior carotid foramen) occupy the same relative positions as in the referred cranium of *Gryposuchus colombianus* (see fig. 6b in Langston 1956). The foramen vagi appears to be divided at depth in IGM 184696, and the hypoglossal foramen is but a single opening.

The thin, ventrally directed flange of the exoccipital, which largely covers the lateral face of the basioccipital in *G. gangeticus,* is narrower and less widely exposed laterally in IGM 184696. It extends ventrad almost to the ventral edge of the basioccipital tuber, a peculiar state shared only with *G. gangeticus* among existing crocodiles. Lateral- and posterior-facing surfaces of this flange are set off obtusely from one another, and posterodistally the flange meets the basioccipital along a sharp, more or less vertical crest that is lacking in *G. gangeticus.* The lateral Eustachian canal emerges in the interface between the exoccipital and basisphenoid bones in UCMP 38358 (fig. 8.7).

BASIOCCIPITAL   The basioccipital (figs. 8.4A, C, D; 8.5A, C, D) is of gavial form but is thinner anteroposteriorly. As in *Gavialis* and in contrast to *Tomistoma* and all other eusuchians, the basioccipital is short dorsoventrally and displays little of the so-called verticalization discussed by Tarsitano (1985) and Tarsitano et al. (1989). The tubera are less expanded and less rugosely marked around the edges than in *Gavialis gangeticus.* The longitudinal axis of the condylar neck is strongly flexed caudoventrally so that the ventral edge of the occipital condyle lies closer to the ventral edge of the basioccipital than in *G. gangeticus.* There is no median crest on the ventral side of the condylar neck, this area being occupied by an excavation that contains paired foramina leading forward and upward into the basioccipital. The right foramen, which is considerably larger than the left one, is 6 mm wide and 8 mm long. One or more, often tiny and variably disposed foramina are usually present in this area in eusuchians and are said to transmit a vein from the substance of the bone (Miall 1878). But in at least some *Crocodylus* and *Alligator* skulls, the passage traverses the basioccipital to open onto the floor of the braincase. In *Gavialis* the foramina are often almost invisible to the unaided eye and large foramina may be lacking. By contrast, in *T. schlegelii* there is usually one large opening. In IGM 184696 the median Eustachian foramen opens into a transverse, slitlike excavation 24 mm wide and 4 mm high, much different from the rounded opening in *G. gangeticus* or *T. schlegelii.* The space between the venous foramina and the opening of the median Eustachian canal is occupied by a transverse bridge of bone 9 mm thick, much less than the corresponding distance in *G. gangeticus.* The basioccipital forms the floor of the foramen magnum.

BASISPHENOID The limited lateral exposure of the basisphenoid (figs. 8.4C, 8.5C, 8.7) is similar to that of *Gavialis gangeticus*. The part of the bone that is interposed between the basioccipital and the pterygoid is apparently very thin. Beneath (anteroventral to) the opening of the median Eustachian canal a transverse lamina 1 mm thick forms the roof of a triangular depression medially at the posterodorsal end of the choanae. Nothing of the sort is present in *G. gangeticus,* but the lamina is probably formed by the basisphenoid. The lateral edges of the triangular depression are probably formed of the pterygoids, which generally appear as a delicate lamella closely applied to the basisphenoid and forming the roof of the choanae on each side of the midline. Like the basioccipital, the basisphenoid is relatively compressed dorsoventrally as in *Gavialis.*

SUPRAOCCIPITAL The supraoccipital (figs. 8.4A, B, D; 8.5A, B, D) is remarkably large in IGM 184696. Paired, irregularly conical postoccipital processes (the right one is broken off) project posteriorly over the occiput and are broadly visible from above. Such postoccipital processes, likely containing an ossified component of the spinalis capitis tendons, occur in large *Gavialis gangeticus* skulls, for example, ZUMZ 100–197 and ZSM 28/1912, but are less developed than in IGM 184696. Where, in *G. gangeticus,* a small V-shaped process of the supraoccipital is insinuated in the skull roof posteromedially between the parietals, there is in IGM 184696 a diamond-shaped median depression 2.8 cm wide and approximately 3.6 cm long. Posteriorly, this depression plunges downward between the postoccipital processes. No sutures have been found within the depression, so its relationship to the supraoccipital is unknown. The edges of the depression are raised and sharp but are confluent laterally with the eavelike lappets of the parietals that overhang the post-temporal fenestrae. Indeed, it appears that the parietals have been extended posteriorly over the top of the nuchal prominence of the supraoccipital. Under this interpretation the supraoccipital would not be exposed in the cranial roof. Support for this idea is found in UCMP 38358 and the isolated supraoccipital IGM 183933. Both specimens display the top of the nuchal prominence, which occurs slightly below the level of the parietal skull roof. In UCMP 38358 the parietal roof appears to be broken away or stripped off of the top of the nuchal promi-

nence. We are not aware of this morphology being reported previously, but it recalls the question of the existence of dermoccipital (=postparietal) bones in eusuchians (see, e.g., Mook 1921; Iordansky 1973). In any event, the supraoccipital is evidently not wedged into the parietal plate in *Gryposuchus colombianus.*

QUADRATE The quadrates of IGM 184696 are lodged in the glenoid fossae of the mandible (figs. 8.4A, B, D; 8.5A, B, D), so the condyles are largely obscured; the quadrates of IGM 184057 are damaged and incomplete. When seen from behind, the axis of breadth of the jaw articulation is inclined ventromesad at an angle of about 22°, almost the same as in a *Gavialis gangeticus* skull 48.6 cm long (TMM 5448). Seen from above, the angle between the axis of articulation and the sagittal plane exceeds 80° compared to about 84° in the gavial skull. The condyles are not deeply divided; the trochlear notch is wider and deeper in IGM 184696 than in IGM 184057 or in *G. gangeticus.* The articular surface of the lateral condyle is slightly more exposed dorsally than the medial one. The opening of the aeriferous canal, which ordinarily lies anterior to the medial condyle on the dorsal side of the quadrate, has not been located in either skull. Seen from below, the quadrate displays various crests (fig. 8.9) for the attachment of the B, B′, and D tendons of Iordansky (1964, 1973). The edge of Crest B is sharp. Crest B′, incompletely preserved, is strongly expressed and forms a lateral boundary for a semiovate excavation in the inferior surface of the quadrate. The D crest is subdued and is identifiable only as the edge of a shallow excavation slightly forward of the medial condyle. Posterolateral to the posterior end of the Crest B′ is a 21-mm-long, finely sculptured tuberosity lying close to the quadrate-quadratojugal suture. This may correspond to Crest A of other eusuchians. Iordansky (1964) indicates that the A and A′ crests are absent in *G. gangeticus* and *T. schlegelii,* but in the largest skulls of both taxa available to us, a small slightly roughened area occurs in the vicinity of the distal part of Crest A as seen in other eusuchians. This discrepancy may be owing to the individual variability in placement of the crests noted by Iordansky. The quadrate plays a greater role in the side of the cranium than in either *G. gangeticus* or *T. schlegelii,* extending to and forming the outer posterior and ventral edge of the trigeminal foramen

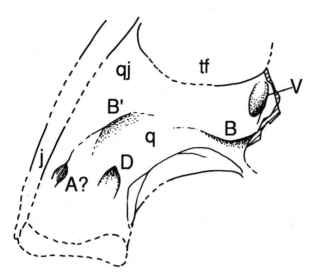

Figure 8.9. *Gryposuchus colombianus*, IGM 184696. Schema of crests on the ventral surface of the quadrate for tendinous attachment of mandibular adductor muscles. Identification of tendon scars follows Iordansky (1964). See text for explanation. Abbreviations: *j*, jugal; *q*, quadrate; *qj*, quadratojugal; *tf*, temporal fossa; *V*, trigeminal foramen. *Hatched area* is broken wall of otic chamber.

in UCMP 38358, and here excluding the prootic from the external surface of the braincase (fig. 8.7). The quadrate also broadly overlaps the ventromedial surface of the postorbital bone anterodorsally. UCMP 38358 is ambiguous because, in contrast to the usual condition in *Gavialis* and *Tomistoma*, the quadrate seems to extend anterodorsally beyond the end of the quadratojugal and, thus, forms the dorsal edge of the lateral temporal fenestra.

Mandible

FORM AND PROPORTIONS The mandible of IGM 184696 is the most complete and least distorted of several examples attributed to *Gryposuchus colombianus* (figs. 8.4B, C; 8.5B). Viewed laterally, the mandible is about as high posteriorly as in *Gavialis gangeticus* and *T. schlegelii* (table 8.2: ratios 13; fig. 8.10). Both the symphysis and retroarticular process are a little shorter relative to the length of the jaw than in *G. gangeticus* but are closer to it in these proportions than to *T. schlegelii* (table 8.2: ratios 14, 15). The width of the jaw posteriorly is close to the upper limit for large *T. schlegelii* and is significantly wider than in the gavial (table 8.2: ratio 18). The rostal symphysis is strongly depressed (fig. 8.11),

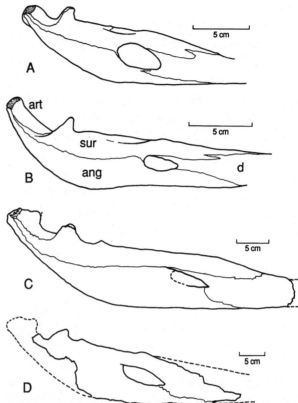

Figure 8.10. Posterior segments of longirostral crocodilian mandibles, lateral aspect. *A. Tomistoma schlegelii*, TMM M6342. *B. Gavialis gangeticus*, TMM M5485. *C. Gryposuchus colombianus*, IGM 184696, left side reversed. *D. Gryposuchus neogaeus*, MLP 26–413, left side reversed. Abbreviations: *ang*, angular; *art*, articular; *d*, dentary; *sur*, surangular.

relatively flatter than in a large example of *G. gangeticus* and much flatter than in large *T. schlegelii* (table 8.2: ratio no. 19). The rostrum of IGM 250480 is even more depressed with a height : width ratio of 0.38. The jaw is sharply expanded transversely at the level of the second dentary tooth (16.0 cm wide in IGM 250712 and 12.9 cm in IGM 184696; see figs. 8.4B; 8.5B). The narrowest part of the symphysis occurs at the interspace between the fourth and fifth teeth. The lateral mandibular fenestra, although crushed dorsoventrally, is best preserved on the left ramus of IGM 184696 (fig. 8.10C). It is much longer relative to its height than in large *G. gangeticus* (table 8.2: ratio 20; fig. 8.10B, C). The ratio in *Gryposuchus neogaeus* (~4.2) is closer to that in *G. colombianus* and, as there, is relatively narrower than in other eusuchians (fig. 8.10D). The posterior end of the

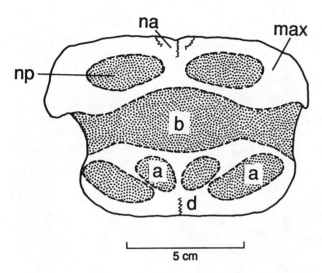

Figure 8.11. *Gryposuchus colombianus,* IGM 184696, semischematic transverse section of rostrum at the level of the seventh maxillary tooth and the tenth to eleventh dental interspace of the mandible. Abbreviations: *a,* alveolus; *b,* buccal space; *d,* dentary; *max,* maxilla; *na,* nasal; *np,* narial passage. *Stipple* indicates matrix.

fenestra was probably more broadly rounded than it now appears, to judge from *G. neogaeus* (MLP 260-413). The surangular is excluded from the edge of the fenestra by a long narrow posterior process of the dentary that meets the angular in the posterodorsal edge of the opening (fig. 8.10C). In *G. gangeticus* the surangular bounds the fenestra posteriorly and forms as much as one-half of its superior edge (fig. 8.10B). The arrangement in *G. colombianus* may be unique among eusuchians; we have found no parallels among existing taxa.

The external surface of the rostrum is marked by strong anteroposterior grooves and scattered foramina, the latter best developed near the labial edge. The sculpture on the lingual surface of the symphysis, marked by finer longitudinal striations and smaller foramina, is coarser than in large *G. gangeticus.* Larger vascular and neural foramina occur medial to the tooth row and approximately opposite the interalveolar spaces.

DENTARY   The dentary is best seen in the large jaw IGM 250712 (fig. 8.12A, B). A deep median cleft separates the tips of the opposing dentary bones as in *Gavialis gangeticus* and to a lesser extent in *T. schlegelii.* This feature accentuates the apparent hypertrophy of the bone around the base of the first tooth on each side of the jaw. Viewed in the horizontal

plane, the dentary displays modest to strong alveolar salients, which gradually disappear caudad. The amplitude of the salients varies with individual size (fig. 8.12 A–C), but in some specimens it is clearly accentuated by dorsoventral crushing. The first dentary alveolus is the largest in the mandible and is markedly wider than long (22 mm compared to 16 mm in IGM 250480). The second alveolus is smaller, and the third is the smallest in the jaw, round with a diameter of 11 mm in IGM 250480. In this example, it lies about 1.1 cm in front of the fourth, and 2.0 cm behind the second. The space between the second and third alveoli is the greatest interalveolar distance in the mandible. The fourth alveolus is transversely oval with principal diameters of 22 and 12 mm in IGM 250480. Except for the first four alveoli, there is not much difference in the size of the sockets, which vary in shape from subcircular to broadly oval, nor in the interalveolar spaces, which, reflecting somewhat wider spacing of the teeth than in large *G. gangeticus,* are about equal to the lengths of the adjacent alveoli. Posteriorly, the tooth row crosses the dentary from the labial edge of the rostrum onto the dorsal side at around the sixteenth or seventeenth tooth. The three postsymphyseal alveoli are bounded superficially on the medial side by the splenial. Shallow occlusal pits for the tips of the upper teeth are present only in the neighborhood of the seventeenth to twenty-second teeth, and their depth and frequency vary in different jaws.

SPLENIAL   The splenial symphysis begins posteriorly at a level midway between the eighteenth and nineteenth teeth in IGM 184696 and even with the nineteenth tooth in IGM 250712. Ventrally, it extends forward to the level of the twelfth tooth. On the lingual side of the rostrum, the splenial symphysis ends anteriorly, even with dental interspace 12–13 in IGM 250712 (fig. 8.12B). The deep conical median pit, which occurs at the posterior end of the splenial symphysis in *Gavialis gangeticus* and *T. schlegelii,* is absent in all gavial jaws from Colombia. In place of the pit there is a tiny foramen, visible from above on each side of the symphyseal suture, in each splenial. In *G. gangeticus* a foramen perforates each splenial deep within the median pit (fig. 5 in Norell 1989), but there are also several additional foramina in the medial wall of the splenial. One, opening within the sagittal suture, has been identified tentatively by Norell as

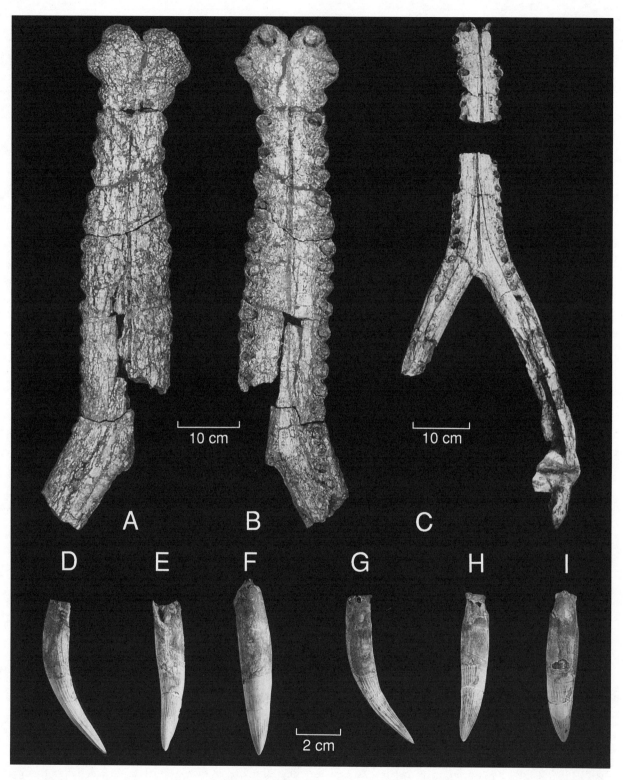

Figure 8.12. *Gryposuchus colombianus*, mandibles and teeth. *A–B.* IGM 250712. *A,* ventral aspect; *B,* lingual aspect. *C.* IGM 250480, lingual aspect. (Note size-related differences in development of anterior end in *B* and *C.*) *D–I.* Two teeth of IGM 184696, fallen, presumably, from the upper jaw before burial (sequential position unknown). *D, G,* lateral aspect; *E, H,* posterior aspect; *F, I,* anterior aspect. *A–C,* scale bars 10 cm. *D–I,* scale bar 2 cm.

WANN LANGSTON, JR., AND ZULMA GASPARINI

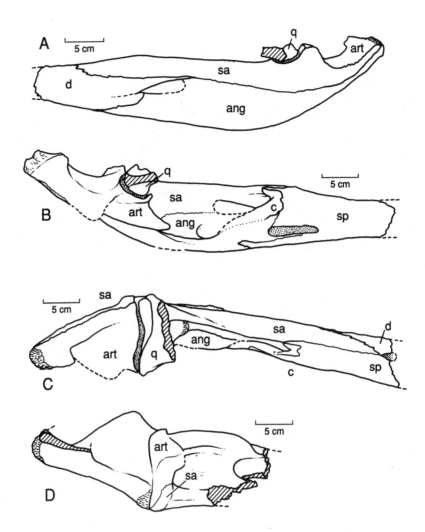

Figure 8.13. *Gryposuchus colombianus.*
A–C. IGM 184696, posterior left ramus of
mandible. *A*, lateral aspect; *B*, medial aspect;
C, dorsal aspect. *D.* IGM 184057, articular
region of right ramus (reversed). In *A–C*,
part of quadrate is lodged in the glenoid
cavity. Abbreviations: *ang*, angular;
art, articular; *c*, coronoid; *d*, dentary;
q, quadrate; *sa*, surangular; *sp*, splenial.

the foramen intermandibularis oralis, which in crocodylids leaves Meckel's canal anteriorly but in alligatorids passes out through the splenial at the same point as the peculiar foramen in *Gavialis* (Norell 1989). Probing of the various foramina in the jaw of *G. gangeticus* suggests an intricate system of anastomoses within the splenial and dentary bones and demonstrates a possible passageway directly from the median pit into the splenial and then, almost immediately, out again into the sutural plane through the foramen intermandibularis oralis of Norell. Moreover, there is a connection within the splenial between this foramen and a small opening in the medial wall of the splenial a short distance behind the symphysis. This indicates that the homologies of the several foramina associated with the posterior part of the splenial symphysis in *Gavialis* need further investigation; but however it may be, there is evidence in IGM 250712 of

one or more foramina in the area pointed out by Norell (1989, fig. 5).

SURANGULAR The surangular extends anterodorsally only as far as the last alveolus (fig. 8.13); unlike *Gavialis* and *Tomistoma*, it is not insinuated between the splenial and dentary medial to the two to four most posterior teeth. The surangular bounds the glenoid fossa laterally and extends posteriorly almost to the end of the retroarticular process as in *T. schlegelii* (fig. 8.10A, C).

ANGULAR The junction of the anterodorsal process of the angular with the dentary above the lateral mandibular fenestra has been noted. The angular forms the ventral margin of the fenestra as well. Posterodorsally, it extends in concert with the surangular almost to the end of the retro-

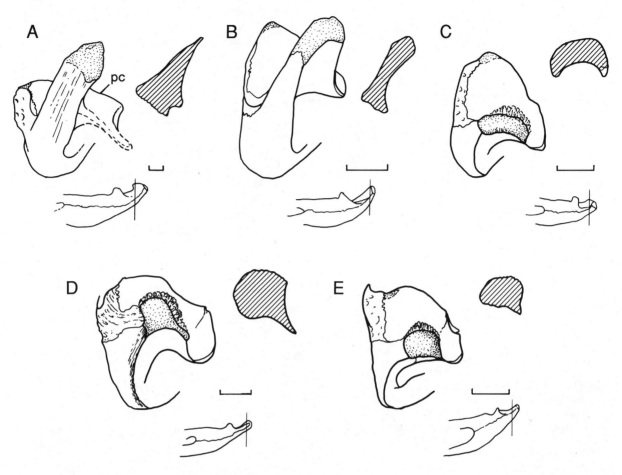

Figure 8.14. Retroarticular processes and associated structures of selected crocodilian mandibles compared: posterior aspects of left mandibles, all drawn approximately to same scale. *Small diagrams* show location of section through retroarticular process, and *hatched shapes* represent shapes of the cross-sections. *A. Gryposuchus colombianus,* IGM 184696. *B. Gavialis gangeticus. C. Tomistoma schlegelii. D. Crocodylus acutus. E. Alligator mississippiensis.* Abbreviation: *pc,* posterior transverse crest of glenoid fossa. *A–E.* Scale bars 1 cm.

articular process where it persists as a very thin longitudinally striated lamina closely applied to the articular (fig. 8.13A).

CORONOID   A well-preserved left coronoid bone is preserved in natural position in IGM 184696. It displays no distinctive features, however.

ARTICULAR   The glenoid fossa resembles *Gavialis gangeticus* more than *T. schlegelii.* The relationship between greatest width and least length in IGM 250480 is approximately 2.3, hence is closer to *Gavialis* (1.5–2.21) than to *T. schlegelii* (~1.5). As usual in gavials, the fossa is shallow and relatively flat, the diagonal intrafossa elevation, usually undeveloped in large *G. gangeticus,* being much less prominent than in

Tomistoma. The lateral one-fourth of the fossa's area (best seen in IGM 184057 and 250480) is formed, in gavial fashion, of the surangular (fig. 8.13D). The posterior transverse crest is low and is less strongly inclined ventromedially than in *G. gangeticus* (fig. 8.14A, B). The triangular posterolateral eminence, formed of articular and surangular in *Tomistoma* and *Gavialis,* is less pronounced than in those taxa (fig. 8.10). The medially expanded flange of the retroarticular process (figs. 8.13D; 8.14A, B) is broad, irregularly triangular, and wider and shorter than in *G. gangeticus.*

The dorsal longitudinal crest of the retroarticular process is constricted transversely in strong contrast to the broad and low crest of *Tomistoma* and most other eusuchians (fig. 8.14). In *G. gangeticus* the crest is elevated as in *Gryposuchus* but is less compressed transversely (fig. 8.14A, B). The crest expands distally into a peculiar, transversely flattened, hooklike process that appears to be characteristic of *G. colombianus* (figs. 8.4 A–C, 8.10 A–C, 8.13A). The end of the process, composed of the articular bone, is knoblike but is less thickened than in other eusuchians except *G. gangeticus* (fig. 8.14). Its principal diameter is inclined about 45° dorsomedially as in *G. gangeticus,* whereas in other eusuchians the axis tends toward the horizontal (fig. 8.14). We are unable to locate the aeriferous foramen that transmits the siphonium into the articular bone. This foramen is readily recognizable in most crocodilians (see illustrations in Iordansky 1973), but its development in *G. gangeticus* is variable. Thus, in a *G. gangeticus* mandible 53.4 cm long (TMM M5485) a foramen so small that it will not pass a bristle occurs on the left articular and no opening is present on the right side; however, large *G. gangeticus* usually have larger foramina. IGM 184696 contains at least three tiny foramina arranged in tandem proximal to and slightly below the plane of the medial flange of the articular. These may be related to the siphonium, but they are situated below the level at which the aeriferous foramen usually occurs.

For comparative measurements of skulls and jaws, see table 8.2 and figure 8.3.

Dentition

The dental formula of IGM 184696 is 4 + 21(22)/23. Thus, there is one less tooth in the premaxilla, and there are as many as three fewer teeth in both the maxilla and dentary than in *Gavialis gangeticus,* but five to six more in the dentary than there are in *T. schlegelii.* Juvenile *Tomistoma* possess five premaxillary teeth, but the second disappears in ontogeny, leaving a complement of four teeth in the adults (Kälin 1933 reported that the second premaxillary tooth of *Tomistoma* is lost at an early age, but it is present in skulls up to a basal length of 35.6 cm in the Texas Memorial Museum collection). As in *G. gangeticus,* there are three postsymphyseal teeth, two less than in *Tomistoma.*

Heterodonty is expressed mainly in differing size. The first two very large premaxillary and dentary teeth of the Greifapparat are noteworthy. The fourth premaxillary and the third dentary teeth are smaller, and the two or three most posterior teeth in both upper and lower jaws are the shortest in the dentition. Although there is some variation, depending on the relative ages of the respective teeth, those posterior to the fourth premaxillary and the third or fourth dentary teeth decrease in size at a fairly uniform and gradual rate posteriorly, the twenty-first dentary tooth, for example, being about two-thirds as large as the fourth.

Upper and lower teeth are similar (fig. 8.12 D–I). The larger teeth are strikingly long and slender and, as in other long-snouted crocodilians, often exhibit a slight sigmoid curvature (fig. 8.12D, G). The crowns bear weak longitudinal fluting anteriorly and delicate longitudinal striae posteriorly. In cross section the crowns are broadly lentoid with the longer axis oriented transversely in the more anterior teeth and slightly more obliquely in those behind. Sharp carinae divide the crowns into almost equal parts as in *G. gangeticus.* A number of (mostly upper) teeth, fallen from their alveoli, were found scattered in the matrix around the two crania IGM 184696 and IGM 184057. Some of these teeth, with intact roots, are fanglike and are up to 7.8 cm long. Crowns are generally a little longer than the roots. The roots are more rounded in section than the crowns.

Size and Proportions of the Species

Gryposuchus colombianus was a large crocodilian although not in a league with some truly gigantic South American Neogene species, for example, *Purrusaurus brasiliensis* and an undescribed longitrostral form from the lower Pliocene of Venezuela. Indeed, IGM 184696 was probably smaller than the largest existing examples of *Gavialis gangeticus,* which according to Martin and Bellairs (1977), citing Bustard 1976, may attain a length of about 7.62 m. Martin and Bellairs conclude, based on a graph by Wermuth (1964, fig. 5), that such an animal should have had a skull length of about 1.15 m, measured from the tip of the snout to the posterior edge of the supraoccipital. With a corresponding skull length of 0.966 m, IGM 184696 would have been around 6.4 m long in life, according to Wermouth's chart cited previously, or approximately 6.6 times the length of the skull to the supraoccipital.

We lack gavial data from which to calculate the mass of IGM 184696 in life. According to Jane Peterson (pers. comm. 1993), a hypothetical *Alligator mississippiensis* with a total length of 6.4 m should have a mass of 1,300 kg. Allow-

Table 8.4. Fossil Gavialidae of South America

Taxon	Source	Age	Material	Current Designation
Rhamphostomopsis neogaeus (Burmeister, 1885)	Argentina (Paraná)	Upper Miocene–Pliocene "Mesopotamian"	Rostral fragments, cranium	*Gryposuchus neogaeus*
Rhamphostomopsis intermedius Rusconi 1935	Argentina (Paraná)	Upper Miocene–Pliocene "Mesopotamian"	Rostral fragment	?*Gryposuchus neogaeus*
cf. *Rhamphostomopsis* (see Langston 1965)	Colombia (Huila)	Middle Miocene Friasian	Several mandibles	*Gryposuchus colombianus*
Leptorrhamphus entrerianus Ambrosetti 1887	Argentina	"Mesopotamian"	Rostral fragment	nomen dubium
Oxysdonsaurus striatus Ambrosetti 1887	Argentina	"Mesopotamian"	Tooth	nomen nudum
Gryposuchus jessei Gürich 1911	Brazil	Upper Miocene–Pliocene[a]	Ends of cranial rostra	Valid taxon
Gavialis colombianus Langston 1965	Colombia (Huila)	?Upper Oligocene–middle Miocene	Several skulls, jaws, and fragments of postcranial bones	*Gryposuchus colombianus*
Ikanogavialis gameroi Sill 1970	Venezuela (Urumaco)	Upper Miocene[b]	Several skulls and jaws	Valid taxon
Hesperogavialis cruxenti Bocquentin Villanueva & Buffetaut 1981	Venezuela (Urumaco)	Upper Miocene[b]	Section of cranial and mandibular rostrum	Valid taxon
Hesperogavialis sp. (Souza Filho et al. 1993)	Brazil	Upper Miocene–Pliocene	Skull	Valid reference
Brasilosuchus mendesi Souza Filho & Bocquentin Villanueva 1989	Brazil (Acre)	Upper Miocene–Pliocene	Premaxillae	Valid taxon
"El gavial de Brasil" Sill 1970	Brazil (Acre)	Upper Miocene–Pliocene	Undescribed skull	?*Gryposuchus* sp.

[a]The date given for the holotype by Gürich (1912, 70) is "jungster Tertier oder Quartär." Buffetaut (1982), based on certain unsubstantiated inferences about a "court" Pliocene and gavial-like characters of *Gryposuchus* (that is, *Rhamphostomopsis* sensu Gasparini) imputes an upper Miocene age to *G. jessei*, but see Frailey et al. (1988) and Langston (1965, notes to tables 3 and 4) concerning the difficulties of dating fossil bones and strata from the lower reaches of some of the larger rivers in South America. The "door-stop" specimen of *G. jessei* (see Gürich 1912, 49) might, improbably, be as old as upper Cretaceous.

[b]Previously dated as mid-Pliocene (Huayquerien). See discussion of correlation and radiometric dating by Bocquentin Villanueva and Buffetaut (1981).

ing for differences in body shape between the alligator and a gavial, whose long narrow rostrum accounts for about one-sixth of the total length, a mass of about 1,000 kg seems reasonable for a *Gryposuchus colombianus* the size of IGM 184696.

DISCUSSION

Fossil longirostrine crocodilians have been known from Cenozoic deposits in South America since Burmeister first reported them in 1885. Several taxa (table 8.4) have been regarded as gavials by most recent authors (but see Hoffstetter 1971). There seems to be general agreement (e.g., Gasparini 1981; Buffetaut 1982) that *Oxysdonsaurus* is a nomen nudum and *Leptorrhamphus striatus* is at best a nomen dubium. But *Ikanogavialis, Hesperogavialis,* and *Gryposuchus* are valid genera, and the recently described *Brasilosuchus mendesi* Souza Filho and Bocquentin Villanueva (1989) may represent a fourth gavial from South America. *Rhamphostomopsis neogaeus* (Burmeister 1885), including *R. inter-*

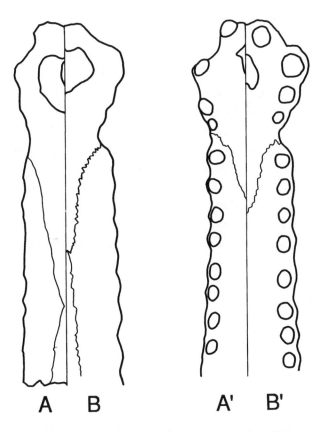

A B   A' B'

Figure 8.15. Cranial rostra compared, drawn to same scale (width at space between last premaxillary tooth and first maxillary tooth). *A, A', Gryposuchus* (= *Rhamphostomopsis*) *neogaeus*, MLP 26-413; *B, B', Gryposuchus jessei*, HM-GI. (*A, A'* modified from Gasparini 1981 and photograph; *B, B'* after Gürich 1912.)

medius Rusconi (1935), has also been widely accepted (e.g., Langston 1965; Gasparini 1968; Sill 1970). But as authors have noted (Langston 1965; Buffetaut 1982), the original description is not very informative and the whereabouts of the type specimen is unknown, nor is it clear whether Rovereto (1912), who discussed it, had ever examined it. Rusconi (1933), however, who coined the name *Rhamphostomopsis,* had not seen it. The type is a segment of a rostrum but whether from the skull or the mandible is not known (contra Gasparini 1968). Gasparini (1968) referred a cranial segment and the tip of a cranial rostrum in the Museo de La Plata (MLP 26-413), previously mentioned by Rusconi (1935), to *Rhamphostomopsis neogaeus,* and Buffetaut (1982) reassigned this specimen to *Gryposuchus.* He was ambivalent, however, about whether it was specifically distinct from *Gryposuchus jessei.* We concur in Buffetaut's attribu-

tion of MLP 26-413 to *Gryposuchus.* The fragmentary nature of the holotype of *G. jessei* (the end of a cranial rostrum) provides only a limited basis for comparisons with other specimens, but Buffetaut noted the close agreement in the relative sizes and arrangement of the four premaxillary alveoli and the general shape of the premaxillary bones between *G. jessei* and MLP 26-413 (fig. 8.15). Similar also in the two specimens is the pattern of the V-shaped maxillary-premaxillary suture in the palate. The width : length ratio of the external nares is 1.26 in *G. jessei* compared with about 1.2 in MLP 26-413. Among differences is the contact between premaxillae and the nasal bones (only the right nasal reaches the posterior tip of the premaxilla in the type of *G. jessei*), which occurs even with the interspace between the sixth and seventh maxillary alveoli in MLP 26-413 and with the posterior edge of the fourth alveolus in *G. jessei.* Another difference, not noted by Buffetaut, is the abrupt transverse expansion of the rostrum at the first maxillary alveolus, which lends a distinctive "shouldered" appearance to the anterior end of the maxillae in the type of *G. jessei* and which is less pronounced in MLP 26-413 and in a recently described *Gryposuchus* rostrum from Brazil (Souza Filho and Bocquentin Villanueva 1989). Additionally, the premaxillae are relatively wider at the second alveoli in the otherwise narrower rostrum of MLP 26-413 than in the type of *G. jessei.* We believe, therefore, that *Gryposuchus jessei* Gürich and *Gryposuchus neogaeus* sensu Gasparini (1968) should be retained, awaiting more complete examples of both taxa from their respective provenances. By assigning *Rhamphostomopsis neogaeus* to *Gryposuchus,* we invalidate Langston's (1965) attribution of two longirostral mandibles from Colombia to cf. *Rhamphostomopsis,* which now are referred to *Gryposuchus* because of their resemblance to jaws in the IGM collection described herein.

An exceptionally well-preserved gavial skull from the upper Miocene/lower Pliocene of Brazil, figured but not described by Sill (1970), has been tentatively referred to *Gryposuchus* by Buffetaut (1982). Nasal bones contact the premaxillae in this individual, but no further information about the specimen is available to us, and we accept Buffetaut's identification.

Buffetaut (1982), while recognizing the inconclusiveness of some evidence (resemblance in degree of flaring of the orbital rim and depression of the interorbital skull roof) and assuming relative primitiveness vis-à-vis Gasparini's *R. neogaeus* specimen, suggested that *Gavialis colombianus* Langston should also be included in *Gryposuchus*. Langston's (1965) description of *Gavialis colombianus* was based on an incomplete skull (UCMP 41136, holotype) lacking most of the rostrum, skull table, and braincase, and a skull table, braincase, and occiput of a second individual of similar size (UCMP 38358). Both specimens are from University of California Locality Coyaima 6, from deposits believed by Stirton (1953) to belong to the upper Oligocene. Comparisons of the specimens at the University of California and the IGM skulls leave little doubt that these individuals were conspecific. A possible difference in geological ages between them may be rendered moot by some recent unpublished work of Richard Madden (pers. comm. 1992), which indicates that the Coyaima 6 deposits are correlative with those of La Venta and, thus, are middle Miocene in age.

COMPARISONS

From the foregoing description it seems clear that *Gryposuchus colombianus* is more like *Gavialis gangeticus* than any other existing crocodilian (fig. 8.8A; table 8.5), including *Tomistoma schlegelii*, which it nevertheless resembles in a number of respects (fig. 8.8B; table 8.5).

Both IGM skulls of *Gryposuchus colombianus* are comparable to the holotype of *Gryposuchus jessei* in size, but the nares of IGM 184696 are significantly wider than in either known specimen of *G. jessei* (width : length ratio = 1.42, compared with 1.26 and 1.13 in the holotype and in a smaller referred rostrum UFAC 1272) (Souza Filho and Bocquentin Villanueva 1991). The striking "shouldered" appearance resulting from the sudden narrowing of the rostrum at the premaxillary-maxillary notch in the holotype of *G. jessei* (fig. 8.15B, B') is less noticeable in IGM 184696 (and also in UFAC 1272). The first premaxillary alveolus in IGM 184696 is slightly larger than that in the holotype of *G. jessei*, but succeeding alveoli are consistently about 10% smaller in the Colombian skull. The length of the rostrum back to the posterior edge of the thirteenth alveolus (ninth maxillary tooth) is greater in *G.*

colombianus (45.4 compared with 39.5 cm), reflecting wider spacing of the teeth in IGM 184696.

In the type of *G. jessei* a contact between the premaxilla and nasal bones exists only on the right side of the skull at the level of the posterior edge of the fourth maxillary tooth, whereas a similar contact occurs on the left side in *G. colombianus*, IGM 184696. Gürich (1912, pl. 1, 1A) shows a considerably wider separation between the left nasal and premaxilla than occurs in IGM 184696, but this scarcely seems significant (in *G. gangeticus* the nasals differ in length beneath the surface of the skull roof).

It is possible that *G. jessei* and *G. colombianus* are conspecific. We believe, however, that stability of nomenclature will be best served, until better examples of *G. jessei* are forthcoming, by recognizing the Brazilian and Colombian examples as distinct species distinguished by the differing proportions of the nares.

Resemblances between *Gryposuchus colombianus* and *Gryposuchus neogaeus* are extensive, and significant differences are few (table 8.5). Both IGM examples of *G. colombianus* are a little larger than *G. neogaeus*, MLP 26-413. The Greifapparat is slightly more robust and the premaxillae are more expanded across the second teeth in IGM 184696, possibly correlated with its larger size. Maxillary teeth one to ten are closer together (only the anterior fifteen teeth and alveoli are known in MLP 26-413) and alveoli are less regularly rounded than in *G. neogaeus*. The dorsal contact between premaxillae and nasals occurs slightly farther forward (even with interdental space 8–9 compared with the tenth tooth in *G. neogaeus*). The failure of the premaxilla to meet the nasal bone on the right side of IGM 184696 is of importance only if future discoveries demonstrate that this relationship exists uniformly in *G. colombianus*. The skull roof is less deeply sculptured in IGM 184696 despite its greater size.

The width of the interorbital skull roof relative to the width of the parietal roof is about the same in IGM 184696 and MLP 26-413, but in the (smaller) holotype of *G. colombianus* the frontal plate is relatively a little wider. The quadrate condyles of IGM 184696 are more strongly differentiated than in *G. neogaeus*, which appear more like the simpler structure of *Gavialis gangeticus*. The supratemporal arcade is wider in dorsal aspect in *G. colombianus* than in MLP 26-413. The posttemporal arcade, comprising parietal and

Table 8.5. Character states in various gavialids and *Tomistoma*

Species (Source of Data)	A	B	C	D	E	F	G	H	I	J	K	L	M	N	O	P	Q	R	S	T	U	V	W	X	Y	Z
Gavialis gangeticus (d, e, f)	0	0	0	0	0	0	0	0	0	0	0	0	0	0	0	0	0	0	0	0	0	0	0	0	—	0
Gryposuchus colombianus (d)	2	1	0	0	3	0	1	0	0	1	1	0	0	0	0	0	0	0	0	0	0	1	0	2	—	0
Gryposuchus neogaeus (d)	—	1	—	0	—	0	1	0	—	—	—	—	0	0	—	0	0	—	—	—	—	—	—	—	0	0
Ikanogavialis gameroi (c, d)	2	—	0	1	3	0	1	1	0	1	0	1	0	0	0	0	0	—	—	—	0	—	—	0	—	0
Eogavialis gavialoides (a, d, g, h)	2	0	1	0	1	2	0	2	0	0	—	?	0	—	0	1	0	—	—	—	—	—	—	—	0	0
Eogavialis africanus (a, d)	2	—	1	2	0	—	0	0	0	0	0	—	—	—	—	0	—	—	—	—	—	—	—	—	0	0
Tomistoma schlegelii (d)	1	0	1	1	1	1	0	1	1	1	1	1	1	1	1	1	0	1	0	1	1	1	1	0	1	1

Sources: a, Buffetaut (1978, 1982); b, Hecht and Malone (1972); c, Sill (1970); d, authors' observations; e, Norell (1989); f, Kälin (1931, 1933); g, Andrews (1906); h, Müller (1927).

Notes: Character states are defined as 0, present; 1, absent; 2, intermediate between *G. gangeticus* and *T. schlegelii;* 3, intermediate between *G. gangeticus* and *E. gavialoides.* Characters A–Z are as given in appendix 8.1.

squamosal, appears broader anteroposteriorly in *G. colombianus* than is depicted in MLP 26-413 (fig. 1 in Gasparini 1968). Partial destruction of the top of the arcade (the upper bar of the "T" as described previously), however, was the basis for Gasparini's (1968) statement that the post-temporal bar is reduced dorsally to a lamina in *"Rhamphostomopsis" neogaeus.*

The occipital condyle is almost round in IGM 184696 (width : height = 1.10), whereas the condyle is relatively wider in MLP 26-413 (width : height = 1.26), and is wider still in UCMP 38358 (width : height = 1.36). These proportions conform to Kälin's (1933) data showing that height of the condyle increases more rapidly than width during ontogeny in *Gavialis gangeticus, T. schlegelii,* and *Alligator mississippiensis.* Possibly because of crushing, the foramen magnum is much lower and relatively wider in IGM 184696 than in either MLP 26-413 or UCMP 38358.

In the mandible the glenoid fossa of IGM 184696 is wider (9.3 cm) than would be expected when compared to *G. neogaeus* (6.5 cm). The posterior rim of the glenoid fossa undulates more than that of *G. neogaeus* in concordance with the better defined quadrate condyles of *G. colombianus.* Although the form and proportions of the lateral mandibular fenestra probably are closer to the natural condition in *G. neogaeus* (fig. 8.10D), the distortion of the opening in all mandibles of *G. colombianus* prevents useful comparisons of the fenestra. We note here that the fenestra of a large mandible from Huila referred to cf. *Rhamphostomopsis* (UCMP

40293) by Langston (1965, fig. 12B) is incorrectly reconstructed.

Some of the foregoing differences are, as noted, likely the result of accidents of preservation and to ontogenetic or other individual variability. Of more importance systematically are some proportional differences. The nares are significantly wider in relation to their length in *G. colombianus* (table 8.2: ratio 16). Supratemporal fenestrae are more sharply trapezoidal in plan in IGM 184696 (but they are somewhat more rounded in the referred skull roof at Berkeley, UCMP 38358, which agrees better with *G. neogaeus:* fig. 8.8D).

The occiput (measured from the ventral edge of the basal tubera at the midline to the top of the skull table) is 13.7 cm high in both IGM 184696 and MLP 26-413, whereas the width of the occiput across the paroccipital processes is considerably greater in IGM 184696 than in MLP 26-413 (33.6 cm compared with ~27.0 cm); the width across the quadrates is 43.0 cm and 30.0 cm, respectively, in the two skulls.

Important differences are found in the retroarticular process. Here, the medially expanded flange of the articular bone is more triangular in *G. colombianus,* beginning its medial expansion some 2.2 cm behind the glenoid, whereas in *G. neogaeus* it arises farther forward. The dorsal ridge that leads to the peculiar retroarticular "hook" in *G. colombianus* is more powerful, more elevated, and more laterally placed in its anterior part than the corresponding spine of *G.*

neogaeus, which arises closer to the centerline of the retroarticular process. The distal end of the process is broken off in MLP 26-413, but what remains suggests that a hooklike process, if present, was less developed than in *G. colombianus.*

We believe that the differences in the shape of the nares and the supratemporal fenestra together with those in the retroarticular process are adequate to distinguish between *G. neogaeus* and *G. colombianus.*

Brasilosuchus mendesi Souza Filho and Bocquentin Villanueva (1989) is known only from the premaxillae. Although clearly a longirostral crocodilian, the end of the cranial rostrum is so unlike that of other gavials that we wonder if it belongs with that group (perhaps it pertains to *Charactosuchus*). The premaxillary rostrum differs markedly from *Gryposuchus* in its slenderness, narrow nares, and the relative size and arrangement of its five premaxillary teeth, absence of an incisive foramen (unique among eusuchians?), and extremely prognathous first teeth.

Besides *G. colombianus,* the most complete skull and jaw of a South American gavial yet described is that of *Ikanogavialis gameroi* Sill (1970). The holotype of *I. gameroi,* although smaller than either IGM example of *G. colombianus,* is longer owing to the exceptionally long rostrum (fig. 8.8C). Besides the long rostrum, the most obvious difference between the two taxa is the presence in *Ikanogavialis* of greatly hypertrophied alveolar collars in the anterior half of the rostrum reminiscent of the extremely everted alveoli in the North African Neogene *Euthecodon* (see Fourtau 1920; Arambourg 1948). The transverse expansion of the premaxillae at the second tooth, however, present in *Gryposuchus,* is but little developed in *Ikanogavialis. Gryposuchus* displays a much wider interorbital space, a more abrupt transition from face to rostrum, and greater elevation of the anterior and lateral rims of the orbit. The postorbital bar (not preserved in the holotype of *I. gameroi* but present in a small skull in the Museum of Comparative Zoology, MCZ field number 95-72V) seems comparable to that of IGM 184696. The powerful, eavelike projection of the postorbital bone that overhangs the orbit posterolaterally in *G. colombianus* is scarcely evident in *Ikanogavialis.* The palatal fenestrae are wider and the palatine bridge is correspondingly narrower in *Ikanogavialis.*

The basioccipital tubera extend farther ventrally in *G. colombianus* than in *Ikanogavialis.* As would be expected in the shorter-skulled *Gryposuchus,* there are six or seven fewer teeth in both the maxilla and the dentary, and the teeth are more closely spaced than in *Ikanogavialis.* An interesting resemblance between *Gryposuchus* and *Ikanogavialis* is the failure of the anterodorsal process of the surangular bone to extend forward between the dentary and splenial bones at the back of the tooth row, the surangular being even shorter in *Ikanogavialis* than in *Gryposuchus* (see Sill 1970). In all respects, therefore, *Gryposuchus* appears more like *Gavialis* than like *Ikanogavialis* (see table 8.5; fig. 8.8A, C).

Hesperogavialis cruxenti Bocquentin Villanueva and Buffetaut (1981) is known from segments of the upper and lower jaws. A briefly described skull from Brazil has recently been referred to *Hesperogavialis* (Souza Filho et al. 1993). The type of *H. cruxenti* is comparable to IGM 184696 in shape and size; the Brazilian skull is a little wider posteriorly and possibly somewhat longer. Failure of the nasal bones to reach the premaxillae distinguishes *Hesperogavialis* from other South American gavials (the state is almost surely a parallelism with the Asiatic gavials). The nasals end asymmetrically in front, the gap between the right nasal and premaxilla being greater than that on the opposite side of the skull, evocative of the asymmetry in *Gryposuchus.* The anterior, wedgelike process of the palatines is shorter, but is broader behind than in *G. colombianus.* The Brazilian skull contains twenty-one alveoli, indicating one less tooth than in *G. colombianus* (if the number reported by Souza Filho et al. represents the full complement). In the holotype of *H. cruxenti* there is one less tooth posterior to the mandibular symphysis, and the teeth, especially in the mandible, are set in from the edge of the jaw; however, in the referred skull the first eight upper teeth are associated with alveolar collars that give an undulating appearance to the edge of the anterior end of the rostrum (which is not preserved in the type of *H. cruxenti*).

Comparisons between *G. colombianus* and several fossil gavials from India are provided by Langston (1956). The only important additional data derived from the IGM material concerns the anterior rostrum and the existence of pterygoid bullae. Excepting the unique shape of the external nares of *G. colombianus,* the length of the nasal bones, and number of teeth, the differences between *Gryposuchus* and the Indian fossils are similar to those noted between *Gryposuchus colombianus* and *Gavialis gangeticus*.

Of greater interest are the older gavials from northern Africa, early attributed to *Tomistoma* (Andrews 1901, 1906; Müller 1927), later to *Gavialis* (Hecht and Malone 1972), and most recently to a new genus, *Eogavialis*, by Buffetaut (1982). *Eogavialis africanus* (Andrews) is from upper Eocene (Priabonian) marine deltaic and lagoonal beds, and *Eogavialis gavialoides* (Andrews) is from slightly younger lower Oligocene (Rupelian) continental deposits in the Fayum (see Gingerich 1992 for a timely discussion of Fayum stratigraphy and depositional environments). Both taxa appear morphologically intermediate between *T. schlegelii* and *Gavialis gangeticus* (Andrews 1901, 1906; Müller 1927; Hecht and Malone 1972; Buffetaut 1978, 1982; unpublished notes of C. C. Mook in W. Langston's possession). In both *Gryposuchus* and *Eogavialis* the cranial profile is depressed and the facial-cranial transition appears intermediate between *Gavialis* and *Tomistoma* in lateral aspect; from above, however, the transition is more abrupt in *Gryposuchus* (fig. 8.8E). Orbits are smaller than the supratemporal fenestrae, and those of *Gryposuchus* are more rounded. The external nares are shorter and relatively wider in *Gryposuchus*. The plan of the skull roof is proportionally similar, but the lateral edge of the supratemporal arcade is more irregular and the supratemporal fenestrae are larger and much more angular than in *Eogavialis*. Posttemporal fenestrae open within a transverse sulcus in the posterior wall of the temporal fossa, and the foramen magnum may be wider than high in both taxa. Premaxillae make contact with the nasals, an anterolateral process of the postorbital overhangs the orbits posterodorsally, and the massive postorbital bar arises more dorsally from the jugal than in nongavials. The supraoccipital is similarly broad, forms a backward-projecting nuchal prominence, and was apparently not exposed in the skull roof. Basipterygoid tubera are gavial-like and the basicranial axis is non-"verticalized." In the palate, the palatine bones extend forward, wedgelike between the maxillae; the palatine-pterygoid suture is rounded posteriorly, the reverse of the *Gavialis gangeticus* condition. Pterygoid bullae are present, but are more gavial-like in shape in *Eogavialis*. The mandibular symphysis is transversely expanded at the second tooth, the surangular is slender, and the lateral mandibular fenestra is elongate and inclined downward anteriorly. The dentition is homodont, and occlusal pits, when present, are near the lateral bend of the ramus.

Differences between *Gryposuchus* and *Eogavialis* seem mostly related to piscivorous habits. Differences in such features as the shape of the supratemporal fenestrae, numbers and arrangement of teeth, relative lengths of rostrum and mandibular symphyses, and interdental pits, for example, are, we believe, relatively trivial matters in taxa having similar feeding adaptations and separated as far in time and space as the African and South American gavials. Other differences, such as the shapes of orbits, flaring of the anterior rami of the jugals, width of the interorbital space, and the narial structures, are perhaps more fundamental. Where differences between *Gryposuchus* and *Eogavialis* are noted, *Gryposuchus* appears closer to *Gavialis* than to *Eogavialis*, but a thorough analysis of possible *Eogavialis-Gryposuchus* relationships should await further study of well-preserved but incompletely described *Eogavialis* material in the Yale Peabody Museum.

Authors have long sought autapomorphies that would define the Gavialidae. The current debate about the possible relationships between *Gavialis* and *Tomistoma* shows that they have not yet succeeded in their endeavor. A noncritical compilation of characters claimed to be distinctive of gavials sensu stricto and of their distribution among the best known South American gavials, *Tomistoma schlegelii* and *Eogavialis*, is presented in table 8.5. Perusal of this table reveals, besides a strong *Gryposuchus–Gavialis gangeticus* resemblance, also complete agreement in available characters between *Gryposuchus colombianus* and *Gryposuchus neogaeus*. Of nineteen states observed in *Ikanogavialis*, fourteen agree with *G. colombianus*. Ten of seventeen available states are shared by species of *Eogavialis* and *Gryposuchus. Tomistoma*, however, agrees with the gavials in only thirteen of the twenty-six gavial states. This comparison is consistent with the assignment of the South American gavials and, probably, *Eogavialis* to the Gavialidae and exclusion of *Tomistoma* from this family on osteological grounds.

Time and space do not permit further analysis of possible relationships between the early gavials and *Tomistoma* and its allies, but we hope that the data presented will be useful in future investigations.

Although the IGM collection provides no new information about times, places, manner of arrival, or dispersal routes of gavials to South America, we cannot resist the temptation to speculate on their possible transoceanic dispersal. Buffetaut (1982) postulates a crossing of the South Atlantic by salt-tolerant gavials during the late Eocene or early Oligocene; however, physical conditions would likely have been more propitious for such a crossing earlier in the Paleogene when distances between Africa and South America were less and sea surface temperatures were higher than later on (Barron and Peterson 1991). Extrapolating from paleogeographic reconstructions by Smith et al. (1981) and Barron et al. (1981), we assume minimum intercontinental distances in the neighborhood of 1,300–6,800 km, depending on the points of departure and landfall. Ocean General Circulation models by Barron and Peterson (1991) indicate powerful surface currents in the order of 19 cm/sec running westwardly between the western bulge of Africa and the northeastern coast of South America during mid-Eocene times. Modeled streamlines suggest that landings anywhere on the northeastern and northern coasts of South America were possible. Under these circumstances, a passive crocodilian might have required about 4.5 months in the open sea to transit the narrowest passage. We find it difficult to imagine even a large stenohaline crocodilian (gavials today are strictly fresh-water animals) surviving such a journey. Taplin and Loverage (1988) found, for example, that small, unfed *Crocodylus niloticus,* normally a fresh-water species, lost body mass at a rate of 1.4% per day and suffered a marked elevation in plasma sodium chloride when exposed acutely to sea water, despite the presence of functioning lingual salt glands. Small individuals of the salt-water crocodile *Crocodylus porosus* showed parallel physiological responses when transferred acutely from fresh to salt water. Neither species appeared to compensate for water loss by drinking salt water. Experiments involving large individuals have not been conducted, but the viability of *C. porosus* in sea water is obvious, no doubt abetted by efficiently excreting lingual salt glands (see also Mazzotti and Dunson 1989), for further discussion of osmoregulation in crocodilians). *Gavialis gangeticus* possesses only minute lingual glands of very low secretory capacity (Taplin et al.

1985), and this is seen as evidence of an ancestral adaptation to salt water (Taplin and Grigg 1989). Presence of *Eogavialis* in deltaic and estuarine deposits, although not conclusive, also suggests such an adaptation.

As yet, Paleogene gavial remains are known in Africa only from the southeastern Mediterranean region. Paleoclimatic reconstructions of Africa in middle Eocene times (Parrish 1993a and references therein) show a "semiarid/semihumid" belt approximating the northern half of the present-day Sahara, extending across the continent. This zone of "moderately low rainfall" may have barred terrestrial dispersal of crocodilians northward or southward at this time, but salt-water-adapted species could presumably have made their way around the nose of Africa in coastal waters. South of the drier latitudes, Parrish's maps indicate more congenial environmental conditions southward for 30° or so, facing comparable regimes on the northeastern and northern coasts of South America.

Although rigors of a transatlantic crossing were probably less within the lower latitudes, other routes may have been open farther south via island hopping. Islands of unknown extent, rising from the western part of the Walvis Ridge, existed during the middle Eocene (Parrish 1993b and references therein). These islands are believed to have submerged by the end of the Eocene, however. Other islands were present on the crest of the Rio Grande Rise off the east coast of present-day Brazil in the Oligocene (Parrish 1993b). Although the existence of islands could have decreased the over-water distances traveled, we believe that a crossing of the middle South Atlantic is less likely to have occurred after the early Eocene because of colder water and less-favorable currents (see Barron and Peterson 1991; Haq 1981).

Once arrived on South American shores, the lineage may be expected to have readapted to fresh-water habitats as Buffetaut has proposed. Indeed, Taplin and Grigg (1989) report that the highly salt-tolerant *Crocodylus porosus* has no osmoregulatory difficulty upon entering fresh water. Moreover, *Crocodylus acutus,* which is regarded as an estuarine specialist in Florida (Mazzotti and Dunson 1988), is widely distributed in fresh-water rivers in South America (Medem 1981). Thus, we believe that Buffetaut's hypothesis best accords with the limited (and somewhat circumstantial) evidence available, but the crossing of the South

Atlantic probably occurred earlier than he suggested, and the immigrants were probably large animals.

An early dispersal would, of course, have allowed more time for diversification of the South American gavial clade evinced by *Gryposuchus* and *Ikanogavialis*. The seemingly *Gavialis*-like *Hesperogavialis* may indicate that more than one species of ancestral gavials made the transatlantic crossing.

Reconciliation of Current Knowledge with Langston's 1965 Study of Colombian Fossil Crocodilians

In 1965 Langston attempted to relate the La Venta crocodilians to adaptively comparable elements in the existing Neotropical fauna and, hence, to various habitats. The new Colombian material discussed above requires few changes in Langston's earlier inferences, which were then and still are imprecise and somewhat conjectural. The variety of crocodilians in the Honda Group seems exceptional when compared with the geographic distribution of the existing Neotropical crocodilian fauna whose niches are generally restricted to longitudinal segments of river systems (see Medem 1981, 1983). Although as many as five taxa are present in, for example, the Orinoco or Amazon basins, it is unusual for more than three taxa to frequent the same stretch of river or a tributary. The variety of fossil taxa in the La Venta area may reflect real distributions, or they may constitute a sampling of populations living within the "Protomagdalena" drainage system, accumulated as cadavers or as still-living individuals displaced by periodic flooding.

The putative "dry-land" crocodile *Sebecus huilensis* is uncommon in the Honda sediments, but fragmentary remains are distributed from just below the Chunchullo Sandstone Beds upward to just above the base of the El Cardón Red Beds in the Villavieja Formation. All specimens are from mudstone and claystone with the exception of TMM 41658-8, which is from a gray sandstone within the Monkey Beds (Baraya Member, Villavieja Formation). The supposedly quiet, pond-dwelling *Mourasuchus atopus*, like *Sebecus* without an existing analogue, is also scarce, but four of the six known examples included some associated postcranial material. This suggests that they may have been buried closer to their native habitats than the *Sebecus* speci-

mens. Stratigraphically, *Mourasuchus* ranges from the base of the section in the Villavieja area upward to the El Cardón Red Beds at the base of the Cerro Colorado Member of the Villavieja Formation (a mandibular fragment from UCMP Locality V4523 comes from about 12 km northeast of Villavieja).

Gryposuchus colombianus remains are more abundant, although only one find includes postcranial elements, and half of the examples are fragments of rostra; indeed, isolated mandibles are the most frequent indicator of this taxon. *G. colombianus* is recognized from at least seven localities, ranging from just below the Chunchullo Sandstone Beds upward to just above the Ferrugenous Beds, in the lower part of the Villavieja Formation. Most specimens are from claystone or from sandstones within thicker claystone packages. Only three specimens of *Charactosuchus fieldsi,* all from the same locality in the Monkey Beds, are known. Langston (1965) assigned *Charactosuchus* to the analogous longirostrine piscivorous niche of *Crocodylus intermedius* in swift rivers. Medem (1983) points out, however, that *C. intermedius* is more at home in deep, low-energy situations, migrating twice a year to remain in this congenial habitat.

The predominant association of crocodilian remains with mudstone and claystone in the Honda sediments was noted by Langston (1965). The IGM collection conforms to this pattern. Only caiman remains seem to occur with any regularity in sandy deposits. Crocodilians seem to be most abundant in the Monkey Beds at the base of the Villavieja Formation with all known La Venta taxa except *Balanerodus logimus* and *Caiman latirostris* (=*Caiman* cf. *C. lutescens* of Langston 1965; see Gasparini and Báez 1975) having been recognized in this unit. A cursory inspection of the stratigraphic sequence in the La Venta area gives the impression that, except for *Caiman latirostris* and *Sebecus* (teeth), crocodilians disappear above the Ferrugenous Beds. Thus, there may be a correlation between the disappearance of crocodilians and the deterioration of suitable habitat caused by drying out of the region and increased seasonality. Both phenomena are evinced by the development of red beds and a sharp increase in the amount of coarse, cyclical clastics toward the upper part of the Villavieja and the overlying Nieva formations.

Additional genera and species of existing South American crocodilian taxa have come to light as fossils in recent years. Currently known are *Melanosuchus* (late Miocene of Venezuela, Medina 1976), *Caiman latirostris* and *Caiman* cf. *C. jacare* (late Miocene and Pliocene of Argentina; Patterson 1936; Gasparini 1981), and *Caiman niteroiensis* (late Miocene–early Pliocene of Brazil, Souza Filho and Bocquentin 1991). Should future discoveries of *Caiman* cf. *C. lutescens* in the middle Miocene of Colombia confirm its attribution to *C. latirostris* (see Gasparini and Báez 1975; Gasparini 1981), it will be the oldest example of an existing species of crocodilian in the fossil record.

Summary

Joint expeditions by INGEOMINAS (IGM) and Duke University paleontologists and geologists in 1985–90 obtained additional fossil crocodilian remains from middle Miocene rocks in the Tatacoa Desert northeast of Villavieja (Huila).

Some of the localities visited had been discovered by University of California (Berkeley) field parties in the 1940s and 1950s, and the fossil crocodilians collected there were treated monographically by Langston in 1965.

Building on previous knowledge, the IGM-Duke collections provide additional data. *Sebecosuchus huilensis* Langston had a higher and narrower premaxillary rostrum than *Sebecosuchus icaeorhinus* from Argentina; its fourth premaxillary tooth is relatively larger; a large premaxilla-maxilla diastema without indentation or perforation is present; external nares are not bounded by a smooth depression as in *S. icaeorhinus;* the internarial bridge is complete, formed of nasals and premaxillae. The caudal vertebrae are amphyplatyan or amphycoelous.

An incomplete mandible of *Charactosuchus fieldsi* Langston adds no new morphological information. The alleged presence of *Charactosuchus* on the island of Jamaica (Berg 1969) and in Florida (Webb and Tessman 1967) is discounted. A purported Brazilian occurrence (*C. sansoai* Souza Filho 1991) is inadequately known.

Mourasuchus atopus (Langston) is found to have existed in the field area for at least 1.5 million years. It is not known if this species possessed the massive bony tumescence seen in skulls of *Mourasuchus nativus* and *M. arendsi,* but not in *M. amazonensis. Mourasuchus* attained substantial size with some skulls (from Venezuela) reaching a length of about 1.25 m.

Balanarodus logimus Langston is now known from both the Deseadan and Friasian, only in Colombia. Its bulbous dental crowns are distinctive, but the animal remains enigmatic.

Other, limited alligatorid remains in the IGM collection are not dealt with in this chapter.

Gryposuchus colombianus (Langston) is represented by excellent cranial material. One skull, greater than 1.0 m long, shares a long list of synapomorphies with the Gavialidae and displays a number of similarities to the late Eocene–early Oligocene *Eogavialis* of Africa. Resemblances to *Tomistoma schlegelii* are less extensive. Formerly believed to be of late Oligocene age, *G. colombianus* now appears to be restricted to the middle Miocene (Friasian), although some isolated teeth in the ?Deseadan Chaparral fauna of Stirton may indicate the presence of gavials in the early Oligocene. The skull of *Gryposuchus* is depressed, nonverticalized sensu Tarsitano and of less angular outline than *Gavialis gangeticus.* In *Gryposuchus,* at least one nasal bone (the left in *G. colombianus,* the right in *G. jessei,* and both in *G. neogaeus*) reaches the premaxilla. The teeth, although numerous, are fewer than in *G. gangeticus.* There are only four teeth in the premaxilla, twenty-one or twenty-two in the maxilla, and twenty-three in the dentary. The nares are wider than long and lack the peculiar bony excrescences and intranarial shelf present in large male *G. gangeticus.* A postorbital spina is lacking, but large elongate pterygoid bullae are present. The bony arches bounding the supratemporal fenestrae are narrow. The retroarticular process ends in a distinctive hooklike expansion unknown in other longirostral crocodilians.

Gryposuchus differs notably from *Ikanogavialis* in its relatively shorter rostrum, lack of outstanding dental salients, and a more gavial-like outline. It differs most importantly from *Hesperogavialis* in the junction between nasals and premaxillae. In all three taxa the surangular ends anterodorsally at or behind the last mandibular tooth, thus differing from other gavials and tomistomas in which the bone sends a narrow process forward between the dentary and the splenial at the posterior end of the lower tooth row.

Of taphonomic interest is that both IGM crania were buried in normal position after some exposure to air or water. When the peridontal ligaments decayed, most of the upper teeth fell out while the lower teeth remained in place;

however, in one of the mandibles the teeth slid downward into their alveoli and were preserved with only part of the crowns emergent.

As Buffetaut (1982) has suggested, the limited and somewhat circumstantial evidence available suggests that gavials reached South America from Africa by crossing the South Atlantic. Paleoceanographic reconstructions, physiological (osmoregulatory) factors seen in existing crocodilians, and the presence of gavials in South America possibly as far back as the early Oligocene suggest that the most propitious time for the gavial dispersal was in the mid-Eocene or earlier.

Correlation between crocodilian remains and stratal types in the Honda Group and geographic distribution of taxa in the Tatacoa badlands suggests mixing of cadavers and living animals from different habitats carried out of their normal ranges by high water. Frequency of occurrence is consistent with the idea that *Sebecus* and *Balanarodus* were allochthonous elements in the La Venta fauna. Only caimanlike forms appear to be associated with coarser clastics with any degree of regularity.

Crocodilian fossils in the Honda Group are most abundant in the Monkey and Fish bed units and diminish upward in the section as red beds and coarser clastic sediments increase in importance. This may be evidence of local extinctions in response to drying out of the environment and increasing seasonality.

If, as suggested by Gasparini (1981), *Caiman* cf. *C. lutescens* of Langston (1965) is actually *Caiman latirostris,* its presence in the Friasian is the oldest occurrence of an existing crocodilian species in the fossil record.

Acknowledgments

We thank the following individuals for their assistance and counsel during this study: M. C. McKenna and J. M. Clark provided helpful comments and review of parts of the chapter. We benefited from our discussions about South American crocodilians with W. D. Sill. J. Peterson generously shared her data on length and mass in *Alligator mississippiensis,* and C. Brochu offered useful comments on a number of points. P. Wellnhofer extended many courtesies to the senior author during his visits to Munich, which he appreciates.

We are grateful to the following colleagues who provided access to specimens in their respective collections: C. Claude, Zoologisches Museum des Universität Zürich; W. Brinkmann, Paläontologisches Institut und Museum der Universität Zürich; and U. Gruber, Zoologische Staatssammlung München. Many data on *Gavialis gangeticus* and *Tomistoma schlegelii* were obtained during visits over many years to the American Museum of Natural History and the National Museum of Natural History by the senior author.

Illustrations are by M. Crisp, D. Tischler, and the senior author, and photographs are by D. Stevens and J. Jaworski. The senior author gratefully acknowledges financial assistance from the University of Texas Geology Foundation for preparation of illustrations.

Appendix 8.1. Cranial character states identified by various authors as distinctively gavialid

See table 8.5.
A. Rostrum—facial transition, sharply defined
B. Premaxillary teeth—five
C. Maxillary and dentary teeth—greatly increased in number (function of long snout), subhomodont
D. Orbits—circular
E. Orbital rim—elevated, jugal flared and everted anteriorly
F. Skull table—wide
G. Supratemporal fenestrae—circular
H. Interorbital space—wide
I. Postorbital bar (pillar)—broad, arises from dorsal side of jugal
J. Postorbital spina—present[1]
K. Subtemporal ramus of jugal—depressed in relation to orbital ramus, quadrangular section
L. Supratemporal fossa—contains distinct slitlike sulcus in posterior wall (confluent with posttemporal fenestra)
M. Supraoccipital—protrudes posteriorly
N. Ectopterygoid—with pronounced posterior ramus
O. Basioccipital—with wide pendulous tubera
P. Axis of jaw articulation—inclined caudomedially
Q. Palatines—extend to a point far in advance of palatal fenestrae
R. Quadratojugal—forms posterodorsal angle of lateral temporal fenestra
S. Quadratojugal—with anteroventral ramus overlapping dorsomedial surface of the jugal

T. Postorbital—with posterodorsal ramus usually meeting anterodorsal process of quadratojugal and isolating the quadrate from the edge of the lateral temporal fenestra

U. Pterygoid bullae—(in adult males), not contacting prefrontal pillar of the palatines anteriorly

V. Prootic—"broadly" exposed on side of braincase

W. Foramen intermandibularis oralis—present in splenial symphysis[2]

X. Subtemporal ramus of jugal—unsculptured

Y. Supratemporal fossae—larger than orbits

Z. Basicranium not "verticalized" sensu Tarsitano 1985

Notes:

1. Several authors as long ago as Kälin (1931) have mentioned a small spina on the postorbital bone in *Gavialis gangeticus.* It is clear from Kälin's language and illustrations (for example, figure 4 in Kälin 1933, erroneously identified as *Tomistoma schlegelii* in the caption) that the spina occurs on the anterior side of the ventral branch of the postorbital bone. This feature appears to be unique in *G. gangeticus.* Authors, for example, Aoki (1976 [1977?], fig. 2) and possibly Norell (1989, fig. 2) among others seem to confuse the prominent crista that crosses the postorbital bar in many crocodilians at the level of the postoribtal-jugal junction with the little spina of Kälin that occurs higher up on the postorbital bar. Absent further clarification, references to this spine in *Tomistoma* and *Alligator,* for example, are ambiguous. Whether the small "bump" on the anterior edge of the postorbital bar of *Bernissartia fagesii* reported by Norell and Clark (1990) is homologous with the spina in *G. gangeticus* is problematical.

2. Norell (1989, 331) states that in *Gavialis gangeticus* the foramen intermandibularis oralis perforates the splenial within the symphyseal surface in the "identical alligatorid position" and that the splenials in crocodylids, "including the disarticulated rami of *Tomistoma,*" lack this foramen. A very small *Tomistoma* (TT M4720) with a splenial bone 66 mm long, however, has such a perforation almost exactly as in a small *Gavialis* jaw (TMM M5487) with a splenial 137 mm long. See the text for additional discussion of this character.

9
Turtles

Roger Conant Wood

RESUMEN

Los Chelonia fósiles del Grupo Honda son abundantes y en ciertos casos, notablemente bien preservados. De las tres familias registradas para la fauna, Chelidae acuáticos y Pelomedusidae son los más comunes, mientras restos de Testudiniidae (tortugas terrestres) son relativamente raros. Los restos más abundantes son los Pelomedusidae del género *Podocnemis*. Entre el nuevo material fósil figura un caparazón adulto, un cráneo, y material de juveniles y recién nacidos. Las especies de *Podocnemis* comparten un caparazón caracterizado por un aplastamiento moderado de arriba abajo, siete huesos neurales (el primero de los cuales es ovoide y elongado), y mesoplastra hexagonales hasta ovoides ubicados lateralmente en el plastron. Dos nuevas especies, *Podocnemis pritchardi* y *P. medemi,* son descritas. La evidencia disponible sugiere que *Podocnemis pritchardi* está relacionado a *Podocnemis lewyana*. *Podocnemis medemi* es más grande que todas las especies vivas. Adicionalmente, se describe un caparazón indistinguible de la especie actual *P. expansa.* Un caparazón de cría es referido al género *Podocnemis* en base a la ausencia completa de fontanelas.

Un espécimen de tortuga terrestre gigante, preservando el plastron y parte del carapacho, gran parte de las cinturas pélvica y pectoral, y huesos de los miembros, es asignado a *Geochelone* (Testudiniidae). Este ejemplar es más grande que las tres especies que se encuentran en la actualidad distribuidas ampliamente en la parte norte sudamericana. Hay al menos cinco o hasta seis tipos diferentes de Chelonia en la fauna de La Venta, la fauna fósil de tortugas más diversa que se conoce hasta ahora para América del Sur. En los llanos de la Orinoquia venezolana se encuentran cuatro especies de Pelomedusidae y dos especies de *Geochelone,* así que éste sería el mejor modelo para las condiciones ambientales en el área de La Venta durante el Mioceno medio.

Fossil turtle remains from Miocene sediments of the upper Magdalena River valley, Colombia, are moderately abundant and, in some cases, remarkably well preserved. Three different families of turtles are found here; aquatic chelids and pelomedusids are common, whereas remains of testudinids (land tortoises) are relatively rare.

Representatives of two of these families—the chelid *Chelus colombianus* (Wood 1976b) and the testudinid *Geochelone hesterna* (Auffenberg 1971)—have already been described. Most abundant of all the Colombian fossil turtles, but until now formally undocumented, are the pelomedusids. These, as well as an intriguing specimen of a giant tortoise whose existence has not previously been reported, are described here.

The fossils that form the subject of this chapter are housed either in the University of California Museum of Paleontology (UCMP) or the Instituto de Investigaciones en Geociencias, Minería y Química, Museo Geológico

Bogotá (IGM). Recent skeletal material used for comparative purposes was examined at the Instituto Roberto Franco, Villavicencio, Colombia (IRF), the herpetology department of the National Museum of Natural History Smithsonian Institution (USNM), and in the collection of P. C. H. Pritchard (PCHP).

Systematic Paleontology

Order Testudines
Suborder Pleurodira
Family Pelomedusidae
Podocnemis

Podocnemis is a genus to which a large number of fossil turtles have been referred, based on a commonly shared typical shell morphology, generally characterized by (1) moderate dorsoventral flattening; (2) seven neural bones, the first of which is oval and elongated, whereas successive ones are coffin-shaped except for the last, which is pentagonal; and (3) the presence of laterally placed, hexagonal to oval mesoplastra on the plastron. As has been noted elsewhere (Wood and Gamero 1971), when better shell material or additional parts of the skeleton other than the shell become known, fossil taxa originally referred to *Podocnemis* often prove to be clearly separable from that genus. Such was the case, for example, with both *"Podocnemis" alabamae* and *"Podocnemis" barberi* from North America (now *Bothremys;* Gaffney and Zangerl 1968), *"Podocnemis" congolensis* (now *Taphrosphys;* Wood 1975) and *"Podocnemis" antiqua* (now *Shweboemys;* Wood 1970), both from Africa, and, most recently, *"Podocnemis" venezuelensis* from northern South America, which, on the basis of a quarry sample collected subsequent to its original description, is now undoubtedly referable to a new genus. It is under study by the author.

Hence, the attribution of the pelomedusids described in this chapter to *Podocnemis*, although quite possibly correct, is provisional and may be subject to revision, especially if associated skull material becomes available. Although pelomedusid remains are rather abundant in the Miocene deposits of the upper Magdalena valley (see apps. 9.1–9.3), many of these also are quite fragmentary and, therefore,

difficult to attribute unequivocally to a particular taxon. Although it is possible that all the pelomedusids described here may eventually prove to be genuine representatives of *Podocnemis,* it is by no means certain at present. A definitive resolution of this question must await the recovery of more and better material.

Podocnemis pritchardi sp. nov.
(fig. 9.1)

TYPE

UCMP 63782, a nearly complete and well-preserved shell, lacking only a small portion of the posterior margin of the carapace.

HYPODIGM

The type and UCMP 40359, the anterior half of a plastron; IGM 1933, a nearly complete right xiphiplastron; IGM 1784, a partial right xiphiplastron; IGM 1866 (part), fragment of a right xiphiplastron; and IGM 1777 (part), a right xiphiplastral tip.

TYPE LOCALITY AND HORIZON

UCMP Locality V4531, La Victoria Formation, between the Tatacoa Sandstone Beds and Chunchullo Sandstone Beds (chap. 2). All fossil pelomedusids from Colombia mentioned in this chapter are from the Honda Group, La Victoria and Villavieja formations (sensu Guerrero, chap. 2), upper Magdalena River valley, Colombia (see Wood 1976b for a more extended discussion).

DIAGNOSIS

Differs from all other South American pelomedusids referred to *Podocnemis* in having very narrow, almost rectangular, laterally placed mesoplastra; outer surfaces of the bridges curving downward and outward; extreme dorsoventral flattening of the shell; only six rather than seven neural bones; ischial scar on xiphiplastron very narrow along its entire length and situated directly adjacent to anal notch.

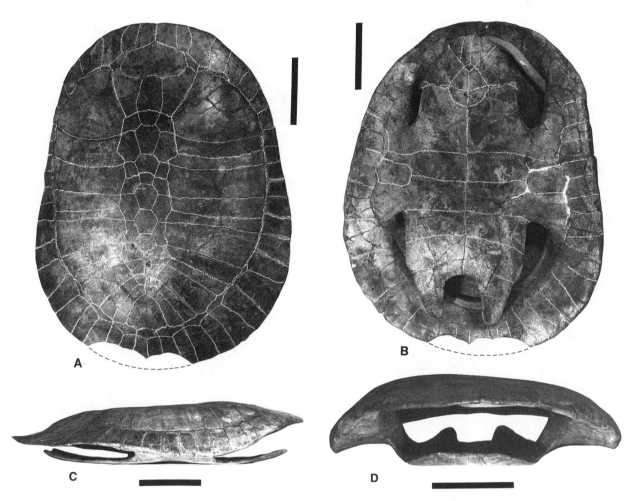

Figure 9.1. Carapace of *Podocnemis pritchardi* sp. nov., type, UCMP 63782. *A*, dorsal view; *B*, ventral view; *C*, lateral view; *D*, anterior view. *A–D*. Scale bars 15 cm.

ETYMOLOGY

Named in honor of my esteemed friend and colleague, Peter C. H. Pritchard, in recognition of his numerous contributions to knowledge of the world's turtles.

DESCRIPTION

The type specimen represents a rather large individual (see table 9.1 for measurements and Royo y Gómez 1946 for photographs of it in the field) in an excellent state of preservation. There is little evidence of crushing, compaction, or distortion of the bone, so the shell's dorsoventral flatness is unquestionably a real character representative of the speci-

men in life rather than an artifact of poor preservation. All bone and scute sutures can be easily traced.

The carapace is oval in outline, attaining its greatest breadth just behind the inguinal notches. Outer portions of the tenth left peripheral, both eleventh peripherals, and the pygal are missing. Otherwise, the posterior peripherals form a smoothly rounded, unserrated rim. The borders of the damaged bones are smooth rather than freshly broken, suggesting that this missing bone represents an injury that healed during the turtle's life. Unlike other species of *Podocnemis*, which almost invariably have seven neural bones extending rearward from the nuchal in uninterrupted sequence, *P. pritchardi* has only six. (The only exceptions known to me among living species of *Podocnemis* are several

Table 9.1. Shell measurements for the type of *Podocnemis*
pritchardi **(UCMP 63782)**

Variable	Measure (cm)
Carapace, midline length:	
As preserved (straight-line distance)	72.9
With posterior margin restored (estimated)	~77.0
Carapace, greatest width (at level of posterior ends of 7th peripheral bones)	60.7
Plastron, midline	51.9
Anal notch:	
Midline depth	11.5
Maximum width (distance between left and right xiphiplastral tips)	9.7
Anterior plastral lobe:	
Width at axial notches	28.1
Width at epiplastral-hyoplastral sutures	22.2
Midline length	~20.0
Posterior plastral lobe:	
Width at inguinal notches	28.2
Midline length (including anal notch)	~24.0
Bridge, anteroposterior minimum length:[a]	
Left side	19.8
Right side	18.8

[a]A plaster-filled crack on the left side largely accounts for the difference in length between the left and right bridges.

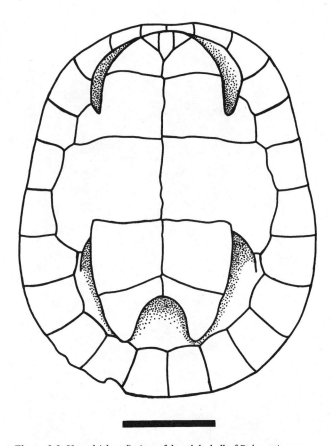

Figure 9.2. Ventral (plastral) view of the adult shell of *Podocnemis erythrocephala*, PCHP 1194. Scale bar 15 cm.

female specimens of *P. lewyana* [IRF 586, IRF 816, and, possibly, also IRF 364], plus an immature specimen of *P. unifilis,* illustrated in Pritchard and Trebbau 1984, all with only six neurals. The *P. unifilis* carapace also has a supernumerary bone intervening between the sixth and seventh left pleurals.) The shape and arrangement, however, is typical of the genus. The first neural is elongate and roughly rectangular. Numbers two through five are hexagonal, being coffin-shaped. The sixth neural is pentagonal. Part of the sixth and all of the seventh and eighth pairs of pleurals meet in the midline behind the neural series. The clearly demarcated scute sulci exhibit no remarkable features.

Superficially, the plastron is somewhat reminiscent of the living *P. erythrocephala* from the Río Negro basin (Mittermeier and Wilson 1974; see also fig. 9.2), only on a gigantic scale. The anal notch is very deeply excavated into a U-shape. The lateral margins of both the anterior and posterior plastral lobes are essentially parallel to the midline axis

of the shell, the maximum width of the posterior plastral lobe being significantly greater than that of the anterior lobe. The entoplastron does not extend rearward of the axillary notches. Uniquely among the pelomedusids, the laterally placed mesoplastra are very narrow and almost rectangular. The pectoral-abdominal scute sulcus crosses the plastron well forward of the anteriormost limits of the mesoplastra. The plastral formula is pectoral > abdominal > femoral > intergular > humeral > anal. The midline length of the anterior plastral lobe is slightly greater, whereas that of the posterior lobe (including the anal notch) is considerably greater, than the anteroposterior length of the bridge. Both the anterior and posterior lobes are slightly upturned at their front and back ends, respectively.

Ischial scars on the visceral surface of the xiphiplastra are distinctive in both shape and position (see fig. 9.4). The attachment of the ischium to the plastron in *P. pritchardi* appears to be weaker than in other species of *Podocnemis.*

The ischial scar is very narrow and only slightly widened at its outermost extent. Moreover, it is situated very close to the border (within approximately 0.5 cm) of the anal notch.

Another unusual feature of this species is the shape of the bridges connecting carapace to plastron. In all other pelomedusids, the outer surfaces of the bridge are essentially flat and angle upward and outward. But in *P. pritchardi* the outer surfaces of the bridge are strongly curved so that the later-almost extremities actually slope downward (fig. 9.1D). This characteristic is clearly not an artifact of preservation for several reasons: (1) the rather weakly developed axillary and inguinal buttresses are perfectly intact, not telescoped downward as they would be if the shell had been dorso-ventrally compressed; (2) none of the sutural contacts between the adjacent bones of the bridge is significantly displaced on either side; and, finally, (3) the curvature of the left and right bridges is essentially symmetrical.

Although not necessarily characteristics differentiating it from all other species of *Podocnemis* and, therefore, not included as part of the formal diagnosis, a number of features of *P. pritchardi* clearly distinguish it from its contemporary, *P. medemi* (see below). These include the deep, U-shaped anal notch; parallel (or nearly so) sides of both the anterior and posterior plastral lobes; marked constriction of the posterior lobe where the sulcus marking the junction between the femoral and anal sucles reaches the outer margin of the plastron; the narrower width of the anterior plastral lobe at the axillary notches compared to the somewhat greater width of the posterior plastral lobe at the inguinal notches, and an entoplastron whose posteriormost extent does not reach a point even with or rearward of the axillary notches.

The sex represented by the type specimen is problematical. In all the living species of *Podocnemis,* adult females attain a considerably larger size than do males (Pritchard 1979). The type of *P. pritchardi* is only slightly shorter in carapace length (by 5 cm) than the record-sized specimen (a female) of *P. expansa,* the largest of the living species of *Podocnemis.* On the basis of mere size, therefore, one might infer that the type of *P. pritchardi* represents a female. Other evidence suggests the contrary, however. A sample of six male shells of *P. lewyana,* examined in the IRF collection, all exhibit distinctive, very deep, broad, U-shaped anal notches very similar in shape and proportions to that of *P. pritchardi.* In a similar-sized sample of female *P. lewyana,* the anal notch is proportionately less wide, less deep, and V-shaped. Given the consistent sexual differences within this particular living species, if one were to ignore size as a factor one might reasonably presume that the type of *P. pritchardi* represents a male. In view of these contradictory kinds of evidence, the question of this specimen's sex cannot be clearly resolved.

Podocnemis medemi sp. nov.
(fig. 9.3)

TYPE

IGM 1863, a nearly complete plastron with an associated anterior half of the carapace.

HYPODIGM

The type and UCMP 125524, a complete right hyoplastron; IGM 1808, an essentially complete left hyoplastron; IGM 1922, a right xiphiplastron; and IGM 2315, a nearly complete left xiphiplastron.

TYPE LOCALITY AND HORIZON

Cuenca de Melgar, Carmen de Apicalá, Honda Group, upper Magdalena River Valley, Colombia.

DIAGNOSIS

Differs from all other pelomedusids referred to *Podocnemis* in having relatively short, rounded anterior and posterior plastral lobes; extremely narrow, V-shaped anal notch; proportionately small, squarish first pair of marginal scutes; enormous size (estimated midline carapace length = 101.5 cm); and position and shape of ischial scar on xiphiplastron (triangular in shape, expanded laterally, and moderately withdrawn from margin of anal notch).

ETYMOLOGY

Named in honor of the distinguished Colombian herpetologist, the late Federico Medem M., in recognition of his outstanding contributions to knowledge of living Colombian turtles.

Figure 9.3. *Podocnemis medemi* sp. nov., type, IGM 1863. *A.* Dorsal view of anterior half of the carapace. *B.* Ventral (plastral) view of nearly complete plastron. Scale bars 15 cm.

DESCRIPTION

The type specimen is represented by a nearly intact plastron, lacking only portions of both xiphiplastra, and somewhat fragmented anterior half of the carapace of a very large individual (see table 9.2 for measurements). All bone and scute sutures are readily discernible.

What remains of the carapace gives the impression that it was more rounded than that of any of the living species of *Podocnemis.* Although largely subjective and necessarily based on fragmentary material, this impression is nevertheless reinforced by several comparable length: width ratios from recent specimens of *P. expansa* (table 9.2).

The extrapolated carapace length of *P. medemi* (table 9.2) is significantly greater (by roughly 20 cm) than that of the largest known recent specimen of *Podocnemis* (a female *P. expansa* measuring 82 cm from front to back in a straight line; Williams 1954). When compared to the mean length of a large sample of adult *P. expansa* (66 cm; P. C. H. Pritchard, pers. comm.) rather than the record, moreover, the difference in size is even more impressive. Because males of all living species of *Podocnemis* are considerably smaller than females, one might reasonably suppose that the enormous size of the type of *P. medemi* indicates that it was a female. It also appears, however, that in at least one recent species of *Podocnemis* (*P. expansa*) males have a more nearly circular carapace than females. In view of its subcircular proportions, therefore, the type of *P. medemi* could instead be interpreted as representing a male. Given the seemingly contradictory nature of the available evidence, the question of the type specimen's sex must remain moot at present.

Aside from overall size, the only other notable feature of that part of *P. medemi*'s carapace which has been preserved is the size and shape of the first pair of marginal scutes, which are rectangular rather than the typical trapezoidal shape and also are proportionately small.

More distinctive than anything else about *P. medemi* are the proportions of its plastron. Both the anterior and posterior lobes are rounded, the midline lengths of both being considerably shorter than the anteroposterior lengths of the bridges. The circularity of the plastron further strengthens the impression that the carapace of this species was likely to have been nearly as broad as long. Contrasting markedly with *P. pritchardi,* the anal notch of *P. medemi* is extremely narrow and V-shaped (fig. 9.4). Moreover, the outer margins of the xiphiplastra are smoothly curved rather than markedly indented as in *P. pritchardi*.

Both the shape and position of the pelvic scars on the visceral surface of the xiphiplastron are comparable to the arrangement typically found in *Podocnemis,* unlike what is seen in *P. pritchardi*. The ischium was solidly fused to the plastron as evidenced by the triangular ischial scar, which is narrow medially and markedly broadens laterally. Further,

Table 9.2. Shell measurements (cm), type of *Podocnemis medemi* (IGM 1863) compared with two recent adult specimens of *Podocnemis expansa* from Bolivia (PCHP 1509 and 1510)

Variable Measured	*P. medemi,* IGM 1863	*P. expansa*[a] PCHP 1510[a]	PCHP 1509[a]
Carapace:			
Midline length, as preserved (to back of 4th neural)	53.4	38.0	31.7
Midline length, total	101.5[b]	71.2	60.7
Width from midline to posterior end of 6th peripheral, left side	46.2	28.0	24.8
Width : length ratio	1:2.2	1:2.5	1:2.5
2d neural, midline length	8.2	6.3	4.6
1st left marginal scute:			
Greatest anteroposterior length	6.1	6.2	6.0
Greatest width	7.5	7.2	6.7
Plastron, midline length	76.5	—	—
Plastral lobe widths:			
Anterior lobe, at axillary notch	~41.0[c]	—	—
Posterior lobe, at inguinal notch	43.0	—	—

[a]Only the carapace has been preserved in these specimens.

[b]This is an estimate based on the assumption that the midline proportions of the carapaces of *P. medemi* and examples of *P. expansa* are roughly comparable.

[c]Because of an anteroposterior crack in the plastron, its width when the turtle was alive would actually have been approximately 0.75 cm less.

the scar itself is not situated right at the edge of the anal notch, as in *P. pritchardi,* but instead sits between 1.0 and 2.0 cm away from its rim.

In other respects as well, the plastron of *P. medemi* exhibits typically podocnemine characters. The laterally placed mesoplastra are hexagonal and roughly as broad as they are long. The pectoral-abdominal scute sulcus traverses the plastron in front of the mesoplastra. Because of damage along the midline of the posterior plastral lobe it is not possible to determine the entire plastral formula. Nevertheless, it is clear that the abdominal midline sulcus is the longest, as it is generally in living species of *Podocnemis,* unlike *P. pritchardi.*

Podocnemis cf. *P. expansa*

Despite the relative abundance of shell material, fossil turtle skulls are rarely preserved in the Miocene sediments of Colombia. Only three are known, none of which is complete. Two of these have previously been referred to the fossil tortoise *Geochelone hesterna* (Auffenberg 1971). The only other skull so far collected from La Venta (IGM

182911, Duke Locality 11, between the Tatacoa Sandstone Beds and Cerbatana Conglomerate Beds, La Victoria Formation; fig. 9.5) is described here for the first time.

The skull is rather large. Aside from some minor dorso-ventral compaction in the cranial region, it is essentially undistorted. Taking into consideration the missing bone along the midline, the snout/condyle length of the skull originally would have been between 12.5 and 13 cm. The width of the skull at its widest part (measured from the midline to the outer rim of the tympanic annulus on the right side) is 5.2 cm; the maximum width of the skull when intact would, therefore, have been between 10.0 and 10.5 cm.

Missing portions of the skull include, on the left side, the snout (premaxilla and prefrontal), the lateral terminus of the outwardly flaring ectopterygoid process, the distal elements of the ear region (quadrate, quadratojugal, and squamosal), and much of the skull roof. Just enough more of the skull roof on the right side has been preserved so that part of its posterior margin can be discerned. This indicates moderate, but not extensive, emargination at the back end of the skull roof. Most of the ear region has been preserved on the

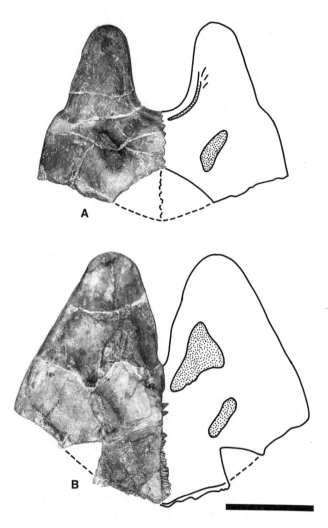

Figure 9.4. Xiphiplastra. *A. Podocnemis pritchardi,* IGM 1933. *B. Podocnemis medemi,* IGM 1922. Mirror image outlines of each of these bones emphasize the difference in anal notch shape. Note also the difference in both shape and position of the ischial scars (represented by *stippled shapes*) on the two specimens. *A–B.* Scale bar 15 cm.

way that the skull has become disarticulated, with its roof separated from the rest of the specimen, a considerable portion of the internal surface of the cranium is readily visible.

As evidenced by the prominently projecting ectopterygoid processes, enlarged carotid canals, and extensive roofing, the skull clearly represents a member of the family Pelomedusidae. Insofar as it has been preserved, in fact, the skull is indistinguishable from the modern species *Podocnemis expansa,* the largest of the six living representatives of this widespread and endemic South American genus. It is conceivable that this skull may be referable to either *Podocnemis medemi* or *P. pritchardi,* but owing to its lack of association with any shell material, there is no basis for confirming one or the other of these alternatives. It is also possible that this skull may represent a third different taxon. If so, however, there is not enough material yet for an adequate diagnosis of it.

Podocnemis, sp. indet.
Hatchling and juvenile shell material

Most remarkable, perhaps, of all the fossil turtles from the Miocene of Colombia is a nearly complete shell (IGM 183823, Duke Locality 67, between the Cerro Gordo Sandstone Beds and Chunchullo Sandstone Beds, La Victoria Formation) of a pelomedusid hatchling. Hatchling shells typically are extremely fragile. The various bones of which a hatchling shell is composed are very thin, only partially ossified, and weakly ankylosed with their neighbors. In view of these circumstances, the shell of a dead hatchling normally disintegrates almost instantaneously in nature (if it has not already been swallowed whole by a predator). To find a fossil hatchling shell so beautifully preserved as the one here described is a very rare circumstance indeed.

A small portion of the anterior rim of the carapace is missing, as are all but the anteriormost wedge of the suprapygal, the entire pygal, adjacent peripheral bones forming the posterior rim, and the lateral ends of the seventh and eighth pleurals from both sides of the shell. Most of the anterior plastral lobe has been lost (both epiplastra, the entoplastron, and the anterior end of the right hyoplastron),

right side of the skull, but all of the palate and snout on this side are missing. Neither the knob of the occipital condyle nor the posterior tip of the supraoccipital crest have been preserved.

Several features of the skull deserve special mention. A pair of moderately well developed, parallel ridges are present on the triturating surface of the maxilla; these clearly extended forward onto the premaxilla. On the roof of the skull, there was a large, elongate, diamond-shaped interparietal scale extending along the midline. The remainder of the scale pattern cannot be determined. Because of the

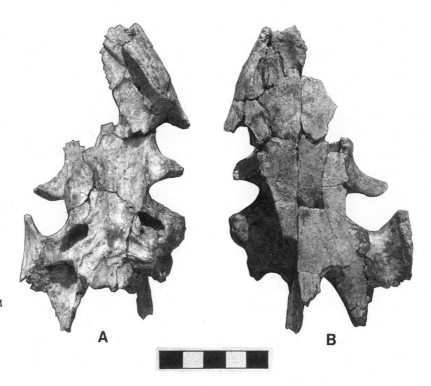

Figure 9.5. *Podocnemis* sp., partial skull, IGM 182911, from Duke Locality 11 between the Tatacoa Sandstone Beds and the Cerbatana Conglomerate Beds, La Victoria Formation. *A,* ventral view; *B,* dorsal view. Scale bar 5 cm.

and the posterior tip of the right xiphiplastron and much of the left xiphiplastron are also lacking.

As preserved, the midline length of the carapace is 8.1 cm. The estimated length of the complete carapace would range between 9.5 and 10 cm. The maximum width of the shell (in the vicinity of the inguinal notches) is 7.7 cm. Some dorsoventral flattening of the shell has occurred owing to breakage in the bridge region on both sides, resulting in a moderate upward displacement of the plastron relative to its original position. Damage to the bridge region also makes it impossible to discern the outlines of the mesoplastral bones. The carapace is nonetheless broadly and symmetrically arched from side to side and also along the antero-posterior axis, showing little evidence of distortion and probably still closely approximating its original shape.

Seven neural bones are present along the midline of the carapace, arranged in a continuous row. The first of these is elongate and roughly rectangular in outline; it abuts directly against the posterior margin of the nuchal bone. Neurals two through six are all hexagonal, becoming relatively shorter and broader toward the posterior end of the series. The seventh neural is pentagonal and is separated from the

suprapygal by the intervention along the midline of the seventh and eighth pairs of neural bones. A discontinuous midline keel is present on the carapace. It extends along that part of the shell which, in a live turtle, would have been covered by the second, third, and fourth vertebral scutes. Although none is particularly protuberant, the keel underlying the fourth vertebral is less prominent than the other two in front of it. In all respects this description fits equally well hatchlings of the four species of *Podocnemis* (*P. erythrocephala, P. expansa, P. unifilis,* and *P. vogli*) described by Pritchard and Trebbau (1984).

Unlike the hatchling-sized shells of most turtles, this one entirely lacks the unossified fontanelles typically present between many of the adjoining carapace and plastron bones. Among all the living South American turtles (no other fossil hatchlings being known from the continent), only hatchlings of the various species of *Podocnemis* are characterized by the complete absence of shell fontanelles (Pritchard and Trebbau 1984; the extent of ossification in hatchling shells of the closely related and monotypic South American pelomedusid *Peltocephalus* has not yet been determined). A small sample of recent *Podocnemis* hatchlings that I examined (*P. unifilis*—PCHP 2266, USNM 50210; and *P.*

vogli—PCHP 1241, 1341, and 1959), all lacking fontanelles, ranged in carapace length from 5.7 to 7.7 cm. Depending on the species, freshly emerged *Podocnemis* hatchlings range in length from 3.6 to 5.5 cm; a captive specimen of *P. expansa* reached 12 cm in carapace length one year after hatching (Pritchard and Trebbau 1984). The La Venta fossil falls in the middle of this spectrum of measurements, and hence probably represents an individual less than one year old.

The lack of fontanelles, coupled with the other features described—the streamlined profile of the shell, the number and arrangement of the neural bones, and the keels of the carapace—clearly indicate that this fossil is referable to the genus *Podocnemis*. Because it has seven rather than six neurals, it presumably does not represent *P. pritchardi*. Unfortunately, the available evidence does not reveal how many neurals characterized the carapace of *P. medemi,* so it cannot automatically be assumed that the fossil hatchling is an example of this species. Other potentially useful taxonomic features—especially the shapes of the mesoplastra, anterior, and posterior lobes—are not adequately preserved in the La Venta hatchling. At present, therefore, attribution to a particular species of *Podocnemis* is premature.

Remains slightly larger but still, nevertheless, of very small fossil pelomedusids have been collected from several regions (Coyaima, Carmen de Apicalá, and Villavieja) in the upper Magdalena valley. These are listed in appendix 9.2. It seems reasonable to suppose that all of this material belongs to a single species. The most complete specimen (IGM 2021) consists of the better part of a plastron (see table 9.3 for measurements) and a somewhat lesser amount of an associated carapace (fig. 9.6). Cataloged under this same number are less complete fragments of a second, comparably small specimen. The description that follows is based on the more complete of these two specimens.

The shell appears to have been fully ossified. The bone is uniformly thick, and there is no evidence that there were ever any fontanelles. This fact alone, as noted, immediately suggests that *Podocnemis* is represented. There were clearly seven neural bones of the usual shape arranged in the usual manner typical of living species of *Podocnemis*. Uncharacteristically for such small examples of *Podocnemis,* the dorsal surfaces of the few neurals that have been preserved all lack keels. Most of the seventh and all of the eighth pairs of

Table 9.3. Plastron measurements for *Podocnemis* sp. indet., juvenile, IGM 2021

Variable	Measure (cm)
Midline length:	
As preserved	9.4
If complete (estimated)	12.3
Transverse width:	
At inguinal notches	5.6
Of right hyoplastron at axillary notch	3.0
Total at axillary notches (estimated)	6.0

Note: IGM 2021 is the most nearly complete of the juvenile specimens referred to *Podocnemis* sp. indet.

pleural bones intervened in the midline of the carapace to separate the last neural bone from the suprapygal. None of the shell's bridge region or the immediately adjoining portions of the plastron are preserved. Consequently, the shape of the mesoplastra cannot be determined.

The posterior plastral lobe differs in several respects from those of both *P. pritchardi* and *P. medemi;* its lateral margins are essentially parallel along their entire lengths, and it terminates in a broad, V-shaped anal notch. The ischial scars on the visceral surface adjacent to the anal notch are roughly comparable in shape and placement to those of *P. medemi* (as well as modern species of *Podocnemis*). But the entoplastron on the anterior plastral lobe appears to have been positioned like that of *P. pritchardi* rather than like that of *P. medemi.* Unlike *P. pritchardi,* however, there were seven neural bones on the carapace rather than six. Thus, the shell material typified by IGM 2021 differs in various ways from both *P. pritchardi* and *P. medemi.* Although it is conceivable, by virtue of considerable ontogenetic change, that these tiny turtles may represent very young growth stages of either *P. pritchardi* or *P. medemi,* there is no compelling reason to believe that this is the case. Equally plausible is the possibility that a third new species is represented. Given the incompleteness of preservation of individual specimens and the small size of the total available sample, however, there is, at present, no unequivocal basis for formal recognition of still another new fossil pelomedusid taxon from the Miocene of Colombia. Until such time as better material may

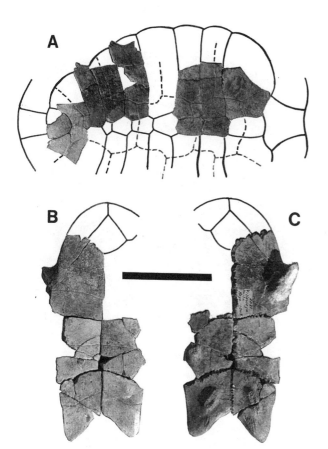

Figure 9.6. Very small juvenile *Podocnemis* sp. indet., IGM 2021. *A*, associated carapace in dorsal view; *B*, plastron in ventral view; *C*, plastron in visceral view. *A–C*. Scale bar 5 cm.

Comparisons with recent *Podocnemis* material indicate that the presence or absence of neural keeling varies from one species to another. Hatchlings of all six of the living species are moderately to strongly keeled. But adult individuals of *P. expansa*, *P. lewyana*, and *P. vogli* (Pritchard and Trebbau 1984; Nicéforo María 1952) invariably lose their carapacial keels as they mature. In *P. erythrocephala* the presence or absence of keels in adults is evidently a sexually dimorphic character; males retain them, whereas the larger females lose them (Pritchard and Trebbau 1984). All adult *P. sextuberculata* in a small sample (IRF 352, 353, and 354) that I examined from Colombia are characterized by a distinct midline keel that extends from the posterior half of the second vertebral scute back onto the anterior half of the fourth. Likewise, adult specimens of *P. unifilis* (IRF 17, 120, 161, 799, and 801; see also Pritchard and Trebbau 1984) all possess very slight midline keels of similar anteroposterior extent as in *P. sextuberculata*, which may be either slightly serrated or continuous.

Given the relatively small size of the fossil neurals described here and the fact that all living species of *Podocnemis* are strongly keeled as hatchlings, and often moderately keeled in immature growth stages as well, it is not possible to determine what taxon (or taxa) may be represented on the basis of neural bone evidence alone. In the absence of any other information, the most parsimonious assumption is that these keeled neurals represent juvenile specimens of *P. medemi*, *P. pritchardi*, or both. Presumably, as is the case for the modern species *P. expansa*, *P. lewyana*, and *P. vogli*, continued shell growth from juvenile to adult size would result in the loss of neural bone keels because adult specimens of both *P. medemi* and *P. pritchardi* lack any suggestion of keeling on their neurals.

be forthcoming, it seems best simply to refer to these fragmentary remains as *Podocnemis*, sp. indet.

Isolated neural bones from various localities (see app. 9.3 for a full listing of these specimens) deserve special comment. Generally modest in size (the largest, UCMP 126525, being 2.96 cm long), these bones have the coffin-shaped outline typical of all but the first and last elements in the neural series of *Podocnemis*. Their common trait is the presence of a moderately developed anteroposterior ridge along the midline axis of the dorsal surface. The proportions of the hexagonal neural bones along the carapacial midline of *Podocnemis* characteristically vary from front to back, with the anteriormost ones being somewhat longer than broad and the more posterior ones being broader than long. The majority of the keeled neurals in the assortment I have examined are broader than long and therefore represent elements from the posterior part of the series.

Suborder Cryptodira
Family Testudinidae
Geochelone, sp. indet.

Most of the fossil turtles from the Miocene of the upper Magdalena River valley are aquatic forms. But tortoises are represented as well, albeit by only a very few specimens.

Several of these have been attributed to a new species, *Geochelone hesterna,* by Auffenberg (1971). The type specimen of *G. hesterna* is the somewhat crushed but essentially complete shell of an adult male (UCMP 40200), having a midline carapace length of 27.8 cm. Within this shell is a poorly preserved skull and lower jaw whose existence was mentioned by Auffenberg, but which were not in any way further described. Only one additional specimen, a very large skull (UCMP 38940), was referred to this species. (Although the specimen's catalog number was misprinted as UCMP 3890 when first mentioned in Auffenberg's original description [1971, 106], it was subsequently correctly cited in the caption for fig. 3 of the same publication.) The snout/condyle length of this skull, which is not associated with any shell material, is 13.2 cm, almost half the entire length of the carapace of the type specimen. It is thus roughly comparable in size to skulls of modern Galapagos tortoises (*G. elephantopus*) and obviously belonged to a tortoise with a very big shell. Although not mentioned by Auffenberg, associated with this large skull is a nearly complete left mandible, covered extensively by matrix.

Contrary to Auffenberg's statement (1971, 106), the two specimens referred to *G. hesterna* are not from the same locality; the type shell is from UCMP Locality V-4536, whereas the large skull is from UCMP Locality V-4524, between the Fish Bed and Ferruginous Beds, Villavieja Formation. Thus, because of its much larger size and its collection from a different locality, there is no compelling reason to associate the referred skull to the same taxon as is represented by the type specimen of *G. hesterna*.

Instead, I think it more probable that this skull belongs to a different, giant, and so far undescribed species of tortoise (here referred to, as a matter of taxonomic convenience, simply as *Geochelone* sp. indet.), which is best represented by a complete plastron and partial carapace (lacking the upper part of its vaulted dome; fig. 9.7), much of the pectoral and pelvic girdles, and an assortment of well-preserved limb bones. It was collected by the Peter Creutzberg family, residents of Bogotá at the time I saw the specimen (in 1970). The only locality data for the specimen is rather imprecise: "from Villavieja." On the sole occasion that I examined this giant tortoise, it was in the Creutzberg home resting on a wooden pedestal to which it was firmly attached; therefore, regrettably, I was unable to examine the external surface of

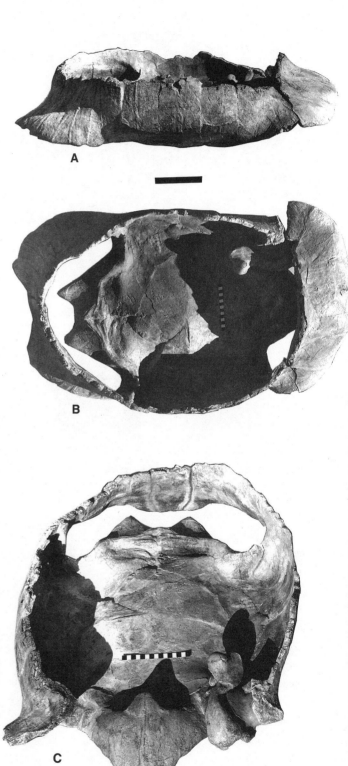

Figure 9.7. *Geochelone* sp. indet., Creutzberg specimen, complete plastron and partial carapace. *A,* right lateral view; *B,* dorsal view with same orientation; *C,* dorsal view from the anterior aspect with the anterior carapace removed. Scale bars 15 cm.

the plastron, an area of considerable potential taxonomic interest. This specimen has subsequently been transferred to the IGM collections but as yet has not been assigned a catalog number (Sánchez Villagra, pers. comm.).

The Creutzberg specimen may well represent a new taxon. It is far larger than the shell of *G. hesterna* or typical representatives of either of the two living South American tortoise species (*G. carbonaria* and *G. denticulata*), to which *G. hesterna* is most closely related. (Both of these living species, incidentally, are broadly distributed across northern South America, having ranges that extend into parts of Colombia.) The midline carapace length of the Creutzberg specimen is 95.0 cm, nearly four times as long as the same measurement (27.8 cm) for *G. hesterna*. Its midline plastron length is 76.5 cm. Typical specimens of both *G. carbonaria* and *G. denticulata* attain maximum carapace lengths in the range of 35–40 cm. The largest known specimen of *G. carbonaria* has a maximum carapace length that barely exceeds 50 cm. A few specimens of *G. denticulata* have attained even greater carapace lengths, in the 60 cm to low 70 cm range. One extraordinary *G. denticulata* specimen has a carapace 82.0 cm long (Pritchard and Trebbau 1984). Thus, the length of the Creutzberg fossil tortoise significantly exceeds that of the sole known shell of *G. hesterna* or typical specimens of both *G. carbonaria* and *G. denticulata*, being more than double the size of the latter two modern species. Even the largest, by far, recorded specimen of *G. denticulata* is still some 13 cm shorter in length than the Creutzberg shell.

Shell size alone, of course, is not a sufficient basis for recognition of a new turtle taxon. Moreover, because of the manner in which it was displayed when I examined it, a detailed morphological description of the shell and associated skeletal material is not possible anyway, so no definitive taxonomic conclusions about the specimen can be reached at present. A more comprehensive analysis than I can give here is currently being undertaken by M. Sanchez Villagra.

Medem et al. (1979) noted the presence of *G. denticulata* in the La Venta fauna. The basis for this conclusion is unclear. A small assortment of giant tortoiselike fragments (see app. 9.4) has been collected; these, however, are not really adequate for detailed taxonomic analysis. Whatever the appropriate formal taxonomic designation might ultimately prove to be for both these fragments, as well as for the Creutzberg shell, they all presumably represent some sort of giant tortoise and indicate the presence (rarely) in the La Venta fauna of two different types of fossil tortoises.

Discussion

The Miocene fauna of the upper Magdalena River valley of Colombia includes representatives of three different families of turtles: the Chelidae, Pelomedusidae, and Testudinidae. All three families are also components of the modern South American turtle fauna as well. Chelids and pelomedusids are strictly aquatic forms; in contrast, testudinids are terrestrial turtles. Chelids occur today in Colombia but, except for *Phrynops dahli* (known only from the vicinity of Sincelejo), are restricted to tropical lowlands east of the Andes. Living testudinids occur both there and in the lower reaches of the Magdalena River Valley. The pelomedusid genus *Podocnemis* is presently found in the Magdalena drainage of central Colombia (where it is represented by a single species, *P. lewyana*, which is restricted to this river alone) and in the eastern lowlands of the country, where three broadly sympatric species (*P. expansa, P. unifilis,* and *P. volgi*) occur. These latter three species represent, respectively, gigantic (89 cm maximum carapace length), rather large (46 cm maximum carapace length), and moderate-sized (36 cm maximum carapace length) turtles (Pritchard and Trebbau 1984).

Although only a single pelomedusid species (*P. lewyana*) today inhabits the Magdalena River of Colombia, during the Miocene, this region—then continuous to the east with the Amazon basin—was inhabited by two, and perhaps even more, different pelomedusid taxa, which also appear to be referable to *Podocnemis*. The two clearly definable fossil species from Colombia were both unusually large; the type of *P. pritchardi* attained an estimated midline carapace length of approximately 77 cm, whereas the comparable dimension of *P. medemi* was in the vicinity of 100 cm. In terms of modern pelomedusid sizes, both fossil species would therefore be considered gigantic. There is no place in the world today where two comparably large pelomedusid species coexist (and, in fact, there is only one living pelomedusid of comparable size, *P. expansa*).

From late Miocene times to the present, pelomedusid species diversity in the geographic area of the upper Magdalena River valley has decreased. The fossil record of fish from this same region indicates a similar taxonomic diminution over the same period (Lundberg et al. 1986). Whereas the region is now an area of arid badlands (Royo y Gómez 1946), paleontological and geological studies (e.g., Stirton 1953b) indicate that, during the middle Miocene, wide expanses of savanna, crossed by meandering streams bordered by dense forests, characterized the scene. The decrease in pelomedusid diversity through time may consequently be correlated with a decrease in the extent (and possibly also variety) of suitable habitat for aquatic turtles.

The Venezuelan llanos today perhaps afford the best model for what conditions may have been like in the upper Magdalena River valley during the late Miocene. Four pelomedusid species (*Podocnemis expansa*, *P. unifilis*, *P. vogli*, and *Peltocephalus dumerilianus*) have overlapping distributions in parts of the llanos. Recent studies of free-ranging, wild populations of two of these taxa (*Peltocephalus dumerilianus*, Pérez Eman 1990; *Podocnemis vogli*, Ramo 1982) have provided substantital new insight into their behavior and ecology; these, together with earlier studies (Alarcón Pardo 1969; Pritchard and Trebbau 1984; S. Maness, pers. comm.) clearly indicate that somewhat differing habitat preferences and behavior patterns serve to minimize or eliminate competition among these species. *Chelus* is the only member of its family known to occur in the llanos, where it is broadly distributed. There is only one record of *Geochelone* from the northern fringe of the llanos; the two species of this genus from northern South America (*G. carbonaria* and *G. denticulata*) are both reported to prefer forested areas rather than extensive open savannas such as the llanos, which are characterized by dramatic annual climatic extremes, ranging from extensive inundation by flood waters to prolonged dry seasons (Pritchard and Trebbau 1984). In some areas, these two species are sympatric (Moskovits 1988; see also Moskovits and Kiester [1987] and Moskovits and Bjorndahl [1990] for further information on the behavior and ecology of these two tortoises).

Podocnemis pritchardi, alone among the Miocene pelomedusids from Colombia, exhibits characters that permit inferences about its relationships to other fossil or living representatives of the genus. The distinctive shape of its anal notch and its number of neurals (six) are both shared only with the living species *P. lewyana*. *P. pritchardi* retains a per-

fectly good suprapygal bone, which is not present in *P. lewyana*, but this is a feature that might well have been lost over the subsequent time, through progressive diminution. Although the shell of *P. pritchardi* is considerably greater in size than that of the largest known *P. lewyana* (~40 cm; Pritchard 1979), this difference does not necessarily preclude the possibility of some kind of close relationship between the two taxa, because size tends to be a much more plastic character than, for instance, the number, shape, and/or arrangement of bones in a skeleton. The narrow, nearly rectangular mesoplastra, combined with the greatly reduced ischial scars and downward curving bridges of *P. pritchardi*, however, are unique among turtles. These characters alone, I believe, are sufficient to preclude any direct ancestor-descendant relationship between *P. pritchardi* and *P. lewyana*. The available evidence suggests that *P. pritchardi* may be related, but not ancestral, to *P. lewyana* (fig. 9.8).

P. medemi is among the largest of any pelomedusids yet known. It is substantially larger than any living species, roughly equal in length to an unnamed carapace from Morocco (Moody 1976), and exceeded in size only by the gigantic *Stupendemys geographicus* (Wood 1976a) from Venezuela and a shell from the Miocene of the upper Río Acre, Brazil, measuring an estimated 120 cm in midline carapace length, which closely resembles the modern species *Podocnemis expansa* (Rancy and Bocquentin Villanueva 1987).

The relationship of *Podocnemis medemi* to other members of the genus is unclear. Neither the roundness of its plastron nor the acuteness of its anal notch are mirrored in any of the known fossil or living species. Moreover, beyond those

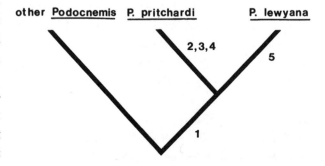

Figure 9.8. Cladogram depicting the possible phylogenetic relationship between *Podocnemis pritchardi* and *P. lewyana*.

comparisons that have already been made with living species, too little is known about the other Colombian fossil pelomedusids to make reasonable inferences about their phylogeny. For the most part, therefore, the relationships of these Colombian fossil pelomedusids to each other and to other species of *Podocnemis* remain obscure.

Four species of fossil turtles at La Venta are well enough represented to justify formal taxonomic designation: *Chelus colombianus, Geochelone hesterna, Podocnemis medemi,* and *P. pritchardi.* The available evidence suggests the likelihood that a second species of tortoise and the possibility that a third species of *Podocnemis* may also be present. Thus, there are almost certainly five, and perhaps as many as six, different kinds of turtles included in the La Venta fauna. The Miocene sediments from the upper Magdalena valley have, in fact, yielded the most diverse fossil turtle fauna yet known from anywhere in South America. The only other fossil turtle fauna on the continent that approaches equal richness is from the late Tertiary Urumaco locality in northernmost Venezuela. There, two pelomedusid taxa (*Stupendemys geographicus* and *"Podocnemis" venezuelensis,* the latter being known from a quarry sample that has yielded skulls, shells, and other other assorted skeletal material) occur besides the extinct chelid *Chelus lewisi* (see Sanchez Villagra 1992 for information on recently discovered material referable to this species). The unexpected presence of soft-shelled (trionychid) turtles is indicated by the presence of a single plastral bone (Wood and Patterson 1973); in addition, a few nondescript scraps of tortoise bone have been found. Five different kinds of turtles are thus known from Urumaco, although only three of these are, so far, well enough preserved to merit formal naming. No other South American fossil localities yet known even closely approach La Venta or Urumaco in terms of chelonian diversity or abundance.

Acknowledgments

A National Science Foundation Science Faculty Professional Development Award provided the time necessary for preparation of the first draft of this manuscript, while the Florida Audubon Society headquarters afforded congenial surroundings in which to prepare it. Travel funds from the National Geographic Society enabled me to examine the specimens described in this chapter. Additional grant support was provided by the Faculty Research Fund of Stockton State College.

For access to fossils under their care, I am grateful to the curators and staffs of the Museo Geológico in Bogotá and the Museum of Paleontology at the University of California, Berkeley. Moreover, I thank R. F. Kay for the opportunity to study recently collected Colombian fossil turtles, and the Creutzberg family of Bogotá for permission to examine the giant Miocene fossil tortoise that they collected. Preparation of the fossil pelomedusid skull here described was skillfully undertaken by Arnie Lewis.

In addition, I thank P. C. H. Pritchard and F. Medem, both for their hospitality and for allowing me to study recent turtles in collections under their care. I thank J. Zug for access to modern hatchling specimens in the collections of the USNM. I am also much obliged to M. Logan for x-ray photographs of hatchling pelomedusid shells, generously loaned by P. C. H. Pritchard. I particularly thank P. C. H. Pritchard, J. H. Hutchinson, M. Sánchez Villagra, and A. Paolillo for helpful reviews of various drafts of this manuscript. I am grateful to R. Vaughn for help with preparation of many of the figures. Finally, I am much obliged to S. Maness, now deceased, for information about the ecology and behavior of *Podocnemis vogli* in Venezuela.

Appendixes 9.1–9.4 list remains from the Miocene of the upper Magdalena River valley, Colombia.

Appendix 9.1. Indeterminate pelomedusid remains

These specimen numbers represent fragments of neurals, pleurals, peripherals, the pelvis, and various elements of the plastron, all of which lack diagnostic features but which clearly were derived from large turtles with flat-arched shells.

UCMP 37894, 37995, 38110, 38112, 38381, 39004, 126526, and 126527.

IGM 1777 (part), 1779 (part), 1812, 1822, 1826, 1828, 1833, 1841 (part), 1843 (part), 1845, 1860, 1863, 1866 (part), 1868, 1884/85, 1899 (part), 1903, 1904, 1909 (part), 1932, 2002, 2085, 2089, 2100 (part), 2130, 2198, 2242 (part), 2275, 2386, 182716, and 182775.

Appendix 9.2. Remains of juvenile pelomedusids referred to *Podocnemis,* sp. indet.

IGM 2021, parts of two shells; IGM 1776, a collective number that includes an isolated neural and a xiphiplastral fragment; IGM 1819, two medial halves of pleurals; IGM 1909, medial half of a single pleural; IGM 2069, hexagonal neural, probably from posterior part of the series; IGM 2100, a collective number including a xiphiplastral fragment; IGM 2120, a single bridge marginal; and IGM 2242, a single posterior peripheral.

Appendix 9.3. Keeled neural bones designated *Podocnemis,* sp. indet.

Most of the catalog numbers listed here are collective ones for multiple fragments. In parentheses is the number of keeled neurals included under the catalog number.

IGM 1841 (1); IGM 1842 (1); IGM 1843 (1); IGM 1889 (3); IGM 2038 (1); IGM 2100 (2); IGM 2108 (2); UCMP 126525 (1); UCMP 126528 (1); and UCMP 38933 (1).

Appendix 9.4. Specimens probably representing giant fossil tortoises

Most of these remains are very fragmentary.

IGM 1812-12, portion from front of carapace; IGM 1812-15, partial entoplastron and indeterminate fragments evidently associated with it; IGM 1822-4, posterior peripheral and other miscellaneous fragments; IGM 2425, miscellaneous pieces of shell and skeleton; IGM 2435, nuchal bone, first pair of peripherals and second right peripheral; UCMP 38013, complete humerus, part of scapula.

10
Birds

D. Tab Rasmussen

RESUMEN

La riqueza y diversidad de la avifauna moderna de la región Neotropical es única y espectacular, especialmente para las zonas tropicales y subtropicales. Aparte de los registros del Terciario de Patagonia, gran parte de la evolución de las aves en América tropical durante el Cenozoico no ha sido muestreada. Hasta ahora, solamente dos aves fósiles han sido descritas para la fauna de La Venta. La parte distal del cúbito izquierdo y la vértebra cervical decimotercera (o duodécima) han sido referidas a *Anhinga* (Anhingidae, Pelecaniformes). *Hoazinoides magdalenae*, reconocido originalmente en base a un cráneo postorbital, es aquí asignado a una nueva familia Hoazinoididae en base a un nuevo esqueleto parcial asociada con especializaciones en el tarsometatarsus que permiten una circumducción del segundo dedo del pie. *Hoazinoides* puede distinguirse de los hoatzínes de la familia Opisthocomidae al poseer el coracoide y esternón no fusionados, el puente supratendinal del tibiotarso no osificado, y la región parietal del neurocráneo cóncava.

Una nueva especie *Galbula hylochoreutes* (Galbulidae, Coraciiformes) se propone en base a la extremidad próxima del húmero derecho más grande y un surco ligamental más pronunciado que las especies actuales de *Galbula*. Una nueva especie *Aramus paludigrus* (Aramidae, Gruiformes) es descrita en base a un tibiotarso izquierdo completo muy semejante pero más grande que la especie actual *Aramus guarauna*. El extremo distal de un tibiotarso izquierdo de cigüeña, con una fosa distinta y redondeada próxima al cóndilo interno, el margen interno del cual forma un labio agudo en el extremo anteromedial, muestra claras afinidades estructurales con *Jabiru mycteria* (Ciconiidae).

Las aves fósiles del Grupo Honda proveen un panorama claro de los ambientes durante el Mioceno medio. Los representantes vivos de *Anhinga, Aramus, Jabiru,* y *Galbula* ocurren en simpatría en pantanos tropicales bajos y boscosos. *Anhinga* sugiere la presencia de aguas calmadas y relativamente profundas. *Aramus* prefiere pantanos boscosos u otra vegetación tupida en los bordes de ciénagas y tierras inundadas. *Jabiru* se encuentra hoy en zonas inundadas y sabanas húmedas. La especie grande de *Galbula* provee evidencia clara para la presencia de bosque tropical húmedo primario o secundario alrededor de zonas abiertas.

Al menos nueve de las once familias de aves sudamericanas pre-Pliocenas con eficientes capacidades para volar, han sido registradas en el archivo fósil norteamericano o de otros continentes antes de la formación del Istmo de Panamá, indicando que la avifauna de América del Sur no evolucionó en aislamiento sino que mantenía un intercambio faunístico importante con otras regiones del mundo.

The living bird fauna of the Neotropics is spectacular for its richness and diversity. Thousands of species belonging to more than twenty orders are represented today in tropical and subtropical habitats of mainland South America. The fossil record of the South American avifauna, however, is regrettably meager. Nearly all known representatives of the continent's Tertiary avifauna come from Patagonia; most of these were studied in the nineteenth century by Ameghino, and they need taxonomic revision (Tonni 1980; Olson 1985a, 1985b). Apart from these Patagonian fossils and a considerable diversity of Pleistocene taxa (Campbell 1976, 1979), there is nearly nothing left. Millions of years of avian evolution within the vast Neotropical region has gone essentially unsampled.

The absence of fossil birds largely reflects the shortage of productive vertebrate localities. Frailey (1986) was able to map only six major vertebrate localities of Tertiary age north of Argentina. Even at these six sites, however, birds have been particularly rare probably because of taphonomic processes and possibly also because avian elements have not been studied or have not been recognized as important. For many years, the only Tertiary bird described from any of these six sites was *Hoazinoides magdalenae* Miller 1953 known from an incomplete cranium from the Miocene La Venta badlands of Colombia.

In recent years a small assemblage of avian fossils from La Venta has finally begun to accumulate as a result of the collecting efforts of Richard Kay, Carlos Villarroel, and their colleagues. Most bird specimens found at La Venta are broken, isolated fragments that are of limited taxonomic value. Several of the bird fossils, however, are of diagnostic elements that do allow precise identification; one specimen is a partial skeleton comprising more than one dozen pieces from a single bird. The fossils were obtained by surface collecting at several productive vertebrate localities within the Honda Group, source of the La Venta fauna (Stirton 1953b; Hirschfeld and Marshall 1976; also chap. 1). The Honda Group contains two formations, the lower La Victoria Formation and the upper Villavieja Formation. Localities within the Honda Group have yielded birds that date to about 13 Ma (chap. 3).

Specimens are housed at INGEOMINAS Instituto de Investigaciones en Geociencias, Minería y Química, Museo Geológico, Bogotá (IGM); the Facultad de Geociencias, Universidad Nacional de Colombia, Bogotá (UNC); and the University of California Museum of Paleontology, Berkeley (UCMP). The fossils were compared to osteological material in the collections of the Los Angeles County Museum of Natural History (LACM); the University of California, Los Angeles (UCLA); the Field Museum of Natural History, Chicago (FMNH); and the National Museum of Natural History, Smithsonian Institution, Washington, D.C. (USNM). The systematic order followed here is that of Olson (1985b).

Systematic Paleontology

Order Cuculiformes
Hoazinoididae fam. nov.

TYPE GENUS

Hoazinoides Miller 1953.

DIAGNOSIS

Cuculiform birds with specialized modifications of the tarsometatarsus allowing circumduction of the second pedal digit. Can be distinguished from all other families of birds by the modified trochlea for the second digit, which is reduced in length, hemispherical, with a stout, curved hooklike process directed medially and ventrally (the hemispherical shape and the hooked process superficially resemble the trochlea for the second digit in the family Tytonidae). Can be further distinguished from hoatzins (Family Opisthocomidae) by the unfused coracoid and sternum, the ossified supratendinal bridge of the tibiotarsus, and by the concave parietal region of the cranium.

Hoazinoides Miller 1953
Hoazinoides magdalenae Miller 1953

HOLOTYPE

UCMP 42823, a postorbital cranium.

TYPE LOCALITY AND HORIZON

UCMP Locality V4517, Royo number 1999, near "middle of the Villavieja–Cerro Gordo section" (Stirton 1951); Monkey Beds, Baraya Member, Villavieja Formation, Honda Group.

REFERRED SPECIMEN

IGM 184249, an associated partial skeleton, consisting of the following fragmentary elements: the shaft of the right coracoid, preserving part of the sternal articulation; the proximal end of the right ulna; the distal end of the right ulna; the distal end of the left radius; the proximal ends of the right and left carpometacarpi; the distal end of the right tibiotarsus; the distal end of the right tarsometatarsus; two nonungual pedal phalanges; two ungual phalanges; miscellaneous small fragments of bone shafts.

None of the elements belonging to IGM 184249 are directly comparable to the holotype of *H. magdalenae* described by Miller (1953), so neither generic nor specific distinction can be demonstrated. The partial skeleton is referred to *H. magdalenae* for the following reasons. Both the holotype and the referred specimens are hoatzin-like cuculiforms, and both also bear some general similarities in shape and proportions to chachalacas (*Ortalis*). The holotypical cranium is similar in size to that of the extant hoatzin, *Opisthocomus hoazin,* whereas the referred postcranial specimen is slightly smaller (78–90% of *O. hoazin* in linear dimensions). The cranium and postcranium are, thus, compatible with being from a single species. The slight difference in relative size between the holotypical cranium and the referred postcranium when compared to modern hoatzins may be the result of individual size variation, or it may indicate that the fossil bird was relatively large-headed compared to the modern hoatzin, which is a notably small-headed species compared to other members of the order Cuculiformes.

LOCALITY AND HORIZON

Duke Locality 75, La Victoria Formation, below the Tatacoa Sandstone Beds.

MEASUREMENTS

Standard measurements were not always possible because of fragmentation. Coracoid: width of shaft at narrowest point, 4.1 mm. Ulna: depth of distal end, dorsal surface to carpal tubercle, 5.8 mm. Tibiotarsus: depth of lateral condyle, 7.1 mm; maximum width from medial to lateral condyle, 8.6 mm. Tarsometatarsus: depth and width of the trochlea for

digit III, 4.0 and 3.6 mm. Ungual phalanx: complete length, 10.3 mm.

DESCRIPTION

Several taxonomically valuable elements are included in the partial skeleton, but most of these are extremely fragmentary. The most useful elements are described here. Extensive comparisons were made between IGM 184249 and nearly all living families of nonpasseriform birds; the brief summary that follows focuses primarily on the members of the order Cuculiformes, the only order that shows consistent resemblances to the fossil.

The damaged tarsometatarsus retains only two trochleae (fig. 10.1). The specimen is broken immediately proximal to the trochlear bases, but, fortunately, a small trace of the rim of the distal foramen is present on the palmar surface. The position of this foramen demonstrates that the preserved trochleae are those for digits II and III, not digits III and IV. The trochlea for digit III is similar to those of cuculiforms and some other bird orders (e.g., Galliformes), perhaps most closely resembling the tarsometatarsus of turacos (Family Musophagidae), such as *Crinifer.* The trochlea for digit II, however, is extremely unusual. It is very short, and the articular surface itself is hemispherical without a groove on the trochlear surface. This allows for circumduction of the proximal phalanx at the trochlear-phalangeal joint. A phalanx belonging to IGM 184249 pivots smoothly and easily at practically any angle on the trochlear surface. A robust, hooked process continues medially and ventrally away from the base of the articular surface.

The structure of the trochlea for digit II resembles that of no bird I have examined except for barn owls (Family Tytonidae), which have a vaguely hemispherical articular surface, and a medio-ventrally directed hooked process. Both *Tyto* and IGM 184249 bear a long, medial ridge (with a corresponding sulcus distal to it) running from the dorsal side of the trochlear base out to near the tip of the hooked process. Similarities between the trochlea of digit II of *Tyto* and the fossil, however, are certainly convergent (presumably related to the ability to rotate and grasp with the second digit) because in other respects the tarsometatarsi of barn owls are very different from those of the fossil. For example,

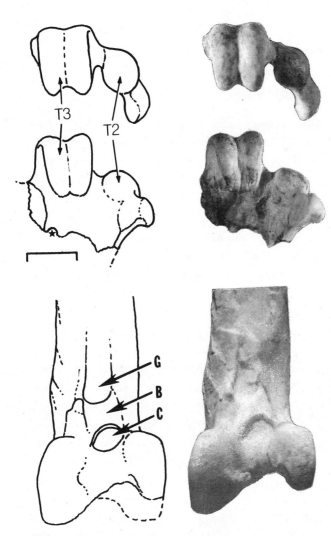

Figure 10.1. *Hoazinoides magdalenae,* IGM 184249, schematic drawings and photographs. *Top.* Broken distal end of the right tarsometatarsus in distal and palmar views. *Bottom.* Distal end of the right tibiotarsus in anterior view. Abbreviations: *B,* supratendinal bridge; *C,* extensor canal; *G,* extensor groove; *T2,* trochlea for digit II; *T3,* trochlea for digit III. The *star* marks the position of the distal foramen on the tarsometatarsus. Scale bar 3 mm.

in *Tyto* the trochlea for digit II extends distally to nearly the same length as the trochlea for digit III, and the structure of the trochlea for digit III is unlike that of the fossil.

The trochlea for digit II in the hoatzin differs from that of the fossil in its relatively larger size and broader shape, its flattish articular surface, and its lack of a prominent hook-like projection directed ventrally and medially. The trochlea of digit II of turacos is somewhat rounded, but not nearly as spherical as in the fossil, and in turacos the trochlea bears

only a small, ventrally oriented projection. Cuckoos and their relatives (Family Cuculidae) have a long trochlea that retains a distinct, spool-like trochlear surface. In cuculids it is the lateral trochlea that has been highly modified for reversed digit function. Outside cuculiforms the fossil shows a general resemblance to species in the cracid genus *Ortalis* in shape and proportions, but there is no special resemblance in the structure of the trochlea for digit II.

The distal end of the tibiotarsus of IGM 184249 includes the lateral condyle, the intercondylar region, and a crushed portion of the shaft (fig. 10.1). It has a fully ossified supratendinal bridge. In its broad extensor groove, the narrow arched bridge, the slight medial placement of the distal opening of the extensor canal, and in proportions of the lateral condyle, the fossil closely resembles some genera of turacos, notably *Tauraco* and *Crinifer,* and, to a lesser extent, some genera of cuculids, such as coucals (*Centropus*). The hoatzin specimens examined by me had unossified, fibrous supratendinal bridges. The distal end of the tibiotarsus of *Ortalis* is quite different from that of the fossil in its narrower, deeper extensor groove, the broad, unarched bridge, the small distal opening of the canal, and the relatively constricted anterior intercondylar fossa. The tibiotarsi of owls (Order Strigiformes) have no extensor grooves or supratendinal bridges, and a different shape of the lateral condyle. The fossil tibiotarsus is more similar to those of cuculiforms than to anything else, but its lack of specific specializations defies placement within any known family.

The coracoid is a poor specimen that retains only a small portion of the sternal facet and the shaft (fig. 10.2). The sternal facet is stepped, with an unusually deep, broad surface or shelf for articulation with the sternum's dorsal lip of the coracoidal sulcus, in addition to the primary articulation with the ventral lip. Similar development of the dorsal shelf is seen in species of *Centropus* and in most other cuculids. Turacos differ from the fossil in having a much greater step between the two shelves and in having a narrower dorsal articulation. The coracoids of the extant hoatzin are firmly fused to the sternum. The sternal facet found in species of *Ortalis* is structurally unlike that of the fossil.

The proximal end of the carpometacarpus of IGM 184249 resembles those of cuculiforms in its combination

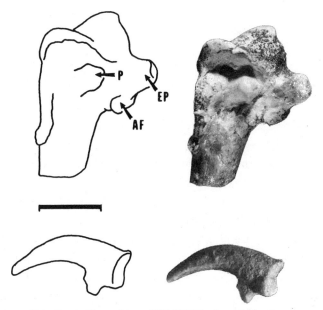

Figure 10.3. *Hoazinoides magdalenae*, IGM 184249, schematic drawings and photographs. *Top.* Proximal end of the left carpometacarpus. *Bottom.* An ungual phalanx. Abbreviations: *AF,* alular facet; *EP,* extensor process of the alular metacarpal; *P,* pisiform process. Scale bar 5 mm.

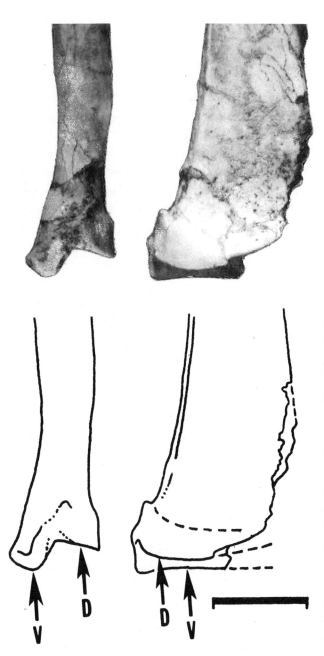

Figure 10.2. *Hoazinoides magdalenae*, IGM 184249, photographs and schematic drawings of the sternal end of the broken coracoid: *left,* medial view; *right,* dorsal view. Abbreviations: *D,* shelf for articulation with the sternum's dorsal lip of the coracoidal sulcus; *V,* ridge for articulation with the ventral lip of the coracoidal sulcus. Scale bar 5 mm.

features there is a particularly close resemblance between the fossil and extant hoatzins.

The distal ulna of IGM 184249 is characterized by a relatively high, arching dorsal condyle that extends distally beyond the ventral condyle and the carpal tubercle. The carpal tubercle is relatively small and bears a very flat ventral surface. These features are typical of cuculiforms. The fossil lacks the unevenly curved rim of the dorsal condyle found in turacos. The fossil differs from species of *Ortalis* in lacking a distally produced ventral condyle and carpal tubercle.

The claws included in IGM 184249 are similar in robustness and curvature to those of extant hoatzins (fig. 10.3). The remaining few fragmentary elements of IGM 184249 are of limited or no diagnostic value.

In Miller's (1953) description of the holotypical cranium, explicit comparisons were made only among the fossil, *Opisthocomus hoazin,* and galliforms, especially species of *Ortalis.* This reflected Miller's belief that the hoatzin was galliform. The ways in which the holotype differed from the extant *Opisthocomus hoazin* (e.g., the concave parietal and frontal areas, and the development of the occipital ridge) are similarities to other cuculiforms, such as *Centropus* spp., as well as to *Ortalis* spp. I suspect that re-

of: a simple alular facet placed well out on the alular metacarpal; the robust pisiform process; and the stout extensor process of the alular metacarpal, which is slightly hooked ventrally and is waisted in proximal view (fig. 10.3). In these evaluation of Miller's specimen may reveal additional generalized cuculiform features.

REMARKS

The fossil clearly belongs to the "basal land bird assemblage" of Olson (1985b), which includes the orders Cuculiformes, Galliformes, Columbiformes, Psittaciformes, and Falconiformes. Among galliforms, superficial similarities exist between IGM 184249 and cracids, particularly species of *Ortalis*. The fossil was presumably a bird of similar size, shape and proportions to modern *Ortalis*, but there is no evidence that the fossil is actually phylogenetically related to gallinaceous birds.

I can find no compelling points of resemblance between IGM 184249 and various members of the order Gruiformes. The family Cariamidae (seriemas) appears to be unmistakably gruiform in character for those elements compared in this study; I detected no additional evidence of hoatzin-seriema resemblances (Olson, 1985b). The fossil provides no evidence for a relationship to the orders Columbiformes (pigeons and doves), Psittaciformes (parrots), or Falconiformes (diurnal raptors). Cracraft (1971) named a new family of hoatzin-like birds, Onychopterygidae, based on a single proximal fragment of a tarsometatarsus. Although there is no reason to believe that Cracraft's Eocene bird from Argentina is the same species as the Miocene bird from Colombia, only future finds can assess the familial relationship between the two.

Throughout the disparate elements of IGM 184249, one detects recurrent cuculiform themes. It is certainly *not* a hoatzin, as indicated by the lack of hoatzin specializations (e.g., fused coracoid and sternum and unossified supratendinal bridge of the tibiotarsus), but it does resemble the three cuculiform clades—turacos, hoatzins, and cuculids—more closely than any other extant birds. With the available evidence it is impossible to place *Hoazinoides* more closely to one of these three groups than another because of the overlapping, mosaic patterns of resemblance. The distal ulna, proximal carpometacarpus, and, probably, the cranium suggest a closer tie with the hoatzin than with turacos or cuckoos. There is, however, currently no rational or empirical basis for determining which characters are primi-

tive for cuculiforms and which, on the contrary, may reflect genealogical affinity.

Olson (1992) has recently diagnosed a new cuculiform family, Foratidae, based on a fossil bird, *Foro panarium*, from the early Eocene of North America that, like *Hoazinoides magdalenae*, resembles in some respects the families Opisthocomidae, Musophagidae, and Cuculidae. *Foro* is morphologically distinct from *Hoazinoides magdalenae*, but both of these extinct species defy more precise taxonomic placement because of their unspecialized, primitive structure, and their mosaic patterns of resemblance to other cuculiforms. Together, they highlight the fact that the basal radiation of the order Cuculiformes was more diverse than would be recognized on the basis of the extant representatives.

The members of the order Cuculiformes have undergone evolutionary experimentation in foot and wing structure, which is, perhaps, related to their generally poor flying abilities. Cuckoos have evolved a fully reversed fourth pedal digit, turacos can facultatively reverse the fourth digit, hoatzins have evolved a distinctive shape of the trochlea for the second digit, and *Hoazinoides* evolved an owl-like configuration of the second digit that probably allowed a wide angle of circumduction or perhaps facultative heterodactyly. In the wing young turacos and hoatzins maintain prehensile alular and major digits that they use for clambering about in trees. Among all known skeletal elements, *Hoazinoides* probably resembles *Opisthocomus* most closely in the structure of the carpometacarpus and the distal ulna; one might speculate that these structural similarities correspond to hoatzinlike locomotor adaptations.

Order **Coraciiformes**
Family **Galbulidae**
Galbula Latham 1790
Galbula hylochoreutes sp. nov.

HOLOTYPE

IGM 250565, proximal end of the right humerus (fig. 10.4).

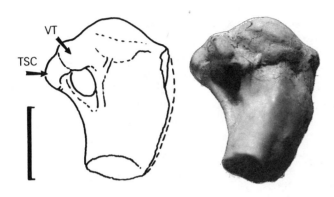

Figure 10.4. *Galbula hylochoreutes,* IGM 250565, schematic drawing and photograph of the proximal end of the right humerus in anconal (caudal) view, tilted slightly so it is viewed from a distal angle. Abbreviations: *TSC,* tubercle for the M. scapulohumeralis caudalis; *VT,* ventral tubercle. Scale bar 5 mm.

LOCALITY AND HORIZON

Duke Locality 113, La Victoria Formation, about 30 m above the Chunchullo Sandstone Beds.

DIAGNOSIS

Differs from all coraciiform families except Galbulidae in the combination of a ventrally produced tubercle for insertion of the M. scapulohumeralis caudalis, the more concave shape of the bicipital crest in palmar or anconal views, and the nearly round, large pneumatic foramen. Differs from all jacamar species but *Galbula dea* and *Jacamerops aurea* in having a slight crest crossing the capital groove from ventral tubercle to humeral head. Further differs from *Jacamerops aurea* in having a more dorsally situated dorsal tubercle and a relatively longer pectoral crest. Further differs from all extant species of *Galbula* (including *G. dea*) in its larger size and in having a more pronounced ligamental groove.

ETYMOLOGY

"Forest dancer," from Greek *hyle,* forest, and *choreutes,* dancer, in reference to the inferred environment in which it lived and the graceful, acrobatic flight of jacamars. The name is a masculine noun in apposition.

MEASUREMENTS

See table 10.1.

DESCRIPTION

All features of the proximal end of the humerus are preserved in fine detail except for a broken pectoral crest. The specimen is especially distinctive in the structure of the bicipital region and in the tubercle for the M. scapulohumeralis caudalis (hereafter abbreviated TSC). This tubercle is very robust, and it projects ventrally to an unusual degree. It is approximately the same size and robustness as the ventral tubercle proper (fig. 10.4). The bicipital crest is extremely concave in either palmar or anconal views, sweeping inward from the extended TSC toward the pneumatic foramen. In palmar view this gives the bicipital surface an unusual triangular shape. The pneumatic foramen is relatively large and round. The median crest descending from the ventral tubercle forms the border of the foramen, and it is continuous with a slight rim formed around the distal border of the foramen. A shallow, triangular fossa (part of the larger pneumatic fossa) lies immediately distal to this rim, bordered by the rim, the bicipital crest, and the main shaft of the humerus. The capital groove is shaped like a straight gutter; a slight fold crosses it between the median crest and the humeral head.

A robust, ventrally extended TSC is characteristic of several birds that exhibit acrobatic flight for aerial insect capture. This is presumably related to the larger size of M. scapulohumeralis caudalis, which serves to rotate the anterior edge of the wing ventrally (Hudson and Lanzillotti 1955; Raikow 1985). For example, a prominent and robust TSC occurs in bee-eaters (Family Meropidae), in some New World flycatchers (Family Tyrannidae), and in jacamars (Family Galbulidae), despite important differences among these groups in other features of the humerus.

In the shape of the pneumatic fossa and the ventral tubercle IGM 250565 resembles coraciiforms, especially members of the New World families Galbulidae (jacamars), Bucconidae (puffbirds), Momotidae (motmots), and Todidae (todies). The fossil humerus is nearly indistinguishable in structure from the humerus of the smaller species, *Galbula dea,* even to details of the distal rim bordering the pneumatic foramen and the triangular fossa distal to that. In size the fossil humerus slightly exceeds *Jacamerops aurea,* the largest extant jacamar, and small species of motmots in the genera *Electron* and *Eumomota* (table 10.1). *J. aurea* (73 g) is

Table 10.1. Measurements (mm) of the proximal humerus of *Galbula hylochoreutes* and of representative jacamars (Family Galbulidae) and motmots (Family Momotidae), arranged in rank order of increasing size

Family	Species	N	Width of Humeral Head[a]		Depth of Proximal Humerus[b]	
			Mean	SD	Mean	SD
G	*Brachygalbula lugubris*	1	1.6	—	5.0	—
G	*Galbula albirostris*	5	1.7	0.1	5.6	0.3
G	*Galbula ruficauda*	4	1.9	0.1	6.0	0.3
G	*Galbula cyanescens*	6	2.0	0.1	6.2	0.1
G	*Galbula dea*	1	2.2	—	6.8	—
M	*Eumomota superciliosus*	5	2.6	0.1	8.3	0.3
G	*Jacamerops aurea*	3	2.6	0.1	8.4	0.2
M	*Electron platyrhynchum*	1	3.0	—	9.0	—
G	*Galbula hylochoreutes*	1	3.1	—	9.3	—
M	*Momotus momota ignobilis*	3	3.5	0.2	10.1	0.4
M	*Momotus momota lessonii*	4	3.6	0.1	10.3	0.2
M	*Baryphthengus martii*	2	3.7	—	11.3	—
M	*Momotus momota momota*	3	3.9	0.3	11.3	0.3
M	*Baryphthengus ruficapillus*	1	4.1	—	11.9	—

[a]Maximum anconal-palmar (cranial-caudal) width.

[b]Distance from the tip of dorsal tubercle to the tip of tubercle for attachment of the M. scapulohumeralis caudalis.

more than three times the weight of typically sized jacamars (e.g., *G. ruficauda,* 22 g) and twice the weight of *G. dea* 34.5 g; Graves and Zusi 1989).

The fossil differs from the humeri of the family Momotidae in having a more robust, projecting TSC. The pneumatic foramen is variable in size and shape among motmots, but generally it tends to be relatively smaller than in jacamars, and in some cases it may consist of several small foramina scattered within the pneumatic fossa (especially in specimens of *Momotus* spp.). The fossil is also fairly similar to puffbirds of the genus *Bucco,* but *Bucco* differs from the fossil in having a notably heavier and longer ventral tubercle, which hangs low over the pneumatic foramen, in having a smaller pneumatic foramen, and in having a different shape of the bicipital crest.

No special resemblances are evident between the fossil and other bird families. Kingfishers (Family Alcedinidae) have a much-reduced TSC, a less-convex bicipital crest, and a larger, more robust ventral tubercle. Bee-eaters (Family Meropidae) have a capital groove that is oriented somewhat more proximodistally, rather than transversely; a small pneumatic foramen; and a slightly less-robust and projecting TSC. Bee-eaters have a humeral morphology that tends toward that of the families Alcedinidae and Coraciidae than toward the families Momotidae and Galbulidae. Rollers (Family Coraciidae) have a very small or absent TSC, and a very broad bicipital surface with a convex bicipital crest. The humerus of wood-hoopoes (Family Phoeniculidae), hoopoes (Family Upupidae) and hornbills (Family Bucerotidae) show no special similarities to the fossil, and they clearly resemble each other more closely than any other coraciiform group. Trogons (Family Trogonidae) have a broader, rounder bicipital region, narrower humeral head, weaker dorsal tubercle, more proximally positioned ventral tubercle, and a crestlike rather than a tuberosity-like TSC. Species of the extinct family Archaeotrogonidae have a prominent TSC, but they have a different shape of the bicipital crest and they lack pneumatic foramina (Mourer-

Chauviré 1980). The fossil differs from members of the order Piciformes (within which the Galbulae used to be classified; Olson 1983; Burton 1985) in numerous respects. For example, the true piciform birds (Suborder Pici) have a peculiar, bladelike shape of the large ventral tubercle.

REMARKS

Except for its large size, the fossil is nearly indistinguishable in structure from the humerus of *Galbula dea,* so there is no justification for diagnosing a new genus. Additional elements will be required to gauge the overall degree of resemblance between the Miocene bird and extant species of *Galbula.* Living jacamars are all active, aerial predators that catch insects on the wing during sorties from an arboreal perch. Extant *G. dea* prefers canopy and forest edges of the Amazon and Orinoco basins. *Jacamerops aurea* inhabits tropical forests in Central and northern South America where it forages in the midlevels and canopies of humid forest and tall second growth. Both species nest in arboreal termitaries (Haverschmidt 1968; Stiles et al. 1989).

Order Gruiformes
Family Aramidae
Aramus Vieillot 1816
Aramus paludigrus sp. nov.

HOLOTYPE

UNC 29-IV-8-4, an entire left tibiotarsus (figs. 10.5, 10.6).

LOCALITY AND HORIZON

Thirty meters west of Cerro Gordo near Duke Locality 33, below the Chunchullo Sandstone Beds.

DIAGNOSIS

Within the order Gruiformes, the fossil resembles only the families Psophiidae and Gruidae. It differs from trumpeters

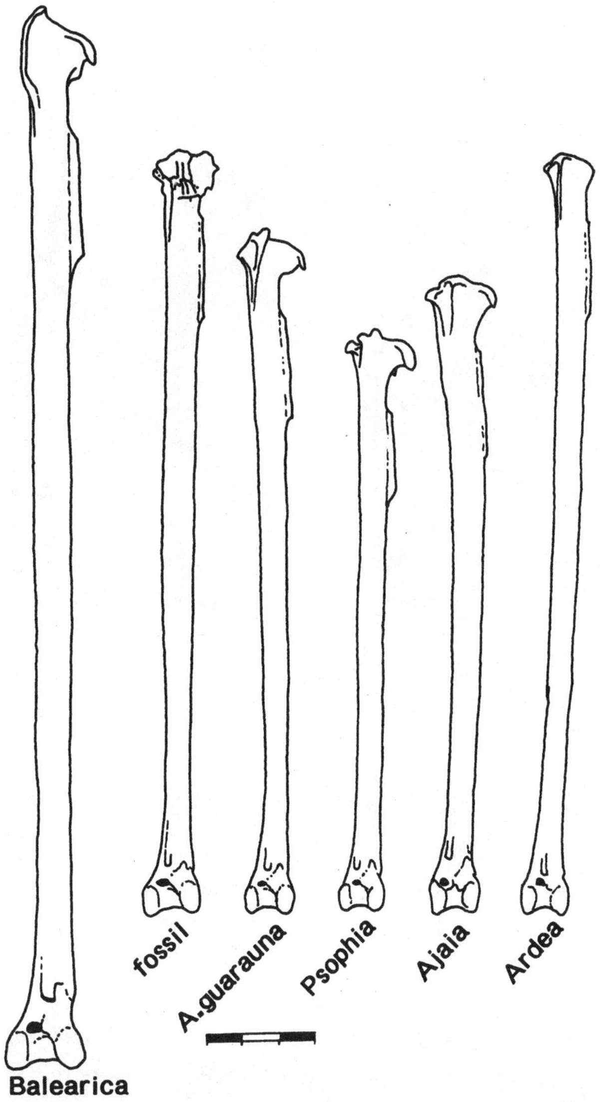

Figure 10.5. Left tibiotarsi in anterior view. *Left to right:* A crowned crane, *Balearica pavonina* (Gruidae); UNC 29-IV-89-4, the holotype of *Aramus paludigrus* sp. nov. (Aramidae); an extant limpkin, *Aramus guarauna* (Aramidae); a trumpeter, *Psophia leucoptera* (Psophiidae); a roseate spoonbill, *Ajaia ajaja* (Plataleidae); and a great egret, *Ardea alba* (Ardeidae). *Aramus paludigrus* is closest to *Aramus guarauna* in proportions and in the detailed structure of the distal end; the proximal end has been crushed and distorted. Osteological specimens are from FMNH. Scale bar 30 mm.

(Psophiidae) in its slightly broader, more inflated lateral condyle; by the relatively long, narrow shaft (specimens of *Aramus* with distal dimensions of the tibiotarsus that are smaller than in *Psophia* still have notably longer bones); by

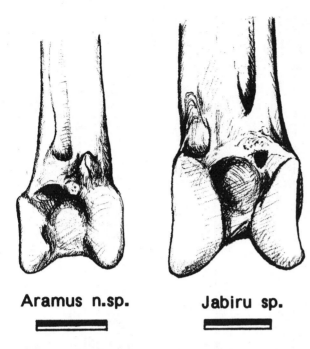

Figure 10.6. Tibiotarsi in anterior view. *Left.* Left tibiotarsus of *Aramus paludigrus* sp. nov., UNC 29-IV-89-4. *Right.* Right tibiotarsus of *Jabiru* sp., IGM 184266. Scale bars 10 mm.

the oval, rather than notched, proximal opening of the extensor canal; and by the smoother bony rim around the medial edge of the proximal articular surface (which in trumpeters is a rough, uneven lip). It differs from cranes (Gruidae), and, in particular, from *Balearica* (crowned cranes) in that the condyles do not spread or splay distally but retain a slightly more narrow, compressed alignment; the posterior rim of the medial condyle is less sharp and less flaring; the median protuberance at the proximal lip of the anterior intercondylar fossa is less robust; the interarticular prominence on the proximal end is smaller; the tuberosity for the restraining retinaculum of the peroneal muscles is shorter. It differs from *Aramus guarauna*, the only extant species of the genus *Aramus*, in its greater size.

ETYMOLOGY

"Swamp crane," based on the genitive stem of Latin *palus*, swamp, and *grus*, a crane, in reference to the preferred habitat of limpkins and their close phylogenetic relationship to cranes. The name is a masculine noun in apposition.

MEASUREMENTS

Total length, 196 mm; depth of lateral condyle, 13.3 mm; depth of medial condyle, 14.2 mm; width across distal condyles, 14.3 mm; minimum width of shaft (~40 mm from distal end), 6.5 mm.

DESCRIPTION

The tibiotarsus is nearly complete, lacking only the fine cnemial crests and other details of the crushed proximal end (figs. 10.5, 10.6). In its structure the fossil tibiotarsus closely resembles that of the extant limpkin, *Aramus guarauna*, and it certainly cannot be distinguished from *Aramus* at the generic level. It is significantly larger than the largest specimens of *A. guarauna* examined by me and exceeds the mean linear dimensions of *A. guarauna* by about 20%.

REMARKS

Cracraft (1973) concluded that two of Ameghino's putative limpkins from Argentina, the Deseadan species *Aminornis excavatus* and *Loncornis erectus*, did not belong to this family. Olson (1985b) suggested that a third species, the Santacrucian *Anisolornis excavatus*, may belong in the family Psophiidae. According to Olson (1985b) also, among the North American fossils identified as limpkins, only *Badistornis aramus* Wetmore 1942, based on a tarsometatarsus from the Oligocene of South Dakota, may be closer to *Aramus* than to a small crane. Because of these previous misidentifications or uncertainties surrounding limpkins and limpkinlike birds, I have been especially cautious in distinguishing *Aramus paludigrus* from cranes and trumpeters. I believe that this species can confidently serve as the first fossil record of limpkins in South America.

The family Aramidae includes only one extant species, *Aramus guarauna*, which is widely distributed in tropical and subtropical wooded marshes, swamps, and mangroves of the New World, including the southeastern United States. Limpkins primarily feed upon large snails (*Pomacea*), but they also consume bivalves, insects, and some plant foods. They are sedentary and, therefore, serve as useful environmental indicators.

Family Ciconiidae
Jabiru Hellmayr 1906
Jabiru sp. aff. *J. mycteria* (Lichtenstein 1819)

REFERRED SPECIMEN

IGM 184266, the distal end of a left tibiotarsus (fig. 10.6).

LOCALITY AND HORIZON

Purchased from local collector; locality unknown.

MEASUREMENTS

Depth of lateral condyle, 24.2 mm; maximum width across distal condyles, 18.1 mm.

DESCRIPTION

The fossil is fairly well preserved except for some crushing and abrasion of the posterior portion of the medial condyle and the loss of the major tuberosities of the anterior intercondylar area. The fossil can be identified as that of a stork by the narrow extensor groove, the broad supratendinal bridge, the anteroposteriorly deep condyles, and the characteristic rounded anterior intercondylar fossa (fig. 10.6). The tibiotarsi of extant stork genera are very similar to each other, but they differ substantially from all other bird families, including Vulturidae (New World vultures) and Balaenicipitidae (shoebill storks).

The fossil resembles the jabiru (*Jabiru mycteria*), but differs from other storks, in having a distinct, rounded fossa just proximal to the internal condyle. The medial edge of the cuplike fossa forms a sharp lip at the anteromedial corner of the tibiotarsus. In other stork genera this fossa is smaller, shallower, elongate rather than round, and placed laterally away from the medial edge of the bone. *Ephippiorhynchus* most closely approaches the condition observed in *Jabiru*. The fossa is considerably less developed in *Mycteria* and *Ciconia*. It is on the basis of this one character that the fossil is allocated to the genus *Jabiru*. The fossil tibiotarsus is smaller than specimens of *Jabiru* that I have examined but not enough to warrant a specific diagnosis.

REMARKS

"Storks have a reasonably good fossil record, although as yet it is not readily comprehended" (Olson 1985b). Many species have been named on the basis of isolated elements, and these are widely scattered in museums around the world. Because the distal ends of the tibiotarsi of most storks are fairly similar to each other, I do not think it is wise to add yet another poorly known, named stork to the many already demanding careful comparison and revision.

In South America only one Tertiary stork has been described, *Ciconiopsis antarctica* Ameghino, a specimen that some authorities believe cannot be accepted as a stork without restudy (Tonni 1980; Olson 1985a, 1985b). *Jabiru mycteria* ranges widely over Central and South America, where it prefers aquatic habitats, but it may occur away from water as well.

Order Pelecaniformes
Family Anhingidae
Anhinga Brisson 1760
Anhinga cf. *A. grandis* Martin and Mengel 1975

REFERRED SPECIMENS

IGM 183485, distal left ulna; IGM 183549, a thirteenth (or twelfth) cervical vertebra.

LOCALITIES AND HORIZON

IGM 183485 was found in the Fish Bed, Baraya Member, Villavieja Formation; IGM 183549 is of unknown provenance within the Honda Group.

REMARKS

The La Venta anhinga was described and discussed by Rasmussen and Kay (1992). Morphologically it is very similar to extant anhingas (Genus *Anhinga*); in size it resembles *A. grandis* from the Miocene of North America (Martin and Mengel 1975; Becker 1987). Like jacanas and limpkins, anhingas are excellent paleoenvironmental indicators. They prefer still bodies of fresh water in warm, wet, tropical and subtropical climates. They hunt by quietly submerging below the surface and stabbing passing fish with their sharp bills.

Discussion

PALEOENVIRONMENT

The La Venta birds present an unambiguous picture of the environment of central Colombia during the middle Miocene. Living analogs of the fossil birds can be selected in the manner of Olson and Rasmussen (1986) to serve as paleoenvironmental indicators. Today, there is one New World species of anhinga (*Anhinga anhinga*), one limpkin (*Aramus guarauna*), one jabiru (*Jabiru mycteria*), and several species of the jacamar genus, *Galbula*. Representatives of these four genera occur sympatrically in modern, forested, lowland Neotropical swamps from Mexico to Argentina (Blake 1977). Anhingas suggest the presence of still, relatively deep water where they submerge themselves to fish. Limpkins prefer wooded swamps or other heavy vegetation in marshes and wetlands. Jabirus occur widely in wetlands and and more open grasslands. The large jacamar from La Venta provides solid evidence of tropical forest, either tall primary canopy forest or lush secondary growth. Some jacamars forage within the forest itself, whereas other species prefer woodland or openings in the forest, just as one might expect, above a swamp or bayou.

BIOGEOGRAPHY

Rasmussen and Kay (1992) pointed out that eight of nine families of pre-Pliocene South American birds known to be efficiently volant have also been found as fossils in North America or elsewhere well before the formation of a land connection. With the addition of the new birds from La Venta, at least nine of the eleven South American families known to be proficient fliers occur on other continents before the Pliocene, the two exceptions being Teratornithidae (Campbell and Tonni 1980) and Galbulidae. This indicates that the South American avifauna did not evolve in isolation but instead had significant faunal exchange with other regions of the world, contrary to the conclusions of Vuillemier (1985) based on generic-level analyses of North and South American avifaunas.

The families Hoazinoididae and Galbulidae are not represented by fossils anywhere outside the La Venta badlands (hoazinoidids are not known to be proficient fliers; if modern hoatzins are to be our guide, they were not). Although it might seem a safe bet to conclude on the basis of modern jacamar distribution that this family has always been exclusively Neotropical, fossils of other small Neotropical coraciiforms (Todidae, Momotidae) have been found in the Paleogene of North America and Europe (Olson 1976; Mourer-Chauviré 1982). Evidence suggests that many, if not most, apparent cases of endemism among Neotropical nonpasseriform birds actually represent examples of relictual taxa surviving from the nearly cosmopolitan tropical and subtropical habitats of the Eocene (Mourer-Chauviré 1982; Olson 1987).

PHYLOGENY

Any general statements about avian phylogeny derived from isolated and broken skeletal elements are bound to be tenuous and highly speculative; nevertheless, the Colombian fossils do suggest several conjectures. The fossil limpkin and the fossil jacamar are each larger than living representatives of their respective families. It is tempting to view the larger size of the Miocene fossils as the primitive condition for each family, Aramidae being derived ultimately from a crane and Galbulidae being derived from a motmot- or roller-sized coraciiform. If this is true, then these two families would constitute exceptions to Cope's rule.

The obvious phylogenetic inference to be drawn from the family Hoazinoididae is that the basal radiation of cuculiform birds was more diverse than previously known. The hoazinoidids may have occupied a chachalaca-like niche before the invasion of cracids into South America during the Pliocene. The new fossils belonging to the families Ciconiidae and Anhingidae provide little insight about phylogeny, other than demonstrating that the evolutionary radiation of these groups occurred on a worldwide stage.

Acknowledgments

I thank R. Kay of Duke University for inviting me to work on the La Venta birds. C. Villarroel of the Facultad de

Geociencias, Universidad Nacional de Colombia, generously granted his permission to work on the limpkin tibiotarsus. I thank R. Madden of Duke for his help with specimen numbers and locality information and for informative discussion. For access to comparative osteological collections I thank K. Campbell (LACM), J. Northern (UCLA), and D. Willard (FMNH). F. Hertel of UCLA helped me with the LACM and UCLA collections and also provided valuable suggestions and advice concerning the comparative osteology of the fossils. I thank S. Olson for the loan of three USNM specimens of *Jacamerops aurea* and for offering me his insights on IGM 184249. Finally, I thank Campbell and Olson for their detailed reviews of the manuscript.

Part Five
Nonprimate Mammals

11
New Clues for Understanding Neogene Marsupial Radiations

Francisco J. Goin

RESUMEN

Los marsupiales exhumados en los niveles del Mioceno medio del Grupo Honda componen una de las faunas neógenas más diversas de América del Sur para este grupo de mamíferos, con representantes de al menos cuatro órdenes y una docena de especies. El rango de tamaños representados por ellos es muy grande, así como también los hábitos alimentarios inferidos. La nueva especie *Pachybiotherium minor* constituye el último representante de este género, caracterizado por la presencia de una rama mandibular fuerte y alta, caninos robustos y de implantación subvertical, molares inferiores bunodontes con los trigónidos angostos y los paracónidos muy reducidos, y m4 sólo moderadamente reducido. La morfología dentaria de esta especie sugiere hábitos alimentarios frugívoro-insectívoros, así como también la persistencia de antiguos linajes adaptados a este tipo de dieta en el Cenozoico tardío de las áreas intertropicales de América del Sur.

Tres de las cuatro especies de Didelphimorphia aquí reconocidas son asignables a los didélfidos Marmosinae, y sugieren que los Didelphoidea constituyen una radiación neógena distinta de didelfimorfios sudamericanos a partir de un grupo basal insectívoro. *Micoureus laventicus* difiere de la especie viviente *M. cinereus* en poseer el cíngulo anterobasal de los molares inferiores poco desarrollado y en el ectoflexo poco profundo y metacristas grandes de los molares superiores. *Thylamys minutus* y *T. colombianus* se reconocen en base a la morfología de la dentición inferior y la compresión anteroposterior de los últimos molares superiores, los cuales presentan metaconos reducidos. *Dukecynus magnus* es el protilacínido más grande hasta ahora conocido, y muestra rasgos craneomandibulares sumamente distintivos, tales como el rostro largo y angosto, sínfisis mandibular fusionada, y una morfología dentaria claramente adaptada a hábitos carnívoros. *Anachlysictis gracilis* posee las sinapomorfías propias de los Thylacosmilidae, si bien la manifestación de estos rasgos no es tan extrema como en *Achlysictis lelongi* (=*Thylacosmilus atrox*). Estas incluyen un cuerpo mandibular recto con una incipiente "pestaña" sinfisaria, premolares inferiores reducidos aunque birradiculados, primer hasta tercer molares inferiores con talónidos reducidos, cóndilo mandibular bajo, cresta masetérica poco desarrollada, y aparentemente sin un proceso ascendente del maxilar. Finalmente, *Hondadelphys fieldsi* presenta un complejo conjunto de rasgos que hacen difícil la interpretación filogenética de los Hondadelphidae.

Marsupials from the middle Miocene Honda Group at Quebrada La Venta and neighboring localities in the La Venta area, upper Magdalena valley, Colombia, have been known for several decades (Stirton 1953b). Subsequent work by Marshall (1976a, 1977) led to the recognition of five species represented in the University of California collections from the Honda Group: *Marmosa* sp., *Marmosa laventica*, *Hondadelphys fieldsi*, *Lycopsis longirostrus*, and a fifth, indeterminate species of large borhyaenid. New and remarkable collections made by Duke University and INGEOMINAS researchers, greatly enhance knowledge of Miocene marsupials from equatorial South America. The purpose of this work is the preliminary description of what now becomes one of the best-represented Neogene South American marsupial faunas.

The answers to many time- and space-related questions

make this study especially interesting: time, because the middle Miocene is a turning point in the decline or radiation of many South American mammalian lineages; space, because equatorial South American faunas are only poorly known in the fossil record of this continent.

As is the case for living armadillos (chap. 13), living South American marsupials, mostly didelphoid opossums, show their greatest diversity in the equatorial tropics. It follows that paleontologic evidence from these latitudes could be crucial for the understanding of their evolution and Neogene radiation. The INGEOMINAS collections from the middle Miocene of Colombia now include representatives of four marsupial orders. This chapter deals with three of them: Microbiotheria, Didelphimorphia, and Sparassodonta. The fourth, the Paucituberculata, is treated separately by Dumont and Bown (chap. 12).

Dental Nomenclature

I, C, P, M = upper incisors, canines, premolars, and molars, respectively; i, c, p, m = lower incisors, canines, premolars, and molars, respectively; L = length, W = width (in lower molars, trigonid width). All measurements are in millimeters.

The generalized marsupial dental formula is presently considered to be I 5/4, C 1/1, P 3/3, M 5/5 (Archer 1978). For the individual dental notation, I have proposed the following scheme (Goin 1991): I 1,2,3,4,5/i 2,3,4,5; C 1/c 1; P 1,2,3/p 1,2,3; M 0,1,2,3,4/m 0,1,2,3,4. Three presuppositions are implicit in this scheme: (1) the first upper incisor (I1) is not homologous to the first lower incisor (i2; see Hershkovitz 1982); (2) there is no reason to assume that the ancestral marsupial dental formula included four premolars instead of three (Archer 1978), and (3) the deciduous jugal tooth in marsupials belongs to the molar "Zahnreihe" (Archer 1978; Goin 1991). I have proposed the notation M/m 0,1,2,3,4 instead of M/m 1,2,3,4,5 for two reasons. First, for practical purposes it is better to designate the four permanent molars of each quadrant as has been done for more than one century—that is, 1,2,3, and 4, instead of 2,3,4, and 5, as would follow if the deciduous jugal tooth of marsupials were a molar. Secondly, the notation M0 or m0 is sufficiently unusual to attract attention to the special (deciduous) nature of these teeth.

A question mark directly precedes a designation that is uncertain. For example, M?1 denotes a molar of a locus that could be identified only tentatively as 1.

Systematic Paleontology

Subclass Metatheria
Superorder Marsupialia
Order Microbiotheria Ameghino 1889
Superfamily Microbiotherioidea (Ameghino 1889) Reig, Kirsch, and Marshall 1985
Family Microbiotheriidae Ameghino 1887
Pachybiotherium Ameghino 1902
Pachybiotherium minor sp. nov.

ETYMOLOGY

[*M*]*inor,* for its small size in relation to the other known species of the genus.

HOLOTYPE

IGM 253051, a left mandibular fragment with roots of m2–3 and almost complete m4 (fig. 11.1A, B).

TYPE LOCALITY

Precise locality unknown. Probably the Honda Group, upper Magdalena valley.

HYPODIGM

The holotype, and IGM 253026, a broken m3 (fig. 11.2A, B); IGM 253027, isolated right trigonid; and IGM 253028, a fragment of upper molar. All specimens are from Duke Locality 22. Measurements are given in table 11.1.

REFERRED SPECIMEN

IGM 253029 an isolated upper molar (M?1; see fig. 11.2C) from Duke Locality 22.

STRATIGRAPHIC RANGE AND GEOCHRONOLOGIC AGE

Duke Locality 22 (San Nicolás Locality) Monkey Beds, Villavieja Formation, Honda Group.

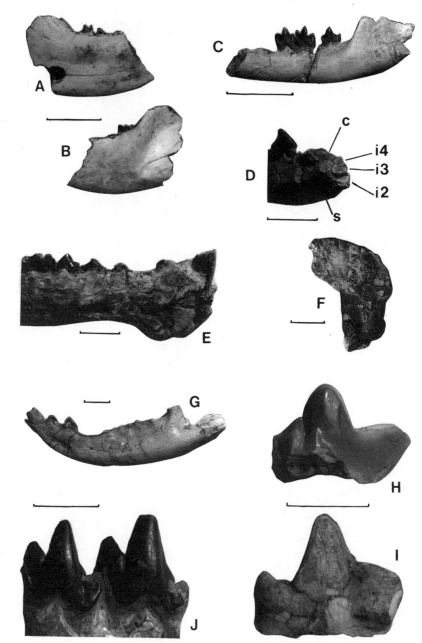

Figure 11.1. *A, B. Pachybiotherium minor* sp. nov., holotype, IGM 253051, left mandibular fragment with m4 complete: *A,* lingual view; *B,* labial view. *C. Micoureus laventicus* (Marshall 1976a), IGM 184151 (tentatively referred specimen), left horizontal ramus, labial view. *D. Hondadelphys fieldsi* Marshall 1976a, IGM 250961, anterior portion of left horizontal ramus with roots of canine (c) and the three incisors (i2–4). *E.* ?Thylacosmilidae gen. et sp. indet., IGM 251108, anterior portion of right mandibular ramus. *F.* Sparassodonta indet., IGM 250475, right maxillary fragment with upper premolar, anterior view. *G.* ?*Hondadelphys,* IGM 184041, left horizontal ramus with p1–2 complete. *H, I. Dukecynus magnus* gen. et sp. nov., holotype, IGM 251149; *H,* isolated M?3, lingual view; *I,* isolated m2, lingual view. *J. Anachlysictis gracilis* gen. et sp. nov., holotype, IGM 184247, left m2–3. Scale bars 10 mm.

DIAGNOSIS

Differs from *Pachybiotherium acclinum* in its smaller size, having hypoconulids closer to entoconids, and in that m4 lacks an entoconid, with the metaconid reduced and the hypoconulid greatly enlarged.

COMMENTS

The phylogenetic affinities of *Pachybiotherium* have been the subject of several revisions since the genus was described in 1902. The type species, *P. acclinum,* comes from the Colhuehuapian (early Miocene) of Patagonia; it was originally

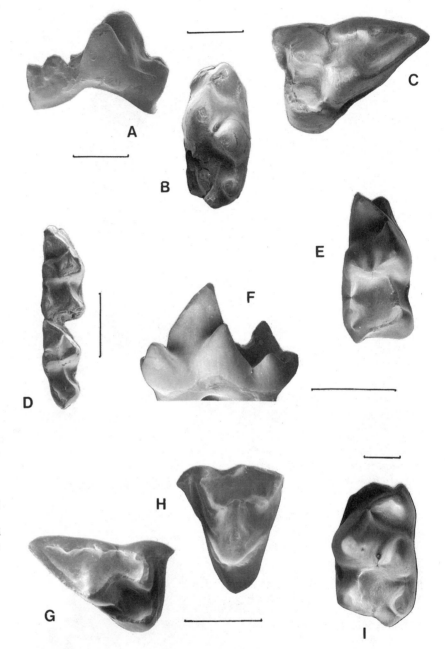

Figure 11.2. *A–C. Pachybiotherium minor* sp. nov. *A, B.* IGM 253026, left m3: *A,* lingual view; *B,* occlusal view. *C.* IGM 253029, referred specimen, left M?1, occlusal view. *D. Thylamys minutus* sp. nov., holotype, IGM 253042, right m3–4, occlusal view. *E, F. Thylamys colombianus* sp. nov., holotype, IGM 251010, isolated right m?2: *E,* occlusal view; *F,* lingual view. *G, H. Micoureus laventicus* (Marshall 1976a). *G,* IGM 251011, isolated right M1, occlusal view; *H,* IGM 250278, isolated left M3, posterolabially broken, occlusal view. *I.* Didelphinae, gen. et sp. indet., IGM 184600, isolated left lower molar, occlusal view. Scale bars 1 mm.

assigned by Ameghino (1902b) to the Microbiotheriidae because of its molar morphology and mandibular structure. Ameghino distinguished it from *Microbiotherium* by its larger size, the greater depth of the horizontal ramus, and its greater curvature near the symphyseal joint. Simpson (1932) and later Ringuelet (1953) regarded it as a junior synonym of *Microbiotherium.* Reig (1955) argued in favor of its recognition as a distinct genus and later (1981) con-

sidered it to be closely allied to *Caluromys.* This position has been maintained in subsequent works by Reig et al. (1985, 1987). In his review of the Microbiotheriidae, Marshall (1982b) excluded *Eomicrobiotherium* and *Pachybiotherium* from the family, arguing that they could have paralleled microbiotheriids (see also Marshall 1982a). Finally, Marshall (1987), Marshall et al. (1989), and Goin (1991) have returned to the view that *Pachybiotherium* should be in-

Table 11.1. Dental measurements (mm) for Honda Group Marsupialia

Specimen	Tooth	Length	Width
Pachybiotherium minor sp. nov.			
IGM 253027	m3	—	1.10
IGM 253026	m3	—	1.00
IGM 253051	m4	2.15	1.10
IGM 253029	M?1	2.70	1.80
Thylamys minutus sp. nov.			
IGM 253051	m1	1.20	0.60
IGM 253038	m?2	—	0.70
IGM 253032	m3	1.50	0.70
IGM 253042	m3	1.50	0.80
	m4	1.40	0.75
IGM 251013	M2	~1.40	~1.45
IGM 250345	M0	1.05	0.65
UCMP 108563[a]	M1	1.45	1.30
	M2	1.40	1.55
	M3	—	1.75
IGM 251232	M3	1.60	1.75
Thylamys colombianus sp. nov.			
IGM 251010	m?2	1.85	1.00
IGM 253034	m?2	1.90	0.95
IGM 253031	m?1	1.80	0.85
IGM 253033	M4	0.90	2.15
IGM 251009	M?3	1.55	—
Lycopsis longirostrus Marshall 1977			
IGM 250974	p2	13.7	5.5
	m4	16.8	7.8
Micoureus laventicus (Marshall 1976a)			
IGM 184336	m2	2.60	1.50
	m3	2.80	1.60
	m4	~2.90	1.50
IGM 184151	m2	2.95	1.80
	m3	~3.00	1.80
	m4	3.20	1.70
IGM 251011	M1	1.90	1.30
IGM 251012	M2	1.90	1.75
IGM 250278	M3	—	2.30
Dukecynus magnus gen. et sp. nov.			
IGM 251149	m2	15.5	7.0
	m4	15.0	8.0
	M?3	16.8	9.0
Anachlysictis gracilis gen. et sp. nov.			
IGM 184247	p2	~6.0	—
	p3	~7.2	—
	m1	~9.2	—
	m2	10.2	5.2
	m3	12.6	6.0
	m4	13.4	6.0

Table 11.1. Continued

Specimen	Tooth	Length	Width
?Thylacosmilidae			
IGM 251108	c	6.1	4.2
	p1	~3.5	1.4
	p2	~6.5	—
	p3	6.5	2.3
	m1	7.4	3.2
	m2	7.7	3.7
	m4	8.7	~4.1
	C	9.2	5.9

[a]Measurements taken on an epoxy resin cast of UCMP 108563; they differ from those reported by Marshall (1976a, 405).

cluded in the Microbiotheriidae. This view appears to be supported by the recent discovery of new remains of *Pachybiotherium acclinum* (Goin 1991) and of the new species *P. minor* described here.

Among the diagnostic features of species of *Pachybiotherium* are the following: (1) very deep and strong horizontal ramus, with mesial flexure anteriorly at the symphysis; (2) mandibular symphysis extends posteriorly to a point immediately posterior to the canine; (3) canine strong, subcircular in cross section, and vertically implanted in the jaw; (4) p3 much larger than p2; (5) lower molars with narrow trigonids, poorly developed anterobasal cingula, very reduced paraconids, metaconids slightly posterior to the protoconids, and hypoconulids slightly separated from the entoconids (see fig. 11.2A, B); and (6) m4 only moderately reduced with respect to m3. The combination of these features clearly distinguishes *Pachybiotherium* from any known Didelphimorphia, and several of these characters (1, 4, 5, and 6) suggest that this genus represents the culmination of a distinct lineage among the Microbiotheriidae. Work in progress favors the hypothesis that *Pachybiotherium, Eomicrobiotherium,* and, probably, *Mirandatherium* represent the sister group of all other known microbiotheriids.

Pachybiotherium minor is more derived than *P. acclinum* in the morphology of its m4, which lacks the entoconid and has a reduced metaconid. However, *P. acclinum* is more derived than *P. minor* in the greater depth and robusticity of the horizontal ramus and in the position of the hypoconulid (slightly more separate from the entoconid), especially in the anterior molars.

I have tentatively referred IGM 253029, an isolated upper molar, to *P. minor* (fig. 11.2C). Its centrocrista is clearly straight, as in all microbiotheres, and the paracone is fused with stylar cusp B, as it is in *Microbiotherium tehuelchum*. Crown size agrees with what would be expected for a first upper molar of *P. minor.* It should be noted, however, that this specimen may correspond to the deciduous molar of a medium-sized didelphoid (M0 in the nomenclature used here). A straight centrocrista and a paracone fused with stylar cusp B, are common features of didelphoid deciduous molars. Similarly, the metacone is clearly larger than the paracone, the metacrista is well developed and the protocone somewhat reduced. These are also common features among M0 of Didelphoidea.

Order Didelphimorphia (Gill 1872)
Aplin and Archer 1987
Superfamily Didelphoidea (Gray 1821)
Osborn 1910
Family Didelphidae Gray 1821
Subfamily Didelphinae (Gray 1821) Simpson 1927
Tribe Marmosini Reig, Kirsch, and Marshall
(in Reig 1981)
Thylamys Gray 1843
Thylamys minutus sp. nov.

ETYMOLOGY

[*M*]*inutus,* for its very small size among marmosine didelphids.

HOLOTYPE

IGM 253042, right mandibular fragment with posterior alveolus of m2 and complete m3–4 (fig. 11.2D).

TYPE LOCALITY

Duke Locality 22, Monkey Beds, Villavieja Formation, Honda Group, Huila Department, Colombia (see chap. 2).

HYPODIGM

From Duke Locality 22, the holotype, and IGM 253032, broken m3; IGM 253045, right mandibular fragment with roots of p2 and m1, and broken p3; IGM 253038, broken right m?2; IGM 250345, right M0. From Duke Localities CVP-8 and CVP-9, both in the Fish Bed, IGM 251013, an isolated right M2, broken labially, and IGM 251232, an isolated right M3. IGM 253052, partial left horizontal ramus with alveoli for p2 and complete p3–m1, from an unknown locality in the Honda Group. Measurements are given in table 11.1.

REFERRED SPECIMEN

UCMP 108563, right maxillary fragment with M1–2 complete and broken M3, from UCMP Locality V4517, in the Monkey Beds (see Marshall 1976a).

STRATIGRAPHIC RANGE

All known localities are in the Baraya Member, Villavieja Formation, Honda Group.

DIAGNOSIS

One of the smallest known species of *Thylamys;* differs from other species of the genus in the following features: metaconid of m4 reduced in relation to m1–3; cristid obliqua ends anteriorly just below the protocristid notch; upper molars with stylar cusp C very compressed labiolingually; stylar cusp A very small but distinct; metacrista poorly developed.

COMMENTS

The first known remains of this species were referred by Marshall (1976a) to "*Marmosa* sp." Splitting "*Marmosa*" into several genera was later proposed by Reig et al. (1985, 1987). This new material from Colombia may be confidently assigned to *Thylamys* and to a new species that differs in several features from other living or extinct species of the genus. Most interesting of these differences are the orientation of the cristid obliqua (see fig. 11.2D) and the persistence of a laterally compressed stylar cusp C. Marshall

(1976a) has already noted the limited development of the metacrista in the upper molars. A tiny upper right molar (IGM 250345) is assigned to this species and is identified as a deciduous molar.

Thylamys colombianus sp. nov.

ETYMOLOGY

[*C*]*olombianus,* for Colombia, from which all known remains of this species come.

HOLOTYPE

IGM 251010 an isolated right m?2 (fig. 11.2E, F).

TYPE LOCALITY

Duke Locality CVP-9, in the Fish Bed, Villavieja Formation, Honda Group, Huila, Colombia (see chap. 2).

HYPODIGM

IGM 253031, isolated right m?2; IGM 253033, broken right M4; IGM 253034, isolated right m?1, badly worn, all from Duke Locality 22. Measurements are given in table 11.1.

REFERRED SPECIMEN

IGM 251009, the labial half of a left M?3 from Duke Locality CVP-9 in the Fish Bed.

STRATIGRAPHIC RANGE

Both localities are in the Baraya Member, Villavieja Formation, Honda Group.

DIAGNOSIS

Similar in size to the living *Thylamys elegans;* differs from other species of the genus in its comparatively larger and slightly longer talonids, entoconids laterally compressed, and paraconids mesially oriented in the trigonid.

COMMENTS

This new species is clearly assignable to *Thylamys* on lower molar morphology and the distinctive features of M4 (very compressed anteroposteriorly and highly reduced metacone). Interestingly, the paraconid is oriented in a relatively mesial position in the trigonid; in most other marmosines (except in the m1 of the ?Pleistocene [not Montehermosan; see Goin 1991] *Zygolestes paranensis*) the paraconid is clearly aligned lingually with the metaconid.

Thylamys colombianus may be related to a new (and still unnamed) species of *Thylamys* from the late Miocene (Huayquerian) of Argentina (see Goin 1991). Despite their general resemblance in size and overall morphology, however, the Huayquerian species shows wider and shorter talonids, stronger hypoconids, and larger hypoconulids.

IGM 251009 is tentatively assigned to *T. colombianus.* The specimen consists of the labial half of a left M?3, including the paracone, metacone, metastylar spur, and stylar shelf. The ectoflexus is deeper than in M1–2 of *Thylamys minutus,* the metacone much higher than the paracone, and the metacrista better developed. Stylar cusp A is small but distinct and coalescent to cusp B, the latter being the larger stylar cusp. Interestingly, stylar cusp C is larger than D. In size and metastylar development, this specimen fits well in what could be expected for a third upper molar of this species.

Micoureus Lesson 1842
Micoureus laventicus (Marshall 1976a) comb. nov.

SYNONOMY

Marmosa laventica Marshall 1976a.

HOLOTYPE

UCMP 39273, partial right mandibular ramus with m1–2, anterior root and posterior alveolus for m3, and m4 complete (see Marshall 1976a, 403–4 and text-fig. 1).

TYPE LOCALITY

UCMP Locality V4936, Monkey Beds, Villavieja Formation, Honda Group. Huila, Colombia (see Marshall 1976a; also chap. 2).

HYPODIGM

IGM 251012, M2 from Duke Locality CVP-9; IGM 250278, fragmentary right M3 from Duke Locality CVP-5 (fig. 11.2H); IGM 251231, metastylar spur of an M?3 from Duke Locality CVP-8; and IGM 251011, an isolated right M1 from Duke Locality CVP-9 (fig. 11.2G), all from the Fish Bed. IGM 184336, a partial right mandibular ramus with m2–4 complete, from Duke Locality 40; IGM 250266, fragments of a lower molar from Duke Locality 34B; and IGM 253035, badly broken upper molar from Duke Locality 22. Measurements are given in table 11.1.

TENTATIVELY REFERRED SPECIMEN

IGM 184151, partial left horizontal ramus with m2–4 complete, from Duke Locality 22 (fig. 11.1C).

STRATIGRAPHIC RANGE

Duke Localities 40 and 34 are in the La Victoria Formation, Honda Group. Duke Locality 22 and Fish Bed localities (CVP-5, 8, and 9) are in the Baraya Member, Villavieja Formation, Honda Group.

REVISED DIAGNOSIS

Slightly larger than the living *Micoureus cinereus;* differs from the living species of the genus in that the anterobasal cingulum of the lower molars is only partially developed and in that the ectoflexa of the upper molars are comparatively shallow and with larger metacrista.

COMMENTS

As was noted by Gardner and Creighton (1989), the correct spelling of this genus is *Micoureus,* not *Micoures. Micoureus laventicus* is assigned to this genus by its comparatively large size, m2–4 trigonids as wide as long, and protoconids only moderately larger than paraconids. At this time, *M. laventicus* is the only known extinct species of the genus. In some features (e.g., in the moderately developed anterobasal cingulum compared with *M. cinereus), M. laventicus* is primitive with respect to the living species of the genus. How-

ever, it shows some specializations not seen in the living species, such as a well-developed postmetacrista (especially on M1) and an anteroposteriorly compressed m4 metaconid.

Marmosini, gen. et sp. indet.

In the IGM collection several marmosine specimens are too fragmentary or damaged for formal assignment. These include four specimens from localities in the Fish Bed (Villavieja Formation): IGM 250343, fragment of an isolated premolar from Duke Locality CVP-8; IGM 250280, isolated premolar from Duke Locality CVP-5; IGM 250344, two isolated premolars from Duke Locality CVP-13C; and IGM 250293, two isolated premolars from Duke Locality CVP-10. Included also are IGM 251214, isolated canine from Duke Locality 132 in the La Victoria Formation, and the following four specimens from Duke Locality 22 in the Monkey Unit, Baraya Member, Villavieja Formation: IGM 250328, isolated incisor; IGM 253041, fragmentary mandible with roots of two premolars; IGM 253040, fragment of an upper molar; and IGM 253046, fragmentary mandible with roots of a premolar or molar.

Didelphinae, gen. et sp. indet.

Among the didelphimorph material are three enigmatic didelphid specimens: IGM 250597, three badly broken mandibular fragments and an isolated lower molar from Duke Locality 32 in the El Cardón Red Beds (Villavieja Formation); IGM 184600, an isolated left lower molar (fig. 11.2I) from an unknown locality; and IGM 253055, a partial left lower molar from an unknown locality. Even though these specimens show some general resemblances with specimens here assigned to *Micoureus laventicus,* they differ in the following features: larger size, talonids comparatively shorter, trigonids much stronger and with prominent protoconids, and paracristids sharp and well developed, indicating carnivorous habits. The three specimens clearly belong to a single taxon. It is my opinion that they may not belong to a marmosine but to a new didelphine didelphid.

Among the didelphid material from Colombia also is a partial skull with an almost complete palate and left jugal dentition, collected by the Kyōto University Expeditions (IGM KU-IV-1). This very interesting specimen may represent a new genus and species of Didelphidae, for it apparently bears some unique features. Its dentition seems to correspond in size and occlusal morphology to the mentioned IGM specimens. In any case, pending a thorough study of the original skull and upper dentition of IGM KU-IV-1, I prefer not to assign these specimens to a new taxon.

Order Sparassodonta (Ameghino 1894)
Aplin and Archer 1987
Superfamily Borhyaenoidea (Ameghino 1894)
Simpson 1930
Family Hondadelphidae Marshall,
Case, and Woodburne 1989
Hondadelphys Marshall 1976a
Hondadelphys fieldsi Marshall 1976a

NEWLY REFERRED SPECIMENS

IGM 253078, two fragments of a left horizontal ramus with p2 and m3–4; IGM 253050, partial right horizontal ramus with talonid of m1 and m2–4 complete; IGM 253079, partial left horizontal ramus with roots of m1, and broken m2–4; and IGM 253049, partial left horizontal ramus with p2–3 and m2–4 complete, collected by Kyōto University at an unknown locality. IGM 250833, fragmentary right mandible with broken m2–4 from Duke Locality 22; and IGM 250961, anterior portion of left horizontal ramus with part of the canine, p1, and broken p2 from Duke Locality 113 (fig. 11.1D). Measurements are given in table 11.2.

COMMENTS

Hondadelphys fieldsi is an enigmatic sparassodont whose systematic position and borhyaenoid affinities are unclear. In his initial description of *Hondadelphys fieldsi,* Marshall (1976a, 416) assumed that it and the didelphine *Thylophorops chapalmalensis* "are more closely related to each other than either is to any other known didelphid, fossil or Recent." Crochet (1980) doubted the didelphine status of

Hondadelphys because of the straight centrocrista in the upper molars. Later, it was considered a specialized didelphine:

> *Hondadelphys* also seems to belong to this group, although Crochet (1980, 96) remarked that it hardly seems to belong to the Didelphinae, mainly on account of its straight centrocrista and reduction of the anterior portion of the stylar shelf on M2. Yet, in most other characters analyzed by us, *Hondadelphys* approaches the group of large-bodied opossums with 2n = 22 chromosomes, and in spite of the peculiar and quite unexpected feature of seemingly predilamdodont molar structure, it is clearly part of the large-bodied group in Wagner trees. (Reig et al. 1987, 73)

In the same year and volume, Aplin and Archer (1987, xxxiv) suggested including *Hondadelphys* among the Sparassodonta: "We would draw attention to the presence in *H. fieldsi* of a straight centrocrista, a poorly developed alisphenoid tympanic process and no mastoid contribution on the hind surface of the skull. However, it lacks the naso-lachrymal contact otherwise ubiquitous in borhyaenids." Finally Marshall, Case, and Woodburne (1989) considered *Hondadelphys* the only known genus of a new family, Hondadelphidae, Superfamily Borhyaenoidea, Order Sparassodonta. To clarify the affinities of *Hondadelphys fieldsi,* several features of this taxon deserve further comment.

In known Didelphoidea, fossil and living, incisor number is not reduced. A decrease of incisor number characterizes many other marsupial lineages, however, reaching extremes in some sparassodont groups (see Marshall 1978; Goin and Pascual 1987; Marshall et al. 1989). In his original description of the holotype of *H. fieldsi,* Marshall (1976a, 412) stated that "$I_1 > I_2 \geq I_3 > I_4$; I_4 is extremely small." More recently, Marshall et al. (1989) maintained that there is secure evidence for only three lower incisors. I have recently examined an epoxy resin cast of UCMP 37960, the holotype of *Hondadelphys fieldsi* (see Goin 1991), and now the referred specimen IGM 250961, which preserves the precanine mandibular structure (see fig. 11.1D). Only three alveoli and incisor roots can be observed; the middle one is more lingually implanted, corresponding to i3, the "staggered" marsupial incisor of Hershkovitz (1982). The marsupial lower dentition lacks i1 (Hershkovitz 1982a); therefore, the three incisors in *Hondadelpys fieldsi* are homologous

Table 11.2. Dental measurements (lengths and widths, mm) for *Hondadelphys*

Specimen	p1–m4	m1–4	p1L	p1W	p2L	p2W	p3L	p3W	m1L	m1W	m2L	m2W	m3L	m3W	m4L	m4W
H. fieldsi																
IGM 253078	—	—	—	—	6.0	2.1	—	—	—	—	8.0	—	7.4	—	—	4.0
IGM 250961	—	—	4.5	1.9	~6.1	2.1	—	—	—	—	—	—	—	—	—	—
IGM 253050	—	~27.2	—	—	—	—	—	—	~7.0	—	6.9	3.4	7.0	3.7	—	4.0
IGM 253079	—	~26.8	—	—	—	—	—	—	~6.8	—	6.8	3.4	7.1	~3.6	~7.8	~4.4
IGM 250833	—	—	~5.0	—	~5.7	—	~6.8	—	~6.6	—	~6.8	—	~6.9	—	~6.8	—
IGM 253049	~50.1	~28.4	—	—	6.3	2.5	6.6	2.6	—	—	6.8	3.4	7.4	3.8	7.5	4.0
?*Hondadelphys*																
IGM 250962	—	—	—	—	—	—	—	—	—	—	—	—	6.5	—	6.7	—
IGM 250364	—	—	—	—	5.2	—	5.7	2.9	~6.1	2.6	—	—	—	—	—	—
IGM 184041	~48.0	26.5	4.6	2.0	5.8	2.4	—	—	—	—	—	—	—	—	—	—

Specimen	P1–M4	M1–4	P1L	P1W	P2L	P2W	P3L	P3W	M1L	M1W	M2L	M2W	M3L	M3W	M4L	M4W
IGM 184041																
Right side	47.5	24.1	—	—	—	—	—	—	6.5	4.6	7.3	5.8	6.5	6.7	~3.1	—
Left side	48.4	—	4.0	1.8	—	—	—	—	—	—	—	6.0	7.0	7.3	3.5	7.3

with i2, i3, and i4 so that Marshall's "I_1" corresponds to i2, his "I_2" to i_3, and his "I3" to i4. There is no evidence of "I_4" (i5) among the studied specimens.

There are several conspicuous differences between the upper molar dentition of *H. fieldsi* and that of the Didelphoidea. (1) A straight centrocrista is clearly visible in M1–4 (this feature is very striking in *H. fieldsi* and not "seemingly predilambdodont" as stated by Reig et al. 1987). Among the Didelphoidea, V-shaped centrocristae can only be clearly observed in marmosines and the Metachirini. In *Caluromys* this feature is very poorly developed and seems to occupy an intermediate position. Among the Didelphinae the centrocrista is straight on the M0 and M4 of most species; it may be straight in M3 of *Lutreolina crassicaudata*, in M2–4 of *Philander opossum*, and, frequently, in M1–2 of *Didelphis albiventris* and *D. marsupialis*. Among fossil taxa the centrocrista is fairly straight on the M1–2 of *Thylophorops chapalmalensis*, at least on M2–3 of *Hyperdidelphys biforata*, and M1–2 of *Didelphis crucialis*. Thus, significant variability can be found. The significant point here is that, in *Hondadelphys fieldsi*, a straight centrocrista is present throughout the molar series, something that does not occur in any Didelphoidea. (2) Para- and metaconules are better developed in *H. fieldsi* than in didelphoid opossums. (3) In *Hondadelphys fieldsi* the preparacrista does not connect with stylar cusp B, but with cusp A (at least on M1–2). This is contrary to the statement of Marshall (1976a) and in contrast to the condition in most Didelphoidea. (4) Finally, in *Hondadelphys fieldsi* the stylar shelf is greatly reduced and does not bear a stylar cusp B.

The lower molars of *H. fieldsi* are peculiar in the following features: (1) the extreme reduction of the metaconid, (2) the great reduction of the anterobasal cingulum, (3) the remarkable development and depth of the talonids, and (4) the absence of molar imbrication in the jugal row (Goin 1988). Characters 1 and 2 are either infrequent or simply do not exist among Didelphoidea, whereas characters 3 and 4 are infrequent or simply do not exist among Borhyaenoidea.

As noted by Marshall (1976a), the basicranium and ear region of *H. fieldsi* have some characters unknown in didelphoids. These include "loss of pars mastoidea and tympanic process of petrosal, development of a large posterior epitympanic sinus, absence of transverse canals, and enlargement of the foramen lacerum medium and posterior carotid foramina, with the latter opening within the basioccipital" (Marshall 1976a, 416).

Aplin and Archer (1987) and Marshall et al. (1989, 439, node 21) assumed that *Hondadelphys fieldsi* lacks a distinct naso-lacrimal contact. Marshall's (1976a) original description, however, is ambiguous. With respect to the sutures at the posterior end of nasals he states: "Nasofrontal contact is similar to *D. marsupialis*. Maxillaries are similar to *D. marsupialis*, as are sutural patterns of jugal and lacrimal" (Marshall 1976a, 413). From this it is not clear if there is a distinct frontomaxillary contact or a nasolacrimal contact. In his discussion on the affinities of *Hondadelphys* Marshall does not mention these contacts as an argument for either didelphoid or borhyaenoid relationships. I suggest that in neither the holotype nor the paratype is there clear evidence of the sutural relationships in the nasal-frontal-lacrimal area. As elaborated here, however, two specimens referred to ?*Hondadelphys* (IGM 250364, see fig. 11.3; and IGM 184041) show two diagnostic sparassodont features; an unfenestrated palate and a distinct nasolacrimal contact.

It seems clear that *Hondadelphys* should be excluded from the Didelphoidea, but the sparassodont affinities of this genus are uncertain. On the basis of some coincident features in the molar arrangement in the jugal row, Goin (1991) recently suggested a possible link between *Hondadelphys* and thylacosmilids. The structure of the lower molars in IGM 251148 (?Thylacosmilidae; see this chap.) makes this suggestion even more interesting.

Hondadelphidae, gen. et sp. indet.
(*Hondadelphys?*)

Six other badly preserved IGM specimens are probably referable to *Hondadelphys*: IGM 250962, partial left ramus, lingually and labially deformed, with roots for m3–4, from Duke Locality 113, La Victoria Formation; IGM 250578, a fragmentary p?3 from Duke Locality CVP-14, Fish Bed, Villavieja Formation; IGM 251215, small portion of right mandible with broken m4, from Duke Locality 132, La Victoria Formation; IGM 250471, lower canine from Duke Locality 106, La Victoria Formation; IGM 250364, facial portion of skull with anterior jugal teeth badly broken (fig. 11.3), a fragment of horizontal ramus with incomplete

COMMENTS

As mentioned, the two new specimens preserving the facial portions of the skull (IGM 250364 and IGM 184041) show two diagnostic features of the Sparassodonta; unfenestrated palate and distinct nasolacrimal contact. IGM 250364 differs from the holotype of *Hondadelphys fieldsi* in its slightly smaller size, distinct canine implantation, and the relative sizes of P2 and P3 (in the holotype of *H. fieldsi* P3 is apparently much larger than P2). In addition, the left femur preserved with IGM 250364 is somewhat longer and more slender than that of the holotype of *H. fieldsi* (Marshall 1976a). Nevertheless, these specimens show several diagnostic features of *Hondadelphys:* m1 bears a minute, vestigial metaconid; the m1 talonid (partially broken) seems to have been wide and deeply basined; and, as can be inferred from the partially preserved right M2, the upper molars have well-developed protocones, the paracones are set clearly apart from the metacones, and the centrocrista is well developed and straight.

IGM 184041, a surface find, consists of remains assignable to three different taxa, including a notoungulate calcaneum (*Miocochilius?*), limb bones of a large sparassodont, and cranial remains here referred to ?*Hondadelphys.* The cranial remains differ from the holotype of *H. fieldsi* in that the ventral edge of the horizontal ramus is not straight but anteriorly bowed, the M4 is narrower, M1–2 protocones are slightly wider, and M2 paracrista is oriented differently. The cheek teeth, however, are very similar to those of *H. fieldsi.*

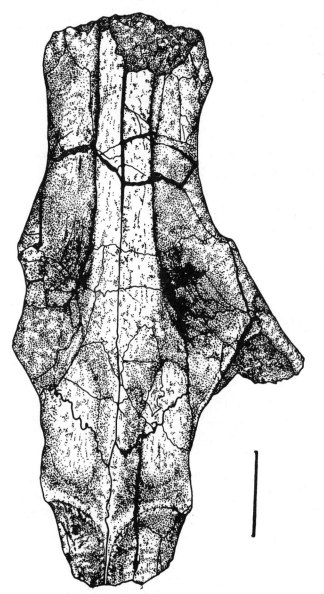

Figure 11.3. ?*Hondadelphys,* IGM 250364, facial portion of skull, dorsal view. Scale bar 10 mm.

c–m1, and associated postcranial bones, most of them very fragmentary, from Duke Locality 11 B, La Victoria Formation; and IGM 184041, posterior and facial parts of a skull with left and right cheek dentition poorly preserved, left horizontal ramus with roots of c, p1–2 complete and alveoli for p3–m4 (fig. 11.1E), and a fragment of right horizontal ramus with part of the canine and p1–2 complete, from Duke Locality 49, La Victoria Formation. Measurements are given in table 11.2.

Family **Borhyaenidae** Ameghino 1894
Subfamily **Prothylacyninae** (Ameghino 1894)
Trouessart 1898
Lycopsis Cabrera 1927
Lycopsis longirostrus Marshall 1977

NEWLY REFERRED SPECIMEN

IGM 250974, right horizontal ramus with p2 complete, alveoli for m1–3, m4 broken, an isolated incisor, and some fragmentary cranial and postcranial bones (fig. 11.4). Measurements are given in table 11.1.

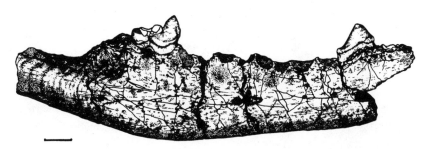

Figure 11.4. *Lycopsis longirostrus* Marshall 1976a, IGM 250974, right horizontal ramus with complete p2 and broken m4, labial view. Scale bar 10 mm.

LOCALITY

Duke Locality 113, at a level between the Chunchullo Sandstone and the Tatacoa Sandstone beds, La Victoria Formation, Honda Group.

COMMENTS

In comparable morphology and measurements, IGM 250974 is almost identical to the holotype of *Lycopsis longirostrus,* UCMP 38061 (see Marshall 1977). Three minor differences, however, can be noted: (1) There are five mental foramina on the labial side of the mandible, instead of the six in the holotype. (2) The mandibular symphysis in IGM 250974 is much more complex and better developed than that of the holotype. (3) The p1 is more obliquely set in the jaw. Some of these differences are probably age related because the holotype belonged to a subadult, as inferred by the incomplete eruption of m4.

Dukecynus gen. nov.

ETYMOLOGY

Duke, in homage to Duke University and Duke University paleontologists who, through years of field work (1985–91) in the upper Magdalena valley of Colombia, were able to make outstanding collections of fossil vertebrates from equatorial South America; *cynus,* dog, in reference to the Prothylacyninae, extinct "doglike" marsupials.

TYPE SPECIES

Dukecynus magnus sp. nov.

REFERRED SPECIES

Only the type species.

GEOCHRONOLOGIC AND GEOGRAPHIC DISTRIBUTION

Middle Miocene of Colombia.

DIAGNOSIS

As for the species, given here.

Dukecynus magnus sp. nov.

ETYMOLOGY

[M]*agnus,* in reference to it being the largest known species of Prothylacyninae.

HOLOTYPE

IGM 251149, heavily cracked and deformed partial skull, anterior fragment of left horizontal ramus, almost complete right mandible with partial dentition badly preserved, isolated m2, m4, and M?3, and associated cranial and postcranial fragments, most of them poorly preserved (fig. 11.5). For measurements, see table 11.1.

TYPE LOCALITY

Duke Locality 140, Polonia Red Beds, Villavieja Formation, Honda Group.

HYPODIGM

The type only.

DIAGNOSIS

Differs from all other prothylacynines in the following combination of characters: elongated and very narrow ros-

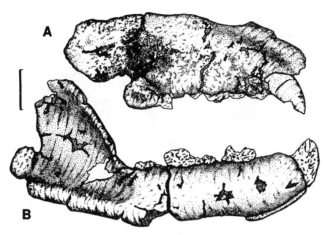

Figure 11.5. Holotype of *Dukecynus magnus* gen. et sp. nov., IGM 251149. *A*. Right side of facial portion of skull. *B*. Right mandible. Scale bar 30 mm.

trum; mandibular symphysis fused but not completely co-ossified; lower molars lack entoconid, m1–3 with moderately reduced talonids, m4 with very reduced or vestigial talonid; upper molars with reduced paracone, well-developed metacrista, protocone apparently reduced and set anterolingually, and paracrista ending at the anterolabial edge of the tooth.

DESCRIPTION AND COMMENTS

The holotype is intensely cracked and laterally compressed and will need further preparation to reveal details of the morphology. Notwithstanding, most of the diagnostic features of the Prothylacyninae are present. The mandibular symphysis extends to a point below p3, a large mental foramen occurs below p2, the canines are only moderately developed, p1 is set obliquely in the jaw, p3 is only moderately larger than p2, and the lower molars have moderately developed talonids and metacristae. The holotype lacks several features usually present in borhyaenines of comparable size—that is, a short snout, strong mandibles, crowded jugal tooth rows with inflated premolar or molar roots, or very large canines with the uppers somewhat vertically implanted.

Dukecynus magnus is the largest known species of Prothylacyninae and bears a unique combination of characters among members of this subfamily of Borhyaenidae. The

mandibular symphysis is tightly interlocking but apparently not completely co-ossified because both rami can be seen at their anterior ventral edges. Below the anterior root of p2 is a large mental foramen, but it is not as large as seen in the Montehermosan *Stylocynus*. The jugal teeth are not crowded in the jaw, and p1 is set only slightly obliquely. The third lower premolar is trenchant and laterally compressed; its posterobasal heel is very reduced. Two isolated lower left molars (m2 and m4) and a left upper molar (M?3) are the best-preserved teeth (fig. 11.1H, I). Talonid development on m2 is intermediate between that of *Pseudothylacynus rectus* (with comparatively reduced talonids) and *Stylocynus paranensis* (with much larger talonids). The m2 hypoconulid is very well developed, and the cristid obliqua is low and does not end in a distinct hypoconid. The m4 talonid was apparently minute or vestigial. The M?3 protocone is broken but was apparently reduced, as in *Prothylacynus patagonicus* (Santacrucian). There is almost no stylar shelf, and the postmetacrista is sharp and well developed. The preparacrista is not aligned with paracone and metacone but ends at the anterolabial edge of the tooth. The ectoflexus is very shallow. Even though badly deformed, it is clear that the skull is elongate (comparatively more than in *Lycopsis longirostrus*) and very narrow anteriorly. P1 is set somewhat obliquely in the maxillary, and the canine projects somewhat anteriorly. The upper cheek teeth are very poorly preserved, except the isolated left M?3 already mentioned. The palate ends anteriorly in a pair of small incisor vacuities.

It is difficult to suggest direct affinities between *Dukecynus magnus* and any other known Prothylacyninae. Only one member of this subfamily is comparable in size, *Stylocynus paranensis*, but differences in mandibular and molar morphology are substantial. In particular, the lower premolars of *Dukecynus* are comparatively smaller, p1 is more obliquely set in the jaw, the lower molars have much more reduced talonids and lack metaconids, the mandibular ramus is stronger with a smaller mental foramen and fused symphysis; the upper molars have comparatively larger postmetacristas and reduced protocones.

Dukecynus magnus is also much larger than other prothylacynines. Compared with *Pseudothylacynus rectus*, it differs in having a less-crowded jugal dentition, better-developed lower molar talonids, reduced posterobasal heels on the lower premolars, and a fused symphysis. *Dukecynus* also differs from species of *Lycopsis* in having reduced hypoconids and, apparently, protocones; no diastemata between the premolars, obliquely set upper and lower first premolars, a fused symphysis and a reduced infraorbital foramen. Compared with *Prothylacynus patagonicus, Dukecynus* has a longer and narrower snout, less-crowded jugal teeth, better-developed talonids, a distinct preparacrista, and more-elongated upper and lower molars. Finally, *Dukecynus* also differs from the Chasicoan *Pseudolycopsis cabrerai* in having a comparatively larger P3, longer and narrower snout, and upper molar paracrista more obliquely set.

Dukecynus magnus seems to have been more carnivorous than omnivorous in its feeding habits. With the exception of its elongated snout, a feature also present in *Lycopsis,* the peculiar specializations of *Dukecynus* are distinct from those of *Lycopsis*. Some general resemblances in upper molar morphology suggest closer affinities with *Pseudolycopsis cabrerai*.

Stirton (1953b), and later Marshall (1976a, 1977), mentioned the existence of a large borhyaenid from Miocene Honda Group deposits in the La Venta badlands. The specimen (UCMP 39250) "is known from a fragmentary skull and associated skeleton. Apart from being a large animal with crowded and obliquely set premolars the specimen is indeterminate" (Marshall 1976a, 418). I have not seen UCMP 39250, but I mention it here because it may eventually be demonstrated to belong to *Dukecynus magnus*.

Family Thylacosmilidae (Riggs 1933)
Marshall 1976b
Anachlysictis gen. nov.

ETYMOLOGY

[A]*na,* from the Greek form that means "from down to up" or "toward"; and *Achlysictis,* the type genus of Thylacosmilidae.

TYPE SPECIES

Anachlysictis gracilis sp. nov.

REFERRED SPECIES

Only the type species.

GEOCHRONOLOGIC AND GEOGRAPHIC DISTRIBUTION

Middle Miocene Honda Group of Colombia.

DIAGNOSIS

As for the type and only known species of the genus, given next.

COMMENTS

In their recent review of the biology and taxonomy of the South American thylacosmilids, Goin and Pascual (1987) concluded that all known remains of this family could be regarded as a single species. Invoking Article 23a of the International Code of Zoological Nomenclature, we proposed to keep the name *Thylacosmilus atrox* Riggs 1933, instead of *Achlysictis lelongi* Ameghino 1891, which would strictly correspond in terms of priority. Our recommendation to recognize a name that has been widely known for more than fifty years had no success (see, e.g., Marshall et al. 1989).

Anachlysictis gracilis sp. nov.

ETYMOLOGY

From the Latin *gracilis,* slender.

HOLOTYPE

IGM 184247, right horizontal ramus with almost complete m2, and broken m3–4; left mandibular fragment with m2–3 complete (fig. 11.1J); fragment of left symphysis; very fragmentary left frontolacrimal portion of the skull, and associated, very fragmentary postcranial material (fig. 11.6). Measurements in table 11.1.

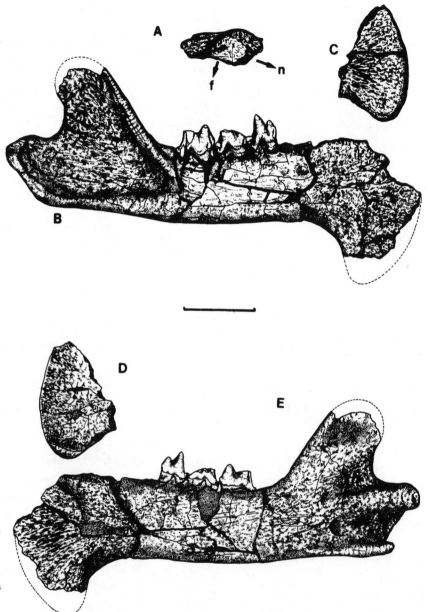

Figure 11.6. Holotype of *Anachlysictis gracilis* gen. et sp. nov., IGM 184247. *A.* Fragment of skull roof, dorsal view (*f*, frontal; *n*, nasal). *B, E.* Right horizontal ramus: *B*, labial view; *E*, lingual view. *C–D.* Fragment of left ramus at the symphyseal region: *C*, lingual view; *D*, labial view. Scale bar 30 mm.

TYPE LOCALITY

Duke Locality 75, at a level between the Chunchullo Sandstone Beds and the Tatacoa Sandstone Beds, La Victoria Formation, Honda Group.

HYPODIGM

The type only.

DIAGNOSIS

Differs from *Achlysictis lelongi* in its smaller size, mandible much more slender and with the symphyseal flange much less developed; lower premolars double-rooted and apparently not conical; lower molars, except m4, with small but distinct talonids; distance from m4 to mandibular condyle comparatively much longer; masseteric fossa much better developed; facial portion of skull apparently flat, with the ascending process of the maxillary very reduced or lacking; postorbital bar absent.

COMMENTS

This remarkable new material sheds much light on our understanding of thylacosmilid evolution. The best-preserved portion of the holotype consists of a right horizontal ramus lacking only the anterior and lower borders of the symphyseal flange and the uppermost portion of the masseteric crest (fig. 11.6). Most of the diagnostic features for the Thylacosmilidae are present in this ramus: (1) large, subvertical symphyseal flange with radially oriented, lingual bony striations; (2) alveolar and ventral edges of the ramus are subparallel and straight; (3) the masseteric crest is low and poorly developed; (4) the poorly inflected angle is oriented posteriorly; (5) the condyle is low in relation to the alveolar plane; and (6) the jugal series is somewhat bowed with its convexity labially oriented (see Goin and Pascual 1987). In some features *Anachlysictis gracilis* differs from *Achlysictis lelongi* only in the lesser development of its thylacosmilid specializations—that is, much less developed symphyseal flange (but typically thylacosmilid in general pattern); smaller, more vertical ramus (but not weakly developed as in *Achlysictis*); and very low condyle (but not as low as in *Achlysictis*). In other features, however, *Anachlysictis gracilis* is clearly different. *Anachlysictis* displays a well-developed masseteric fossa so that the last molar is comparatively much more distant from the condyle than in *Achlysictis;* the premolars are double-rooted and probably not conical as in *Achlysictis;* the lower molars (except m4) have small but distinct talonids; and the last molars are slightly imbricated (fig. 11.1J).

IGM 184247 includes a small portion of the skull roof at the frontonasal contact. From this small fragment it can be established that (l) the skull roof is flat, at least in the orbital area; (2) there is no postorbital bar; (3) there is no evidence of an ascending process of the maxillary following the canine implantation; and (4) the frontal crests converge posteriorly from the postorbital process to the sagittal crest. None of these features is present in *Achlysictis,* and features 1 and 3 suggest that the intraalveolar extention of the upper canine was much less than in *Achlysictis,* resulting in a differently shaped facial portion of the skull (fig. 11.6).

In *Achlysictis* the presence of a postorbital bar, together with the peculiar dorsal profile of the skull, the high and salient occipital condyles, and the intraalveolar extention of the upper canines, has been correlated with a distinct posture of the skull and predation strategy (Goin and Pascual 1987). On the basis of its known remains, the masticatory mechanics of *Anachlysictis* was almost certainly unlike *Achlysictis.* Jugal teeth of *Achlysictis lelongi* are not crowded but set in an elongated row, so that the last molar lies very near the mandibular condyle. *Anachlysictis gracilis,* on the contrary, displays some degree of molar imbrication (Goin 1988), so that its m4 is set closer to the midpoint of the horizontal ramus. Crowding of the jugal teeth and molar imbrication are typical sparassodont adaptations to maximize bite forces at the midpoint of the jaw, where the largest tooth of the molar series (m4) is usually positioned.

Anachlysictis gracilis makes an ideal ancestor for *Achlysictis lelongi;* it is geologically older and all comparable features are relatively more primitive.

Borhyaenoidea, fam. et gen. incertae sedis (Thylacosmilidae?)

REFERRED SPECIMEN

IGM 251108, partial left horizontal ramus with alveoli for c–m1, right horizontal ramus with partially broken canine and m4 and almost complete p1–m2, right maxillary fragment with the canine partially broken and roots for p1, and associated poorly preserved and fragmentary postcranial bones, from Duke Locality 132 in the La Victoria Formation (figs. 11.1E, 11.7) For measurements, see table 11.1.

COMMENTS

This very interesting specimen is badly damaged and will need more preparation before any conclusion is reached about its phylogenetic affinities or adaptations. Nevertheless, I mention it here because it clearly represents a new taxon for the Miocene of Colombia and because some of its peculiar features may prove to be thylacosmilid synapomorphies. At first sight it resembles a *Cladosictis*-like sparassodont, but both the mandible and the dentition show unique features. The lower canines are not subcircular in cross

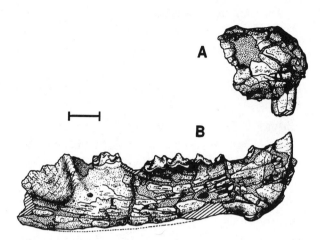

Figure 11.7. ?Thylacosmilidae, gen. et sp. indet., IGM 251108. *A.* Anterior portion of skull, lateral view. *B.* Right mandible, lateral view. Scale bar 10 mm.

section but are laterally compressed and obliquely set in the jaw. There is no evidence for incisors or alveoli anterior to the canines. There are wide diastemata between the canine and p1, and between p1 and p2. The first lower premolar is very small; p2 and p3 are subequal in length, although p3 seems to be larger in width. The lower molars increase in size from front to back, lack entoconids, and m2 has a well-developed talonid and anterobasal cingulum. The talonids on m2–4 are peculiar in that their basins are labially open—that is, they are not bounded by the cristid obliqua, which is extremely low and forms part of the talonid floor.

The mandible shows an incipiently developed symphyseal flange that is not as extended as in *Anachlysictis* or, of course, *Achlysictis,* but it presents a peculiar profile unmatched in any other sparassodont. The anteroventral margin of the horizontal ramus is flattened in the coronal plane, and in lateral view the anterior margin is colinear with the canine alveolus and prolonged inferiorly to form an incipient flange.

The preserved right maxillary is also oddly blunt at the anterior edge. The upper canine, disproportionately larger than the lower canine, is subvertically implanted. The enamel is clearly thicker labially than lingually. The first upper premolar is very small and lingually implanted in relation to the canine. There is no evidence of a premaxillary-maxillary suture, nor of incisor teeth or alveoli.

Although the lower postcanine dentition of IGM

251108 agrees in general terms with other hathlyacynid sparassodonts, upper and lower canine morphology, the symphyseal area of the mandible, and the preserved maxillary strongly suggest affinities with the Thylacosmilidae. A case could be made, however, that it represents a distinct borhyaenoid lineage that paralleled thylacosmilids in its facial adaptations, as has occurred in several taxa of placental sabertooths.

Sparassodonta, fam. et gen. indet.

Two other INGEOMINAS specimens can be assigned to the Sparassodonta, despite the fact that they are fragmentary and that it would be futile to attempt to establish more precise phylogenetic affinities on the basis of only these remains.

IGM 250475 is a right maxillary fragment with the roots and crown of P1, from Duke Locality 106. The large size of this specimen and the strong, inflated P1 roots suggest that it may be a large borhyaenine. P1, however, is implaced in-line with the tooth row, an uncommon feature among the group.

IGM 184041, part of the hind skeleton of a medium- to large-sized borhyaenid, is from Duke Locality 49. As already mentioned, IGM 184041 consists of bone remains of at least three different taxa; a notoungulate, ?*Hondadelphys,* and a larger borhyaenid. The bones of this larger borhyaenid are presently being prepared. They clearly do not belong to *Hondadelphys* nor to *Dukecynus.* In size and general morphology, these remains may correspond to *Lycopsis* or a *Lycopsis*-like prothylacynine.

Discussion

The marsupial fauna from the middle Miocene Honda Group of Colombia represents one of the most diverse marsupial fossil assemblages known from the Neogene of South America. Four marsupial orders and at least twelve species (two of them indeterminate) are represented (table 11.3). In addition, two other sparassodont species may also be represented among the INGEOMINAS collections (IGM 250475 and IGM 251196, here considered as Sparassodonta, fam. et gen. indet.).

This faunal assemblage includes a diverse range of body

Table 11.3. The twelve lowest identifiable taxa of marsupials in the middle Miocene Honda Group of Colombia

Order	Lower Taxon
Microbiotheria	*Pachybiotherium minor* sp. nov.
Didelphimorphia	*Thylamys minutus* sp. nov.
	Thylamys colombianus sp. nov.
	Micoureus laventicus
	Didelphidae, gen. et sp. indet.
Sparassodonta	*Hondadelphys fieldsi*
	Lycopsis longirostrus
	Dukecynus magnus gen. et sp. nov.
	Anachlysictis gracilis gen. et sp. nov.
	?Thylacosmilidae, gen. et sp. indet.
Paucituberculata	*Pithiculites chenche*
	Hondathentes cazador

Note: The two Paucituberculata species are described in chapter 12.

sizes, from the tiny marmosine didelphoid *Thylamys minutus* to the giant prothylacynine borhyaenoid *Dukecynus magnus*. The inferred feeding habits of each of these marsupials indicates a remarkable variety as well, including the frugivorous-insectivorous *Pachybiotherium* and Caenolestoidea, the insectivorous-carnivorous Marmosinae, omnivorous-carnivorous Didelphidae indet., and the predominantly carnivorous Sparassodonta.

In general terms, three important evolutionary events are indicated by the middle Miocene marsupials from Colombia: first, the persistence, in the South American tropics, of old frugivore-insectivore lineages known in the continent since the early to middle Tertiary (microbiotheriids and caenolestoids); second, the probable origin of a didelphimorphian Neogene radiation from a basically insectivorous stock; and third, the differentiation of new carnivorous lineages among the Sparassodonta. Each of these inferred events will be considered further here.

Work in progress suggests that *Pachybiotherium minor* is the last surviving member of this genus and of a distinct clade among the Microbiotheriidae that also includes *Eomicrobiotherium* and, probably, *Mirandatherium*. It is interesting to note that, although there is no evidence for the presence of *Pachybiotherium* in Friasian deposits of Patagonia, there are records, in the Friasian, of species of *Microbiotherium* (the other microbiotheriid lineage). It

seems no coincidence that *Dromiciops* (a clear member of the *Microbiotherium* lineage) presently lives in the wet-temperate forests of southern Chile and Argentina. At least two features present in *Pachybiotherium* suggest primatelike adaptations: strong and deep horizontal rami, slightly bowed anteriorly (producing a somewhat rounded chin); and low-crowned molars with bunoid cusps (fig. 11.2A, B). These features, together with its small body size, suggest that *Pachybiotherium minor* had frugivorous-insectivorous habits, whereas the *Microbiotherium* clade is characterized by dental and mandible features that imply a more insectivorous diet.

Didelphids have been largely thought to be present in continental faunas of the New World as early as the late Cretaceous. The consideration of some supposedly conservative living taxa (e.g., *Didelphis marsupialis*) as "living fossils" has been common in the literature for the last few decades, and much work dealing with their physiology, locomotion, and masticatory mechanics has attempted to reflect the "primitive metatherian pattern" or even the most primitive mammalian pattern as well. It is my opinion that this dangerous mythology has obscured much thinking on marsupial origins and early radiations. I have recently proposed the use of a more restricted concept for the superfamily Didelphoidea in a sense that excludes all pre-Neogene didelphimorphian opossums (Groin 1991). This change reflects didelphimorph phylogeny more clearly, while accounting for the basic structural unity and monophyly of Neogene didelphoid opossums. I also suggested that marmosine opossums may represent the plesiomorphic sister-group of all other known didelphids. Interestingly, three of the four didelphimorph species from the middle Miocene of Colombia are clearly assignable to marmosine genera. Of them, *Micoureus* may well represent a structural ancestor for all didelphoid opossums. Living members of this genus show primitive states for several characters, including body size, otic region, dental morphology, reproductive tract, and chromosome number (Reig et al. 1987).

Apart from *Micoureus laventicus*, two other marmosine species are herein recognized in the La Venta fauna: *Thylamys minutus* and *T. colombianus*. Living members of this genus may enter into daily torpor or hibernation and have incrassated tails, features that indicate tolerance to seasonal variation in environmental conditions. Morton (1980) suggested that caudal fat storage in marsupials is a response to seasonal shortages in insect availability.

Finally, several unrelated sparassodont taxa can be distinguished in the La Venta fauna. Of these, the only species clearly related to an earlier taxon is the prothylacynine *Lycopsis longirostrus*. From Marshall's (1977, 1979b) studies, it seems clear that *L. longirostrus* is derived with respect to the Santacrucian species *L. torresi* from southern Argentina. Both species are dolichocephalic and were probably omnivorous. Another sparassodont, *Hondadelphys fieldsi,* shows omnivore adaptations in the dentition (large talonids and protocones). The phylogenetic affinities of this genus are far from clear. None of the pre-Friasian sparassodonts that are known make a plausible structural ancestor to *Hondadelphys.* Available information about living South American carnivores suggests that omnivory (or, at least, some vegetable component in the diet) is related to seasonal variation in food availability (Armesto et al. 1987).

Dukecynus magnus is the largest known prothylacynine. Despite some superficial similarities relating to facial elongation of the skull, *Dukecynus* shows no direct affinities with the *Lycopsis torresi*–*L. longirostrus* lineage. The cranial and dental morphology of *Dukecynus* indicate a more strictly carnivorous feeding habit.

The earliest record of unquestionable thylacosmilids is another novelty of the remarkable faunal assemblage from the Miocene Honda Group of Colombia. *Anachlysictis gracilis* is highly derived in dental and mandibular morphology and makes an ideal ancestor for the extremely specialized *Achlysictis lelongi,* from Huayquerian-Chapadmalalan levels in Argentina. Comparisons between *Anachlysictis* and *Achlysictis* suggest, however, that their predatory strategy and masticatory mechanics may have been quite different.

As is the case for the Cingulata, a faunal correlation between the La Venta fauna and the Friasian age faunas of Patagonia age would be impossible on the basis of the marsupial taxa alone (see also Marshall 1976a). Friasian marsupials of Colombia comprise a unique assemblage among marsupial faunas of equivalent age. L. G. Marshall (1990) suggested that the marsupial fauna from the type Friasian Age at Alto Río Cisnes, Chile, may be of Santacrucian Age. As far as the didelphimorphs and sparassodonts are concerned, one constraint on the age of the La Venta fauna is clear: it is a post-Santacrucian assemblage. Several recorded taxa most probably represent new Neogene radiations (marmosines, thylacosmilids). Others (*Hondadelphys, Dukecynus*) are more difficult to interpret, given the present state of knowledge. In any case, the fossil marsupials from Colombia constitute remarkable clues for the understanding of marsupial evolution in South America. As can be deduced from many other vertebrate groups, Miocene life in the tropics was as diverse and surprising as it is today, or even more so.

Acknowledgments

I am indebted to R. F. Kay and R. H. Madden (Duke University Medical Center, USA) and to A. López Reina (INGEOMINAS, Colombia), who made the collection of marsupials from the Miocene of Colombia available to me for study. M. Bond kindly oriented me and discussed several aspects of this work. G. Scillato-Yané, S. Vizcaíno, and M. Bond critically reviewed the manuscript. M. Lezcano prepared figures 11.3–11.6. Many thanks to P. Sarmiento for her help in the preparation of specimens for the scanning electron photomicroscopy. Finally, many thanks to R. H. Madden for helping me with the English version of this work.

12
New Caenolestoid Marsupials

Elizabeth R. Dumont and Thomas M. Bown

RESUMEN

En base a material mandibular de Caenolestoidea, se describe un nuevo género de Palaeothentidae y una nueva especie del Abderitidae *Pithiculites* para el Grupo Honda. Estos registros extienden la distribución geográfica y temporal de estas dos familias. El nuevo Palaeothentidae, *Hondathentes cazador,* con dientes jugales exodinodontes, el troncal generalizado de este linaje, presenta una mezcla de rasgos primitivos y derivados que sugieren un período largo de aislamiento de otros linajes de la familia. Entre éstas características se señalan un p3 inferior de corona alta y simple y un primer molar inferior de cúspides bajas, trigónido acotado anteroposteriormente, paracrístida oblícua y sin paracónido bifurcado. *Pithiculites chenche,* la especie más grande de un género reconocido también en depósitos del Colhuehuapense en Patagonia, se identifica en base a un primer molar inferior en forma de hoja de navaja con tres dentículos y pocas rayaduras. Ninguno de estos dos taxa nuevos para la fauna de La Venta diverge marcadamente de taxa antes considerados como frugívoros o faunívoros, de manera que estos nuevos caenolestoideos probablemente tuvieron dietas semejantes.

Although living caenolestoid marsupials represent a single family confined to the high altitudes of western South America, the Tertiary fossil record illustrates that caenolestoids once exhibited remarkable geographic and taxonomic diversity (Marshall 1976; Patterson and Marshall 1976; Bown et al. 1990; L. G. Marshall 1990; Bown and Fleagle 1993). Representatives of three families are known from Tertiary deposits in Argentina, Chile, and Bolivia. The discovery of new caenolestoid species from the Honda Group of Colombia further expands the geographic and temporal ranges of two families. Through analogies to reconstructed dietary adaptations and diversity patterns of abderitid and palaeothentid caenolestoids from Argentina, the occurrence of these new taxa in this tropical fauna offers insights into the paleoenvironment and paleoecology of the Honda Group during the Miocene.

Abbreviation: IGM, prefix for specimens housed at the Instituto de Investigaciones en Geociencias, Minería y Química, Museo Geológico Bogotá (INGEOMINAS).

Systematic Paleontology

Superorder Marsupialia
Order Paucituberculata
Superfamily Caenolestoidea
Family Abderitidae
Pithiculites Ameghino 1902
Pithiculites chenche sp. nov.

TYPE SPECIMEN

IGM 250941, fragmentary right dentary with P_3 through M_3 (fig. 12.1); only known specimen.

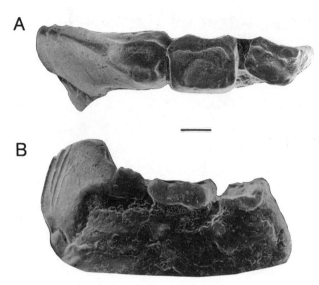

A

B

Figure 12.1. *Pithiculites chenche,* IGM 250941, type specimen, fragment of right ramus with P_3, M_1–M_3: *A,* occlusal view; *B,* lingual view. Scale bar 1 mm.

Table 12.1. Measurements (mm) of cheek teeth of *Pithiculites chenche* **sp. nov. and** *Hondathentes cazador* **gen. et sp. nov.**

Specimen	Tooth	Length	Breadth
Pithiculites chenche			
IGM 250941 (type)	M_1	4.2	1.9
	M_2	2.1	1.7
	M_3	1.5	1.3
Hondathentes cazador			
IGM 250440 (type)	P_3	1.3	1.1
	M_1	2.8	1.5
	M_2	2.3	1.6
IGM 251008	M^1	2.0	2.5
IGM 250342	M^3	1.5	1.8

HORIZON AND LOCALITY

La Victoria Formation, unit between Tatacoa Sandstone Beds and Chunchullo Sandstone Beds, Duke Locality 113.

DISTRIBUTION

Known only from the type locality; Miocene, Colombia.

ETYMOLOGY

Named for a mythical person of the Grand Tolima region of Colombia who decorates himself in jewelery of pure gold and emeralds and seeks sanctuary in the Magdalena River.

DIAGNOSIS

Larger than *Pithiculites minimus* and *P. rothi.* M_1 bladelike, exhibiting three denticles (as does *P. minimus*) and four striations. Denticles are apically positioned, lending trigonid a more squared outline than in *P. minimus.* M_1 talonid uniquely possesses three mesiodistally aligned cuspules, but the lingual edge of M_2 and, possibly, M_3 talonid is composed of three small cuspules as in *P. minimus* (measurements in table 12.1).

DESCRIPTION

The type is a fragmentary right dentary containing P_3 through M_3. All of the teeth are unbroken but somewhat worn. P_3 is vestigial and incorporated into a basal notch at the mesiolingual corner of M_1. The M_1 is enlarged and bladelike, exhibiting four striations both buccally and lingually. The anterior three striations end in small denticles separated by notches along the crest of the trigonid. The most distal striation is heavily worn and shows no evidence of having ended in a denticle. Overall, the outline of the trigonid has a squared appearance. Buccally, the postvallid is steep and smooth and lacks a hypoflexid. In contrast, the lingual edge of the trigonid slopes more gently toward the talonid and is separated from it by a slight talonid notch. The lingual edge of the talonid is uniquely decorated with three small cuspules that are aligned to form a mesiodistal ridge; the buccal edge of the talonid is smooth. The lingual portion of the postcristid is strongly inflected mesially, leaving isolated the small postcingulid.

The M_2 is worn relatively flat so that the lingual cusps are only slightly higher than the buccal cusps. One lingual cusp is visible on the trigonid. Two buccal cusps are represented by two conflated dentine islands. Three small, mesiodistally aligned cuspules form the lingual edge of the talonid. The hypoconid is large and slightly extended distally. Unlike the condition on M_1, the M_2 postcristid is relatively straight and there is no postcingulid.

The M_3 has been artificially shifted out of the axis of the tooth row. All of the cusps are heavily worn and are represented by dentine islands. As on M_2, the trigonid carries one lingual and two buccal cusps. In addition, although the tooth is very worn, it appears that two large and one small cusp form the lingual border of the talonid. The hypoconid is large and relatively more distally expanded than on M_2.

COMPARISONS

The new specimen of *P. chenche* can clearly be referred to *Pithiculites* based on its possession of three apical denticles and relatively few striations on M_1 and its small size relative to species of *Abderites*. Other abderitids are known from Colhuehuapian through Santacrucian deposits of southern Argentina and Frisian deposits of Río Cisnes, Chile. *P. chenche* is similar to *P. minimus* in having a lingual talonid boundary, formed by three distinct cuspules, on M_{2-3}. However, the possession of three distinct cuspules on the lingual edge of the M_1 talonid, the squared shape of the M_1 trigonid, and its large size demonstrate that *P. chenche* is clearly distinct from *P. minimus*. Similarities between *P. chenche* and *P. rothi* include an M_3 that is more rectangular than square and a relatively large M_2. *P. chenche* can be distinguished from *P. rothi* by the presence of three lingual talonid cusps on M_{2-3}, the lack of M_2 and M_3 postcingulids, poorly differentiated trigonids and talonids, and larger size. Overall, *P. chenche* shares many more features with *P. minimus* than with *P. rothi*.

Given that *P. rothi* is the most primitive known species of *Pithiculites* (L. G. Marshall 1990), the post-M_1 dentition of *P. minimus* and *P. chenche* must be viewed as equally derived. The unique features of *P. chenche* relative to those of *P. minimus* are aspects of M_1 morphology. Because the M_1 of *P. rothi* is unknown, it is not possible to determine whether *P. chenche* or *P. minimus* is the more derived taxon.

Family **Palaeothentidae**
Subfamily incertae sedis
Hondathentes gen. nov.

TYPE

IGM 250440, fragmentary right dentary with P_3 through M_2 (fig. 12.2A, B).

Figure 12.2. *Hondathentes cazador. A–B.* IGM 250440, type specimen, fragment of right ramus with P_3, M_1–M_2: *A,* occlusal view; *B,* lingual view. *C.* IGM 250342, isolated right M^1, occlusal view. *D.* IGM 251008, isolated right M^3, occlusal view. Scale bar 1 mm.

HYPODIGM

The type specimen, as well as IGM 250342, isolated right M^1 (Duke Locality CVP-8, Fish Bed, Villevieja Formation) (fig. 12.2C); IGM 251008, isolated right M^3 (Duke Locality CVP-9, Fish Bed, Villevieja Formation) (fig. 12.2D).

HORIZON AND LOCALITY

Villavieja Formation, El Cardón Redbeds, Duke Locality 32, Colombia.

DISTRIBUTION

Miocene, Colombia.

ETYMOLOGY

Named for the Honda Group of Colombia with the suffix *thentes* (Greek, hunter).

DIAGNOSIS

Small palaeothentid about the size of *Palaeothentes minutus*. Differs from *Palaeothentes* and resembles *Acdestis* in lacking an anterobasal cuspule on P_3, in having an M_1 with anterolingually (not anteriorly) oriented paracristid, and in lacking bifurcation of the M_1 paraconid. Differs from *Acdestis* and resembles *Palaeothentes* in having the P_3 crown as tall as the M_1 paraconid, a very short, lingually directed paracristid on M_1, and peripherally dispersed lingual cusps. Differs from both *Palaeothentes* and *Acdestis* in having a posteriorly broad-basined P_3, a relatively transverse anterolingually posteriolabially oriented M_1 paracristid, M_1 and M_2 that are exodaenodont and lack an entoconid notch, and an M^1 postprotocrista that is confluent with the anterior margin of the hypocone platform (measurements in table 12.1).

Hondathentes cazador sp. nov.

TYPE

As for the genus.

HORIZON AND LOCALITY

As for the genus.

DISTRIBUTION

As for the genus.

ETYMOLOGY

Named for El Cazador, a mythical hunter of the Grand Tolima region of Colombia.

DESCRIPTION

The type specimen is a fragmentary right dentary with P_3 through M_2. The P_3 is premolariform and equal in height to the M_1 trigonid. The protoconid is simple and lacks an anterior cuspule. Distally, a small talonid shelf is worn so that it is slightly wider lingually than labially.

The M_1 trigonid is well defined by a strong paralophid and protocristid. The protocristid and cristid obliqua are heavily worn. As in other palaeothentids, the paraconid is mesial and the metaconid is distal to the protoconid. The protoconid base carries a very small mesiobuccal swelling. The cristid obliqua is prominent and extends from the hypoconid to intersect the postvallid wall approximately one-third of the distance buccally from its lingual edge. On the talonid the entoconid is slightly distally expanded relative to the hypoconid. In addition, the entoconid carries a small distal shelf near its base. There is no hypoconulid.

The morphology of M_2 differs from that of M_1 in several respects. The M_2 trigonid is relatively shorter with the protoconid and metaconid transversely aligned and the paraconid shifted distally and buccally. The anterobuccal swelling on the base of the protoconid is more pronounced than on M_1. In addition, the intersection of the cristid obliqua and the postvallid wall is shifted buccally and is less acute. The entoconid and hypoconid are approximately equal in size and are transversely aligned. The distobasal shelf around the entoconid is more well developed than on M_1, but it is still excluded from the talonid.

The hypodigm includes one specimen each of M^1 and M^3. Each tooth is deeply basined and exhibits a generalized, tribosphenic pattern. M^1 has a linguobuccally short talon and a small distinct protoconule. The hypocone platform is primitively narrow. The confluence of the postprotocrista with the anterior margin of the hypocone platform is a unique feature. M^3 exhibits a slight labially directed fold at the anterior margin of the metacone.

COMPARISONS

The new palaeothentid *Hondathentes cazador* possesses a unique dental morphology within the family as it shares some features with *Palaeothentes*, others with *Acdestis*, and others that it possesses alone. Other palaeothentids are known from the Deseadan, Colhuehuapian, and Santacrucian of southern Argentina, the Deseadan of Bolivia, and the Friasian of Chile. *Hondathentes*, despite its young age, appears to be the most generalized known palaeothentid, based on the combination of the following features: tall and uncomplicated P_3; M_1 with a low crown and antero-posteriorly short trigonid that lacks a bifurcated paraconid and presents an anterolingually posterobuccally oriented paracristid; and possession of relatively less trenchant molar cusps than found among other palaeothentids. It is advanced in being the only palaeothentid with exodaenodont cheek teeth.

Examination of character polarities in all Caenolestoidea indicate that, excepting molar exodaenodonty, nearly all of the diagnostic dental characters in *Hondathentes* are generalized for Palaeothentidae. Two possible exceptions are the lack of an anterobasal cusp on P_3 (which is of unknown polarity) and the lack of an entoconid notch on M_1 and M_2. If the lack of the anterior P_3 cusp is generalized for the family, *Hondathentes* is an excellent morphotype for the Palaeothentidae. If, instead, the absence of that cusp is derived, *Hondathentes* is a suitable antecedent for *Palaeothentes*. In either case, the known dentition of *Hondathentes cazador* appears to be near that of the stem palaeothentid.

Discussion

Pithiculites chenche and *Hondathentes cazador* extend both the geographic and temporal ranges of abderitid and palaeothentid marsupials, respectively. Abderitidae are otherwise restricted to the Colhuehuapian and Santacrucian of Chubut and Santa Cruz provinces in southern Argentina (Marshall 1976b), and the Friasian of Chile (L. G. Marshall 1990). Palaeothentids are temporally and geographically more diverse than abderitids. Members of the family are recorded from the Descadan, Colhuehuapian, and Santacrucian of southern Argentina. One species (*Palaeothentes boliviensis*) is known from the Deseadan of Bolivia (Patterson and Marshall 1976), and three species are known from the Friasian of Chile (L. G. Marshall 1990).

The presence of *Pithiculites chenche* and *Hondathentes cazador* in the Miocene of Colombia illustrates that caenolestoids had a more cosmopolitan distribution than previously known. In *H. cazador* the presence of both primitive and moderately derived features suggests a long period of isolation from Argentine and Chilean centers of pa-

laeothentid evolution. The geographical and temporal seclusion of *P. chenche* from its apparent close relative, *Pithiculites minutus* (known only from the Colhuehuapian of Argentina), likewise suggests a long period of isolation for the Colombian fauna.

Paleoenvironmental reconstructions of the Pinturas Formation and of the Santa Cruz Formation at Monte Observación in southern Argentina, from which both palaeothentids and abderitids are best known (Marshall 1976b; Bown et al. 1990), suggest that caenolestoid localities there represent humid, upland forest and lowland coastal plain environments respectively (Bown and Larriestra 1990; Bown et al. 1990). Abderitid and palaeothentid marsupials are abundant and diverse components of the mammalian fauna at both of these localities. The presence of *H. cazador* and *P. chenche* in the La Venta fauna suggests that perhaps a similar humid paleoenvironment existed in that region as well.

Dietary reconstructions for the Santacrucian caenolestoids suggest that abderitids were primarily frugivorous, whereas palaeothentids encompassed a wider range of adaptations, from faunivory to frugivory and possibly folivory (Strait et al. 1990). The dental morphology of the caenolestoids from Colombia is not widely divergent from that seen in Santacrucian forms, suggesting that these taxa were occupying similar dietary niches. In contrast to the abundance of caenolestoids in Santacrucian faunas of southern Argentina, La Venta caenolestoids are apparently rare.

Pascual and Ortiz Jaureguizar (1990) noted the low diversity and rarity of abderitids and palaeothentids associated with the occurrence of small primates such as *Neosaimiri fieldsi* in the La Venta fauna. The authors suggested that abderitids and palaeothentids may have occupied dietary niches currently filled by small-bodied primates, a proposal that seems plausible in view of the distribution of those groups in the Santacrucian of Argentina. Bown et al. (1990) reported that abderitids and palaeothentids exhibit opposite patterns of abundance and diversity at two geographically distinct Santacrucian localities: the inland Pinturas Formation located in the Río Pinturas valley and the coastal Santa Cruz Formation at Monte Observación. Whereas abderitids were more abundant and diverse at Pinturas, palaeothentids exhibited higher diversity and abundance at Monte Observación. The pattern of primate diversity and abundance at these localities mirrored that of abderitids (Fleagle 1990). Therefore, it seems possible that the inverse pattern of abundance and diversity among palaeothentids and primates may indicate competition between these taxa for similar niches. It is, however, currently difficult to discern whether the comparative rarity of all caenolestoids in the Honda Group is the result of environmental, temporal, or taphonomic factors, either singly or in combination. It is hoped that with additional collecting and analysis it will be possible to control some of these variables better and gain a broader understanding of caenolestoid adaptation and evolution in the Miocene of Colombia.

Acknowledgments

We thank D. Krause for comments on earlier versions of this manuscript. This research was supported by National Geographic Society grants 2964-84 and 3292-86 and National Science Foundation grants BSR 86-14133 and BSR 89-18657 to R. F. Kay.

13
Armored Xenarthrans: A Unique Taxonomic and Ecologic Assemblage

Alfredo A. Carlini, Sergio F. Vizcaíno, and
Gustavo J. Scillato-Yané

RESUMEN

La Venta es una de las pocas localidades fosilíferas de Sudamérica que ha provisto una diversidad tan grande de Cingulata. Todos los armadillos y gliptodontes (Xenarthra, Cingulata) representan nuevos géneros (excepto *Asterostemma*) y especies, con relaciones filogenéticas relativamente distantes con los cingulados coevos de Patagonia. Durante el Mioceno medio en el cono austral sudamericano se registran solamente Peltephilinae, Euphractini, Eutatini, y Stegotheriini. Por el contrario, en el Mioceno medio de Sudamérica ecuatorial se registran Astegotheriini, Dasypodini, y Tolypeutinae.

Nanoastegotherium prostatum, el más pequeño armadillo de La Venta, se reconoce en base a osteodermos de morfología variable, entre aquellos con figura central lageniforme y los que poseen enormes forámenes en su superficie. La figura central lageniforme y los escasos forámenes pilíferos lo asemejan a algunos Astegotheriini del Casamayorense de Patagonia. Es el registro más moderno de la Tribu. Las figuras periféricas laterales y anteriores de los osteodermos anticipan la probable condición estructural ancestral de los Dasypodini.

Anadasypus hondanus, el más antiguo Dasypodini conocido, se asigna a esta Tribu en base a las figuras periféricas de las placas de las bandas móviles, que sugieren la presencia de escamas córneas epidérmicas triangulares con el ápice dirigido hacia atrás y cubriendo dos osteodermos adyacentes, así como por los anillos caudales con doble hilera de placas. Por sus caracteres arcaicos, como, por ejemplo, las placas fijas con la figura central lageniformes y sin figuritas periféricas laterales y posteriores, *Anadasypus hondanus* es el más primitivo Dasypodini conocido.

Pedrolypeutes praecursor es el más antiguo y primitivo Tolypeutinae. La presencia de caracteres primitivos tales como osteodermos con la figura central bien desarrollada y tubérculos incipientes, cuarto metacarpiano poco reducido, ileon poco comprimido y sinsacro amplio, sugieren que los Tolypeutinae pueden haberse originado a partir de armadillos generalizados como los Euphractinae.

Neoglyptatelus originalis, un primitivo Glyptatelinae, aparece en el Grupo Honda mucho después de la desaparición de la Subfamilia en altas latitudes. Sus osteodermos, pentagonales a hexagonales, con la figura central contra el margen posterior, son más similares a los de los primitivos armadillos que aquéllos de *Glyptatelus.* A diferencia de otros gliptodontes, las vértebras dorsales no se fusionan en un tubo.

Dos especies de Propalaeohoplophorinae (*Asterostemma gigantea* and ?*A. acostae*) están presentes en el Grupo Honda. Estos taxa muestran algunos caracteres derivados compartidos con los posteriores Sclerocalyptinae: osteodermos con figura en roseta, sin figuras periféricas adicionales, y molariformes mandibulares posteriores simplemente trilobados.

Los requerimientos ambientales de los Cingulata del Grupo Honda sólo pueden ser supuestos en base a las exigencias ecológicas de taxa estrechamente relacionados (especies de *Dasypus* y *Tolypeutes* actuales) y de inferencias

213

generales sobre las posibles adaptaciones de los extinguidos Astegotheriini y Glyptodontidae. El variado conjunto de taxa que se registran en el Grupo Honda indica una heterogeneidad ambiental. Los Glyptodontidae y Pampatheriidae pueden haber ocupado áreas abiertas y pastizales, mientras que los Astegotheriini y Dasypodini pueden haber habitado formaciones vegetales más cerradas y bosques. Los Dasypodini, los Tolypeutinae, y tal vez los Glyptatelinae ocuparon áreas ecotonales.

Most of the South American Tertiary armored edentates are known by remains collected in Argentina. Furthermore, for the period between the middle Paleocene and middle Miocene almost all the known material comes from localities that are restricted to southern Argentina, or Patagonia.

Living armadillos of the family Dasypodidae attain their greatest diversity in the forests and savannas of the equatorial tropics (Wetzel 1982, 1985a, 1985b). For this reason it would seem an admissible hypothesis that most of the evolutionarily significant events, such as the origin of higher taxonomic groups, occurred within the tropics and subtropics of South America (Scillato-Yané 1986). Our initial analysis of the available paleontologic evidence confirms this conjecture (Vizcaíno et al. 1990).

During the Tertiary, especially the early and middle portion, tropical and subtropical conditions extended farther to the south than they do today. This would have resulted in the introduction of greater diversity in the armadillos of Patagonia. Nevertheless, we believe that even in these favorable times the present territory of Argentina was marginal to the central events of the phyletic history of the Cingulata. In this context the Tertiary fossil-bearing deposits of equatorial South America have increased significance as potential sources of new taxa and evidence for these fundamental cladogenetic events.

The Miocene fauna of La Venta and neighboring localities in the Magdalena River valley in Huila Department, Colombia, has been known for several decades. Stirton (1953b) tentatively correlated this fauna with the middle Miocene Friasian Land Mammal Age, and his suggestion has been generally accepted.

Until now, the fossil record of armored edentates in the La Venta fauna has been curiously poor; only one genus of

Pampatheriidae (*Kraglievichia* Castellanos following Porta 1962, probably *Vassallia* Castellanos following Robertson 1976, and a new genus following Edmund and Theodor, chap. 14) and one (*Asterostemma* Ameghino) or two genera (*Asterostemma* and *Propalaeohoplophorus* Ameghino) of Glyptodontidae, according to different authorities (Porta 1962; Krohn 1978; Villarroel 1983). It is remarkable that not one Dasypodidae had been identified although the presence of at least two genera in the La Venta fauna was recognized by Savage (1951b).

Today this panorama has become notably enlarged because of the rich collections made by researchers from Duke University and from INGEOMINAS in Colombia (Carlini, Scillato-Yané, et al. 1989). All the Dasypodidae and Glytodontidae included in these collections are new to science (for the Pampatheriidae, see chap. 14). The remains are very numerous and some specimens are very complete. More detailed descriptions are being prepared for future publication.

The purpose of this chapter is to make known the new taxa and to propose some new phylogenetic, paleobiogeographic, and paleoenvironmental hypotheses about their evolutionary history.

Systematic Paleontology

Order Cingulata Illiger 1811
Superfamily Dasypodoidea Bonaparte 1838
Family Dasypodidae Bonaparte 1838
Subfamily Dasypodinae Bonaparte 1838
Tribe Astegotheriini Ameghino 1906
Nanoastegotherium gen. nov.

ETYMOLOGY

From the Latin *nanus,* and this from the Greek *nanos,* dwarf, small; and from *Astegotherium,* type genus of Astegotheriini.

TYPE SPECIES

Nanoastegotherium prostatum sp. nov.

Figure 13.1. *A. Nanoastegotherium prostatum* gen. et sp. nov., holotype, IGM 183912, dorsal carapace scutes. *B. Neoglyptatelus originalis* gen. et sp. nov., holotype, IGM 183153, partial dorsal carapace. *C. Asterostemma gigantea* sp. nov., IGM 250499, ventral view of a partial caudal tube. *D–E. Asterostemma gigantea* sp. nov., holotype, IGM 250928: *D*, detailed view of the scutes; *E*, partial dorsal carapace. Scale bars 1 cm.

REFERRED SPECIES

Only the type species.

GEOCHRONOLOGIC AND GEOGRAPHIC DISTRIBUTION

Middle Miocene, Colombia.

DIAGNOSIS

Small size, somewhat smaller than *Dasypus hybridus* (Desmarest). Subrectangular fixed scutes, from 4.0 to 7.0 mm long and from 3.0 to 4.5 mm wide; lageniform central

figure reaching the posterior margin as in *Astegotherium* Ameghino and *Prostegotherium* Ameghino; chief figure two-thirds to three-fourths of total scute length. Four to five anterior and lateral peripheral figures separated from the chief figure by a sulcus that is wider and deeper than in other Astegotheriini. Generally, two large foramina in the sulcus at each side of the anterior portion of the chief figure (positioned as in *Prostegotherium notostylopianum* Ameghino). Other fixed scutes with even larger foramina, which on some fixed scutes occupy nearly the entire surface. Without foramina on the hind margin or with few small foramina as in *Astegotherium* and *Prostegotherium* and differing from *Pseudostegotherium* Ameghino.

Nanoastegotherium prostatum sp. nov.
(fig. 13.1A)

ETYMOLOGY

From the Greek *prostatum,* to be before.

HOLOTYPE

IGM 183912. Numerous isolated dorsal carapace scutes and one from the caudal tube extracted from a coprolite.

TYPE LOCALITY

Duke Locality 70, at a level between the Chunchullo Sandstone Beds and the Tatacoa Sandstone Beds, La Victoria Formation, Honda Group, Huila Department, Colombia (see chap. 2).

HYPODIGM

The holotype, IGM 183127, isolated scutes in coprolites, from Duke Locality 31; and IGM 184703, isolated scutes in coprolites, from Duke Locality 65.

STRATIGRAPHIC RANGE AND GEOCHRONOLOGIC AGE

See figure 13.4 and chapter 2.

DIAGNOSIS

The same as the genus by monotypy.

COMMENTS

All the remains were extracted from probable crocodilian coprolites. There are a few pieces of the skeleton, one scute from the caudal tube and numerous fixed scutes (from the pelvic shield?). Curiously, scutes from the mobile bands have not been found. There is a morphological gradation from scutes with lageniform figure to those with enormous foramina; we suppose the latter occupied some particular region of the carapace (middle axis? middle pelvic region?). Another feature of several of the scutes is a remarkable widening of the radial sulcus separating the two anterior peripheral figures, the radial sulcus develops into a depression as large as a peripheral figure.

Originally, the Astegotheriini were separated from Stegotheriini as distinct families by Ameghino (1906) but this scheme was not adopted by later authors (Simpson 1948; Hoffstetter 1958; Patterson and Pascual 1968; Scillato-Yané 1976a, 1980, 1986b). Vizcaíno (1990, 1994) revived Ameghino's systematic hypothesis and included the genera *Astegotherium, Prostegotherium,* and *Pseudostegotherium* in the tribe Astegotheriini.

The Astegotheriini are known from the Itaboraian of southern Brazil and from Riochican to Colhuehuapian localities in Patagonia but are not known from deposits representing a very significant temporal hiatus comprising the Mustersan and Deseadan. The record of new Astegotheriini in Colombia extends the biochron of this tribe significantly.

Some features of the scutes (the lageniform chief figure, scarce piliferous foramina) are similar to those of some Casamayoran Astegotheriini such as *Astegotherium dichotomus* Ameghino and *Prostegotherium notostylopianum* Ameghino and differ from those of the Colhuehuapian *Pseudostegotherium glangeaudi* Ameghino.

The presence of anterior and lateral peripheral figures heralds the condition characteristic of the Dasypodini (fig. 13.2A). The lateral figures suggest the presence of epidermal scales covering partially two adjacent scutes, independently of the scale covering the chief figure. In recent Dasypodini (*Dasypus* Linné) the peripheral figures completely surround the chief figure and epidermal scales cover part of two or three neighboring scutes.

Tribe Dasypodini Bonaparte 1838
Anadasypus gen. nov.

ETYMOLOGY

From the Greek *ana,* "from down to up" or "toward"; and *Dasypus,* type genus of living Dasypodini.

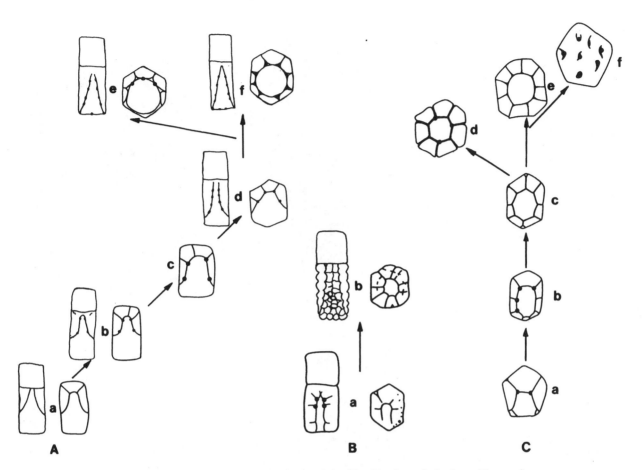

Figure 13.2. Hypothetical scute design morphocline in middle Miocene Colombian Cingulata and related taxa. Not to scale.

A. Dasypodinae, mobile and fixed scutes. *a–c.* Astegotheriini, as follows: *a, Astegotherium; b, Prostegotherium; c, Nanoastegotherium* gen. nov. *d–f.* Dasypodini, as follows: *Anadasypus* gen. nov.; *e, Propraopus; f, Dasypus.*

B. Tolypeutinae, mobile and fixed scutes, as follows: *a, Pedrolypeutes* gen. nov.; *b, Tolypeutes.*

C. Glyptodontidae. *a, b.* Glyptatelinae, as follows: *a, Neoglyptatelus* gen. nov.; *b, Glyptatelus. c–f.* Representatives of other glyptodontid subfamilies, as follows: *c, Asterostemma* (Propalaeohoplophorinae); *d, Glyptodon* (Glyptodontinae); *e. Sclerocalyptus* (Sclerocalyptinae); *f, Doedicurus* (Doedicurinae).

TYPE SPECIES

Anadasypus hondanus sp. nov.

REFERRED SPECIES

The type species and another as yet unnamed species from the Miocene of Ecuador (Carlini, Vizcaíno, et al. 1989, 23).

GEOCHRONOLOGIC AND GEOGRAPHIC DISTRIBUTION

The middle Miocene of Colombia and the middle to late Miocene of Ecuador.

DIAGNOSIS

Medium-size Dasypodini, similar in size to *Dasypus novem-cinctus.* At least six movable bands (seven in *Propraopus,* six to eleven in *Dasypus*). Mobile band scutes 25–33 mm long and 6–9 mm wide with a subrectangular central figure; with three to five holes in each sulcus delimiting the central figure (three to four in *Propraopus,* five to eight in *Dasypus*) and one to two holes in the posterior margin (generally three in *Propraopus* and four in *Dasypus*). Quadrangular to hexagonal scutes 6.5–9.5 mm long and 6.0–7.5 mm wide with relatively lageniform central figure (subcircular in *Pro-praopus* and *Dasypus,* in which it never reaches the posterior margin) and with three anterior peripheral figures (ante-

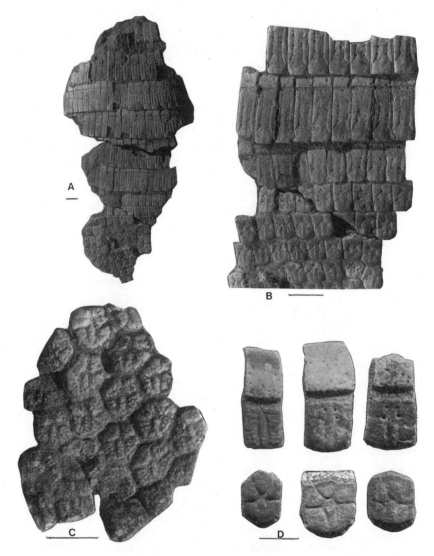

Figure 13.3. *A–B. Anadasypus hondanus* gen. et sp. nov., holotype, IGM 183499: *A,* partial dorsal carapace with six mobile bands and nine fixed bands of the pelvic buckler; *B,* detailed view of the scutes. *C. Pedrolypeutes praecursor* gen. et sp. nov., holotype, IGM 182640, partial pelvic buckler. *D. Pedrolypeutes praecursor* gen. et sp. nov., IGM 184673, dorsal carapace scutes. Scale bars 1 cm.

rior, lateral, and posterior peripheral figures in *Propraopus* and *Dasypus*). Caudal tube comprised of rings with two rows of scutes without peripheral figures (peripheral figures present in the anterior row of each caudal ring in *Propraopus* and *Dasypus*).

Anadasypus hondanus sp. nov.
(fig. 13.3A, B)

ETYMOLOGY

[*H*]*ondanus,* from the Honda Group.

HOLOTYPE

IGM 183499, partial dorsal carapace with six movable bands, one semimovable, and nine fixed bands from the pelvic shield.

TYPE LOCALITY

Duke Locality 59, near where the Fish Bed crosses the road near the headwaters of Quebrada Pachingo, at a stratigraphic level just below the Fish Bed, Villavieja Formation, Honda Group (see chap. 2).

HYPODIGM

The holotype and IGM 183320 numerous partially articulated scutes from the ?pelvic shield, from Duke Locality 41; IGM 183448, partial caudal tube with nine rings, from Duke Locality 52; IGM 183722, partial dorsal carapace and miscellaneous postcranial bones, from Duke Locality 62; and IGM 183862 isolated scutes, partial maxilla with two teeth, and miscellaneous postcranial bones, from Duke Locality 68.

STRATIGRAPHIC RANGE AND GEOCHRONOLOGIC AGE

See figure 13.4 and chapter 2.

DIAGNOSIS

Mobile band scutes with a slight longitudinal keel on the central figure (absent in the species from Ecuador); three to four foramina in the sulcus surrounding the central figure (four to five in the species from Ecuador). Fixed scutes with a more lageniform central figure than that of the species from Ecuador; and with an important longitudinal keel on the central figure (absent in the species from Ecuador).

COMMENTS

The identification of these remains as Dasypodini is indicated by two features in particular: (1) the peripheral figures of the movable band scutes suggest the existence of triangular epidermal scales with a posteriorly directed apex partially covering two adjacent osteoderms; and (2) the caudal tube rings are made up of a double line of scutes which differs from the condition found in all other Dasypodidae (Vizcaíno 1990). *Anadasypus* gen. nov. is the oldest Dasypodini recorded. Until now, the phyletic history of the tribe was only known through two genera: *Dasypus* Linnaeus (Pliocene-Recent) and *Propraopus* Ameghino (?Pliocene-Pleistocene). Furthermore, *Anadasypus* gen. nov. shows the greatest number of relatively primitive features of all known Dasypodini, so it could be regarded as the "structural ancestor" of the other genera of the tribe. The

fixed scutes do not have the lateral and posterior peripheral figures that are characteristic of the later genera *Propraopus* and *Dasypus*. The central figure is relatively lageniform and resembles some Astegotheriini, especially those of the Casamayoran (early Eocene) and *Nanoastegotherium*.

Subfamily Tolypeutinae Gray 1865
Pedrolypeutes gen. nov.

ETYMOLOGY

In honor of Pedro Bondesio of the Museo de Ciencias Naturales de La Plata, and from *Tolypeutes*, the living genus of Tolypeutinae.

TYPE SPECIES

Pedrolypeutes praecursor sp. nov.

REFERRED SPECIES

Only the type species.

GEOCHRONOLOGIC AND GEOGRAPHIC DISTRIBUTION

The middle Miocene of Colombia.

DIAGNOSIS

One-third larger than *Tolypeutes* Illiger but with more numerous and proportionally smaller scutes. Scute surface with numerous tubercles, smaller than those of *Tolypeutes*. Firm lateral joint between adjacent scutes; scute margin more serrate than in other Dasypodidae but less than in *Tolypeutes*. Subrectangular mobile scutes with two deep and parallel sulci delimiting a single central and two lateral figures (the homologous sulci in *Tolypeutes* are shallow and divergent). Imbricated area with a smooth external surface. Semimobile scutes with two anterior large foramina placed at each side of the central figure; small foramina in the posterior and external lateral margin. Subrectangular, pentagonal, or hexagonal fixed scutes, isodiametric or slightly elongate; central and peripheral figures clearly delimited by a deep sulcus; the central figure reaching the posterior margin of the scutes (in *Tolypeutes* the central figure is completely surrounded by peripheral figures and the sulci are shallow). The fourth metacarpal is less reduced than in *Tolypeutes.* The anterior portion of the ilium and synsacrum are less compressed than in *Tolypeutes;* the neural spines of the synsacrum are weakly developed.

Pedrolypeutes praecursor sp. nov.
(fig. 13.3C, D)

ETYMOLOGY

From the Latin *praecursor,* precursor.

HOLOTYPE

IGM 182640 an articulated portion of the pelvic shield and isolated mobile scutes; mesocervical bone and the remaining cervical vertebrae, the first three thoracic vertebrae; the posterior articulated lumbar vertebrae; the anterior portion of the right ilium fused with the synsacral vertebrae; the proximal portion of the right humerus and distal portion of the right femur.

TYPE LOCALITY

Duke Locality 4, in the Honda Group near Coyaima, Tolima Department. Duke Locality 4 corresponds to University of California Museum of Paleontology Vertebrate Locality 4411 (see Stirton 1953b), where nearly all of the Coyaima fauna was collected.

HYPODIGM

The holotype and IGM 183218, several isolated scutes, from Duke Locality 33; IGM 183663, isolated scutes and a partial maxilla with three teeth, from Duke Locality 58; IGM 183855, a few isolated scutes, from Duke Locality 32;

IGM 184240, a few isolated scutes, from Duke Locality 85; and IGM 184682, several isolated scutes, from Duke Locality 79.

STRATIGRAPHIC DISTRIBUTION AND GEOCHRONOLOGIC AGE

The type locality (Duke Locality 4) corresponds with UCMP Locality 4411, one of the Coyaima localities of Stirton (1953b). Stirton (1953b) assigned the Coyaima fauna to the Colhuehuapian (conventionally late Oligocene) based on the presence there of the interathere *Cochilius.* The hypodigm material was collected from Duke Localities 32, 33, 58, 79, and 85. All localities are within the Honda Group and extend from the La Victoria Formation upward to the El Cardón Red Beds, a biostratigraphic range spanning nearly the entire Honda Group stratigraphic column in the upper Magdalena valley (see chap. 2).

DIAGNOSIS

The same as the genus by monotypy.

COMMENTS

Until now, the Tolypeutinae were known only from the Pliocene (at Monte Hermoso, in Buenos Aires Province, and from the highest levels of the subandean Tertiary) and the Quaternary (Buenos Aires Province) of Argentina (Scillato-Yané 1980). Moreover, until now, all known fossil remains pertained to the living genus *Tolypeutes.* The carapace scutes of *Pedrolypeutes* have well-developed principal figures and weakly developed tubercles (fig. 13.3C, D). Moreover, *Pedrolypeutes* has the fourth metacarpal less reduced, the ilium less compressed, and a more expanded synsacrum than *Tolypeutes.* These characteristics suggest that *Pedrolypeutes* could be a structural ancestor of the modern Tolypeutinae (fig. 13.2B) preserving primitive features such as the scute figures and proximally expanded ilium and synsacrum (Scillato-Yané et al. 1991).

Superfamily Glyptodontoidea Burmeister 1879
Family Glyptodontidae Burmeister 1879
Subfamily Glyptatelinae Castellanos 1932
Neoglyptatelus gen. nov.

ETYMOLOGY

From the Greek *neos,* recent, and from *Glyptatelus,* the type genus of Glyptatelinae.

TYPE SPECIES

Neoglyptatelus originalis.

REFERRED SPECIES

Only the type species.

GEOCHRONOLOGIC AND GEOGRAPHIC DISTRIBUTION

From the middle Miocene Honda Group of Colombia.

DIAGNOSIS

Smaller than *Glyptatelus tatusinus* Ameghino and *Glyptatelus malaspinensis* Ameghino and much smaller than *Clypeotherium* Scillato-Yané. Small scutes, 13–20 mm long, 13–16 mm wide, and 6–9 mm thick; pentagonal to hexagonal in outline and with straight-sided central figure (rounded in *Glyptatelus*). Smooth scute surface as in *Glyptatelus* (more rugose in *Clypeotherium*). One to three large foramina in the principal sulcus just at the intersection with the radial sulcus. Caudal ring scutes with proportionally much smaller foramina than *Clypeotherium*.

Neoglyptatelus originalis sp. nov.
(fig. 13.1B)

ETYMOLOGY

From the Latin *originalis,* original, from origin.

HOLOTYPE

IGM 183153, a partial dorsal carapace and isolated osteoderms.

TYPE LOCALITY

Duke Locality 31, near the railway culvert over Quebrada Balsillas, Municipality of Villavieja, Huila Department, Colombia, in the Monkey Beds, Villavieja Formation, Honda Group (see chap. 2).

HYPODIGM

The holotype and IGM 183944, a partial caudal ring, from Duke Locality 46; IGM 184254, numerous isolated scutes and miscellaneous postcranial bones, from Duke Locality 75; IGM 184296, one scute and miscellaneous postcranial bones, from Duke Locality 6; IGM 250659, numerous isolated scutes and miscellaneous bones, from Duke Locality 118; IGM 250841, a partial carapace, from Duke Locality 22; IGM 250853, one scute and one metapodial, from Duke Locality 22.

STRATIGRAPHIC RANGE AND GEOCHRONOLOGIC AGE

See figure 13.4 and chapter 2.

DIAGNOSIS

The same as the genus by monotypy.

COMMENTS

The morphology of the carapace scutes in *Neoglyptatelus* is quite different from that of *Glyptatelus*. Other than in size and the caudal rings, *Neoglyptatelus* is more similar to *Clypeotherium*.

Until now, the record of Glyptatelinae was restricted to the middle Eocene (Mustersan) to Oligocene (Deseadan) of Argentina and the Oligocene (Deseadan) of Bolivia. The presence of *Neoglyptatelus originalis* in the middle Miocene of Colombia represents a remarkable prolongation of the family's known temporal range.

Among the remains available for study is a partial maxilla and mandible with lobate teeth (IGM 250697) of a subadult of a small glyptodont. Unfortunately, these dental remains are associated with only one poorly preserved scute with no diagnostic features. We include this specimen in the hypodigm with doubt, postulating that it belongs in this taxon only because there are no other small glyptodonts at the same stratigraphic levels.

Subfamily Propalaeohoplophorinae Ameghino 1889
Asterostemma Ameghino 1889

TYPE SPECIES

Asterostemma depressa Ameghino 1889, from the Santacrucian (early Miocene), Santa Cruz Province, Argentina.

REFERRED SPECIES

The type species and *Asterostemma barrealense* Rusconi (?middle Miocene, San Juan Province, Argentina). *Asterostemma venezolensis* Simpson (middle Miocene, eastern Venezuela), *Asterostemma ?acostae* Villarroel (middle Miocene, Huila Department, Colombia).

GEOCHRONOLOGIC AND GEOGRAPHIC DISTRIBUTION

Early and ?middle Miocene of Argentina; middle Miocene of Colombia.

EMENDED DIAGNOSIS

Medium to large Propalaeohoplophorinae, ranging in size from *Propalaeohoplophorus australis* (Moreno) to *Eucinepeltus crassus* Scott. Posterior region dorsal scutes between 44 and 48 mm long, 35–48 mm wide, and 9–15 mm thick. Scutes arranged in more regular transverse bands than in other Propalaeohoplophorinae. Generally, a single series of peripheral figures in each scute. The peripheral figures of two or three adjacent scutes forming composite figures. Scutes

from the posterior region of the caudal tube are slightly imbricated, more similar to the caudal tube in Dasypodidae than to that of *Propalaeohoplophorus* Ameghino or *Cochlops* Ameghino.

Asterostemma gigantea sp. nov.
(fig. 13.1 C–E)

ETYMOLOGY

From the Latin *gigantea,* gigantic.

HOLOTYPE

IGM 250928, including a portion of the mid-dorsal carapace of a mature individual preserving the ventral impression of the articulation with the synsacrum and the forewings of both iliums, the last synsacral vertebra joined to the ischium, the left astragalus, the proximal portion of the left scapula, the distal portion of the left humerus and the anterior portion of the right mandible preserving the alveoli of five molariform teeth.

TYPE LOCALITY

Duke Locality 113, 4 km west-northwest of El Libano School, Municipality of Baraya, Huila Department, Colombia, about 50 m stratigraphically above the Chunchullo Sandstone Beds in the La Victoria Formation, Honda Group (see chap. 2).

HYPODIGM

The holotype and IGM 250499, isolated scutes from the carapace and caudal rings, partial caudal tube (fig. 13.1C), miscellaneous bones and nine partial teeth, from Duke Locality 108; IGM 250588, one scute from the carapace, from Duke Locality 32; IGM 250637, a few scutes from the carapace, from Duke Locality 114; and IGM 250718, a few scutes from the carapace, one vertebral body, and a partial left mandible with seven partial teeth, from Duke Locality 121 West.

STRATIGRAPHIC RANGE AND GEOCHRONOLOGIC AGE

See figure 13.4 and chapter 2.

DIAGNOSIS

The largest species of the genus, a little larger than *Eucinepeltus petestatus* Ameghino, as large as *Eucinepeltus crassus* Scott (the largest known Propalaehoplophorinae), and one-fourth smaller in linear dimensions than the sclerocalyptine *Sclerocalyptus ornatus* (Owen). The central figure of the scutes with a slight depression on the posterior third. The central figure of the scutes from the posterior region of the carapace are more convex than in *Asterostemma venezolensis* and ?*Asterostemma acostae*. The central figure of the anterior and lateral dorsal scutes is nearly planar, whereas it is planar to slightly concave in *A. depressa* and *A. barrealense*. The mid-dorsal scutes are proportionally much broader than in *A. venezolensis*. The apex of the caudal tube is blunter than in *A. depressa* but less blunt than in *Propalaeohoplophorus australis*. The second lower molariform tooth tends to trilobate form and is more complex than that of *A. venezolensis* (which has rounded lingual sides with no sulcus) and much more complex than all Patagonian Propalaeohoplophorinae, except *Propalaeohoplophorus minor* Ameghino (perhaps also *Asterostemma;* see Simpson 1947, 9).

COMMENTS

The presence of propalaehoplophorine glyptodonts in the Miocene of Colombia was first recognized by Jaime de Porta (1962) who ascribed isolated scutes to *Asterostemma* cf. *A. venezolensis* and *Propalaeohoplophorus* sp. The same author (Porta 1969) added *Asterostemma depressa* to the La Venta fauna list, but the occurrence of this species cannot be confirmed. Villarroel (1983) included every formerly mentioned taxon from the La Venta fauna in the synonymy of a new species doubtfully assigned to *Asterostemma*, ?*A. acostae*. We agree with Villarroel's doubt and add the gigantic species described herein provisionally to *Asterostemma*. As a matter of fact, we believe that *Asterostemma gigantea* is clearly different from species of *Propalaeohoplophorus* Ameghino, *Cochlops* Ameghino, *Eucinepeltus* Ameghino, and *Metopotoxus* Ameghino. Regrettably, the carapace of the type species *A. depressa* from the early Miocene of Patagonia is very poorly known.

Like Villarroel (1983), we cannot completely discredit the idea that the Propalaehoplophorinae from northern South America, including *A. venezolensis* Simpson, could belong to a new genus distinct from *Asterostemma*. In this case, *Asterostemma* would have a geographic distribution limited to Patagonia. Into this new genus we would include the species described herein. Nevertheless, given our present state of knowledge about the family, we prefer not to add a sixth genus until new and more complete remains, especially of the type species of *Asterostemma*, allow us to do so more convincingly.

Discussion

Mammal evolution in the South American Tertiary is known fundamentally through fossil remains from Argentina, and our knowledge of the more common Argentine armored edentates is rather complete (Scillato-Yané 1986). Nevertheless, we have had some doubts about whether the pattern of cingulate evolution in the fossil record of Argentina would be duplicated in more equatorial latitudes of South America. The great geographic distance and the fact that environmental conditions in the equatorial tropics would have been less affected by the gradual global climatic zonation evident in southern South America during the Miocene (Pascual et al. 1985) led us to predict, a priori, quite different patterns of evolutionary change (Scillato-Yané 1986). This appears to be substantiated by the fossils from Colombia.

The Honda Group provides numerous and diverse vertebrate remains, and these fossils allow us to establish a relatively complete scenario for the Colombian middle Miocene, the most complete record of Tertiary mammals from intertropical South America. The Cingulata occurring in the La Venta fauna are quite different from those from southern South America. It must be emphasized that all the armadillos and glyptodonts described herein and almost all recorded up to now (except ?*A. acostae* and *A. gigantea*, which might represent a new genus) are new species and genera. Most of these taxa have relatively distant phyletic relationships with Cingulata from the classic Friasian localities of Patagonia. The correlation of the La Venta fauna to the Friasian would be impossible on the basis of the Cingulata alone. No Friasian guide fossils among the Cingulata (*Epipeltephilus* Ameghino) occur in the La Venta fauna nor do any Cingulata with first appearances or last occurrences in the Friasian SALMA of Patagonia (Bondesio et al. 1980).

The armadillos and glyptodonts in the middle Miocene La Venta fauna comprise a very peculiar assemblage for which we propose the following evolutionary, paleobiogeographic, and paleoenvironmental interpretations. With the possible exception of the Deseadan locality of Quebrada Fiera in Mendoza Province, Argentina (Gorroño et al. 1979; Scillato-Yané 1988) we believe that there are only a very few (or no) fossil localities with such brief temporal representation (chap. 3) wherein Cingulata of such diverse phylogenetic histories are recorded together.

Nanoastegotherium is the latest record of the tribe Astegotheriini now known to have had a long temporal (Itaboraian-Friasian) and latitudinal (Patagonia-Colombia, 45°) range. Anatomical features point to this group as the probable structural ancestors of the Dasypodini. It is remarkable that *Nanoastegotherium,* the Astegotheriini that is most closely related to later Dasypodini, occurs at the same stratigraphic level as *Anadasypus,* the oldest and most primitive known Dasypodini (fig. 13.4). This fact supports the hypothesis of an intertropical origin for the Dasypodini (Vizcaíno 1990; Vizcaíno et al. 1990).

Pedrolypeutes and *Anadasypus* are representatives of two living groups, the Tolypeutini and Dasypodini, respectively, both with a heretofore poorly known fossil record but now having their oldest and most primitive representatives recorded in Miocene sediments of equatorial South America. The skeleton of *Pedrolypeutes* shows some features that suggest the Tolypeutinae may have originated from a generalized Dasypodidae such as the Euphractinae.

In a recent comparison of middle Miocene armadillos from Colombia, Ecuador, Chile, and Argentina, we pointed out the following significant contrast (Vizcaíno et al. 1990). In the Friasian of southern South America the Euphractinae are the dominant group of Dasypodidae. In the Friasian of Patagonia the Euphractini and Eutatini occur along with the specialized carnivorous Peltephilinae and myrmecophagous Stegotheriini. By contrast, in equatorial South America no Euphractinae, Peltephilinae, or Stegotheriini occur. By contrast, in Ecuador and Colombia the Astegotheriini, Dasypodini, and Tolypeutinae are recorded, and none of these groups are known from the Friasian of Patagonia.

The new Glyptatelinae represent a vestige of a group that until now was known only from older rocks at higher latitudes, that is, from the middle Eocene-Oligocene of Patagonia and the Oligocene of Bolivia. This is another example of taxa that disappeared from higher latitudes and reappeared at later times in the tropical region (McKenna 1956; Marshall et al. 1983; Cifelli and Guerrero 1989). *Neoglyptatelus* is a primitive Glyptatelinae displaying more plesiomorphic character states than the other Glyptodontidae. The sculpture of the osteoderms in *Neoglyptatelus* is more similar to that of armadillos than are those of *Glyptatelus.* Among the available material are some dorsal vertebrae from an adult that do not form a tube as in all other Glyptodontidae. It is not possible to say whether or not this is a feature shared among all Glyptatelinae. This feature of *Neoglyptatelus* probably represents the primitive character state for the Glyptodontidae. Among Propalaeohoplophorinae vertebral fusion is noticeable, and in the Glyptodontinae, Sclerocalyptinae, and Doedicurinae vertebral fusion is even greater, developing a continuous tube.

The Propalaeohoplophorinae are represented by two taxa in the Honda Group: *Asterostemma gigantea* (apparently a specialized autapomorphous terminal taxon) and ?*A. acostae.* ?*Asterostemma acostae* displays some shared derived features with the Sclerocalyptinae, including osteoderms with rosette sculpture and without the additional peripheral figures (such as are seen in *Propalaeohoplophorus* and *Cochlops*) and simple trilobate (without additional reentrant folds) posterior mandibular teeth (such as are seen in the Huayquerian Sclerocalyptine *Hoplophractus* Cabrera). We announced the occurrence of a new genus of Glyptodontidae from the Miocene of Ecuador that we consider to be the most primitive known sclerocalyptine although it preserves some propalaeohoplophorine features in its dentition (Carlini, Vizcaíno, et al. 1989).

The coexistence in the Honda Group of two very closely related but morphologically quite distinct glyptodonts is not unique for the fossil record of South America. In the Deseadan, as in the Honda Group, a very large, morphologically specialized glyptodont (the gigantic *Cly-*

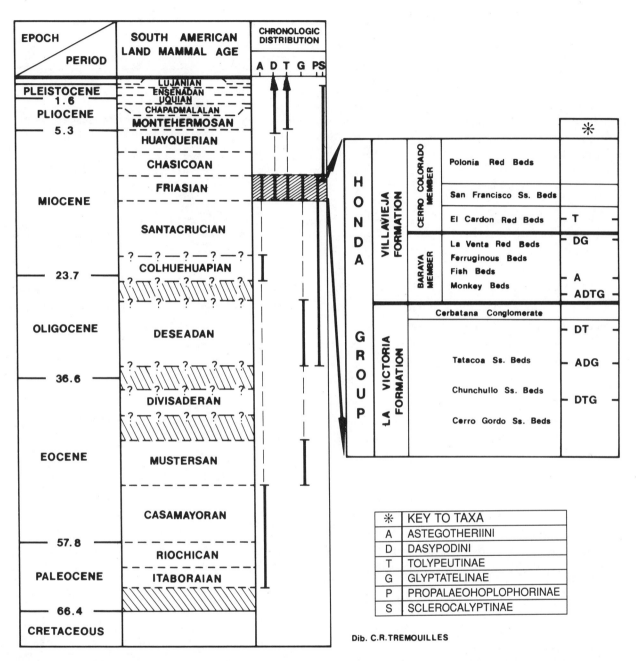

Figure 13.4. Cenozoic chronologic distribution of the major groups of Dasypodidae and Glyptodontidae from the middle Miocene of Colombia, and their stratigraphic range within the Honda Group. Abbreviations: *Ss.,* sandstone; others are defined in the key. Numbers in the "Epoch/Period" column refer to age (Ma).

peotherium Scillato-Yané) coexists with a very closely related but much smaller as yet unnamed form (the oldest known propalaeohoplophorine) that apparently shares more derived features of morphology with later taxa.

The environmental preferences of the Honda Group Cingulata (table 13.1) can only be speculated based upon

the ecological preferences of closely related living cingulates (*Dasypus* and *Tolypeutes*) and very general inferences about the extinct Astegotheriini and Glyptodontidae.

Honda Group Cingulata are represented relatively uniformly throughout the Honda Group stratigraphic column in the La Venta area. The association of such diverse taxa

Table 13.1. Inferred body size and ecological requirements of the middle Miocene Cingulata of Colombia

Taxon	Body Size	Feeding Habits	Habitat
Dasypodidae:			
Nanoastegotherium	Small, <2 kg	Insectivorous	Humid forest
Anadasypus	Medium, <5 kg	Omnivorous-insectivorous	Humid forest, ecotonal areas, and savannas
Pedrolypeutes	Medium, <4 kg	Insectivorous	Dry forest, ecotonal areas, and grasslands
Glyptodontidae:			
Neoglyptatelus	Small, <15 kg	Herbivorous	Savannas, ecotonal areas, and grasslands
Asterostemma	Medium to large, 50–100 kg	Herbivorous	Savannas and grasslands
Pampatheriidae:			
Scirrotherium[a]	Small, <40 kg	Herbivorous	Savannas and ecotonal areas

[a]See chapter 14.

indicates environmental heterogeneity. The Astegotheriini and Dasypodini were probably humid forest dwelling, the Glyptodontidae and Pampatheriidae may have occupied grasslands, and the Tolypeutinae may have occurred in dryer shrub formations. With respect to vegetation density, the Astegotheriini and Dasypodini may have inhabited closed formations, whereas the other groups are generally associated with more open environments. The Dasypodini, Tolypeutinae, and, perhaps, the Glyptatelinae could have occupied ecotonal areas such as those where *Tolypeutes* occurs today in northwestern Argentina.

The environment required by the diverse Cingulata is in general agreement with that inferred by Stirton (1953b) based upon sedimentology, that is, widespread savannas and a meandering stream system with closed marginal forest and by Bondesio and Pascual (1977), that is, wide and relatively shallow flood plains. We believe that the lateral displacement of a meandering river channel would have resulted in a succession of different plant communities in the same area at an ecological time scale.

Acknowledgments

We are indebted to R. F. Kay and R. H. Madden (Duke University Medical Center) and to A. L. Reina (INGEOMINAS) who made the collection of Cingulata available to us for study. R. Pascual, M. G. Vucetich, E. Ortiz Jaureguizar, and M. Bond critically reviewed the Spanish manuscript. E. Aragón reviewed the English manuscript. The illustrations are by C. R. Tremouilles, Museo de Ciencias Naturales de La Plata staff illustrator.

14
A New Giant Pampatheriid Armadillo

Gordon Edmund and Jessica Theodor

RESUMEN

Numerosos especímenes permiten describir un nuevo género y especie de Pampatheriidae, *Scirrotherium hondaensis,* del Grupo Honda. *Scirrotherium hondaensis* es el más antiguo pampatérido conocido hasta ahora. El tamaño y la alta proporción de osteodermos hexagonales con pocos hoyos foliculares grandes reunidos por un surco son características diagnósticas que lo distinguen de *Kraglievichia paranensis. Scirrotherium hondaensis* y *Vassallia minuta* muestran dentaduras primitivas en las cuales los primeros cuatro dientes son cilíndricos hasta elípticos y no bilobados. Aunque cercana a *Holmesina floridanus* en tamaño, *Scirrotherium* difiere en tener las dos carillas de la articulación calcanoastragalar bien separadas. Características derivadas en común sugieren que *Scirrotherium* está más estrechamente relacionado a *Vassallia minuta.*

P orta (1962) described three groups of pampathere osteoderms from the upper Magdalena valley of Colombia and assigned them to *Kraglievichia paranensis* (Ameghino 1883) Castellanos, *K.* cf. *K. paranensis,* and *Kraglievichia* sp. Porta's illustrations and descriptions are more than adequate, but the assignment to *Kraglievichia* is incorrect, apparently made from the literature rather than from comparison with original specimens. Collections had been made in 1944, 1949, and 1950 by field parties from UCMP and INGEOMINAS, but the skeletal material (in UCMP)

was not seen by Porta. The identification of *Kraglievichia* from Colombia has been widely accepted in the literature (Hirschfeld and Marshall 1976; Scillato-Yané 1980; Carroll 1988).

In addition to the UCMP material, the authors have studied the Duke University–INGEOMINAS collections of 1985–89. Comparisons were made with all known pre-Pleistocene pampathere specimens from North and South America, including the types and other material of *Kraglievichia paranensis, Vassallia* spp., *Plaina* spp., and *Holmesina floridanus.* Unfortunately, the association of skeletal, dental, and carapacial material is tenuous for all but the last species.

Few differences between the skeleton of *S. hondaensis* and the other taxa were noted, except for absolute size. The osteoderms of *Scirrotherium,* however, are distinct, resembling those of *Vassallia minuta* and *Holmesina floridanus* more than those of *Kraglievichia, V. maxima,* or *Pampatherium.* Dimensions of skeletal elements and osteoderms vary somewhat throughout the section, but no size trend was noted. The geologically oldest specimen, a partial skull and carapace, IGM 250533, is among the smallest but is probably a subadult. Because of the general uniformity in size and morphology, all of the pampathere material from the Honda group is here considered to represent a single taxon.

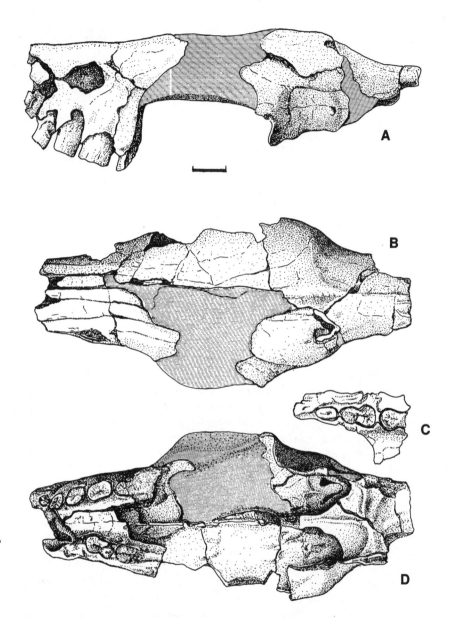

Figure 14.1. *Scirrotherium hondaensis,* UCMP 40201, holotype, skull: *A,* left lateral view; *B,* dorsal view; *C,* fragment with left posterior teeth, palatal view; *D,* ventral view. Scale bar 1 cm.

Systematic Paleontology

Order Xenarthra Cope 1889
Infraorder Cingulata Illiger 1811
Superfamily Dasypodoidea Cabrera 1929
Family Pampatheriidae Edmund 1987
Scirrotherium gen. nov.

TYPE AND ONLY KNOWN SPECIES

Scirrotherium hondaensis.

HOLOTYPE

UCMP 40201. Incomplete skull with most of the snout and teeth M1 to M5 on the left side; isolated fragment from the left side, including teeth M7 to M9, part of the palate and base of the jugal process (fig. 14.1), partial left mandible with teeth m3 to m9.

HYPODIGM

The type and an incomplete right mandible with most of the teeth, UCMP 40056 (fig. 14.2). Anterior end of a

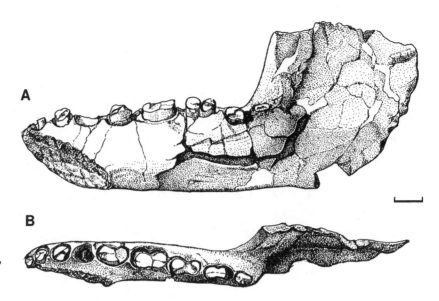

Figure 14.2. *Scirrotherium hondaensis,* UCMP 40056, right mandibular ramus: *A,* medial view; *B,* occlusal view. Scale bar 1 cm.

weathered skeleton with part of the skull, mandibles, vertebral column, and associated osteoderms, IGM 250533. Osteoderms UCMP 38066 and 37924, and IGM 182192 (fig. 14.3 A–C). Numerous cranial and postcranial elements and osteoderms from the type area in the collections of IGM and UCMP, to be described in a future paper, are not formally included in the hypodigm.

DIAGNOSIS

A small pampathere, slightly larger than *Vassallia minuta,* but smaller than *Kraglievichia paranensis.* General features of skull resemble those of other pampatheres. Anterior four upper teeth generally cylindrical. M4 elongate-ovate. Teeth M5 to M8 clearly bilobate, with emarginations on the medial side in all, also on the lingual side in some specimens; M9 varies, often reniform or egg shaped. Mandible relatively shallow dorsoventrally, elongate; ascending ramus broad anteroposteriorly. Lower dentition resembles upper. Skeletal elements almost identical with those of other pampatheres. Two facets of the astragalo-calcaneal articulation well separated. Osteoderms approximately the size of *V. minuta;* half are hexagonal in outline. Typical buckler osteoderm has small number of well-spaced follicular pits connected by distinct channel. Submarginal band continuous, confluent posteriorly with elongate central figure.

ETYMOLOGY

Greek, *skiros,* a hard covering, and *therion,* a wild beast.

AGE AND REMARKS

Specimens of *S. hondaensis* are found throughout the fossiliferous stratigraphic units of the Honda Group. The type skull is from UCMP Locality V4503 just below the Tatacoa Sandstone Beds. Mandible 40056 and osteoderm 37924 are from the Monkey Beds, Locality V4519. Osteoderm 38066 is from Locality V4530 below the Chunchullo Sandstone Beds.

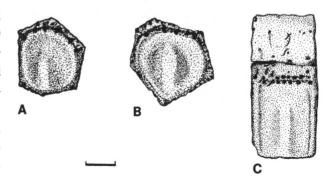

Figure 14.3. *Scirrotherium hondaensis,* osteoderms. *A.* Buckler osteoderm, UCMP 38066. *B.* Buckler osteoderm, UCMP 37924. *C.* Imbricating osteoderm, IGM 182194. Scale bar 1 cm.

DESCRIPTION

Skull

Four skulls are known, all incomplete. The holotype UCMP 40201 is laterally compressed but retains most of the dentition and extends caudad to the beginning of the braincase (fig. 14.1). Skull IGM 184702 is the least distorted, slightly flattened, with much of the palate and left side. Specimen 182941 is badly crushed and skewed but preserves many sutures. IGM 250533 is much weathered but retains some of the right side including the occiput. Little remains of the premaxilla in any specimen, but the suture with the maxilla is present in UCMP 40201, characteristically passing through the second alveolus. The nasals are weakly sutured in the midline but join the frontals in a highly serrated suture. The maxillae form the palate back to the anterior edge of M8 at which point the palatines assume this function, extending back to the choana. Where preserved, the choanal margin is sinuous. In life the pterygoids probably formed the choanal margin, but as with most other pampathere specimens they have been lost post mortem.

The maxilla is highest at the junction with the nasal and frontal at the anterior margin of M7. The jugal process is pneumatic and well developed, with the infraorbital canal extending from the anterior edge of M7 to the anterior edge of M8. The lachrymal forms the dorsal part of the jugal process and the anterior part of the orbit. The lachrymal foramen is located near the midpoint of the lachrymal anteroposteriorly. The frontals are united in the midline by a closed suture and are strongly inflated by sinuses above the orbits and the lachrymals. The cribriform plate is missing but apparently filled a large space at the orbital constriction. Skull IGM 250533 has an estimated length of 143 mm, including a braincase length of 62 mm.

The first four upper teeth are cylindrical. Cross sections through the first three vary from nearly circular to irregularly ovate. The fourth is quite variable from elliptical to reniform. Teeth five through eight are bilobate with emarginations on the lingual side in all specimens and also on the labial in some. Tooth nine is quite variable, most commonly elongate-elliptical in section.

Mandible

No complete mandible is known. Our description is based on the following: UCMP 40201, 40056, 40058, 39842, 38066, and 38880; and IGM 183749, 183769, and 250533. The horizontal ramus is long and shallow with the ventral margin almost straight from the rear of the symphysis to well behind the level of m9 (fig. 14.2). The ascending ramus is broad anteroposteriorly, and arises at approximately 80° to the alveolar plane. Two mental foramina are present near the anterior margins of m4 and m5.

Teeth four to nine lie in a straight line, with teeth one through three curving gently mediad. The occlusal surfaces of the teeth are distinctly stepped in the better-preserved specimens, the anterior of each being approximately 2 mm higher than the posterior margin of the preceding tooth. This complements the arrangement of the upper dentition, providing a strong mechanical limit to the forward excursion of the mandible. Trituration was accomplished by transverse movement of the pterygoid and masseter apparatus as shown by striae on the occlusal surfaces of many of the teeth. In occlusal view the teeth resemble those of their upper counterparts. The first three are generally ovate, whereas m4 varies from reniform to elongate-elliptical. Teeth m5 to m8 are bilobate as in the uppers, whereas m9 varies from reniform to pear-shaped.

Vertebrae

The only articulated vertebrae are those of IGM 250533. As with all known pampatheres, the atlas is free, whereas cervicals two to five are fused. In this possibly immature individual the separate centra can still be recognized. The sixth cervical is free, as in other pampatheres, but its centrum is as robust as those of the other cervicals. In Pleistocene species it is reduced to a thin bar. Although the area is not well preserved, it appears that the seventh cervical is not fused to the first two thoracics, the condition found in Pleistocene species of *Holmesina* and *Pampatherium*. An isolated third thoracic, UCMP 39842, lacks the single disc-shaped central facets found in other parts of the column. Rather, the articulation consists of a large median surface flanked by two smaller dorsolaterally situated facets on both the anterior and posterior ends. This condition, which must have restricted movement, is also found in the anterior thoracics of other known pampatheres. None of the posterior thoracics or lumbars, with their multiple articulating facets, has yet been collected. The caudal vertebrae are mostly represented by isolated centra. The central facets are almost circular and slightly concave. These facets are parallel, but the anterior facets are set markedly more dorsal than the posteriors, thus producing a decidedly ventrad curve to the postsacral vertebrae. The shapes of the centra along with the high proportion of hexagonal osteoderms suggest that the carapace may have been rather high-domed.

Osteoderms

The typical division of the carapace of pampatheres into anterior and posterior bucklers separated by three rows of imbricating osteoderms is seen in specimen IGM 250533. In addition, this specimen retains many of the smaller osteoderms of the cephalic casque and for the first time in pampatheres indicates the presence of small osteoderms in the skin of the neck.

The shape of osteoderms is important to the identification of pampathere taxa. Each lot of nonmarginal buckler osteoderms is sorted for percentages of rectangulars, hexagonals (regular and irregular), and polygonals. *Scirrotherium* is unique in having an average of more than 50% (48-65%) nonmarginal buckler osteoderms being hexagonal. Comparison with other taxa revealed closest agreement with the series of *V. minuta* mentioned previously, which had 47% hexagonals. In *Pampatherium* close to 50% of the osteoderms are rectangular as confirmed by the mounted carapace of *P. mexicanum*, IGMM 56 501, in which the posterior buckler is composed almost entirely of rectangular osteoderms. *Holmesina*, including *H. floridanus*, exhibits a preponderance of polygons and some hexagons but few rectangles. The morphology of osteoderms of *Kraglievichia paranensis* is distinctly different from those of *Scirrotherium* or *Vassallia minuta* although many osteoderms of *Kraglievichia* are hexagonal.

Area of the buckler osteoderms is expressed as anteroposterior length multiplied by the width at right angles. Because the shapes are irregular, this cannot be given in standard units such as mm², and the measurement is merely referred to as *area*. The areas of *Scirrotherium* specimens ranged from 358 to 525, with the oldest specimen, IGM 250533, being 425. *Vassallia minuta* (MLP 69-xii-26-17) measured 488, well within the range of *Scirrotherium*. The wide variation in the *Scirrotherium* series may be attributed at least in part to age and sex differences. On the basis of outline and area *Scirrotherium* appears to be closest to *V. minuta*.

The surface ornamentation of the nonmarginal osteoderms of *Scirrotherium* is unique. Most bear a single row of follicular pits at the anterior end. These are relatively few, large, and connected by a distinct channel (fig. 14.3), a feature not seen in any other pampathere. As seen in ground sections, the pits extend into the cancellous bone of the interior of the osteoderm. The submarginal band is usually clearly developed, forming a narrow ridge posterior to the row of follicular pits and continuing around the entire osteoderm. The central figure is somewhat variable in shape, but it is usually rounded and elongate. It is always separated anteriorly from the submarginal band but is confluent with it posteriorly. The sulcus is broad and horseshoe-shaped. The imbricating osteoderms (fig. 14.3C) have an ornamented area similar to that of the bucklers, but are more elongated, bear more follicular pits, and lack the submarginal band posteriorly. The small ossicles of the neck are subcircular in outline and radiate from a slightly raised center.

Conclusions

Scirrotherium hondaensis compares most closely with *Vassallia minuta*. The type of the latter, a nearly complete mandible with a tooth row length of 75 mm, is slightly smaller than that of *Scirrotherium* (range, 85–92 mm; mean, 88 mm). In both species the four anterior teeth are ovate to elliptical, the most primitive state yet known. The dentition of *Scirrotherium* is smaller and more primitive than that of *Kraglievichia paranensis* or *Vassallia maxima*. Unfortunately, the anterior dentition of *Holmesina floridanus* is not adequately preserved for comparison.

The osteoderms of *Scirrotherium* resemble most closely those of *V. minuta* as represented by MLP 69-xii-26-17. In

Figure 14.4. Restoration of *Scirrotherium hondaensis,* formerly used as a restoration of *Holmesina septentrionalis.* The osteoderm pattern conforms more closely with that of *S. hondaensis.*

size the latter fall within the range of *Scirrotherium* and contain a high percentage with hexagonal outlines. The ornamentation is similar, but *Scirrotherium* is unique in having a small number of large follicular pits joined by a possibly vascular channel. Comparison was made with series of osteoderms of *Vassallia maxima.* These are much larger, proportionately thinner, and have a broad central figure and very shallow sulcus. Buckler osteoderms of *Kraglievichia* are often rectangular although hexagons are common. They differ from *Scirrotherium* in being larger, thicker, and with a distinctive ornamentation (fig. 1C in Edmund 1987).

The skeleton of *Scirrotherium* is abundantly represented in collections and differs little from that of other pampatheres. The most interesting feature noted in preliminary study was in the astragalo-calcaneal articulation in which the two facets are distinctly separated. In *Holmesina floridanus* (Pliocene, Florida), approximately the same size as *Scirrotherium,* the two facets are united, becoming separated during the evolution toward *H. septentrionalis* in the Pleistocene (fig. 14.4). Because it is unlikely that these facets would unite then separate later, *Scirrotherium* is proba-

bly not ancestral to *H. floridanus* and its descendants. The status of this articulation is unknown for *V. minuta,* but the facets are well separated in *V. maxima.*

In comparison with other pampatheres, the size, dentition, and osteodermal characters most closely resemble those of *Vassallia minuta* and are at an evolutionary level expected for a pampathere of the mid-Miocene.

Acknowledgments

We give special thanks to the curators and collection managers of the many museums in North and South America who made available the series of pampatheres in their care. The manuscript was greatly enhanced by the critical reviews of S. Vizcaíno, MLP, C. Ray, USNM, and E. Lundelius, University of Texas, Austin. Travel and other research funds were provided to the senior author from Grant A 7937 and to the junior author from Grant 3-641-111-40, both from the Natural Sciences and Engineering Research Council of Canada.

15
Xenarthrans: Pilosans

H. Gregory McDonald

RESUMEN

La fauna de La Venta del Mioceno medio contiene un conjunto diverso de ocho especies de perezozos (Pilosa, Xenarthra) dominado por los Mylodontidae. *Pseudoprepotherium confusum,* el perezozo más grande de la fauna, demuestra caracteres milodontinos tales como el lacrimal y forámen lacrimal reducidos, la superficie de la carilla cuboide del astrágalo convexa, una apófisis lateral en el tercer metatarso, y calcis tuber ensanchado, indicando que la superficie inferior entera estaba en contacto con la tierra. Un paladar parcial, rama mandibular, y esqueleto postcraneal incompleto permiten identificar el Mylodontinae *Glossotheriopsis pascuali* para la fauna de La Venta. *Glossotheriopsis* demuestra unos caracteres algo más derivados de los Mylodontinae tales como el primer diente jugal caniniforme. La carilla articular distal de la tibia no modificada, vértebras lumbares no fusionadas, y el calcáneo aplanado de arriba abajo y no engrosado, sugieren que los Mylodontinae no habían alcanzado aún una rotación pedal ni las características para soportar grandes pesos corporales de las formas más tardías de la subfamilia. Elementos del cráneo y esqueleto no asociados establecen la presencia de *Neonematherium flabellatum* en la fauna de La Venta, aquí asignado a la subfamilia Scelidotheriinae por presentar la carilla cuboide del astrágalo cóncava.

En los Megalonychidae de La Venta, se distingue la especie grande de la pequeña por el tamaño del astrágalo. Los Nothrotheriinae están representados por dos taxa que demuestran características típicas del astrágalo, como, por ejemplo, el borde interno de la polea extendida hasta una altura más allá del borde lateral y la presencia, aunque rudimentaria, de una apófisis odontoide. La presencia en la fauna de un megaterio Megateriinae está indicada por un extremo superior de tibia y peroné fusionado. El material disponible no permite una determinación más precisa de estos taxa, y la presencia de *Hapalops* en la fauna sigue siendo incierta.

La gran variedad de morfología craneana y dental, y de proporciones corporales y tamaños, sugiere que los perezozos llenaban una variedad de nichos herbívoros. Los dientes molariformes de *Neonematherium flabellatum, Glossotheriopsis pascuali,* y *Pseudoprepotherium confusum* se caracterizan por poseer dos superficies triturantes que forman una cresta transversal cortante semejante a *Choloepus,* el perezozo folívoro actual. La diversidad de los perezozos se mantiene a lo largo de la columna estratigráfica con los mismos taxa registrados tanto para la Formación La Victoria como para la Formación Villavieja. La presencia en la fauna de *Glossotheriopsis* (fósil guía del Colloncurense) y *Neonematherium* (también reconocido en el Mayoense) apoya a la correlación de la fauna de La Venta al "Friasense."

Our current knowledge of the sloths of the La Venta fauna of Colombia is based on the work of Hirschfeld (1985). Additional work in the region by Kay and colleagues since 1985 has led to the discovery of sloth taxa not previously recorded for the fauna, plus additional material of the less-common taxa.

Based on the number of specimens recovered, megalonychids, nothrotheres, and megatheres are present in the fauna but are not as prominent as the mylodonts. This is in marked contrast to Santacrucian-age faunas of Patagonia in which these groups are dominant and the mylodonts compose a minor component (Scott 1903–1904).

Systematic Paleontology

Family Mylodontidae
Subfamily Mylodontinae
Pseudoprepotherium confusum Hirschfeld 1985

REFERRED SPECIMENS

Duke Locality 6: IGM 184284, right third metacarpal, carina of metapodial; IGM 182926, right ulna. Duke Locality 12: IGM 182720, molariform; IGM 182719, molariform. Duke Locality 17: IGM 182757, left humerus; IGM 182759, right lower first molariform. Duke Locality 23: IGM 182863, upper fifth molariform; IGM 182868, right lower fourth molariform. Duke Locality 25: IGM 184481, two fragmentary teeth. Duke Locality 26: IGM 250392, molariform. Duke Locality 33: IGM 183221, upper ?fifth molariform; IGM 184224, first lower molariform. Duke Locality 42: IGM 183384, juvenile molariform. Duke Locality 50: IGM 183439, tooth fragment. Duke Locality 51: IGM 183442, mandible with fourth molariform. Duke Locality 52: IGM 183475, right upper fifth molariform. Duke Locality 59: IGM 183683, right navicular, two proximal phalanges, second phalanx. Duke Locality 71: IGM 183923, two molariforms, second phalanx, caudal. Duke Locality 74: IGM 183961, right lower third molariform; IGM 183962, molariform. Duke Locality 90: IGM 184348, first upper molariform. Duke Locality 91: IGM 184413, molariform. Duke Locality 98: IGM 184557, teeth and rib fragments; IGM 184569, distal left radius. Duke Locality 100: IGM 184576, molariform, proximal end left and right humerus, partial right femur. Duke Locality 121E: IGM 250807, proximal humerus minus epiphysis, distal left tibia. Duke Locality 136: IGM 251141, left ulna, left radius, left unciform. Fish Bed: IGM 183501, upper first molariform; IGM 183502, first and fifth upper molariforms; IGM 184130, first molariform; IGM 184478, left distal humerus.

The osteology of *Pseudoprepotherium* was described in detail by Hirschfeld (1985) based on an extensive sample. Although the specimens listed here complement Hirschfeld's work, they provide no new information about the anatomy of the animal.

Hirschfeld (1985) placed *Pseudoprepotherium* in the subfamily Mylodontinae, indicating a closer relationship to such genera as *Glossotherium, Glossotheriopsis, Thinobadistes, Lestodon,* and *Mylodon* than to the subfamily Scelidotheriinae, which includes *Nematherium, Neonematherium, Proscelidodon, Catonyx,* and *Scelidotherium.* She did not discuss the characters that distinguish the two subfamilies or the characters present in *Pseudoprepotherium* that support this placement. Although I agree with Hirschfeld's conclusion, it should be noted that *Pseudoprepotherium* shares a number of features with the scelidotheres that are best assessed as retained primitive features from the common ancestor of the two subfamilies. Among these shared features retained by *Pseudoprepotherium* and present in scelidotheres are a long, narrow skull with parallel tooth rows and no anterior expansion of the rostrum, the first cheek tooth not modified into a caniniform; the predental spout of the mandible narrower than the width of tooth rows and dorsally inflected; and the humerus with an entepicondylar foramen.

Recognition of *Pseudoprepotherium* as a mylodontine rather than a scelidothere is based primarily on four skeletal features. In the skull the lachrymal bone is reduced as is the lachrymal foramen; in scelidotheres both the bone and foramen are large and prominent. In the pes are three features. On the astragalus the articular surface for the cuboid is convex as in the other mylodontines, whereas in scelidotheres this area is concave. A lateral process is developed on the third metatarsal, overlapping the fourth metatarsal; this process is lacking in scelidotheres. In addition, the third metatarsal in scelidotheres is short, blocky, and more closely resembles that of a megalonychid. Finally, the tuber calcis of the calcaneum is expanded, permitting the entire ventral surface of the calcaneum to be in contact with the ground. In scelidotheres, only the posterior part of the tuber calcis is expanded and in contact with the ground, with the articulations for the astragalus positioned off the ground, giving the pes a more arched appearance.

Using *Pseudoprepotherium* as a primitive mylodontine, *Glossotheriopsis* as a more derived mylodontine, and *Neonematherium* as an early representative of the Scelidotheriinae permits the recognition of a suite of features most likely common to the mylodonts in general and present in the ancestor of both subfamilies. Many of these features are modified in later members of both lineages, but more so in the mylodontines. In both *Pseudoprepotherium* and *Glossotheriopsis* most of the molariforms are circular in cross section, except the last molariforms, or (as in *Glossotheriopsis*) when the anterior dentition is modified into a caniniform. It is only in later mylodontines that the other molariforms become rectangular and develop lobes. The development of lobes on the molariforms appears to occur earlier in the scelidotheres than in the mylodontines because the teeth of both *Nematherium* and *Neonematherium* show some lobation.

In both *Pseudoprepotherium* and *Glossotheriopsis* the molariforms develop a single transverse loph on the occlusal surface. In late Pleistocene mylodontines such as *Glossotherium* the structure of the first molariform, with two occlusal surfaces separated by a single transverse loph, is presumably retained from this earlier feature. In *Neonematherium* the molariforms also initially develop a single transverse loph that, in some individuals, is eventually lost and replaced by a flat occlusal surface. Transverse cutting lophs on the molariforms are also present in the Santacrucian scelidothere, *Nematherium*.

The development of transverse lophs on the teeth of sloths is functionally related to the position of the mandibular condyle relative to the tooth row (McDonald 1977) and the relative position of the upper and lower teeth to each other (Naples 1990). The teeth of both megalonychids and megatheres are characterized by a pair of transverse cutting lophs and are functionally bilophodont as in tapirs and kangaroos. In both groups the mandibular condyle is elevated above the occlusal plane of the tooth row. In *Pseudoprepotherium* the mandibular condyle is elevated above the tooth row as it is in *Nematherium* and *Neonematherium* although it is not as pronounced. This portion of the mandi-

ble is not preserved in any specimen of *Glossotheriopsis* but, based on tooth morphology, it probably also was elevated. In later mylodontines and scelidotheres the mandibular condyle is in the same plane as the tooth row. The molariforms develop occlusal surfaces perpendicular to the long axis of the tooth and lack transverse lophs.

The long, narrow predental spout of the mandible of *Pseudoprepotherium* and scelidotheres is similar in structure to that of *Hapalops* and other Santacrucian nothrotheres and megalonychids. The modification of this part of the mandible so that it is shorter and wider in late mylodontines (such as *Glossotherium* and *Lestodon*) is probably related to the development of anteriorly divergent tooth rows. Another primitive feature for mylodonts is the presence of parallel tooth rows with no diastemas.

The recent collection of additional specimens of *Pseudoprepotherium* slightly extends the stratigraphic range of this taxon beyond that described by Hirschfeld (1985). Although previously known from both the La Victoria and Villavieja formations, it is now known from below the Chunchullo Sandstone Beds in the La Victoria Formation (fig. 15.1).

Glossotheriopsis pascuali Scillato-Yané 1976

REFERRED SPECIMENS

Duke Locality 16: IGM 183243, upper right second molariform. Duke Locality 21: IGM 184297, partial palate with left three molariforms, base of right first molariform and second and third molariforms, isolated fifth upper molariform, and fragment of left mandible with first lower molariform. Duke Locality 40: IGM 184354, molariform from sandstone above Duke Locality 40. Duke Locality 42: IGM 183384, right upper fifth molariform. Duke Locality 54: IGM 250839, partial left mandible bearing caniniform and third and fourth molariforms. Duke Locality 67: IGM 184063, molariform. Duke Locality 87: IGM 184310, left upper molariform. Duke Locality 102: IGM 184605, posterior part of right mandible bearing partial m3 and m4 with broken occlusal surface. Duke Locality 103: IGM 184644, right upper molariform. Duke Locality 132: IGM

	Fossil Level	Stratigraphic Unit	*Pseudo-prepotherium*	*Neonem-atherium*	*Glosso-theriopsis*	Megalonychid Large	Megalonychid Small	Nothrothere Large	Nothrothere Small	Megathere
Villavieja Formation		San Francisco Sandstone Beds								
	45	El Cardón Red Beds	O							O
	44		X							
	43		XO							
	42	La Venta Red Beds								
	41					X				
	40		O	XO		XO				
	39	Ferruginous Red Beds	O							
	38		XO					O		
	37	Fish Bed	XO			X				
	36		XO							O
	35	Monkey Beds	XO	XO	X	XO		X	X	O
La Victoria Formation	34	Cerbatana Conglomerate Beds								
	33		X					X		
	32									
	31		O							
	30									
	29									
	28					X				
	27									
	26									
	25	Tatacoa Sandstone Beds								
	24		O	O		O				
	23		X		X		X			
	22									
	21		X							
	20		X						XO	
	19									
	18							X		
	17		X							
	16			X			X			
	15									X
	14		O	X	X			X		
	13			X	X		X			
	12	Chunchullo Sandstone Beds								
	11			X	O					
	10		X	X	X	X		XO		
	9									
	8									
	7		X	X				X		
	6			X	X	X		X		
	5							X		
		Cerro Gordo Sandstone Beds								
	4									
	3			X				X		
	2			X						
	1									
		Coyaima								

Figure 15.1. Stratigraphic distribution of sloths in the Honda Group. Abbreviations: *X*, Duke Locality; *O*, UCMP Locality. *Thick line* indicates contact between upper Villavieja Formation and lower La Victoria Formation.

251213, partial skeleton with palate bearing left caniniform and first molariform. UCMP Locality V5046: IGM 252928, mandible fragment with m2 and m3, left humerus minus proximal epiphysis, right unciform, two unguals, medial half of distal epiphysis of right femur, left tibia with medial half of distal epiphysis, left calcaneum minus epiphysis of tuber calcis, left cuboid, two cervicals, centrum of thoracic, sacral, and caudal vertebrae, articulated five posterior thoracics and first lumbar.

The partial palate and mandible, IGM 184297 (fig. 15.2A, B), and a fragmented palate with part of the postcranial skeleton, IGM 251213 (fig. 15.2E), can be directly compared to the holotype (fig. 15.3) and provide the best evidence for the presence of *Glossotheriopsis* in the La Venta fauna. Reference of other elements to this taxon is based on their similarity to these specimens.

The tooth rows of IGM 184297 diverge anteriorly, giving the palate a triangular appearance similar to that seen in later mylodonts such as *Glossotherium*. The tooth rows are continuous and lack diastemas. The first cheek tooth is functionally caniniform and tapers to a sharp point. It is roughly an equilateral triangle in cross section with one apex directed anteriorly. The opposite side develops an occlusal surface that is positioned obliquely to the long axis of the tooth. This occlusal surface develops from contact with the anterior edge of the first lower cheek tooth. The first cheek tooth is strongly curved along its anteroposterior long axis and each side is grooved.

The second tooth in the series is oval in cross section with its greatest diameter parallel to the axis of the tooth row. Shallow longitudinal grooves are present along the midline on both the lateral and medial sides. The tooth is curved along its long axis, but the degree of curvature is not as great as the tooth anterior to it. The occlusal surface is divided into two wear surfaces of which the anterior is larger than the posterior. The two surfaces meet at a right angle. Contact between the two areas forms a prominent transverse ridge. The general configuration of the tooth is similar to that seen in later mylodontines such as *Glossotherium*.

The third and fourth cheek teeth are small and about half the size of the second. In cross section the third cheek tooth is a rounded triangle with one apex positioned laterally; the cross section of the fourth cheek tooth is circular. The third cheek tooth has a shallow groove on its medial size. Unlike

later mylodontines in which the occlusal surface is flat and develops at a right angle to the long axis of the tooth, the occlusal surface of these teeth is formed by two wear facets that meet at an angle. As in the second cheek tooth, this results in the formation of a transverse ridge extending from the lateral to medial side of the tooth at a right angle to the long axis of the tooth row. Commonly, the softer internal core of osteodentine will be more worn than the outer harder shell, resulting in two "cusps" positioned on the lateral and medial sides. Curvature of the long axis of the third cheek tooth is less than that of the second; the fourth cheek tooth is straight.

An isolated tooth found associated with the palate is here interpreted as the fifth upper cheek tooth. Its long axis is straight, and it has a longitudinal groove on the lateral side. As in the other teeth, there are two occlusal surfaces, but these are offset in a steplike fashion. The anterior of these two surfaces shows more wear than the posterior. The posterior wear facet is formed as a narrow band on the posterior edge of the tooth and is confined to the hard outer layer of osteodentine. A second specimen (IGM 183384) preserves all the same details of configuration.

The partial palate preserves the zygomatic process of the maxilla. This is positioned above the third cheek tooth as in other mylodontines such as *Glossotherium*.

Only a small fragment of the left mandible of IGM 184297 is preserved. The specimen is broken just posterior to the caniniform and through the alveolus of the first molariform. The caniniform is triangular in cross section. The medial side of the tooth is grooved, giving the tooth a lobate appearance in cross section. The occlusal surface is formed by two wear facets. The anterior facet, formed by occlusion with the upper caniniform, is oriented more acutely to the long axis of the tooth than the second wear facet. The posterior wear facet is the larger of the two and contacts the anterior part of the second upper cheek tooth. The two wear surfaces meet at an angle, forming a prominent mediolateral transverse ridge.

The three other specimens of *Glossotheriopsis,* IGM 184310, a first upper cheek tooth (=caniniform), and IGM 183243 and IGM 184644, second upper cheek teeth (=first molariforms), agree with that just described for the same teeth in IGM 184297 in all details. Measurements of IGM 184297 and IGM 251213 are in table 15.1. The other two

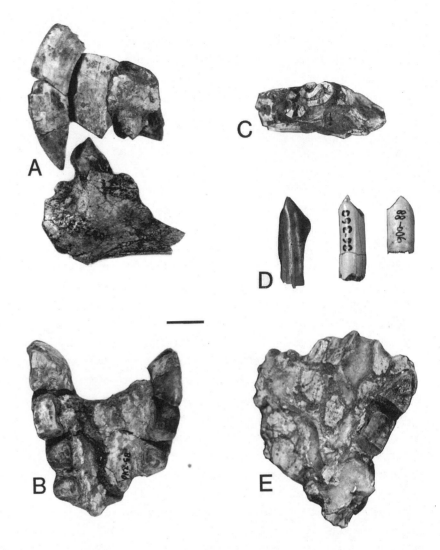

Figure 15.2. *Glossotheriopsis pascuali.* *A–B.* IGM 184297: *A,* left lateral view of maxilla and mandible; *B,* palatal view. *C.* IGM 184605, posterior mandible, occlusal view. *D.* Isolated molariforms: *left,* IGM 184354; *center,* IGM 183384; *right,* IGM 184063. *E.* IGM 251213, partial palate, palatal view. Scale bar 10 mm.

isolated molariforms, IGM 184063 and IGM 184354, both have the double occlusal surface and a circular cross section, but only IGM 184063 has a groove on one side.

The best preserved mandibles are IGM 250839 and IGM 184605, each of which has a complete lower fourth molariform. The outline of the last lower molariform in the partial mandible (IGM 184605; fig. 15.2C) of *Glossotheriopsis* is similar to that in *Neonematherium.* The two genera can be distinguished by the greater anterolabial elongation of the anterior lobe in *Neonematherium* and narrower widths of the anterior and posterior lobes relative to their lengths. In *Glossotheriopsis* the last lower molariform is less elongated and more compact. Although it is broken, enough of the third molariform is preserved to indicate a tooth with a circular cross section, distinguishing it from the

same tooth in *Neonematherium* which is rhomboidal in cross section.

Specimen IGM 250839 has a general configuration markedly similar to that of *Orophodon hapaloides* from the Deseadan illustrated by Hoffstetter (1954a). The general appearance of both specimens is that the mandible is shorter and more massive than that of other sloths. In both, the first and second cheek teeth (M_2 and M_3 of Hoffstetter) are circular, and the last tooth in the series is bilobate with a prominent transverse ridge, or step, on the anterior lobe. The known specimens of *Orophodon* do not preserve the anteriormost tooth, so it is not known if it was circular or modified into a caniniform as in *Glossotheriopsis.*

The lower caniniform in *Glossotheriopsis* is large and well developed. It is triangular in cross section with one apex

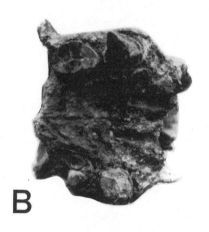

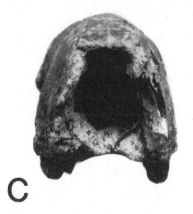

Figure 15.3. *Glossotheriopsis pascuali*, holotype, MLP 76-VIII-30-1, anterior part of skull and lower caniniform: *A*, right lateral view; *B*, palatal view; *C*, anterior view. Scale bar 10 mm.

Table 15.1. Dental measurements (mm) for
Glossotheriopsis pascuali

Variable Measured	IGM 184297	IGM 251213
Skull, anterior width across caniniforms	~43	~51.3
Alveolar length:		
Upper tooth row	~50	—
Posterior cheek teeth	~40	—
Caniniform:		
Anteriorposterior length	9.7	9.1
Mediolateral width	9.4	7.7
M1:		
Anteroposterior length	12.9	11.1
Mediolateral width	7.3	6.9
M2:		
Anteroposterior length	8.4	—
Mediolateral width	7.4	—
M3:		
Anteroposterior length	7.3	—
Mediolateral width	7.5	—
M4:		
Anteroposterior length	7.9	—
Mediolateral width	6.8	—

directed anteriorly. There are two occlusal surfaces, positioned anteriorly and posteriorly, which meet at an acute angle to form a prominent transverse ridge. The posterior occlusal surface is larger than the anterior. Unlike that of some later mylodontines, the caniniform is not separated from the cheek teeth by a diastema.

The last cheek tooth of *Glossotheriopsis* has a more bilobate appearance than in *Orophodon* with a more pronounced narrowing of the isthmus between the anterior and posterior lobes. The osteodentine core of this tooth is also proportionately larger and more distinct in *Glossotheriopsis* than in *Orophodon*. Despite these differences, however, the taxa are markedly similar. Whether these similarities reflect close phylogenetic relationship or merely convergence cannot be determined by available material. At the moment there are no known taxa from the Colhuehuapian or Santacrucian with the appropriate morphology to link *Orophodon* and *Glossotheriopsis*.

The only association of postcranial elements with a partial mandible is IGM 252928. The humerus is gracile in proportions and is not as robust as in later mylodontines. The humerus still possesses an entepicondylar foramen in contrast to later members of the lineage in which it is lost.

Like the humerus, the tibia is slender for its length and does not exhibit the robustness characteristic of later mylodontines such as *Glossotherium*. Distally a single tendonal groove is recessed into the posteromedial edge of the shaft. The presence of a single tendonal groove is one of the features that distinguishes this group of mylodonts from the scelidotheres, which have two tendonal grooves located on a process on the medial side of the distal end. Enough of the distal epiphysis is preserved to determine that the medial articular surface for the astraglus is flattened, not concave as in later mylodontines. The concave medial articular surface of the tibia reflects the modification of medial trochlea of the astragalus into a prominent raised odontoid process that is characteristic of later mylodontines such as *Lestodon*, *Mylodon*, and *Glossotherium*. Modification of the medial trochlea into this process is associated with the rotation of the foot so that the animal's weight was supported on the fifth metatarsal (Stock 1917). The unmodified nature of the distal articular surface of the tibia and the structure of the calcaneum suggests that the mylodontines had not yet achieved the rotation of the pes that characterizes later members of the lineage. In contrast, the feet of *Pseudoprepotherium* and *Neonematherium* appear to have already undergone this pedolateral rotation because their astragali have the medial trochlea modified into an odontoid process.

The tuber calcis of the calcaneum of *Glossotheriopsis* is mediolaterally expanded but not to the degree seen in *Neonematherium*. It is dorsoventrally flattened and does not exhibit the thickening and modification seen in later mylodontines for weight bearing related to the rotation of the anterior part of the pes. There is a prominent sustenaculum, and the two astragalar articulations are separate. A process on the anterolateral edge has a prominent tendonal groove. In later mylodontines this groove is recessed into the body of the calcaneum.

The two unguals associated with IGM 252928 resemble those of megalonychids in being mediolaterally compressed and triangular in cross section. In later mylodontines the unguals are flattened dorsoventrally and semicircular in cross section.

In later mylodontines such as *Glossotherium* the lumbar vertebrae coossify with the sacrum. The one preserved lumbar of *Glossotheriopsis* is distinct, and there is no indication of fusing with the lumbar posterior to it.

At present, most records of *Glossotheriopsis* are from low in the La Victoria Formation. Duke Localities 21, 40, 42, and 67 and UCMP Locality V5046 are just below the Chunchullo Sandstone Beds, whereas Duke Localities 102, 103, and 132 are just above these sandstone beds. Duke Locality 87 is located higher in the La Victoria Formation just below the Tatacoa Sandstone Beds. The specimens from Duke Localities 16 and 54, which are in the Monkey Beds, mark the only records of this taxon from the Villavieja Formation.

Subfamily Scelidotheriinae
Neonematherium flabellatum Ameghino 1904

REFERRED SPECIMENS

Duke Locality 5: IGM 183997, distal right tibia. Duke Locality 27: IGM 183276, fragment left mandible with m3, m4. Duke Locality 28: IGM 183423, proximal right femur, proximal right humerus, distal right humerus, proximal end of right ulna, left astragalus. IGM 184359, partial skeleton: right maxilla with M2–M4, right side braincase, humerus fragments, fragment right femur, distal tibia, patella, three unguals, proximal calcaneum. Duke Locality 33: IGM 183171, fragment of left mandible with m3, m4. IGM 183174, right astragalus. Duke Locality 40: IGM 183292, palate. IGM 183294, right mandible fragment with m3. IGM 184373, right femur minus proximal end. Duke Locality 42: IGM 183333, right mandible with m3, m4. IGM 183342, proximal end left ulna, right astragalus, right third metatarsal. Duke Locality 48: IGM 250522, partial right mandible. Duke Locality 63: IGM 184243, partial skeleton: skull, mandible, right humerus, ulna and radius, right femur, proximal end both tibiae, proximal calcaneum, metapodials, phalanges, ribs, axis, and cervicals. Duke Locality 66: IGM 183791, left humerus. Duke Locality 70: IGM 183922, left m4. Duke Locality 74: IGM 183963, left third metatarsal. Duke Locality 79: IGM 184668, left mandible with m1–m4, IGM 184689, left maxilla with M2–M4, right mandible with m2–m4. IGM 184690, posterior part of left mandible with m4. IGM 184050, partial skeleton. Duke Locality 85: IGM 184227, left maxilla with m2–m4. Duke Locality 88: IGM 184315, left m4. Duke Locality 113: IGM 250966, left mandible with m2–m3, distal right scapula, proximal right humerus, distal right radius, proximal ends of both tibiae, distal left tibia, ribs, vertebrae. Duke Locality 132: IGM 251218, distal right tibia, right astragalus, ungual.

Before the recovery of the large sample of *Neonematherium* from La Venta, our knowledge of this taxon has been restricted to the type specimen from Río Fénix, Santa Cruz Province, Argentina. Unlike many of the La Venta sloths, which are known primarily from isolated skeletal elements that cannot be compared to the types of many taxa, *Neonematherium* is known from partial skeletons. The preservation of associated cranial and postcranial material permits a direct comparison with the type of *Neonematherium flabellatum* (the anterior part of a skull with right fourth and fifth and left third, fourth and fifth cheek teeth, MACN 11628; Ameghino 1904) and, in turn, makes it possible to identify with greater certainty many of the isolated bones of this taxon.

One of the major features that distinguishes scelidotheres from mylodonts is the configuration of the cuboid facet of the astragalus. In mylodonts this facet is convex (as is typical of most ground sloths), but in scelidotheres the facet is concave. No astragali referred to the earliest scelidothere, *Nematherium,* have been described, so it is not known if this character had developed by the Santacrucian. Kraglievich (1928b) illustrated an astragalus of *Neonematherium* that shows this feature, and it is present in astragali of *Neonematherium* in the La Venta fauna.

As previously discussed, morphology of the calcaneum is different between mylodonts and scelidotheres. In *Proscelidodon* (from the Huayquerian) and later scelidotheres the distal end of the tuber calcis is expanded and its dimensions mediolaterally and anteroposteriorly are roughly equal. Only the distal portion is in contact with the ground. This is also true of *Neonematherium,* but its proportions are different from those of later scelidotheres. This part of the calcaneum is flattened anteroposteriorly but expanded mediolaterally so that its overall configuration more closely resembles a megalonychid rather than other scelidotheres.

Another feature that has been used to distinguish

scelidotheres from mylodontines is the coossification of the proximal and second phalanges of the third digit of the pes. These two bones never fuse in mylodontines. In *Neonematherium* these two bones are still separate based on a partial pes associated with IGM 184050.

Hirschfeld (1985) listed *Neonematherium* from only two localities, both in the Baraya Member of the Villavieja Formation. The eleven additional localities and larger sample in this study extends the range of this genus to the lowest part of the La Victoria Formation as represented by Duke Locality 28 at Cerro Chacarón (fig. 15.1).

Family Megalonychidae
Gen. et sp. indet.
Large species

REFERRED SPECIMENS

Duke Locality 8: IGM 183973, proximal left femur. Duke Locality 31: IGM 183122, right first metacarpal, left and right second metacarpal, proximal right third metacarpal, left first metatarsal, distal left calcanum, left navicular, phalanges, unguals. IGM 183118, distal right femur. Duke Locality 40: IGM 184385, partial right mandible with second molariform. Duke Locality 42: IGM 183380, left astragalus. Duke Locality 66: IGM 253264, parts of skeleton, including right ulna and radius, proximal end right femur, right fibula, left patella. Duke Locality Fish Bed: IGM 183936, left third metacarpal. Duke Locality at a level between the La Venta Red Beds and Ferruginous Beds: IGM 184118, proximal right femur.

The larger of the two species of megalonychid present in the fauna is probably the same taxon described and illustrated by Hirschfeld (1985). None of the newly collected material permits any definite generic assignment. The major impediment to the identification of these specimens is the lack of materials comparable to those known for previously described taxa. Currently, only two megalonychids, *Eucholoeops* and *Diellipsodon,* are known from the Friasian. Both of these taxa are based on cranial and mandibular material. Santacrucian genera such as *Analcimorphus* and *Megalonychotherium,* which could be present in the Friasian, are also known only from cranial material. The genus *Protomegalonyx,* based on distal femora, is known from the

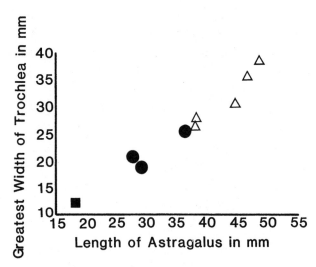

Figure 15.4. Scatter diagram of greatest width of trochlea of the astragalus plotted against the length of the astragalus for nothrotheres and megalonychids in the La Venta fauna. Symbols: *square*, small megalonycid; *circle*, large megalonycid; *triangle*, cf. *Hapalops.*

Chasicoan (Scillato-Yané 1977a), but the La Venta specimens cannot be referred to this taxon. In *Protomegalonyx* the articular surfaces of the distal condyles are separated from the patellar groove, but in the large La Venta megalonychid these surfaces are connected.

The stratigraphic range of the large megalonychid is from Level 6 to Level 40/41, that is, from below the Chunchullo Sandstone Beds in the La Victoria Formation to just below the La Venta Red Beds in the Baraya Member of the Villavieja Formation (fig. 15.1).

Megalonychidae, gen. et sp. indet.
Small species

REFERRED SPECIMENS

Duke Locality 36: IGM 183237, right astragalus. Duke Locality 70: IGM 183905, distal half of right femur. Duke Locality 136: IGM 251172, left second metacarpal, right astragalus, phalanges, unguals.

In addition to the larger species of megalonychid recognized by Hirschfeld (1985) from La Venta, a second smaller species is present in the fauna. Based on the astragalus, the species was half the size of the larger megalonychid in

the fauna (fig. 15.4), making it similar in size to many of the Santacrucian genera. This taxon ranges from Level 13, just above the Chunchullo Sandstone Beds, to Level 23, just below the Tatacoa Sandstone Beds, La Victoria Formation.

Family Megatheriidae
Subfamily Nothrotheriinae
cf. *Hapalops* Ameghino 1887
Gen. et sp. indet.
Large species

REFERRED SPECIMENS

Duke Locality 12: IGM 182715, left astragalus. Duke Locality 28: IGM 250896, proximal end of left humerus. Duke Locality 33: IGM 183167, proximal left ulna. Duke Locality 39: IGM 183264, fragment left mandible, distal right scapula, proximal left femur. Duke Locality 40: IGM 184370, right astragalus. IGM 184369, right astragalus. Duke Locality 42: IGM 183378, right second metacarpal and ungual. Duke Locality 57: IGM 183650, molariform. Duke Locality 67: IGM 183817, left humerus. Duke Locality 92 below sandstone: IGM 184450, fragments of mandible. Duke Locality 100: IGM 184574, proximal right humerus, distal left humerus, proximal left radius, fragment proximal left femur, and distal left femur, distal right fibula. Duke Locality 134: IGM 251119, left astragalus.

Hirschfeld (1985) described a nothrothere she referred to as cf. *Hapalops* from two localities. The larger sample now available consists of isolated bones, which are referred to this taxon based on their relative size. It is possible that the material cited here consists of two or more species but, lacking associated cranial and postcranial material, finer resolution is not possible. The astragali preserved do permit the recognition of the animal as a nothrothere, based on the modification of the medial trochlea into an enlarged odontoid process similar to that in later nothrotheres such as *Pronothrotherium* from the Huayquerian, and can be distinguished from the similar-sized megalonychid in which the medial trochlea is not modified.

The stratigraphic range of this taxon is from Level 3, below the Cerro Gordo Sandstone Beds of the La Victoria Formation, to Level 38, below the Ferruginous Beds of the Baraya Member, Villavieja Formation (fig. 15.1).

Nothrotheriinae, gen. et sp. indet.
Small species

REFERRED SPECIMENS

Duke Locality 31: IGM 183640, right distal tibia, right distal fibula, right astragalus, left and right calcaneum, distal metapodial, proximal phalanx, centrum of thoracic. IGM 183200, synsacrum, right femur, proximal right tibia, left proximal second metatarsal, left third metatarsal, left fourth metatarsal, left and right navicular, right second metatarsal, proximal and second phalanx, thoracic, four caudals.

This is the smallest sloth in the fauna and in linear dimensions is about half the size of the smaller of the two megalonychids. The specimens cited are probably the same taxon described by Hirschfeld (1985) from UCMP Locality V4935.

The available sample of this small sloth comes from two collections made at Duke Locality 31. The two samples complement each other with regard to portions of the skeleton preserved, and there is no duplication of bones. It is likely that the two collections represent the same individual.

This taxon is considered to be a nothrothere based on the morphology of the astragalus. The medial trochlea extends higher than the lateral trochlea and has the rudimentary development of the odontoid process that is so prominent and characteristic of later nothrotheres, such as *Nothrotherium* and *Nothrotheriops*.

This taxon is currently known from only two localities, Duke Locality 31 in the Monkey Beds (Level 35), which is in the lowest part of the Baraya Member of the Villavieja Formation, and UCMP Locality V4935 (Level 20), between the Chunchullo and Tatacoa Sandstone Beds of the La Victoria Formation.

Family Megatheriidae
Subfamily Megatheriinae
Gen. et sp. indet.

REFERRED SPECIMEN

Duke Locality 139: IGM 251145, right tibia-fibula.

Hirschfeld (1985) described a number of specimens of a small megathere from the La Venta fauna. Three genera of megatheriines are currently recognized from the Friasian: *Eomegatherium, Megathericulus,* and *Promegatherium*. Assignment of the La Venta material to any one of these is not possible, and the small additional sample does not permit resolution of this problem. IGM 251145 does, however, provide some information regarding the evolutionary development of one feature of this subfamily. One characteristic of megatheriines that distinguishes them from the other sloths is the fusion of the proximal end of the fibula to the tibia, a feature noted by Owen (1860). Although Hirschfeld described a megathere tibia from La Venta, the fibula was not preserved. The fibula preserved with IGM 251145 shows that this pattern of coossification of the two bones was established in the taxon it represents. In *Planops*, a Santacrucian genus considered to be a megatheriine (Hoffstetter 1961a), the tibia and fibula are separate.

Previously, all known occurrences of this megathere at La Venta were confined to the Villavieja Formation. The newly collected specimen extends the presence of a large megathere into the La Victoria Formation.

Biostratigraphy

Except for the rarer taxa such as the small megalonychid, small nothrothere, and the megathere, which are known only from two or three localities, the stratigraphic distribution of sloth taxa at La Venta is similar (fig. 15.1). Based on current samples, there does not appear to be any turnover in the fauna or replacement of one type of sloth by another.

Marshall et al. (1983) consider *Glossotheriopsis* to be one of the guide fossils for the Friasian faunas of Argentina, as the only records of the genus are from deposits considered to be Friasian in age. The La Venta fauna has been considered to be Friasian in age (Marshall et al. 1983) based on the similarity of many of its constituent taxa to the type Friasian of Chile. The addition of *Glossotheriopsis* to the La Venta fauna strengthens this interpretation. Although the type of *G. pascuali* is from the "Colloncurense," the earliest part of the Friasian (Scillato-Yané 1976b), current information on the temporal distribution of the taxon is limited and its stratigraphic range may not be limited to this part of the Friasian. The presence of this species in the La Venta fauna does, however, provide another taxon common to the Friasian age faunas of Argentina and Colombia.

The La Venta fauna of Colombia differs from the earlier Santacrucian faunas by the presence of *Neonematherium, Pseudoprepotherium,* and *Glossotheriopsis,* none of which is present in the Santacrucian. The type of *Neonematherium* was recovered from the Río Fénix, which is currently considered to be Mayoan or late Friasian in age. The only genus of sloth common to the Santacrucian of Patagonia and the La Venta fauna of Colombia is, possibly, *Hapalops.* Given the need for a complete taxonomic revision of *Hapalops* and the current lack of diagnostic specimens of both megalonychids and nothrotheres in the La Venta fauna, the presence of *Hapalops* in the La Venta fauna is uncertain.

Paleoecology

The wide variety of skull morphology, dentition, body proportions, and size present in the La Venta sloths suggests that members of this group had diversified to fill a variety of niches as herbivores. Although interpretations regarding dietary preference and ecology of some fossil sloths have been proposed (Stock 1925), these interpretations tend to describe each animal's ecology in only a general way, with little discussion regarding how the conclusions were reached. In many cases fossil sloth dietary preference has been inferred from taxonomy. Megalonychids, nothrotheres, and megatheres are considered to be browsers, while mylodonts (mylodontines and scelidotheres) are considered to be grazers (Stock 1925). The lack of precision in interpreting fossil sloth ecology reflects the marked differences in the anatomy between sloths and other herbivores and the lack of close modern analogs. Even the extant tree sloths are too specialized to provide good models for reconstructing many aspects of fossil sloth ecology. Yet despite the many pronounced anatomical differences, there are functionally analogous features, particularly in the skull but also in the postcranial skeleton, that make it possible to more precisely interpret their paleoecology. Rather than relying on a single modern species as a model, it is necessary to use features common to a variety of modern herbivores with similar ecological requirements.

The absence of teeth in the premaxillae of sloths makes this part of the skull functionally similar to that of ruminant artiodactyls. Following work by Solounias et al. (1988) and Janis and Ehrhardt (1988) on artiodactyls, the shape of the premaxillae can be used to infer the dietary category (brows-

er, grazer, or intermediate) to which a sloth belonged. The shape of the premaxillae in sloths is also reflected in the shape of the predental spout of the mandible so that even an isolated jaw can be used in making this assignment.

One feature common to the earliest representatives of each sloth lineage and still present in most of the sloths in the La Venta fauna is the structure of the anterior part of the mandible or predental spout. In *Pseudoprepotherium* and *Neonematherium,* for which this part of the mandible is preserved, it is elongate and narrower than the tooth rows. This morphology is similar to that of early nothrotheres, such as *Hapalops,* and to later nothrotheres such as *Pronothrotherium* (Huayquerian) and *Nothrotherium* (Lujanian). If one views the shape and proportions of the premaxillae and predental spout of the mandible in sloths as functionally equivalent to ungulates, as described by Janis and Ehrhardt (1988), then sloths with long, narrow predental spouts can be interpreted as selective browsers. This mandibular morphology suggests that the earliest mylodonts, both mylodontines and scelidotheres as represented by *Pseudoprepotherium* and *Neonematherium* in Colombia and the earlier Santacrucian genera *Analcitherium* and *Nematherium* in Patagonia, were not grazers but browsers.

Glossotheriopsis differs from the other mylodonts in the La Venta fauna in having an anteriorly flared muzzle and related wider mandibular symphysis, which is typical of later genera of mylodonts. A wide muzzle in ungulates has been used by Janis and Ehrhardt (1988) as an indication of grazing. This feature suggests that although earlier mylodonts may have been adapted for browsing, *Glossotheriopsis* probably represents the beginning of an adaptive shift to grazing that characterizes the later mylodontine mylodonts and that this group of sloths shifted to feeding on plants or plant parts lower in nutrients. A related modification was that, like many ungulate grazers, they evolved a dentition for crushing and grinding grasses. This suggests a shift in habitat preference from woodland to savanna and from browsing to grazing in the lineage of mylodontines that included *Glossotheriopsis.* We have good evidence that at least one member of this group, *Mylodon darwini,* used this habitat. As shown by Moore (1978), the dung of *Mylodon darwini* from the late Pleistocene is composed entirely of Cyperaceae and Gramineae, including species indicative of cool, wet sedge-grasslands.

The dentition of post-Friasian mylodonts, both mylodontines and scelidotheres, is characterized by flat occlusal surfaces that develop at right angles to the long axis of the molariforms. This crushing type of dentition is in marked contrast to that seen in the three mylodonts in the La Venta fauna, *Pseudoprepotherium, Glossotheriopsis,* and *Neonematherium* in which the molariforms are characterized by the presence of two occlusal surfaces positioned anteriorly and posteriorly and meeting at an acute angle to form a single transverse crest or loph. The position of the occlusal surfaces and the development of the transverse loph is similar to that seen in the extant tree sloth, *Choloepus,* as described by Naples (1982). This type of occlusal surface is best described as slicing vegetation and is functionally similar to that of megalonychids, nothrotheres, and megatheres, except that the teeth of the latter three groups have two transverse lophs instead of one.

Because mastication breaks food into smaller pieces and, thus, increases surface area so that digestive fluids act more quickly on the material, the functional shift in the dentition from slicing to crushing suggests a major change in the types of plants used by mylodonts. Although later mylodonts retain some shearing capability, as in the anterior dentition, the development of a crushing mode of mastication in the posterior dentition is characteristic of post-Friasian mylodonts. Because sloths are monogastric caecalids (terminology after Guthrie 1984), the shift from slicing to crushing would greatly affect the possible nutrient uptake during digestion. Monogastric caecalids are adapted to eating high-fiber, low-protein plants (Guthrie 1984). Because browsers feed on plant parts containing more nutrients and less fiber, their chewing does not need to be as efficient as that of grazers. The crudely chopped plants in the dung of the browsing ground sloth *Nothrotheriops* reflect these differences (Hansen 1978).

Crushing of vegetation rather than slicing results in the compression of the vegetation between the teeth and greater destruction of cell walls, thus making more nutrients available for digestion (Rensberger 1973). The greater degree of commutation or crushing of vegetation before swallowing increases the available surface area for digestion and, thus, efficiency in nutrient uptake and allows an animal to use plants or parts of plants that are relatively nutrient poor.

As noted by Marshall and Cifelli (1990), the Colhuehuapian, Santacrucian, and Friasian are dominated by mammalian groups, suggesting savanna and open woodland habitat. Considering the morphology of the premaxillae and predental spout on the mandible for those sloths from Argentina and Colombia for which skull and mandible structure and dentition are known, it is most likely that the open woodland component was probably the habitat used by the earliest sloths. Use of this habitat was continued in later nothrotheres, megatheres, and megalonychids but apparently not by the mylodonts. *Glossotheriopsis* represents the earliest stage of the adaptive shift to grazing, a shift that would allow later mylodontines to move into grassland savanna habitats.

Summary

The middle Miocene La Venta fauna contains a diverse assemblage of sloths dominated by members of the family Mylodontidae, including *Pseudoprepotherium confusum, Neonematherium flabellatum,* and *Glossotheriopsis pascuali. Glossotheriopsis,* the earliest known mylodontine mylodont, is recorded in the La Venta fauna of Colombia for the first time. Although previously known from the fauna, *Neonematherium flabellatum* is more common than previously thought. Megalonychids and nothrotheres are each represented by two taxa and megatheres by one, but available material from the fauna does not permit definite identification. Sloth diversity is stable throughout the stratigraphic section with the same taxa present in both the La Victoria and Villavieja formations.

Acknowledgments

I thank R. F. Kay for the opportunity to study the sloths of the La Venta fauna. During my visit to South America, R. Pascual of the Museo de Ciencias Naturales de La Plata and J. Bonaparte of the Museo Argentino Ciencias Naturales graciously provided access to the collections in their care. Travel in South America was partially funded by National Sciences and Engineering Research Council of Canada grant A 7973 to A. G. Edmund.

16
Locomotor Adaptations in Miocene Xenarthrans

Jennifer L. White

RESUMEN

Los perezozos (Pilosa, Xenarthra) han sido divididos tradicionalmente en dos grupos, las formas arborícolas de la actualidad, y las formas terrestres del archivo fósil. Faltando evidencias paleontológicas, sigue siendo oscuro el orígen de los perezozos arborícolas actuales y sus adaptaciones altamente especializadas para suspensión arbórea. En realidad, los perezozos fósiles de supuestos hábitos terrestres demuestran una diversidad amplia de morfologías postcraneanas que sugieren una comparable diversidad locomotora, hasta inclusive hábitos arbóreos.

La relación entre comportamiento y morfología en animales vivos nos permite interpretar las adaptaciones de animales extinguidos en base a índices de la forma de los elementos del esqueleto. Para poder averiguar el comportamiento locomotor de los posibles ancestros miocénicos de los perezozos terrestres gigantescos extinguidos y los perezozos arborícolas actuales, se examinó un número de carácteres morfológicos indicadores entre diversas especies actuales de Xenarthra. Dieciocho dimensiones lineares fueron medidas en el esqueleto de muchas especies de armadillos, perezozos, y vermilinguas actuales. Fueron escojidos dimensiones que tuvieron una relación con el comportamiento locomotor en otros mamíferos. De estas mediciones, diez índices morfométricos fueron formulados para poder permitir comparaciones independientemente de las diferencias en forma que se pueden atribuir a distintos tamaños corporales absolutos.

En base a éstos resultados, los Nothrotheriinae del Mioceno medio de Colombia se ubican dentro de la distribución de valores para taxa arbóreos actuales para casi todos los índices. La pequeña especie de Megalonychidae presenta valores para curvatura y compresión en las falanges ungulares del pie comparables a los Xenarthra trepadores actuales. El extremo inferior o distal del fémur de la especie grande de Megalonychidae es profundo de arriba abajo, dentro de la distribución de valores para Xenarthra terrestres actuales. *Neonematherium* demuestra una mezcla de adaptaciones tanto arbóreas como terrestres. *Pseudoprepotherium* demuestra características de hábitos terrestres, como por ejemplo, andar sobre el miembro inferior y sobre la cara lateral del pie. Las semejanzas entre los Nothrotheriinae del Santacrucense y los de La Venta, y entre *Nematherium* y *Neonematherium*, indican una reciente herencia común o una convergencia morfológica en adaptaciones para la vida arbórea. La gran diversidad de las adaptaciones locomotores entre las especies de La Venta indica que las condiciones ambientales durante el Mioceno medio de Colombia deberían haber sido diversas.

Sloths have traditionally been dichotomized into the extant tree sloths, sharing some of the most unique locomotor adaptations among mammals, and the fossil "ground" sloths, popularly represented by the huge, robust Pleistocene forms (Owen 1842; Simpson 1945; Hoffstetter 1954b; Romer 1966). As a result of the apparent absence of transitional forms, the origins of the tree sloths and their highly specialized suspensory adaptations have remained elusive. This often presumed phylogenetic and functional distinction between extant, small, arboreal sloths and their extinct, large, seemingly terrestrial relatives raises two problematic issues that require more attention.

First, extant sloths have been united by limb features, yet phylogenies of fossil sloths are based primarily on cranial features. The possibility of a diphyletic origin of the tree sloths has been proposed on the basis of certain cranial similarities of each extant genus with a different extinct family (Patterson and Pascual 1968; Webb 1985a; Webb and Perrigo 1985; Pascual et al. 1990). If this scenario is correct, two separate lineages of putative sister taxa to *Bradypus* and *Choloepus,* respectively, may have evolved in South America during the Tertiary.

Second, the temporally diverse and geographically widespread fossil sloths, often lumped as "ground" sloths, in fact exhibit a broad range of postcranial morphologies suggesting considerable locomotor diversity, including arboreality. Assertions in the literature that certain fossil sloths may have been semiarboreal have not been rigorously quantified and documented (e.g., Anthony 1916; Matthew and Paula Couto 1959; Romer 1966; Hirschfeld and Webb 1968; Paula Couto 1971; Mayo 1980; Scillato-Yané 1981a; Hirschfeld 1985; Webb 1985a), with the exception of Coombs (1983) who compared limb indices among several fossil sloth taxa, and Sherman (1984) who performed a detailed morphometric analysis of forelimb features. Inferring a diversity of locomotor habits among fossil sloths would substantiate speculations about locomotor habits and further discredit the traditional "ground" versus "tree" distinction.

In this chapter I investigate skeletal correlates of locomotor behavior in a variety of extant xenarthran genera (sloths, vermilinguas, and armadillos) and fossil sloth genera from two different faunas in South America, the middle Miocene Santacrucian of Argentina and the middle Miocene La Venta fauna of Colombia.

Sloth Radiations

The Santacrucian Land Mammal Age of Argentina heralds a great diversification in xenarthran evolution (Scillato-Yané 1986a). Recent radioisotope dates place the Santacrucian in the middle Miocene at about 18–15 Ma (Marshall, Drake, et al. 1986; MacFadden 1990). It is geographically vast and contains abundant fossil material (Marshall, Hoffstetter, et al. 1983). Extensive collections have been amassed, and the fauna appears to have been phylogenetically diverse (Marshall, Drake, et al. 1986). The three major families of fossil sloths seem to be well represented (Marshall, Hoffstetter, et al. 1983) although distinctions between the three are very subtle. Climatic conditions are presumed to have been warm and moist, supporting both subtropical and temperate forested areas and grasslands (Webb 1978; Scillato-Yané 1986a; Pascual and Ortiz 1990).

The taxonomic status of many members of the rich assemblage of Santacrucian fossil sloths is unresolved, yet the lineage(s) leading to the extant sloths may have arisen within this diverse group (Scott 1903–1904; Hoffstetter 1961a; Hirschfeld and Webb 1968; Paula Couto 1979; Marshall, Hoffstetter, et al. 1983; Frailey 1988). The genus *Hapalops* is of particular interest for several reasons. *Hapalops* has been identified as a possible generalized ancestor to later Tertiary megatheriid and megalonychid sloths (Hirschfeld and Webb 1968; Frailey 1988), and it is well represented in the fossil record. Furthermore, many species of *Hapalops* are relatively small and gracile compared to their larger, more robust relatives, and, therefore, they may provide clues about the origin and evolution of the arboreal adaptations of the tree sloths, *Bradypus* and *Choloepus*. In overall size the complete skeleton of *Hapalops longiceps* YPM-PU 15523 is similar to that of the terrestrial vermilingua (*Myrmecophaga*). Twelve lengths and diameters of limb elements of this specimen yield a mean estimated body weight of 26 kg in Gingerich's BODYMASS program (1990). A similar estimated body weight of 25.5 kg for *Hapalops* was obtained by Sherman (1984) based on a regression against log scapular length. Estimates through regression analyses of extant xenarthrans and *H. longiceps,* however, range up to approximately 40 kg based on femoral head diameter.

The phylogenetic position of the numerous other genera of small-bodied Santacrucian sloths is still in question. *Analcimorphus, Pelecyodon, Schismotherium, Eucholoeops,* and *Hyperleptus* are generally placed with *Hapalops* in the subfamily Nothrotheriinae (Paula Couto 1979; Barrio et al. 1984), but it is unclear whether this group belongs with the

Megatheriidae (e.g., Hoffstetter 1969; Scillato-Yané 1978; Paula Couto 1979; Barrio et al. 1984; Scillato-Yané et al. 1987) or the Megalonychidae (Scott 1903–1904; Cartelle and Fonseca 1983; Marshall, Hoffstetter, et al. 1983). In addition, the morphological variation between some genera and species of this group is substantial, and it is likely that not all of these genera belong in the same subfamily. The persistent tendency to recognize one cohesive subfamily in some classifications probably results from the rather unspecialized nature and small size of all of these taxa although it is not wise to base a phylogenetic classification on such criteria. For the sake of brevity, when referring to several of these genera together I will use the term *nothrotheres*.

The only Santacrucian large-bodied sloths (see size bars in figs. 16.2–16.4 to compare sizes of various limb elements in different xenarthran taxa) that are present in any abundance are *Planops* and *Prepotherium* of the subfamily Planopsinae. Material is still scarce, however, and it is unclear whether these really are separate genera (Hoffstetter 1961a; pers. obs.). They are placed within the Megatheriidae (Simpson 1945; Paula Couto 1979), but show affinities with some genera commonly referred to as nothrotheres even though they are much larger (Scott 1903–1904; Hoffstetter 1961a; Paula Couto 1979; Hirschfeld 1985). The family Mylodontidae is represented in the Santacrucian by two fairly small genera, *Nematherium* and *Analcitherium*. The predominance of relatively small sloths in the Santacrucian can be related to climatic conditions. Large-bodied sloths were prevalent in the Oligocene when Patagonia was becoming cooler and drier and grasslands provided grazing opportunities. Webb (1978) noted that in the late Oligocene and early Miocene, conditions were much wetter evidenced by the presence of marine beds; therefore, during the Santacrucian, moist subtropical and temperate forests were supported and grazing opportunities declined. This accounts for the scarcity of large mylodontids, which were likely grazers (Webb 1978; Coombs 1983). Later in the middle Miocene, the rise of the major part of the Andes Mountains initiated another cooling and drying trend, and large-bodied grazers again became abundant.

The middle Miocene La Venta fauna occurred between 13.5 and 11.5 Ma, as discussed by Flynn et al. (chap. 3). The fossils are found in the Honda Group; the Villavieja and La Victoria formations are exposed in the Magdalena River valley between the Eastern and Central Cordilleras of Colombia and contain the most varied and abundant assemblage of mammalian fossils from equatorial South America (Marshall, Hoffstetter, et al. 1983; Hirschfeld 1985). Although climatic conditions during the Tertiary in this region are poorly understood (Pascual and Ortiz 1990), one possible explanation for the diversity of taxa represented in the Miocene of Colombia is that wet tropical forests and woodland savannas had persisted for a long time (Webb 1978).

The fossil sloth taxa represented in the Colombian Miocene fauna include some types that are similar to the Santacrucian nothrotheres in being small and relatively unspecialized. There are some very small nothrotheres and some larger ones that may be referred to *Hapalops*. Megalonychids are scarce but also fall into small and large size categories. The La Venta fauna includes some larger scelidotheres (*Neonematherium*) and mylodonts (*Pseudoprepotherium*) that are not present in the Santacrucian fauna. The former genus appears to be closely related to the Santacrucian *Nematherium* (Engelmann 1985; Hirschfeld 1985). *Pseudoprepotherium* apparently manifests some resemblances to *Orophodon*, a large Oligocene mylodont; perhaps its distinction from the other Miocene Colombian taxa reflects retained similarities to earlier large mylodontids that were abundant before the Miocene in South America (Webb 1978; Engelmann 1985; Hirschfeld 1985). Large members of the family Mylodontidae dominate the xenarthran fauna of the Colombian Miocene as they dominate the southern continent in the early Tertiary. In contrast, in the Santacrucian of Argentina, mylodontids are much rarer whereas small nothrotheres are abundant. As demonstrated here, a comparison of postcranial morphologies of the two assemblages reveals a corresponding variability in locomotor adaptations.

One way of inferring locomotor behaviors from skeletal features of extinct organisms is by performing detailed functional character analyses. Extant xenarthrans exhibit a wide array of locomotor and corresponding morphological specializations, including suspension, fossoriality, and arboreal and terrestrial quadrupedalism. Such diversity within an order offers an opportunity to examine the relative contribution of heritage and habitus to morphological evolution. In this study comparative functional analyses of a set of morphometric data from both the hind limb and forelimb of extant xenarthrans provide a basis for assessing locomotor adaptations in fossil sloths.

Materials and Methods

The three behaviorally diverse clades of extant xenarthrans provide a comparative taxonomic sample for this study. Taxa included (with approximate body weights) are the two genera of sloths (*Bradypus*, 4–5 kg; and *Choloepus*, 5–7 kg), the three genera of vermilinguas (*Myrmecophaga*, 25–40 kg; *Tamandua*, 4–5 kg; and *Cyclopes*, 240 g), and two genera of armadillos (*Euphractus*, 3.5–6.5 kg; and *Dasypus*, 3–4 kg). *Bradypus*, *Choloepus*, and *Cyclopes* are fully arboreal; the sloths are suspensory as well (Miller 1935; Von Hagen 1939; Nowak and Paradiso 1975; Mendel 1985a; Taylor 1985). *Tamandua* is both an arboreal and terrestrial quadruped and a powerful digger (Taylor 1978; Barlow 1984; Hildebrand 1985). *Myrmecophaga* is a terrestrial quadruped and digs as well (Nowak and Paradiso 1975; Shaw et al. 1985; Taylor 1985; Burton 1987). Both *Dasypus* and *Euphractus* are terrestrial quadrupeds with fossorial habits (Nowak and Paradiso 1975; Barlow 1984; Redford and Wetzel 1985). In the cases of the sloths and *Tamandua*, species were combined within genera because, for the purposes of this study, there are no significant locomotor differences between species within a genus (Mendel 1985a, 1985b; Taylor 1985). The other vermilingua genera comprise only one species each. Armadillos do manifest substantial interspecific variation, so the study was confined to *E. sexcinctus* and *D. novemcinctus*; these species are relatively abundant in museum collections.

Eighteen skeletal measurements were taken on between five and forty-one individuals of each extant taxon, depending on availability of the appropriate bone or difficulty of measurement (app. 16.1). These particular measurements were chosen based on their demonstrated relevance to locomotor behavior in other mammalian groups such as primates, carnivores, and bovids. The specimens are housed at the American Museum of Natural History in New York, the U.S. National Museum at the Smithsonian Institution in Washington, D.C., and the Field Museum of Natural History in Chicago.

Ten morphometric indices were constructed from the

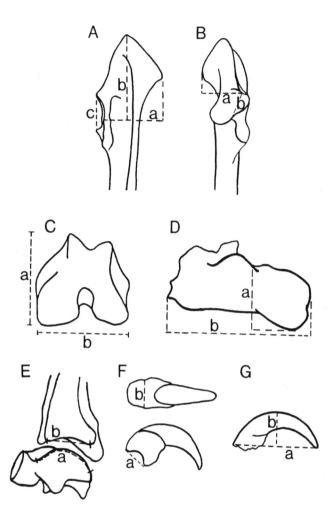

Figure 16.1. Pictorial representations of the morphological indices analyzed in the text: *A*, dorsal (*a*/*c*) and proximal (*b*/*c*) olecranon projections; *B*, medial olecranon projection (*a*/*b*); *C*, distal femur shape (*a*/*b*); *D*, volar calcaneal tuberosity (*a*/*b*); *E*, astragalar/tibial mobility (*a*/*b*); *F*, ungual phalanx compression (*a*/*b*); *G*, ungual phalanx curvature (*a*/*b*).

eighteen skeletal measurements; each index is represented pictorially in figure 16.1. There is some debate over the appropriateness of using ratios, or indices, in multivariate analysis because they are often not normally distributed and because it is difficult to determine whether the numerator or denominator is changing (Atchley et al. 1976; Albrecht 1978). Most of the data (indices) in this study are, in fact, normally distributed, and when they are not, nonparametric tests were used in statistical analyses. In the type of analysis presented here, data are most meaningful and functionally interpretable when viewed as morphometric

Table 16.1. Fossil taxa included in this study

Taxon	Specimen[a]	Locality/Horizon[b]	Graph on Which Specimen Appears[c]
		SANTACRUCIAN, ARGENTINA	
Analcimorphus giganteus	MACN (Ameg.) 6502	La Cueva	MU
Eucholoeops fronto	AMNH 9241	Río Gallegos	DF, PU
Eucholoeops ingens	FMNH 13125	Fairweather	AS, MU
Pelecyodon arcuatus	AMNH 9240	Río Gallegos	DOL, MOL, DF, PU
	FMNH 12062	Río Gallegos	PU
Hapalops longiceps	YPM-PU 15523	Coy Inlet	DOL, MOL, POL, DF, AS, VC, MU, PU
Hapalops elongatus	YPM-PU 15160	Coy Inlet	VC
Hapalops sp.	YPM-PU 15594	Killik Aike	AS, VC
	MACN (Fleagle) SC23.69B	Monte Obs./69	AS
Schismotherium fractum	MACN (Ameg.) 6445-6470	La Cueva	DOL, MOL, POL, DF
cf. *Schismotherium*	MACN (Fleagle) SC22.69E	Monte Obs./69	MU
Prepotherium potens	YPM-PU 15345	Killik Aike	DOL, MOL, POL, DF
Planopsine	MLP 4-334	?	PU
Planops cf. *P. martini*	MLP 267	?	VC
Nematherium angulatum	FMNH 13129	Fairweather	DOL, MOL, POL
	FMNH 13258	Río Gallegos	VC
cf. *Nematherium*	MACN (Ameg.) 6740, 6744, 6754, 6757, 6765	?	VC
Nothrothere	MLP 4-496	?	MU
	MLP 4-291, 4-320	?	MU
	MLP 4-79, -173, -174, -179, -183, -185, -188, -189, -190, -191	?	DOL, POL, MOL
		HONDA GROUP, COLOMBIA	
Pseudoprepotherium confusum	UCMP 38011	V-4523	DF
	UCMP 38014	V-4523	AS, VC
	UCMP 38881	V-4528	DOL, POL, MOL
	UCMP 39957	V-5045	DOL, POL, MOL, DF, MU
	UCMP 40465	V-5044	AS
	UCMP 41137	V-4522	DF
	IGM 250528	110/1	PU
	IGM 250807	121E/3	DF
	IGM 251141	136/3	DOL, POL, MOL
Neonematherium	UCMP 39275	V-4932	AS
	UCMP 39330	V-4517	MU, PU
	IGM 183342	042/2	DOL, POL, MOL
	IGM 183423	028/2	DOL, POL, MOL
	IGM 184243	063/3	DF
	IGM 184373	040/2	DF
	IGM 251218	132/3	AS
Large nothrothere, cf. *Hapalops*	IGM 183167	033	DOL
	IGM 184574	100	DF

Table 16.1. Continued

Taxon	Specimen[a]	Locality/Horizon[b]	Graph on Which Specimen Appears[c]
Large nothrothere, cf. *Hapalops*	UCMP 39949	V-4935	DOL, MOL, POL, DF, AS, VC
Small nothrothere	IGM 183640	031	AS
Large megalonychid	IGM 183118	031/4	DF
Small megalonychid	IGM 251172	136	PU

[a]AMNH: American Museum of Natural History; FMNH: Field Museum of Natural History; ING: Instituto Nacional de Investigaciones en Geociencias, Mineria y Química, Museo Geológico Bogotá; MACN: Museo Argentino de Ciencias Naturales (Ameghino Collection or Fleagle Collection); MLP: Museo de La Plata; UCMP: University of California Museum of Paleontology; YPM-PU: Princeton Collection at Yale. Specimen numbers beginning with SC are my own.

[b]Monte Obs. is Monte Observación; Fairweather is Angelina Ranch, Cape Fairweather. For IGM specimens, locality numbers denote Duke localities, horizons 1–3 are La Victoria Formation, and horizon 4 is Villavieja Formation. For UCMP specimens, locality numbers denote UCMP localities.

[c]AS: astragular/tibial mobility; DF: distal femur shape; DOL: dorsal olecranon projection; MOL: medial olecranon projection; MU: manual ungual phalanx shape; POL: proximal olecranon projection; PU: pedal ungual phalanx shape; VC: volar calcaneal tuberosity.

indices. Indices provide a method for locally scaling shape differences among taxa of disparate absolute sizes. An incomplete knowledge of the constancy of either the numerator or denominator does not detract from the overall patterns of discrimination elucidated by the shape variables. Moreover, using a standard size variable such as the geometric mean as the denominator is impractical in this study, because fossil specimens are so often represented by single bones.

Indices were first tested for normality and homogeneity of variances. Unplanned comparisons among index means were performed for normal, homogeneous data using the parametric GT2 method. Heterogeneous data were tested for equality of means by the Games and Howell method. Differences in locations of means for non-normal data were tested by using Kruskal-Wallis tests. All statistics were performed with the BIOM package of statistical programs (Rohlf 1988). Means or individual values for each index are displayed in univariate plots with the index description and its value along the abscissa and taxa along the ordinate.

Because most fossil sloth taxa have been described on the basis of crania, it is often extremely difficult to identify isolated limb elements to genus, and in most cases it is unwise to combine such individuals into a single taxon. Most of the individuals included in this study consist of single elements or partial skeletons. For this reason, this study necessarily suffers from small sample sizes for fossils. I have included a sample of fossil individuals on each univariate plot (see table 16.1); the choice of taxa depended upon the ability to obtain the measurements required for any particular index. In a few cases that will be discussed I have combined a group of individuals. Because this is not a study of systematics, I will refer to an individual that is not positively identified by a more general term (e.g., large nothrothere).

A principal coordinates analysis was performed on all standardized means of morphometric indices for all the extant taxa and one fossil individual. Principal coordinates analysis is an ordination technique based on a Euclidean distance matrix; it reduces the dimensionality of the data to, usually, two or three axes that account for most of the variation among taxa (Sneath and Sokal 1973; James and McCulloch 1990). This multivariate technique is advocated instead of principal components analysis when the number of characters exceeds the number of taxa (Sneath and Sokal 1973). A minimum spanning tree was superimposed on the plot of the first two principal coordinate axes to reveal which taxa are most similar when all axes are considered (James and McCulloch 1990). These analyses were accomplished by using NTSYS-PC (Rohlf 1989).

A single, complete specimen of *Hapalops longiceps* (YPM-PU 15523; for this and other abbreviations, see table 16.1) could be included for multivariate analysis. A single individual was used because the measurements in this study are scattered throughout the limb elements, and it is rare to have use of a complete skeleton of a single fossil sloth specimen. Although the genus *Hapalops* has been split perhaps unjustifiably into numerous species (Scott [1903–1904] lists thirty-seven species) and is in need of taxonomic revision (Scillato-Yané 1986), I did not feel secure enough to combine other *Hapalops* individuals with YPM-PU 15523 to obtain a mean for each measurement. No other Miocene taxon included in the study was complete enough to include in the multivariate analysis.

Results and Discussion

A brief description of indices of skeletal shape included in this work, their functional significance, and their utility as discriminators of habitat utilization among extant taxa follows; the values of these indices for Santacrucian and Colombian sloths are interpreted on this basis. Features that were examined are represented by line drawings of both extant genera and representative fossil taxa; indices are quantified graphically in univariate and bivariate plots. Vertical lines with open bars are means and standard deviations, squares are individual values, and horizontal lines connecting squares represent a range of variation for a taxon.

OLECRANON PROCESS

The size, shape, and orientation of the olecranon process of the ulna has been associated with variations in locomotor behavior in primates and carnivores. The olecranon process serves primarily as the insertion point for the triceps brachii, which extends the forearm at the elbow. The olecranon projects dorsally in terrestrial animals that habitually hold the forearm in full extension (fig. 16.2A: MY); this arrangement maximizes triceps leverage (Jolly 1972b; Bown et al. 1982; Van Valkenburgh 1987). The olecranon is proximally extended (in line with the shaft of the ulna) in climbers and especially in diggers (fig. 16.2A: TA, DA, EU); this maximizes triceps leverage while the elbow is flexed as it usually is during climbing and digging (Miller 1935; Bown et al. 1982; Coombs 1983; Rose and Emry 1983; Fleagle and Meldrum 1988; Ford 1990). The olecranon is reduced in both dorsal and proximal dimensions in suspensory animals that habitually hang by their arms (fig. 16.2A: BR, CH) (Mendel 1979; Taylor 1989; Ford 1990). Two indices were constructed from measurements of the olecranon process to quantify these differences (fig. 16.1A). Dorsal olecranon projection is the maximum dorsal or posterior projection from the center of the trochlear notch divided by the length of the trochlear notch. Proximal olecranon projection is the maximum proximal extent divided by the length of the trochlear notch.

A plot of dorsal olecranon projection is presented in figure 16.2C. Among vermilinguas there are significant (p = 0.01) differences between the terrestrial *Myrmecophaga*, the semiarboreal *Tamandua*, and the fully arboreal *Cyclopes*; each has a progressively lower index than the last. The tree sloths, *Bradypus* and *Choloepus* are distinguished from all other taxa (p = 0.01) and represent an extreme form of the tendency exhibited by the arboreal *Cyclopes*. The armadillos fall within the range of *Cyclopes*, but the proximal projection index indicates how differently shaped their olecrana are.

The range of variation among Santacrucian taxa is not great (fig. 16.2A, C); *Prepotherium* (PR★) has the greatest dorsal expansion, whereas *Nematherium* (NE★) and the nothrothere taxa (PE★, HL★, SC★) exhibit reduced dorsal projections and ulnar shafts that curve ventrally. Only *Pelecyodon* (PE★) falls near the range of the tree sloths. For this and the next two indices, the combined group of Santacrucian nothrotheres consists of ten ulnae of highly similar shapes and sizes that were collected together; because the standard deviation is so low, I felt confident enough to include them under one taxon name. The range of variation for Colombian taxa is slightly larger and the values are higher (fig. 16.2A, C); *Pseudoprepotherium* (PS+) has the greatest dorsal expansion of any of the fossils, yet the nothrotheres (NO+), which are probably *Hapalops*, have comparable values to the Santacrucian nothrotheres and *Hapalops*. *Neonematherium* (NN+) has an intermediate value; its shaft is straight yet it has no substantial dorsal projection. Overall, none of the fossil taxa falls entirely

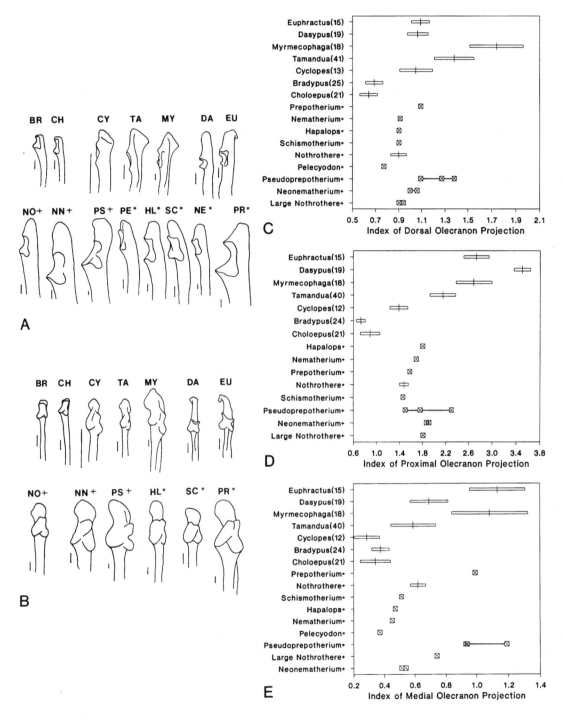

Figure 16.2. Forelimb indices. *A.* Lateral view of proximal left ulnae. *B.* Anterior view of proximal left ulnae. *C.* Plot of dorsal olecranon projection. *D.* Plot of proximal olecranon projection. *E.* Plot of medial olecranon projection. *A–B.* Scale bars 1 cm. Colombian specimens are indicated by *plus signs*, Santacrucian by *asterisks*. Abbreviations for extant xenarthrans: *BR, Bradypus; CH, Choloepus; CY, Cyclopes; DA, Dasypus; EU, Euphractus; MY, Myrmecophaga; TA, Tamandua*. Abbreviations for fossil sloths: *HL★, Hapalops longiceps; NE★, Nematherium; PE★, Pelecyodon; PR★, Prepotherium; SC★, Schismotherium; NO+, large nothrothere; NN+, Neonematherium; PS+, Pseudoprepotherium*. Plots include sample sizes for extant taxa (*in parentheses*), means (*vertical lines*), standard deviations (*open bars*), individual values (*squares*), and ranges of variation (*horizontal lines*). For nothrothere★ in *C–E*, n = 10.

within the range of the most extreme arboreal or terrestrial extant taxa.

In the plot of proximal olecranon projection (fig. 16.2D), the digging armadillos and *Myrmecophaga* lie significantly (p = 0.01) above the arboreal taxa. *Myrmecophaga* has a fairly large proximal projection as a corollary of its dorsal enlargement (fig. 16.2A). The armadillos have greatly expanded olecrana in the proximal direction but no dorsal projection in accordance with the flexed position of their elbows during digging activities. The suspensory tree sloths manifest a significantly (p = 0.01) reduced proximal expansion because their arms are usually hanging and extended by gravity.

This index demonstrates that some fossil taxa (e.g., *Prepotherium*, PR★) exhibit olecrana that are shaped completely differently than those of any extant taxa (fig. 16.2A). Among Santacrucian taxa, *Hapalops* (HL★) and *Nematherium* (NE★) have reduced olecrana and straight ulnar shafts, suggesting that they did not rely on powerful forearm extension (fig. 16.2A, D). *Pelecyodon* (PE★, not measured), *Schismotherium* (SC★), and the nothrotheres also have very reduced olecrana and ulnar shafts that curve ventrally, which is seen in animals that habitually flex the forearm. The value for *Prepotherium* falls between the values of the Santacrucian taxa that exhibit reduced olecrana; unlike these other taxa, however, it has a substantial dorsal projection. Among Colombian taxa, *Neonematherium* (fig. 16.2A: NN+; fig. 16.2D) manifests a very proximally extended, straight olecranon, falling between semiarboreal and arboreal vermilinguas. Proximal olecranon values for *Pseudoprepotherium* encompass those of *Neonematherium* and the large nothrothere and indicate that the olecranon is not greatly extended proximally (fig. 16.2D). This reflects, in part, the relatively large trochlear notch that would have borne heavy loads at the elbow. There is a great amount of variation in size of the olecranon in *Pseudoprepotherium*, but it is generally bulky (fig. 16.2A: PS+); this enlarged insertion area may be suggestive of digging adaptations (see the discussion of the medial olecranon projection that follows). Again, no fossil sloth manifests the extreme values of the digging armadillos or suspensory tree sloths.

The olecranon has a substantial medial expansion in the digging and terrestrial taxa (fig. 16.2B: TA, MY, DA, EU); these animals rely on powerful extension of the forearm for moving substrate and propulsion, respectively (Hildebrand 1985). This expansion is correlated with an enlarged medial epicondyle of the humerus; both these features may increase the area of origin of the carpal flexors (Taylor 1978; Hildebrand 1985). Another possible scenario is that the triceps insertion creates lateral torque on the olecranon that is countered by epitrochleoanconeus, a muscle passing between the medial aspect of the olecranon and the medial epicondyle (Taylor 1978). Epitrochleoanconeus is a relatively large muscle in *Tamandua*; its size and orientation contribute both to extension and abduction of the elbow (Taylor 1978). Dorsoepitrochlearis is a muscle that extends from the latissimus dorsi tendon to the medial olecranon and the extensor side of the proximal forearm (Taylor 1978). This muscle contributes to adduction and extension of the elbow and is relatively large in diggers and climbers (Miller 1935; Hildebrand 1985). At the opposite extremes are the tree sloths and *Cyclopes* (fig. 16.2B: BR, CH, CY); in these animals elbow extension is not so important, so neither the stabilization of epitrochleoanconeus nor the extension power of dorsoepitrochlearis are crucial (Miller 1935).

The arrangement of extant taxa in the plot of the medial olecranon projection (fig. 16.2E) is essentially consistent with the pattern observed for the proximal olecranon. In this case, however, all the arboreal taxa are statistically indistinguishable from each other and significantly different (p = 0.01) as a group from digging and terrestrial taxa. The Santacrucian fossils manifest a range of values comparable to that of the extant taxa (fig. 16.2B, E); *Prepotherium* (PR★) has a great medial expansion comparable to that in *Myrmecophaga* and *Euphractus*, whereas *Pelecyodon* is well within the range of the suspensory forms. *Schismotherium* (SC★) and the group of nothrotheres have values comparable to the semiarboreal, digging vermilinguas, whereas *Hapalops* (HL★), *Pelecyodon*, and *Nematherium* manifest little medial expansion, suggesting minimal reliance on elbow extension. Perhaps their elbows were continually in a flexed position. The range of variation is just as great in the Colombian radiation (fig. 16.2B, E). There is clear separation between *Neonematherium* (NN+) with a very small medial expansion, the large nothrothere (NO+), and *Pseudoprepotherium* (PS+) with a greatly expanded medial olecranon like that of the diggers.

DISTAL FEMUR SHAPE

Terrestrial quadrupeds tend to have anteroposteriorly deep distal femora (fig. 16.1C); this increases the leverage arm for extension of the leg by quadriceps femoris (Taylor 1976; Kappelman 1988). This expansion is evident in the terrestrial *Myrmecophaga* and the armadillos (fig. 16.3A: MY, DA, EU). Arboreal animals that continually flex the hind limb have broader, shallower distal femora (fig. 16.3A: BR, CH, CY). The tree sloths (*Bradypus* in particular) and *Cyclopes* manifest this shallow condition. *Bradypus* tends to climb and cling to vertical substrates more frequently than *Choloepus* (Miller 1973); this may explain the slightly deeper distal femur of the latter. In the plot of distal femur shape (fig. 16.3D), the location of the vermilinguas clearly illustrates the utility of this index in discriminating locomotor mode because of the clear statistical separation between the three genera. *Myrmecophaga*, along with the digging *Dasypus*, has by far the deepest distal femora (p = 0.01), and all genera practicing terrestrial locomotion have significantly greater values than the fully arboreal genera.

The distal femur index reveals some striking differences between the two radiations. All the Santacrucian taxa exhibit relatively shallow distal femora (fig. 16.3A, D), perhaps because of extreme flexion of the knee while climbing in the taxa with relatively symmetrical condyles (e.g., *Schismotherium*: SC★). Alternatively, this shallow morphology could result from lateral stresses imposed by an abducted posture of the femur. An anteroposteriorly flattened distal femur, along with an abducted posture of the knee with a larger weight-bearing surface on the medial side, is thought to be associated with arboreality and the potential for a close approximation of the two hind feet in some primates (Jungers 1976; Fleagle and Meldrum 1988; Ruff 1988). *Prepotherium* (fig. 16.3A: PR★) falls within the range of arboreal forms for this index, but its extremely enlarged medial condyle, the high degree of bicondylar angulation of the articular surface, and values for other indices indicate that it was not characterized by the arboreal adaptations of some of the other Santacrucian genera; rather, this morphology may reflect walking on abducted hind limbs.

The Colombian taxa exhibit a much wider range of variation for this index (fig. 16.3A, D). *Pseudoprepotherium* (PS+) falls within the range of extant terrestrial taxa with

extremely deep distal femora, and it is also characterized by a huge medial condyle. This morphology suggests heavy loading while walking on an abducted knee. The large megalonychid also has a value in the terrestrial range. *Neonematherium* (NN+) is clearly distinguishable from *Pseudoprepotherium* and the large megalonychid; it has a moderately deep distal femur with relatively symmetrical condyles, perhaps indicating adaptation to both climbing and terrestrial walking on nonabducted limbs. One of the large Colombian nothrothere individuals (NO+) has a shallower distal femur than any other fossil taxon.

VOLAR CALCANEAL TUBEROSITY

The shape of the calcaneal tuberosity is determined by the relative development of its posterior (proximal), volar, and medial dimensions; this index describes the volar extent (fig. 16.1D). The tuberosity is similarly elongated posteriorly in the armadillos, *Myrmecophaga,* and *Tamandua* (fig. 16.3B: DA, EU, MY, TA). In contrast, it shows a marked volar expansion relative to length in the fully arboreal *Cyclopes* and the tree sloths, especially *Choloepus* (fig. 16.3B: BR, CH, CY; fig. 16.3E). Whereas *Bradypus* descends trees rear first, *Choloepus* descends head first (Goffart 1971; Mendel 1985b) and, therefore, has a special need for plantarflexion of the pes to grip the substrate (Jenkins and Krause 1983). Increasing the volar extent of the calcaneal tuberosity may increase the lever arm for plantarflexion of the pes by soleus and gastrocnemius (Mendel 1981a). It seems more likely, however, that the extended tuberosity increases the lever arm for quadratus plantae and flexor digitorum brevis, which flex the pedal digits (Mendel 1981a). Quadratus plantae has been observed to be continuous with soleus in some specimens of *Choloepus* (Mendel 1981a). The muscle originates on the calcaneal tuberosity, and is also closely associated with flexor hallucis longus and flexor digitorum longus; it may help to direct the pull of the latter two muscles (Lewis 1962).

In the graph of volar calcaneal tuberosity (fig. 16.3E) *Choloepus* lies significantly above all other taxa (p = 0.01). *Cyclopes* also exhibits a high mean for this index although the variance is high. Digital flexion is clearly important for these fully arboreal animals (Flower 1882; Mendel 1981a). *Bradypus* is not statistically distinguishable from *Myr-*

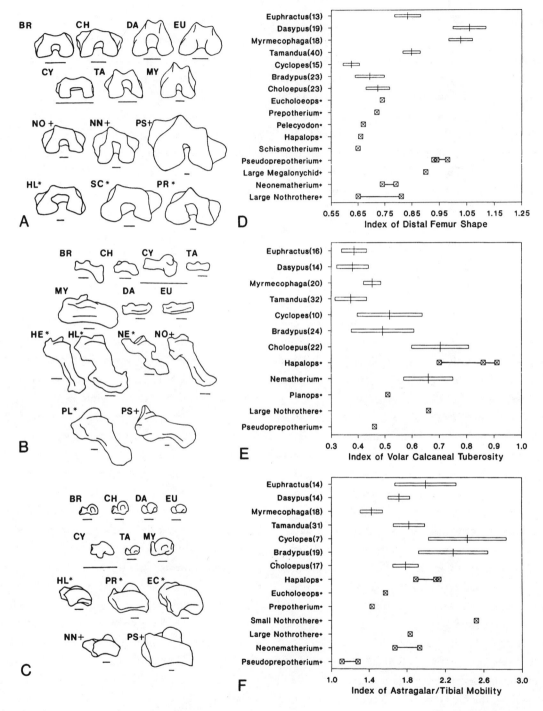

Figure 16.3. Hind limb indices. *A*. Distal view of left femora. *B*. Lateral view of left calcanei. *C*. Lateral view of left astragali. *D*. Plot of distal femur shape. *E*. Plot of volar calcaneal tuberosity. *F.* Plot of astragalar/tibial mobility. *A–C*. Scale bars 1 cm. Colombian specimens are indicated by *plus signs*, Santacrucian by *asterisks*. Abbreviations for extant xenarthrans: *BR, Bradypus; CH, Choloepus; CY, Cyclopes; DA, Dasypus; EU, Euphractus; MY, Myrmecophaga; TA, Tamandua*. Abbreviations for fossil sloths: *EC*★, *Eucholoeops; HE*★, *Hapalops elongatus; HL*★, *H. longiceps; NE*★, *Nematherium; PL*★, *Planops; PR*★, *Prepotherium; SC*★, *Schismotherium; NO*+, large nothrothere; *NN*+, *Neonematherium; PS*+, *Pseudopropotherium*. Plots include sample sizes for extant taxa (*in parentheses*), means (*vertical lines*), standard deviations (*open bars*), individual values (*squares*), and ranges of variation (*horizontal lines*). For *Nematherium*★ in *E*, n = 6.

mecophaga, but this is not surprising because *Bradypus* is able to flex only its distal phalanges owing to extensive fusion of the pes and, therefore, would not benefit from added leverage of the aforementioned muscles.

Santacrucian taxa exhibit a substantial range of variation for this index (fig. 16.3B, E). *Hapalops* and some other "nothrothere" taxa are characterized by remarkably expanded calcaneal tuberosities and manifest values comparable to or greater than that of *Choloepus* (fig. 16.3B: HL★, HE★). The tuberosity of *Hapalops longiceps* is greatly expanded both volarly and medially; this may be suggestive of flexed pedal digits along with an inverted position of the pes. *Nematherium* (NE★) has an easily recognizable, very distinctive calcaneal morphology, and I have combined six specimens for the plot. The calcaneus of the large Santacrucian *Planops* (PL★) is similar in general morphology to the "nothrotheres," but the tuberosity is more robust and less volarly expanded.

The two Colombian taxa in the plot have extremely dissimilar calcaneal morphologies. The large nothrothere calcaneus (NO+) exhibits a shape comparable to that of the Santacrucian *Hapalops* but slightly less volarly expanded; it probably is *Hapalops* material. In contrast, the calcaneus of *Pseudoprepotherium* (PS+) is unlike that of any other fossil sloths represented here (including the Santacrucian mylodont *Nematherium*) and is actually more similar to that of *Myrmecophaga* (MY). Its volar surface appears to be designed to be placed fully on the ground because (1) it is expanded and rugose mediolaterally but lacks the bladelike volar expansion seen in the nothrotheres, and (2) it rests alone in a position suitable for articulation with the astragalus (and, therefore, the tibia), whereas the volarly expanded calcanei must be externally supported in such a position (pers. obs.; Hirschfeld 1985).

ASTRAGALAR/TIBIAL MOBILITY

The length of the curved surface of the astragalar trochlea relative to the anteroposterior depth of the distal tibial articular surface (fig. 16.1E) reflects the degree of movement possible at this joint (Szalay and Decker 1974). The lateral trochlea was measured because the medial trochlea is reduced in sloths. A relatively long trochlear arc articulated with a shallow distal tibia suggests a great range of potential flexion and extension at this joint. This is crucial for climbers (fig. 16.3C: BR, CH, CY); the values for *Bradypus* and *Cyclopes* fall significantly above most other taxa (p = 0.05) in the astragalar curvature plot (fig. 16.3F). The discrepancy between values for the tree sloths might be partially explained by the observation that *Bradypus* climbs more frequently than *Choloepus* (Miller 1973), which falls within the range of the digging xenarthrans. The high value for *Euphractus* is likely related to extreme flexion and extension of the pes during digging; scratch diggers use the hind and forefeet together when digging (Hildebrand 1974). Having more equally sized articular surfaces suggests restricted movement of the ankle as in the fully terrestrial *Myrmecophaga*, whose value lies significantly below the values of all other taxa (p = 0.01).

Similar patterns in talocrural joint morphology hold true in some other groups of mammals. The fully arboreal kinkajous and tree squirrels manifest nearly 180° of curvature in the astragalar trochlea, and extreme plantarflexion is possible at this joint (Jenkins and McClearn 1984). Langdon (1984) noted that among primates, terrestrial cercopithecids have a more restricted range of motion at this joint than do suspensory hominoids, which rely on extreme plantarflexion. Meldrum (1991) found that arboreal cercopithecines exhibit a great degree of plantarflexion when the pes is released from the substrate and a proximodistally extended trochlear surface, whereas terrestrial cercopithecines demonstrate limited plantarflexion, an abbreviated trochlear articular surface, and a general reduction in ankle mobility. As a caveat, animals with very different locomotor patterns may have similar values for a given index for different reasons. Some terrestrial cursors such as the horse and coyote, for example, also have fairly high values for this index (1.83 and 1.84, respectively; pers. obs.). These values, however, are not as high as those of the fully arboreal xenarthrans. In any case, no extant cursors are included in this study, and it is unlikely that fossil sloths were cursorial; therefore, there is little danger of misinterpreting the astragalar/tibial index in this manner.

Among Santacrucian taxa *Hapalops* (HL★) falls within the range of the extant taxa with the most highly mobile ankle joints, whereas *Eucholoeops* (EC★) and *Prepotherium* (PR★) have much lower values (fig. 16.3C, F). *Prepotherium* reveals an incipient stage of the trend toward the development of a medial odontoid process on a very short, wide astragalus, a morphology that is highly developed and probably convergent in later megatheres and mylodonts (Owen 1860; McDonald 1977; Hirschfeld 1985). There is little anteroposterior movement between the astragalus and the anteroposteriorly deep tibia; most movement is rotational around the odontoid process, which is consistent with a posture in which weight was borne on the massive lateral side of the pes and the knee was habitually abducted. (Stock [1917] describes seven characteristics of a pedolateral pes.)

The range of variation in this index is much greater among Miocene Colombian sloths (fig. 16.3F). The small nothrothere exhibits an extremely high value for this index, indicating a potential for a significant amount of flexion and extension of the ankle. The large nothrothere and *Neonematherium* exhibit moderately high ranges of movement at the talocrural joint. The astragalus of *Neonematherium* displays an incipient form of a few characteristic scelidothere features such as a concave cuboid facet and an elevated and anteroposteriorly reduced medial trochlea (fig. 16.3C: NN+). In overall morphology and size, however, it resembles that of *Nematherium* and other Santacrucian taxa more closely. *Pseudoprepotherium* (PS+) has the lowest value of any taxon on the plot; its astragalar and tibial surfaces are nearly congruent in length and the astragalus manifests a more highly developed stage in the evolution of the odontoid process than the unrelated *Prepotherium* (PR★).

UNGUAL PHALANX SHAPE

The shape of the ungual phalanx is often indicative of a particular mode of locomotion or behavior, and the overlying claw mirrors the shape of the ungual phalanx (Van Valkenburgh 1987). Two aspects of ungual shape were measured on the manual and pedal ungual phalanges of digit III. Ungual phalanx compression equals the base height divided by the base width (fig. 16.1F); ungual phalanx curvature is the length of the phalanx divided by the maximum height of curvature (fig. 16.1G). Values for individuals are depicted on bivariate scattergrams of both compression and curvature for manual ungual phalanges (fig. 16.4C) and pedal ungual phalanges (fig. 16.4D).

UNGUAL PHALANX COMPRESSION

A high value for ungual phalanx compression indicates lateral compression of the claw; this is seen in suspensory forms and climbers that sustain dorsoventral stresses in the claws (Preuschoft 1970; Mendel 1981b; Van Valkenburgh 1987) (fig. 16.4A, B: CY, BR, CH). Fossorial (especially scratch digging) and terrestrial animals (DA, EU, MY, TA) have relatively flat and wide claws to facilitate plantigrade walking and digging (Preuschoft 1970; Coombs 1983; Van Valkenburgh 1987; Taylor 1989). *Myrmecophaga, Dasypus*, and the tree sloths, have similar manual and pedal indices (fig. 16.4C, D). The climbing vermilinguas tend to have more laterally compressed pedal unguals than manual; this is likely because *Cyclopes* often hangs by its feet, and *Tamandua* has relatively flattened foreclaws that reflect digging adaptations.

UNGUAL PHALANX CURVATURE

A low value for ungual phalanx curvature indicates a highly curved claw, which is characteristic of suspensory and climbing animals; hooklike claws facilitate hanging, grasping, and piercing bark (Cartmill 1974; Mendel 1981b; Coombs 1983; Van Valkenburgh 1987; also fig. 16.4A, B: CY, BR, CH). Although the climbing vermilinguas have more highly curved claws than *Myrmecophaga* (especially in the pes), all vermilinguas tend to have curved claws, probably owing to nest-ripping behavior. The terrestrial, fossorial armadillos exhibit relatively straight claws (DA, EU).

The Santacrucian *Hapalops, Schismotherium, Pelecyodon*, and *Analcimorphus* manifest quite laterally compressed manual ungual phalanges, but they are also rather straight (fig. 16.4A: HL★, SC★; fig. 16.4C). This combination of features might be consistent with a scansorial life style that included both arboreal and terrestrial quadrupedalism; the lateral compression suggests that dorsoventral stresses were placed on these claws during climbing activities, whereas the lack of curvature may indicate plantigrady (as opposed to curling the phalanges under the palm as in *Myrmecophaga*) while the manus was placed on the substrate. Among the extant taxa, the climbers with the most extreme curvature values demonstrate additional roles for the claws that exag-

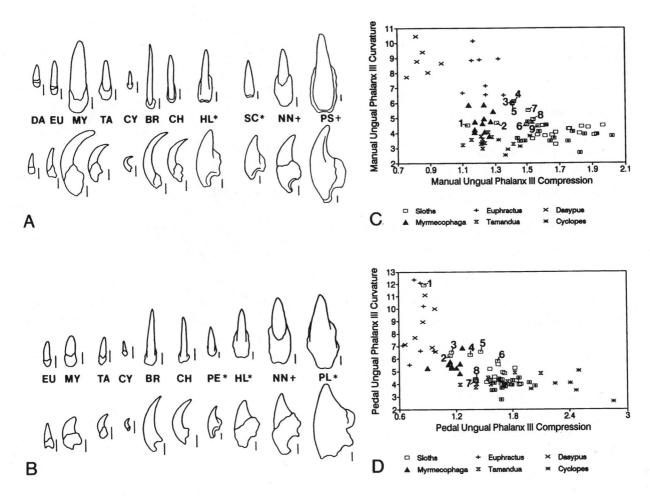

Figure 16.4. Ungual phalanx indices. *A.* Third manual ungual phalanges, dorsal view (*top row*) and lateral view (*bottom row*). *B.* Third pedal ungual phalanges, dorsal view (*top row*) and lateral view (*bottom row*). Scale bars 1 cm. Colombian specimens are indicated by *plus signs,* Santacrucian by *asterisks.* Abbreviations for extant xenarthrans: *BR, Bradypus; CH, Choloepus; CY, Cyclopes; DA, Dasypus; EU, Euphractus; MY, Myrmecophaga; TA, Tamandua.* Abbreviations for fossil sloths: *HL*, Hapalops longiceps; PE*, Pelecyodon; PL*,* planopsine; *SC*, Schismotherium; NN+, Neonematherium; PS+, Pseudoprepotherium. C.* Bivariate scattergram of individual values for manual ungual phalanx curvature versus manual ungual phalanx compression. Extant sloths: *dotted squares, Choloepus; open squares, Bradypus.* Numbered fossil sloths: *1, Neonematherium+; 2, Pseudoprepotherium+; 3, Schismotherium*; 4, Hapalops longiceps*; 5, Analcimorphus*; 6, Eucholoeops*; 7, 8, 9,* nothrothere*. *D.* Bivariate scattergram of individual values for pedal ungual phalanx curvature versus pedal ungual phalanx compression. Extant sloths: *dotted squares, Choloepus; open squares, Bradypus.* Numbered fossil sloths: *1, Pseudoprepotherium+; 2,* planopsine*; *3, Neonematherium+; 4, Hapalops longiceps*; 5, 6, Pelecyodon*; 7, Eucholoeops*; 8,* small megalonychid+.

gerate the curvature: suspension and nest ripping. The three Santacrucian nothrothere individuals and *Eucholoeops* cluster within the lower range for curvature and lateral compression of the tree sloths; such curved, narrow claws are indicative of climbing activities (fig. 16.4C).

Both Colombian mylodontid sloths on the manual plot have much more dorsoventrally compressed ungual phalanges than any of the Santacrucian taxa, and they both fall

in the range of the terrestrial *Myrmecophaga* (fig. 16.4A: NN+, PS+; fig. 16.4C). Mylodontids are often characterized by having flat claws, but these are not as flat as some of the armadillos. The manual claws of these individuals are also much more curved than those of the armadillos; in fact, they are more curved than those of the Santacrucian *Hapalops, Schismotherium,* and *Analcimorphus* (fig. 16.4C). The curved nature of the mylodontid manual claws could

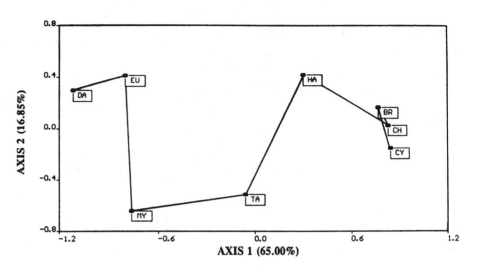

Figure 16.5. Axis 1 and axis 2 of principal coordinates analysis of the ten shape indices for extant xenarthran genera and the fossil sloth *Hapalops longiceps*, with superimposed minimum spanning tree. Abbreviations: BR, *Bradypus*; CH, *Choloepus*; CY, *Cyclopes*; DA, *Dasypus*; EU, *Euphractus*; MY, *Myrmecophaga*; TA, *Tamandua*; HA, *Hapalops longiceps*.

be an adaptation for ripping up roots, an activity that might require a similar ungual morphology to that of *Myrmecophaga*, a nest ripper. Coombs (1983) suggested that mylodonts concentrated the weight of the forelimb on the ulnar side of the manus like the giant vermilingua, which turns its long, curved claws inward. Alternatively, large, curved claws could be associated with bipedal browsing behavior; limb proportions, tooth morphology, and dung composition, however, all indicate that mylodonts were grazers rather than browsers (Coombs 1983). The other indices included in this study and analyses of limb proportions by Coombs (1983) preclude an explanation of arboreality for the curved nature of the Colombian mylodont claws. Thus, these claws were most likely used either for tearing and ripping, for digging in a manner unlike that of armadillos, or for an unrelated activity such as defending themselves (Coombs 1983).

On the pedal plot most Santacrucian phalanges cluster around compression values for the semiarboreal *Tamandua* and curvature values for *Myrmecophaga* (fig. 16.4B: PE★, HL★; fig. 16.4D). The most highly curved pedal claws are seen in *Eucholoeops*. The planopsine exhibits a curvature value similar to that of the other Santacrucian pedal phalanges, yet its pedal ungual phalanx is much more dorsoventrally compressed (fig. 16.4B: PL★; fig. 16.4D). Among Colombian sloths on the pedal plot, the small megalonychid falls within the range of the arboreal extant xenarthrans for curvature and in the range of the semiarboreal *Tamandua* for compression. In contrast, *Neo-*

nematherium has pedal values between those of *Myrmecophaga* and the armadillos for both curvature and compression, and similar to those of the Santacrucian planopsine (fig. 16.4B: NN+; fig. 16.4D). *Pseudoprepotherium* falls within the range of the flattest, straightest armadillo claws on the pedal plot with more extreme values for both indices than any other included fossil taxon (fig. 16.4D).

PRINCIPAL COORDINATES ANALYSIS

The ten morphometric indices reported here are summarized in a principal coordinates analysis for all extant genera and *Hapalops longiceps* YPM-PU 15523, the only individual included in this study on which all measurements could be obtained.

In figure 16.5 the first two axes of the analysis account for 81.85% of the variation among the taxa under consideration. The third axis, not displayed, accounts for only 9.07% of the variation, the functional interpretation of which is unclear. Matrix comparison reveals that this plot corresponds to the original distance matrix upon which it was based with a correlation coefficient of 0.983 (normalized Mantel statistic Z).

The first principal coordinate axis accounts for the majority of variation among the eight taxa (65.00%). Seven of the ten variables are highly correlated with this axis (highly correlated is defined here as a Pearson correlation coefficient > 0.70, and p < 0.05; see table 16.2). Positively loading variables are pedal and manual ungual compression, astraga-

Table 16.2. Pearson correlation coefficients and, below them, probabilities for correlations between the ten indices used in this study and for correlations between indices and the first three principal coordinate axes

	DOL	MOL	POL	DF	AS	VC	MUCU	MUCO	PUCU	PUCO
MOL	0.67									
	0.07									
POL	0.63	0.74								
	0.10	0.03								
DF	0.65	0.70	0.84							
	0.08	0.05	0.01							
AS	−0.59	−0.60	−0.57	−0.83						
	0.12	0.11	0.14	0.01						
VC	−0.56	−0.57	−0.64	−0.63	0.21					
	0.15	0.14	0.09	0.10	0.61					
MUCU	0.03	0.56	0.76	0.53	−0.26	−0.32				
	0.95	0.15	0.03	0.18	0.54	0.43				
MUCO	−0.60	−0.58	−0.97	−0.77	0.45	0.66	−0.69			
	0.11	0.13	0.00	0.02	0.27	0.07	0.06			
PUCU	0.12	0.67	0.78	0.52	−0.24	−0.40	0.98	−0.68		
	0.77	0.07	0.02	0.18	0.56	0.33	0.00	0.06		
PUCO	−0.27	−0.79	−0.77	−0.70	0.56	0.41	−0.88	0.62	−0.90	
	0.52	0.02	0.03	0.05	0.15	0.31	0.00	0.10	0.00	
PC1	−0.62	−0.86	−0.97	−0.89	0.65	0.67	−0.77	0.88	−0.81	0.87
	0.100	0.007	0.000	0.003	0.078	0.072	0.025	0.004	0.016	0.005
PC2	−0.70	−0.11	0.00	−0.28	0.42	0.25	0.63	0.03	0.58	−0.35
	0.050	0.788	0.995	0.504	0.297	0.546	0.096	0.941	0.131	0.401
PC3	0.10	−0.12	0.12	−0.15	0.60	−0.59	−0.02	−0.29	0.03	0.25
	0.809	0.784	0.780	0.723	0.119	0.127	0.969	0.472	0.938	0.552

Notes: Taxa included in this analysis are all extant genera and *Hapalops longiceps*. DOL: dorsal olecranon projection; MOL: medial olecranon projection; POL: proximal olecranon projection; DF: distal femur shape; AS: astragular/tibial mobility; VC: volar calcaneal tuberosity; MUCU: manual ungual curvature; MUCO: manual ungual compression; PUCU: pedal ungual curvature; PUCO: pedal ungual compression; PC1: first principal coordinate axis; PC2: second PC axis; PC3: third PC axis.

lar/tibial mobility, and volar calcaneal tuberosity. Negatively loading variables are distal femur shape, pedal and manual ungual curvature, and proximal and medial olecranon projection. Among the extant taxa, this is a terrestrial/arboreal axis. The terrestrial armadillos and *Myrmecophaga* have low values, the fully arboreal tree sloths and *Cyclopes* have high values, and the semiarboreal *Tamandua* occupies an intermediate position.

The second principal coordinate axis explains 16.85% of the variation. Highly loading variables include dorsal olecranon projection and manual ungual curvature, and pedal ungual curvature is a moderately loading variable.

Although the interpretation of this axis is not as straightforward as that of Axis 1, taxa with dorsally projecting olecrana occupy the lower half of the plot, whereas those with a smaller dorsal projection occupy the upper half.

Each extant clade can be defined by at least some of these ten variables; therefore, each clade's morphological integrity is partially revealed by this plot. It is clear, however, that the distribution of taxa in this space can be explained functionally as well, especially because Axis 1 is determined by a suite of characters that are highly correlated with each other and clearly distinguish arboreal forms from terrestrial forms.

On Axis 1 *Hapalops* has positive values toward those of the tree sloths and *Cyclopes* but not overlapping with them; the latter three occupy the extreme of Axis 1 as fully arboreal animals. The minimum spanning tree connecting the taxa reveals that *Hapalops* lies between the semiarboreal *Tamandua* and the fully arboreal taxa. Although it is unlikely that *Hapalops* was suspensory in the sense that the tree sloths are, it is characterized by a functional complex of features associated with arboreality or climbing abilities. Although *Hapalops* is clustered with the tree sloths relative to other taxa, phylogeny is clearly not the only reason; the three genera of vermilinguas are distributed across the entire axis. In addition, the minimum spanning tree places all sloth genera between *Tamandua* and *Cyclopes*.

Size is another factor that might be invoked as the primary contributor to the first principal coordinate or component (James and McCulloch 1990); in a general sense, size often does differentiate arboreal and terrestrial forms (Eisenberg 1981). It is true that some features that appear more "terrestrial" are seen in the large *Pseudoprepotherium*, whereas apparent adaptations for arboreality seem to be more common in the smaller fossil sloths. Although size probably does have some influence here, because the extant taxa range from 240 g (*Cyclopes*) to 25–40 kg (*Myrmecophaga*), this principal coordinates plot demonstrates that it is clearly not the main determinant of the distribution of taxa in multivariate space. *Myrmecophaga* is the largest animal and *Cyclopes* is the smallest, yet the tree sloths are larger than the armadillos, and *Hapalops* is comparable in size to *Myrmecophaga*.

Conclusions

Hapalops and other Santacrucian "nothrotheres" have been posited to be generalized ancestors to later megalonychids and megatheriids (Hirschfeld and Webb 1968; Frailey 1988), whereas the Santacrucian *Nematherium* and the Colombian *Neonematherium* and *Pseudoprepotherium* are purportedly ancestral to later more specialized scelidotheres and mylodonts (Hirschfeld 1985). To investigate the locomotor behavior of the assemblage of possible ancestors to later Tertiary sloths and the extant tree sloths, it is necessary to examine the functional implications and distribution of a large set of morphological characters among extant

xenarthrans. The analyses of the shape indices presented here demonstrate that skeletal morphology successfully differentiates terrestrial, fossorial, climbing, and suspensory locomotor habits among extant xenarthrans. This correlation between locomotor behavior and the development of morphological traits in extant taxa permits adaptations of fossil taxa to be interpreted from skeletal shape indices.

Based on this subset of characters, many of the Santacrucian sloths appear to have been semiarboreal. *Hapalops* exhibits primarily arboreal, or climbing, adaptations in the univariate plots; this interpretation is supported both by the principal coordinates analysis presented here and the general conclusions of Sherman (1984) from morphometric studies of the forelimb. Other Santacrucian taxa such as *Pelecyodon, Schismotherium, Analcimorphus,* and the unidentified nothrotheres exhibit similar climbing adaptations for all the indices (for which they are represented) except, perhaps, manual ungual phalanx curvature. *Schismotherium* and *Pelecyodon* are closest in morphology to the suspensory tree sloths, especially in the olecranon and distal femur indices. *Eucholoeops* is intermediate in morphology in distal femur shape and astragalar/tibial mobility; it is difficult to interpret its highly curved claws without further analysis. *Nematherium* also exhibits intermediate values in the olecranon and calcaneal indices. This mylodont is more robust and, perhaps, was more terrestrial than the Santacrucian nothrotheres, but it is quite gracile compared to other mylodonts. The relatively rare planopsines appear to have been terrestrial for all indices except proximal olecranon projection, and the morphology of the distal femur and astragalus suggests that they walked on abducted hind limbs and the lateral side of the pes.

There are fewer definitively identified sloth taxa represented in the Miocene of Colombia owing to a paucity of material, but the range of size classes and morphological variation exceeds that of the Santacrucian radiation. The gracile nothrotheres, including a small form and a larger form that is probably *Hapalops,* fall within the range of arboreal extant taxa for most indices with values comparable to those of the Santacrucian "nothrothere" taxa. The small megalonychid manifests curvature and compression values of the pedal ungual phalanges comparable to those of extant climbing xenarthrans. The large megalonychid is only represented by a distal femur, but it has quite a deep one in the range of extant terrestrial values. *Neonematherium* exhibits a mixture of features suggestive of both arboreal and terrestrial adaptations, as does *Tamandua*. It has intermediate values for most of the indices, along with fairly curved manual claws. Although the younger *Neonematherium* is slightly more specialized than the presumably closely related Santacrucian *Nematherium* (at least in astragalar morphology), neither mylodontid manifests the derived features of later Tertiary and Pleistocene mylodontids. To the contrary, these taxa are rather generalized in many aspects like the Santacrucian "nothrotheres" and are quite unlike the other mylodont, *Pseudoprepotherium*. The abundant *Pseudoprepotherium* exhibits a suite of features indicative of terrestrial habits and walking on abducted hind limbs and the lateral side of the pes. This genus is definitely the most terrestrial looking of either radiation for all indices although some of these characters (such as limb robusticity and odontoid morphology of the astragalus) are not as extremely developed as those seen in some later Tertiary and Pleistocene terrestrial sloths. It is difficult and unwise to make definitive conclusions regarding the inferred locomotor repertoire of a given fossil taxon based on a small subset of measurements, especially when partial skeletons limit data collection. Based on comparisons to extant xenarthrans, however, the diversity in skeletal correlates of locomotion in these fossil sloth taxa is clear.

It is apparent that some Colombian taxa resemble Santacrucian taxa, whereas *Pseudoprepotherium* represents a common Colombian morphotype that is not present in the Miocene of Argentina. The phenetic similarity between the Santacrucian and slightly younger Colombian "nothrotheres" and between *Nematherium* and *Neonematherium* could be interpreted as phylogenetic relatedness; if this were an accurate assessment, either or both radiations had fairly recent common ancestors, or some Santacrucian taxa expanded northward perhaps to avoid the progressive cooling and drying of Patagonia. Alternatively, the similarities may represent morphological convergences in taxa adapted to forested areas. Patagonia was apparently wet and forested in Santacrucian times with a diversity of potential habitats, but large, open, arid grasslands that would have supported large-bodied sloths were not dominant until the later Miocene (Webb 1978; Pascual et al. 1985; Pascual and Ortiz 1990). Although little is known about environmental conditions of the Tertiary in Colombia, the area must have been characterized by a greater diversity of habitats (including grasslands) for a relatively longer period of time than in Patagonia to have supported the locomotor diversity seen in the La Venta fossil sloths.

In nearly every index, the range of morphological variation is greater for the Colombian radiation. The greater diversity of locomotor adaptations among the La Venta taxa could be associated with a number of phenomena: environmental conditions may have been more varied in Colombia because of the equatorial latitude, consequently supporting more adaptive niches; the large-bodied *Pseudoprepotherium* may represent the survival of a pre-Miocene lineage that was not ecologically suited to the environmental conditions (e.g., the scarcity of grazing opportunities) of the earlier Miocene of Patagonia (Webb 1978; Hirschfeld 1985); or certain morphological trends (e.g., the cuboid facet concavity in scelidothere astragali and the development of an odontoid process in megatheriid and mylodontid astragali) may simply be more developed in the Miocene of Colombia than in the Miocene of Argentina.

In relation to the extant taxa the range of variation (in some indices) for all fossil sloths is nearly comparable to that of all extant xenarthrans as a group; this indicates that Miocene sloths alone manifested substantial variation in locomotor habits. In certain indices there is more variation than in others, suggesting that phylogenetic constraints played a role in the evolution of certain features in sloths. Some fossil taxa seem to exhibit a mosaic of features; that is, a fossil does not necessarily fall on the "terrestrial" or the "arboreal" side of the graph for every index. This result substantiates the observation that there may not be an exact modern analog for any extinct sloths because the overall composition of morphological features for a given fossil taxon is unlike that of any modern taxon. Particularly noteworthy is the fact that none of the Miocene sloths appears to have been suspensory like the tree sloths. The extensive morphospace occupied by the Miocene sloths encompassed a range of behaviors, including arboreality, terrestriality, and possibly digging activities, but it did not overlap the limited suspensory/climbing niche of the extant sloths.

The results of this study indicate that (1) there is quantifiable evidence for arboreality among fossil "ground" sloths, at least in the Miocene radiations of both Argentina and Colombia; (2) although postcranial morphologies are diverse, generic and familial distinctions are subtle in the Miocene of Colombia and especially in the Santacrucian, and extreme specializations for terrestriality are not evident; and (3) if *Hapalops* and other "nothrotheres" are actually directly ancestral to one or both tree sloths, it is not unreasonable to suspect that arboreality might be an ancestral condition in sloths because arboreal adaptations were evident quite early in the evolution of the group.

Acknowledgments

I thank M. A. Turner (Yale Peabody Museum), G. Musser (American Museum of Natural History), J. H. Hutchison (University of California Museum of Paleontology), and R. H. Madden (Duke University) for the loan of material; W. Fuchs (AMNH), L. Gordon and R. W. Thorington, Jr. (U.S. National Museum), J. Flynn, W. Simpson, and W. Stanley (Field Museum of Natural History), G. McDonald (Cincinnati Museum of Natural History), R. Pascual and G. J. Scillato-Yané (Museo de La Plata), and J. Bonaparte (Museo Argentino de Ciencias Naturales) for facilitating access to collections; W. Jungers, R. Sokal, F. J. Rohlf, and J. Fleagle for helpful discussions; and J. Fleagle, R. Kay, R. Cifelli, and an anonymous reviewer for commenting on and helping to improve drafts of the manuscript. Funding for this project was provided by J. Fleagle, the American Museum of Natural History (Theodore Roosevelt Memorial Fund), the Field Museum of Natural History (Thomas J. Dee Fund), the U.S. National Museum at the Smithsonian (Short-Term Visitor Award), and grant BSR 89-18657 to R. F. Kay (Duke University).

Appendix 16.1. Skeletal measurements taken for this study

Indices constructed from these measurements are discussed in text and depicted in figure 16.1.

1. Maximum dorsal or posterior projection of the olecranon from the center of the trochlear notch, perpendicular to the shaft of the ulna (fig. 16.1A: a)
2. Maximum proximal extent of the olecranon from the center of the trochlear notch to the tip of the olecranon, in line with the shaft of the ulna (fig. 16.1A: b)
3. Maximum medial projection of the olecranon from the most anterior, proximal point of the trochlear notch, perpendicular to the ulnar shaft (fig. 16.1B: a)
4. Length of the trochlear notch from the most anterior, proximal point to the junction of the radial notch and coronoid process (fig. 16.1A: c; fig. 16.1B: b)
5. Anteroposterior depth of distal femur from the plane of both condyles to the tip of the patellar groove (fig. 16.1C: a)
6. Mediolateral bicondylar width of distal femur (fig. 16.1C: b)
7. Volar extent of calcaneal tuberosity from proximal edge of the astragalar articular surface (when oriented horizontally), perpendicular to the long axis of the bone (fig. 16.1D: a)
8. Maximum length of calcaneus in the same orientation as in (7) (fig. 16.1D: b)
9. Arc length of the articular surface of the lateral trochlea of the astragalus (fig. 16.1E: a)
10. Anteroposterior depth of the distal tibial lateral articular surface (fig. 16.1E: b)
11. Length of manual ungual phalanx III from dorsal aspect of base to tip (fig. 16.1G: a)
12. Length of pedal ungual phalanx III from dorsal aspect of base to tip (fig. 16.1G: a)
13. Maximum height of curvature of manual ungual phalanx III measured perpendicular to length cord (fig. 16.1G: b)
14. Maximum height of curvature of pedal ungual phalanx III measured perpendicular to length cord (fig. 16.1G: b)
15. Base height of manual ungual phalanx III (fig. 16.1F: a)
16. Base height of pedal ungual phalanx III (fig. 16.1F: a)
17. Base width of manual ungual phalanx III (fig. 16.1F: b)
18. Base width of pedal ungual phalanx III (fig. 16.1F: b)

17
Paleobiology and Affinities of *Megadolodus*

Richard L. Cifelli and Carlos Villarroel

RESUMEN

Nuevo material de *Megadolodus molariformis,* incluyendo la mayoría de la dentición juvenil y adulta, y mucho del esqueleto postcraneal, aumenta significativamente el conocimiento de una especie previamente conocida sólo por un molar y medio de la mandíbula inferior.

Los dientes del frente, muy seguramente caninos, fueron modificados a colmillos. Aunque los molariformes son presumiblemente primitivos en caracteres tales como la baja altura de la corona y la ausencia de crestas, *Megadolodus* tiene especializaciones dentales observadas en los Litopterna, específicamente en los Proterotheriidae. Estos caracteres no se encuentran en la familia Didolodontidae (Condylarthra), a la cual el género había sido referido previamente. El esqueleto es altamente distintivo entre los ungulados suramericanos, y tiene varias características que solamente comparte con los Litopterna. Las articulaciones suplementarias entre los procesos transversales de las vértebras lumbares posteriores, que restringen la movilidad lateral del tronco posterior, están presentes, como apófisis lumbares de enlace. Los carpos, y especialmente los tarsos, presentan numerosas similaridades con los Litopterna. Los pies son tridáctilos, con los dígitos laterales muy reducidos. Estas características en los tarsos, carpos, y esqueleto axial corroboran la evidencia dental de una estrecha relación del género con los Proterotheriidae.

La dentición y el esqueleto postcraneal representan un curioso mosaico de caracteres primitivos y derivados. En las proporciones de sus extremidades, *Megadolodus* es comparable con otros proterotéridos, y difiere en algunos aspectos con ungulados recientes. Sin embargo, en muchos caracteres del esqueleto, tales como la construcción robusta y sus brazos y piernas cortos, *Megadolodus* es comparable con Suoidea vivientes. Los grandes molariformes bunodontes, con gruesas capas de esmalte, sugieren que *Megadolodus* fue omnívoro, o más probablemente se alimentó de frutos duros. En todas las características dentales y postcraneales que pueden ser comparadas entre *Megadolodus* y ungulados vivientes, el género colombiano es similar a taxa con preferencias selváticas o boscosas y difiere de la mayoría de los que se encuentran en hábitats abiertos. Se concluye que el hábito preferencial de *Megadolodus* fue cerrado, y más probablemente selvático.

Megadolodus representa una rama diferente de los Proterotheriidae. La pobre representación de esta estirpe en el Terciario de Suramérica tropical, puede estar relacionada con un inadecuado registro fósil. La ausencia de ellos de las faunas bien conocidas de Patagonia se debe muy probablemente a factores paleobiogeográficos desconocidos, que quizás incluyen una dependencia en especies vegetales restringidas a los trópicos.

The Didolodontidae comprise a poorly known assemblage of archaic ungulates restricted to South America. Although they have long been considered to be allied to the Litopterna, an ungulate order indigenous to South America (Scott 1913), didolodonts are comparable to

North American early Tertiary "condylarths" and are among the most plesiomorphic of known South American ungulates. For this reason, they are integral to an understanding of the origin and diversification of mammals in the Neotropics (e.g., Simpson 1948; Cifelli 1983a, 1983b). Didolodontids, represented primarily by dentulous jaw fragments, were long known only from Paleocene and Eocene faunas of South America. The description of a putative didolodontid, *Megadolodus molariformis*, from the Miocene La Venta fauna (of presumed Friasian age; Stirton 1953b), thus represented a significant extension in temporal range for the family (McKenna 1956). Because the sequence of South American mammalian faunas is largely restricted to the southern part of the continent, where the record is relatively more complete (Simpson 1940; Patterson and Pascual 1972; Marshall et al. 1983), the occurrence of the archaic Didolodontidae in the Miocene of Colombia has been forwarded as evidence of biogeographic zonation on the continent (e.g., Simpson 1980) and of the tendency for primitive taxa to survive in the Tropics (McKenna 1956). The type and hitherto only known specimen of *Megadolodus molariformis*, consisting of a jaw fragment bearing the first molar and half of the last premolar, was collected by a University of California expedition directed by R. A. Stirton. Renewed paleontological investigations in the Honda Group of the Magdalena valley, Colombia, has produced additional materials, including most of the dentition and skeleton, substantially improving the basis for assessing the relationships and paleobiology of this enigmatic species.

Systematic Paleontology

Order Litopterna Ameghino 1889
Family Proterotheriidae Ameghino 1887
Megadolodinae subfam. nov.

TYPE GENUS

Megadolodus McKenna 1956.

INCLUDED GENERA

The type only.

DISTRIBUTION

Middle Miocene, Colombia.

SUBFAMILIAL DIAGNOSIS

Distinct from all other Proterotheriidae in having upper and lower canines developed into tusks and a large hypocone present on M^3. Limbs differ from Proterotheriidae for which the skeleton is known in being more robust and in having relatively shorter distal elements (tibia, metatarsals, first phalanges). Cheek teeth differ from those of other proterotheriids in being relatively larger.

DISCUSSION

As developed below, numerous similarities, best regarded as synapomorphies, characterize *Megadolodus* and Proterotheriidae, leading us to refer the Colombian genus to this family with little doubt. However, *Megadolodus* differs strikingly from members of both proterotheriid subfamilies, Proterotheriinae and Anisolambdinae. Because the Colombian genus shares no known synapomorphies with either subfamily and has numerous autapomorphies of its own, we refer it to a new subfamily. As recognized on the basis of the type and only genus, Megadolodinae differs from other proterotheriids in having more bunodont cheek teeth. Although significant for a differential diagnosis, differences relating to bunodonty in Megadolodinae have been omitted from the diagnosis because they appear to represent plesiomorphies.

Megadolodus molariformis McKenna 1956

TYPE SPECIMEN

UCMP 39270, part of left lower jaw with labial half of P_4, complete M_1, and roots of M_2.

NEWLY REFERRED MATERIAL

IGM 183282, mandibular rami and fragments thereof, including right P_1, P_2 (still in the crypt), broken right dP_{3-4}, M_1, broken M_2, and badly broken left dP_{3-4} and M_1; upper jaw fragments and teeth including left P³, right P⁴ (both unerupted), and right M²; canine fragments, and various postcranial scraps, including the proximal part of phalanx 1 from manual or pedal digit III; IGM 183544 and 183916, partial skeleton, jaw fragments, and isolated teeth, including ?dC¹, upper and lower ?deciduous canine fragments; right lateral incisor or, less probably, P¹, right dP³, broken right dP⁴, lingual halves of left M¹⁻², left P₁, broken right dP₃, right M₁, broken right M₂, and left M₂; IGM 184019, right P₃; IGM 250400, right M₃; TATAC-1, fragment of right maxilla with well-worn P⁴⁻³; and TATAC-2, left mandibular ramus with P₂–M₂.

HORIZONS AND LOCALITIES

The type was collected from UCMP Locality V4932, Monkey Bed, Villavieja Formation (Fields 1959); IGM 183544/183916 were collected from Duke Locality 54 in the same unit. IGM 183282 and 184019 were collected from Duke Localities 28 and 61, respectively, which lie stratigraphically between the Chunchullo Sandstone Beds and the Cerro Gordo Sandstone Beds, lower La Victoria Formation (Guerrero 1990). IGM 250400 was collected from Duke Locality 90 in the upper La Victoria Formation. Interestingly, primates, which are generally rare, are represented at all of these sites except Locality 61. TATAC-1 and TATAC-2 were collected from Honda Group rocks in the Tatacoa Desert; their exact provenance and stratigraphic position is uncertain. Both the Villavieja and La Victoria formations are placed in the Honda Group (cf. Wellman 1970). Although the stratigraphically lowest (level of Duke Locality 28) and highest (Monkey Bed) occurrences of the species are separated by more than 500 m of section, there is no geochronological evidence to suggest that this represents a notably long temporal range for the species (cf. chap. 3).

DESCRIPTION

Dentition

Both IGM 183282 and 183544/183916 represent juvenile individuals. In the former specimen, both dP³⁻⁴ and dP₃₋₄, although heavily worn, remained in place (the nondescript remnants of the upper deciduous teeth were removed to expose the underlying P³⁻⁴); wear is light on P₁, and the M¹ and M₁ are almost unworn. In IGM 183544/183916, the deciduous teeth (dP³⁻⁴, dP₃) are lightly worn, wear on P₁, M¹, and M₁ is almost indistinct, and M² and M₂ appear to have been unerupted. TATAC-1 represents a fully adult individual; the teeth are well worn, although the coronal patterns are still visible. P⁴ has been somewhat displaced labially and M¹ lingually; enamel is broken and missing from the posterolingual angle of P⁴, the labial part of M², and the posterolabial part of M³. TATAC-2 preserves P₂–M₂ complete although all of the teeth are fractured.

Only fragments of the anterior dentition are preserved, and for this reason there is considerable uncertainty about the nature of the incisors, canines, and anterior upper premolars. A crown of one upper ?canine is known for IGM 183544/183916. Because dental eruption and wear otherwise indicate this to be a young individual and because of the extreme thinness of enamel on this tooth, we believe it to be deciduous. The tooth is large, tusklike, and greatly resembles the upper canine of the extant suoid *Tayassu*. It is laterally compressed with sharp edges defining its anterolingual and posterolingual borders; the lingual face is nearly flat while the labial face is gently convex anteroposteriorly. From its base the canine broadens somewhat and curves posteriorly to a sharp point. A well-developed, flat wear facet for occlusion with the lower canine extends from the tip for about half of the crown height on the anterolabial surface. Fragments of one or both of the lower deciduous ?canines are preserved with IGM 183544/183916. The lower ?canine was much larger than the upper and, like that tooth, broadened from the base, was laterally compressed, posteriorly recurved, and ended in a sharp point.

IGM 183544/183916 includes a jaw fragment with a small tooth of uncertain identity (fig. 17.1J). A large diastema was present on one side of it, 11 mm of which are preserved on the fragment (which lacks indication of an alveolus at its extremity) itself. The tooth is almost certainly an incisor because the lower premolars are already known and the jaw fragment is slender, tapering to an edge at the tooth row, which would be unexpected for a fragment of the maxilla posterior to C¹. The tooth is low-crowned, bulbous, and oval in cross section. A single cusp is present,

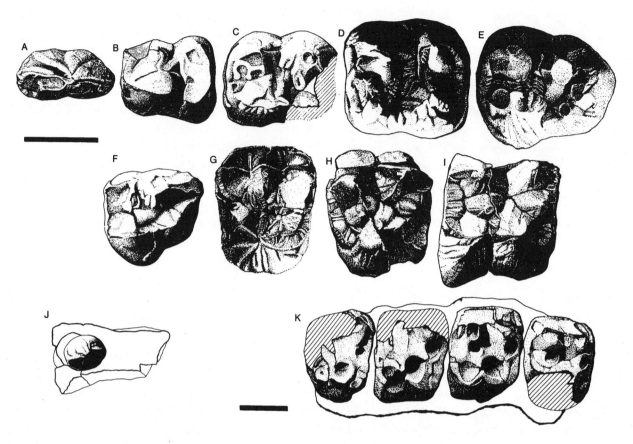

Figure 17.1. Dentition of *Megadolodus molariformis*. *A,* left P$_1$, IGM 183544; *B,* right P$_3$, IGM 184019; *C,* right dP$_3$, IGM 183544; *D,* left M$_2$, IGM 183544; *E,* right M$_3$, IGM 250400; *F,* right dP^3, IGM 183544; *G,* left P^3, IGM 183282; *H,* right P^4, IGM 183282; *I,* right M^2, IGM 183282; *J,* ?right I^3 or P^1, IGM 183544; *K,* right P^4–M^3, TATAC-1. *A–I, J–K.* Scale bars 1 cm.

with crests extending from it to the anterior and posterior edges of the crown, respectively.

The dP^3 (fig. 17.1F) is anteroposteriorly elongate compared to the molars. The crown bears two labial cusps, paracone and metacone, of which the former is slightly larger. The paracone and metacone are tightly appressed and bear weak, irregular ridges on their lingual faces; faint paracone and metacone folds are present on the labial side of the ectoloph, which is otherwise featureless. The preparacrista and postmetacrista descend to small parastylar and metastylar spurs at the antero- and posterolabial corners of the tooth, respectively. The protocone is low and is placed anterolingual to the paracone. From it a crest descends anterolingually to the parastyle. The hypocone is large, forming a prominent posterolingual bulge in the tooth crown. It is connected to the apex of the protocone by a crest; a weak ridge lies in the lingual sulcus between the two cusps. A crest descends the posterior slope of the hypocone to join the postcingulum, which is weakly developed. A stronger crest, the metaloph, extends labially from the hypocone to the metacone. Midway between the two cusps, the crest is expanded into a metaconule; there is otherwise no indication of conules on the tooth. DP^4 is poorly preserved. The paracone and metacone are more widely separated than in dP^3, and a small mesostyle is present between them. A weak labial sulcus is present at the base of the ectoloph. Other features of the crown are too indistinct to be interpreted.

The adult cheek teeth are low crowned, large relative to the size of known parts of the skeleton, and have relatively thick enamel. The enamel is strongly crenulated. The upper cheek teeth increase in size from P^3 to M^2; M^3 is similar in size to P^4 (fig. 17.1K). The major axis of both premolars and molars is transverse. All cheek teeth are brachyodont and have robust, short roots. The crowns are bunodont to slightly bunolophodont. On all teeth, it is possible to distinguish a faint crest extending from parastyle to metastyle and connecting paracone to metacone; similarly, a faint crest runs from the parastyle to paraconule, protocone, and (in the case of the molars), hypocone, terminating at a connection with the postcingulum.

P^{3-4} (fig. 17.1G, H) are molariform, with the major cusps and conules well developed; the most important distinction from the molars is the lack of a hypocone. Owing to the curvature of the lingual faces of P^{3-4}, these teeth are subtriangular in occlusal view. The molars, by contrast, are quadrangular with a well-developed hypocone projecting lingually. On the posterior premolars the paracone is situated anterior and slightly lingual to the metacone, whereas on the molars the paracone is anterior and labial to the metacone. On P^{3-4}, the protocone is somewhat more massive than the paracone and metacone, which are subequal in size. A small mesostyle is present on the molars, but the centrocrista retains an essentially linear, anteroposterior configuration.

The hypocone, present on all three molars, is posterior and slightly lingual to the protocone. On M^{1-2}, the hypocone is nearly as large as the protocone, whereas on M^3 the latter cusp is noticeably larger (figure 17.1I, K). The hypocone is separated from the protocone by a lingual sulcus, in which a small pillar is variably developed.

The paraconule and metaconule are well developed on both molars and posterior premolars. The paraconule is located more or less equidistant between paracone and protocone, and is slightly anterior to the axis of those cusps. The metaconule is slightly closer to the metacone than the protocone and is somewhat posterior to the axis of those cusps. Whereas the paraconule is joined by crests to the protocone and anterior cingulum, the metaconule is isolated, lacking crests.

The buccal cingulum is represented by a thin, crenulated rim extending across the entire labial margin of the upper molars (with variable disruption labial to the paracone and metacone) and posterior upper premolars from the parastyle to metastyle. These two styles are developed as small prominences of the cingulum. The anterior cingulum is thick and isolated, circumscribing the anterior face of the tooth. The posterior cingulum, also thick, is crenulated; it

connects labially with the metastyle and lingually with the crest uniting paraconule, protocone, and hypocone.

The only specimen that preserves the alveolar border of the mandibular symphysis, TATAC-2, is somewhat broken, obscuring details of incisor size and morphology. The canine alveoli are distorted so that no exact dimensions may be determined. It is clear, however, that the adult canines were very large; the space between the canine alveoli is short, indicating that the incisors were small. The mandibular rami are massive. The only lower deciduous tooth sufficiently well preserved to show coronal morphology is dP_3 of IGM 183544/183916 (fig. 17.1C). It is molariform but not as quadrate as the molars, being longer than wide; the trigonid is somewhat broader than the talonid. A paraconid is lacking; protoconid and metaconid are approximately equal in height and are connected by low, faint crests. The talonid was apparently tri-cusped, although this cannot be determined with certainty owing to breakage of the specimen. Small, irregular enamel ridges are present in the trigonid and talonid basins.

The adult cheek teeth increase in size from P_1 to M_2 (fig. 17.1 A–D). In occlusal view, P_{1-2} are ovoid and somewhat pointed anteriorly. P_1 is narrower labiolingually than P_2; both teeth are structurally simple, possessing a single cusp (situated slightly anterior to the center of the oval) and a small posterior spur, the latter of which is better developed on P_2. A faint crest runs longitudinally along each premolar, beginning at the anterior base of the crown and extending to the talonid spur. The protoconid of P_1 is sharp in the unworn condition, although the cusp is robust and has somewhat of an inflated appearance. On P_2, a faint lingual bulge ("paraconid") is present anterior to the protoconid.

P_3 differs from the anterior premolars in being well molarized. The trigonid is narrower but longer than the talonid. The paraconid is lacking with the paracristid terminating low on the anterior face of the metaconid; protoconid and metaconid are connate and subequal in size. On the talonid of P_3, the hypoconid is more massive than the entoconid; the hypoconulid is absent. Small cingula are present on the anterior and posterior faces of the tooth.

P_4 is similar to the molars, differing from them principally in lacking a hypoconulid and in having a narrower trigonid relative to talonid. The lower molars are bunodont, with low, somewhat inflated, conical cusps and weak

Table 17.1. Dental measurements (mm) of *Megadolodus molariformis*

Variable Measured	UCMP 39270	IGM 183282	IGM 183544, 183916	IGM 184019	TATAC 1	TATAC 2	IGM 250400
P¹ length	—	—	6.0	—	—	—	—
width	—	—	4.9	—	—	—	—
P³ length	—	14.2	—	—	—	—	—
width	—	17.5	—	—	—	—	—
P⁴ length	—	14.5	—	—	13.8	—	—
width	—	17.5	—	—	18.6	—	—
dP³ length	—	—	13.5	—	—	—	—
width	—	—	11.8	—	—	—	—
dP⁴ length	—	—	—	—	—	—	—
width	—	—	—	—	—	—	—
M¹ length	—	—	—	—	14.4	—	—
width	—	—	—	—	19.9	—	—
M² length	—	17.2	—	—	15.1	—	—
width	—	—	—	—	—	—	—
M³ length	—	—	—	—	13.7	—	—
width	—	—	—	—	18.6	—	—
P₁ length	—	13.0	13.4	—	—	—	—
width	—	7.5	6.8	—	—	—	—
P₂ length	—	—	—	—	—	14.4	—
width	—	—	—	—	—	9.4	—
P₃ length	—	—	—	13.6	—	12.8	—
width	—	—	—	12.2	—	12.9ᵃ	—
P₄ length	15.3ᵃ	—	—	—	—	14.6	—
width	—	—	—	—	—	15.7ᵃ	—
dP₃ length	—	16.3ᵃ	15.5ᵃ	—	—	—	—
width	—	12.0ᵃ	11.8	—	—	—	—
dP₄ length	—	15.4ᵃ	—	—	—	—	—
width	—	13.2	—	—	—	—	—
M₁ length	16.4	15.1	15.6	—	—	14.8ᵃ	—
width	15.5	16.4ᵃ	16.0	—	—	—	—
M₂ length	—	16.8	17.8	—	—	15.6	—
width	—	17.7ᵃ	16.6	—	—	15.9	—
M₃ length	—	—	—	—	—	—	19.1
width	—	—	—	—	—	—	15.7

ᵃFigure is approximate owing to specimen breakage.

connecting crests. M_{1-2} are subquadrate; the trigonids and talonids are comparable in length and width although the trigonid is, on occasion, wider than the talonid. The paraconid is absent. The protoconid and metaconid, subequal in size, are connected anteriorly by a thick, crenulated paracristid and posteriorly by a weaker but similarly thick and crenulated protocristid. The hypoconid is the best developed of the talonid cusps and is similar to the two trigonid cusps in size and height. The smaller entoconid and hypoconulid are subequal in size; the hypoconulid is situated slightly posterior to the other talonid cusps and is closer to the entoconid than to the hypoconid. Two crenulated crests descend from the hypoconid, the anterior of which (cristid obliqua) terminates at the lingual side of the protocristid. The posterior crest is weaker and runs to the posterobasal part of the hypoconulid. The anterior cingulum is represented by a weak, crenulated rim that runs across the entire anterior face of the lower molars. The posterior cingulum is very weak and ends at the middle of the posterior face of the teeth. Measurements are given in table 17.1.

Axial Skeleton

The neck and tail are not known, but IGM 183544/183916 preserves most of the axial skeleton of the trunk region in articulation (fig. 17.2). The posterior six thoracic vertebrae, identified by the presence of attaching ribs, are represented; the last thoracic is also the anticlinal vertebra (Slijper 1946). Anteriorly, four additional ribs are present on the left side of the specimen. The anterior ribs are short and broad, the first being the shortest and broadest. The first rib is also more sharply curved, suggesting that it occupied a very anterior position in the series, probably rib 1 or 2. Thus, *Megadolodus* had at least ten but probably no more than eleven thoracic vertebrae. The prezygapophyses of all presacral vertebrae flare dorsally; on the posterior thoracics and all lumbars, the prezygapophyses turn inward, enveloping the postzygapophyses of the preceding vertebrae, as in

Artiodactyla. The neural spines of the thoracic series are broken, but the preantepenultimate (T.7 or T.8) appears to have been the longest. Posterior to this, the neural spines of the thoracics become progressively shorter and more erect, as is typical in terrestrial mammals (Slijper 1946). The lumbosacral transition is difficult to determine because the posteriormost preserved vertebra is partly disarticulated and rotated, and because the preserved morphology of the sacrolumbar region is unusual. The last preserved vertebra is interpreted as the first sacral because it lacks a neural spine and well-developed transverse processes (thus distinguishing it from the preceding lumbar vertebrae), and because it bears articular surfaces on the anterior faces of the transverse processes (which would not be predicted for succeeding sacral vertebrae). Thus, five lumbar vertebrae were present in *Megadolodus*. The transverse processes of the lumbars, which project slightly downward, increase in length and anterior inclination from L.1 to L.4. The transverse processes of L.4 are anteroposteriorly broad and bear prominent knobs, with apparent articular surfaces, on their posterior faces. Corresponding knobs are present on the anterior faces of L.5 transverse processes, as are similar, but

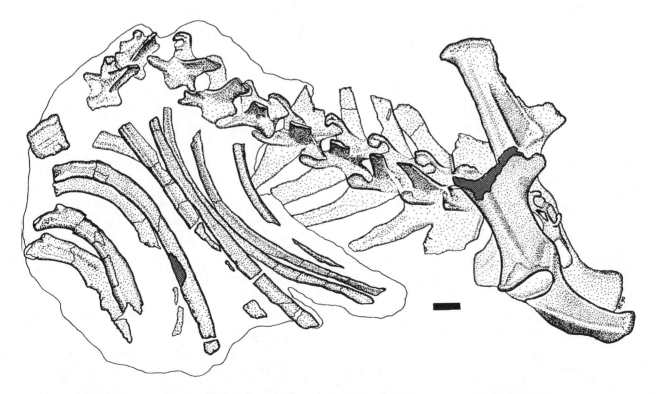

Figure 17.2. *Megadolodus molariformis,* IGM 183544, vertebral column and pelvis (the hind limbs were removed for study). Scale bar 1 cm.

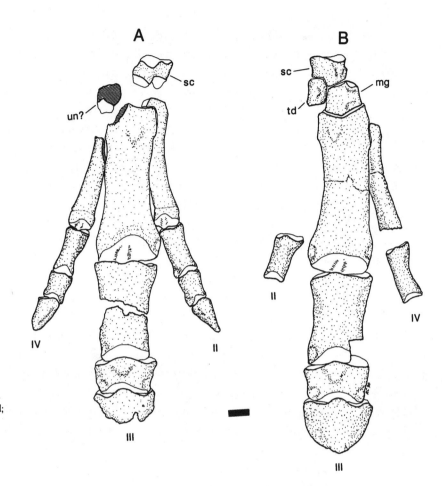

Figure 17.3. *Megadolodus molariformis,* IGM 183544, hands, anterior view: *A*, right hand; *B*, left hand. Abbreviations: *sc,* scaphoid; *td,* trapazoideum; *un,* unciform; *mg,* magnum; *roman numerals,* digit rays. Scale bar 1 cm.

larger, knobs on the posterior faces for articulation with presumed S.1. Disarticulation between L.5 and S.1 shows that the anterior knobs are convex and the posterior ones concave. Articular surfaces between lumbar transverse processes are unusual among mammals; they occur in some perissodactyls such as *Tapirus* and *Equus.* The strong development of these articular surfaces indicates that lateral (side to side) mobility in the lower trunk would have been greatly restricted, if not altogether precluded. The neural spines of the lumbars achieve successively stronger anterior inclinations posteriorly in the series. The neural spine of L.4 is the longest; L.5 has a noticeably smaller spine.

Forelimb

Fragments of the humerus, radius, and ulna are represented in IGM 183544/183916, but they are too incomplete to

merit description. The epiphyses of the proximal humerus and ulna, at least, were unfused. Parts of both hands are preserved in IGM 183544/183916; together, most of the manus is represented (fig. 17.3). All epiphyses of the hands are fused, with the exception of the distal epiphyses of metacarpals II and IV. The scaphoid, the only element known for the proximal row of carpals, has a large, concavoconvex radial facet. The scaphoid is broadest dorsally (unlike the condition seen in perissodactyls), where it is convex; the facet slopes ventrad posteriorly before sweeping upward at the plantar margin of the bone. A prominent beak is developed on the plantar side of the scaphoid. In dorsal view, the distal end of the scaphoid shows a distinct "step." Medially, a flat articular surface is developed for the proximal end of the trapezoideum; this is confluent with an oblique facet, forming the side of the "step" with which the trapezoideum articulates. On the lateral side of the distal scaphoid, a lower, slightly convex facet is present for artic-ulation with the proximal end of the magnum. Toward the plantar side of the scaphoid, on the inferior part of the plantar beak, a small, laterally oriented facet is present for articulation with the trapezium; thus, the trapezium lay palmar rather than distal to the scaphoid. The trapezoideum has a large, angled proximal facet, the medial part of which articulates with the medial scaphoid "step"; the lateral, the side of the "step." The medial side of the trapezoideum bears a vertically oriented, flat facet with which the tra-pezium articulated; the trapezium was thus nested between the trapezoideum and the scaphoid on the medial side of the wrist. An angled, flat facet on the lateral side of the trapezoideum articulated with the proximomedian surface of the magnum. The magnum is distinctly larger than the trapezoideum. At least two facets were present on the prox-imal side of the magnum. These are angled with respect to each other; the medial facet articulates with the scaphoid and the lateral, the lunate. Breakage on the lateral side of the only magnum preserved, the left, leaves open the question of whether the element articulated also with the cunei-form, as in litopterns (Cifelli 1993a). One poorly preserved wrist element is tentatively identified as the right unciform. Distally, it bears a convex, posteriorly narrowing facet for metacarpal IV; part of a small, obliquely oriented facet, for articulation with metacarpal III, is preserved on the medial side of the element. Facets for the magnum and cuneiform are not preserved. The ?unciform bears a prominent plantar beak, which terminates in a rounded knob. This knob bears an apparent articular surface, perhaps for a sesamoid.

The manus is tridactyl and mesaxonic, with the lateral cheiridia being markedly smaller than digit III. Except for the proximal articulation, metacarpal III is bilaterally sym-metrical. This element is anteroposteriorly compressed; it is shorter and more robust (L = 82 mm, midshaft W = 21 mm) than that of *Tapirus,* although it does not broaden as much distally as in that genus. The main articular surface on the proximal end, for the magnum, is gently concave and is higher medially than laterally. A small, flat facet, for meta-carpal II, is present on the medial side, proximal end, of metacarpal III. A thin, straplike facet, also for metacarpal II, extends ventrally a short distance along the medial side of metacarpal III. On the proximolateral side of metacarpal III, a small, oblique facet is present for articulation with the

unciform. On the plantar side of metacarpal III, a pro-tuberance is present, inferior to the magnum facet and set to the lateral side of the axis of the bone. This bears a small, vertically oriented facet for articulation with metacarpal IV. A vertically oriented facet, for the fourth metacarpal, ex-tends a short distance down the lateral side of metacarpal III. On the distal end of metacarpal III, the median keel for articulation with the first phalanx is pronounced. The first phalanx is long (L = 43 mm) relative to the metacarpal of digit III, in comparison to homologous elements of ar-tiodactyls and perissodactyls. The second phalanx is broader than long; the terminal phalanx is developed into a broad, noncloven hoof. The lateral digits, II and IV, are subequal in length, although metacarpal II extends more distally than metacarpal IV because their proximal relation-ships with the carpus differ. The proximal end of metacar-pal II is laterally extended so that it sits, in part, atop meta-carpal III and is interlocked with that element. By contrast, the proximal end of metacarpal II is well below that of metacarpal III, with the unciform occupying a position lateral to metacarpal III. The phalanges of the lateral digits are of typical ungulate morphology; the terminal phalanges bear pointed hooves. The degree of reduction of the lateral digits, in terms of both length and robusticity, appears to be only slightly less than in the Santacrucian proterotheriid litoptern, *Diadiaphorus.*

Hind Limb

The pubes and parts of the iliac blades (fig. 17.2) are missing from the pelvis of IGM 183544/183916. The acetabulum is deep and is bordered by strong anterior and posterior rims. The blade of the ilium flares sharply dorsad about one-quarter of the distance from acetabulum to tip. The sacroiliac contact, at the base of this dorsal flare, is thus more posteriorly located than is typical of most living ungulates. Anteriorly, the iliac blade flares laterally and bears a marked prominence on its anteroventral tip, perhaps marking the termination of origin of the iliacus muscle. The ischium, which is slightly less than one-third the length of the ilium, extends posteriorly from the acetabulum, flaring dorsally into an expanded prominence (ischial tuberosity) near its tip, where the biceps femoris and semitendinosus muscles took origin.

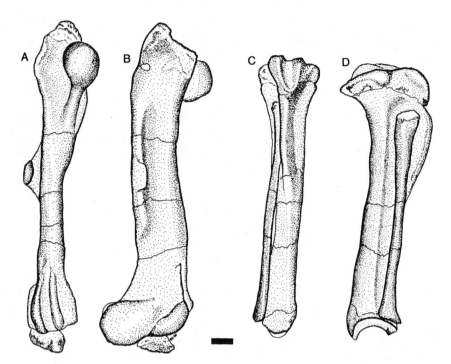

Figure 17.4. *Megadolodus molariformis*, IGM 183544. *A–B.* Right femur/patella: *A*, anterior view; *B*, lateral view. *C–D.* Tibia/fibula: *C*, anterior view; *D*, lateral view. Scale bar 1 cm.

The entire right hind limb, including all phalanges, sesamoids, and patella, are preserved in almost complete articulation, although the long bones (fig. 17.4) are somewhat crushed transversely; parts of the left hind limb and most of the foot were also recovered. The femur (length from head to condyles = 192 mm) is robustly built, being more reminiscent of modern suoids than of more lightly built artiodactyls such as antelopes. The head is hemispherical, lacking the semicylindrical appearance seen in many Recent ungulates. Nonetheless, the anterior and posterior parts of the articular surface are strongly developed, suggesting the capacity for considerable hip excursion in the anteroposterior plane. The greater trochanter is robust and is broadly developed laterally and dorsally, extending well beyond the head, suggesting strong development of the moment arms of the gluteal muscles. No second trochanter is present as such, an unusual absence among terrestrial mammals, but a long, well-defined crest extends ventrally from the base of the head. The trochanteric fossa is deep. A third trochanter is present on the lateral side of the femur, about halfway down from head to condyles and connected to the greater trochanter by a strong lateral ridge. Although not dorsoventrally elongate, the third trochanter is very

well developed laterally, terminating in an anteriorly directed flare. The distal femur is strongly developed anteroposteriorly, as is common among ungulates, and the patellar groove is deep, even accounting for crushing. The tibia is somewhat shorter than the femur (177 mm) and, like that bone, is fairly robust. As with the distal femur, the proximal tibia is well developed anteroposteriorly. The proximal epiphysis, although preserved in place, appears to have been unfused. A strong groove for the patellar tendon is present; inferiorly, the cnemial crest is likewise strong, even accounting for transverse compression of the bone. The distal tibia is transversely narrow. No medial malleolus is present; stability of the upper ankle joint apparently was provided by the complementing, deeply trochlear surfaces of the tibia and astragalus. The crurotarsal articulation is deeply concave anteroposteriorly, with deep medial and lateral grooves and a salient median ridge. The fibula, by comparison, is a frail element. As preserved, it is shorter (total length = 137 mm) than the tibia; the proximal end appears to have been poorly ossified and thus was not preserved. The distal fibula articulates with the tibia anterolaterally; the fibula terminates at a point adjacent to the dorsal rim of the lateral crurotarsal groove. A small articular

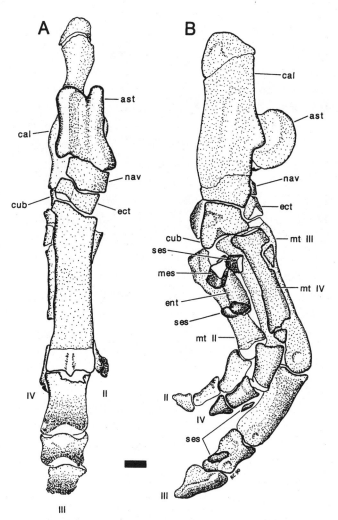

Figure 17.5. *Megadolodus molariformis,* IGM 183544, right foot, shown as preserved: *A,* anterior view; *B,* lateral view. Abbreviations: *ast,* astrangulus; *cal,* calcaneus; *nav,* navicular; *cub,* cuboid; *ect,* ecto-cuneiform; *ses,* sesamoid bones; *mt,* metatarsal; *roman numerals,* digit rays. Scale bar 1 cm.

surface is developed; articulation with the calcaneus was present but not extensive.

The astragalus and calcaneus (fig. 17.5) are of typical litoptern morphology (Cifelli 1983a) and complete description is unnecessary. The astragalar body is spool-like, with strong, parallel medial and lateral ridges and a deep trochlea. The crurotarsal articulation extends far anteroinferiorly, to the neck of the astragalus, and posteroinferiorly, nearly to the astragalocalcaneal facets, where it is confluent with the groove for the digital flexor tendons, suggesting a great capacity for flexion and extension at the proximal

ankle joint. The astragalar head is subcylindrical, with its axis subparallel to that of the crurotarsal joint; the navicular facet is extensively developed on the dorsal and inferior surfaces of the head. On the calcaneus, the sustentacular facet is anteroposteriorly elongate and concave; the ectal facet is strongly convex and bears a marked medial orientation. The calcaneal neck (that part anterior to the astragalocalcaneal facets) is elongate, terminating in an oblique, concave cuboid facet. The tip of the tuber calcis, although preserved in place, was unfused to the body of the calcaneus. The navicular is by far the largest of the remaining tarsals. Its proximal surface, for articulation with the astragalus, is deeply concave, with its major axis of articular curvature being transverse. An enormous, hooklike plantar tuberosity, as long as the remaining anteroposterior width of the bone, is present. On its distal end, the lateral four-fifths of the navicular articulates with the ectocuneiform, which is the dominant element of the distal tarsals. A small part of the distal surface, at its medialmost extent, articulated with the mesocuneiform. The cuboid is transversely narrow, with a limited exposure in dorsal aspect, but like the navicular, has a strong, beaklike plantar tuberosity. The cuboid extends distally past the end of the ectocuneiform so that it articulates distally with both metatarsal IV and III, the former on its medial side. The mesocuneiform, the largest of the three distal tarsals, occupies most of the proximal articulation of metatarsal III. The mesocuneiform, which is very small, articulated with the ectocuneiform laterally, the navicular dorsally, and metatarsal II distally. Metatarsal II thus interlocks metatarsal III, articulating with its dorsal surface and with the medial surface of the ectocuneiform. The entocuneiform, which is an oblong, nondescript element, articulated with the medial side of the proximal end of metatarsal II.

The pes is mesaxonic and tridactyl, with no remnants of digits I and V present. The distal epiphyses of the metatarsals, although present, are unfused. The pes is dominated by metatarsal III (total length = 81 mm), which is bilaterally symmetrical and has the appearance of that of a perissodactyl or litoptern. It is anteroposteriorly compressed, although this appearance may in part be the result of postmortem crushing. Metatarsal III is smaller than but otherwise similar to that of *Tapirus*. The median keel on the distal end is pronounced. Phalanx 1 is long (42 mm) relative to the metatarsal, in comparison to homologous elements of most living perissodactyls and artiodactyls. The second phalanx is short, broad (L = 22 mm; W = 25 mm), and anteroposteriorly compressed. The terminal phalanx, which bears small nutrient foramina, is bilaterally symmetrical and unguliform; a median notch is lacking. The lateral metatarsals are subequal in length (metatarsal II = 62 mm; metatarsal IV = 59 mm) and are considerably shorter and more slender than metatarsal III. They were tightly appressed to the median metatarsal and articulated with it through most of their lengths. Digits II and IV bear first and second phalanges, which are, respectively, similar. The proximal phalanx of each digit (~22 mm long) is longer than the second (~15 mm long). The terminal phalanges are developed as small, pointed hooves strongly reminiscent of the Santacrucian proterotheriid *Diadiaphorus*. In degree of reduction of the lateral digits of the foot *Megadolodus* was similar also to that genus.

Ecological Morphology: Limb and Body Comparisons between *Megadolodus* and Other Ungulates

Because of their obvious close relationship to substrate, the structure and proportions of the limbs can provide important evidence about modes of posture and locomotion in terrestrial mammals (e.g., Gregory 1929; Howell 1944; Maynard Smith and Savage 1959). The relatively complete hind limb known for *Megadolodus molariformis* affords a basis for comparison of the species' build and limb proportions with that of other ungulates. In this section, we compare the hind limb of *Megadolodus* with that of some Recent ungulates. We also estimate the body mass of the Colombian genus, based on programs developed by Gingerich (1990) and on regression coefficients provided by Damuth and MacFadden (1990), and compare aspects of femoral morphology to living ungulates based on the work of Kappelman (1988).

PROPORTIONS OF THE HIND LIMB

Methods

We selected twenty-two species (listed in the caption to fig. 17.6) of living ungulates from six families (Tapiridae, Sui-

dae, Tayassuidae, Tragulidae, Cervidae, and Bovidae) for comparison with *Megadolodus molariformis*. Sampling, constrained by availability of specimens, was designed to include species of various body types, habitat preferences, and degrees of perceived cursorial modification. The sample also includes a wide range of body sizes (from ~1 to 200 kg), although extremely large species were excluded, as were those with obviously irrelevant modifications correlative with high level browsing, aquatic habitat preference, or graviportality. We also included measurements for two proterotheriids, *Diadiaphorus majusculus* and *Thoatherium minisculum*, based on published data and illustrations (Scott 1910). We point out that there are several deficiencies in this sample and its data. The sample of Recent taxa is small and is based on one individual of each species. Some Recent taxa are represented by zoo specimens; measurements of the Santacrucian proterotheriids were not taken first hand. Thus, although we subject the data from this sample to statistical analyses, we restrict ourselves to qualitative interpretation of the results.

Five postcranial measurements were selected for this study, defined as follows: femur length (FEM L), length of femur from head to medial condyle; midshaft diameter of femur (FEM W), least diameter of femur at the midpoint in the shaft defined by FEM L; tibia length (TIB L), maximum length of tibia (including medial malleolus, if present); metatarsal length (MT3 L), length of longest (third) metatarsal; and longest first phalanx length (PH1 L), maximum length of the first phalanx of tarsal digit III. FEM W was taken as a mimimum because of differences among taxa in the development of ridges and crests for muscular insertions. The only complete femur of *Megadolodus molariformis*, IGM 183544, is somewhat flattened in the midshaft region, as a result of postmortem crushing. We estimated FEM W by measuring the circumference at midshaft, and subtracting 8% (the difference by which a sample of comparably sized artiodactyls departed from circularity in cross section of the femur at midshaft) from the circular diameter. The limits of the original diameter are constrained by its preserved maximum and minimum; an average of these gave similar results. For some of the suoids, a plantar beak is present on the proximal end of metatarsal III; our measurement extends between articular surfaces only and does not include this protuberance. To aid in interpretation of the influence of

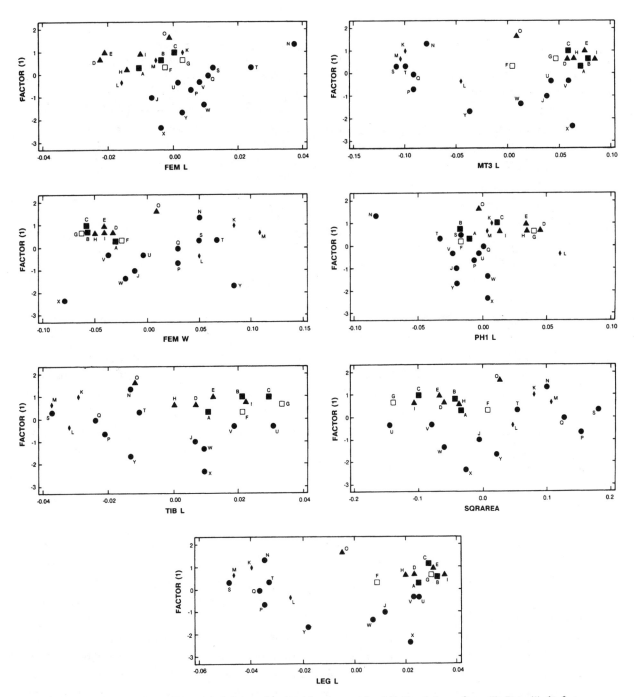

Figure 17.6. Bivariate plots of standardized residuals for one dental and five postcranial variables in relation to factor (1). *Factor (1),* the first unrotated axis derived from a principal components analysis of the six original variables (FEM L, FEM W, TIB L, MT3 L, PH1 L, SQRAREA), is assumed to represent "size" and was the independent variable used for least squares regression of each variable (slopes and Y-intercepts for these regressions are given in table 17.2). *LEG L* is a secondary variable representing the sum total of TIB L, MT3 L, and PH1 L. Symbols denote taxonomic (fossil species) or habitat preference (extant species) groupings as follows: *filled diamonds,* Proterotheriidae; *filled circles,* closed habitats; *filled squares,* mixed habitats; *open squares,* rocky or mountainous terrain; *filled triangles,* open habitats. Habitat preferences for extant species are based on Nowak and Paradiso (1983), Janis (1988), and Kappelman (1988). Letters denote species: *A, Cervus nippon; B, Odocoileus virginianus; C, Tragelaphus spekei; D, Antilope cervicapra; E, Gazella granti; F, Capra aegagrus; G, Rupricapra rupricapra; H, Saiga tatarica; I, Antilocapra americana; J, Muntiacus muntjak; K, Diadiaphorus majusculus; L, Thoatherium minisculum; M, Megadolodus molariformis; N, Tapirus* sp.; *O, Equus asinus; P, Tayassu tajacu; Q, Tayassu pecari; R* (not plotted because of missing data), *Potamochoerus porcus; S, Hylochoerus meinertzhageni; T, Babyrousa babyrussa; U, Moschus berezovskii; V, Mazama americana; W, Pudu mephistopheles; X, Tragulus javanicus; Y, Hyemoschus aquaticus.*

size on proportions, we included a dental variable in the analyses. M_2 was selected because it was the best represented among taxa and appeared to be the least influenced by varying specializations among the taxa sampled. To minimize the effect of shape differences, we multiplied maximum length by maximum width of M_2 and used the square root of the product (SQRAREA) as a linear variable in the analyses. All data were converted to their natural logarithms to lessen the effect of outlier taxa on the analyses (cf. Smith 1984). Analyses were performed on the SYSTAT statistical package (Wilkinson 1990).

We tested two methods to account for size-related differences in limb proportions. In the first, we employed least-squares regression to determine the relationship between an independent variable, selected from the pool, and each of the remaining variables. To negate the influence of differences in absolute value of data between variables, the residuals were divided by their expected values so that all data were normalized to values between 1 and −1. We performed several principal components analyses on these data, using a correlation matrix and a varimax rotation. The analyses were performed on different data sets in which we varied the independent variable against which the remaining variables were regressed to obtain the residuals. The independent variables chosen were FEM W, which Gingerich (1990) has shown to be the most highly correlated (of the available pool) with body mass; FEM L, which we predicted would be the limb length least affected by contrasting postcranial specializations; and SQRAREA, which is a dental variable and thus not directly related to limb proportions. When regression against one of the postcranial measurements (FEM W or FEM L) formed the basis for the standardized residuals subjected to analysis, other variables loaded on the first component in order of their physical relationship to the independent variable (e.g., when the data was regressed against FEM W, highest loadings for factor 1 were, in order, FEM L, TIB L, MT3 L, and PH1 L), only SQRAREA had a high loading on principal component 2, and postcranial variables loaded on component 3 in the reverse order as they did on component 1. When the dental variable (SQRAREA) was used as the independent variable in the regressions, factors 1 and 3 loaded as before except for FEM W, which had a high loading only on component 2. Thus, the associations of variable loadings

Table 17.2. Correlation of variables with "size," and regression parameters, based on a sample of twenty-two Recent ungulates and three proterotheriid litopterns

Variable	Factor 1	Y-intercept	Slope
FEM L	0.973	5.173	0.301
FEM W	0.918	2.744	0.342
TIB L	0.929	5.267	0.296
MT3 L	0.682	4.777	0.320
PH1 L	0.959	3.583	0.377
SQRAREA	0.793	2.453	0.297
LEG L (TIB L + MT3 L + PH1 L)	—	5.862	0.314

Note: "Size" = component loadings on factor 1 of principal components analysis based on log-converted data.

appear to have been related to their physical association or proximity in the body rather than to any obvious model of form/function correlation. Plotting of the case scores on the principal conponents revealed no intelligible taxonomic, habitat preference, or functional groupings. These results, coupled with evidence for systematic variability that cannot be attributed to size among all variables (fig. 17.6), led us to reject this approach for size-adjusting our data set. We point this out because caution appears to be warranted in using this technique and because similar approaches have been used in other studies (see, e.g., Radinsky 1981).

Because most or all of the variables are highly correlated with body size (see, e.g., Gingerich et al. 1982; Legendre 1986; Gingerich 1990), we elected to base our regressions on case scores for principal component 1 of an analysis on the entire set of logged (but otherwise unadjusted) data. In this case, principal component 1 was assumed to represent an average size component for all variables, and for this reason we did not employ a varimax or other rotation. All six variables had high loadings on the first principal component (table 17.2). Residuals were calculated from the parameters resulting from least-squares regression of each variable against factor 1, the independent variable. The residuals were standardized by dividing them by expected values. (We tested another method of standardization, dividing case values by the maximum observed for each variable, and found that results of subsequent principal components analyses were similar.) Plots of the standardized residuals for each variable against factor 1 are shown in

Table 17.3. Loadings of transformed variables and variance explained by components, based on principal components analysis of standardized data, on first three rotated axes

Variable	Factor 1	Factor 2	Factor 3
COMPONENT LOADING			
TIB L	−0.982	0.006	0.052
SQRAREA	0.922	0.265	−0.244
MT3 L	−0.898	−0.327	−0.231
FEM W	0.877	0.196	0.319
PH1 L	−0.113	−0.951	0.159
FEM L	0.226	0.905	0.255
VARIANCE EXPLAINED BY ROTATED COMPONENTS[a]			
All 5 variables	3.455	1.939	0.307
	(57.585)	(32.322)	(5.124)

[a]Percentage of total variance that the explained variance represents is given in parentheses.

figure 17.6. The standardized residuals were then subjected to principal components analysis (based on the correlation matrix and using a varimax rotation). Component loadings and factor scores for principal components 1–3 of this analysis are shown in table 17.3 and figure 17.7, respectively. Principal component 1, which accounted for 58% of the variance, had high positive loadings for SQRAREA and FEM W, and strongly negative loadings for TIB L and MT3

L. Principal component 2 explained 32% of the total variance. Only two variables loaded strongly on this factor, PH1 L, which was highly negative, and FEM L, which was strongly positive. Factor 3 explained only 5% of the variance, and no variable loading exceeded an absolute value of 0.32.

Results

As shown in figure 17.7, factor 1 clearly separated living suoids, which have large teeth, robust femora, and short distal limb elements (fig. 17.6), from more cursorial artiodactyls, in which the reverse conditions pertain. All three proterotheriids, *Diadiaphorus*, *Thoatherium*, and *Megadolodus*, were similar to suoids in this respect; *Megadolodus*, in fact, scored highest on factor 1 because of its combination of a very robust femur with a short tibia and metatarsal III (fig. 17.6). Other Recent taxa with positive scores on principal component 1 include the tapir (*Tapirus* sp.), which has large molars and short distal limb elements (fig. 17.6); the donkey (*Equus asinus*), which has a short tibia and robust femur compared to most of the ruminants in the sample; and the water chevrotain (*Hyemoschus aquaticus*), which has a very robust femur and a short tibia. By contrast, the mouse deer (*Tragulus javanicus*), the only other tragulid included in the sample, fell within the cluster of remaining artiodactyls because of its extremely slender femur and somewhat longer than average distal limb elements. Principal component 1 does not clearly separate remaining artiodactyls (Cervidae, Bovidae) according to habitat preference or lo-

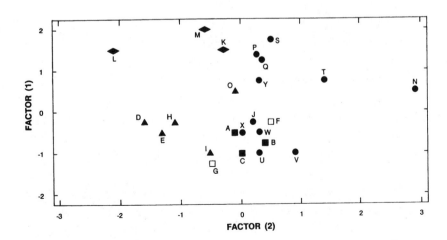

Figure 17.7. Plot of case scores on first two rotated factors of principal components analysis. Symbols and letters are as given in the caption to figure 17.6.

comotor category, although most of the small, forest-dwelling deer scored negatively on this axis because of their slender femur, long tibia and metatarsal III, and small teeth.

The tapir (*Tapirus* sp.) scored highest on principal component 2 because of its unusually long femur and short first phalanx. The suoids scored moderately high on principal component 2 for the same reason, as did most of the small, forest-dwelling deer. Among suoids, the babirusa (*Babyrousa babirussa*) was an outlier because of its longer femur and shorter first phalanx, as it was on component 1 owing to its somewhat smaller teeth than other suoids. Among Recent taxa, the lowest-scoring species on component 2 are all open-country cursors noted for their speed (*Antilope cervicapra*, *Saiga tatarica*, *Gazella granti*). All proterotheriids also had relatively low scores on component 2. *Thoatherium minisculum*, which scored lowest of all taxa on axis 2, has a much shorter femur and longer first phalanx than would be predicted on the basis of its size.

BODY MASS

Regardless of the exact placement of *Megadolodus* among indigenous South American ungulates, the genus unarguably is not closely related to any living mammal. For this reason we employed a variety of regression equations, based on both postcranial and dental measurements, in estimating body mass for *Megadolodus molariformis*. Gingerich (1990) studied the relationship between postcranial element size (length and parasagittal midshaft diameter) and body mass in thirty-six species of living mammals, based on individual and multiple regressions. P. D. Gingerich generously supplied us with a copy of his BODYMASS program (Gingerich 1990), and the results for eight postcranial measurements of *Megadolodus* are given in table 17.4A. The predictions have poor consistency, with those derived from most of the length measurements (especially tibia length) having values less than the geometric mean, and those for limb diameters exceeding the mean. Thus, compared to the "average" mammal (based on the sample of thirty-six species), *Megadolodus* had short (especially the tibia) and robust (especially the third metacarpal and tibia) limb bones, results that are reasonably consistent with the comparisons based strictly on limb proportions. Significantly, the sample of taxa upon which the regressions are based includes no

Table 17.4. Body mass predictions for *Megadolodus molariformis*

A. Predictions based on postcranial measurements of IGM 183544 and on the program BIOMASS

Variable Measured	Actual Measure (mm)	Predicted Body Mass (kg)		
		Mean	95% Min.	95% Max.
Length of:				
Metacarpal III	81.5	46.3	7.1	301.7
Femur	192	20.4	5.6	74.7
Tibia	177	15.3	4.1	56.3
Metatarsal III	81	30.5	4.5	207.2
Diameter of:				
Metacarpal III	10.5	95.3	38.3	237.4
Femur	25	78.7	41.2	150.6
Tibia	25.2	89.7	34.9	230.7
Metatarsal III	9.0	44.1	11.5	169.2
All (N = 8), geometric mean	—	43.9	41.2	56.3

B. Predictions based on anteroposterior diameters of leg bones of IGM 183544 and on regression coefficients

Regression Sample	%SE[a]	Mean %PE[a]	Predicted Body Mass (kg)
Suoids:			
Femur (=25 mm)	33	21	65.9
Tibia (=25.2 mm)	31	21	64.2
All ungulates:			
Femur	25	17	35.4
Tibia	25	16	115.7

Continued on next page

suoids or other species that appear to be comparable to *Megadolodus* in limb proportions. Because the diameters of limb bones, especially the proximal elements (humerus and femur), have been shown to be most accurate in predicting body mass (Gingerich 1990; Scott 1990), the figure for femur diameter (78.7 kg) and, to a lesser extent, that for the tibia (89.7 kg) is taken to be the most reasonable approximation.

Scott (1990) studied the relationship between body mass

Table 17.4. Continued

C. Predictions based on dental measurements (mean lengths) and regression coefficients for nonselenodonts

Structure Measured	%SE	Mean %PE	Predicted Body Mass (kg)
M_1	40	27	67.7
p^3	37	23	64.0
P_3	43	28	58.1
P_4–M_2	39	24	61.9
Lower postcanine tooth row	31	22	92.0

Sources: **A.** Calculations based on program by Gingerich (1990). **B.** Calculations based on regression coefficients of Scott (1990). **C.** Dental measurements (mean values) used in this analysis are from table 17.1 and regression coefficients are from Damuth (1990).

Notes: Min. = minimum; Max. = Maximum; SE = standard error of the estimate; PE = prediction error.

[a]cf. Smith (1981, 1984).

and postcranial measurements in a large sample (160 species) of living ungulates. Based on the regression coefficients for her all-ungulate sample, the anteroposterior femur and tibia measurements for *Megadolodus* yielded predicted body masses of 35.4 and 115.7 kg, respectively (table 17.4B). The discrepancies between these figures and those obtained previously from the all-mammal sample emphasize the facts that ungulates differ from other mammals in their body proportions, and that (as indicated by the comparisons presented here) *Megadolodus* and other litopterns are simply unlike other ungulates in their limb proportions. Among living ungulates, *Megadolodus* appears to be most structurally similar to suoids in general build, although they differ in many details (as shown, e.g., in fig. 17.6). Based on the regression coefficients presented by Scott (1990) for a sample of living suoids, body mass estimates for *Megadolodus* are 64.2 kg (tibia diameter) and 65.9 kg (femur diameter), results which are reasonably consistent with those obtained from the all-mammal sample (table 17.4A, B).

Because the modern ungulate fauna is dominated by grazing selenodonts that have narrow cheek teeth in comparison to archaic ungulates such as *Megadolodus*, weight estimates for such fossil taxa will tend to be high when based on equations incorporating cheek tooth widths and on

samples including selenodonts (Damuth 1990). Furthermore, *Megadolodus* appears to have abnormally wide molars, even in comparison to living nonselenodonts (J. Damuth, pers. comm.), and we accordingly restrict ourselves to estimates based only on cheek tooth lengths. J. Damuth, who reviewed an earlier version of this manuscript, calculated body mass estimates for *Megadolodus* and kindly permitted us to publish the results (table 17.4C). These estimates, based on the coefficients of Damuth (1990) for cheek tooth lengths of nonselenodont ungulates, range from 58.1 to 92 kg, with an arithmetic mean of 68.8 kg. Except for the high outlying value, these results are consistent with those obtained for femur and tibia diameters, based on the suoid sample of Scott (1990) and the all-mammal sample of Gingerich (1990), but not the all-ungulate sample of Scott (1990).

The wide range in body mass estimates for *Megadolodus* (35.4–115.7 kg) reflects structural diversity within the Mammalia and the lack of entirely comparable living models for this extinct ungulate. Based on the results of table 17.4, however, a reasonable estimate of body mass for *Megadolodus* is 60–80 kg.

FEMORAL MORPHOLOGY

Kappelman (1988) has demonstrated that, among living bovid artiodactyls, femoral morphology correlates closely with locomotor pattern and habitat preference. Kappelman showed that four femoral characters (extracted from measurements reflecting femoral head shape, shaft proportions, and knee structure) may be used to distinguish species preferring open, intermediate, or closed habitats. The apparent independent evolution of these characters among several bovid lineages and among cursorial carnivores (Kappelman 1988) prompted us to evaluate their applicability among some nonbovid ungulates. We collected data for *Megadolodus*, two perissodactyls, a forest-dwelling cervid, and two tayassuids (table 17.5), based on the methods described by Kappelman (1988).

Femoral head shape score (FHSS), which approximates the cross-sectional area of the femoral head in a coronal plane, measures the degree of lateral taper or sphericity of the femoral head. Kappelman (1988) found that, among

Table 17.5. Femoral characters of *Megadolodus molariformis* and several extant ungulate species

Species	FHSS	PSR	MPLR	PGR
Megadolodus molariformis	7.48	0.895	1.51	1.23
Tapirus sp.	5.45	0.653	1.63	1.10
Equus caballus	−1.03	0.704	1.57	0.826
Moschus berezovskii	4.281	0.956	1.107	1.059
Tayassu tajacu	1.121	0.847	—	—
Tayassu pecari	1.250	0.816	—	—

Notes: Figures based on one specimen of each taxon. Data collected following methods described by Kappelman (1988). Mean values obtained by Kappelman (1988, 124) for Bovidae, by habitat group, are as follows (95% confidence intervals given in parentheses): FHSS—open, −0.80 (−1.40 to −0.20); intermediate, 3.03 (2.13 to 3.93); closed, 9.07 (8.14 to 10.00). PSR—open, 0.962 (0.950 to 0.974); intermediate, 1.011 (0.993 to 1.029); closed, 1.061 (1.035 to 1.087). MPLR—open, 1.532 (1.520 to 1.544); intermediate, 1.477 (1.463 to 1.491); closed, 1.411 (1.393 to 1.429). PGR—open, 1.129 (1.107 to 1.151); intermediate, 1.058 (1.042 to 1.074); closed, 1.031 (1.009 to 1.053).

bovids, open-country species tend to have femoral heads of cylindrical shape that, in these presumably more cursorial taxa, tends to limit abduction and axial rotation, restricting movement to the parasagittal plane. The femoral head of *Megadolodus* is hemispherical, with an FHSS fitting within the 95% confidence limits for Kappelman's closed habitat preference group. Similarly, most of the other forest-dwelling taxa included in this study (table 17.5) scored high on FHSS (the two peccaries have somewhat lower scores than would have been expected), whereas the perissodactyl *Equus caballus* easily fell within the 95% confidence interval for open-country bovids. Thus, the limited data presented here for living ungulates corroborates the relationship of femoral head shape to habitat preference and locomotor pattern.

The relationship of three other femoral ratios to habitat preference among nonbovid ungulates is less clear. Proximal shaft ratio (PSR) indicates the relationship of anteroposterior to mediolateral shaft width in the proximal femur, providing an estimate of loading forces through this part of the element. Kappelman (1988) found that bovids preferring open habitats have a proximal femur that is relatively wider mediolaterally than that of forest-dwelling species, hypothesizing that greater mediolateral bending mo-

ment is generated in open-country taxa by femoral head morphology and orientation. We found that *Megadolodus* and perissodactyls are unlike bovids in having the lesser trochanter developed as a long ridge, and in having a strong lateral ridge descending from the greater trochanter to the third trochanter. These two features combine to greatly inflate the mediolateral shaft width of the proximal femur. The PSR of both perissodactyls is less than that of open-country bovids; even without the trochanteric ridges, the proximal femur of the forest-dwelling tapir is much greater in mediolateral than anteroposterior width. This anomaly cannot be attributed to ordinal relationships, because the horse (*Equus caballus*) has a higher value than the tapir (*Tapirus* sp.), the opposite of what would be predicted. Anomalously low values were also obtained for the other Recent taxa. Because we encountered difficulty in consistently defining comparable measuring points and because only one specimen of each species was measured, we have little confidence in the data. In view of these results, we are unable to evaluate the significance of the unusually low PSR in *Megadolodus*.

The medial patellar lip ratio (MPLR) measures the development of the distal femur anterior to the condyles, reflecting the relative mechanical advantage of the leg extensor group. Kappelman (1988) found that highly cursorial, open-country bovid species are characterized by a higher MPLR than are those from closed habitats. Neither MPLR nor the following measurement of the distal femur, patellar groove ratio, could be reliably measured for the species of *Tayassu* because the patellar lips are ill defined. The musk deer included in this study (*Moschus berezovskii*), which has the most bovidlike femur of the sample, had a value falling into the range for closed-habitat bovids. Both perissodactyls in this study, however, proved to have MPLR values fitting within the open-country habitat group established for bovids; enigmatically, the ratio for the tapir slightly exceeded that for the horse (table 17.5). These data suggest that the MPLR categories established for bovids may not be universally applicable and that this ratio may not vary according to habitat preference in all ungulate groups. Given this, we are hesitant to ascribe significance to the value determined for *Megadolodus*.

The patellar groove ratio (PGR) is a measure of the

symmetry of the medial and lateral margins of the patellar groove on the femur. Kappelman (1988) found asymmetry to be higher, with the medial side longer, in open-country bovid species. The mechanical significance of this difference is unclear to us, although Kappelman (1988) attributed the greater degree of asymmetry in cursorial bovids to an increased moment of rotation at the knee joint. As with PSR and MPLR, the results for PGR in this study are ambiguous. For the perissodactyls, *Tapirus* scored much higher than did *Equus*, so that prediction of habitat based on scores obtained for bovids would be the reverse of the actual situation (table 17.5). PGR for the musk deer is somewhat higher than would be predicted on the basis of its habitat preference. As with PSR and MPLR, these ambiguous results suggest that the model developed for bovid artiodactyls is not universally applicable. The significance of the relatively high value obtained for *Megadolodus* is unclear.

Discussion

RELATIONSHIPS

McKenna (1956) compared *Megadolodus molariformis* to Holarctic "condylarths" but referred the species without question to the Didolodontidae. Within that family he distinguished two lineages, and he allied *Megadolodus* with taxa such as *Ernestokokenia* and *Asmithwoodwardia* on the basis of molarization of P_4. Didolodontidae, in turn, have generally been considered to be related to Litopterna. Simpson (1948, 95) remarked that "the Litopterna are, in effect, no more than advanced condylarths surviving in South America long after this general structural grade had been replaced elsewhere by still more progressive condylarth derivatives, the perissodactyls and artiodactyls." The Didolodontidae and early Tertiary Litopterna were reviewed by Cifelli (1983b), who defined the latter order on the basis of postcranial rather than dental specializations. Although nondental remains of Didolodontidae are rare, several early Tertiary members of the family, at least, lack these distinctively litoptern postcranial specializations; conversely, other taxa, although dentally indistinguishable from didolodontids (that is, they are primitive in this regard), bear litoptern synapomorphies of the tarsus and can be

referred to that order. Based on several lines of evidence, we are able to refer *Megadolodus molariformis* to the Litopterna with some confidence. First, *Megadolodus* shares the complex of tarsal synapomorphies characteristic of Litopterna but not known Didolodontidae; second, despite the general primitiveness of the dentition in this Colombian genus (bestowed on it largely by bunodonty), it shares litoptern dental synapomorphies; and third, *Megadolodus* shares both dental and postcranial synapomorphies with an advanced litoptern family (Proterotheriidae) but not other members of the order or, where known, Didolodontidae.

In the dentition, *Megadolodus* shares with Litopterna, but not Didolodontidae, advanced molarization of the last premolars. P^4, which is distinguished from the molars only by lacking a hypocone, has broadly spaced, subequal paracone and metacone, and both conules are well developed. P_4 is fully molariform, with a broadly developed talonid. As in Litopterna, the cristid obliqua of the lower molars achieves its anterior attachment to the lingual side of the trigonid, at the base of the metaconid. On the upper molars, a mesostyle is developed; this is characteristic of the advanced litoptern families Proterotheriidae, Macraucheniidae, and Adianthidae and apparently was acquired independently in *Didolodus* (Cifelli 1983b). In the postcranial skeleton, all litoptern tarsal synapomorphies are present (Cifelli 1983a; Cifelli and Guerrero 1989). These features form a complex as distinctive and diagnostic as that of artiodactyls or perissodactyls and, as in those ungulates, represent modifications presumed to be related to terrestrial locomotion. The postcranial structure of *Megadolodus* is litoptern-like in many other respects. Although these resemblances represent derived features within the Mammalia, they cannot be cited as certain synapomorphies diagnosing the Litopterna because their distribution among Didolodontidae is unknown (although the presence of some or all of them in didolodonts seems unlikely). Some of these characters are related to the mode and degree of reduction in the lateral digits: both manus and pes are mesaxonic and tridactyl. In the vertebral column the anterior sacral and last two lumbar vertebrae have large articular surfaces on the transverse processes. These articulations would have permitted flexion and extension but would have restricted lateral movement in the lower trunk. Similar but smaller articulations are found in perissodactyls (Slijper 1946), although they are usually restricted to the last lumbar and first sacral vertebrae. In addition, in the lumbar region of the body axis, the vertebrae are interlocking: the postzygapophyses are cylindrical and the respective prezygapophyses wrap upward, inward, and downward to enclose them. Elsewhere this feature is rare among the Mammalia, a notable exception being Artiodactyla. Presumably these articulations are related to stabilization and mobility restriction in the trunk, although their exact function is not obvious (see, e.g., Slijper 1946). In the wrist, the scaphoid has a prominent plantar beak and a distinct "step" in anterior view, separating the facets for trapezoideum and magnum. Both of these features, unusual among mammals but found in litopterns (Scott 1910), are of uncertain functional significance. The distinctive articulations of the scaphoid with elements of the distal carpal row are probably related to reduction of the lateral digits in litopterns, although they are unlike those seen in other, adaptively similar, ungulate mammals. In the hind limb, the second trochanter of the femur is absent as such, being instead developed as a low, thin ridge, as it is in perissodactyls. Judged by the condition in early Tertiary ungulates and in most living mammals, the lesser trochanter primitively was developed as a knoblike protuberance. In the tarsus, both cuboid and navicular have salient plantar projections on them, another presumed synapomorphy (of uncertain functional significance) of *Megadolodus* with Litopterna.

A special relationship of *Megadolodus* to litopterns is supported by synapomorphies shared by that genus and Proterotheriidae to the exclusion of Macraucheniidae (and, where known, Adianthidae). Placement of *Megadolodus* within the Didolodontidae would therefore require independent evolution of these characters. Some derived resemblances of *Megadolodus* to proterotheriids appear to be associated with advanced reduction of the lateral digits. Characters necessarily so associated include the complete loss of pedal digit V (and, possibly, digit V of the hand as well, although Scott [1910] reported a small, possible remnant of metacarpal V in *Diadiaphorus*), extreme reduction of digits II and IV, and increase in relative size and importance of digit III. In both manus and pes, *Megadolodus* and proterotheres have an enlarged, equiform ungual phalanx on digit III, not seen in macraucheniids (Scott 1910). Other

characters appear to be sufficient but not necessary consequences of digital reduction because they are not general to other mammalian groups undergoing loss or reduction of the lateral digits. In the manus, Proterotheriidae and *Megadolodus* share the following synapomorphies: scaphoid no longer overlies trapezium in anterior view; magnum enlarged and greatly exceeding the trapezoideum in size; metacarpal II with prominent lateral projection on its proximal end, resting on top of metacarpal III; and metacarpal III with plantar projection, bearing a second proximal articulation for metacarpal IV. In the hind limb, the fibula is reduced as in proterotheres but not macraucheniids; in the pes, *Megadolodus* and proterotheres are distinctive in the development of an articular contact between the cuboid and metatarsal III, and in the extreme reduction of the mesocuneiform. Although the dentition of *Megadolodus* appears to be primitive in its bunodonty and lack of strong crest development, such plesiomorphies are not unparalleled among the Proterotheriidae (e.g., *Prothoatherium*; Cifelli and Guerrero 1989). Furthermore, *Megadolodus* shares several apomorphies with proterotheres, to the exclusion of Didolodontidae, Protolipternidae, and the advanced litoptern families Macraucheniidae and Adianthidae. The third premolars are molarized: P3 has subequal, broadly spaced paracone and metacone, and both conules are present; P_3 has a lophate talonid. A significant feature of the upper molars (and posterior upper premolars) is that the metaconule has lost its primitive attachment to the protocone and the metacone via the postprotocrista, so that it is an isolated cusp, as is characteristic of Proterotheriidae. The only known contradiction to a relationship of *Megadolodus* with Proterotheriidae is the presence of a hypocone on M3 of the Colombian genus. This feature is developed in didolodontids and in the litoptern families Macraucheniidae and Adianthidae, but not in known Proterotheriidae (Cifelli 1983b). Thus, placement of *Megadolodus* within the Litopterna as a relative of Proterotheriidae implies independent acquisition of this character.

More specific relationships of *Megadolodus* are enigmatic. As currently conceived, the Proterotheriidae comprises two subfamilies, each a monophyletic group (Cifelli 1983b): Anisolambdinae (Itaboraian to Deseadan) and Proterotheriinae (Deseadan to Chapadmalalan). The only complete foot known for Anisolambdinae is that of

Deseadan *Protheosodon*, figured by Loomis (1914). *Megadolodus* is similar to Proterotheriinae in its more advanced degree of reduction in the lateral digits, and in having contact between the cuboid and metatarsal III; these are derived conditions not seen in *Protheosodon*. However, this latter taxon (and other Anisolambdinae) resembles Proterotheriinae, to the exclusion of *Megadolodus*, in its more lophate dentition. At present, we place *Megadolodus* as a sister taxon to both subfamilies of Proterotheriidae, with the three joined in an unresolved trichotomy. *Megadolodus* may be autapomorphous with respect to Proterotheriidae in the development of canine tusks if, in fact, the canines are correctly identified. The distinctiveness of *Megadolodus* may eventually warrant recognition of a separate family to include it, but pending more detailed investigations on the phylogeny of Proterotheriidae, we more conservatively retain the Colombian genus in that family.

With the removal of *Megadolodus* to the Litopterna, the occurrence of other post-Eocene Didolodontidae is called into question. Hoffstetter and Soria (1986) referred a species from the Miocene of Colombia to the family. Better materials of the taxon, including the upper dentition and parts of the skeleton, led to its referral to the Proterotheriidae as a species of *Prothoatherium*, otherwise known from the Colhuehuapian (Cifelli and Guerrero 1989). Patterson and Pascual (1972) considered the aforementioned *Protheosodon*, from the Deseadan, to be a didolodontid, but comparative study of the dentition and foot suggest that it, too, is a proterotheriid (Cifelli 1983a, 1983b). The only remaining post-Eocene taxon referred to the Didolodontidae is Deseadan *Salladolodus deuterotheroides*, known by two upper molars (Soria and Hoffstetter 1983). We are unable to assess the relationships of *Salladolodus* based on these materials alone (although the coronal morphology is reminiscent of primitive Macraucheniidae); thus, there is no unambiguous evidence to suggest survival of the Didolodontidae later than Casamayoran.

PALEOBIOLOGY

Many of the postcranial specializations of *Megadolodus* appear to reflect commitment to terrestrial locomotion. The vertebral column is specialized in having interlocked posterior thoracics and lumbars (also seen in Artiodactyla) and articulations between transverse processes of the posterior lumbars and first sacral (also seen in Perissodactyla), both of which would appear to restrict lateral mobility of the posterior trunk. The greater trochanter of the femur is strongly developed (as in ungulates generally), reflecting strong development of the gluteal musculature, and the pronounced ischial tuberosity likewise indicates that the hamstring muscles were well developed. The articular surfaces of the hip joint shows emphasis on flexion and extension at the hip although, as discussed below, the shape of the femoral head suggests capability for abduction and axial rotation. Both hands and feet have strongly reduced lateral digits and well-developed hooves on digit III. The carpus and tarsus are compact, and the modifications in size, morphology, and articular relationships of wrist and ankle elements probably are associated with reduction of the lateral digits, restriction of lateral mobility, and increased capacity for flexion and extension. Given these presumed specializations for terrestriality in *Megadolodus*, it is then relevant to compare the Colombian taxon with Recent ungulates.

The first factor of the principal components analysis (on which FEM W and SQRAREA had highly positive loadings and TIB L and MT3 L had strong negative loadings) tended to segregate heavily built ungulates, which generally have large teeth and short distal limb elements. For primates Kay (1981, 1985) has suggested that enlarged cheek tooth area may reflect ingestion of extremely hard food items, specifically hard fruits. Suoids, which have the largest molars of the sample of living taxa studied, are generally regarded as opportunistic omnivores. Kiltie (e.g., 1981a, 1981b, 1982) found that, in the South American rain forest, peccary species are major predators on hard fruits and seeds, which comprised a significant part of their diet. The extent to which other suoid species use these food resources, and the implications of such use on tooth size and structure remain to be investigated. Among other Recent taxa included in the analysis, both perissodactyls had large teeth, with those of the tapir exceeding those of the donkey in size. Tapirs are generally considered to be browsers (e.g., Janzen 1983), although they ingest fruits and are known to be seed predators (e.g., Janzen 1981; Bodmer 1990). Of the artiodactyls, the relatively largest value for SQRAREA was in the water chevrotain, *Hyemoschus aquaticus*, which feeds almost exclusively on fruits (Dubost 1978). All of the proterotheriids included have very large cheek teeth; those of *Megadolodus* are the largest and are comparable in this respect to suoids. As in the extant peccaries and pigs, cheek teeth of *Megadolodus* are bunodont and very low crowned, suggesting an omnivorous diet (chap. 18). The large size of these teeth, coupled with the relatively thick enamel and robust mandible, suggests that hard fruits may have formed a component of its diet.

No simple relationship of tibia length to habitat preference emerged from the analysis (see also Scott 1985), although suoids have a relatively short tibia, as do the perissodactyls *Tapirus* and *Equus* and the water chevrotain (*Hyemoschus aquaticus*). The fastest cursors have a tibia of moderate length, whereas the longest tibia tended to occur in forest- and mountain-dwelling artiodactyls (fig. 17.6). As with suoids, proterotheriids have a very short tibia, with that of *Megadolodus* being relatively the shortest.

Suoids also had the relatively shortest metatarsals among living taxa included in the analysis, with those of the tapir and water chevrotain also being quite short. Metatarsal length has a clearer relationship to habitat preference and running ability than does tibia length (cf. Scott 1985). As shown in figure 17.6, open-country cursors have the longest metatarsals, with forest-dwelling taxa having shorter elements (the position of the mouse deer, *Tragulus javanicus*, appears anomalous in this respect).

The second principal component, which had high and inverse loadings for FEM L and PH1 L, more clearly segregates living taxa according to habitat preference and degree of cursoriality than does the first component, with highly cursorial, open-country species having a relatively short femur and long first phalanx. The proterotheriids scored surprisingly high on this component (fig. 17.7), especially *Thoatherium minisculum*, which has an extremely long first phalanx. Relative elongation of distal limb elements has long been recognized to be a trait of cursorial mammals (e.g., Howell 1944; Maynard Smith and Savage 1959), suggesting that the elongation of the proximal phalanges among the Proterotheriidae might represent a functional alternative to elongation of the metatarsus. Consideration of the total hind limb length distal to the femur (i.e., TIB L + MT3 L + PH1 L) does not bear this out (fig. 17.6). All proterotheres have extremely short legs, comparable to the

tapir and the suoids included in the sample, and far shorter than any cursorial species. It is also of interest to note that, in contrast to published accounts (Nowak and Paradiso 1983), the leg of the water chevrotain (*Hyemoschus aquaticus*) is also short.

Megadolodus is highly similar to other proterotheriids in the proportions of its hind limbs and relative molar size. Proterotheriids are characterized by a combination of limb proportions differentiating them from Recent ungulates; the most unusual characteristic of these litopterns is the relatively elongate first phalanx. Among extant ungulates, *Megadolodus* is most comparable to suoids, especially in such features as relatively large molars, robust femur, and short distal limb elements. Although living suoids are fairly labile in their habitat requirements, all (except the wart hog, *Phacochoerus aethiopicus*) are generally associated with closed, forested habitats. Furthermore, notwithstanding the uniqueness of *Megadolodus* and other proterotheriids in certain respects, the Colombian genus shares distinctive limb proportions with forest-dwelling taxa (regardless of familial affinities) and in these same respects differs from all ungulates preferring open habitats.

The femur of *Megadolodus* bears a hemispherical head. In extant cursors, the articular surface of the head is medio-laterally developed so that it is subcylindrical, which restricts movement to the parasagittal plane (Kappelman 1988). Quantitative data indicates that femoral head shape in *Megadolodus* corresponds more closely to forest dwelling than to open country Bovidae, although its hemispherical development is not approximated by any member of that family. This joint morphology would have permitted hip mobility in the form of abduction and axial rotation. The survey of Recent ungulates presented herein corroborates the suggestion of Kappelman (1988) that such morphology correlates with preference for a closed habitat, where the implied mobility would be well suited to coping with a more three-dimensional substrate. In the other characteristics of the femur studied by Kappelman (1988), *Megadolodus* is not comparable to living Bovidae. The utility of these femoral characters as indicators of habitat preference is presumably limited by constraints imposed by phylogenetic background and by the fact that simple mechanical considerations alone cannot always predict biological role. The proportions of the shaft of the proximal femur in *Mega-*

dolodus and Perissodactyla, for example, are not comparable to those seen in Artiodactyla because litopterns, like perissodactyls, have a strongly developed third trochanter and a long, ridgelike second trochanter. In perissodactyls, at least (associated hind limbs for early Tertiary litopterns are unknown), these features are temporally archaic. Although the functional significance of this unusual trochanteric morphology is uncertain (see, e.g., Howell 1944), it clearly influences the shape of the proximal femur. The results obtained for characters of the distal femur are more obscure still. Kappelman (1988) indicated that the relatively high scores seen among open-country bovids reflect the need in cursorial species to increase mechanical advantage for the knee extensors (mm. vasti) by increasing moment of rotation at the knee. The Recent taxa included in this study do not fit this pattern, however, with both perissodactyls having anomalously high values, and the forest-dwelling tapir exceeding the open-country horse for both indices (MPLR and PGR). As Kappelman (1988) pointed out, the relative moment of the mm. vasti in extending the leg depends on numerous factors. Highly cursorial taxa have greatly elongated distal limb elements, particularly the metatarsals (Gregory 1929; Howell 1944; Scott 1985). This effectively increases the distance between the point of rotation (knee) and load (substrate) so that, other factors being equal, the moment of rotation would be lower than in comparable but noncursorial species. Thus, within a group such as Bovidae, the anteroposterior expansion of the distal femur in highly cursorial taxa could be related to mechanical exigencies imposed by elongation of the limb distal to the knee, rather than to an absolute increase in moment for the mm. vasti (thus, the two characters of the distal femur, MPLR and PGR, would be predicted to correlate highly with metatarsal length). The perissodactyls and suoids included in this study not only have shorter distal limb elements than the ruminant artiodactyls, but also are more robustly built. We speculate that the relatively great anteroposterior expansion of the distal femur in these taxa may be related to a high absolute moment for the leg extensors, implying more powerful but slower extension capability at the knee joint than in cursors (this would explain the otherwise enigmatic results obtained for the tapir and horse, which are the reverse of what would be expected if anteroposterior expansion of the distal femur were related to cursorial modifica-

tion per se). This hypothesis could be tested by developing means of obtaining anatomically comparable and functionally relevant measurements of the distal femur which are applicable to suoids as well as ruminant artiodactyls. In any event, the robust build and short distal limb elements of *Megadolodus*, coupled with its anteroposteriorly expanded distal femur, are more suggestive of powerful extension capability than of great speed.

Consistency among the various postcranial predictors of body mass is poor for *Megadolodus molariformis* because this species differs in body proportions from those living taxa upon which the regression coefficients were based. Estimates based on dental measurements and the most accurate postcranial indicators suggest, however, that a body mass of 60–80 kg represents a reasonable estimate for *Megadolodus*. In general terms, *Megadolodus molariformis* was somewhat larger than the living species of *Tayassu* and within the weight range of most living species of Suidae.

Megadolodus represents a morphologically distinctive clade of Proterotheriidae. Because the group is monotypic and is currently known only from the Miocene of Colombia, the antiquity of the clade is difficult to assess. However, the lack of either anisolambdine or proterotheriine synapomorphies in *Megadolodus*, together with the apparent primitiveness of its cheek teeth, suggests that Megadolodinae have been distinct since the Paleogene. Insofar as the record of South American Tertiary mammals stems almost entirely from the southern extreme of the continent, the lack of additional reports of Megadolodinae from the Tropics is unsurprising. Nonrepresentation of the group in the well-known faunas from Patagonia, however, suggests that *Megadolodus* and allied taxa were rare or absent in that area. Dental and postcranial evidence presented herein indicates that *Megadolodus*, at least, was a suoid-like ungulate that was restricted to closed (and probably heavily forested) habitat. Although increasing aridification characterized much of the Tertiary in Patagonia, a combination of geological, floral, and faunal evidence suggests the presence of closed habitat in that region before and including the Santacrucian (e.g., Pascual and Odreman Rivas 1971; Webb 1978). Nonrepresentation of Megadolodinae in Patagonia cannot, therefore, be explained strictly on the basis of habitat differences. If absence of the group from the southern part of South America reflects a true distribution pattern rather than an artifact of the fossil record, the paleo-biogeographic reason(s) for this remain unknown. A possible but speculative explanation for the distribution of Megadolodinae is that the group specialized on certain plant species, perhaps, but not necessarily, including fruit-bearing taxa that were restricted to the Tropics.

Acknowledgments

We appreciate the efforts of M. McKenna, F. Szalay, and J. Damuth, who reviewed an earlier version of this chapter; we are especially grateful to Damuth for his help and for permission to publish calculations made by him. We thank D. Savage, Museum of Paleontology, University of California, for providing us with a cast of the type specimen of *Megadolodus molariformis*; D. F. Schmidt, Smithsonian Institution, for providing us with measurements of Recent ungulate postcrania; H. Thewissen, Duke University, for information; and G. Schnell, University of Oklahoma, for advice on statistics. Figure 17.1 was prepared by S. Hansen. Partial support for this research was provided by NGS grants 2964-84 and 3292-86 and by NSF grants BSR8614533 and BSR8918657 to R. F. Kay.

18
Litopterns

Richard L. Cifelli and Javier Guerrero

RESUMEN

El Grupo Honda en Colombia contiene una diversa asociación de especies de un grupo de ungulados fósiles exclusivamente sudamericanos, que se incluyen en el orden Litopterna. Se encuentran al menos seis especies, de las cuales cinco tienen afinidades con la familia Proterotheriidae y la otra es un Macraucheniidae.

La familia Macraucheniidae está representada con una especie por identificar del género *Theosodon*. El primer registro tropical de este género, extiende aun más una distribución muy amplia temporal y geográficamente. Los taxa con afinidades en la familia Proterotheriidae incluyen *Megadolodus* y *Prothoatherium* (ya descritos), una nueva especie de *Prolicaphrium*, *Villarroelia* (un nuevo género y especie), y otro género y especie por definir. Los litopternas del Grupo Honda son taxonómicamente y morfológicamente arcaicos. *Prothoatherium* y *Prolicaphrium* estaban restringidos al Colhuehuapense, y *Megadolodus* y el género y especie por identificar son estructuralmente más similares a taxa del Paleógeno. Además, los Litopterna del Grupo Honda también se caracterizan por tener molariformes de baja corona.

Juzgando la distribución por altura de corona de ungulados vivientes, los proterotéridos colombianos probablemente representan una variedad de formas de alimentación, incluyendo omnívoros, frugívoros, y herbívoros. Todos preferían un hábitat relativamente cerrado y ninguno se alimentaba predominantemente de pastos. En altura de corona, los proterotéridos colombianos son sim-ilares a los del Colhuehuapense de Patagonia, y tienen coronas más bajas que los del Santacrucense de altas latitudes.

Después del Santacrucense, la distribución de proterotéridos parece haberse restringido a causa del inicio de climas más áridos y desérticos en altas latitudes de Suramérica. La presencia de una diversa asociación de proterotéridos arcaicos y relictuales en el Mioceno medio de Colombia, junto con un régimen dietario que no dependía de pastos, y la preferencia probable por hábitats cerrados, apoya la sugerencia de que la distribución de la familia estaba estrechamente ligada a cambios climáticos en el Terciario tardío de Suramérica, y que en general, sus taxa estaban restringidos a comunidades de selva o sabana arbórea.

The Litopterna rank second to Notoungulata as South America's most successful and diverse indigenous ungulate order. All three advanced families (Proterotheriidae, Macraucheniidae, and Adianthidae) appear to have originated by the Eocene (Cifelli 1983b) and survived through most of the remaining Cenozoic; adianthids apparently became extinct during the Miocene, proterotheriids at the end of the Pliocene, and macraucheniids in the Pleistocene. Although litoptern clades were, in general, morphologically stereotyped after the Oligocene and were not exceedingly abundant in most known local faunas, they nonetheless constitute important elements in South American land mammal assemblages of the Tertiary. Unlike their notoungulate contemporaries, most litopterns retained a relatively conservative dentition and low-crowned cheek teeth, features that presumably were not well suited to the increasingly dry, open savannas that characterized the later Tertiary of Patagonia (e.g., Pascual and Odreman Rivas 1971; Webb 1978). Thus, the distribution, diversification, and extinction of litopterns would be predicted to be closely associated with environmental conditions (e.g., Bianchini and Bianchini 1971). Litopterns may be useful climatic indicators and important sources of information on the paleobiogeography of South America.

Litopterns have long been noted to occur in the terrestrial rocks of the Honda Group (Stirton 1953b). With the exception of *Prothoatherium* (Cifelli and Guerrero 1989) and *Megadolodus* (McKenna 1956; also chap. 17), however, Colombian litopterns have neither been described nor even tentatively identified, most faunal lists simply including the families Macraucheniidae and Proterotheriidae in the La Venta faunal list (e.g., Stirton 1953b; Hirschfeld and Marshall 1976; Marshall et al. 1983). The present account has two purposes: to present a systematic and descriptive account of the Litopterna of the Honda Group, and to compare the Colombian litopterns with those of the well-known Patagonian assemblages.

Systematic Paleontology

Order Litopterna Ameghino 1889
Family Proterotheriidae Ameghino 1887

With the exception of Colhuehuapian and Mesopotamian taxa, fully revised by Soria (1981) and Bianchini and Bianchini (1971), respectively, the systematics of Neogene proterotheres are in a state of virtual chaos. In most cases the literature on later proterotheres consists of little more than cursory descriptions published near the turn of the century, with synonymies (particularly at the specific level), individual variation, and relationships remaining to be established. Under these circumstances, the basis for establishing the identity of the Colombian taxa must be clarified.

Two proterotheriid genera, *Prolicaphrium* (monotypic) and *Prothoatherium* (two species), are known from the Colhuehuapian, each apparently representing a major clade in

the family (Soria 1981). By contrast, Scott's (1910) classic monograph of the Litopterna from the succeeding Santacrucian lists six genera (*Diadiaphorus, Licaphrium, Proterotherium, Tetramerorhinus, Tichodon,* and *Thoatherium*), collectively including a seemingly impossible twenty-eight species. *Tetramerorhinus* species are probably referable to *Proterotherium* (Scott 1910); *Tichodon* is of uncertain status and has been omitted from subsequent synopses (e.g., Simpson 1945; Marshall et al. 1983). Scott and subsequent authors (e.g., Bianchini and Bianchini 1971) have noted that Santacrucian species referred to *Proterotherium* and *Licaphrium* are probably not generically separable. The type of the former genus, *P. cervioides,* is not, however, from the Santacrucian, but is a member of the Pliocene Mesopotamian fauna. Bianchini and Bianchini (1971) have shown that *Proterotherium cervioides* is generically distinct from all Santacrucian species. Thus, reference to *Proterotherium* should be restricted to the type species alone; we follow convention (e.g., Bianchini and Bianchini 1971; Soria 1981) in referring to the Santacrucian species (which, by implication, probably belong in *Licaphrium*) as *"Proterotherium."* Because the circumscription of most Santacrucian species is not clear, we refer simply to genera in our comparisons here.

Kraglievich (1930b) described *Proterotherium berroi* from beds in Uruguay that he believed to be late Miocene or Pliocene in age. Insofar as the type and only specimen is a distal humerus, we do not believe that the validity of the species can be addressed at present. The most diverse and best understood assemblage of later Proterotheriidae is that of the Huayquerian, which includes Mesopotamian taxa. Bianchini and Bianchini (1971) recorded five Mesopotamian genera, including *Proterotherium, Brachytherium, Licaphrium, Epitherium,* and *Thoatherium.* Of these, Simpson (1945) considered *Epitherium* to be a junior synonym of *Brachytherium*; in apparent agreement, the former genus was omitted from the Huayquerian faunal list of Marshall et al. (1983). Bianchini and Bianchini (1971), however, maintained *Epitherium* as a distinct genus. We follow the usage and definition of these two genera as given by Bianchini and Bianchini (1971). Additional Huayquerian records of the Proterotheriidae include ?*Diadiaphorus, Eoauchenia* (a possible synonym of *Epitherium*; Pascual 1966), and *Epecuneia*; the status of the last-named genus is uncertain because it is recognized mainly on the basis of postcranial characters and size (Pascual 1966). The only remaining member of the family from post-Santacrucian faunas is *Diplasiotherium* from the Pliocene Monte Hermoso Formation (Rovereto 1914), represented by a single, inadequately known species.

At least four members of the Proterotheriidae are known from the Honda Group; a possible fifth, *Megadolodus molariformis,* was originally referred to the Didolodontidae (McKenna 1956) and is now considered to be a member or close relative of the Proterotheriidae (chap. 17). *Megadolodus* will not be treated in this section, nor will *Prothoatherium colombianus,* a proterotheriid from the Honda Group, which has also been described elsewhere (Hoffstetter and Soria 1986; Cifelli and Guerrero 1989). Of the remaining three taxa, we herein describe two as new; the third is represented by inadequate materials, and its identity is uncertain.

Prolicaphrium Ameghino 1902
Prolicaphrium sanalfonensis sp. nov.

TYPE SPECIMEN

IGM 182852, symphysis and right ramus of mandible with right I_2 and P_2–M_3.

HYPODIGM

The type specimen, and IGM 184499, right mandibular ramus with P_2–M_3; IGM 183246 and 183620, fragment of maxilla with left M^{1-2} and associated right P^3; IGM 250873, right M^3; UCMP 39254, isolated broken teeth and fragments; and UCMP 39256, associated but badly broken and well-worn upper and lower cheek teeth.

HORIZONS AND LOCALITIES

The type was collected at Duke Locality 21, below the Chunchullo Sandstone Beds, La Victoria Formation. The remaining specimens are from UCMP locality V4523, Fish Bed, Baraya Member, Villavieja Formation; Duke Locality 16, Monkey Bed, Baraya Member, Villavieja Formation; and Duke Locality 95, unit between the Chunchullo and Tatacoa Sandstone beds, La Victoria Formation (stratigraphic nomenclature follows that of Guerrero 1990; also chap. 2).

ETYMOLOGY

Named for the town of San Alfonso, Huila Department, Colombia, which is near the type locality for the species.

DIAGNOSIS

The smaller of the two species referred to the genus. Differs from *P. specillatum,* the type species, in having slightly lower crowns on the cheek teeth, a stronger connection between protocone and metaconule on the upper molars, and P^3 with a better-developed mesostyle and anterocingulum. Differs from species of *"Proterotherium,"* *Thoatherium,* *Diadiaphorus,* *Brachytherium,* *Villarroelia,* *Epitherium,* and *Licaphrium* in having lower-crowned cheek teeth.

DESCRIPTION AND DISCUSSION

The cheek teeth (fig. 18.1) are remarkable for their low crown height relative to other Oligocene and later proterotheriids, being comparable only to *Prothoatherium* in this regard. Some fragments of the anterior upper dentition are preserved with UCMP 39254 and 39256, but these are uninformative. The upper molars and posterior upper premolars have weak styles as in *P. specillatum* and in contrast to species of *"Proterotherium,"* *Diadiaphorus,* *Brachytherium,* *Thoatherium,* *Villarroelia,* and *Licaphrium.* As in *Proterotherium cervioides* and species of *Prolicaphrium,* *Villarroelia,* and *Epitherium,* noticeable paracone and metacone ribs are apparent on the ectoloph of P^3–M^3; these are lacking in *"Proterotherium,"* *Licaphrium,* *Diadiaphorus,* *Thoatherium,* and *Brachytherium.* P^3 (fig. 18.1C) bears a strong anterior cingulum. The posterior cingulum produces a notable bulge at the posterolingual corner of the tooth, but a hypocone as such is lacking. In species of *"Proterotherium,"* *Diadiaphorus,* *Brachytherium,* *Thoatherium,* and *Licaphrium,* a hypocone is well developed and attaches to the protocone near the apex of that cusp with a sweeping crest; this crest attaches low to the protocone on P^3 of *Prolicaphrium sanal-*

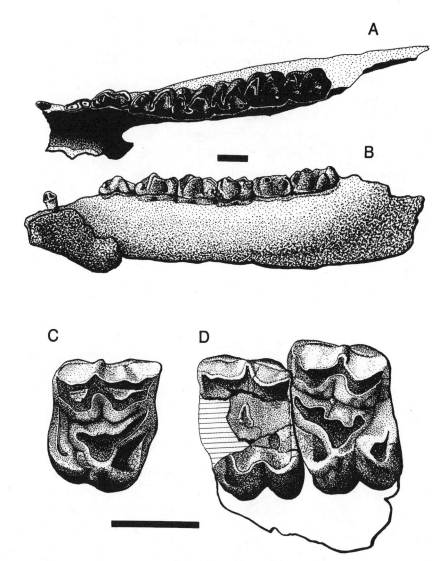

Figure 18.1. *Prolicaphrium sanalfonsensis* sp. nov. *A–B.* IGM 182852, type specimen, right I$_2$ and P$_2$–M$_3$: *A,* occlusal view; *B,* lingual view. *C.* IGM 183620, right P^3, occlusal view. *D.* IGM 183246, left M$^{1-2}$ (same individual as IGM 183620), occlusal view. Scale bars 1 cm.

fonsensis. The metaconule is large and, with wear, attaches to the protocone, forming part of a V-shaped arrangement of trigonal crests. The development and relationships of the metaconule are similar on P^4–M^3; P^4 differs from P^3 in having a well-developed hypocone. On P^4 and succeeding teeth, the metaconule is distinct and is more connected to the protocone than to any other cusp. Bianchini and Bianchini (1971) described two major coronal patterns on upper molars of Proterotheriidae: in one, typified by *"Proterotherium,"* the metaconule is isolated and small or lacking; in the other (e.g., *Thoatherium*), the metaconule is shifted lingually and is incorporated into the postcingular crest, and thus has an attachment to the hypocone. Although

Prolicaphrium specillatum apparently fits the former pattern (cf. Soria 1981), the V-shaped arrangement seen on upper molariform teeth of *P. sanalfonsensis* does not fit readily into either model. Because this arrangement appears to be primitive for tribosphenic mammals, because intraspecific variability is poorly documented, and because the difference between the species is slight, the significance of this minor difference is problematic.

The lower dentition (fig. 18.1A, B) is comparable to that of *Licaphrium* and *Prolicaphrium* species, except that it is more gracile. The trigonid of P$_2$ and, to a lesser extent, P$_3$, is more anteroposteriorly elongate than in other taxa, with the exception of *P. specillatum*. Development of an ento-

Table 18.1. Dental measurements (mm) of *Prolicaphrium sanalfonsensis* sp. nov.

Specimen	P³ L	P³ W	P⁴ L	P⁴ W	M¹ L	M¹ W	M² L	M² W	M³ L	M³ W
IGM 183246	12.0	14.3	—	—	10.6	16.9	12.0	19.2	—	—
UCMP 39254	—	—	11.8	—	—	—	—	—	—	—
UCMP 39256	—	—	—	—	—	—	—	—	9.7	13.1
IGM 184499	—	—	—	—	—	—	—	—	—	—
IGM 182852	—	—	—	—	—	—	—	—	—	—

Specimen	P₂ L	P₂ W	P₃ L	P₃ W	P₄ L	P₄ W	M₁ L	M₁ W	M₂ L	M₂ W	M₃ L	M₃ W
IGM 183246	—	—	—	—	—	—	—	—	—	—	—	—
UCMP 39254	—	—	—	—	—	—	—	—	—	—	—	—
UCMP 39256	—	—	14.5	8.3	—	—	12.5	8.7	12.3	—	14.6	9.8
IGM 184499	13.1	4.9	14.8	8.7	12.9	9.8	12.1	—	12.1	9.5	14.7	9.0
IGM 182852	13.4	5.6	14.5	8.9	13.0	10.4	11.1	9.5	11.7	10.0	15.6	9.6

conid is variable on P_3, being present on one specimen (IGM 184499) and absent on another (IGM 182852). On succeeding teeth the entoconid is a strong, independent cusp, as it is in *Prolicaphrium specillatum* and species of *Brachytherium* and *Licaphrium,* and in contrast to other advanced proterotheriids, in which the entoconid is merged into the talonid crescent. The posterior lower premolars and all molars lack paraconids, with the paracristid terminating before it reaches the lingual margin of the tooth. The hypoconulid is a small cusp only noticeable on unworn teeth. On M_3, the hypoconulid is developed as a third lobe, clearly separated from the hypoconid by a sulcus on the labial side of the tooth (unlike the condition seen in *Diadiaphorus* and *Thoatherium*). Measurements are given in table 18.1.

Prolicaphrium sanalfonsensis differs from species of most described genera in a number of dental features, as elaborated previously. Because of the poor status of understanding regarding the variability and polarity of these characters, we have omitted these from the diagnosis, pending more adequate definition of advanced proterothere genera. Referral to *Prolicaphrium* is, faute de mieux, based on overall similarity rather than on the sharing of uniquely derived morphology with the type species.

Villarroelia gen. nov.

TYPE SPECIES

Villarroelia totoyoi sp. nov.

INCLUDED SPECIES

The type only.

DISTRIBUTION

As for the type and only species.

DIAGNOSIS

As for the type and only species.

ETYMOLOGY

For Carlos Villarroel, in recognition of his contributions to knowledge of South American fossil mammals.

Villarroelia totoyoi sp. nov.

TYPE SPECIMEN

IGM 250965, skull lacking the anterior rostrum, with roots of right P^1 and with right P^2–M^3, left P^3–M^3, and miscellaneous fragmentary postcranial elements.

HYPODIGM

The type, and UCMP 39970, fragmentary skull with broken right dM^4, P^4 in the crypt, M^1 in early wear, and M^2 in eruption, and left M^1 (broken) and M^2; IGM 183407, partial palate with right and left M^{2-3}; GM-31, right P^3; GM-01, associated maxillary, mandibular, and postcranial fragments including right M^3, P_4, M_1, and M_3, broken left M_3, two cervical vertebrae, two thoracic vertebrae, proximal right scapula, ends of the right radius, fragmentary right femur, proximal left femur, left tibia, broken left calcaneus, and several phalanges; DGP 7-V-89–2, fragment of right maxilla with broken P^2 and P^3–M^1 intact; IGM 250500, associated but broken upper and lower teeth, including parts of P^3–M^2 and M_3 from both sides, premolar fragments, and right I_2; IGM 250502, distal femur and fragment of mandibular ramus with left P_4; IGM 184036, right P_3; IGM 184051, isolated teeth, including right P^{1-4} and left P^2 and M^{2-3}; IGM 183998, postcranial fragments and right mandibular ramus with dM_{3-4}, M_1, and M_2 (in eruption); IGM 183999, associated skeletal parts including mandibular rami with left dM_{1-4}, M_1, and M_2 in the crypt, maxillary fragment with right dM^{3-4}, and various postcranial fragments; IGM 250956, broken postcranial elements and fragment of left mandibular ramus with left M_2; IGM 184661, associated proximal scapula and left mandibular ramus with broken left M_{2-3}; IGM 184653, associated left P_{1-2}, M_{2-3}, right P_2, M_1, and M_3, and fragments of the vertebral column; IGM 183628, left M_1; IGM 183051, left M_2; and IGM 184501, fragment of right mandibular ramus with right P_3 and broken P_4.

ETYMOLOGY

Named for Totoyó, chief of the indigenous tribe that, in the sixteenth century, inhabited the banks of the river bearing the same name (now the Río Villavieja).

HORIZONS AND LOCALITIES

Duke Localities 49, below the Cerro Gordo Sandstone Beds; 79, between the Cerro Gordo Sandstone and Chunchullo Sandstone beds; 96, 113, between the Chunchullo Sandstone and Tatacoa Sandstone beds; and 108, between the Tatacoa Sandstone Beds and the Cerbatana Conglomerate Beds (all in the La Victoria Formation). DGP 7-V-89-2 was collected several m below the Tatacoa Sandstone Beds, in the middle of the La Victoria Formation.

DIAGNOSIS

Proterotheriid roughly the size of *Licaphrium mesopotamiense* and with generally similar dental morphology, differing from that species and from all comparable taxa in the complete molarization of P^{3-4}, with salient conules and with fully developed, lingually placed hypocone.

DESCRIPTION AND DISCUSSION

Several partial crania and a number of fragmentary postcranial elements are represented in the hypodigm; these do not differ materially from specimens of *"Proterotherium"* or *Licaphrium* as described by Scott (1910), and their description is unwarranted. The anterior deciduous molars of the upper dentition are not known. DM^{3-4} are generally similar to those of *"Proterotherium,"* dM^4 differing from that of Santacrucian taxa in having a crest connecting the protocone and metaconule. The adult cheek teeth are comparable in height to those of *Licaphrium* and *"Proterotherium."* The upper premolars are, in general, more transversely expanded than in other proterotheres. P^1 is similar to the corresponding tooth in *Prolicaphrium, Diadiaphorus,* or *"Proterotherium,"* differing in having a more lingually developed protocone, which forms a prominent bulge at the posterolingual corner of the tooth. As in most other proterotheriids, labial folds on the ectoloph of P^1 are faint or absent. P^2 (fig. 18.2) is also similar to that of other proterotheriids but is distinct in having a more lingually expanded protocone (as with P^1) and in having a distinct paracone fold on the ectoloph. Both parastylar and metastylar spurs are present on the ectoloph, as in other Proterotheriinae. P^3 is much larger than P^2, being comparable

Figure 18.2. *Villarroelia totoyoi* gen. et sp. nov., IGM 250965, P^2–M^3 of type specimen, occlusal view. Scale bar 1 cm.

to M^1 in size. This, coupled with the fact that P^3 is completely molariform, strongly distinguishes *Villarroelia* from all other Proterotheriidae, in which the premolars form a relatively graded series. The mesostyle is prominent on P^3 as it is on all succeeding teeth; parastyle and metastyle are strong and subequal on P^3, whereas the parastyle is distinctly more prominent on succeeding teeth. Both paracone and metacone have distinct labial ribs, as they do on all succeeding teeth except M^3, in which the metacone fold is lacking. The cusp pattern of P^3 is molariform and, except for the differences just noted, isolated teeth from this locus could readily be confused with M^1. The hypocone is developed as a strong cusp, subequal in height to the protocone and projecting an equivalent distance lingually. The metaconule is also strong on P^3 and on all succeeding teeth. Early in wear, the metaconule connects with the protocone so that a V-shaped pattern is formed with those two cusps and the paraconule. By contrast, the metaconule on upper molars and posterior premolars of *"Proterotherium," Licaphrium,* and *Diadiaphorus* is generally small and remains independent. In *Proterotherium cervioides* and *Thoatherium,* which we regard as otherwise specialized, the metaconule is lingually placed, nearly merging with the protocone (cf. Bianchini and Bianchini 1971). P^4 is larger than either P^3 or M^1 and is molariform, differing from the molars only in having a more strongly developed anterior cingulum. On P^4 and succeeding teeth, the hypocone is deeply separated from the protocone by a lingual sulcus, connecting with the latter cusp only in very advanced stages of wear. The coronal pattern of M$^{1-2}$ (fig. 18.2) is as described for the posterior premolars. On M^3 the metaconule is a large cusp, but it remains unconnected to the protocone. Although a faint bulge is present at the lingual extremity of the postcingulum, M^3 lacks any hint of a hypocone, unlike the condition seen in *Prolicaphrium, Licaphrium, "Proterotherium," Thoatherium,* and *Brachytherium.*

DM$_1$ is a small, lanceolate tooth with lingually concave crests descending from the protoconid; there is no development of other cusps or a talonid. DM$_2$ is much larger than dM$_1$. A small paraconid is developed at the anterior terminus of the anterior trigonid crest; two crests descend posteriorly from the protoconid, the labial of which forms a small talonid crescent. DM$_3$, the longest of the lower deciduous teeth (table 18.2), is bicrescentic, has a broader talonid than trigonid, and bears a distinct paraconid. DM$_4$ resembles the molars, except that it is longer relative to width and retains a small paraconid. P$_{1-2}$ (fig. 18.3A, B) resemble their deciduous predecessors (except that they are larger) and imbricate slightly, as in *Prolicaphrium* (Soria 1981). P$_{3-4}$ (fig. 18.3C, D) resemble the corresponding teeth of *Licaphrium, "Proterotherium,"* and *Prolicaphrium.* The lower molars (fig. 18.3E, F) are also highly similar to those of *Licaphrium* and *Prolicaphrium.* On M$_{2-3}$, the paralophid terminates in a median rather than lingual position, a point of difference from species of *Diadiaphorus, Thoatherium,* and some species of *"Proterotherium."* A small entoconid is present on M$_{1-2}$; on M$_3$, this cusp is large and distinctly separated from the hypoconulid. The hypoconulid projects as a third lobe on M$_3$, as it does in most Proterotheriidae. Measurements are given in table 18.2.

Villarroelia totoyoi is obviously distinctive in its advanced molarization of the posterior upper premolars, particularly P^3. The affinities of the species are enigmatic. In most other respects *Villarroelia* is dentally similar to *Licaphrium* and *Prolicaphrium.* Because character distribution and polarity among Proterotheriinae are not well understood, the significance of these similarities is unclear. We interpret most such characters in *Villarroelia* (e.g., the presence of paracone and metacone ribs on the ectolophs of upper molars and the lack of *Thoatherium*- or *"Proterotherium"*-like relationships of metaconule to protocone) as primitive retentions, which merely serves to underscore the problems of interpreting the relationships of the Colombian genus.

Proterotheriidae, gen. et sp. indet.

IGM 183233 (fig. 18.4A) is a right upper molar, probably M^2, of an unidentified proterotheriid. The specimen was collected at Duke Locality 33, which lies stratigraphically

Table 18.2. Dental measurements (lengths and widths, mm) of *Villarroelia totoyoi* gen. et sp. nov.

Specimen	dM³ L	dM³ W	dM⁴ L	dM⁴ W	P¹ L	P¹ W	P² L	P² W	P³ L	P³ W	P⁴ L	P⁴ W	M¹ L	M¹ W	M² L	M² W	M³ L	M³ W
IGM 183999	14.7	13.7	13.2	14.9	—	—	—	—	—	—	—	—	—	—	—	—	—	—
IGM 183407	—	—	—	—	—	—	—	—	—	—	15.9	22.9	—	—	15.5	19.8	13.3	20.8
GM-01	—	—	—	—	—	—	—	—	—	—	—	—	—	—	—	—	13.6	17.5
DGP 7V89-2	—	—	—	—	—	—	—	—	13.3	19.1	13.9	20.4	14.5	19.9	—	—	—	—
UCMP 39970	—	—	—	—	—	—	—	—	—	—	—	—	14.6	20.6	15.6	19.8	—	—
IGM 250965	—	—	—	—	—	—	12.3	15.5	13.9	19.7	15.0	21.5	13.8	20.4	15.1	20.9	13.5	21.1
DGP-31	—	—	—	—	—	—	—	—	14.2	18.9	—	—	—	—	—	—	—	—
IGM 184051	—	—	—	—	12.7	9.7	13.1	13.0	13.9	18.1	14.7	20.0	—	—	14.9	21.3	13.6	20.9

Specimen	dM₁ L	dM₁ W	dM₂ L	dM₂ W	dM₃ L	dM₃ W	dM₄ L	dM₄ W	P₁ L	P₁ W	P₂ L	P₂ W	P₃ L	P₃ W	P₄ L	P₄ W	M₁ L	M₁ W	M₂ L	M₂ W	M₃ L	M₃ W
IGM 183999	9.5	4.3	13.8	6.2	17.1	9.1	14.9	9.7	14.3	—	—	—	—	—	—	—	14.3	10.0	—	—	—	—
IGM 184661	—	—	—	—	—	—	—	—	—	—	—	—	—	—	—	—	—	—	14.2	9.9	18.1	9.1
IGM 250956	—	—	—	—	—	—	—	—	—	—	—	—	—	—	—	—	—	—	15.7	11.0	—	—
IGM 184653	—	—	—	—	—	—	—	—	10.2	4.5	13.8	6.8	—	—	—	—	13.6	10.1	14.9	10.3	17.9	9.6
DGP-01	—	—	—	—	—	—	—	—	—	—	—	—	—	—	13.9	10.5	13.5	10.5	—	—	16.2	9.8
IGM 183998	—	—	—	—	18.3	10.2	15.1	12.1	—	—	—	—	—	—	—	—	13.9	9.9	—	—	—	—
IGM 250502	—	—	—	—	—	—	—	—	—	—	—	—	—	—	14.0	11.8	—	—	—	—	—	—
IGM 184501	—	—	—	—	—	—	—	—	—	—	—	—	15.5	9.7	—	—	—	—	—	—	—	—
IGM 184036	—	—	—	—	—	—	—	—	—	—	—	—	—	—	12.8	10.1	—	—	—	—	—	—

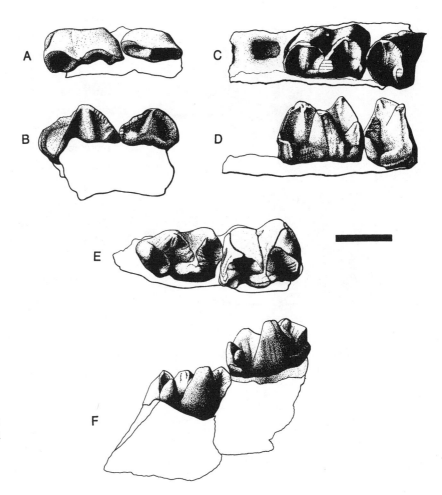

Figure 18.3. *Villarroelia totoyoi* gen. et
sp. nov. *A–B*. IGM 184653, left P$_{1-2}$:
A, occlusal view; *B,* lingual view. *C–D.* IGM
184501, right P$_3$: *C,* occlusal view; *D,* lingual
view. *E–F.* IGM 184653, left M$_{2-3}$:
E, occlusal view; *F,* lingual view. Scale bar
1 cm.

below the Chunchullo Beds in the lowermost La Victoria
Formation. The tooth is intermediate in size (preserved L =
9.7, W = 13.9) between comparable molars of the two
smallest proterotheriids otherwise known from the Honda
Group, *Prothoatherium colombianus* and *Prolicaphrium sanalfonsensis*. In both size and morphology the tooth cannot
belong to either of these taxa. IGM 183233 probably represents a new genus and species, but the available material is
insufficient to diagnose it satisfactorily. The tooth is low
crowned with weak cresting, features that we interpret as
primitive. The mesostyle is salient, but parastylar and, probably, metastylar spurs are weak (the posterolabial corner of
the tooth is missing). Both paracone and metacone ribs are
present on the labial face of the ectoloph. The paraconule
and metaconule are strong and subequal in development;
the latter cusp lacks attachment to either protocone or hypocone. The hypocone, although shorter than the pro-

tocone, is a tall, independent cusp separated from the protocone by a deep lingual sulcus.

Most characteristics of this puzzling specimen appear to
be primitive, being more reminiscent of Paleogene Litopterna, such as *Anisolambda* (see Simpson 1948; Cifelli
1983b), than Neogene taxa. We know of no comparable
proterotheriids from Deseadan or later faunas.

Family Macraucheniidae Gervais 1855
Theosodon Ameghino 1887
Theosodon, sp. indet.

REFERRED SPECIMENS

UCMP 38088, associated phalanx and fragment of right
maxilla with M$^{1-3}$; IGM 252921, right dM$_4$; IGM 184407

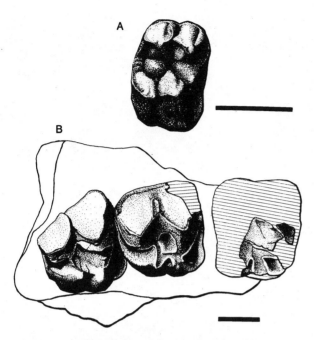

A

B

Figure 18.4. *A.* Proterotheriidae, gen. et sp. indet., IGM 183233, right M$^{?2}$, occlusal view. *B. Theosodon,* sp. indet., UCMP 38088, right M$^{1-3}$, occlusal view. Scale bars 1 cm.

and 184408, left P^4 and fragment of maxilla with left M$^{2-3}$; IGM 250458, left C^1 or I^3, left M^2 lacking the hypocone, and right M^1; IGM 183498, anterior half of M$_1$ and associated fragments of the skeleton; and IGM 183074, right calcaneus.

HORIZONS AND LOCALITIES

UCMP Locality V4534, between Tatacoa Sandstone Beds and Chunchullo Sandstone Beds, La Victoria Formation; unnumbered locality, Fish Bed, Baraya Member, Villavieja Formation; Duke Locality 106, between the Cerbatana Conglomerate and the Tatacoa Sandstone Beds, La Victoria Formation; and Duke Locality 91, below the Chunchullo Sandstone Beds, lowest La Victoria Formation.

DISCUSSION

Theosodon is surely present in the Honda Group, but, owing to the general disrepair of macraucheniid systematics, the species cannot be surely identified or established as new. *Theosodon* has been reported from Colhuehuapian (Simpson 1935; Soria 1981) through Chasicoan (Pascual 1966)

rocks of South America. The Colhuehuapian species, ?*T. frenguellii* Soria 1981, is distinct from all well-established Santacrucian species; its status with respect to poorly represented taxa, such as *T. karaikensis,* is uncertain. Scott (1910) listed a somewhat unlikely seven species from the Santacrucian; of these, the status of *T. fontanae, T. patagonicus,* and *T. karaikensis,* at least, is uncertain, either because their fossil representation is poor (*T. karaikensis,* for instance, is apparently known only by one specimen each of M^3 and M$_3$; cf. Ameghino 1904) or their distinctness not well established (e.g., *T. fontanae;* cf. Scott 1910, 151). Another species of *Theosodon* was added to the roster from the Santacrucian by Kraglievich and Parodi (1931). In the Friasian, Marshall et al. (1983) list both *Theosodon* and *Phoenixauchenia.* The presence of the former genus in this Land Mammal Age is based on Kraglievich's (1930a) assignment of a specimen from the Río Frías to *T.* aff. *gracilis,* an otherwise Santacrucian species. *Phoenixauchenia* (monotypic; the type is *P. tehuelcha*) was described by Ameghino (1904) on the basis of an astragalus, said to be smaller than that of *T. gracilis,* from the Río Fénix. Kraglievich (1930a) listed ?*P. tehuelcha* from the Río Fénix and Laguna Blanca fauna, based on a calcaneus reported to be smaller than that of *T. gracilis.* As Kraglievich noted, the status of *P. tehuelcha* is itself open to debate. The last reported occurrence of *Theosodon* is in the Chasicoan (Pascual 1966), based on the description of *T. hystatus* from the Arroyo Chasicó Formation (Cabrera and Kraglievich 1931). No figures were published of specimens belonging to this species, the type of which is a mandibular ramus bearing teeth in advanced wear. Cabrera and Kraglievich (1931) also described *Cullinia levis,* said to be *Theosodon*-like except that it lacks P$_1$, from the Chasicoan; this genus has been included in some compilations (e.g., Simpson 1945) but omitted from more recent syntheses (Pascual 1966; Marshall et al. 1983).

Under the forgoing circumstances identification of the Colombian taxon is somewhat less than straightforward. *Theosodon* sp. from the Honda Group differs from Colhuehuapian ?*T. frenguellii* in having stronger labial cingula on the lower molars and is notably smaller than Santacrucian *T. lallemanti, T. fontanae, T. lydekkeri, T. pozzii,* or *T. patagonicus.* The Colombian species is also probably smaller than Chasicoan *T. hystatus.* It is similar to *T. garretorum,* especially resembling that species in the relatively reduced

Table 18.3. Dental measurements (lengths and widths, mm) of *Theosodon*, **sp. indet.**

Specimen	P^4 L	P^4 W	M^1 L	M^1 W	M^2 L	M^2 W	M^3 L	M^3 W	M_1 L	M_1 W
UCMP 38088	—	—	21.8	21.5	21.3	19.6	19.0	18.1	—	—
IGM 250458	—	—	—	—	18.8	19.3	—	—	—	—
IGM 184407	—	—	—	—	—	19.8	17.6	18.8	—	—
IGM 184408	15.2	16.0	—	—	—	—	—	—	—	—
IGM 252921	—	—	—	—	—	—	—	—	22.1	11.5

M^3 (fig. 18.4B), but appears to be smaller than that Santacrucian species as well. IGM 252921, a dM_4 of the species from the Honda Group, bears a small crest descending posteriorly from the protoconid. This feature is, to our knowledge, not found in other species of *Theosodon* for which the tooth is known, but in the absence of a sample documenting intraspecific variation, we are unable to evaluate its significance. Noting especially the unusual characteristics of dM_4, we suspect that a hitherto undescribed species is represented by specimens from the La Victoria and Villavieja formations. Based on published descriptions, measurements, and illustrations, however, the Colombian species cannot at present be distinguished from *Theosodon gracilis, T. karaikensis, "Cullinia" levis,* or *"Phoenixauchenia" tehuelche.* Measurements are given in table 18.3.

A peculiarity of *Theosodon* that has not received attention, to our knowledge, lies in the sequence of premolar eruption. P4 in Santacrucian *Theosodon* (e.g., FMNH P13175, assigned to *T. garretorum*) is always more worn than the preceding tooth, P3; some specimens (e.g., FMNH P13292) suggest that P4 erupted before P3 and may, in fact, have been the first of the series to erupt. Such an eruption pattern was first observed by Scott (1910, 112), who remarked that "in one individual with milk-dentition still complete, the fully formed, but as yet rootless, crown of P^4 is nearly ready to displace dP^4, and it, therefore, seems likely that that tooth is the first of the second series to erupt." Judged by figures and available specimens, this pattern applies not only to species of *Theosodon* (including Col-

huehuapian *T. frenguellii*) but also to *Cramauchenia* and, possibly, *Macrauchenia.* Adianthids, like proterotheriids, retain the sequence presumed to be primitive (with P4 being the last of the premolar series to erupt). Premolar eruption sequence may prove to be a synapomorphy of or within Macraucheniidae when distribution of the character becomes better understood.

Discussion

With few exceptions, post-Eocene Litopterna were morphologically anachronistic with respect to other herbivores (such as Notoungulata) on the South American continent, being distinguished by their generally brachydont cheek teeth. As is well known (e.g., Patterson and Pascual 1972), the record of South American fossil mammals stems mainly from the southern extreme of the continent. Santacrucian faunas and floras of Patagonia, which slightly predate deposition of the Honda Group, suggest the presence of a remarkably diverse habitat, including well-watered woodlands and well-vegetated, open park savanna (Webb 1978). Although conditions in Patagonia during the preceding Colhuehuapian appear to have been somewhat wetter than during the Santacrucian (Webb 1978), succeeding ages were characterized by progressive aridification through the remainder of the Miocene and Pliocene with the establishment of treeless savannas and, ultimately, desert grasslands (Pascual and Odreman Rivas 1971). Because litopterns, particularly Proterotheriidae, retained brachydont teeth generally associated with browsing and with more closed habitats, it might be predicted that their distribution and success would be strongly influenced by such changes in environment and habitat.

Although revision of Neogene Litopterna is urgently needed in order to gain a true appreciation of their diversity, there can be no doubt that, in Patagonia at least, Proterotheriidae flourished and enjoyed a considerable radiation during the Santacrucian. By contrast, Litopterna from the preceding Colhuehuapian age were neither diverse nor abundant and were morphologically closer to Paleogene taxa than to their Miocene successors (Soria 1981). Litoptern diversity and abundance, particularly that of Proterotheriidae, experienced a strong decline at high latitudes subsequent to the Santacrucian (Webb 1978). Marked climatic zonation characterized the continent by this time, however, and this appears to have affected litoptern distribution as well as that of many other terrestrial mammals (Pascual et al. 1985) and of the flora (Menendez 1971). In the Mesopotamian (Huayquerian) fauna of the Entre Ríos Formation, for instance, Bianchini and Bianchini (1971) record a peculiar mixture of litoptern taxa. The survival of relictual taxa such as *Licaphrium* and other genera typical of the Santacrucian of Patagonia suggests that wetter, more moderate conditions persisted in northern Argentina, as they do today, well after the onset of aridification in Patagonia.

The Litopterna of the Honda Group are of some interest in this regard, insofar as they date from 13.5 to 11.8 Ma (chap. 3), approximately contemporaneous with (or slightly younger than) Friasian faunas of Patagonia (Marshall et al. 1977) and, therefore, postdating the climatic and faunal disruption that followed the Santacrucian in Patagonia (Bondesio et al. 1980). The macraucheniid *Theosodon* was widely distributed geographically and temporally. Its presence in the Honda Group represents, as far as we are aware, a substantial northward range extension (into the Tropics), but little more can be said of this occurrence. Of the Proterotheriidae and proterotheriid-like taxa, most are morphologically and/or taxonomically archaic. *Prothoatherium* and *Prolicaphrium* are genera otherwise restricted to the Colhuehuapian (Soria 1981); *Megadolodus* and the unidentified taxon bear strongest structural similarity to truly archaic forms not recorded in Patagonia after the Eocene.

Villarroelia, known only from the Honda Group of Colombia, is advanced in its strongly molariform posterior upper premolars, although it is otherwise similar to taxa of the Colhuehuapian and Santacrucian. *Villarroelia* resembles both *Prolicaphrium sanalfonsensis* and *Epitherium* in lacking specializations associated with either of the coronal patterns seen in advanced proterotheriids (Bianchini and Bianchini 1971). It is possible that these three taxa are related, but such a grouping remains to be documented by unambigous synapomorphy.

Even within the context of the primitiveness general to the family Proterotheriidae, most of the Honda Group proterotheres are notable for their extremely low-crowned cheek teeth; in the case of two taxa (*Megadolodus molariformis* and *Prothoatherium colombianus*), at least, the cheek teeth are arguably bunodont, with their crowns consisting simply of low cusps connected by poorly defined crests. Crown height of ungulate cheek teeth has long been assumed to be related to diet, with relatively higher crowns characterizing species that ingest greater proportions of grass (e.g., Simpson 1950). Recent work indicates that crown height is related to a number of factors associated with dental wear (Fortelius 1985); Janis (1988) has suggested that habitat preference may be more important than dietary preference in determining the degree of hypsodonty. Although simple indices of hypsodonty can be misleading when comparisons are made between taxa with differing types of molar construction (Webb 1983), Janis (1988) was able to categorize most living ungulates to dietary and/or habitat preference groups, based on a hypsodonty index (HI; height divided by width) for M_3.

The confused state of proterotheriid systematics and the nature of available samples preclude rigorous morphometric treatment; however, qualitative comparison with the data presented by Janis (1988) reveals some points of interest. Table 18.4 lists hypsodonty indices (based on measurements of unworn or slightly worn teeth) for proterotheriids from the Honda Group and, where available, Patagonia, were calculated following the methods described by Janis (1988). We emphasize that these data are preliminary (based in some cases on published illustrations and in others on poorly preserved specimens) and should be taken as suggestive rather than indicative. As with the systematics section of this chapter, we hope this work will help

Table 18.4. Hypsodonty indices of Proterotheriidae from the Honda Group, compared to those of selected Santacrucian and Colhuehuapian taxa

Taxon	Specimen	HI
Friasian, Colombia		
Megadolodus molariformis	IGM 250400	0.57
Prothoatherium colombianus	IGM 183116	0.89
Prolicaphrium sanalfonsensis	IGM 182852	0.79
Villarroelia totoyoi	IGM 184653	1.14
Santacrucian, Patagonia		
"*Proterotherium*" *mixtum*	AMNH unnumbered	1.14
"*Proterotherium*" sp. indet.	FMNH P13191	1.11
	FMNH P13189	1.16
Thoatherium miniusculum	YPM-PU 15719	1.00[a]
Licaphrium proximum	YPM-PU unnumbered	1.11[a]
Colhuehuapian, Patagonia		
Prolicaphrium specillatum	FMNH P13306	0.88
Prothoatherium lacerum	MACN A52-247	0.83

Notes: Hypsodonty index, HI = M_3 height divided by width. Based on data from 127 living species, Janis (1988, 384) found the following distribution of hypsodonty indices among dietary/habitat preference groups (observed ranges given, with means in parentheses): grazers, 1.7–8.7 (5.18); mixed feeding in open habitat, 1.5–5.3 (3.80); fresh grass grazers, 1.9–4.1 (3.2); selective browsers, 1.2–3.5 (2.06); mixed feeding in closed habitats, 1.1–3.0 (2.05); regular browsers, 1.2–2.3 (1.68); high-level browsers, 0.8–2.2 (1.31); omnivores, 1.0–1.3 (1.16).

[a] Calculated from illustrations and data in Scott (1910).

to stimulate further study of the evolutionary history of Litopterna.

As shown in table 18.4, cheek teeth of all Honda Group taxa are relatively low crowned, easily fitting within the brachydont category (HI < 1.5) as defined by Janis (1988). Of living ungulates, the lowest-crowned cheek teeth are found among omnivorous suoids. *Megadolodus molariformis* may have had a similar dietary preference, although it has notably lower-crowned cheek teeth than any living ungulate. Postcranial anatomy suggests that, like most living suoids, *Megadolodus* was a forest-dweller (chap. 17). The extremely low-crowned, bunodont teeth with thick enamel, coupled with a massive jaw, suggest that fruits (particularly hard fruits) may have been a significant part of diet in *Megadolodus* (cf. Janzen and Martin 1982). This speculation might be tested through study of dental microwear features. *Prothoatherium colombianus* and *Prolicaphrium sanalfonsensis* also had extremely low-crowned, bunodont or sub-

bunodont cheek teeth. Based on the data presented by Janis (1988), these species may have had an omnivorous, frugivorous, or browsing diet (see also Cifelli and Guerrero 1989). Although habitat preference cannot be directly established, no living ungulate found in open environs has cheek teeth anywhere near as low crowned, regardless of diet. We infer that *Prothoatherium colombianus* and *Prolicaphrium sanalfonsensis* preferred a forested habitat. Although comparable data are not available for the unidentified proterotheriid from the Honda Group, its upper molar is similar in evolutionary "grade" to those of *P. colombianus* and, in fact, is somewhat lower crowned still. A similar habitat preference may be inferred for this species.

Villarroelia totoyoi, by far the largest of Colombia's Miocene Proterotheriidae, has more specialized, higher-crowned cheek teeth than the preceding taxa. By comparison to Recent ungulates, the HI of *V. totoyoi* (1.14) is compatible with those of either mixed feeders in closed habitats, high-level browsers, or omnivores, being slightly less than the observed ranges for "regular" browsers or selective browsers (categories as defined by Janis 1988). Although the postcranial skeleton of *Villarroelia totoyoi* is poorly known, it is similar in known respects to those of Santacrucian Proterotheriidae, which lack specializations for high-level browsing. Despite their low crowns, the lophodont cheek teeth of *V. totoyoi* contrast with the more bunodont teeth of living ungulate omnivores, and we therefore suggest that this species was, most likely, a mixed feeder in closed habitats. The strong molarization of the posterior upper premolars in *V. totoyoi* represents continuation of a general trend in the evolution of proterotheriids, which early in their history acquired some molariform characteristics (Cifelli 1983b). Molarization of the premolars effectively increases the volume of the cheek tooth battery and is a specialization seen especially in Recent hindgut fermenters (Janis 1988). We speculate that proterotheriids were hindgut fermenters and that the enlarged premolars of *Villarroelia* may be related to ingestion of a generally coarser diet than that of other Colombian proterotheres, perhaps enabling the species to better exploit local grass resources. Nonetheless, the HI for *V. totoyoi* is far lower than that of any living ungulate whose diet is primarily grass.

In correlation with the aforementioned aridification of

Patagonia during the Miocene and Pliocene, available materials of Santacrucian proterotheriids indicate that hypsodonty was greater than in the preceding Colhuehuapian (we suspect that later proterotheriids from Patagonia, particularly those of Friasian age, will show that this trend continued and increased). The Colombian taxa are comparable to those from the Colhuehuapian (of Patagonia) in hypsodonty; with the exception of *Villarroelia,* they are noticeably lower crowned than the sample of Santacrucian taxa (note that, because suitable materials were unavailable, the highest crowned of Santacrucian genera, *Diadiaphorus,* is omitted from table 18.4). Based on relative height of M_3, all Santacrucian proterotheres could have been mixed feeders in closed habitats, high-level browsers, or omnivores. As with *Villarroelia,* it is unlikely than any of the Santacrucian species was a high-level browser. Although it is also unlikely that any of the Santacrucian proterotheres preferred open habitat or was committed to a dietary regimen of grazing, their generally higher-crowned cheek teeth suggest either a coarser diet, somewhat more open habitat, or both, than the Colombian species.

Acknowledgments

We thank C. Villarroel, D. Savage, M. McKenna, and J. Flynn for permission to study specimens in their care, and C. Janis for helpful comments on an earlier draft of the manuscript. Partial support for this research was provided by NGS grants 2964-84 and 3292-86, and by NSF grants BSR8614533 and BSR 8918657 to R. F. Kay.

19
A New Leontiniid Notoungulate

Carlos Villarroel and Jane Colwell Danis

RESUMEN

Huilatherium pluriplicatum (Leontiniidae, Notoungulata) del Mioceno medio del Grupo Honda en Colombia fue originalmente descrito en base a un espécimen juvenil. Se presenta una definición revisada y una descripción más completa del cráneo, mandíbula, dentición, y elementos de pies y manos, basadas en el estudio de las colecciones del Museo de Paleontología de la Universidad de California, y de gran cantidad de nuevo material recientemente descubierto.

Las características diagnósticas incluyen $I^{2?}$ e I_3 procumbentes y alargados, dentición anterior reducida, molariformes hipsodontes, molariformes superiores con cuencas rodeadas de esmalte, y un hipolofúlido alargado en los molares inferiores.

Un análisis filogenético de 23 caracteres dentales revela que *Huilatherium* (del Mioceno medio de Colombia) y *Taubatherium* (del Deseadense de Brasil) comparten colmillos hipertrofiados, depresiones de los premolares superiores con cuencas dobles incompletamente separadas, pérdida del cíngulo anterolingual de los molares superiores, y pérdida del cíngulo labial de los molares inferiores. Algún grado de diferenciación filogenética y gran parte de la historia de la familia Leontiniidae pudo haber ocurrido en Suramérica tropical.

The family Leontiniidae is a group of medium to large herbivorous toxodonts. The family attained its greatest diversity during the Deseadan (Oligocene) in the southern part of South America. Of the seven known genera, only two survived beyond the Oligocene: *Colpodon* known from the Colhuehuapian (early to middle Miocene) and *Huilatherium* from the middle Miocene Honda Group in the upper valley of the Magdalena River, Colombia.

Around 1920, Brother Ariste Joseph (Mauricio Rollot), of the Brothers of the Christian Schools and Institute of La Salle in Bogotá, recovered the first fossil leontiniid from Colombia (which he mistakenly identified as an astrapothere), from Tolima Department. That specimen, a fragment of the left maxilla with M^{1-2} of *Huilatherium pluriplicatum*, is now in the collection of the Smithsonian Institution (USNM 414735). In the late 1930s, George R. Hyle discovered two upper premolars of leontiniid while mapping the geology of the area west of the Magdalena River in the vicinity of Santa Elena between Neiva and the Municipality of Aipe. The specimens were later identified by R. A. Stirton. Years later, in 1940, José Royo y Gómez, traveling through the Tatacoa Desert between Cerro Gordo and the Quebrada Chunchullo, collected some fossils, among them the left P^4 of a notoungulate, which he mentioned and illustrated in 1942 (fig. 5.2 in Royo y Gómez 1942b) and again in 1945. This P^4 can also be referred to *Huilatherium pluriplicatum*.

Stirton (1953b) reported the presence of Leontiniidae in two distinct faunal assemblages in the upper valley of the Magdalena, in the La Venta fauna, and in the Coyaima fauna attributed by him to the late Oligocene. C. L. Gazin communicated to Stirton the existence of two upper molars of leontiniid from Coyaima (in Stirton 1953b). As discussed here, the "Coyaima" leontiniid mentioned by Stirton is probably the previously mentioned specimen recovered by Brother Ariste Joseph (USNM 414735).

The most complete description of the leontiniid from the La Venta fauna is that of Colwell (1965), a master's thesis in paleontology, University of California, Berkeley. In this unpublished thesis, which constitutes, in large part, the basis of the present work, Colwell proposed *"Laventatherium hylei"* as a new genus and species of leontiniid. Colwell's thesis remained unpublished, however, and in later revisions of the La Venta fauna (Hirshfeld and Marshall 1976; Marshall et al. 1983) the presence of leontiniids was reported without further details.

Villarroel and Guerrero (1985), without knowing of Colwell's study, proposed a new genus and species of leontiniid, *Huilatherium pluriplicatum,* based upon a fragment of the left maxilla from the La Venta area. New collections from this area, made by a team under the direction of R. F. Kay, through a cooperative agreement between INGEO-MINAS and Duke University, led to the discovery of material that demonstrates irrefutably that the type and, at that time, the only known specimen of *Huilatherium pluriplicatum* correspond to a juvenile with dM^3 and dM^4. Based upon all the available collections, only a single monospecific genus of leontiniid occurs in the Honda Group.

Systematic Paleontology

Subclass Eutheria
Order Notoungulata Roth 1903
Suborder Toxodontia Owen 1858 (see Mones 1988)
Family Leontiniidae Ameghino 1895
Huilatherium Villarroel and
Guerrero 1985, p. 36

SYNONYMY

= *"Laventatherium"* Colwell, thesis, University of California, Berkeley, 1965, p. 10.

TYPE SPECIES

Huilatherium pluriplicatum Villarroel and Guerrero 1985.

EMENDED DIAGNOSIS

Dental formula 1.0.4.3/1.0.3.3; $I^{2?}$ and I_3 enlarged, procumbent; cingulum on incisors lacking; $I^{1?}$, I^3, I_1, I_2, and canine lacking; conspicuous incisor-premolar diastema, even in individuals with deciduous dentition; upper and lower premolars and molars hypsodont; premolars progressively molariform from P^{1-4}/P_{2-4}; small accessory ridge parallel to parastyle on P^3; presence of single basin on anterior surface at lingual corner (leontiniid depression) on P^{2-4}; posterior arm of hypofossette on M^{1-2} well developed; mesial (second) crista on M^{1-2} well developed; first crista weakly developed; enamel-lined pits constant and numerous in upper premolars and molars; protoconid well developed, paraconid and metaconid weakly developed on P_{2-4}; paraflexid progressively developed from M_{1-3}; postfossettid U-shaped; hypolophulid lobe lengthened on M_{1-3}. Juvenile dM^3 and dM^4 are molariform and mesodont, possess well differentiated and long roots, weak protoloph, deep postfossette positioned mesiolingually; lingual margin of the mesial crista with several small plications, few enamel-lined pits; lingual fold separating protocone and hypocone very deep; mesolingual cingulum weak and restricted to the lingual one-half; internal cingulum lacking; dM_1 small; dM_2 smaller but similar to P_2, its cingula are more developed; dM_4 molariform with crown structure similar to M^1 but with a shorter trigonid and well-developed lingual cingulum.

NOTE OF CLARIFICATION

The original diagnosis proposed by Villarroel and Guerrero (1985) was based upon a fragment of maxilla from an individual thought to be adult. With the discovery of new material, especially IGM 151144, it can be shown that the type specimen corresponds to a juvenile individual in which dM^3 and dM^4 are still in place. Based upon this new evidence, the original diagnosis is complemented by material with the permanent dentition. This discovery also permits us to refer *Huilatherium* to the family Leontiniidae without any of the doubt expressed in the original publication.

Huilatherium pluriplicatum Villarroel and Guerrero 1985

SYNONYMY

= *"Laventatherium hylei"* Colwell, thesis, University of California, Berkeley, 1965, p. 10.

HOLOTYPE

DGP ICN-P-328, a fragment of the left maxilla with dM³ and dM⁴. The specimen is now housed at the Departamento de Geociencias, Universidad Nacional de Colombia, Santa Fe de Bogotá.

TYPE LOCALITY

Tatacoa Desert, northeast of Villavieja, Huila Department, Colombia.

STRATIGRAPHIC HORIZON AND AGE

The extensive collection of specimens of *H. pluriplicatum* come from the La Victoria Formation and the Villavieja Formation, Honda Group (see Guerrero 1990; also chap. 2). The available collections show that the species occurs from the base of the La Victoria Formation to the top of the Baraya Member of the Villavieja Formation. Leontiniids have not been reported in the Cerro Colorado Member of the Villavieja Formation.

The only reference to leontiniids outside the area of the Tatacoa Desert in Colombia is that of Stirton (1953b, 611), who reported "two upper molars in United States National Museum" that according to information from C. L. Gazin came from the area of Coyaima (Tolima Department), 60 km north of the Tatacoa. Stirton included this specimen in the Coyaima fauna. This specimen (of which we have seen a drawing in occlusal view made by R. Madden) can be referred to *Huilatherium pluriplicatum* of the La Venta fauna.

HYPODIGM

The type and the following material: UCMP 39961, a nearly complete cranium, slightly distorted. Palate (partially restored in plaster) with alveoli of right and left I² and P¹; crowns of right and left P² mostly destroyed; left P³–M¹ partly restored and M²⁻³ complete; palate and postpalatine areas essentially complete. UCMP 40276, right and left mandibles broken off posterior to M₃; most of left ramus and symphysis restored; right I₃ and P₃–M₃, left I₃ and P₂–M₃ complete; right P₁ and part of left P₁ isolated. UCMP 40042, left M². UCMP 40278, left I²? complete, base of crown and part of root of right I²?; left M₁; fragments of upper molars and premolars; fragment of mandible with roots of M₂ and M₃; left metacarpal II; ?distal end and part of proximal end of left metacarpal IV; ?distal end of metacarpal IV; ?proximal end of metacarpal IV; proximal phalanx digit II; distal phalanx digit II; proximal phalanx digit III; proximal phalanx digit IV; median phalanx digit IV; five podials (indeterminate); one caudal vertebra; end of a second caudal vertebra; two sesamoids. UCMP 39977, parts of badly fractured right and left mandibles with right P₂–M₂ and left P₂₋₄ (teeth except P₂ and P₃ heavily worn and fractured). UCMP 38052, left metatarsal IV. All above-mentioned specimens from UCMP Leontiniid Locality V4525 (for more information about UCMP localities, see Fields 1959).

UCMP 34288, left P³ and P⁴ moderately worn, from the Santa Elena Locality V3902. UCMP 38949, five upper molars and premolars and parts of two lower incisors, minutely fractured and most encrusted with calcium carbonate, from Cerro Gordo Locality V4530. UCMP 40280, right P¹, parts of two lower incisors, right and left P₃, right M₃, left P₄–M₃, from Cerro Gordo Locality V4532. UCMP 39636, right and left M² in early stages of wear, from the Railway Bridge Locality V4932. UCMP 39634, left P³–M³, left P₄–M₃ well worn with parts missing; right P₄; UCMP 39272, right P⁴, part of ectoloph of M², most of M³, left M²⁻³, right M₁₋₃, left P₂ and M₂, two proximal phalanges; both specimens from the San Nicolas Locality V4937. UCMP 38096, left mandible with symphysis, roots of incisors in place; P₂–M₃, molars partly broken away; and UCMP 37907, left M³; both from the Monkey Locality V4517. UCMP 38966, parts of left P³, part of M¹, part of left M_2; UCMP 37938, lingual half of right M^2; UCMP 37988, parts of left and right M^2; all three specimens from the Toxodont Locality V4519. UCMP 37989, lower incisors, left M^2, and right P_4 surface leached, well-preserved left P_2 and P_4; UCMP 38010, left M_3 surface leached; both from the "Unit between Fish Bed and Monkey Unit," UCMP Locality V4522. UCMP 38043, right M_2, podial, from the "Unit above Fish Bed," UCMP Locality V4524. UCMP 38229, right mandible with P_2–M_3 and most of horizontal and ascending ramus and condyle; UCMP 38065, left mandible with P_2–M_3, most of horizontal and ascending ramus restored, but condyle preserved, right M^2; and UCMP 40063, fragment of left mandible with M_2 in place; all three specimens from the "Unit between Upper and Lower Red Beds," UCMP Locality V4527. UCMP 38230, left mandible with base of ascending ramus (partly restored), P_3–M_3 weathered; right mandible with root of I_3, P_3–M_3; also UCMP 38916, lingual half of left P^4, and UCMP 38884, right P^2, anterolabial corner of M^3, anterior part of right M^3; all three specimens from the "Upper Red Beds," UCMP Locality V4528.

IGM 182755, edentulous ramus, from Duke Locality 14. IGM 182903, anteroexternal portion of left $M^{1?}$, from Duke Locality 24. IGM 182983, left $M_{2?}$ without the distal extremity of the roots; IGM 183052, right $P_{3?}$ lacking the distal extremities of both roots, moderate wear; and IGM 250901, right M^3, little worn; all three specimens from Duke Locality 28. IGM 184269, unworn right P_2, well preserved; IGM 184596, left P^1, unworn, lacking the distal extremity of the root; both from the Tatacoa Desert, locality unknown. IGM 184287, right $M^{2?}$, enamelless crown, from Duke Locality 6. IGM 184432, right $M_{2?}$, well-worn crown, from Duke Locality 92. IGM 184540, right P^4, well-worn and anteroposteriorly compressed, from Duke Locality 96. IGM 184514, left P^1 with moderate wear, lacking the distal portion of the root, from Duke Locality 97. IGM 184616, partial left $M_{1?}$, trigonid and anterior portion, from Duke Locality 102. IGM 250372, right P_2 well preserved and little worn, from Duke Locality 11B. IGM 250448, $M_{2?}$ very worn and fractured, from Duke Locality 106. IGM 251144, partial right maxilla with dM^{3-4}, P^{3-4} encrypted, a fragment of horizontal ramus with a molar broken at the alveolus, and other associated poorly preserved bone fragment, from Duke Locality

138. IGM 251207, left I_3, from Duke Locality 132. IGM 251192, right P_3; IGM 251091, mandible; and IGM 251099, tooth fragments; all three specimens from Duke Locality 80. IGM 251181, tooth and tooth fragments, from Duke Locality 136. IGM without number, symphysis and left mandibular ramus with dM_2, dM_4 and M_1, juvenile individual from the La Venta area, locality unknown.

DGP LV-82-1, right P^2–M^3, little worn and well preserved, from the La Venta Red Beds. DGP LV-37-1, left P^2, little worn; DGP LV-16-VI-87-2, right P_4; DGP TATAC-47, partial right maxilla with P^4–M^3, well preserved and much worn; DGP TATAC-26, right P_2; DGP TATAC-45, posterior portion of the right horizontal ramus with M_{2-3}; DGP TATAC-50, left P_3; DGP TATAC-52, left P3; and DGP TATAC-53, right P_3; all eight specimens from the La Venta area, locality unknown. DGP 19-VII-88-1, $I^{2?}$, distal extremity of the crown; DGP 30-IV-85-1, two M_3, left and right sides, well preserved; both specimens collected northwest of Baraya.

USNM 414735, partial left maxilla with M^{1-2}. Although the loose specimen label written by Brother Ariste Joseph reads "Neiva, 1920, Huila Department," Stirton (1953b) reported that Charles Lewis Gazin claimed the specimen came from the area of Coyaima. A blue, circular label on the teeth reads "Astrapothere Tolima Dept." (Purdy, pers. comm.).

DIAGNOSIS

As for the genus.

COMPARATIVE DESCRIPTION

Cranium

The dorsal region of the single available cranium UCMP 39961 has not been preserved and the ventral portion is somewhat distorted by crushing (fig. 19.1). The bone is minutely cracked and fractured and the position of the sutures are obscure. In overall aspect the cranium is similar to those of *Leontinia* and *Scarrittia* but with a longer snout and less-expanded zygomatic arches than in *Ancylocoelus*.

Along the dorsal surface of the premaxillaries are thick rugose ridges that Loomis (1914) stated were also present in

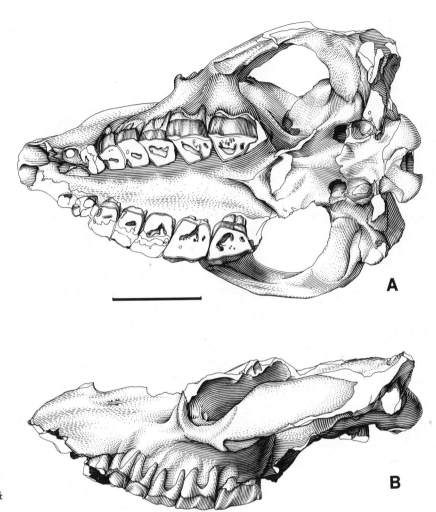

Figure 19.1. *Huilatherium pluriplicatum,* UCMP 39961, cranium, slightly distorted and partially restored: *A,* palatal view; *B,* left lateral view. Scale bar 10 cm.

Leontinia and appeared to continue dorsally in life as a cartilagenous nasal septum. The anteroposterior length of the nasal ridge is relatively longer than in *Leontinia* and considerably longer than in *Ancylocoelus* in which only a shortened ridge is developed. The premaxillary-maxillary boundary is near the anterior root of P⁴. The maxillary extends posteriorly to form the anterior part of the zygomatic arch. The major part of the broad heavy arch is formed by the jugal. Extending from the most anterior region of the zygomatic arch lateral to the infraorbital canal is a heavy flattened preorbital ridge that is better developed and extends more dorsoventrally than in *Leontinia*. It is similar in position to that in *Ancylocoelus* but is slightly thinner and less rounded. The most anterior part of the zygomatic arch is gently concave as in *Leontinia*. The ventral edge is more or less straight as in *Leontinia* and lacks the sharply convex anteroventral bulge seen in *Ancylocoelus*. The posterior part of the arch is broad and flattened. The infraorbital canal is relatively small and extends only about 20.0 mm internal to the preorbital ridge.

A large supraorbital foramen lies just posterior to the orbit slightly anterior to the median region of the zygomatic arch. Its exact shape and size are indeterminate because of crushing. The optic foramen and anterior lacerate foramen lie close together in the intraorbital region ventral and posterior to the supraorbital foramen and lateral to the probable pterygoid region. Both are round with the anterior lacerate foramen slightly posteromedial to the orbital foramen. The foramen rotundum, equal in size to the orbital and anterior lacerate foramina, lies posteromedial to them and medial to the most posterior region of the zygomatic arch. A "subsquamosal" foramen lies in the internal dorsal part of the most posterior region of the zygomatic arch.

The most anterior part of the palate is sharply concave and narrow. It becomes broader and more flattened as the tooth rows diverge. Rugose strips lateral to the internal nares indicate the probable position of reduced pterygoid, however, the bones themselves have been lost. Postero-lateral to the internal narial openings and just anterior to the tympanic bulla is a large oval foramen. The bone in the area of this foramen is thin and badly fractured; it is impossible to determine the homologies of this foramen precisely. A smaller channel that is assumed to be for the Eustachian tube occurs on the posterior surface. Because of the position of this foramen and of some identifiable foramina in the vicinity, it is assumed that the large opening may include the foramen ovale and the median lacerate foramen.

Only the basal part of the tympanic bullae is preserved. The basal portion is round, however, extensions of spongy bone interomedial to the bulla appear to represent an extension of the bulk of the bulla into an oval or pear-shaped form similar to the condition seen in *Leontinia*. The preserved portion of the bulla is separated into two chambers by a partition of bone. Just lateral to the median part of the bulla is a transverse slitlike stylomastoid foramen. Just posterior to the stylomastoid is the small round tympanohyal pit. Along the posteromedial edge of the bulla is the small posterior lacerate foramen. Posterior and medial to this foramen is the slightly larger hypoglossal foramen.

The paroccipital process lies posterior and somewhat lateral to the bulla. It has a broad triangular base and becomes progressively more slender toward the tip. The postglenoid process is not preserved; however, the base is slightly larger than that of the paroccipital process, and transversely it is oval to reniform. Set back under the overhang of the posteromedial part of the postglenoid process is the postglenoid foramen. Slightly posterior to the postglenoid foramen is the external auditory meatus. Because of crushing and fracturing, it is impossible to trace the canal structure in this area to compare it with *Leontinia* and *Scarrittia*. Median to the postglenoid foramen and the external auditory meatus beneath the transverse edge of the paroccipital

process is another foramen, but it is poorly preserved. The occipital condyles are similar to those in *Leontinia*.

Mandible

The mandible is deep and robust (fig. 19.2). The ventral surface is horizontal, rising sharply at the symphysis. The region housing the tooth row is of about equal depth throughout. A large mental foramen lies ventral to the anterior rooth of P_3 midway down the mandible. A smaller foramen lies dorsal and posterior to the mental foramen. The ascending ramus rises at a width equal to that of the tooth row, becoming narrow posteriorly. A deep fossa lies on the internal surface ventral to the anterior region of the ascending ramus. The condyle lies dorsal to the tooth row

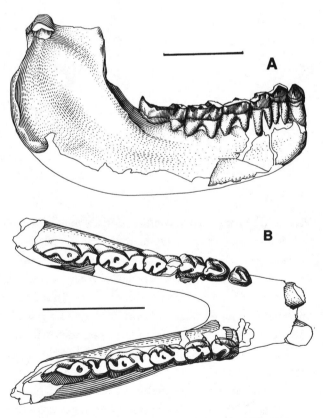

Figure 19.2. *Huilatherium pluriplicatum,* mandible. *A.* UCMP 38065, left mandible, with P_2–M_3, most of horizontal and ascending ramus restored, lateral view. *B.* UCMP 40276, right and left mandibles, most of left ramus and symphysis restored, occlusal view. Scale bars 10 cm.

and is long and dorsally flattened. None of the mandibular material is complete or well preserved enough to give accurate information on other processes.

Upper Teeth

Huilatherium possesses only a single pair of upper incisors I²?. I¹?, I³, and the upper canines are absent, leaving a conspicuous diastema between I²? and P¹. In *Leontinia, Scarrittia, Ancylocoelus, Colpodon,* and *Taubatherium* from Brazil, I¹⁻³ are present. In *Leontinia, Scarrittia,* and *Taubatherium,* the upper canine is also retained. No incisor-premolar diastem occurs in these genera.

In *Huilatherium*, I²? is known from three isolated teeth, UCMP 40278, UCMP 37989, and DGP 19-VII-88-1. They are large, procumbent, tusklike teeth that lack a cingulum. In *Leontinia*, I¹ and I³ possess a cingulum, and a cingulum is seen in all of the upper incisors in *Scarrittia* and *Ancylocoelus*. The absence of a cingulum in known specimens of *Colpodon* may be attributed to wear.

P¹ has a nearly circular basal outline; with a thick triangular root; a single cusp dominates the crown (fig. 19.3B). A high cingulum, set off from the main cusp by a deep trench, borders the cusp anteriorly. A second, weaker cingulum (the anterior cingulum of Patterson [1934b, 93, fig. 10A]) is seen posterolabially, and a third cingulum connects posterolingually to an oblique trenchant cusp (protocone of Patterson 1934b) in UCMP 40278. In UCMP 40278 and IGM 184596, a shallow basin, defined by a short lingual spur and the labial "metaloph" of Patterson (1934b), lies between the trenchant cusp and the posterior base of the main cusp. In *Leontinia* this basin opens freely into a lesser posterolabial depression.

P² has three roots; that of the paracone is the best developed. In occlusal view the outline of P² is ovoid and oriented transversely. The ectoloph is positioned longitudinally; the labial surface is very convex and near the base of the crown displays three or four accessory ridges that are independent of the robust and crenulate labial cingulum. On the anterior part of the crown of DGP LV-82-1 and DGP LV-37-1 can be seen a valley, delimited anteriorly by a weakly developed anterolingual cingulum and covered by a thin layer of enamel. This valley is the serial homologue of the leontiniid depression of P³ and P⁴. A mesial fossette is not present.

P³ and P⁴ are nearly rectangular in occlusal aspect, al-

Figure 19.3. Crown views of isolated teeth of *Huilatherium pluriplicatum. A,* UCMP 34288, left P³; *B,* UCMP 40278, left P¹; *C,* UCMP 40280, right M₃; *D,* UCMP 39636, left M². Scale bar 1 cm.

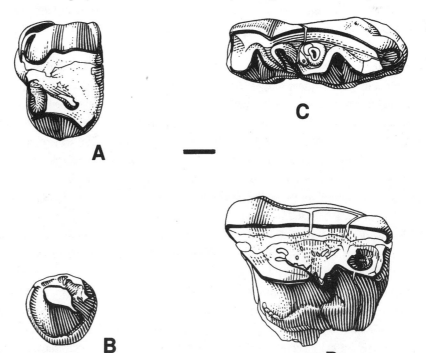

Table 19.1. Upper premolar measurements (lengths and widths, mm) for
Huilatherium pluriplicatum

Specimen	Side	P1		P2		P3		P4	
		L	W	L	W	L	W	L	W
IGM 184596	Left	21.0	23.0	—	—	—	—	—	—
IGM 184514	Left	20.0	23.0	—	—	—	—	—	—
DGP LV-82-1	Right	—	—	25.0	31.0	—	—	27.0	35.0
DGP LV-37-1	Left	—	—	24.0	28.0	27.0	35.0	—	—
DGP TATAC-47	Right	—	—	—	—	—	—	30.0	42.0
UCMP 39961	Right	—	—	—	—	28.0	38.0	28.0	43.0
	Left	—	—	—	—	27.0	39.0	29.0	44.0
UCMP 34288	Right	—	—	—	—	26.0	34.0	—	—

though the ectoloph is strongly curved lingually in each. In P3 the labial cingulum meets the parastyle and metastyle, and a small accessory ridge occurs on the parastyle parallel to and partly fused with it (UCMP 34288). The paracone and metacone ridges on the ectoloph tend to diminish toward the base of the crown. The protoloph and metaloph converge near the protocone to isolate a mesial fossette (fig. 19.3A). In early stages of wear (as in UCMP 34288 and DGP LV-82-1) the mesial fossette is prolonged anterolabially by a narrow anterolingual valley. The leontinid depression (Colwell 1965, 20), characteristic of the Leontiniidae, is separated from the mesial fossette by a weaker crest and is emarginated mesially by the anterolingual cingulum. A suggestion of a cingulum lingual to the protoloph occurs in UCMP 39634 but is not present in UCMP 39961 and DGP LV-82-1. The labial cingulum is poorly developed or absent in *Huilatherium* but is well developed in *Leontinia* (Patterson 1934b). Numerous small pits in the dentine of the occlusal surface of the P3 (especially the left P3 of UCMP 39961) may be homologous to the enamel-lined pits in the molars. The premolar pits lack enamel lining and occur on both sides of the mesial fossette.

P4 is larger than P3 (table 19.1), but the parastyle and metastyle are less prominent. The labial cingulum is reduced or absent, and a basal lingual cingulum is lacking. In P4 the mesial fossette and the leontiniid depression are more widely separated. Posterolingual to the mesial fossette is a labially prolonged sulcus in DGP LV-82-1 and DGP TATAC-47, which disappears with wear. In *Leontinia* a

cingulum descends to the lingual end of the metaloph (Patterson 1934b).

The upper molars differ from the premolars in that the molar ectoloph is relatively and absolutely longer, becoming progressively longer from M1 to M3 (table 19.2). The upper molars lack a labial cingulum. The leontinid depression, conspicuous anterolingually in the premolars, is absent in the molars. The relief of the anterolingual enamel border of the anterior arm of the hypoflexus is much higher than on P4 and becomes even more pronounced in well-worn M1–3.

In contrast to *Leontinia, Ancylocoelus,* and *Colpodon,* the mesial crista (second crista of Patterson 1934b) is well developed on the molars of *Huilatherium* (fig. 19.3D). Moreover, in this genus the anterior extremity of the anterior arm of the hypoflexus is forked. The anterolingual cingulum that is strongly developed in *Leontinia,* and present but less well pronounced in *Scarrittia* (Chaffee 1952), is absent in *Huilatherium*. In some specimens (DGP LV-82-1) is a spur that may be the equivalent of the pillar that separates the anterior and lingual cingulums at the anterolingual corner of the molars. The pillar is also seen in *Colpodon*.

The M1 of *Huilatherium,* in occlusal outline, differs from the premolars and is similar to M2 as in *Leontinia* and *Scarrittia*. The posterior arm of the hypoflexus is slightly longer than in *Leontinia* and *Scarrittia* and much longer than in *Ancylocoelus* and *Colpodon*. The occlusal depression between the median ridge and the outer enamel ridge of the ectoloph in *Leontinia* (Patterson 1934b) is absent in *Huila-*

Table 19.2. Upper molar measurements (mm) for _Huilatherium pluriplicatum_

Specimen	Side	M^1		M^2		M^3	
		L	W	L	W	L	W
USNM 414735	Left	40.0	37.0	47.0	38.0	—	—
UCMP 39961	Right	42.0	36.0	51.0	49.0	56.0	47.0
	Left	41.0	36.0	52.0	50.0	56.0	47.0
UCMP 37007	Right	—	—	50.0	47.0	—	—
UCMP 39636	Right	—	—	54.0	51.0	—	—
	Left	—	—	53.0	51.0	—	—
UCMP 39272	Right	—	—	51.0	42.0	54.0	44.0
	Left	—	—	—	—	54.0	44.0
UCMP 39634	Left	—	—	—	—	59.0	49.0
UCMP 40278	Left	36.0	37.0	—	—	—	—
DGP LV-82-1	Right	44.0	41.0	50.0	41.0	52.0	40.0
DGP TATAC-47	Right	43.0	41.0	52.0	44.0	58.0	43.0

therium. A postfossette occurs posterior to the tip of the posterior arm of the hypoflexus; in DGP LV-82-1 it is oval, and in DGP TATAC-47 it is U-shaped. The hypocone extends more lingually than on M$^{2-3}$, and, consequently, the posterior half of M^1 is relatively wider than in the last two molars. Three small pits occur in the dentine on the occlusal surface labial to the hypofossette on the right M^1 of UCMP 39961; these pits lack an enamel lining.

The ectoloph length of M^2 exceeds the transverse width of the tooth (fig. 19.3D). The parastyle is more prominent and more widely separated from the labial ridge opposite the paracone than in _Leontinia_. The hypoflexus opens lingually, and the anterior and posterior arms are separated by the mesial crista. The anterior arm of the hypoflexus is persistently forked (UCMP 39636, DGP LV-82-1, DGP TATAC-47) and is distally threadlike; proximally, its labial edge is convolute. The posterior arm of the hypofossette is open lingually; the protoloph curves abruptly posteriorly and extends parallel to the enamel edge of the ectoloph. The postfossette is nearly square but narrows slightly across the labial side. Anterior and posterior cristae invade the postfossette. The anterior crista is rapidly lost with wear, and, in later stages, the fossette becomes crescentic (DGP TATAC-47). The labial enamel surface is convex, and the lingual surface is bulbous basally with a weak cingulum in some specimens. Although the inner edge of the hypocone occurs more lingually than on M^3, it does not extend as far as on M^1. The anterior surface of M^2 is nearly flat dorso-

ventrally and directed anterolabially to the parastyle, as in M^3.

Between the hypoflexus and the enamel border of the ectoloph, the occlusal surface of M^2 is dotted with numerous enamel-lined pits of different sizes. They vary in size, configuration, enamel thickness, and the depth to which they penetrate the dentine. The dentine is slightly elevated around the enamel of each pit. The pits are more numerous in _Huilatherium_ than in the other known genera of Leontiniidae.

In ectoloph length M^3 exceeds that of M^2. The parastyle, paracone, and wide metacone ridges resemble those on M^2. Distinct from M$^{1-2}$, the M^3 hypoflexus is only present in the earliest wear stages (as on IGM 150901) and the hypofossette becomes isolated in early wear stages. In _Leontinia_ and _Scarrittia_ the hypoflexus is developed more nearly to the base of the crown than in _Huilatherium_. A short inflection extends the postfossette posteriorly just lingual to the metastyle on M^3. The occlusal surface between the hypoflexus and the enamel edge of the ectoloph has as many as twelve enamel-lined pits, resembling those of M^2.

Lower Teeth

Huilatherium lacks I$_1$, I$_2$, the lower canine, and P$_1$. A conspicuous diastema separates I$_3$ and P$_2$. I$_3$ is a large, procumbent, tusklike tooth in UCMP 40276 and UCMP 37989. The crown has a rounded labial surface and a flattened

Table 19.3. Lower premolar measurements (mm) for *Huilatherium pluriplicatum*

Specimen	Side	P₂		P₃		P₄	
		L	W	L	W	L	W
UCMP 37989	Right	—	—	—	—	27.0	19.0
	Left	20.0	14.0	—	—	27.0	19.0
UCMP 38065	Left	21.0	17.0	24.0	20.0	26.0	22.0
UCMP 38229	Right	22.0	19.0	25.0	21.0	26.0	22.0
UCMP 39272	Left	22.0	17.0	—	—	—	—
UCMP 40280	Right	—	—	22.0	18.0	—	—
	Left	—	—	23.0	18.0	27.0	21.0
UCMP 40276	Right	—	—	24.0	19.0	25.0	21.0
	Left	19.0	18.0	23.0	20.0	26.0	20.0
UCMP 38096	Left	—	—	24.0	21.0	25.0	22.0
UCMP 38230	Right	—	—	25.0	22.0	29.0	23.0
	Left	—	—	25.0	19.0	—	—
IGM 183052	Right	—	—	22.0	19.0	—	—
IGM 184269	Right	—	—	25.0	19.0	—	—
DGP 16-VI-87-2	Right	—	—	—	—	25.0	22.0
DGP TATAC-26	Right	20.0	18.0	—	—	—	—
DGP TATAC-50	Left	—	—	25.0	19.0	—	—
DGP TATAC-52	Left	—	—	23.0	16.0	—	—
DGP TATAC-53	Right	—	—	—	—	24.0	19.0

lingual surface and is D-shaped in occlusal view. I_3 gradually flattens transversely into a bladelike tip from occlusion against I^2?, and the tooth lacks a cingulum. The root is nearly circular in cross section and carries little evidence of the angular boundary that separates the labial and lingual surfaces of the crown. In UCMP 40276, in which I_3 occurs in place, the incisor diverges upward at an angle of approximately 35°.

P_2 is nearly square in basal outline although it is slightly longer than wide (see table 19.3). The tooth appears double rooted lingually, but the broad labial root presents a smooth, rounded, continuous surface. The protoconid is prominent. From the protoconid a ridge extends posterolabially almost to the labial cingulum. Posterior and parallel to this ridge is a deep groove analogous to the hypostriad on the molars. A crest extends posterolingually from the protoconid and connects with the well-developed lingual cingulum. Posteriorly on the crest are one or more small cuspules, seen on UCMP 39272; these are quickly obliterated with wear. A well-developed ridge, nearly parallel to the posterolabial corner of the tooth, branches from the crest. The labial cingulum diminishes anteriorly. The lin-

gual cingulum is continuous to an anteromedian crest. In *Leontinia* both the labial and lingual cingula are more prominent anteriorly than in *Huilatherium*. *Huilatherium* lacks the development of an accessory cusp at the base of the anteromedian crest seen in *Leontinia* and *Scarrittia*.

P_3 is more molariform than P_2. P_3 has two distinct roots corresponding to the trigonid and talonid. The trigonid and talonid are separated labially by a deep hypostriad and lingually by a less-pronounced metastriad. The trigonid consists of a large protoconid and smaller paraconid and metaconid crests. A prominent posterolabial ridge at the anterior border of the hypostriad arises from the protoconid. A vertical anterior ridge sometimes continuous with the labial cingulum extends to the base of the enamel. The presence or absence of the labial cingulum is individually variable, but a lingual cingulum is always present. *Huilatherium* lacks the deep lingual striad seen in *Leontinia*, *Scarrittia*, and *Ancylocoelus* and possesses smaller paraconid and metaconid crests than in those genera.

P_4 is more molariform than P_3. The trigonid and talonid are nearly equal in size and are separated by a hypostriad labially and a deep metaflexid lingually. In early stages of

Table 19.4. Lower molar measurements (mm) for *Huilatherium pluriplicatum*

Specimen	Side	M$_1$ L	M$_1$ W	M$_2$ L	M$_2$ W	M$_3$ L	M$_3$ W
UCMP 38065	Right	35.0	21.0	44.0	24.0	60.0	24.0
UCMP 40276	Right	35.0	20.0	44.0	21.0	61.0	19.0
	Left	34.0	21.0	44.0	20.0	60.0	19.0
UCMP 40280	Right	—	—	—	—	54.0	19.0
	Left	35.0	19.0	40.0	20.0	59.0	21.0
UCMP 40063	Left	—	—	43.0	20.0	—	—
UCMP 38096	Left	—	—	43.0	24.0	61.0	22.0
UCMP 38230	Left	—	—	44.0	23.0	60.0	22.0
	Right	—	—	44.0	23.0	63.0	23.0
UCMP 37989	Left	—	—	48.0	24.0	—	—
UCMP 38043	Right	—	—	44.0	20.0	—	—
UCMP 39272	Right	—	—	43.0	21.0	58.0	22.0
	Left	—	—	45.0	23.0	—	—
UCMP 38018	Right	—	—	—	—	63.0	24.0
IGM 182983	Left	—	—	42.0	19.0	—	—
IGM 184432	Right	—	—	41.0	19.0	—	—
IGM 150448	Right	44.0	20.0	—	—	—	—
IGM unnumbered	Left	37.0	17.0	—	—	—	—
DGP 30-IV-85-1	Left	—	—	—	—	58.0	19.0
DGP TATAC-45	Right	—	—	40.0	22.0	57.0	21.0

wear a poorly developed paraflexid is present, and a small paralophid lobe lies just anterior to the protolophid. A lingual crest extends along the anterior portion of the paralophid to the well-developed lingual cingulum. A labial crest extends dorsoventrally to the base of the enamel. The labial cingulum is absent. The entoconid dominates the talonid and becomes a flattened shelf with wear. The paralophid lobe is not as well developed in *Huilatherium* as in comparable stages of wear in *Leontinia*, *Scarrittia*, and *Ancylocoelus*. In *Huilatherium* the metaflexid lacks the complex anterior and posterior arms seen in *Ancylocoelus*. In *Leontinia* and *Scarrittia* metaflexid arms are present but are slightly less developed.

The lower molars are long and narrow (fig. 19.3C), and are progressively more so from M$_1$ to M$_3$ (table 19.4). In comparison to the premolars, the molars are relatively and absolutely longer. The ridges that descend from the lingual and labial corners of the anterior surface of the lower premolars are not seen on the molars. The molar hypostriad, especially pronounced on the labial surface of the anterior premolars, is shallow, and the entoflexid, absent in the pre-

molars, is prominent on the molars. A U-shaped postfossettid is developed on the anterior region of the molar talonids. This postfossettid, in little-worn molars (such as in both M$_3$s of DGP 30-IV-85-1), is lingually elongate and opens onto the lingual side of the crown. The hypolophulid lobe is well differentiated. The lingual cingulum is reduced or absent, and the labial cingulum is absent.

The paraconid of M$_1$ is stronger than on P$_4$, giving the trigonid a more nearly square basal outline than the premolar. The anterior ridge present on P$_4$ is lacking on M^1. The metalophid is well developed. A distinct labial hypoflexid is absent from M$_1$, but the shallow hypostriad persists to the base of the enamel. The metaflexid is more strongly developed and extends more anterolabially than on P^4. A U-shaped posterofossettid occurs on the anterior part of the talonid but is obliterated with extensive wear in M^1. A weak entoflexid occurs in UCMP 40276 and UCMP 39634. Posterior to the entoflexid lies a strongly curved transverse hypolophulid lobe. The labial cingulum is lacking from M$_1$, and the lingual cingulum is reduced to an irregular row of cuspules along the lingual surface. The metaflexid is relatively shorter in *Huilatherium* than in *Leontinia, Scarrittia,* or *Ancylocoelus,* owing to the sharp curvature of the hypolophulid. In the related genera just mentioned the hypolophulid lobe is less strongly curved and is, as a result, more nearly parallel to the anteroposterior axis of the tooth. A postfossettid also occurs in *Leontinia, Ancylocoelus,* and *Colpodon*; it is oval in *Leontinia* and circular in *Ancylocoelus* and *Colpodon.*

On M_2 the paraflexid is more strongly developed than on M_1 so that the metalophid is more nearly isolated in early stages of wear. The U-shaped postfossettid is better developed than on M_1 and is present on more heavily worn teeth. M_2 also exhibits a stronger and more nearly circular entoflexid that forms a small isolated pit with wear. The hypolophulid lobe is better developed than on M_1 and less curved transversely (fig. 19.2B). As on M_1 the labial cingulum is lost on M_2, and the lingual cingulum is reduced or absent. The hypolophulid lobe in *Huilatherium* resembles the lobe on M_1 in *Leontinia* in position and orientation.

On the trigonid of M_3 the paraflexid is better developed, and the metaflexid is more nearly vertical than on the anterior molars. On the talonid of M_3 the U-shaped postfossettid is large and well developed (fig. 19.3C). On one specimen (UCMP 39272) a tiny enamel-lined pit occurs inside the U-shaped postfossettid. The entoflexid widens to form a long, deep, lingual depression that extends almost to the base of the enamel. A large rounded pit appears n this depression with extensive wear. The hypolophulid lobe is greatly enlarged and almost parallels the anteroposterior axis of the tooth. The most posterior part of M_3 narrows to a slightly rounded bladelike edge. The labial cingulum is absent, and the lingual cingulum is reduced or absent.

Manus and Pes Elements

Three Honda Group localities have yielded foot bones that are referred to *Huilatherium.* UCMP 40278 includes a left scaphoid, left ?trapezoid, ?pisiform, left metacarpal II, the distal end of metacarpal III, the ?proximal and distal ends of metacarpal IV, the proximal and distal phalanges of digit II, the proximal phalanx of digit III, proximal and median phalanges of digit IV; and from the pes, a left navicular, two partly preserved indeterminate podials, and two sesamoids.

Carpals

The scaphoid has been weathered, and the facets on the ventral side are incompletely preserved. The anterior surface is more sharply convex on the lunar side than in *Scarrittia* but becomes gradually flattened transversely. The anteroproximal edge is rounded. The radial facet is almost circular in outline, convex anteriorly, and gradually flattened posteriorly with a sharp upturning of the posterior edge. The edge of the facet opposite the lunar facet is rounded, whereas the lunar edge is flattened. The surface of the lunar facet is flattened and ovoid in outline, lengthened anteroposteriorly. As in *Scarrittia,* the facet extends distally, leaving a roughly concave gap below. Distally, the bone has a small flattened surface that represents all that remains of the magnum facet. The remainder of this facet and the surface of articulation with the trapezoid are not preserved.

The ?trapezoid has been weathered, and the articular facets are not distinct. It is slightly less than half the size of the scaphoid. The outline of the proximal and distal surfaces are triangular. The external surface is concave. The flattened weathered surfaces seem to correspond generally to articular facets described for *Scarrittia* (Chaffee 1952).

The ?pisiform is dumbbell shaped in dorsal outline. The proximal end is oval, lengthened slightly dorsoventrally. A slightly concave facet on the dorsal surface is assumed to articulate with the unciform. Ventral to it is a smaller, flatter facet thought to articulate with metacarpal IV. The shaft is concave dorsally and constricted transversely. The distal portion is a rough knob, somewhat flattened posterodistally.

Metacarpals

Parts of metacarpals II, III, and IV are preserved; however, only metacarpal II is complete. The proximal end is weathered, and, consequently, smooth articular facets are not preserved. On the dorsal side is the concave, somewhat triangular trapezoid facet. The remainder of the proximal surface is occupied by the magnum facet. The proximal cross section of the shaft is flattened transversely. A rough tuberosity occurs distally on the ventral surface. The cross section of the middle part of the shaft is more nearly circular than either end. The distal part of the shaft is flattened at an angle to the transverse plane, giving the appearance of twisting to the shaft. The sharply protruding tuberosities on either side of the distal condyle for the attachment of the lateral ligament do not occur at a corresponding position on each side; instead, the one on the side of metacarpal III is much more proximally located. This differs somewhat from the condition in metacarpals III and IV. In metacarpal III one tuberosity is only slightly more proximally located, and in metacarpal IV they are almost equal distance from the distal end. On the distal end the rounded dorsal articulation is convex both proximodistally and transversely with no evidence of the groove seen in *Leontinia* and *Scarrittia*. The proximal edge of the articulation is slightly concave proximodistally on the side closest to metacarpal II and is sharply rounded proximally and becomes gradually rounded toward the outer distal surface. The shallow groove, better developed in *Leontinia* and *Scarrittia*, appears on the rounded distal surface and extends along the outer side of the keel on the distal and ventral articular surfaces. A shallower groove occurs along the inside of the keel. The ventral edge of the articular surface is more rounded than in metacarpal III or IV.

The preserved distal portion of metacarpal III includes the distal end and some of the shaft. The shaft is considerably more flattened dorsoventrally than in metacarpals II and IV. The most proximal edge of the dorsal articulation is semicircular. The dorsal part of the facet is slightly convex proximodistally and flat transversely. A single deep groove occurs on the distal end, and no keel is apparent; however, there is an indication of abnormal wear or weathering. The ventral articular surface is worn and has been broken away; however, it appears to be wider than in metacarpals II and IV. The ventral surface of the distal shaft is slightly crushed but is flattened like the dorsal surface.

The proximal section of metacarpal IV is preserved with most of the proximal facets. The proximal part of the shaft is more nearly rounded in cross section than in metacarpal II, with only slight dorsal flattening and apparently only a slight development of the rough transverae tuberosities so prominent on metacarpal II. On the proximal surface is a transverse elongate facet that articulates with the unciform. The ventral edge of the facet is sharply concave, and the dorsal surface is slightly concave. A broadly triangular facet, transverse to the unciform facet, is the surface which articulates with metacarpal III. No trace of metacarpal I was preserved, so there is a possibility that this structure was not

retained in *Huilatherium*. A small rectangular facet on the dorsal surface opposite the articular surface of metacarpal III, however, indicates a possible contact with a greatly reduced metacarpal I. The distal end of metacarpal IV is more flattened transversely than metacarpal II, but less than metacarpal III. On the dorsal side the facet is strongly convex proximodistally and more gradually convex transversely. The dorsal edge of the facet is more nearly semicircular than in metacarpal II. The groove and keel on the distal end is like that of metacarpal II.

Phalanges

The proximal phalanges are broad and rather flattened distally, with somewhat bulbous proximal ends. The proximal end has the rounded concave facet for articulation with metacarpal II. The dorsal surface is irregular, with no development of the flattened distal region seen in metacarpals III and IV. The distal articulation is ovoid with a gentle median groove running dorsoventrally.

The proximal phalanx of digit III is lengthened transversely as is the metacarpal. The proximal articulation is a lengthened oval transversely. The bulbous tuberosities on the ventral surface are more extensively developed than in metacarpals II and IV. The distal portion of the phalanx is flattened transversely. The distal articulation is weathered but appears similar to the same bone in digit II.

The proximal phalanx of digit IV is more similar to digit II than to digit III in appearance. The proximal articulation is slightly smaller and the ventral tuberosities less equally developed. The dorsal surface is slightly more regular and flattened transversely than in digit II. The distal articulation is ovoid but is more continuously concave, with no development of a groove as seen in digit II.

The median phalanx of digit IV is considerably shorter proximodistally than the proximal phalanges and less flattened distally. The proximal and distal articulations are of almost equal size and shape; they are oval and lengthened slightly transversely. The dorsal and ventral surfaces between the facets are concave.

The ungual phalanx of digit II is hoof-shaped. The proximal articulation is ovoid, lengthened transversely. The bone is asymmetrical, with the side closest to digit III smaller and less extended, similar to that of *Nesodon* (Toxodontidae; see Scott 1912). Unlike the ungual phalanges of *Leontinia* and *Scarrittia,* there is no tendency for the phalanx to be flattened or spatulate and no evidence of development of a median cleft.

Tarsals

The ?navicular is similar in general appearance to specimens of *Scarrittia* (Chaffee 1952) and to specimens of *Leontinia* (Gaudry 1906). An oval concave facet on the proximal surface is assumed to articulate with the astragalus. Posterior to this facet is a small proximodistally flattened surface for articulation with the mesocuneiform. The dorsal surface is almost trapezoidal, whereas the ventral surface is triangular. The distal surface is another concave facet similar in shape and size to that on the proximal surface but set at an angle to it.

Podials

From UCMP Locality V4525 also is an isolated metatarsal IV (UCMP 38052). The proximal end is a single convex surface that articulates with the cuboid. Transverse to the ventral part of the cuboid facet is a proximodistally flattened surface that may be the contact area with metatarsal III. The shape of the shaft is generally similar along its entire length, slightly flattened on the dorsal surface and convex ventrally. The shaft curves slightly laterally. The distal articular surface is convex. The dorsal edge of that surface is rounded. A median keel occurs on the ventral surface with a shallow groove on the side next to metatarsal III and a deeper groove on the opposite side.

Two proximal phalanges (UCMP 39272) collected with fragments of teeth of *Huilatherium* from Locality V4937 are referred to this genus. One is the proximal phalanx of digit II of the carpus. It is slightly larger than UCMP 40278 but otherwise very similar. The second phalanx is minutely fractured and weathered, especially on the distal half, and is questionably identified as the proximal phalanx of digit II of the tarsus. It is larger overall than the carpal phalanges. The ledge of bone just distal to and surrounding the proximal articular surface is better developed around the entire perimeter than in the phalanges of the manus. The proximal articulation is D-shaped with the ventral surface straight,

unlike those in the manus that are slightly concave. Dorsoventrally, the articulation is concave, and transversely, it is flattened. The distal area is poorly preserved.

From Locality V4524 a single podial (UCMP 38043) was collected with teeth of *Huilatherium*; however, this specimen is thus far indeterminate.

DISCUSSION

Given the poor quality of the material available for some genera (such as *Henricofilholia*) and the lack of detailed descriptions for most of the other genera, no stable and comprehensive taxonomy is available for the family at this time. The following brief review of the taxonomic history of the group is a necessary preamble to a discussion of the phylogenetic position of *Huilatherium* within the family.

In 1985 Ameghino referred nine genera to the family: *Colpodon* Burmeister 1885, *Baenodon* Ameghino 1897, *Ancylocoelus* Ameghino 1895, *Leontinia* Ameghino 1895, *Loxocoelus* Ameghino 1895, *Rodiotherium* Ameghino 1895, *Scaphops* Ameghino 1895, *Senodon* Ameghino 1895, and *Stenogenium* Ameghino 1895. To these Ameghino later added two additional genera: *Hedralophus* Ameghino 1901 and *Henricofilholia* Ameghino 1895. Loomis (1914) reconsidered the taxonomic status of some of these genera and reduced the family to the following genera: *Leontinia* (=*Senodon,* =*Stenogenium,* =*Scaphops*), *Ancylocoelus* (=*Rodiotherium*), and perhaps *Loxocoelus,* whose validity was put in doubt. *Henricofilholia* was removed from the family and placed in the Isotemnidae. Simpson (1934) introduced a new genus, *Scarrittia,* and in 1945 recognized only four genera: *Ancylocoelus, Leontinia* (including *Stenogenium* and *Scaphops*), *Colpodon,* and *Scarrittia.* Chaffee (1952) described *Scarrittia* in greater detail in his study of the Scarritt Pocket Deseadan fauna. Chaffee recognized five genera, *Leontinia, Colpodon, Ancylocoelus* (=*Rodiotherium*), *Henricofilholia* (in doubt), and *Scarrittia,* and left the taxon *Scaphops grypus* in doubt. Since that time, two new genera have been described: one by Colwell (1965) in an unpublished thesis (which corresponds to *Huilatherium* Villarroel and Guerrero 1985) and *Taubatherium* from Tremembé in Brazil (Soria and Alvarenga 1989).

Table 19.5. Character/taxon matrix for Leontiniidae (Notoungulata) and
Pleurostylodon **(Isotemnidae) used in the phylogenetic analysis**

Taxon	Character																						
	1	2	3	4	5	6	7	8	9	10	11	12	13	14	15	16	17	18	19	20	21	22	23
Taubatherium	?	?	2	?	1	1	0	0	0	0	0	0	0	1	1	?	0	0	0	2	0	0	0
Colpodon	1	0	0	?	2	1	1	1	2	0	1	0	1	0	0	?	?	?	0	1	0	1	1
Ancylocoelus	1	0	1	1	2	0	0	0/1	?	0	1	1	1	1	1	1	0	0	0	1	2	1	1
Leontinia	0	0	1	0	3	0	1	1	1	0	1	0	1	0	2	1	1	1	0	2	1	1	1
Scarrittia	0	0	1	1	2	0	?	1	1	0	?	0	1	0	2	1	1	1	0	2	2	1	1
Huilatherium	2	1	2	0	1	0/1	1	2	2	1	2	0	0	1	2	0	2	2	1	3	2	0	0
Pleurostylodon	0	0	1	0	0	1	1	2	0	1	1	1	1	0	0	1	0	0	0	0	1	0	1

Note: Characters 1–23 and their states are defined in appendix 19.1.

Soria and Bond (1988), in a brief abstract, proposed to remove *Colpodon* from the Leontiniidae and refer it to the Notohippidae, arguing that *Colpodon* possesses hypsodont cheek teeth and unhypertophied incisors/canines, that is, "las especializaciones dentarias propias de los Leontínidos." This interpretation of Soria and Bond (1988) contradicts the family diagnosis established by Chaffee (1952). Although Chaffee's (1952) family diagnosis is based on characters of the dentition, a comparison of the postcranial skeleton of *Scarrittia* with other Toxodontia led him to propose that the Leontiniidae possess characters intermediate between homalodotheriids and toxodontids, with greatest similarity to the homalodotheres. Chaffee (1952, 550) concluded that although the evidence is inconclusive, "the Leontiniidae are a trifle closer to the Homalodotheriidae than to the Toxodontidae and even farther removed from the Notohippidae."

Our assessment of the phylogenetic relationships of *Huilatherium* is based exclusively on an analysis of dental characters among the six best-known genera of Leontiniidae (*Leontinia, Scarrittia, Ancylocoelus, Taubatherium, Colpodon,* and *Huilatherium*) (see app. 19.1). Chaffee (1952), Simpson (1967), and Paula Couto (1983) agree that the ancestry of the family is probably with the Isotemnidae, although as Simpson noted, no single taxon serves as a satisfactory ancestor for the Toxodontia (Leontiniidae, Toxodontidae, Homalodotheriidae, and Notohippidae). In our analysis, the well-known Casamayoran isotemnid *Pleurostylodon* was selected as the outgroup.

Maximum parsimony analysis was performed by PAUP Version 3.0s (Swofford 1991) on the dental characters described in appendix 19.1. The exhaustive search algorithm found a single maximum parsimony tree of forty-two steps (fig. 19.4). Although a more stable taxonomy should incorporate the cranial, mandibular, and postcranial characters, the following can be established from the dental characters (table 19.5).

Taubatherium and *Huilatherium* share four synapomorphies, hypertrophied incisors, double-basined and incompletely separated leontiniid depressions on the upper premolars, loss of the upper molar anterolingual cingulum,

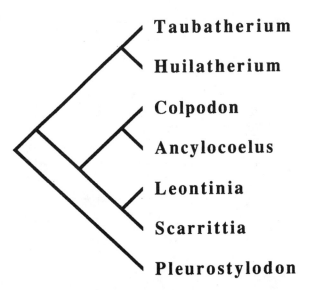

Figure 19.4. Maximum parsimony tree for the family Leontiniidae based upon characters of the dentition (see app. 19.1 and table 19.5).

and the loss of the lower molar labial cingulum. Thus, the maximum parsimony tree suggests some degree of phylogenetic differentiation of leontiniids in the tropical zone of South America. With the possible exception of *Taubatherium* from Brazil, which may antedate the Oligocene (see Simões Ferreira, in Paula Couto 1983), leontiniids are known only from Deseadan (Oligocene) and later deposits in South America. The diversity of the family in the Oligocene and the absence of Leontiniidae in the Mustersan suggests that the differentiation of the family occurred sometime in the Mustersan-Deseadan hiatus (middle Eocene to Oligocene), a period of time that is very poorly known throughout South America. The presence of *Taubatherium* in Brazil and *Huilatherium* in Colombia suggests that an important part of the history of the family may have occurred in the tropics as pointed out by Paula Couto (1983).

Acknowledgments

We express our sincere appreciation to D. E. Savage, R. F. Kay, R. H. Madden, and R. A. Stirton for their invaluable cooperation. Without their assistance this work would not have been possible. To others who have contributed to this project in different ways, S. E. Benson, J. E. Mawby, W. Langston Jr., R. C. Fox, C. R. Harrington, R. G. Chaffee, W. Turnbull, and J. Guerrero, we extend an affectionate thank you. The Department of Paleontology of the University of California and the Facultad de Ciencias de la Universidad Nacional de Colombia (Santa Fe de Bogotá) contributed monetarily. A. Lucas made the illustrations.

Appendix 19.1. Dental characters and character states observed in genera of Leontiniidae

Numbers in parentheses refer to character states observed in the various genera as scored in table 19.5.

1. Dental formula: (0) complete; (1) lacking upper and lower canines; (2) lacking upper and lower canines and P_1.

2. Diastema: (0) absent; (1) present.
3. $I^{2?}$: (0) not hypertrophied; (1) slightly hypertrophied; (2) hypertrophied.
4. Cingulum on the hypertrophied incisor: (0) absent; (1) present.
5. Leontiniid depression on the upper premolars: (0) simple; (1) doubled, basins incompletely separated; (2) doubled, basins well separated.
6. Upper premolar labial cingulum: (0) present on P^{2-4}; (1) present on P^3 but absent or weakly developed on P^4.
7. Enamel-lined pits on the upper premolars: (0) absent; (1) present.
8. Posterior arm of the upper premolar hypofossette: (0) not developed; (1) weakly developed; (2) well developed.
9. Cheek teeth: (0) brachydont; (1) mesodont; (2) hypsodont.
10. Upper molar mesial crista: (0) weakly developed; (1) well developed.
11. Enamel-lined pits on the upper molars: (0) absent; (1) present and numerous.
12. Upper molar lingual cingulum: (0) weakly developed; (1) well developed.
13. Upper molar anterolingual cingulum: (0) absent; (1) present.
14. Anterolingual enamel border of the upper molar hypoflexus: (0) not conspicuous with wear; (1) conspicuous in relief with wear.
15. I_3: (0) not hypertrophied; (1) somewhat hypertrophied; (1) hypertrophied.
16. Lower incisor cingulum: (0) absent; (1) present.
17. Lower premolar metaflexid: (0) large; (1) moderate; (2) small.
18. Bifurcation of the lower premolar metaflexid: (0) present, with anterior and posterior branches well developed; (1) present, with the branches weakly developed; (2) without observable bifurcation.
19. Lower premolar labial cingulum: (0) present on P_{2-4}; (1) present on P_2 but absent on P_4.
20. Configuration of the lower molar postfossettid: (0) round; (1) ovate; (2) U-shaped.
21. Lower molar entoflexid: (0) narrow with sides sharply curved; (1) broad with sharply curving posterior edge; (2) broad with gently rounded sides.
22. Lower molar lingual cingulum: (0) reduced or absent; (1) present.
23. Lower molar labial cingulum: (0) absent; (1) present.

20
Enamel Microstructure in Notoungulates

Mary C. Maas

RESUMEN

La estructura microscópica del esmalte en unas especies nominales de Notoungulata fue examinada para determinar relaciones filogenéticas dentro del orden, y en especial, la relación de los notoungulados primitivos "Notioprogonia," la validez de Hegetotheria, y las relaciones entre varias familias de Hegetotheria y Typotheria. Se examinó la microestructura del esmalte en los molares de nueve especies, entre ellos *Henricosbornia* (Henricosborniidae), *Notostylops* (Notostylopidae), *Archaeohyrax* (Archaeohyracidae), *Hegetotherium* (Hegetotheriidae), *Rhynchippus* (Notohippidae), *Trachytherus* (Mesotheriidae), *Notopithecus* (Interatheriidae) y dos especies de la fauna de La Venta, *Miocochilius* (Interatheriidae) y *Pericotoxodon* (Toxodontidae). Los dientes fueron seccionados longitudinalmente en el plano transversal, la superficie de la sección anterior fue examinada para determinar el patrón de decusación de los prismas. Para determinar la forma en corte transversal, el empaquetado de los prismas, y para observar el esmalte en profundidades distintas, se hizo pulir en tres profundidades una carilla tangencial a la superficie externa.

Tanto rasgos cualitativos como variables cuantitativos fueron observados, como, por ejemplo, el arreglo de los prismas en corte transversal, el grado de perfeccionamiento de los márgenes de los prismas, la cantidad relativa de esmalte prismático e interprismático, y la forma y el tamaño de los prismas. Valores promedios para cada especie fueron usados en un análisis de agrupación para

poder discriminar patrones de distribución entre estas especies.

Los resultados indican que esmalte superficial de Tipo 1, esmalte profundo de Tipo 2, y zonas de decusación no muy nítidas son características primitivas para el orden Notoungulata al igual que en el condilártro mioclénido *Litaletes*. Se distinguen las especies hipsodontes (proto- y euhipsodontes) por una serie de características derivadas del esmalte, entre otras un "esmalte radial modificado" caracterizado por láminas distintas y gruesas entre hileras de prismas y cristalitos entre hileras de prismas orientados perpendicularmente con respecto a los cristalitos prismáticos. Se vieron patrones de semejanza no muy claros entre miembros de distintos grupos monofiléticos de Notoungulata y hasta patrones distintos entre miembros del mismo grupo monofilético. Por ejemplo, *Archaeohyrax* y *Hegetotherium* no comparten ninguna característica del esmalte que los distinga como grupo aparte de los Typotheria, y se distinguen entre sí en el patron de decusación. Las características de la estructura microscópica del esmalte no apoyan a una relación estrecha entre Mesotheriidae y Interatheriidae ni entre Mesotheriidae y Hegetotheria. La estructura del esmalte en *Trachytherus* es semejante al esmalte en los Toxodontidae.

The Notoungulata were a distinctive order of Cenozoic ungulates confined almost entirely to the South American continent. They first appeared in the Riochican

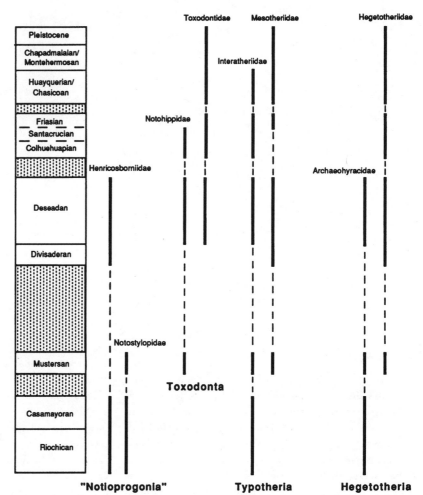

Figure 20.1. Approximate stratigraphic ranges by suborder of the eight notoungulate families included in this study: Henricosborniidae and Notostylopidae ("Notioprogonia"), Notohippidae and Toxodontidae (Toxodonta), Interatheriidae and Mesotheriidae (Typotheria), Archaeohyracidae and Hegetotheriidae ("Hegetotheria"). *Dashed lines* indicate gaps in the fossil record of a family. The column at left refers to Paleocene through Pliocene South American land mammal ages; *stippled areas* indicate hiatuses in the fossil record. (Redrawn from Cifelli 1985.)

(late Paleocene), reached their greatest familial diversity during the Deseadan (early Oligocene), and gradually declined in diversity throughout the rest of the Tertiary until their disappearance in the Pleistocene (fig. 20.1). Simpson (1945, 1948, 1967) distinguished four suborders within the Notoungulata: "Notioprogonia," Toxodonta, Hegetotheria, and Typotheria. The "Notioprogonia," currently considered to be paraphyletic, include the most primitive members of the order. The Toxodonta are generally considered a long-distinct, monophyletic group. The two remaining suborders, Hegetotheria and Typotheria, include representatives of a variety of small, superficially rodentlike and rabbitlike families whose relationships are more uncertain.

Simpson (1945, 1967) argued for separate subordinal status for Typotheria (Oldfieldthomasiidae, Archae-

opithecidae, Mesotheriidae, and Interatheriidae) and Hegetotheria (Archaeohyracidae and Hegetotheriidae) based on cranial characters, particularly features of the ear region. More recently, Cifelli (1985) concluded that the dental resemblance between some archaeohyracid hegetotheres and mesotheriid typotheres, on the one hand, and the cranial resemblance between archaeohyracid and hegetotheriid hegetotheres, on the other hand, support the inclusion of Hegetotheria within Typotheria. As further support of hegetothere affinities to typotheres, he noted similarities between cheek teeth of archaeohyracid hegetotheres and those of primitive typotheres (oldfieldthomasiids, archaeopithecids, and notopithecine interatheriids). Cifelli (1985) also pointed out derived dental characters linking mesotheriid typotheres and hegetotheres, although Patterson (1934a) argued that mesotheriids have a closer relationship to interatheriid typotheres.

Unresolved issues concerning the subordinal classification of notoungulates, therefore, include (1) the affinities of members of the paraphyletic "Notioprogonia," (2) the subordinal validity of Hegetotheria, and (3) relationships of the "hegetotheres" and the typothere families Mesotheriidae and Interatheriidae. Current phylogenetic hypotheses are based largely on cranial characters and morphology of the anterior dentition, features that are not known for all taxa. Analyses of gross dental characters are typically a mainstay of phylogenetic analyses but in notoungulates are frequently confounded by wear-related changes in tooth morphology. Enamel microstructure represents a potentially informative character complex that is independent of the gross morphological characters previously used in phylogenetic analysis of these animals. Enamel structure also is readily accessible for most taxa. The goal of this study is to examine and describe the enamel microstructure of representative notoungulates and to explore its potential for clarifying subordinal relationships within the group.

Enamel Microstructure

Mammalian tooth enamel is a highly mineralized tissue composed largely of groups of similarly oriented hydroxyapatite crystallites (prisms and interprismatic material). When viewed in cross sections perpendicular to prism long axes, the arrangement of enamel prisms and interprismatic enamel has been shown to differ in a consistent manner among groups, particularly at familial and higher levels of taxonomic resolution (e.g., Boyde 1971; Krause and Carlson 1986; Martin et al. 1988). Boyde (1964, 1971) recognized three basic prism cross-sectional packing arrangements in mammalian enamel:

1. Pattern 1, in which prisms are roughly circular in cross-section and are distinguished from interprismatic enamel by complete "boundaries" formed by discontinuities in crystallite orientation. Boyde described pattern 1 enamel as characterized by a hexagonal packing arrangement of prisms when viewed in cross section. In hexagonal packing, prisms in adjacent horizontal rows are offset from one another (fig. 20.2).

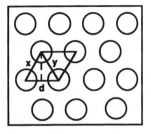 

Pattern 1 Model   **Pattern 2 Model**

Figure 20.2. Cross-sectional shapes and packing arrangements of pattern 1 prisms (hexagonally packed) and pattern 2 prisms (arranged in apicocervical columns). The sides x, y, and d of the triangles represent distances between the centers of three contiguous prisms. The *dashed lines* represent the height (h_d for pattern 1, h_y for pattern 2) of the triangle. The pattern 1 model also applies to pattern 3 enamel. Distances x, y, and d are used to calculate parameters of prism cross-sectional shape and packing arrangement (see table 20.2 and text for explanation).

2. Pattern 2, in which prism boundaries are incomplete cervically (toward the root of a tooth), prisms are arranged in apicocervically oriented columns when viewed in cross section, and columns of prism cross sections are separated by sheets of interprismatic enamel (interrow sheets) (fig. 20.2).
3. Pattern 3, in which prism boundaries are incomplete cervically, as in pattern 2 enamel, but prism cross sections are arranged hexagonally as in pattern 1.

Boyde (1964) demonstrated the developmental basis of these three prism patterns and argued that their distribution had phylogenetic implications.

Features of enamel that have proved useful in distinguishing taxa include the arrangement of prism cross sections either in apicocervically oriented columns or in horizontal rows (prism packing patterns), completeness of prism boundaries, relative amounts of prismatic and interprismatic enamel, and size and shape of prisms. Quantification of these parameters, first applied to human enamel (e.g., Fosse 1968), more recently has been used to distinguish taxa in such disparate groups as Multituberculata (Carlson and Krause 1985; Krause and Carlson 1986) and Artiodactyla (Grine et al. 1986, 1987).

Both qualitative and quantitative studies of enamel must consider variability in the intrinsic structure of enamel and apparent variability that represents the effects of specimen preparation. Intrinsic variability has been demonstrated by

Table 20.1. Specimens used in analysis of notoungulate tooth enamel structure

Suborder and Family	Species	Provenance	Specimen Number	Tooth Type[a]
Mioclaenidae	*Litaletes disjunctus*	Torrejonian, Montana	Uncataloged	Right M_2
"NOTIOPROGONIA"				
Henricosborniidae	*Henricosbornia lophodonta*	Casamayoran, Patagonia	AMNH 28966	Right M_x, right M_3
Notostylopidae	*Notostylops pendens*	Casamayoran, Argentina	AMNH 28495	Right M_x, left M_x
TOXODONTA				
Notohippidae	*Rhynchippus* sp.	Deseadan, Bolivia	UF 91400	M_x
	Rhynchippus sp.	Deseadan, Bolivia	UF 20680	M^x
Toxodontidae	*Pericotoxodon platignathus*	Honda Group, Colombia	Uncataloged	M_x, M^x
TYPOTHERIA				
Interatheriidae	*Notopithecus adapinus*	Casamayoran, Patagonia	AMNH 15901	Left M^x, right M^x
	Miocochilius anamopodus	Honda Group, Colombia	Uncataloged	Right M_x, right M_x
Mesotheriidae	*Trachytherus* sp.	Deseadan, Bolivia	UF/FSM 121803	M_3
	Trachytherus sp.	Deseadan, Bolivia	UF/FSM 121804	M_3
Hegetotheriidae	*Hegetotherium* sp.	Friasian, Chile	Uncataloged	M^x, M^x
Archaeohyracidae	*Archaeohyrax* sp.	Deseadan, Bolivia	Uncataloged	M^x, M_x

Note: All specimens represent Order Notoungulata, except *Litaletes disjunctus* (Order "Condylarthra").

[a] x = tooth position uncertain.

quantitative studies that have shown prism size, shape, and packing to vary significantly according to depth of enamel relative to the tooth surface and enamel dentine junction (EDJ). In addition, however, cross-sectional shapes of prisms may appear to differ significantly according to the plane (relative to prism long axes) in which a specimen is sectioned (Fosse 1968; Krause and Carlson 1986). This apparent variation occurs because prisms do not follow a straight course from the EDJ to the surface, and, therefore, a plane of section may intercept long axes of different prisms at various angles. Although preparation-dependent variability must be considered and evaluated, however, it has been demonstrated that when depth-related variation is taken into account, prism morphology can clarify phylogenetic relationships among some groups of mammals (e.g., Krause and Carlson 1986; Martin et al. 1988).

Prism decussation is another aspect of enamel microstructure of potential taxonomic importance (Boyde 1969). Decussation refers to the coordinated change in orientation of long axes of adjacent prisms between the EDJ and the surface of the tooth. In decussating enamel, long axes of one group of adjacent prisms are roughly parallel to one another but at an angle to long axes of prisms in adjacent groups.

Each group of parallel prisms is designated a decussation zone. Patterns of prism decussation differ among taxa in several respects: the difference in orientation of prism long axes in adjacent zones may be large or small; the change in orientation of prism long axes between zones may be gradual or abrupt; decussation zones may extend from the EDJ to the surface or only occur in part of the enamel (Kawai 1955; Boyde 1969).

In this chapter I describe prism morphology, packing patterns, and decussation among notoungulate species representing the four suborders defined by Simpson (1967). A critical aspect of the analysis is documentation of levels of variability in both qualitative and quantitative features of notoungulate enamel microstructure.

Materials and Methods

Enamel microstructure was examined in molar teeth of nine notoungulate species, encompassing all four suborders and including known toxodonts and typotheres from Colombia (table 20.1). Three of the nine species, *Henricosbornia*, *Notostylops* ("Notioprogonia"), and *Notopithecus* (Typotheria), have low-crowned (brachydont) cheek teeth; the

Table 20.2. Equations for calculation of prism shape variables for pattern 1 and pattern 2 models

Variable	Pattern 1 Model	Pattern 2 Model
ASA	$\frac{1}{2}\sqrt{4d^2\gamma^2 - (d^2 + \gamma^2 - x^2)^2}$	$\frac{1}{2}\sqrt{4\gamma^2 x^2 - (\gamma^2 + x^2 - d^2)^2}$
h_d	$\dfrac{\frac{1}{2}\sqrt{4d^2\gamma^2 - (d^2 + \gamma^2 - x^2)^2}}{d}$	—
h_γ	—	$\dfrac{\frac{1}{2}\sqrt{4\gamma^2 x^2 - (\gamma^2 + x^2 - d^2)^2}}{\gamma}$
K	$\dfrac{\frac{1}{2}d\sqrt{3}}{h_d}$	$\dfrac{h_\gamma}{\frac{1}{2}\gamma\sqrt{3}}$
CD	$\sqrt{\dfrac{2dh_d}{\sqrt{3}}}$	$\sqrt{\dfrac{2\gamma h_\gamma}{\sqrt{3}}}$

Source: Grine et al. (1987).

Notes: Pattern 1 equations also apply to pattern 3 enamel. *ASA:* average ameloblastic cross-sectional secretory area; h_d and h_γ: height between prism centers in adjacent rows; *K:* prism compression/distention value; *CD:* distance between prism centers.

others are hypsodont or hypseledont. Samples comprised two individuals for each taxon. Additionally, a single tooth of the North American Paleocene mioclaenid "condylarth" *Litaletes disjunctus* was examined. *Litaletes,* which is thought to be close to the ancestry of several South American ungulate groups, including Notoungulata (Cifelli 1983b), provides a basis for assessing polarity of enamel characters in notoungulates.

All specimens were prepared for examination in a scanning electron microscope (SEM). The individual teeth were sectioned longitudinally in a buccolingual plane. The cut surface of the mesial section of each tooth was polished with increasingly fine grades of abrasive (to 0.05-μm alumina grit) for documentation of prism decussation patterns. The distal section of each tooth was used to examine prism cross-sectional shape and packing arrangement. A polished facet was prepared oriented tangentially to the outer enamel surface. The tangential facet was located approximately midway down the buccal surface of each tooth. Each facet was polished in three steps to expose enamel at three different depths from the outer surface (superficial, intermediate, and deep). First, the total thickness of enamel was measured, and a facet was ground and polished to a depth less than one-third the original enamel thickness (superficial enamel). After SEM examination of the superficial enamel, the facet was polished twice more: for inter-

mediate enamel to a depth one-half the original enamel thickness and for deep enamel to a depth approximately two-thirds the original enamel thickness.

Preparation for SEM examination of polished longitudinally sectioned specimens and tangentially polished specimens (for each of the three depths) included acid-etching (5% HCl for 10 sec or 0.5% H_3PO_4 for 30 sec) and sputter-coating with gold palladium. All specimens were examined in a JEOL 35C SEM at 10 kV accelerating voltage in secondary electron mode. To document prism cross-sectional morphology on tangential facets, two representative micrographs were taken for each enamel depth at magnifications of 2,000 times. Micrographs of the entire enamel thickness (from EDJ to surface) of polished longitudinally sectioned specimens recorded prism decussation patterns for both buccal and lingual enamel midway between the apex and cervix of the sectioned tooth.

Analysis of enamel structure included qualitative descriptions of enamel (prism packing and prism decussation patterns) and quantitative assessment of prism size and shape. For each enamel depth the predominant prism cross-sectional packing patterns were classified following the scheme of Boyde (1964, 1971). Prism decussation was classified qualitatively according to three criteria: (1) presence or absence of prism decussation, (2) the change in orientation of prism long axes between decussation zones

Table 20.3. Decussation characteristics and predominant prism patterns for superficial, intermediate, and deep enamel

Taxon	Prism Packing Pattern (1, 2, or 3)			Decussation	
	Superficial	Intermediate	Deep	Zones[a]	Extent[b]
"Notioprogonia":					
Henricosbornia	1	2/1	2/1	Weak	Partial
Notostylops	1	2/1	2/1	Weak	Partial
Toxodonta:					
Rhynchippus	2/1	2	2	Absent	
Pericotoxodon	1/2	2	2	Strong	Complete
Typotheria:					
Notopithecus	1	2/1	2/1	Weak	Partial
Miocochilius	1/2	2	2	Strong	Complete
Trachytherus	1/2/3	2	2	Absent	
"Hegetotheria":					
Archaeohyrax	1/2	2	2	Weak	Partial
Hegetotherium	2/1	2	2	Strong	Partial

[a]*Absent:* no evidence of decussation. *Weak:* zones occasionally present, but poorly defined, with gradual transition (transition zone three to five prisms wide) in prism orientation between zones. *Strong:* zones regular and well defined with abrupt transition (change in orientation with no "transition" zone) in prism orientation between adjacent zones.

[b]Extent of decussation from enamel-dentine junction (EDJ) to tooth surface. *Partial:* nondecussating enamel present at EDJ and/or at surface. *Complete:* decussation zones extend completely from EDJ to surface with no discernable nondecussating enamel zone.

(abrupt or gradual transition), and (3) the extent of decussation from the EDJ to the surface.

To quantify prism cross-sectional size and shape, prism centers were marked on acetate overlays of micrographs. Lines x, y, and d were drawn connecting prism centers. These lines form the sides of a series of contiguous triangles (fig. 20.2). For each specimen measurements included x, y, and d for ten contiguous triangles and prism area (*PA,* the area delineated by a single prism boundary) for the prism at the apex of each triangle. The variables x, y, d, and *PA* were recorded using the Sigmascan Scientific Measurement System (Jandel Scientific). Descriptive statistics were calculated for each of the four variables at each enamel depth.

Values of x, y, d, and *PA* were used to calculate four parameters of prism shape and packing: (1) *CD* (distance between adjacent prism centers on a horizontal plane); (2) *K* (apicocervical compression or distention of prisms); (3) *ASA* (average ameloblast secretory area, the area of prismatic and interprismatic enamel laid down by a single cell during development); and (4) *PA/ASA* (a measure of the

relative amounts of prismatic and interprismatic enamel) (table 20.2).

Species averages for each parameter were used as variables in cluster analyses to explore the distributions of prism shape and size characters among taxa. Two methods of cluster analysis were applied to each variable and to combinations of standardized variables (deviation from the variable mean in standard deviation units). The first method, unweighted pair-groups using mathematical averages (UPGMA), yields a hierarchical arrangement of clusters based on a data matrix of average Euclidean distances. Optimality of clusters produced by UPGMA can be assessed by cophenetic correlation, a measure of goodness of fit between the original Euclidean distance matrix and a similarity matrix derived from the resultant phenogram (Sneath and Sokal 1973). The second method, the *k*-means system, is a nonhierarchical approach that delineates a predefined number of groups based on their mean vectors (centroids) (Sneath and Sokal 1973). In this study the *k*-means clustering algorithm (Wilkinson 1990) was run for

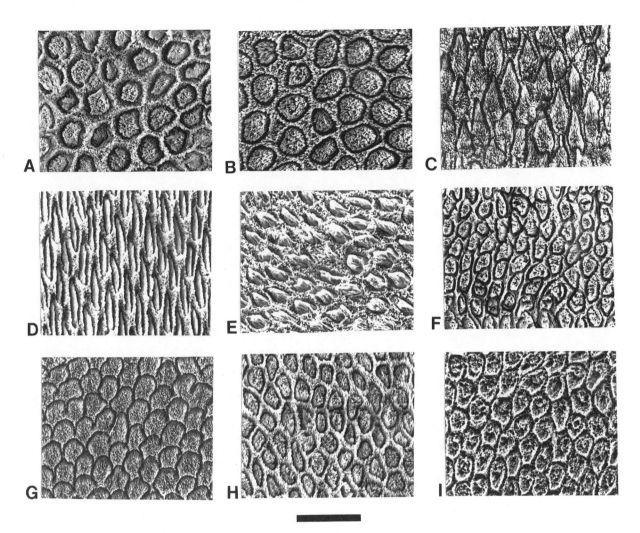

Figure 20.3. Tangential sections of superficial enamel: *A, Henricosbornia lophodonta; B, Notostylops pendens; C, Rhynchippus* sp.; *D, Pericotoxodon platignathus; E, Notopithecus adapinus; F, Miocochilius anamopodus; G, Trachytherus* sp.; *H, Archaeohyrax* sp.; *I, Hegetotherium* sp. Scale bar 10 μm.

three and for four clusters, representing numbers of groups in the alternative schemes of notoungulate subordinal classification.

Results and Discussion

Prism packing patterns differ according to enamel depth for each of the nine notoungulate species (table 20.3). Superficial enamel of all species is characterized for the most part by pattern 1 prisms (fig. 20.3). Although pattern 1 prisms predominate throughout the superficial enamel for *Henricosbornia, Notostylops, Notopithecus,* and *Archaeohyrax,* pat-

tern 2 and pattern 3 prisms are found in some regions of superficial enamel of the other forms. It is noteworthy that the pattern 1 prisms in most taxa are arranged in apicocervically oriented longitudinal rows, an arrangement more typical of pattern 2 than pattern 1 enamel, where prisms are hexagonally packed.

Both intermediate and deep enamel of all nine notoungulate species is characterized by a predominance of pattern 2 prisms (table 20.3). Prisms with complete boundaries (cf. pattern 1), however, are found in some regions of both the intermediate and deep enamel of all three brachydont forms (*Henricosbornia, Notostylops, Notopithecus*), although not the hypsodont species. Furthermore, there is

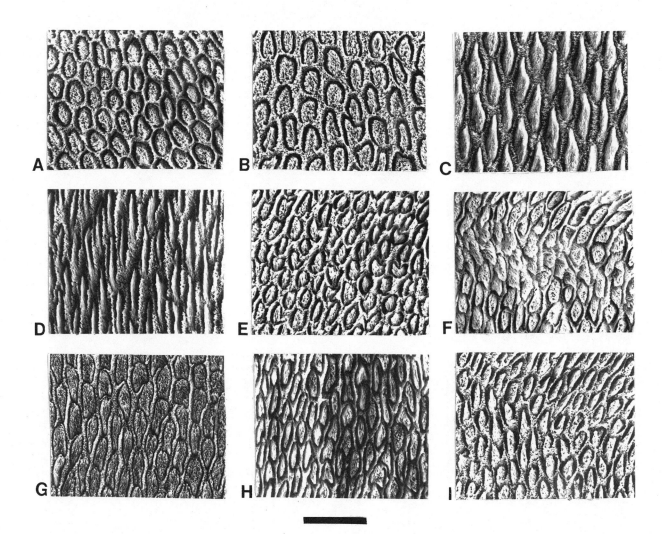

Figure 20.4. Tangential sections of intermediate enamel: *A, Henricosbornia lophodonta; B, Notostylops pendens; C, Rhynchippus* sp.; *D, Pericotoxodon platignathus; E, Notopithecus adapinus; F, Miocochilius anamopodus; G, Trachytherus* sp.; *H, Archaeohyrax* sp.; *I, Hegetotherium* sp. Scale bar 10 μm.

considerable variation in shapes of the pattern 2 prisms and in the prominence of interrow sheets of enamel (figs. 20.4, 20.5).

Interrow sheets are most distinct in deep enamel for all species but are particularly pronounced in the six hypsodont forms (*Rhynchippus, Pericotoxodon, Miocochilius, Trachytherus, Archaeohyrax, Hegetotherium*) (fig. 20.5). In these species interrow sheets are thick and interrow crystallites are nearly perpendicular to prismatic crystallites. This type of enamel has been termed "modified radial enamel" and is considered an adaptation to the biomechanical requirements of hypsodont teeth (Pfretzschner 1993). A similar configuration is found in the deep enamel of the

hypsodont artiodactyls *Ovis* and *Capra* (Grine et al. 1986, 1987).

Polarity of prism packing patterns in Notoungulata was assessed by comparison with enamel structure of the North American mioclaenid condylarth *Litaletes*, a form thought to be close to the ancestry of Notoungulata (Cifelli 1983b). Like the notoungulates, the superficial enamel of *Litaletes* is characterized by pattern 1 prisms, and its deep enamel is predominantly pattern 2, although interrow sheets are not as well developed as in notoungulates. The intermediate enamel of *Litaletes* also is predominantly pattern 2 but unlike the notoungulates has some regions of pattern 3 prisms. My preliminary studies show that this distribution of prism

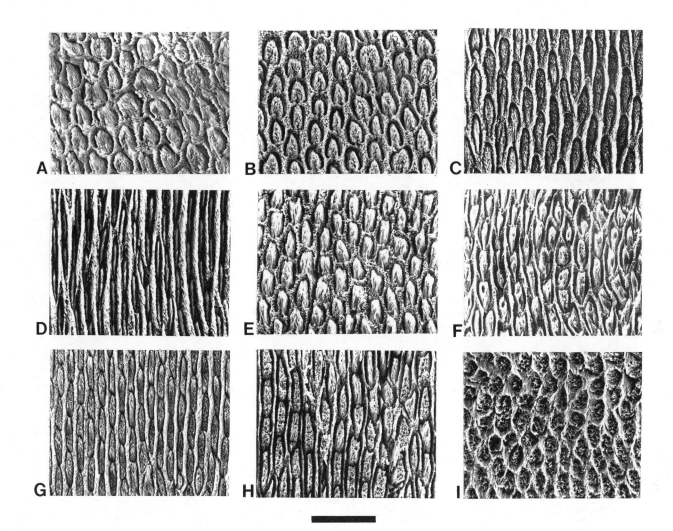

Figure 20.5. Tangential sections of deep enamel: *A, Henricosbornia lophodonta; B, Notostylops pendens; C, Rhynchippus* sp.; *D, Pericotoxodon platignathus; E, Notopithecus adapinus; F, Miocochilius anamopodus; G, Trachytherus* sp.; *H, Archaeohyrax* sp.; *I, Hegetotherium* sp. Scale bar 10 μm.

types is typical of a variety of other North American condylarths as well. The *Litaletes* configuration suggests that pattern 1 superficial enamel and pattern 2 intermediate and deep enamel is primitive for Notoungulata.

Like prism packing patterns, prism decussation varies considerably within the Notoungulata (fig. 20.6; table 20.3). In some forms, decussation is pronounced, with abrupt changes in prism orientation between adjacent decussation zones, whereas in others the change in prism orientation between zones is gradual, and zones are poorly defined. Likewise, in some cases decussation zones extend from the EDJ to the surface, whereas in others the outer enamel (and sometimes the innermost enamel) is nondecussating.

Much of the variation in decussation patterns occurs within suborders. Among typotheres, *Miocochilius* shows pronounced decussation, *Notopithecus* lacks decussation in some regions and is characterized by weakly defined decussation zones in other regions, and *Trachytherus* shows no indication of prism decussation. Likewise, within the Toxodonta *Rhynchippus* is distinguished by nondecussating enamel, whereas decussation is pronounced in *Pericotoxodon*. Hegetotheres also show differences in prism decussation: *Hegetotherium* enamel is strongly decussating, but *Archaeohyrax* has very poorly defined decussation zones or lacks decussation altogether. Among "Notioprogonia," *Notostylops* and, particularly, *Henricosbornia* display only weakly defined zones. In these brachydont forms, as in

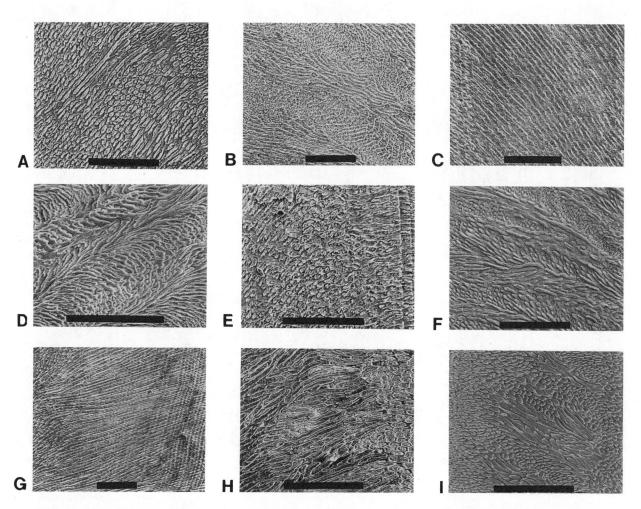

Figure 20.6. Longitudinal sections in a buccolingual plane through notoungulate molars: *A, Henricosbornia lophodonta; B, Notostylops pendens; C, Rhynchippus* sp.; *D, Pericotoxodon platignathus; E, Notopithecus adapinus; F, Miocochilius anamopodus; G, Trachytherus* sp.; *H, Archaeohyrax* sp.; *I, Hegetotherium* sp. Scale bars 100 μm.

Notopithecus, decussation zones are irregular in width and in their extent from the EDJ, and the change in orientation of prism long axes between zones is gradual. Interestingly, *Archaeohyrax, Trachytherus,* and *Rhynchippus,* all of which are hypsodont and either lack decussation or have poorly defined zones, appear to be characterized by distinct inner zones of modified radial enamel.

Decussation is found in the mioclaenid condylarth *Litaletes* as well as the primitive forms *Henricosbornia* and *Notostylops,* suggesting that prism decussation may be primitive for notoungulates. Differences, however, in decussation patterns within suborders and the absence of decussation in some toxodonts and typotheres give some support to the alternative possibility that prism decussation may

have been independently acquired in several notoungulate lineages. Prism decussation is found among many, although not all, families of North American Paleocene condylarths, particularly those of large body size (Von Koenigswald et al. 1987), and in representatives of other orders of South American ungulates (Astrapotheria, Litopterna, Pyrotheria) (Fortelius 1985). The demonstration that decussation confers the biomechanical property of resistance to crack propagation (e.g., Pfretzschner 1986) and its wide distribution within Mammalia indicate that it has developed convergently several times during mammalian evolution (e.g., Von Koenigswald and Pfretzschner 1987; Von Koenigswald et al. 1987), and this is likely to have been the case within the Notoungulata as well.

Table 20.4. Prism size and shape variables (x, y, and d) and prism area (PA) for three depths of enamel relative to the outer tooth surface

Taxon	x Mean	SD	SE	CV	y Mean	SD	SE	CV	d Mean	SD	SE	CV	PA Mean	SD	SE	CV
DEEP ENAMEL
Archaeohyrax	3.96	0.63	0.14	15.8	5.41	0.58	0.13	10.7	3.77	0.39	0.21	10.5	6.47	0.88	0.20	13.6
Henricosbornia	3.91	0.42	0.09	10.8	4.00	0.32	0.07	7.9	4.66	1.16	0.26	25.0	6.21	0.98	0.22	15.7
Hegetotherium	3.69	0.55	0.12	14.9	3.93	0.64	0.14	16.3	3.28	0.33	0.07	10.1	5.94	1.37	0.31	23.0
Miocochilius	3.71	0.55	0.12	14.9	5.02	0.83	0.18	16.6	3.50	0.46	0.10	13.1	5.33	0.86	0.19	16.1
Notopithecus	3.80	0.38	0.09	10.0	3.72	0.48	0.11	13.0	4.55	0.51	0.12	11.3	5.76	1.03	0.23	17.8
Notostylops	3.70	0.59	0.13	15.8	4.09	0.41	0.09	10.0	2.91	0.44	0.10	15.0	5.15	0.78	0.18	15.2
Pericotoxodon	5.06	1.51	0.33	29.7	7.37	0.93	0.21	12.7	4.80	0.98	0.22	20.4	5.53	1.28	0.28	23.1
Rhynchippus	4.64	0.87	0.19	18.7	7.48	1.11	0.25	14.9	4.63	0.59	0.12	12.8	8.88	1.66	0.37	18.7
Trachytherus	5.35	1.10	0.25	20.7	8.22	1.02	0.23	12.4	4.62	0.96	0.22	20.8	7.43	1.26	0.29	16.9
INTERMEDIATE ENAMEL
Archaeohyrax	3.52	1.22	0.27	34.6	4.79	0.93	0.21	19.5	3.34	0.43	0.10	12.9	5.37	0.94	0.21	17.6
Henricosbornia	3.99	0.41	0.09	10.3	4.14	0.30	0.07	7.3	3.44	0.32	0.07	9.2	5.69	1.17	0.26	20.5
Hegetotherium	4.05	0.83	0.18	20.4	5.36	1.56	0.35	29.0	3.41	0.98	0.22	28.8	8.37	3.23	0.72	38.6
Miocochilius	3.79	0.33	0.07	8.6	4.28	0.68	0.15	15.9	2.60	0.25	0.06	9.8	4.02	0.59	0.13	14.6
Notopithecus	3.79	0.38	0.09	9.9	3.11	0.38	0.09	12.3	3.68	0.45	0.10	12.3	3.94	0.82	0.18	20.8
Notostylops	3.61	0.59	0.13	16.4	4.97	1.74	0.39	34.9	3.33	0.60	0.13	17.9	5.34	0.92	0.21	17.1
Pericotoxodon	4.87	0.90	0.20	18.5	8.85	1.50	0.33	17.0	5.08	0.99	0.22	19.5	6.82	1.17	0.26	17.2
Rhynchippus	4.26	0.43	0.10	10.1	7.33	0.47	0.11	6.4	4.18	0.56	0.12	13.5	7.55	1.46	0.33	19.3
Trachytherus	4.61	0.59	0.14	12.7	6.46	0.48	0.11	7.5	4.25	0.81	0.19	19.0	8.46	1.85	0.42	21.8
SUPERFICIAL ENAMEL
Archaeohyrax	3.53	0.61	0.14	17.3	3.46	0.53	0.12	15.4	3.94	0.59	0.13	15.0	4.59	0.74	0.17	16.1
Henricosbornia	4.86	0.56	0.12	11.5	4.70	0.34	0.08	7.3	6.32	1.16	0.26	18.4	11.58	1.37	0.31	11.8
Hegetotherium	5.88	1.20	0.26	20.4	5.08	0.72	0.16	14.1	5.24	0.78	0.17	14.9	14.38	2.50	0.56	17.4
Miocochilius	2.94	0.66	0.15	22.3	3.21	0.24	0.05	7.6	2.84	0.31	0.07	11.0	3.35	0.58	0.13	17.3
Notopithecus	4.51	0.44	0.10	9.8	4.09	0.56	0.12	13.7	4.24	0.68	0.15	16.0	4.46	0.60	0.13	13.5
Notostylops	4.02	0.39	0.09	9.7	4.09	0.32	0.07	7.8	3.80	0.31	0.07	8.1	4.63	1.06	0.24	22.8
Pericotoxodon	6.15	0.54	0.12	8.8	11.08	0.71	0.15	6.4	5.94	0.58	0.13	9.8	8.78	1.95	0.44	22.2
Rhynchippus	4.75	0.59	0.13	12.5	7.67	0.69	0.15	8.9	4.75	0.79	0.18	16.6	12.68	1.85	0.42	14.6
Trachytherus	4.56	1.27	0.28	27.9	5.06	1.40	0.31	27.7	4.19	0.56	0.13	13.4	9.40	2.88	0.64	30.6

Notes: Sample statistics are based on ten prisms from each of two individuals for each taxon (i.e., N = 20). Means, standard deviations (SD), and standard errors (SE) are given here in micrometers for x, y, and d (μm² for *PA*). Coefficient of variation, CV = 100 · SD/mean.

Table 20.5. Average prism shape and size parameters

Taxon	CD	K	ASA	PA/ASA
	DEEP ENAMEL			
Archaeohyrax	4.07	0.58	14.38	0.46
Hegetotherium	3.55	0.85	11.04	0.58
Henricosbornia	4.04	1.03	14.30	0.45
Miocochilius	3.68	0.60	11.86	0.46
Notopithecus	3.90	1.14	13.23	0.44
Notostylops	3.40	0.70	10.05	0.52
Pericotoxodon	5.07	0.49	22.71	0.27
Rhynchippus	4.70	0.42	19.47	0.49
Trachytherus	4.97	0.38	22.13	0.38
	INTERMEDIATE ENAMEL			
Archaeohyrax	3.53	0.58	10.93	0.51
Hegetotherium	3.72	0.60	12.30	0.78
Henricosbornia	3.80	0.85	12.59	0.45
Miocochilius	3.28	0.61	9.39	0.44
Notopithecus	3.45	1.27	10.38	0.39
Notostylops	3.46	0.60	10.54	0.53
Pericotoxodon	4.73	0.29	19.79	0.38
Rhynchippus	4.11	0.34	14.82	0.53
Trachytherus	4.61	0.53	18.64	0.48
	SUPERFICIAL ENAMEL			
Archaeohyrax	3.54	1.10	10.92	0.44
Hegetotherium	5.16	1.06	23.31	0.62
Henricosbornia	4.98	1.13	21.49	0.54
Miocochilius	2.91	0.85	7.45	0.51
Notopithecus	4.20	1.09	15.36	0.29
Notostylops	3.93	0.93	13.37	0.35
Pericotoxodon	5.51	0.25	26.37	0.33
Rhynchippus	4.89	0.42	20.85	0.63
Trachytherus	4.14	0.93	15.35	0.64

Notes: CD: distance between prism centers. *K:* apicocervical compression/distention. *ASA:* average ameloblast secretory area (μm^2). *PA/ASA:* ratio of prism area to *ASA*. See table 20.2 and text for explanation.

In addition to qualitative differences in notoungulate prism packing patterns and decussation, notoungulate enamel is characterized by considerable metrical variation (table 20.4). Distributions of prism cross-sectional variables *x, y, d,* and *PA* for each species and enamel depth were tested for deviations from normality. In almost every case the Kolmogorov-Smirnov test statistic (Sokal and Rohlf 1981) was highly significant. Moreover, Levene's test of equality of variances (Van Valen 1978) demonstrated that each variable was significantly heteroscedastic. Coefficients of variation for each variable typically range between 10 and 20% but are even higher in some cases. Interestingly, forms with the most pronounced prism decussation (*Pericotoxodon, Miocochilius, Hegetotherium*) show no more variability than forms with nondecussating enamel (*Rhynchippus, Trachytherus*), indicating that preparation-dependent variability is no more of a factor for decussating than for nondecussating enamel. It is likely, nevertheless, that some component of variation is, in fact, the result of differences in angle of intercept between prism long axes and the tangential facet, owing to the sinuous course of prisms from the EDJ to the surface, even in forms without decussation.

Average values of prism cross-sectional shape variables *ASA, CD, K,* and *PA/ASA* are reported in table 20.5 for each species and each enamel depth. These values are the basis for cluster analysis of notoungulate enamel characters. Analyses for superficial enamel variables yielded only weakly defined clusters and are not considered further. Results of the nonhierarchical *k*-means cluster analysis for each variable and for the combination of standardized variables are shown in tables 20.6 and 20.7. Results of the hierarchical cluster analysis of combined standardized variables for the nine species are depicted as phenograms (fig. 20.7). Cophenetic correlation coefficients were 0.85 and 0.80 for deep and intermediate enamel trees, respectively, confirming the goodness of fit of the phenograms to the original Euclidean distance matrix.

Although the clustering patterns for intermediate and deep enamel are more consistent among variables than clusters based on superficial enamel, clusters for the two depths differ from one another, and in neither case does cluster membership reflect the subordinal classification based on cranial and gross dental characters. For example, the toxodonts *Pericotoxodon* and *Rhynchippus* form a discrete cluster only according to variable *K* for intermediate enamel. Surprisingly, the mesotheriid typothere *Trachytherus* clusters consistently with either *Rhynchippus* or *Pericotoxodon* according to all variables but *PA/ASA* for deep enamel. The

Table 20.6. Nonhierarchic classification of notoungulates based on variables of deep enamel and *k*-means cluster analysis

Variable(s)[a]	Cluster 1	Cluster 2	Cluster 3	Cluster 4
ASA				(Cluster 2)[b]
	Archaeohyrax	*Pericotoxodon*	*Hegetotherium*	*Rhynchippus*
	Henricosbornia	*Trachytherus*	*Miocochilius*	
	Notopithecus		*Notostylops*	
CD				(Cluster 2)[b]
	Archaeohyrax	*Pericotoxodon*	*Hegetotherium*	*Rhynchippus*
	Henricosbornia	*Trachytherus*	*Miocochilius*	
	Notopithecus		*Notostylops*	
K				(Cluster 3)[b]
	Pericotoxodon	*Henricosbornia*	*Archaeohyrax*	*Hegetotherium*
	Rhynchippus	*Notopithecus*	*Miocochilius*	
	Trachytherus		*Notostylops*	
PA/ASA				(Cluster 1)[b]
	Archaeohyrax	*Pericotoxodon*	*Hegetotherium*	*Trachytherus*
	Henricosbornia		*Notostylops*	
	Notopithecus			
	Miocochilius			
	Rhynchippus			
ASA, CD, K, PA/ASA (standardized)				(Cluster 2)[b]
	Archaeohyrax	*Pericotoxodon*	*Henricosbornia*	*Rhynchippus*
	Hegetotherium	*Trachytherus*	*Notopithecus*	
	Miocochilius			
	Notostylops			

[a]Variables are as defined in table 20.5.

[b]Membership based on three-cluster classification.

two toxodonts and *Trachytherus* form a distinct cluster for all deep enamel variables and combinations of variables, except *PA/ASA*. Moreover, even within a single family variation is pronounced: the two interatheriids, *Miocochilius* and *Notopithecus*, are not closely linked in any analysis. Instead, *Notopithecus* consistently clusters with another brachydont form, *Henricosbornia*.

Cluster analysis does not substantiate a close relationship between the two "hegetotheres" *Archaeohyrax* and *Hegetotherium*. In addition, the indications from gross dental similarities of affinity between mesotheriids and some archaeohyracids (Cifelli 1985) are not supported by cluster analyses of enamel variables of *Trachytherus* and *Archaeo-*

hyrax. Likewise, *Archaeohyrax* shows no particular similarity in enamel structure to the primitive interatheriid *Notopithecus* although the two share some gross dental characters (Cifelli 1985).

Conclusions

Three unresolved issues concerning notoungulate relationships were considered in this study: the relationships of primitive notoungulates ("Notioprogonia"), the validity of Hegetotheria, and relationships among hegetothere and typothere families. Neither of the primitive notoungulates, *Henricosbornia* and *Notostylops*, show a consistent pattern of

Table 20.7. Nonhierarchic classification of notoungulates based on variables of intermediate enamel and *k*-means cluster analysis

Variable(s)[a]	Cluster 1	Cluster 2	Cluster 3	Cluster 4
ASA	*Archaeohyrax* *Hegetotherium* *Henricosbornia* *Notopithecus* *Notostylops*	*Pericotoxodon* *Trachytherus*	*Rhynchippus*	(Cluster 1)[b] *Miocochilius*
CD	*Archaeohyrax* *Hegetotherium* *Henricosbornia* *Notopithecus* *Notostylops*	*Pericotoxodon* *Trachytherus*	*Rhynchippus*	(Cluster 1)[b] *Miocochilius*
K	*Archaeohyrax* *Hegetotherium* *Miocochilius* *Notostylops* *Trachytherus*	*Notopithecus*	*Henricosbornia*	(Cluster 1)[b] *Pericotoxodon* *Rhynchippus*
PA/ASA	*Archaeohyrax* *Rhynchippus* *Trachytherus*	*Hegetotherium*	*Notopithecus* *Pericotoxodon*	(Cluster 1)[b] *Henricosbornia* *Miocochilius*
ASA, CD, K, PA/ASA (standardized)	*Archaeohyrax* *Henricosbornia* *Miocochilius* *Notostylops* *Rhynchippus*	*Hegetotherium*	*Notopithecus*	(Cluster 1)[b] *Pericotoxodon* *Trachytherus*

[a]Variables are as defined in table 20.5.

[b]Membership based on three-cluster classification.

similarity to any of the more derived notoungulates nor to one another. Both are characterized by pattern 1 superficial enamel, by pattern 2 deep enamel, and weakly defined decussation zones, all of which are apparently primitive for the order. It is noteworthy that all of the brachydont forms examined in this study, including the brachydont interatheriid *Notostylops*, are characterized largely by primitive enamel features, whereas hypsodont forms are distinguished by a variety of apparently derived enamel features. The subordinal validity of Hegetotheria (Archaeohyracidae and Hegetotheriidae) is not supported by the details of

enamel structure examined here. *Archaeohyrax* and *Hegetotherium* share no enamel characters that consistently distinguish them from typotheres, and the two hegetotheres also differ markedly from one another in prism decussation patterns. The final issue concerns the affinities of mesotheriid typotheres, hegetotheres, and interatheriids. Characters of enamel structure do not support a close relationship between mesotheriids and interatheriids as suggested by Patterson (1934a) nor between mesotheriids and hegetotheres as postulated by Cifelli (1985). Instead, the enamel structure of the Deseadan mesothere *Trachytherus* is

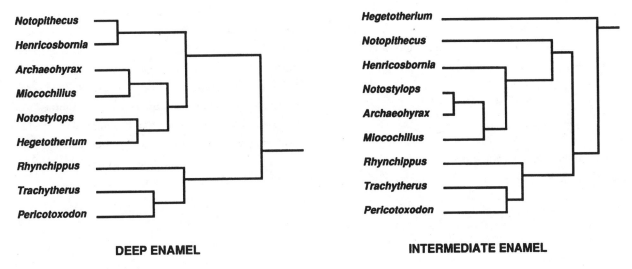

DEEP ENAMEL   **INTERMEDIATE ENAMEL**

Figure 20.7. Phenograms of nine species of notoungulates, based on the method of unweighted pair groups using arithmetic averages (UPGMA) for four standardized parameters *(ASA, CD, K, PA/ASA)* of deep and intermediate enamel. The lengths of the bars reflect relative Euclidean distance.

more similar to toxodont enamel than to that of any typothere or hegetothere.

The failure of the prism variables analyzed here to clarify these phylogenetic relationships within Notounugulata largely reflects the structural complexity of enamel in this group. It has been pointed out (e.g., Von Koenigswald and Pfretzschner 1987; Clemens and Von Koenigswald 1991) that phylogenetic studies must address enamel structure at several levels of complexity, from that of the individual crystallite to the whole dentition. In notoungulates enamel complexity is expressed in varied patterns of crystallite orientation within prisms and interprismatic enamel (e.g., modified radial enamel), in depth related variation in prism packing patterns, and in varied patterns of prism decussation. Similarities in enamel structure are found among members of different notoungulate clades, and different structural patterns occur among members of the same clade. The question then arises as to the significance of this complex pattern of structural variation. Does the apparent lack of phylogenetic information in notoungulate enamel reflect real limitations on the influence of phylogeny on these characters, or does it reflect the interaction of phylogenetic and functional factors?

It has become increasingly clear that a key aspect of enamel structure is the relationship between the physical properties of enamel and dental biomechanics. Two important properties are differential resistance to wear and re-

sistance to fracture. Both are related to the anisotropic structure of mammalian enamel (e.g., Rensberger and Von Koenigswald 1980; Boyde 1984; Stern et al. 1989; Maas 1991). Critical structural features include the regular pattern of differently oriented crystallites that comprise prismatic and interprismatic enamel, the orientation of individual prisms, and the orientation of groups of prisms (decussation zones). Although the latter has received the most attention in relation to dental mechanics (e.g., Rensberger and Von Koenigswald 1980; Fortelius 1985; Pfretzschner 1986; Von Koenigswald et al. 1987), the orientation of individual prisms also is an important influence on gross wear patterns and, thus, dental function. For example, changes in prism orientation between the EDJ and tooth surface have been shown to be related to differential resistance to wear by leading and trailing edges of wear surfaces in herbivorous marsupials (Young et al. 1987; Ferreira et al. 1989) and in notoungulates (Maas 1989). Similarly, among some hypsodont notoungulates enamel on buccal and lingual sides of teeth show different patterns of prism decussation and prism orientation (Maas 1989). Other studies have shown how analysis of microstructural characters within a functional context may elucidate phylogenetic questions (e.g., Von Koenigswald 1982a; Pfretzschner 1993). This study has documented the pattern of variation in enamel structure within the Notoungulata. More detailed studies of the distribution of enamel structural characters at all levels of complexity will be important in understanding functional aspects of the adaptive radiation of this group and the phylogenetic relationships among its members.

Acknowledgments

Thanks are due R. L. Cifelli (Oklahoma Museum of Natural History), R. F. Kay and R. H. Madden (Duke University), B. J. MacFadden and G. Morgan (University of Florida / Florida State Museum), and R. H. Tedford (American Museum of Natural History) for loan of specimens for SEM analysis and to R. L. Cifelli and R. H. Madden for discussions and access to unpublished work concerning notoungulate relationships. The comments of F. E. Grine, R. F. Kay, and W. Von Koenigswald helped to improve this chapter significantly. This research was supported by a grant from the Duke University Research Council and NSF BNS 9020788.

21
A New Toxodontid Notoungulate

Richard H. Madden

RESUMEN

Un nuevo género y especie de Toxodontidae, *Pericotoxodon platignathus,* es distinto de todos los demás Toxodontidae al poseer una combinación de caracteres: entre otros, el margen inferior de la rama horizontal extendido inferolateralmente en el adulto, pliegue lingual bifurcado y persistente y pliegues accesorios muy efímeros (de corta duración con el desgaste) en los molares superiores, pequeño pliegue o plicación distolingual en los primeros dos molares superiores, pliegue distolingual variable en el tercer molar superior, pliegue lingual del cuarto premolar superior y fosétido accesorio de los molares inferiores efímeros, y pliegue ento-hipoconúlido del tercer molar inferior persistente.

Pericotoxodon no está relacionado filogenéticamente con ninguna de las subfamilias supuestamente monofiléticas de Toxodontidae que han sido reconocidos tradicionalmente, sino comparte una serie de sinapomorfías con *Gyrinodon, Plesiotoxodon,* y *Dinotoxodon,* las que sirven para diagnosticar una nueva subfamilia, Dinotoxodontinae. Estas características derivadas en común o sinapomorfías incluyen el margen inferior extendido de la rama horizontal, pliegue lingual bifurcado variable y contacto entre el metalofo y protolofo en los molares superiores, y el alargamiento del tercer molar superior con pliegue distolingual reducido. Material proveniente de la Formación Biblián del Mioceno temprano del Ecuador y asignado al mismo género sugiere que la divergencia de este grupo monofilético pudiera haber ocurrido antes de la divergencia de las subfamilias Haplodontheriinae y Toxodontinae. Los registros de Dinotoxodontinae están restringidos a la zona baja tropical y subtropical de América del Sur.

La masa corporal de *Pericotoxodon platignathus* estimada en base a diversas dimensiones del esqueleto es comparable a la de una jirafa, el rinoceronte negro de Africa, el alce de América del Norte, y el rinoceronte de Sumatra. Las características del rostro y de la dentadura anterior, más un énfasis en los filos cortantes de los dientes jugales, sugieren hábitos pastores para la especie. Por primera vez entre los Toxodontidae se registra un dimorfismo en la forma y el área del corte transversal de los colmillos que sugiere una tendencia de formar grupos sociales o reproductores polígamos, como en diversas especies de ungulados vivientes.

The Toxodontidae are medium to large terrestrial, herbivorous Notoungulata easily recognized by their specialized anterior dentition, ever-growing incisor tusks, and distinctive high-crowned molars. The family is abundant in Miocene faunas throughout South America, being represented in Amazonian Brazil (Paula Couto 1956, 1981, 1982), eastern Peru (Spillmann 1949; Willard 1966), southern Ecuador (Repetto 1977), Venezuela (Hopwood 1928), Bolivia (Hoffstetter 1977, 1986; Villarroel 1977), Patagonia (Scott 1912), and Colombia (Madden 1990). Past studies of the family have dealt most extensively with material from Argentina (Pascual 1965; Pascual et al. 1966).

This chapter describes a new genus of Toxodontidae from Colombia. Toxodonts were among the first fossil vertebrates to be collected from the Miocene Honda Group of Colombia. Brother Ariste Joseph (Mauricio Rollot) collected a toxodontid left P⁴ (USNM 10883) in 1920 near "La Union" in Huila Department. Joint expeditions of the University of California Museum of Paleontology and the Colombian Servicio Geológico National between 1945 and 1950 recovered much additional material of Toxodontidae from the Villavieja area, including a partial skull (UCMP 40199) discovered in 1950 by José E. Périco, which forms the holotype of the new genus described here. Prospecting by investigators at Duke University and the Instituto Nacional de Investigaciones en Geociencias, Mineria y Química (INGEOMINAS) in the Miocene Honda Group has yielded much additional material. These combined collections permit a rather full delimitation of dental and mandibular variation in this new genus.

Notoungulate dental morphology and crown nomenclature used to describe Toxodontidae in this chapter are illustrated in figure 21.1.

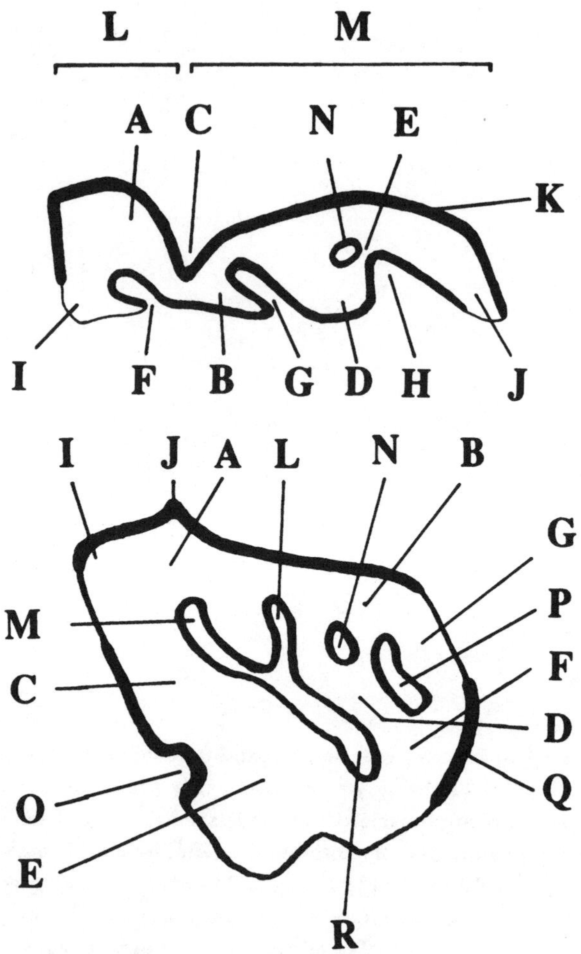

Systematic Paleontology

Order Notoungulata Roth 1903
Suborder Toxodontia Owen 1853
Family Toxodontidae Gervais 1847
Dinotoxodontinae subfam. nov.

TYPE GENUS

Dinotoxodon Mercerat 1895.

INCLUDED GENERA

Gyrinodon Hopwood 1928, *Plesiotoxodon* de Paula Couto 1982, and the new genus described here.

SUBFAMILIAL DIAGNOSIS

Medium-sized advanced Toxodontidae with inferolaterally deep and flaring mandible ramus beneath the molar row in the adult, upper molar F2 fold when present unaffected by wear, upper molar protoloph with enameless lingual col-

Figure 21.1. Dental morphology and terminology in the lower and upper molars of Toxodontidae. *Top*, lower molar features: *A*, anterior crescent, protoconid; *B*, metaconid; *C*, buccal enamel fold; *D*, entoconid, entolophid; *E*, pillar; *F*, anterior fold, sulcus or inflection; *G*, meta-entoconid fold; *H*, ento-hypoconulid fold; *I*, paraconid; *J*, hypoconulid; *K*, buccal enamel; *L*, anterior crescent, trigonid; *M*, posterior crescent, talonid; *N*, accessory fossettid. *Bottom*, upper molar features: *A*, paracone, part of ectoloph; *B*, metacone, part of ectoloph; *C*, paraconule, part of protoloph; *D*, metaconule; *E*, protocone, lingual column, part of protoloph; *F*, part of metaloph; *G*, fourth crest, part of metaloph; *I*, parastyle; *J*, paraconid ridge; *L*, F2 fold; *M*, F1 fold; *N*, F3 fold or fossette; *O*, anterior cingulum; *P*, F4 fold or fossette; *Q*, posterior cingulum; *R*, primary lingual enamel fold.

umn, upper molar metaloph and protoloph in contact lingual to the F2 fold, distally tapering M¹ and M² metalophs with slight lingual enamel indentation, and mesiodistally elongate upper third molars.

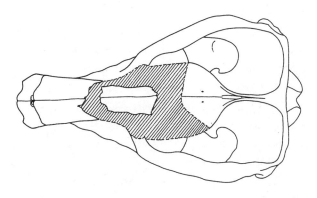

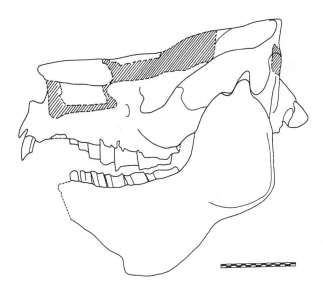

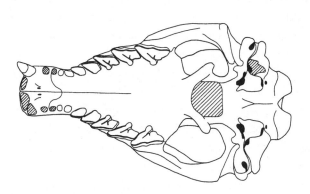

Figure 21.2. Skull of *Pericotoxodon platignathus:* composite reconstruction based on UCMP 40199 (a partial cranium with dentition), UCMP 38002, 39549, and 38980 (mandibles), IGM 184536 (frontal), and IGM 184773 (nasals, sagittal/lambdoid crests). *Top,* superior view; *middle,* lateral view; *bottom,* basal view. Scale bar 15 cm.

Pericotoxodon gen. nov.
(fig. 21.2)

TYPE SPECIES

Pericotoxodon platignathus sp. nov.

GENERIC DIAGNOSIS

Pericotoxodon platignathus is distinct from other Toxodontidae in having an adult mandibular ramus inferolaterally deepened beneath M_2; buccolingually broad lower molar entolophids; individually variable (persistently bifurcate or simple) upper molar primary lingual enamel fold; very ephemeral upper molar accessory lingual enamel folds or fossettes; M^1 and M^2 with distolingual enamel plication; M^3 distolingual fold variably present; ephemeral P^4 lingual enamel fold; ephemeral lower molar accessory fossettid; buccolingually broad P_2 talonid; and persistently open M_3 ento-hypoconulid fold.

Pericotoxodon platignathus sp. nov.

ETYMOLOGY

The genus is named for José Espiritu Périco Araque, discoverer of the holotype. The species name is a combination of two Greek components, *platy,* flat, broad, and *gnathus,* jawed.

HOLOTYPE

UCMP 40199, a partial cranium with left I^{1-2}, P^{2-4}, M^{1-3}, and right P^{3-4}, M^{1-3}, with a portion of the right premaxilla and the internasal margin of the right nasal.

TYPE LOCALITY

The holotype was collected from a concretionary sandstone thought to have been within the "Cerbatana Gravels" but now recognized as the Monkey Beds, Villavieja Formation at UCMP Locality V5043, "La Gaviota," approximately 1 km N65°W of Gaviota Station House on the Neiva-Girardot Railway and approximately 3 km NNW of San

Nicolás house, Municipality of Villavieja, Huila Department, Colombia.

HYPODIGM

The holotype and the following material: IGM 184710, a partial cranium preserving right P^4, M^{1-3}, and left P^{3-4}, M^{1-3}, and associated mandibular rami with left I_{1-3}, P_4, M_1, M_2 (part), and right I_{1-2}, P_4 (Duke Locality 65). UCMP 39549, a partial left mandible ramus with P_4, M_{1-3} (UCMP Locality V4932). UCMP 39845, a partial juvenile cranium preserving left dP^{3-4}, M^{1-2}, and right dP^4, M^{1-2}, and partial mandibular rami with right dP_{2-4}, M_{1-2}, and left dP_{3-4}, and M_2 (UCMP Locality V4937). UCMP 38980, a mandible ramus with left P_{2-4} and M_3 (part) (UCMP Locality V4529). IGM 184367, a partial mandible preserving left I_3, M_{1-2}, and right I_1 (part), I_2, P_4, M_2 (Duke Locality 28). IGM 250903, mandibular rami preserving right I_{1-3}, and left I_{1-3}, C_1, P_{2-4}, M_{1-2} (Duke Locality 116). UCMP 39845, a partial juvenile rostrum and palate with left dP^{3-4}, M^{1-2}, and right dP^4, M^{1-2}, and partial horizontal rami preserving left dP_{3-4}, M_2, and right dP_{2-4}, M_{1-2} (UCMP Locality V4937). IGM 184399, a left mandibular ramus preserving P_3, dP_4, P_4 (encrypted), broken M_{1-3} crowns (Duke Locality 91). IGM 184773, an adult premaxilla, partial left and right nasals, left paroccipital, part of the sagittal and lambdoid crests and other associated cranial fragments, and associated right P^4 and left M^1 or M^2 (Duke Locality 67). UCMP 39632, co-ossified adult premaxillaries with I^1 and I^2 (UCMP Locality V4519). UCMP 39633, a premaxilla with I^1 and I^2 (UCMP Locality V4536). UCMP 39631, left and right unfused premaxillaries with I^1 and (d?)I^2 (UCMP Locality V4519). IGM 183940, a partial mandible of a very young individual preserving the alveoli of right dI_{1-3}, d/c, and the base of the crowns of dP_{1-3} (Duke Locality 73).

LOCALITIES AND HORIZON

The holotype is from a locality in the Monkey Beds, the lowest named subdivision of the Villavieja Formation. The hypodigm material comes from exposures of the Honda Group in the La Venta area at the following localities: UCMP Localities V4519, V4529, V4536, V4932, V4937,

and V5043, and Duke Localities 28, 65, 67, 73, 91, and 116. Less-well-preserved material referred to *Pericotoxodon platignathus* has been collected from numerous additional localities in the Honda Group of the La Venta area. A nearly complete listing of these specimens can be found in Madden (1990). Dental material referrable to this same species also has been recovered in the Polonia Red Beds (Duke Locality 126) in the uppermost Villavieja Formation. These UCMP and DU localities range stratigraphically from a level just above the Cerro Gordo Sandstone Beds (La Victoria Formation) near the base of the section upward through the Polonia Red Beds (Villavieja Formation) (chap. 2). (For the temporal span of this section, see chap. 3.)

DIAGNOSIS

As for the genus.

DESCRIPTION

The *Pericotoxodon* cranium is about the size of that of the Sumatran rhinoceros *Dicerorhinus sumatrensis*. Among Toxodontidae the cranium of *Pericotoxodon* appears unspecialized and presents more similarites with middle Miocene *Palyeidodon* and *Nesodon* than with later, more divergent forms such as *Paratrigodon*, *Trigodon*, *Toxodon*, *Xotodon*, and *Posnanskytherium*.

In superior view the *Pericotoxodon* skull is pyriform, resembling *Nesodon* and *Toxodon* and contrasting with the narrow skulls of *Posnanskytherium* and *Xotodon*, the triangular skulls of *Paratrigodon* and *Trigodon*, and the ovate skulls of *Proadinotherium muensteri* and *Adinotherium*. The temporal fossae in *Pericotoxodon* have a deep and regularly concave posterior margin, similar to the condition seen in *Nesodon* and unlike the angular margin of the fossae in *Palyeidodon* and *Trigodon*. The sagittal crest in *Pericotoxodon* is anteroposteriorly short. Unlike *Nesodon*, the frontal of *Pericotoxodon* appears to have been expanded and the anteriorly divergent temporal crests are short and relatively weakly developed, resembling *Palyeidodon*. The nasals of *Pericotoxodon* are narrow and do not expand laterally as they do in most other Toxodontidae.

In lateral view the inferior margin of the orbit is low on the face; the lateral margin of the nasal aperture flares laterally posterior to the premaxillary tuberosity, which is visible in lateral view. The available material suggests that the nasal aperture was directed forward and the superior profile of the nasals was flat from rhinion to nasion. The frontal is elevated or dome-shaped. The superior profile is more like *Palyeidodon* than *Nesodon*, *Adinotherium*, or *Proadinotherium muensteri* wherein the frontal is depressed and the nasals convex anteroposteriorly from rhinion to nasion.

In *Pericotoxodon* the sharp superior and lateral margins of the infraorbital foramen mark the ventral margin of the zygomatic process as in *Posnanskytherium* and *Trigodon* and unlike *Nesodon* and *Palyeidodon* in which the infraorbital foramen is widely separated from the zygomatic process of the maxilla. The zygomatic arch is dorsoventrally deep and the dorsal margin ascends rather steeply posterior to the inferolateral margin of the orbit, producing a superior profile unlike that of other Toxodontidae. The zygomatic process of the maxilla is positioned above M^3 in *Pericotoxodon* as in most Toxodontidae (except *Posnanskytherium*, where its anterior root extends above M^2). The occipital condyles do not project as far backward and downward as they do in *Nesodon*. The plane of the occiput is perpendicular to the basicranial axis, not sloped forward as in *Nesodon* or backward as in *Adinotherium*.

In vental view the palate of *Pericotoxodon* is broadest at M^3 and narrows sharply to P^2. Anterior to P^2, the palate is elongate and the left and right alveolar margins are subparallel. Upper I^3, C, and P^1 are separated from each other and from the cheek-tooth series by short diastemata. By contrast, in *Nesodon* the palate is more nearly triangular, the tooth rows converge anterior to M^3, the premolar size gradient from P^4 to P^1 is more gradual and diastemata only occur between canine and I^3 and between I^3 and I^2. Like *Palyeidodon*, *Pericotoxodon* displays a swollen, spherical, anteroposteriorly compressed and laterally projecting postglenoid process.

The most distinctive feature of the adult mandibular ramus in *Pericotoxodon* is the inferolateral expansion of the inferior margin beneath the molars (fig. 21.3). In lateral profile the inferior and alveolar margins of the horizontal ramus begin to diverge anteriorly from a point just beneath M_3 and beneath the anteriormost point of the masseter insertion. The horizontal ramus attains greatest dorsoventral depth beneath M_2. From beneath M_2 the inferior and al-

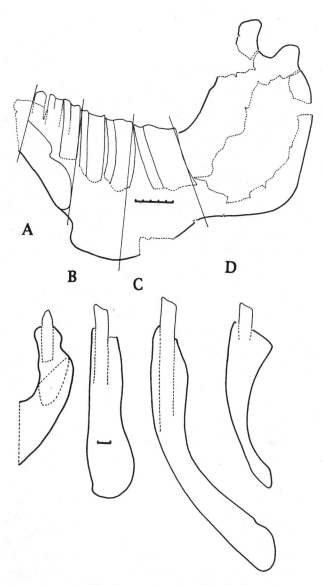

Figure 21.3. Mandible of *Pericotoxodon platignathus* adult. Composite reconstruction based on UCMP 38002, 39549, and 38980 from the Villavieja Formation. *Top:* Lateral view with premolar and molar crowns and midline symphysis profile indicated. Scale bar 5 cm. *Bottom,* four transverse cross sections of the horizontal ramus: *A,* just anterior to P_2; *B,* between P_4 and M_1; *C,* between M_2 and M_3; *D,* posterior to M_3; all sections oriented as viewed from the front, with lateral to the right, medial to the left. Scale bar 1 cm.

veolar margins converge anteriorly to the symphysis. In lateral profile the inferolateral expansion is rounded. In UCMP 39549 the projecting inferior margin extends anteriorly to the base of the symphysis where it forms a slight anterior salient before joining the steep upward slope of the symphysis. The rather pronounced anterior profile in

UCMP 39549 contrasts with the less-well-developed and smoothly rounded projection seen in most other specimens (e.g., UCMP 38980 and IGM 184367). The border of the inferolateral process is rounded and somewhat roughened on the external and inferior surfaces.

Among Toxodontidae a comparable inferolaterally projecting mandible ramus occurs in adult *Dinotoxodon paranensis* (MLP 39-XII-2-8) from the late Miocene of Argentina and other known Dinotoxodontinae. The inferolateral projection or process in *Dinotoxodon* is not as deep or pronounced as in *Pericotoxodon*.

In *Pericotoxodon* both the juvenile and adult mandible is fully synostosed at the symphysis. In transverse cross section viewed anteriorly, the symphysis is rounded inferiorly beneath P_4. From the level of P_2 anteriorly the inferior margin is angled. In cross section beneath P_1 and P_2 the lateral profile of the symphysis is expanded outward to accommodate the shaft of the I_3 tusk. In transverse cross section the symphysis is rounded internally and forms a smooth trough that gradually widens anteriorly. In midsagittal cross section the symphysis extends posteriorly to beneath P_4. The posterior end of the symphysis is bluntly rounded and curves upward anteriorly into a well-developed posterodorsal depression. Above this depression the symphysis attains its greatest dorsovental thickness. Anteriorly, the thickness of the symphysis gradually decreases. The symphyseal cross section in adult *Pericotoxodon* seems to be unique among Toxodontidae. Unlike *Pericotoxodon*, the pronounced posterodorsal depression in *Toxodon* and *Xotodon* is directed more posteriorly. Primitively, the depression is directed ventrally as in *Adinotherium*, *Proadinotherium leptognathum*, and *Nesodon*.

The horizontal ramus of *Pericotoxodon* undergoes progressive inferolateral deepening with age (fig. 21.4). The onset of development of the inferolateral projection appears to coincide with the shedding of dP_4, the last deciduous tooth to be replaced. In general, the very young juvenile mandible of *Pericotoxodon* (e.g., IGM 183940) compares in shape with the very young juvenile mandible of *Nesodon* (e.g., YPM-PU 15746), but with a more keeled ventral midline and fully synostosed symphysis. The inferior margin of the older juvenile mandible (IGM 184399) is smoothly rounded in lateral view.

No single specimen of *Pericotoxodon* preserves a complete

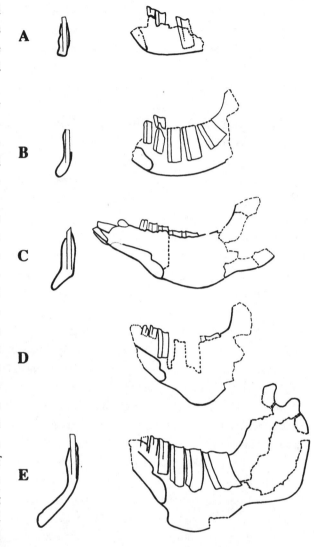

Figure 21.4. Hypothetical growth series of the mandible of *Pericotoxodon platignathus*. *Right:* Horizontal ramus in lateral view with midline profile of the symphysis. *Left:* Transverse sections at the level of M_2, oriented lateral to the left, medial to the right. *A,* UCMP 39845, juvenile; *B,* IGM 184399; *C,* IGM 250903; *D,* UCMP 38980; *E,* composite reconstruction of UCMP 38002, 39549, and 38980. See figure 21.2 for approximate scale.

upper dentition intact. The premaxilla of *Pericotoxodon* is composed of dense alveolar bone surrounding very high-crowned incisors. The upper central incisor crown is scalpriform, strongly curved longitudinally, triangular in cross section, and very high crowned. Enamel covers only the labial side of the crown. The I^1 crown gradually tapers toward the root. In some specimens the pulp cavity is closed. Although enamel deposition was of limited dura-

RICHARD H. MADDEN **341**

tion, the crown height of this tooth in *Pericotoxodon* is significantly greater than I¹ of *Nesodon* or *Palyeidodon*. The lateral
profile of the crown between the occlusal surface and the
site of enamel formation describes an arc greater than 90°.

The lateral incisor (I²) is a longitudinally curved, ever-
growing tusk, triangular in cross section and wears to an
apex at the mesiobuccal corner. The arc of longitudinal
curvature is constant and extends to more than 90° between
crown apex and open pulp cavity. In transverse cross section
only the mesiobuccal angle is covered with enamel. The
enamel covers the full length of the buccal surface and only
half of the mesial surface. There is no enamel on the distal or
lingual surfaces.

The alveolar bone in the area of the premaxillary-
maxillary suture in *Pericotoxodon* is rarely preserved. The
presence of an I³ is suggested by what may be a small
alveolus in the premaxilla of UCMP 40199 although no
crowns can be confidently identified. Isolated teeth (IGM
183028 and IGM 182912) similar in morphology to the
upper canines of *Nesodon* described by Scott (1912) are
identified as upper canines.

No example of P¹ can be identified with confidence. P²
and P³ are generally cylindrical in cross section and are ever-
growing. P⁴, and to a lesser degree, P³, are elongate labially,
giving the tooth a tear-drop shape in cross section (fig.
21.5). A longitudinal concavity or shallow parastyle sulcus
is usually present on the external (labial) surface. The external surface is covered with a broad enamel band that wraps
around the anteroexternal corner. A second enamel band
occurs on the mesiolingual aspect. Both enamel bands
decrease in width and the mesial band is obliterated in
advanced stages of wear. In general, the occlusal surface of
P⁴ is featureless; however, two specimens (IGM 250501 and
IGM 183943) preserve an ephemeral lingual enamel fold at
the occlusal surface.

The upper molars are ever-growing structures with the
occlusal surface complicated by the presence of a lingual
enamel fold and a distolingual enamel inflection. The protoloph supports a lingual column. The upper molar lingual
fold may be simple (or single, as in UCMP 40176) or bifurcate (as in UCMP 39612). The distal branch of the bifurcate
fold may be present or absent at different tooth positions
within a single tooth row (e.g., UCMP 38981). The enamel on the lingual aspect of the metaloph supports a

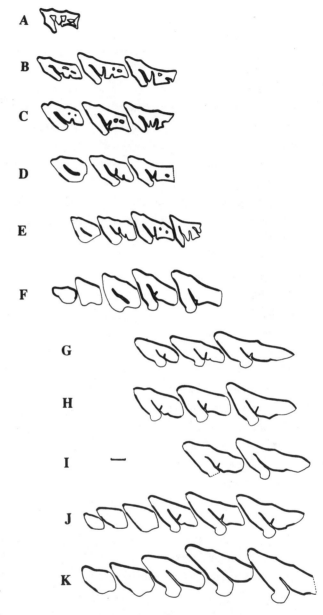

Figure 21.5. Age/wear-related change and individual variation in the
deciduous and permanent upper dentition of *Pericotoxodon platignathus*.
Tooth orientation is buccal side up, mesial side to the left. *A*, UCMP
40047, dP³: *B*, IGM 182753, dP³⁻⁴, M¹; *C*, UCMP 39883, dP³⁻⁴, M¹;
D, UCMP 39883, dP³⁻⁴, M¹ (base of the crowns); *E*, UCMP 39845,
dP³⁻⁴, M¹⁻² (reversed); *F,* UCMP 38882, P², dP³⁻⁴, MP1⁻² (reversed);
G, IGM 183396, right M¹⁻³ (reversed); *H*, IGM 184764, right M¹⁻³
(reversed); *I*, UCMP 38981, left M²⁻³; *J*, UCMP 40199, left P²⁻⁴,
M¹⁻³; *K*, UCMP 40176, left P³⁻⁴, M¹⁻³. Scale bar 1 cm.

distolingual inflection. The expression of this enamel feature is variable, ranging from a slight concavity (IGM
184764) to an enamel inflection or fold (IGM 183396).

The three upper molars increase in size from M¹ to M³. M² differs from M¹ in having a proportionally longer metaloph. The M¹ and M² metaloph usually tapers distally although some specimens (IGM 184049 and IGM 182877) show a slight increase in the transverse breadth of the metaloph distally. M³ displays a mesiodistally elongate crown, more concave ectoloph, and a tapering distal extremity. The distally elongate lingual enamel on M³ supports a variable F4 fold, ranging from a pronounced fold (as in UCMP 39612) to a slight inflection (UCMP 38981).

In the lower tooth row a short diastema occurs between I_3 and the lower canine, and a longer diastema between the canine and the most anterior premolar. I_1 and I_2 are very high crowned, strongly curved longitudinally, and may have been ever-growing. The crowns are triangular in cross section, but cross sectional shape changes notably with wear. Lingual and buccal enamel bands are present in relatively unworn specimens. The lingual band diminishes in width and thickness with wear and disappears before the buccal band. The mesial aspect is covered to varying extent by the buccal enamel band.

As in other Toxodontidae, the lower lateral incisor (I_3) of *Pericotoxodon* is a procumbent, longitudinally curved, ever-growing tusk. Triangular in cross section, the mesial side is shortest, the buccal and lingual sides are long and reach an acute angle laterally. A thin enamel band covers the lateral two-thirds of the buccal face and wraps around the lateral angle of the crown to overlap onto the lingual enamel band. The thin lingual enamel band covers the lateral three-quarters of the lingual face.

The lower canine is diminutive, single-rooted, low crowned, and bladelike. In occlusal view, the crown is crescent-shaped, concave lingually. Judging from crown wear, occlusion between the upper and lower canines occurred along a single plane with mesiodistal strike and buccal dip. Accordingly, the buccal enamel of the lower canine is higher (from cemento-enamel junction to occlusal surface) than the lingual enamel.

No single specimen of mandible preserves more than three lower premolars. The three premolars are similar in morphology and increase in crown height and cross-sectional area posteriorly (fig. 21.6). The P_2 crown is bicrescentic with a buccal enamel fold separating subequal mesial and distal crescents in less-worn specimens. With moderate

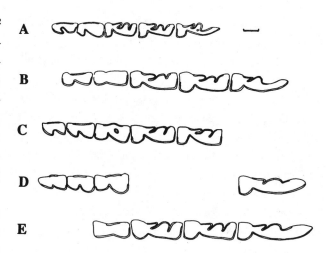

Figure 21.6. Lower dentition of *Pericotoxodon platignathus*, arranged by dental eruption sequence, increasing mandible depth, and increasing M_3 crown length. Tooth orientation is buccal side down, mesial side left. *A*, UCMP 39845, right dP_{2-4}, M_{1-2}; *B*, IGM 184399, left P_3, dP_4, M_{1-3}; *C*, IGM 250903, left P_{2-4}, M_{1-2}; *D*, UCMP 38980, left P_{2-4}, M_3; *E*, UCMP 39549, left P_4, M_{1-3}. Scale bar 1 cm.

wear the mesial crescent becomes longer than the distal crescent. With additional wear the buccal enamel fold is obliterated, the crescents become indistinct, and the lingual enamel band diminishes and is eventually obliterated. At advanced stages of wear the crown becomes ovoid in cross section with a single root (IGM 183616 and IGM 182923).

P_3 and P_4 are very high crowned and rootless (ever-growing). The P_3 crown is slightly concave lingually and distally, whereas P_4 is slightly concave lingually and mesially. Unlike P_2, there is little change in the relative length of the buccal enamel crescents with wear. Some isolated lower premolars preserve an enamel-lined fossettid in the position of the meta-entoconid fold, which becomes obliterated in relatively early wear stages.

Pericotoxodon's lower molars are bicrescentic and ever-growing. M_1 and M_2 have nearly the same dimensions so that identifying the tooth position of isolated specimens is always uncertain. All three lower molars have both meta-entoconid and ento-hypoconulid folds in the lingual enamel band. In general, both folds are trenchant and are reflected mesially in M_1 and M_2. The ento-hypoconulid fold of M_1 becomes an isolated enamel-lined fossettid with advanced wear. In very old individuals, when the ento-hypoconulid fossettid becomes obliterated, the meta-entoconid fold becomes isolated as a fossettid and the buccal enamel fold opens into a shallow concavity. This pattern of obliteration of the lower molar lingual folds in *Pericotoxodon* occurs as well in *Palyeidodon* and is unlike the pattern seen in *Trigodon* or *Stereotoxodon*.

The M_3 of *Pericotoxodon* has an elongate posterior crescent with an open ento-hypoconulid fold. The degree to which the M_3 ento-hypoconulid fold opens is correlated with increasing mesiodistal crown length, which evidently continues throughout the life span of the individual as the tooth grows upward from its base and is worn away at the top. In unworn specimens small, closely appressed, enamel-lined fossettids appear on the entoconid lobe in the position of Loomis's "bay 3" (Loomis 1914). These fossettids are rapidly obliterated with wear.

The deciduous dentition of *Pericotoxodon* is characterized by (1) the absence of evergrowing tusks, both above and below; (2) a shallow mandible with a rounded anterior dental arcade and imbricating lower incisor crowns; (3) rooted upper and lower deciduous cheek teeth with occlusal surfaces complicated by persistent enamel folds and fossettes; and (4) thinner enamel compared with the permanent teeth.

The first upper deciduous incisor crown is larger than dI^2. The crown of dI^1 is mesiodistally elongate, spatulate, and tapers to a single root. The occlusal surface changes from crescentic to triangular with wear. The enamel is continuous around the crown apex in the least-worn specimen but is rapidly worn away from the mesiolingual corner first and then from the lingual surface, thereafter becoming confined to the buccal surface.

The deciduous second upper incisor crown is triangular in outline. In little-worn specimens the enamel wraps continuously around the circumference but with wear disappears first from the mesiolingual corner and then from the lingual and mesial surfaces. The deciduous anterior teeth are arrayed transversely in the premaxilla and form a continuous enamel crest or blade. No maxillary deciduous canine or deciduous first or second premolars can be identified confidently in the available collections.

The deciduous third upper premolar is hypsodont and molariform. With wear the occlusal surface changes from quadrilateral (with no parallel sides) to trapezoidal (with subparallel mesial and distal sides) to irregularly tear-shaped in outline. A distinct parastyle is a persistent feature of the dP³ crown. Enamel initially envelops the dP³ crown but with wear is rapidly lost from the mesiobuccal corner and the lingual end of the protoloph to isolate the mesial enamel band. The enamel is lost from the lingual surface when the lingual fold becomes an isolated fossette. The dP⁴ crown is longer mesiodistally and has a less-pronounced parastyle than dP³. The occlusal surface of dP⁴ is more complex than dP³, displaying a posterointernal fossette. The crown height of the deciduous cheek teeth in *Pericotoxodon* appears to be greater than that of juvenile Santacrucian *Nesodon* and *Adinotherium*.

The deciduous lower incisors and canine were procumbent and no diastemata appear between these tooth positions. The crown of dI_3 in *Pericotoxodon* is nearly identical to that of *Nesodon* except that it lacks any trace of the median buccal ridge or buttress of *Nesodon*.

By contrast with the lower permanent molars, the deciduous premolars form distinct roots. In the unworn state dP_2 displays one buccal and two shallow lingual enamel folds. The meta-entoconid fold is the deeper of the two and persists as a closed fossettid through intermediate wear stages. In more worn examples the lingual folds and fossettids are completely obliterated. The unworn crown of dP_3 is more molariform than dP_2, higher crowned, and has two lingual enamel folds. Lightly worn specimens show a buccolingually narrow trigonid with open anterior sinus and two fossettids in the position of the meta-entoconid and ento-hypoconulid folds of lower molars. The ento-hypoconulid fossettid is the first to become obliterated by wear. The dP_4 crown is fully molariform, that is, the trigonid or anterior crescent is buccolingually broad, there is no trace of an anterior sinus, and there are two lingual enamel folds in the positions of the meta-entoconid and ento-hypoconulid folds. With wear the ento-hypoconulid fold becomes an isolated fossettid before the meta-entoconid fold.

VARIATION

The available material of *Pericotoxodon* reveals considerable metric and morphologic variation in the permanent teeth (table 21.1). Metric variation in *Pericotoxodon platignathus* is comparable to the total variation presented by Santacrucian

Table 21.1. Tooth measurements (cm) for *Pericotoxodon platignathus*

Tooth	Measurement		*Pericotoxodon platignathus*					*Nesodon imbricatus*	
		N	Mean	Minimum	Maximum	SD	CV	N	CV
I²	Maximum diameter	18	2.51	1.80	3.01	0.430	17.02	0	—
	Breadth	19	1.96	1.63	2.13	0.160	8.14	0	—
P³	Maximum diameter	6	2.18	1.79	2.57	—	—	0	—
	Breadth	6	1.76	1.45	2.17	—	—	0	—
P⁴	Maximum diameter	5	3.25	2.97	3.43	—	—	0	—
	Breadth	5	2.01	1.75	2.22	—	—	0	—
M¹	Ectoloph length	7	4.02	3.23	4.83	0.547	13.61	12	7.35
	Maximum breadth	7	2.19	1.61	2.60	0.392	17.95	0	—
M²	Ectoloph length	8	4.60	3.65	5.33	0.544	11.82	10	5.91
	Maximum breadth	7	2.42	1.79	2.72	0.347	14.35	0	—
M³	Ectoloph length	15	5.42	4.81	6.06	0.454	8.38	12	13.35
	Trigonid breadth	12	2.42	2.09	2.87	0.249	10.27	0	—
I₃	Length	10	2.71	2.00	3.21	0.460	16.80	0	—
	Breadth	9	1.49	0.75	1.87	0.350	23.31	0	—
P₂	Length	4	1.57	1.52	1.69	—	—	0	—
	Maximum breadth	4	0.98	0.94	1.07	—	—	0	—
P₃	Length	4	1.75	1.70	1.83	—	—	0	—
	Maximum breadth	4	1.09	1.05	1.12	—	—	0	—
P₄	Length	11	2.21	2.09	2.41	0.109	4.95	0	—
	Maximum breadth	12	1.33	1.03	1.88	0.285	21.50	0	—
M₁	Length	10	3.01	2.44	3.24	0.253	8.43	11	6.68
	Trigonid breadth	7	1.19	1.02	1.36	—	—	10	10.22
	Talonid breadth	10	1.03	0.79	1.29	0.152	14.80	0	—
M₂	Length	9	3.11	2.70	3.52	0.305	9.81	10	4.84
	Trigonid breadth	9	1.24	0.95	1.45	0.155	12.48	10	10.66
	Talonid breadth	9	1.00	0.75	1.17	0.129	12.93	0	—
M₃	Length	11	4.95	3.90	5.59	0.519	10.50	9	13.37
	Trigonid breadth	12	1.34	1.21	1.48	0.092	6.86	9	8.35

Note: Coefficients of variation (CV = 100 · SD/mean) for homologous dimensions in a restricted sample of *Nesodon imbricatus* (from between about 8 and 19 km south of Coy Inlet, Santa Cruz Formation, Patagonia) are provided for comparison.

Nesodon imbricatus. Much of the metric variation in the dentition of *Pericotoxodon* is wear related as can be documented for the lower cheek teeth by virtue of the eruption sequence and correlated increase in mandible depth and M_3 crown length.

The sorts of wear-related changes seen in the size and morphology of the more complex tooth crowns of *Nesodon imbricatus* are comparable to the changes in size and morphology seen in the simpler and ever-growing teeth of *Pericotoxodon platignathus*. The most variable characters of the upper molars of both *Nesodon imbricatus* and *Pericotoxodon platignathus* are the diameter of the lingual column and the length of the lingual enamel folds. Coefficients of varia-

tion for lower molar crown length and upper molar ectoloph length are similar in these two species. For the least-variable crown dimensions (P^4, P_4, and M^3 crown length, and M_3 crown breadth) no apparent size change in this species can be demonstrated through the 900-m-thick fossiliferous Honda Group section. Evidently, most of the observed metric and morphological variation in *Pericotoxodon* is individual, ontogenetic, or wear related.

Two distinct morphologies are represented in the available sample of *Pericotoxodon* I² tusks: (1) triangular with a rounded medial projection that lengthens the medial and lingual surfaces and (2) triangular with equilateral sides (fig. 21.7). A bivariate plot of I² maximum crown length (mesial

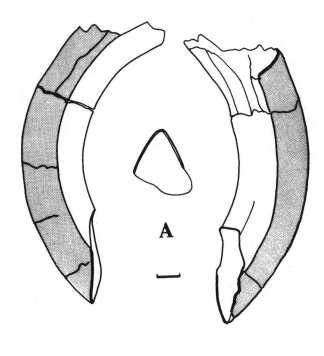

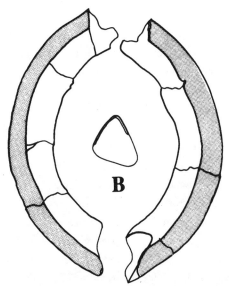

Figure 21.7. Dimorphism in the *Pericotoxodon platignathus* I² tusk: *A,* IGM 182810, "male" right I²; *B,* IGM 250474, "female" right I². Orientation: *left,* mesial aspect; *right,* distolingual aspect; *center,* cross section at occlusal surface (mesial to the right). Scale bar 1 cm.

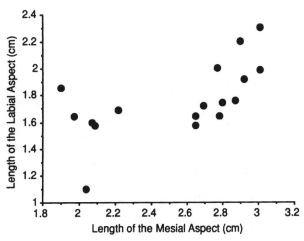

Figure 21.8. Bivariate plot of *Pericotoxodon platignathus* I² tusk crown dimensions.

PHYLOGENETIC AFFINITIES

For character analysis forty-two dental and mandible characters were scored for *Pericotoxodon,* ten other genera of Toxodontidae, and three non-toxodontid notoungulates—*Rhynchippus* (Notohippidae), *Pleurostylodon* (Isotemnidae), and *Simpsonotus* (Henricosborniidae) (app. 21.1). For each genus included in the phylogenetic analysis the holotype and hypodigm material that was examined along with a brief discussion of the taxonomic status is presented elsewhere (Madden 1990). Brief descriptions of the characters and character states (with a single genus exemplary of that condition indicated parenthetically) are given in table 21.2. Problems of homology and variation for these characters and character states are discussed in some detail elsewhere (Madden 1990). It should be noted that patterns of continuous ontogenetic variation in the dentition (such as observed in *Pericotoxodon, Nesodon, Adinotherium, Palyeidodon,* and *Hyperoxotodon*) are not comparably documented for many genera of Toxodontidae.

On this character data maximum parsimony analysis was performed using PAUP 3.0s (Swofford 1991). The heuristic search algorithm with random addition sequence (using preset "factory" settings; i.e., ten replications, one starting seed, TBR branch swapping, MULPARS, etc.) was run assuming Fitch parsimony for all characters, that is, characters were treated as unordered transformation types. Repeated heuristic searches retained eight minimum-length

aspect) and perpendicular breadth (labial aspect) illustrates a marked discontinuity (fig. 21.8). This discontinuity in size and morphology is interpreted to indicate sexual dimorphism in *Pericotoxodon.*

Table 21.2. Character/taxon matrix for genera of Toxodontidae and Notoungulata outgroups used in the phylogenetic analysis

Taxon	1	2	3	4	5	6	7	8	9	10	11	12	13	14	15	16	17	18	19	20	21	22	23	24	25	26	27	28	29	30	31	32	33	34	35	36	37	38	39	40	41	42
Posnanskytherium	3	0	2	2	0	0	1	0	0	1	2	0	2	1	0	1	2	4	2	0	3	2	3	2	3	2	3	?	3	?	0	1	2	3	0	?	0	1	1	1	1	1
Xotodon	2	?	?	2	0	0	0	0	0	1	2	0	2	0	?	?	?	?	?	?	2	?	3	1	3	2	2	1	0	1	2	0	2	3	0	4	0	1	?	1	?	1
Hyperoxotodon	2	0	?	2	?	1	?	0	0	0	2	0	2	?	?	?	?	?	?	?	?	2	0	1	0	2	0	1	1	1	?	0	1	2	0	3	2	0	?	?	0	1
Palyeidodon	1	?	?	2	1	1	2	0	1	0	1	1	1	0	0	0	1	1	0	2	2	2	1	1	1	2	0	1	0	2	0	0	2	2	1	4	2	0	?	1	0	1
Gyrinodon	?	?	?	2	?	?	1	?	0	?	0	0	1	?	?	?	?	?	?	?	?	?	?	0	2	2	2	1	0	?	?	?	?	?	0	3	0	0	?	?	?	1
Toxodon	3	0	2	2	0	0	1	0	0	0	0	1	1	1	0	1	2	3	2	2	3	2	?	1	2	2	2	2	0	2	1	1	2	3	0	2	0	0	1	?	1	1
Trigodon	2	?	?	3	1	0	0	0	0	0	0	1	2	1	?	3	2	3	0	?	3	?	3	1	2	2	1	0	2	?	3	1	2	3	0	4	0	0	?	?	?	1
Adinotherium	3	0	1	2	1	2	3	0	2	1	1	0	0	0	1	0	1	1	1	0	2	2	1	1	0	2	0	1	1	1	1	0	2	1	1	0	3	1	0	0	0	0
Pericotoxodon	2	0	2	2	1	0	1	0	0	0	0	1	1	0	0	1	2	2	2	2	2	1	1	1	2	0	1	1	2	4	0	2	3	1	3	1	0	0	0	0	1	1
Nesodon	2	0	1	2	1	2	2	1	2	0	1	0	0	0	1	0	1	2	1	0	2	1	2	1	1	2	0	1	1	2	0	0	2	1	1	0	2	1	0	0	0	0
Proadinotherium	2	0	1	2	1	1	{2/3}	1	2	0	1	0	0	0	0	?	?	1	1	0	?	0	2	1	0	1	0	1	1	0	1	0	1	1	1	0	3	1	0	?	0	0
Rhynchippus	0	1	1	1	1	1	1	1	?	0	2	0	1	0	0	2	0	0	0	3	1	1	3	1	0	2	2	0	1	3	1	0	1	2	1	1	?	1	0	0	0	0
Pleurostylodon	?	1	0	0	1	3	3	1	1	1	2	0	0	0	1	2	0	0	0	3	1	0	?	1	0	0	0	0	0	0	?	0	0	0	2	0	?	0	0	0	0	0
Simpsonotus	?	?	?	0	1	?	?	1	1	1	2	0	0	?	1	?	?	?	?	?	?	0	0	0	1	0	1	0	0	0	0	0	0	0	0	0	0	?	0	?	0	0

Note: Integer-coded character states for each genus are given in order of the forty-two characters listed in appendix 21.1.

trees of 137 steps. A strict consensus solution of this eight-tree set is the basis for evaluating patterns of character change and the following discussion of phylogeny (fig. 21.9).

To explore the consequences for tree length of characters relating to crown height, characters 1, 3, 22, and 26 were predefined as ordered (Wagner parsimony) character types. In addition, hypselodonty, or rootless, ever-growing crowns (characters 39, 40, 41, and 42) and tooth loss (characters 5 and 32) were considered irreversible (Camin-Sokal parsimony) with polarity either up or down depending on the condition of the outgroup taxa. With this new character-type set a PAUP heuristic search with random addition sequence (at factory settings) found ten minimum length trees of 141 steps. The strict consensus solution is identical for both the Wagner/Camin-Sokol and the Fitch parsimony character sets.

A polytomy at the basal node for the family Toxodontidae involves the unresolved plesiomorphic group "Nesodontinae" (including *Nesodon, Adinotherium,* and *Proadinotherium*) and a clade comprising all other toxodonts. Four features of the anterior dentition, broad triangular I^1, enlarged triangular I^2 tusk, triangular I$_1$ and I$_2$, and pro-cumbent lower incisor row with enlarged I$_3$ are unambiguous synapomorphies for the family.

In the strict consensus tree a polytomy also occurs among *Palyeidodon, Hyperoxotodon,* and a monophyletic clade of "advanced" Toxodontidae (including *Pericotoxodon*). *Palyeidodon* and *Hyperoxotodon* share two unambiguous synapomorphies with *Pericotoxodon* and other advanced genera, high-crowned I^1 and rootless M$_1$. Some degree of cheek-tooth crown simplification also characterizes this clade, involving the loss of upper premolar pre- and postcingula, loss of the M^2 parastyle, reduced P^4 parastyle, loss of M^1 distolingual features, and rootless P$_2$. The two genera of middle Miocene toxodonts from Patagonia, *Palyeidodon* and *Hyperoxotodon,* show increased crown height and more simplified crowns compared with contemporaneous *Adinotherium* and *Nesodon*. Both *Palyeidodon* and *Hyperoxotodon,* however, are relatively primitive in dental morphology compared with *Pericotoxodon*. For example, *Pericotoxodon* upper molar lingual folds never isolate into closed fossettes, upper premolar lingual enamel folds are more ephemeral, and upper molar lingual fold is less persistently bifurcate.

Pericotoxodon shares one unambiguous synapomorphy in

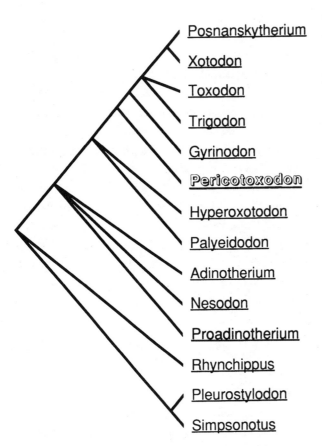

Posnanskytherium

Xotodon

Toxodon

Trigodon

Gyrinodon

Pericotoxodon

Hyperoxotodon

Palyeidodon

Adinotherium

Nesodon

Proadinotherium

Rhynchippus

Pleurostylodon

Simpsonotus

Figure 21.9. Strict consensus maximum parsimony tree depicting the hypothetical phylogenetic relationships of *Pericotoxodon* to other Toxodontidae (fourteen taxa, forty-two characters, tree length = 137). Consensus tree statistics: component information (consensus fork) = 7 (normalized = 0.636); Nelson-Platnick term information = 31; Nelson-Platnick total information = 38; Mickevich's consensus information = 0.583; Colless weighted consensus fork (proportion of maximum information) = 0.494; Schuh-Farris levels sum = 116 (normalized = 0.406); Rohlf's CI_1 = 0.554; Rohlf's $-\ln CI_2$ = 21.574; CI_2 = 4.27e − 0.10.

the dentition with five genera (*Toxodon, Trigodon, Xotodon, Posnanskytherium,* and *Gyrinodon*), that is, ever-growing or rootless P⁴s. Additional synapomorphies for this clade include characters associated with crown simplification, including featureless P², a simple P⁴ lingual fold, loss of the P⁴ parastyle, and a reduced distolingual enamel fold on M³.

Pascual (1965; Pascual et al. 1966) has argued, in effect, that the evolutionary radiation of the Toxodontidae involved well-defined monophyletic subfamilies with long and independent phyletic histories. Pascual (Pascual et al. 1966) typified the subfamilies Toxodontinae, Haplodon-

theriinae, and Xotodontinae by the cranial and dental features of the well-known terminal taxa, *Toxodon, Trigodon,* and *Xotodon,* respectively. In contrast, the consensus tree is polychotomous with respect to these three traditionally recognized subfamilies. Review of the middle Miocene genera *Hyperoxotodon* and *Palyeidodon* based upon more complete material has cast some doubt on their supposed relationships with later Miocene taxa. *Hyperoxotodon* (including *"Nesodonopsis"*) is not a toxodontine and *Palyeidodon* is not a very convincing haplodontheriine (see Madden 1990). More significantly for this discussion, *Pericotoxodon* does not fit the morphological definition of either the Toxodontinae or the Haplodontheriinae. The cranium and mandible of *Pericotoxodon* are unlike either *Toxodon* or *Trigodon,* and, moreover, *Pericotoxodon* shares no dental synapomorphies with either of these genera or with *Xotodon.* Thus, neither the dichotomous relationships among the three subfamilies of "advanced" Toxodontidae nor the phylogenetic affinities of *Pericotoxodon* are resolved by parsimony analysis based upon this set of dental characters.

Pericotoxodon is distinct from middle Miocene Toxodontidae from Patagonia (*Palyeidodon, Hyperoxotodon, Adinotherium,* and *Nesodon*) in the following combination of autapomorphous features: (1) inferolaterally deepened adult mandible ramus; (2) buccolingually broad lower molar entolophids; (3) variably bifurcate or simple, persistent, upper molar primary lingual enamel fold; (4) very ephemeral upper molar accessory lingual enamel folds/fossettes; (5) M¹ and M² with distolingual enamel plications; (6) reduced M³ distolingual fold; (7) simple, nonbifurcate and ephemeral P⁴ lingual enamel fold; (8) ephemeral lower molar accessory fossettid; (9) buccolingually broad P₂ talonid; and (10) the persistently open M₃ ento-hypoconulid fold.

Pericotoxodon can also be distinguished from most other late Miocene Toxodontidae from Argentina. For example, compared with the less-well-known late Miocene *Ocnerotherium, Pericotoxodon* displays neither the upper molar posterior pillar (buccolingually broadened metaloph) nor the relatively indistinct protoloph lingual column. Compared with *Stereotoxodon, Pericotoxodon* has a prominent lingual column on the upper molar protoloph, reduced M³ distolingual fold, distally tapering upper molar crowns, less-persistent lower molar ento-hypoconulid fold, and more persistent meta-entoconid fold. Compared with the late Miocene *"Palaeotoxodon nazari"* (of problematical taxonomic status and known only from juvenile material), *Pericotoxodon* has transversely broader lower molar entolophids and deep horizontal mandibular ramus.

Although parsimony analysis reveals that *Pericotoxodon* does not share a special relationship with any of the better-known genera of Toxodontidae, *Pericotoxodon* resembles three of the more poorly known genera: (1) *Dinotoxodon paranensis* Mercerat 1895 from the late Miocene–early Pliocene Mesopotamian fauna from the Ituzaingo Formation of northeastern Argentina; (2) *Gyrinodon quassus* Hopwood 1928 from the "Miocene" of Falcón State, northern Venezuela; and (3) *Plesiotoxodon amazonensis* Paula Couto 1982 (AMNH 55805) from the Upper Juruá River in Brazil. As argued here, these morphological resemblances are of phylogenetic significance.

In an unpublished doctoral dissertation at the Museo de Ciencias Naturales de La Plata, Zetti (1972) called attention to the inferolaterally deepened mandible of *Dinotoxodon,* contrasting it with the shallow ramus of contemporary *"Palaeotoxodon"* (= *"Pisanodon"* Zetti) and *Alitoxodon* Rovereto. The mandibular ramus of *Gyrinodon* is similarly described as "very deep, in order to accommodate the persistently growing, rootless molars. The greatest depth is immediately below the prolophid of the third molar. From this point the depth diminishes both before and behind" (Hopwood 1928, 574). The inferolaterally deepened mandibular ramus of *Pericotoxodon, Dinotoxodon,* and *Gyrinodon* is a synapomorphy linking these three genera in a monophyletic clade.

Other synapomorphies can be found in the upper dentition, including (1) variably bifurcate or simple upper molar lingual fold, (2) metaloph-protoloph contact, and (3) reduced M³ distolingual fold. The range of morphological variation seen in the M³ morphology of *Pericotoxodon* encompasses the morphology seen in both *Gyrinodon* and *Dinotoxodon,* which have simplified M³s with relatively elongate and buccolingually narrow crowns. Similarly, whereas both *Gyrinodon* and *Dinotoxodon* display simple, single upper molar lingual folds, in *Pericotoxodon,* the upper molar lingual fold is either simple or bifurcate. The upper molar F2 fold (persistent with wear, but variably present in *Pericotoxodon*) was lost in later *Gyrinodon* and *Dinotoxodon.*

Plesiotoxodon resembles *Pericotoxodon* in these same relatively derived features of upper molar morphology.

Thus, for these four genera, a new subfamily of "advanced" Toxodontidae, the Dinotoxodontinae, is herein proposed. The Dinotoxodontinae comprise a monophyletic clade encompassing *Plesiotoxodon, Gyrinodon, Dinotoxodon,* and *Pericotoxodon* and characterized by the following synapomorphies: (1) inferolaterally deep mandible ramus, (2) variably present but persistent F2 fold in the upper molars, (3) metaloph/protoloph contact, (4) distally tapering M¹ and M² metalophs, and (5) mesiodistally elongate and buccolingually narrow upper third molars. The subfamily Dinotoxodontinae is best characterized by features of the genus *Pericotoxodon*. These features are contrasted with those of the other monophyletic subfamilies of Toxodontidae in table 21.3.

ADAPTATIONS

A mean estimate of the body mass of *Pericotoxodon platignathus* was obtained from several least-squares regression equations derived from samples of taxonomically diverse living ungulates (table 21.4). Three separate estimates were obtained from dental measures. For two dental dimensions, M_1 length and P_4–M_2 length, regression equations for nonselenodont ungulates are used (Damuth 1990). Nonselenodont ungulates are a taxonomically heterogeneous group including perissodactyls, hippos, some pigs (but excluding *Phacochoerus* and *Hylochoerus*), and hyraxes. In addition, assuming that lower postdiastema tooth row length in toxodontids is equivalent to lower postcanine tooth row length in selenodont artiodactyls, the regression terms for living selenodont nonbrowsing artiodactyls are used for this dimension (Damuth 1990, 245). In addition, six estimates of the body mass (kg) of *Pericotoxodon platignathus* were obtained for cranial and mandibular measures using regression equations for the living Perissodactyla/hyracoid sample of Janis (1990, table 16.8). Finally, one estimate of the body mass of *Pericotoxodon* was obtained from the distal femur, using a regression equation derived from an "all-ungulate" sample (table 16.7 in Scott 1990).

From these ten dental, cranial, mandibular, and postcranial estimates, a mean estimate of the body mass of *Pericotoxodon platignathus* of 798 kg is obtained. This mean

Table 21.3. Principal features of the five subfamilies of Toxodontidae

Feature	Nesodontinae	Xotodontinae	Toxodontinae	Haplodontheriinae	Dinotoxodontinae
Mandibular symphysis	Short, with a shallow digastric depression	Keeled "chin"	Very elongate and horizontal	Not known	Elongate, with a deep digastric depression
Mandibular ramus	Alveolar and inferior margins parallel	Alveolar and inferior margins diverge anteriorly	Alveolar and inferior margins converge anteriorly	Alveolar and inferior margins parallel posteriorly, convergent anteriorly	Inferior margin greatly deepened inferolaterally beneath the molars
Mandible maximum depth	Not applicable	Below P_3	Below M_3	Below M_2	Below M_2
Mandible angle	Undifferentiated	Well differentiated	Undifferentiated	Undifferentiated	Undifferentiated
I_{1-2}	Right-triangular, rooted	Obtuse-triangular, ever-growing, procumbent	Ever-growing, buccolingually flat, mesiodistally broad ovoid, procumbent	Laterally compressed ovoid	Ever-growing, right-triangular
Diastema	None to slight	Moderate	Long	Short	Short
Lower molars	Two lingual folds, persistent accessory fossettid	Two lingual folds, reduced paraconid	Two lingual folds	One lingual fold (no meta-entoconid fold)	Two lingual folds, ephemeral accessory fossettid
Upper premolars	P^{3-4} with lingual folds/fossettes	P^{3-4} without lingual folds	P^{3-4} with lingual folds	P^{3-4} without lingual folds	P^{3-4} with lingual folds
Upper molars	Bifurcate fold with persistent accessory fossettes, protoloph and metaloph confluent with wear	Buccolingually narrow, simple lingual fold, no lingual column on protoloph	Bifurcate lingual fold, metaloph does not contact protoloph, lingual column present on protoloph	Evolutionary tendency to lose bifurcation, lingual columns on protoloph and metaloph	Bifurcate lingual fold, metaloph contacts but never confluent with the protoloph, lingual column on protoloph
Skull shape in sagittal view	Pyriform	Narrow, parallel sided	Pyriform	Triangular	Pyriform
Premaxilla	Short and broad, with a weakly developed dorsal protuberance	Short and narrow	Long and broad, with a well-developed dorsal spine	Short and broad	Long and broad, with a well-developed dorsal protuberance
Nasals	Low, long and horizontal, aperture directed forward	Low, long and horizontal, aperture directed forward	High and short, aperture directed upward	High and short, aperture directed forward	Low, long and horizontal, aperture directed forward
Occiput	Erect, or inclined slightly backward	Erect	Inclined slightly backward	Inclined forward	Erect
Frontal	Depressed glabella, rugose surface, prominent temporal ridges	Smooth, expanded, globose	Flat, triangular	Elevated basal protuberance for horns	Flat
Dimorphism	Not known	Not known	? Tusks	? Horns	Yes, tusks

Table 21.4. Body mass estimates for *Pericotoxodon platignathus*

Variate Measured	Measure (cm)	Body Mass Estimate (kg)
M$_1$ length	3.01	532.2
P$_4$–M$_2$ length	8.32	363.7
Postdiastema tooth row length	16.58	654.9
Total skull length	49.70	544.1
Posterior length of skull	26.60[a]	889.1
Palatal width	12.00[a]	2,970.2
Length of masseteric fossa	19.30	325.7
Posterior jaw length	10.40	333.0
Depth of mandibular angle	22.00	648.6
Distal femur transverse diameter	10.75	722.2
All variates (N = 10)		798.4 (248.1)[b]

[a]Estimated.

[b]Mean (SE).

estimate compares with that of the giraffe *Giraffa camelopardalis* (750 kg), the black rhino *Diceros bicornis* (816 kg), the moose *Alces alces* (800 kg), and the Sumatran rhino *Dicerorhinus sumatrensis* (800 kg).

Features of the rostrum and anterior dentition suggest a grazing habitus for all Toxodontidae; these include an elongate, distally broadened rostrum as seen in grazing Bovidae (Janis and Ehrhardt 1988) and *Ceratotherium* (Kingdon 1979), and extreme crown height in the central upper incisor and two central lower incisors such as in the living *Vicugna* (see Miller 1942).

The cheek teeth of *Pericotoxodon* display the vertical enamel bands typical of middle Miocene and later Toxodontidae with ever-growing teeth. These enamel bands appear on the occlusal surface as concave shearing crescents. As judged by the orientation of wear striations and occlusal surface relief (sharp-crested, crescent-shaped enamel blades, transverse ridges, and dentin pits), occlusion probably involved a transverse power stroke.

Enamel microstructure, decussation and prism-packing patterns have been described for *Nesodon* and *Toxodon* (Fortelius 1985) and *Pericotoxodon* (chap. 20). Rensberger (1973) has suggested that these and other aspects of the

design of high-crowned cheek teeth may be related to resisting mechanical failure. Two unique features of the upper molars in Toxodontidae suggest a strong emphasis on shearing. The lingual pillar on the protoloph of the toxodontid maxillary molar buttresses the shearing blade formed by the lingual enamel fold. Similarly, crown imbrication in the Toxodontidae, whereby the distal corner of one cheek tooth is overlapped by the antero-external corner (or parastyle) of the next posterior tooth, may also be associated with resistance to mechanical failure by providing lingual support for the mesial enamel shearing blade of the protoloph of the adjacent tooth. Although the evolutionary trend in Miocene Toxodontidae is clearly toward less total shearing (such as between *Nesodon* and *Pericotoxodon*), increasing crown imbrication may compensate, in part, by increasing both the mesiodistal length of the crown and the number of blade sets brought into function during each masticatory cycle.

The developmental timing of the appearance of the inferolateral projection of the mandibular ramus in *Pericotoxodon* coincides with the complete eruption of the adult dentition, that is, after dP$_4$ is shed. This postjuvenile onset of development suggests that this structure may have been ornamental. Ornamental structures along the inferior margin of the mandible are rare in living ungulates. Protuberances and flanges occur on the inferior margin of the horizontal ramus in a wide variety of fossil mammals, however, including entelodonts (*Archaeotherium*), uintatheres (*Uintatherium, Bathyopsis*), merycoidodonts (*Merycoidodon*), Camelidae (*Oxydactylus*), suoids (*Platygonus*), pilosans (*Megatherium*), and also in typotherian notoungulates (*Interatherium*). In all these cases, no muscle insertion scars or rugosities are evident.

Sexual dimorphism in the I^2 tusk has been demonstrated for *Pericotoxodon*. The sexually dimorphic I^2 tusk of *Pericotoxodon* may have affected the appearance of the snout as viewed from the front. As Scott (1912) pointed out, the appearance of the snout of *Nesodon imbricatus* changes dramatically with individual age and the continuous eruption and wear of the upper incisors. Although the degree of sexual dimorphism can predict mating systems ranging from monogamy to extreme polygyny in living ungulates, slight sexual dimorphism occurs in ungulates of diverse grouping tendencies, reproductive behaviors, and habitat preferences (Janis 1982). The complex relationship between resource availability, food item dispersion, antipredator behavior, within habitat visibility, and male-male display on female grouping patterns and male mating strategy (Jarman 1983), makes inferences about herding behavior from tusk dimorphism in *Pericotoxodon* speculative at best. By analogy with large living ungulates of similar body weight (around 800 kg) the dimorphism ratio of 1.55 in male/female tusk cross-sectional area of *Pericotoxodon* suggests possibly polygynous grouping tendencies.

DISCUSSION

The evolutionary acquisition of ever-growing (rootless) tooth crowns in the Toxodontidae occurred sometime during the early and middle Miocene. At least part of the evolutionary transformation in one lineage (Nesodontinae) was sufficiently slow to be detected in the Miocene fossil record of Patagonia, which documents a measurable increase in cheek tooth crown height from Deseadan *Proadinotherium leptognathum* to Colhuehuapian *P. muensteri* to Santacrucian *Adinotherium* and *Nesodon*. It is not clear, however, whether the acquisition of ever-growing cheek teeth occurred once, at the base of the radiation of "advanced" toxodontids (i.e., a synapomorphy) or several times independently.

Some evidence suggests that this transformation may have occurred in parallel in several distinct lineages. This evidence includes (1) the first appearance of hypselodont Xotodontinae (*Hyperoxotodon*) and ?Haplodontheriinae (*Palyeidodon*) in the Santacrucian of Patagonia, (2) the occurrence of another unnamed hypselodont taxon in the Río Frías Formation with possible affinities with *Adinotherium,* and (3) the discovery in the early Miocene of Ecuador of a relatively primitive but hypselodont unnamed species of *Pericotoxodon* (Madden 1990). These "advanced" Toxodontidae span a considerable range in size (from small *Hyperoxotodon* to large *Palyeidodon*) and morphology, being distinct in dental and cranial characters unrelated to increasing crown height. Moreover, the temporal range of these four taxa overlaps, at least in part, the temporal range of the relatively low-crowned Colhuehuapian-Santacrucian Nesodontinae in southern Patagonia. In addition, their widespread geographic distribution and rather abrupt first

appearance in the fossil record of Patagonia without clear evolutionary affinities with Nesodontinae suggests that these "advanced" groups acquired hypselodonty outside southern Patagonia.

An alternative to the independent acquisition of ever-growing teeth in several lineages is that simplified and ever-growing molars represent a synapomorphy linking these four early to middle Miocene taxa with all post-Santacrucian Toxodontidae in a single monophyletic clade descended from a single taxon that also had ever-growing molars. To evaluate these two alternatives, PAUP heuristic searches using the random addition sequence were run using two different character-type sets: one in which hypselodonty is treated in Dollo parsimony (ordered, irreversible, and evolutionarily unique) and another in which hypselodonty is treated in Wagner parsimony (unordered). The resulting trees (a strict consensus tree for the Wagner character-type set and a single maximum parsimony tree for the Dollo character-type set) have identical topology, that is, both interpretations of character-type parsimony are consistent with the branching pattern depicted in figure 21.9.

If *Palyeidodon, Hyperoxotodon,* and *Pericotoxodon* represent a monophyletic group, their last common ancestor, with ever-growing and simplified molars, must antedate the earliest known occurrence of "advanced" Toxodontidae, that is, before about 21.0 Ma or sometime in the Deseadan-Colhuehuapian interval. The only toxodontid so far known at this antiquity, *Proadinotherium* (known from Colombia, Bolivia, and Patagonia), did not have ever-growing molars.

Another feature of "advanced" Toxodontidae is the loss of some or all upper and lower molar lingual enamel folds and fossettes. It has been suggested that in mammals the loss of enamel-lined fossettes (or fossettids) may be correlated with ever-growing (rootless) crowns by developmental and mechanical constraints on the geometry of the enamel-forming organ (Von Koenigswald 1982b). If this constraint is universal (see also Webb and Hulbert 1986), characters of crown simplification in Toxodontidae should be correlated with increasing crown height and would not, therefore, constitute independent synapomorphies for clades of "advanced" Toxodontidae. The developmental pattern of crown formation in the M¹ of *Adinotherium* and *Pericotox-*

odon, as revealed by the study of wear stages, demonstrates that enamel-lined fossettes are formed (and lost) in both rooted and rootless teeth.

The Miocene fossil record in South America suggests that the Dinotoxodontinae had their initial evolutionary radiation and most of their subsequent geographic distribution restricted to lowland tropical or subtropical environments. The earliest occurrence of the subfamily is in the Biblián Formation in Cañar Province, southern Ecuador. Based upon additional material, the M³ identified as *"Prototrigodon"* by Simpson (in Repetto 1977) from the early Miocene Biblián Formation of southern Ecuador, should be referred to *Pericotoxodon* sp. (Madden 1990). In the middle Miocene *Pericotoxodon* is known from the Honda Group of Colombia. In the late Miocene the subfamily is represented in Acre Province, Brazil by the still inadequately known genus *Plesiotoxodon amazonensis* Paula Couto 1982, from the Upper Juruá River. Additional, but poorly preserved, material of probable Dinotoxodontinae (USNM 205350 from Bassler Vertebrate Fossil Locality 56, Playa Mapuya, Río Inuya, Loreto Department, and USNM unnumbered material labeled "Bassler Locality 26, Shapaja, Chasuta, Mishquiyacu) suggest the presence of closely related toxodonts in adjacent areas of eastern Peru. The Dinotoxodontinae are also recorded in northern Venezuela (*Gyrinodon*) and in the Ituzaingo Formation of Entre Ríos Province, Argentina (*Dinotoxodon*).

Summary

Pericotoxodon platignathus is distinct from other middle to late Miocene Toxodontidae in the following combination of features: adult mandible ramus inferolaterally deepened beneath M_2; lower molar entolophids buccolingually broad; upper molar lingual fold bifurcate or simple, persistent with wear; upper molar accessory F3 or F4 lingual enamel folds or fossettes very ephemeral with wear; M¹ and M² with distolingual enamel plication; M³ distolingual F4 fold variably present; P⁴ lingual enamel fold simple, nonbifurcate, ephemeral with wear; lower molar accessory fossettid ephemeral; P_2 talonid buccolingually broad; M_3 ento-hypoconulid fold persistently open.

Pericotoxodon is not related phylogenetically with any of the traditionally recognized monophyletic groups (sub-families) of "advanced" Toxodontidae. It does share a suite of synapomorphies with *Gyrinodon, Plesiotoxodon,* and *Dinotoxodon,* which serves to diagnose a new subfamily, Dinotoxodontinae, of the notoungulate family Toxodontidae. These relatively derived features include an inferiorly deepened mandibular ramus, variably bifurcate or simple upper molar lingual fold, metaloph-protoloph contact, and reduced distolingual fold on the mesiodistally elongate M³.

Material of *Pericotoxodon* from the early Miocene Biblián Formation in southern Ecuador suggests that the clade Dinotoxodontinae antedates the appearance of all other groups of "advanced" Toxodontidae. The Dinotoxodontinae are geographically widespread, known from western Brazil, northern Venezuela, Colombia, southern Ecuador, and northeastern Argentina, a distribution largely restricted to lowland equatorial South America.

A mean body mass estimate of 798 kg for *Pericotoxodon platignathus* compares with that of the living giraffe, black rhino, moose, and Sumatran rhino. Features of the rostrum and anterior dentition and the emphasis on shearing features in the cheek teeth suggest a grazing habitus. Dimorphism in tusk cross-sectional shape and area suggests possibly polygynous grouping tendencies.

Acknowledgments

I acknowledge the authorities of the INGEOMINAS, Bogotá, Colombia, who both authorized fossil collecting and entrusted their fossil vertebrates to R. F. Kay for scientific study. I thank these gentlemen and D. E. Savage and the authorities of the UCMP for permission to study and describe the fossil Toxodontidae from Colombia. I also thank J. Ostrom of the YPM, F. Jenkins of the MCZ at Harvard University, J. Flynn of the FMNH, M. C. McKenna of the AMNH, M. Coombs and the authorities of the Pratt Museum (Amherst College Museum), C. Ray of the USNM, and R. Pascual of the MLP, La Plata, Argentina, for facilitating comparative work in their collections. Financial support for this endeavor was provided by the Graduate School of Duke University and the Fellowship Committee of the Field Museum of Natural History. The fieldwork in Colombia was supported by National Geographic Society grants 2964-84 and 3292-86 and NSF grants BSR 86-14533 and NSF BSR 89-18657 to R. F.

Kay. J. Guerrero of INGEOMINAS and the Department of Geology, Duke University, provided the stratigraphic framework. I thank R. L. Cifelli, University of Oklahoma, and R. F. Kay, Duke University, for valuable advice during all stages in the preparation of this chapter, and S. D. Webb for many helpful suggestions to improve the final manuscript. Figure 21.2 was drawn by L. D. Woods. Figure 21.7 was drawn by S. Wong.

Appendix 21.1. Forty-two dental and mandible characters, and their states, for the family Toxodontidae

Operational taxonomic units (OTUs) exemplifying each character state are indicated parenthetically.

1. Length between I^2 and the upper cheek tooth row as a proportion of the anteroposterior length of the closed upper cheek tooth dentition. Character states: (0) no diastema (*Rhynchippus*, Notohippidae); (1) x ≤ 0.2 (*Palyeidodon*); (2) 0.2 < x < 0.3 (*Trigodon*); (3) x ≥ 0.3 (*Toxodon*).

2. I^1 crown shape in cross section. Character states: (0) triangular (*Nesodon*); (1) oval (mesiodistally elongate) to rounded (*Pleurostylodon*, Isotemnidae).

3. Height of the I^1 labial enamel from the occlusal surface to the cementoenamel junction or the enamel secretory front. Character states: (0) low crowned (less than mesiodistal length) (*Pleurostylodon*, Isotemnidae); (1) moderately high crowned (greater than mesiodistal length to 2 × mesiodistal length) (*Nesodon*); (2) high crowned (greater than 2 × mesiodistal length) (*Posnanskytherium*).

4. I^2 crown cross section. Character states: (0) ovoid (*Pleurostylodon*, Isotemnidae); (1) rounded triangle; (2) triangular (*Pericotoxodon*); (3) laterally compressed (*Trigodon*).

5. I^3. Character states: (0) absent (*Posnanskytherium*); (1) present (*Nesodon*).

6. P^2 primary occlusal surface features. Character states: (0) featureless (*Pericotoxodon*); (1) simple or irregular enamel fold/fossette (*Hyperoxotodon*); (2) bifurcate enamel fold/fossette (*Nesodon*); (3) complex enamel folds/fossettes (*Pleurostylodon*, Isotemnidae).

7. P^4 primary occlusal surface features. Character states: (0) featureless (*Xotodon*); (1) simple or irregular enamel fold/fossette (*Pericotoxodon*); (2) bifurcate enamel fold/fossette (*Nesodon*); (3) complex enamel folds/fossettes (*Pleurostylodon*, Isotemnidae).

8. P^{2-4} lingual precingulum (or anterointernal fossettes). Character states: (0) present (*Nesodon*); (1) absent (*Palyeidodon*).

9. P^{2-4} posterior lingual cingulum. Character states: (0) absent (*Pericotoxodon*); (1) present (*Palyeidodon*); (2) present on P^4 only (*Adinotherium*).

10. P^{2-4} crown imbrication. Character states: (0) absent (*Nesodon*); (1) present (*Xotodon*); (2) present only between P^3 and P^4 (*Trigodon*).

11. The lingual pillar on the M^{1-3} protoloph. Character states: (0) well developed on M^{1-3} (*Pericotoxodon*); (1) weakly developed on M^{1-3} (*Nesodon*); (2) absent (*Xotodon*).

12. The posterior lingual pillar on the M^1 and M^2 metaloph. Character states: (0) absent (*Posnanskytherium*); (1) weakly developed (*Palaeotoxodon*); (2) present, well-developed (*Ocnerotherium*).

13. M^2 ectoloph. Character states: (0) distinct parastyle, paracone, and metacone (*Adinotherium*); (1) straight ectoloph (*Palyeidodon*); (2) concave ectoloph (*Xotodon*).

14. Adult lower dental formula. Character states: (0) complete, 3.1.4.3 (*Nesodon*); (1) reduced, 3.0 or 1.3 or 4.3 (*Toxodon*).

15. Relative enamel crown height of the I_1 and I_2 labial enamel (occlusal surface to the cementoenamel junction). Character states: (0) I_1, I_2 equal (*Proadinotherium leptognathum*); (1) I_2 greater than I_1 (*Adinotherium*).

16. I_1 cross section at the occlusal surface. Character states: (0) triangular (*Palyeidodon*); (1) mesiodistally elongate ovoid triangle (*Posnanskytherium*); (2) spatulate, buttressed lingually (*Pleurostylodon*, Isotemnidae); (3) buccolingually elongate oval (*Trigodon*).

17. Lingual enamel on I_1 and I_2. Character states: (0) present, ephemeral in the adult (*Adinotherium*); (1) present, persistent in the adult (*Nesodon*); (2) absent (*Posnanskytherium*).

18. I_2 labial enamel. Character states: (0) convex to rounded labially (*Pleurostylodon*, Isotemnidae); (1) dihedral, two distinct surfaces (*Proadinotherium leptognathum*); (2) flat to slightly convex (*Nesodon*); (3) smoothly convex labially (*Toxodon*); (4) slightly concave labially (*Posnanskytherium*).

19. I_3 angle of implantation. Character states: (0) greater than 30° (*Trigodon*); (1) about 30° (*Nesodon*); (2) less than 30° (*Toxodon*).

20. The relative width of the I_3 labial and lingual enamel bands; Character states: (0) labial band wider than lingual band (*Nesodon*); (1) labial and lingual bands equal (*Xotodon*); (2) lingual band wider than labial band (*Toxodon*); (3) enamel unbanded (*Rhynchippus*, Notohippidae).

21. P_2 crown shape. Character states: (0) weakly developed heel (*Simpsonotus*, Henricosborniidae); (1) posterior crescent present (*Pleurostylodon*, Isotemnidae); (2) posterior crescent distinct, persistent, equal in breadth to the anterior crescent (*Nesodon*); (3) anterior and posterior crescents indistinct (*Posnanskytherium*).

22. P_4 buccal enamel crown height (x = buccal enamel crown height/mesiodistal crown length). Character states: (0) x <

1.0 (*Pleurostylodon,* Isotemnidae); (1) x between 1.0 and 1.5 (*Rhynchippus,* Notohippidae); (2) x > 1.5 (*Adinotherium*).

23. P_3 lingual enamel folds/fossettids. Character states: (0) anterior fold, and meta-entoconid fold/fossettid (*Simpsonotus,* Henricosborniidae); (1) anterior fold, meta-entoconid and ento-hypoconulid folds/fossettids (*Adinotherium*); (2) meta-entoconid fold only (*Proadinotherium leptognathum*); (3) no folds/fossettids (*Xotodon*).

24. Lower molar buccal enamel fold. Character states: (0) closed, appressed fold (*Gyrinodon*); (1) open, angular fold (*Xotodon*); (2) open indentation (*Posnanskytherium*).

25. M_1 and M_2 talonid (posterior crescent). Character states: (0) smoothly convex (*Adinotherium*); (1) convex to straight externally, rounded right angle posteriorly (*Pericotoxodon*); (2) convex to flat externally, rounded posteriorly (*Toxodon*); (3) flat to concave externally, obtuse angle posteriorly (*Xotodon*).

26. Lower first molar buccal enamel crown height (x = crown height/mesiodistal crown length). Character states: (0) x < 0.49 (*Pleurostylodon,* Isotemnidae); (1) 0.5 < x < 0.99), rooted (*Proadinotherium leptognathum*); (2) x ≥ 1.0 (*Rhynchippus,* Notohippidae).

27. Most persistent lingual enamel fold(s) on M_1. Character states: (0) meta-entoconid fold/fossettid (*Pleurostylodon,* Isotemnidae); (1) ento-hypoconulid fold (*Trigodon*); (2) meta-entoconid and ento-hypoconulid folds (*Gyrinodon*); (3) meta-entoconid fold (*Posnanskytherium*).

28. Relative buccolingual breadth of the lower M_1 trigonid and entolophid (x = 100 × trigonid breadth/entolophid breadth). Character states: (0) trigonid and entolophid of approximately equal breadth (*Pleurostylodon,* Isotemnidae); (1) trigonid broader than entolophid (84 < x < 97) (*Nesodon*); (2) trigonid much broader than entolophid (x < 83) (*Paleotoxodon*).

29. M_3 lingual enamel folds/fossettids. Character states: (0) meta-entoconid and ento-hypoconulid folds (*Pleurostylodon,* Isotemnidae); (1) meta-entoconid and ento-hypoconulid folds, accessory fossettid (*Rhynchippus,* Notohippidae); (2) no meta-entoconid fold, ento-hypoconulid fold only (*Trigodon*); (3) meta-entoconid fold only, no ento-hypoconulid fold (*Posnanskytherium*).

30. Width of the M_1 lingual enamel band expressed as a percentage of total mesiodistal crown length. Character states: (0) x > 80% (*Proadinotherium leptognathum*); (1) 65% < x < 79% (*Xotodon*); (2) 51% < x < 64% (*Toxodon*); (3) x ≤ 50% (*Rhynchippus,* Notohippidae).

31. The inferior and alveolar margins of the adult horizontal ramus (between the angle and the symphysis) in lateral profile. Character states: (0) inferior and alveolar margins parallel (*Palyeidodon*); (1) mandible shallows continuously from the angle to the symphysis (*Toxodon*); (2) mandible deepens continuously from the angle to the symphysis (*Xotodon*); (3) inferior margin deepens to P_4, then shallows anteriorly (*Palaeotoxodon*); (4) inferior margin greatly deepened beneath the molars (*Pericotoxodon*).

32. Upper canine. Character states: (0) absent (*Posnanskytherium*); (1) present (*Pericotoxodon*).

33. P^2 parastyle and parastyle sulcus. Character states: (0) P^2 parastyle prominent, persistent angular sulcus (*Pleurostylodon,* Isotemnidae); (1) P^2 parastyle present, sulcus ephemeral (*Hyperoxotodon*); (2) P^2 parastyle absent (*Toxodon*).

34. P^4 parastyle and parastyle sulcus. Character states: (0) prominent parastyle, persistent sulcus (*Pleurostylodon,* Isotemnidae); (1) parastyle weak, sulcus persistent (*Proadinotherium leptognathum*); (2) parastyle weak, sulcus ephemeral (*Adinotherium*); (3) parastyle absent (*Toxodon*).

35. Upper molar lingual primary fold. Character states: (0) simple (*Xotodon*); (1) bifurcate (*Pericotoxodon*); (2) complex (*Pleurostylodon,* Isotemnidae).

36. M^1 lingual enamel distal fold(s)/plication/indentation. Character states: (0) two distolingual folds/fossettes (*Adinotherium*); (1) one distolingual fold/fossette (*Rhynchippus,* Notohippidae); (2) distolingual plication (*Toxodon*); (3) weak or ephemeral distolingual inflection (*Hyperoxotodon*); (4) no distolingual features (*Xotodon*).

37. M^3 distolingual enamel features. Character states: (0) absent (*Xotodon*); (1) distolingual inflection (*Pericotoxodon*); (2) distolingual plication (*Hyperoxotodon*); (3) two accessory folds/fossettes (*Adinotherium*).

38. M_1 paraconid (or anterior fold). Character states: (0) absent/indistinct (*Toxodon*); (1) present (*Xotodon*).

39. I^1 roots. Character states: (0) roots formed (*Pleurostylodon,* Isotemnidae); (1) rootless (*Posnanskytherium*).

40. P_2 roots. Character states: (0) rooted (*Rhynchippus,* Notohippidae); (1) rootless (*Posnanskytherium*).

41. P_4 roots. Character states: (0) rooted (*Pleurostylodon,* Isotemnidae); (1) rootless (*Pericotoxodon*).

42. M_1 roots. Character states: (0) rooted (*Pleurostylodon,* Isotemnidae); (1) rootless (*Pericotoxodon*).

22
Uruguaytheriine Astrapotheres of Tropical South America

Steven C. Johnson and Richard H. Madden

RESUMEN

Se registran dos géneros de Astrapotheriidae Uruguaytheriinae en el Grupo Honda de Colombia. *Xenastrapotherium* y *Granastrapotherium* son los últimos sobrevivientes de la familia Astrapotheriidae. La forma pequeña, *Xenastrapotherium kraglievichi*, está conocida ahora en base a dentaduras superiores e inferiores en asociación que permiten una diagnosis enmendada para este género. Una nueva especie de *Xenastrapotherium* de la Formación Biblián del Mioceno temprano del Ecuador es descrita aquí. Un maxilar parcial proveniente del Río Juruá en Brasil, originalmente descrito por Paula Couto, sirve del holotipo para otra nueva especie referida al mismo género. Una cuarta especie de *Xenastrapotherium* es descrita en base a material proveniente de la Formación Tuné, cerca a Chaparral, Tolima, Colombia.

Granastrapotherium snorki, un género y especie nueva, el más grande astrapotérido del Grupo Honda, era un megaherbívoro de miembros graviportales y grandes caninos en forma de defensas; probablemente estaba provisto de una trompa. Otro material proveniente de depósitos de edad desconocida en el oriente del Perú es referido al mismo género, extendiendo así la distribución del género hasta incluir la Amazonia.

Material de astrapotérido proveniente de Quebrada Honda en Bolivia puede ser asignado a la subfamilia Uruguaytheriinae. Los Uruguaytheriinae comparten una serie de sinapomorfías dentales, como, por ejemplo, un pozo anterointerno bien desarrollado en el primer molar superior, pérdida del hipofléxido de los molares inferiores, parastilo de los molares superiores reducido, metafléxido del molar inferior encerrado por un cíngulo anterointerno, y el gran surco lingual del tercer molar superior encerrado por una cresta formada por un cíngulo que conecta el protolofo con el metalofo. La Edad Deseadense de los Uruguaytheriinae más antiguos que se conocen hasta ahora sugiere que su divergencia debería haberse ocurrido durante o antes del Oligoceno.

Astrapotheriidae del Oligoceno y aun más recientes demuestran cambios morfológicos importantes que les separan como grupo monofilético de los astrapotéridos del Casamayorense y Mustersense. Entre estas características derivadas se nota el canino alargado en forma de colmillo y de crecimiento continuo, molar superior con posfoseta, primer molar superior con pozo anterointerno, hipocónido del molar inferior con una cresta anterointerna, y paralófido del molar inferior bien desarrollado.

Astrapotheriidae are large archaic terrestrial herbivores of the South American Cenozoic and are conspicuous elements of most Eocene through middle Miocene fossil assemblages from that continent. The family first appears in the early Eocene (Casamayoran) and reaches its highest diversity in Patagonia during the early and middle Miocene (Colhuehuapian-Santacrucian South American Land Mammal Ages [SALMA]). The last occurrence of the family in Patagonia is recorded in middle Miocene local faunas of Friasian SALMA (Kraglievich 1930a).

Oligocene and later Astrapotheriidae from outside Patagonia are generally referred to the subfamily Uruguaytheriinae Kraglievich (1928b), and differ from Miocene astrapotheres in Patagonia by the shared derived loss of the lower molar hypoflexids. The Uruguaytheriinae are less well known than the astrapotheriids from Patagonia. Until now, *Xenastrapotherium* has been known only from isolated mandibles collected from Miocene deposits in Venezuela (Stehlin 1928) and Colombia (Cabrera 1929). In 1947 R. A. Stirton described fragmentary uruguaytheriine astrapotheriid material from the ?Deseadan Chaparral fauna in Tolima Department, Colombia. In 1956 L. I. Price and G. G. Simpson collected a fragment of a maxilla with a broken molar crown, an isolated upper premolar, and lower molar talonid fragments from deposits of uncertain age along the Juruá River in Acre Province, Brazil. The Acre material along with the six fragmentary teeth from Chaparral were included in the hypodigm of a new genus and species, *Synastrapotherium amazonense* by Paula Couto (1976). Fragmentary remains of uruguaytheriines now at the Smithsonian Institution, Washington, D.C., from eastern Peru are reported by Willard (1966). Most recently, Hoffstetter (1977) and Frailey (1987) describe uruguaytheriine material from middle Miocene deposits at Quebrada Honda in Bolivia.

In this paper we describe the uruguaytheriine Astrapotheriidae from the rich middle Miocene Honda Group of the upper Magdalena River valley in Colombia. These remains are the youngest securely dated remains of that order in South America. The Honda Group sample includes the largest assemblage of uruguaytheriine remains anywhere in South America. The following descriptions are based upon material collected by the University of California Museum of Paleontology (UCMP) and the Colombian Servicio Geológico Nacional (now INGEOMINAS) in the late 1940s and early 1950s, and originally studied by Johnson (1984). Additional material from the Honda Group collected by joint Duke University and INGEOMINAS expeditions between 1985 and 1991, and in the collections of the Smithsonian Institution and the Yale Peabody Museum have also been studied. In addition, we have examined material from the Honda Group in the collections of the Departamento de Geociencias, Universidad Nacional de Colombia, Bogotá, the Museo Geológico of INGEOMINAS, and the Museo Paleontológico del Instituto Huilense de Cultura in Villavieja and Neiva, Huila Department, Colombia.

Two distinct genera of uruguaytheriine astrapothere occur in the Honda Group of Colombia. For the smaller species, *Xenastrapotherium kraglievichi,* formerly known only by the type specimen, we now have associated upper and lower dentitions. Based upon this new material, an emended diagnosis for the genus is provided. We describe a new species of *Xenastrapotherium* from the Miocene Biblián Formation of southern Ecuador. The type of *Synastrapotherium amazonense* de Paula Couto (1976) from the Río Juruá in Acre Province, Brazil, is herein assigned to the genus *Xenastrapotherium.*

A morphologically distinct and larger new genus and species from the Honda Group is described from a partial cranium, a complete palate, nearly complete mandibles, several upper and lower dentitions including juveniles, and numerous isolated teeth. Astrapothere material in the Smithsonian Institution from eastern Peru is referred to this new genus. Following the description, we present our views on the probable adaptations of this interesting astrapothere.

Finally, we discuss the results of a phylogenetic analysis of astrapotheres based on the cranium, dentition, and postcranium and discuss the phylogenetic position of the two Colombian middle Miocene astrapotheres.

Systematic Paleontology

Order **Astrapotheria** Lydekker 1894
Family **Astrapotheriidae** Ameghino 1887
Subfamily **Uruguaytheriinae** Kraglievich 1928b
Xenastrapotherium Kraglievich 1928b

TYPE SPECIES

Xenastrapotherium (*"Astrapotherium"*) *christi* (Stehlin 1928) Kraglievich 1928b.

EMENDED GENERIC DIAGNOSIS

Xenastrapotherium differs from all other known astrapotheres by possessing two lower incisors on each side and an M^3 posterolingual cingulum that joins the protoloph with the metaloph and closes the median valley lingually. In addition to these characters, *Xenastrapotherium* differs from other known uruguaytheriine astrapotheres in possessing lower canine tusks with pronounced longitudinal curvature and anteroinferior and posteroinferior grooves in cross section; shallow lower molar metaflexids bounded by a variably developed anterolingual cingulum; and mesiodistally narrow and rounded metalophids.

DISCUSSION

The type species, *Xenastrapotherium christi,* is known from a nearly complete mandible described by Stehlin (1928) from Venezuela. New material of *Xenastrapotherium kraglievichi* from the Honda Group includes associated upper and lower dentitions and permits extending the diagnosis of the genus to features not preserved in the type species, in particular, features of the lower canine, lower incisors, and upper M^3.

The lower molar hypoflexid, which appears as a faint trace on the M_{2-3} of the *X. kraglievichi* type specimen, is completely lacking on other specimens of the species. Lower incisors, which are present only as weakly developed root alveoli on the mandible of the type species *X. christi* Stehlin, are variably expressed in *X. kraglievichi*. For example, UCMP 38115 presents two, well-developed, bilobed lower incisor crowns, whereas the mandibular symphysis of IGM 182786 preserves only two very small incisor alveoli (as in *X. christi*). The ever-growing, lower canine tusks in *Xenastrapotherium* are more horizontally emplaced in the mandible and the longitudinal curve of the lower canine extends the exposed portion anterolaterally. This more horizontal and lateral disposition contrasts with the lower tusks of *Astrapotherium, Parastrapotherium,* and *Astrapothericulus,* which display an upward or dorsal curvature. By comparison with these same three genera, the lower canines in *Xenastrapotherium* are more rounded in cross section with the greatest cross-sectional diameter horizontal, that is, parallel with the base of the ramus. In cross section the lower canine tusks display anteroinferior and posteroinferior longitudinal grooves. These two longitudinal concavities

distinguish this genus from other Uruguaytheriinae for which the lower canines are known.

Xenastrapotherium kraglievichi Cabrera 1929
(fig. 22.1)

HOLOTYPE

MLP 12-96, an adult left mandible fragment preserving M_1 (nearly complete), M_2, and M_3.

HYPODIGM

UCMP 38063, right and left upper canines, fragmentary partial upper dentition including right P^{3-4}, M^{1-3}, left P^{3-4}, M^{1-3}, from UCMP Locality V4527; UCMP 38847, lower canine, left lower incisor, P^3, P^4, M^1 and M^3, and right P^3, P^4, and M^3, from UCMP Locality V4528; UCMP 38860, left P^{3-4}, right P^3 and M^1, from UCMP Locality V4528; UCMP 38115, partial symphysis with right I_{1-2}, left I_1, and left ramus with P_4, M_{1-3}, from UCMP Locality V4528; UCMP 38884, right lower canine, M^1, M^2, and both mandibular rami with M_{1-3}, from UCMP Locality V4528.

REFERRED MATERIAL

UCMP 34287, right mandible fragment with dP_{3-4}, M_1, from UCMP Locality V3902; UCMP 38925, partial M^2, left and right M^3, from UCMP Locality V4524; UCMP 38058, a lower incisor, from UCMP Locality V4526; UCMP 38885, right M_2 and M_3, from UCMP Locality V4527; UCMP 38922, two lower incisors, left upper canine, right M^3, left mandible fragment with M_3, from UCMP Locality V4528; UCMP 39820, right upper canine, from UCMP Locality V4536; UCMP 39823, right M^{1-2}, left P^4, and M^3, from UCMP Locality V4536; UCMP 39818, three lower incisor crowns, partial right upper canine, and left and right lower canines, left P^4, from UCMP Locality V4932; IGM 183551, partial left M^3, locality unknown; IGM 184570, right M^3, from Duke Locality 100; IGM 250579, lower incisor, from Duke Locality 32; IGM 250891, a partial right M^1, from Duke Locality 126; IGM 251162, left P^4 and lower incisor, from Duke

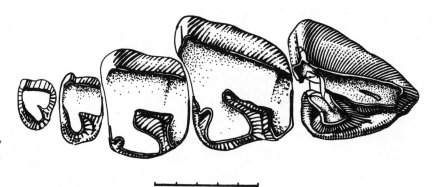

Figure 22.1. Left maxillary cheek tooth row of *Xenastrapotherium kraglievichi,* based upon a composite of UCMP 39823, UCMP 38847, and TATAC-48. Scale bar 5 cm.

Locality 141; IGM 251059, from Duke Locality 130; IGM 182784, a partial symphysis with left lower canine, and the alveoli of left I_{1-2}, from Duke Locality 19; YPM-PU 21936, associated left and right mandibular rami preserving the symphysis, left M_1 (partial), M_2, M_3 (broken) and right M_{2-3}; USNM 10884, left P_4 (see later for locality); TATAC-48, a partial left maxillary dentition with P^4, M^{1-3}, from an unknown locality; MPV-189, a partial left maxilla with M^{2-3}, from an unknown locality; MPV-10, a partial right maxilla with M^1 (broken), M^{2-3}, from an unknown locality; MPV-55, left ramus fragment with M_{1-2}, from an unknown locality; MPV-188, a right ramus fragment with M_{2-3}, from an unknown locality.

GEOLOGIC HORIZON AND LOCALITIES

UCMP localities in the Villavieja Formation containing *Xenastrapotherium kraglievichi* include V3902, Santa Elena, 5 km west of Guacirco, Baraya Member; V4524, Fish Bed, Baraya Member; V4526, Red Bed 2, La Venta Red Beds, Baraya Member; V4527, Red Bed 3, Unit between the La Venta and El Cardón Red Beds," Baraya Member; V4528, Red Bed 4, El Cardón Red Beds, Cerro Colorado Member; V4536, San Nicolás, Monkey Beds, Baraya Member; and V4932, Railway Bridge, Monkey Beds, Baraya Member. More detailed geographic descriptions of these localities can be found in Fields (1959) and on file at the University of California Museum of Paleontology. Duke University localities in the Villavieja Formation containing *Xenastrapotherium kraglievichi* include Duke Locality 32 in the El Cardón Red Beds, Cerro Colorado Member; Duke Localities 126 and 141 in the Polonia Red Beds, Cerro Colorado Member; and Duke Localities 19 and 130 in the

Monkey Beds, Baraya Member. Only one specimen of *Xenastrapotherium kraglievichi* has been recovered in the La Victoria Formation at Duke Locality 100, stratigraphically the highest fossil vertebrate locality in the La Victoria Formation.

YPM-PU 21936 was collected 6–8 km northeast of Villavieja. This general vicinity corresponds to UCMP Locality V4517 and Duke Locality 6 and is probably within the Villavieja Formation. USNM 10876 was collected by Brother Ariste Joseph from somewhere near Neiva, and USNM 10884 was collected near the village of Villavieja in September 1920. All TATAC and MPV specimens were collected by amateurs from unknown localities in the La Venta area.

EMENDED DIAGNOSIS

Xenastrapotherium kraglievichi differs from *X. christi* in lacking or having only a very weak remnant of the hypoflexid depression positioned opposite the paraflexid on M_2 and M_3 (in *X. christi* the weak M_2 and M_3 hypoflexid is positioned opposite the metalophid) and having P_4 implanted more in line with the molar row. *Xenastrapotherium kraglievichi* differs from *X. aequatorialis* in lacking any trace of the posterior pillar and in having a much-reduced lower molar lingual cingulum. *Xenastrapotherium kraglievichi* differs from all other known species of the genus by having significantly smaller cheek teeth.

DESCRIPTION

The dental formula of *Xenastrapotherium kraglievichi* is 0.1.2.3/2.1.1.3. Upper incisors are absent as in all known

Table 22.1. Tooth measurements (cm) for _Xenastrapotherium kraglievichi_

Tooth	Measurement[a]	N	Mean	SD	SE	Observed Range	CV
			LOWER DENTITION				
C	Max. diameter	3	4.43	0.420	0.240	4.10–4.90	—
	Trans. diameter	3	2.72	0.450	0.260	2.30–3.20	—
P_4	Length	6	2.51	0.280	0.110	2.01–2.88	—
	Breadth	6	1.90	0.200	0.080	1.72–2.19	—
M_1	Length	9	3.95	0.322	0.107	3.53–4.48	7.98
	Breadth	9	2.33	0.260	0.090	1.93–2.69	11.32
M_2	Length	9	5.19	0.330	0.110	4.56–5.60	6.43
	Breadth	10	2.64	0.180	0.060	2.35–2.89	6.67
M_3	Length	10	5.89	0.350	0.110	5.34–6.38	5.88
	Breadth	10	2.57	0.310	0.100	2.24–3.30	12.20
			UPPER DENTITION				
C	Max. diameter	3	4.13	0.780	0.450	3.50–5.00	—
	Trans. diameter	3	2.62	0.300	0.170	2.30–2.90	—
P^3	Length	1	2.05	—	—	—	—
	Breadth	1	2.34	—	—	—	—
P^4	Length	4	2.03	0.030	0.020	1.99–2.07	—
	Breadth	4	3.00	0.180	0.090	2.86–3.26	—
M^1	Length	3	4.04	0.090	0.050	3.95–4.12	—
	Breadth	4	4.30	0.210	0.110	4.00–4.46	—
M^2	Length	6	6.01	0.460	0.190	5.30–6.53	—
	Breadth	5	4.71	0.450	0.200	4.20–5.20	—
M^3	Length	12	6.29	0.420	0.120	5.50–6.80	6.73
	Breadth	12	4.34	0.300	0.090	3.83–4.75	6.87

[a]Maximum diameter = major axis of cross section. Transverse diameter = minor axis of cross section. Coefficient of variation, $CV = 100 \cdot SD/mean$.

Astrapotheria. As in most Astrapotheriidae (but see later), the lower incisors in _Xenastrapotherium kraglievichi_ are procumbent, bilobed, and have robust roots. The upper and lower canines are enlarged, curved, ever-growing tusks. The upper premolars are simple and nonmolariform, and both upper and lower molars display a lophodont crown pattern superficially reminiscent of the black rhinoceros (_Diceros_). For tooth measurements, see table 22.1.

The upper canines are enlarged, ever-growing, and downward-curving tusks, with a triangular cross section. The upper canines resemble those of _Astrapotherium magnum_. Enamel covers the two ventral surfaces, whereas dentine is exposed on the dorsal surface (anterior face). The enamel is longitudinally fluted and grooved. There is a beveled rounded-triangular wear facet for contact with the lower canine. This facet truncates the tusk, producing an acute, chisel-like point. The anterior face of the tusk above the wear facet is polished and shows evidence of abrasion.

As in the Miocene Astrapotheriidae from Patagonia, there are two upper premolars in _Xenastrapotherium kraglievichi_. The upper premolars of _Xenastrapotherium_ are relatively more simple in structure with a less-pronounced parastyle and distostyle and weaker lingual cingulum. P^4 is larger than P^3 and has a more pronounced distolingual cingulum, a somewhat more pronounced parastyle, and a small metaloph extending lingually from the ectoloph. Both P^3 and P^4 have two roots.

All the available specimens of M^1 are worn. As in _As-_

trapotherium and other Miocene astrapotheriids, the worn M^1 crown is roughly quadrate in occlusal outline, with a well-developed ectoloph. Compared with other Miocene astrapotheriids, the protoloph displays a persistent, more angular anterolingual pocket or enamel fold. There is no evidence of a crista. By way of contrast to the M^1 of *Astrapotherium,* which has a pronounced buccal cingulum, the M^1 of *Xenastrapotherium* shows no buccal cingulum.

M^2 is the largest tooth in the molar series. As in M^1, the M^2 ectoloph shows no buccal cingulum. Compared with *Astrapotherium* and other Miocene astrapotheres from Patagonia, the *Xenastrapotherium* M^2 displays a weak parastyle, and in advanced wear stages, the protoloph and metaloph become joined by a well-developed posterolingual cingulum that closes off the shallow median valley lingually.

M^3 is roughly triangular in occlusal outline. The ectoloph displays a weak parastyle and no buccal cingulum. A secondary crista and medifossette appears lingual to the ectoloph in moderate wear stages. A well-developed mesiolingual pocket is surrounded by a broad rounded cingulum that wraps around the mesial aspect of the protoloph. *Xenastrapotherium kraglievichi* differs from all other Miocene astrapotheres in the configuration of the M^3 median valley, which is closed lingually by the distal continuation of the broad, rounded lingual cingulum that connects the base of the protoloph with the metaloph. In all other Miocene astrapotheres the deep median valley of M^3 opens distolingually.

A number of isolated lower incisor crowns (UCMP 38058 and 38922; IGM 250579 and 251162) and a partial symphysis with three incisor crowns and the root of a fourth (UCMP 38115) are known. These specimens indicate that *Xenastrapotherium kraglievichi,* like *X. christi,* had two incisors in each mandible. The Patagonian Miocene genera *Parastrapotherium, Astrapotherium,* and *Astrapothericulus* have three lower incisors in each side, and the crowns of each have distinct proportions. In these genera the I_2 crown is highest and has the stoutest root, the I_3 crown is broadest mesiodistally, and the I_1 crown is smallest. The lateral incisor in *Xenastrapotherium kraglievichi* is high-crowned, suggesting that it is the I_3 that has been lost in this genus. The lower incisor crowns in *X. kraglievichi* are spatulate and bilobed in the unworn condition and have both lingual and buccal basal cingula. With wear, the incisor crowns become

more nearly oval in crown view. In preserved complete lower incisors the roots are robust and in length measure at least twice enamel crown height. The lower incisor root alveoli, however, are very small in some specimens (e.g., IGM 182784), indicating that the lower incisors are either variably developed or may have worn out and been shed.

The lower canines are enlarged, ever-growing tusks. In cross section they form a rounded triangle with a broadly rounded base distally and with a somewhat rounded apex mesially. The superior (dorsal) and inferior (ventral) surfaces are covered with enamel. A wide shallow groove on the upper surface extends longitudinally along the leading edge. The enamelless distal surface is smoothly convex. In cross section the lower canine tusks display anteroinferior and posteroinferior longitudinal grooves. The lower canine tusk has a strong longitudinal curve and is implanted horizontally in the symphysis with the greatest cross-sectional diameter horizontal, that is, parallel with the base of the ramus. The horizontally implanted lower canines of *Xenastrapotherium* contrast with the more vertically implanted lower tusks of *Astrapotherium, Parastrapotherium,* and *Astrapothericulus.* A large, beveled, distally facing contact facet for the upper canine occurs near the apex of the crown. As in *Astrapotherium,* a series of worn areas or grooves formed by abrasion appear on the anterior or mesial edge of the lower canines.

The single lower premolar has two roots, and the crown is bicrescentic and lophodont, displaying distinct paralophid, metalophid, and hypolophid. Unlike *Xenastrapotherium christi* in which P_4 is rotated, the P_4 of *X. kraglievichi* is implanted in line with the molar row. As in other Miocene Astrapotheriidae, the lower molars of *Xenastrapotherium* are two-rooted, bicrescentic, and lophodont and increase in crown length from M_1 to M_3. The paralophid, metalophid, and hypolophid are distinct and separated by a shallow paraflexid and deep, persistent entoflexid. The paraflexid is usually enclosed by a cingular ridge descending from the distal aspect of the paralophid. In occlusal view, when relatively less worn, the lophids are round but broaden basally. As was first noticed by Kraglievich (1928b), in the lower molars of *Xenastrapotherium* the prehypocristid does not extend lingually. This results in a very weak hypoflexid and a mesiodistally narrow, rounded metalophid. In *X. kraglievichi* the hypoflexid is indistinct on M_1, appears

Table 22.2. Mandibular ramus depth and thickness (cm) for Honda Group Astrapotheriidae

Specimen	M₂		Mandible Depth at M₂	Mandible Thickness at M₂
	Length	Breadth		
Xenastrapotherium				
MLP 12-96	4.80	1.90	8.10	—
YPM–PU 21936	4.96	2.57	6.36	6.20
UCMP 38115	5.60	2.70	—	—
Granastrapotherium				
UCMP 40187	6.26	3.74	11.90	6.03
UCMP 40408 (L)	6.98	3.79	11.96	5.35
IGM 182778	6.20	3.30	10.70	3.73
UCMP 40017 (L)	6.64	3.30	9.70ᵃ	5.3ᵃ
Genus incertae sedis				
UCMP 34287	Encrypted	Encrypted	8.25	3.85
USNM 10876	Erupting	Erupting	5.63	3.50
UCMP 39820	Np	Np	7.01	4.60

Notes: (L): specimen from left side. Np: not preserved.

ᵃEstimated.

only as a slight indentation on M_2, and as a slight depression opposite the paraflexid in M_3. With wear, the occlusal surface of the lower molars increases in buccolingual breadth. Whereas in M_1 and M_2, crown length decreases toward the base, in M_3 crown length increases. The lower molars show no basal cingulum buccally or distolingually.

Three adult partial mandibles are known, the type MLP 12-96, UCMP 38115, and YPM-PU 21936. These mandibles are similar to those of other Miocene Astrapotheriidae. The condyle is not as high relative to the tooth row as in *Astrapotherium magnum,* and the horizontal ramus is long and shallow dorsoventrally and quite thick transversely (table 22.2). The symphysis is fused, elongate, and narrow and troughlike on the dorsal surface, widening anteriorly near the incisor alveoli. The ventral surface of the symphysis is rugose, pitted, and shows numerous small vascular foramina. There are two prominent mental foramina, one ventral to the premolar and another about 3 cm anterior and somewhat ventral to the first.

DISCUSSION

Xenastrapotherium kraglievichi is the smaller of the two astrapotheres in the Honda Group and has less-specialized lower canine morphology. Unfortunately, *X. kraglievichi* is known only from partial dentitions and is not as common as the larger genus in the Honda Group. The upper dentition and symphyseal region is now known, and these regions permit a more comprehensive emended diagnosis for the genus and species. No cranial material or associated postcranial remains are yet known, however, and juvenile material is difficult to allocate.

With the exception of a single specimen, stratigraphic occurrences are restricted to the Villavieja Formation. The exceptional specimen was collected at the highest fossiliferous level of the underlying La Victoria Formation.

Xenastrapotherium aequatorialis sp. nov.
(fig. 22.2)

HOLOTYPE

EPN(Q) 4280, a portion of the right mandibular ramus preserving the posterior root of P_4, M_{1-2} nearly complete, and the anterior alveolus of M_3.

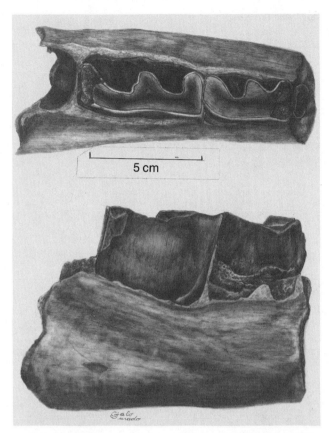

Figure 22.2. *Xenastrapotherium aequatorialis* sp. nov., holotype, EPN(Q) 4280, partial right mandibular ramus preserving the posterior root of P$_4$, M$_{1-2}$, and the anterior wall of the alveolus of M$_3$: *top,* occlusal view; *bottom,* lateral view. Scale bar 5 cm.

HORIZON AND LOCALITY

EPN(Q) 4280, purchased in August 1985, was reported to have been collected along the banks of the Río Burgay near the junction of Qda. de Shunshi, approximately 1 km south of Biblián, Province of Cañar (2°45′ S, 78°55′ W), corresponding to point 354989 on the Azogues topographic sheet (Instituto Geographico Militar [Quito] CT-NV-EI, 3885-IV, 1969). Here, the Río Burgay cuts through the Miocene Biblián Formation in the core of the Biblián Anticline. The Biblián Formation outcrops extensively in the 4-km² area surrounding the site where the type specimen was reported to have been collected.

GEOCHRONOLOGIC AGE

Freshwater invertebrates from the Biblián Formation are similar to those from the richly fossiliferous overlying Loyola Formation, regarded as Miocene in age (Bristow and Parodiz 1982). Bristow (1973) and Bristow and Parodiz (1982) described andesitic lavas at or near the contact between the Biblián and Loyola formations and reported two ^{40}K-^{40}Ar whole rock determinations of 19.7 ± 0.5 Ma and 21.0 ± 0.6 Ma from one of these lavas outcropping near Descanso on the east side of the Cuenca basin. The Descanso andesite is considered to correspond to the uppermost Biblián Formation (Bristow and Parodiz 1982).

ETYMOLOGY

The specific name *aequatorialis* refers to the terrestrial equator, which bisects the country of origin.

DIFFERENTIAL DIAGNOSIS

Xenastrapotherium aequatorialis differs from other known species of the genus (*X. christi* and *X. kraglievichi*) in showing a remnant of either the internal pillar or a short, lingually extended prehypocristid (anterior horn of the posterior crescent) fused to the distal aspect of the metalophid in M$_2$, a rugose, continuous external cingulum on M$_1$, an anteroexternal cingulum on M$_2$, and a well-developed, "beaded" lingual cingulum continuous around the metalophid on M$_1$ and M$_2$.

DESCRIPTION

The type specimen, a portion of the right mandibular ramus, preserves the posterior root of P$_4$, the nearly complete crowns of M$_1$ and M$_2$, and the anterior alveolus of M$_3$. Radiographs reveal no developing teeth in the body of the mandible, supporting their identification as permanent teeth. The corpus of the mandible is shallow and thick as in *X. kraglievichi.* The ramus is strongly convex buccally, and the submandibular fossa is concave lingually. The mental foramen is situated beneath the posterior margin of M$_2$.

P$_4$ was fully erupted as judged by the impression of the distal aspect of the crown on matrix between P$_4$ and M$_1$. Judging from the shape of the matrix impression and the orientation of the posterior root as viewed from the occlusal perspective, P$_4$ was positioned in line with M$_1$ and M$_2$ as in *X. kraglievichi* and unlike the P$_4$ of *X. christi.*

The first lower molar is a lophodont, bicrescentic, and rooted tooth. The tooth shows no external vertical depression or hypoflexid. The M_1 metalophid is rounded superiorly and flattened basally. The entoflexid is deeply folded into the tooth and would have persisted as a distinct enamel fold into advanced wear stages. A buccal cingulum is present as a discontinuous line of irregular rugosities. A prominent and continuous beaded internal cingulum descends from the anterior aspect of the paralophid, extending distally along the base of the lingual side of the metalophid to reascend the posterior aspect of the entolophid.

M_2 is longer and broader than M_1. Only a shallow indentation anteroexternally indicates the position of the hypoflexid. A basal anterolingual cingulum descends from the paralophid and reaches the metalophid but is discontinuous around the metalophid. The metalophid is rounded superiorly and flattens basally. The most distinctive feature of this tooth is the preservation of a remnant of an internal pillar fused to the distal aspect of the metalophid. The entoflexid is deeply folded and would persist in a well-worn tooth. The M_2 presents only a trace of an anteroexternal cingulum. A thick posterolingual cingulum encloses the base of the entoflexid. The third lower molar is not preserved but was apparently fully erupted as evidenced by the interproximal wear facet on the distal aspect of the M_2 crown.

DISCUSSION

As judged from lower molar dimensions, *Xenastrapotherium aequatorialis* is about the same size as *X. kraglievichi* and *X. christi*. *X. aequatorialis*, however, differs from the other two species by showing an internal talonid pillar fused to the metalophid. In this feature the Ecuadorean species resembles Santacrucian *Astrapotherium*, Colhuehuapian *Astrapothericulus*, and Colhuehuapian-Deseadan *Parastrapotherium*. Also like Oligocene and Miocene genera from Patagonia, the lower molars of the *X. aequatorialis* display both buccal and lingual cingulums. By contrast, the lower molars of *X. kraglievichi* and *X. christi* display no buccal cingulum and only a reduced lingual cingulum distally.

Xenastrapotherium aequatorialis is possibly as much as 7 million years older than *X. kraglievichi* from the Villavieja Formation in Colombia and is contemporaneous with Col-

huehuapian faunas in Patagonia containing *Parastrapotherium*, *Astrapothericulus*, and *Astrapotherium* (Marshall et al. 1983).

Xenastrapotherium chaparralensis sp. nov.
(figs. 18–23 in Stirton 1947b)

HOLOTYPE

UCMP 37838, a left M_3.

TYPE LOCALITY, GEOLOGIC HORIZON, AND AGE

The holotype and hypodigm material was collected at UCMP Locality V4406, in the Tuné Formation, Municipality of Chaparral, Tolima Department, Colombia. Stirton (1947b, 1953b) assigned the associated mammal fauna to the Deseadan Land Mammal Age, now calibrated to between 34 and 21 Ma (Marshall, Cifelli, et al. 1986).

HYPODIGM

UCMP 38733, lower canine; UCMP 37839, lower incisor; UCMP 37835, right M^1 or M^2; UCMP 37836, right dP^4; UCMP 37837, right M_1 or M_2; and UCMP 37834, left M^3; all from the same locality.

DIFFERENTIAL DIAGNOSIS

Xenastrapotherium chaparralensis differs from *X. kraglievichi* in displaying an upper molar buccal cingulum and a more lingually directed M^{1-2} crista. *Xenastrapotherium chaparralensis* differs from *X. kraglievichi*, *X. christi*, and *X. aequatorialis* in having greater lower molar area (table 22.3).

DESCRIPTION

The astrapothere remains from UCMP Locality V4406, originally described by Stirton (1947b), include a lower incisor (UCMP 37839) with robust roots and crown dimensions (2.2 cm mesiodistal length, 1.8 cm buccolingual breadth) comparable with *Xenastrapotherium kraglievichi*. The dP^4 (UCMP 37836) differs from uru-

Table 22.3. Tooth measurements (cm) for species of *Xenastrapotherium*

Sample or Variable	P$_4$ Length	P$_4$ Breadth	M$_1$ Length	M$_1$ Breadth	M$_2$ Length	M$_2$ Breadth	M$_3$ Length	M$_3$ Breadth	M^3 Length
X. kraglievichi									
MLP 12-96, type	—	—	3.53	2.00	4.93	2.39	5.42	2.55	—
Honda Group sample:	(N = 6)	(N = 6)	(N = 9)	(N = 9)	(N = 9)	(N = 10)	(N = 10)	(N = 10)	(N = 12)
Mean	2.51	1.90	3.95	2.33	5.19	2.64	5.89	2.57	6.29
Observed minimum	2.01	1.72	3.53	1.93	4.56	2.35	5.34	2.24	5.50
Observed maximum	2.88	2.19	4.48	2.69	5.60	2.89	6.38	3.30	6.80
X. aequatorialis									
EPN(Q) 4280, type	—	—	3.90	2.00	5.50	2.50	—	—	—
X. christi									
Basel Aa-21, type	1.75[a]	1.55[a]	3.90[a]	2.15[a]	5.00[a]	2.25[a]	6.00[a]	2.25[a]	—
X. chaparralensis									
UCMP 37837	—	—	4.50	2.40	—	—	—	—	—
UCMP 37838	—	—	—	—	—	—	6.95	3.10	—
UCMP 37834	—	—	—	—	—	—	—	—	6.45
X. amazonense									
DGM 574-M, type	—	—	—	—	—	—	—	—	8.30[b]

[a]Estimated from a cast at the UCMP.

[b]Estimated from figure 1A in Paula Couto (1976).

guaytheriine dP⁴s from the Honda Group in lacking the mesiolingual pocket. In UCMP 37836 a mesiolingual cingulum extends along the mesial aspect of the protoloph without forming a distinct "pocket."

The M¹ or M² (UCMP 37835) displays a weakly developed labial cingulum and a lingually oriented crista, and, in these features the upper molar resembles those of *Parastrapotherium*. Unlike *Astrapotherium* or *Parastrapotherium*, however, the upper molar of *X. chaparralensis* displays a weaker parastyle and shallow lingual valley. The mesial and lingual aspects of this crown are not preserved.

UCMP 37834, a left M³, is similar in overall morphology to the M³ of *X. kraglievichi* although in a less-worn condition than any known material of that species. The broad, rounded ridge connecting the protoloph with the metaloph is a synapomorphy shared with *X. kraglievichi* that contrasts with the absence of this structure in *Parastrapotherium*, *Astrapotherium*, and the other genus from the Honda Group.

The distalmost tip of the lower canine tusk (UCMP 37833) was tentatively identifed by Stirton (1947b, fig. 15) as a toxodont. The relatively unworn specimen, however, compares in most aspects of its morphology with comparably worn lower canine tusks of *X. kraglievichi*.

The two lower molars (UCMP 37837 and 37838) display only a weak trace of the hypoflexid. The M₁ entoflexid is enclosed by a cingular ridge descending from the protolophid.

DISCUSSION

Stirton (1947b) did not assign the Chaparral material to any known genus of uruguaytheriine astrapothere, even though he noted the absence of a lower molar hypoflexid. Stirton also noted that the quadrate M³ (UCMP 37834 is mistakingly identified as an M² by Stirton) displays a lingual connection between the protoloph and metaloph. The M³ of *X. kraglievichi* (described previously) and *X. amazonense* (see here) show the same feature. On the basis of the connection between the protoloph and metaloph, we refer the material from UCMP Locality V4406 to *Xenastrapotherium*. The larger size of the V4406 lower molars, especially the length of the M₃ (the holotype, UCMP 37838; see table

22.3) is the principal diagnostic character by which the Chaparral material can be recognized as a distinct species.

Stirton's original Deseadan age assignment of the Chaparral fauna was based upon the associated mammals of which, aside from *X. chaparralensis*, only one (*Proadinotherium*) can be confidently identified to genus. *Proadinotherium* is known from both the Deseadan and Colhuehuapian of Patagonia and also occurs in the early Santacrucian Pinturas Formation. The astrapothere material from Chaparral clearly represents *Xenastrapotherium*, a genus otherwise known only from dated deposits of Miocene age. In lacking a basal cingulum and any trace of the pillar on the metaloph, the lower molars of *X. chaparralensis* appear more advanced than those of *X. aequatorialis* from the Biblián Formation of Ecuador. All this suggests that the Chaparral local fauna may be Miocene in age.

Xenastrapotherium amazonense (Paula Couto 1976)
(figs. 1, 2 in Paula Couto 1976)

SYNONYMY

Synastrapotherium amazonense Paula Couto 1976.

HOLOTYPE

DGM 574-M, a maxillary fragment with complete M³, including part of the jugal and inferior margin of the orbit.

TYPE LOCALITY

DGM Locality 28, Río Juruá, left bank at Pedra Pintada, between the mouth of the Tejo River, at Taumaturgo, and the mouth of the Breu River, in Brazil, at the Peruvian border. From "Puca" type deposits at the Pedra Pintada exposure (for more details, see Paula Couto 1976).

DIFFERENTIAL DIAGNOSIS

Xenastrapotherium amazonense differs from all other species of the genus *Xenastrapotherium* by having a substantially larger M³. M³ ectoloph length for *X. amazonense* falls outside the observed range for *Xenastrapotherium kraglievichi* and is 30% greater than that of *X. chaparralensis* (see table 22.3).

DISCUSSION

G. G. Simpson and L. Price collected astrapothere remains from the Tertiary red beds outcropping along the banks of the Río Juruá in northwestern Brazil. On the basis of a partial maxilla (the holotype DGM 574-M), Paula Couto (1976) described a new genus and species, *Synastrapotherium amazonense.* Into this taxon Paula Couto (1976) referred a left P^3 and the fragmentary lower molar material also collected from along the Río Juruá and the astrapothere material collected by Stirton near Chaparral in Colombia.

Paula Couto (1976) identified the upper molar of the holotype (DGM 574-M) as an M^3. The fragmentary maxilla is identified as DGM 574-M in the holotype description and as DGM 611-M (see fig. 1A in Paula Couto 1976, 246). The holotype specimen includes the proximal portion of the jugal and a portion of the inferior margin of the left orbit, also the complete crown of an upper molar. Judging by the position and triangular outline of the crown, DGM 574-M is correctly identified as a left M^3.

In the diagnosis and description Paula Couto emphasized seven features of the M^3 of *Xenastrapotherium* (=*Synastrapotherium*) *amazonense.* Four of these features, (1) a sinuous ectoloph, (2) a strong protoloph, (3) metaloph and postfossette absent, and (4) crista strong and oriented parallel to the ectoloph are characteristic of astrapotheriids generally but do not distinguish one genus from another. The three other features, M^3 size, cingular ridge connecting protoloph and metaloph, and anterolingual pocket, are differentially diagnostic.

The M^3 dimensions fall well outside the observed range for *Xenastrapotherium kraglievichi* (see table 22.3) and the length of the M^3 ectoloph in DGM 574-M is 30% greater than the M^3 of *X. chaparralensis* (table 22.3).

The median valley is approximately parallel to the ectoloph, ending near the posterior end of the crista. It is separated from the lingual side of the crown by a strong, high, convex cingulum that connects the base of the protoloph with the posterior end of the ectoloph. In this feature DGM 574-M resembles *X. chaparralensis* and *X. kraglievichi.* All three species of *Xenastrapotherium* display a broad, rounded cingulum or ridge connecting the protoloph with the ectoloph, herein considered a shared derived feature and diagnostic of the genus. This feature is distinct from that seen in

Astrapotherium, Astrapothericulus, and *Parastrapotherium* and the large uruguaytheriine from the Honda Group wherein the M^3 median valley cuts deeply to the lingual side of the crown.

The anterolingual pocket seen in *Xenastrapotherium amazonense* is a prominent feature of the upper molars of *X. kraglievichi* and the large uruguaytheriine from the Honda Group. This synapomorphy clearly aligns DGM 574-M with the two uruguaytheriine genera from the Honda Group in Colombia. Unfortunately, this aspect of the protoloph is not preserved in *X. chaparralensis.* Based upon the cingular ridge connecting protoloph to ectoloph, DGM 574-M is referred to the genus *Xenastrapotherium.* On the basis of its large size DGM 574-M is retained as a distinct species, *X. amazonense.*

The lower molar fragments (AMNH 55737 and 55738) do not come from the same locality as the holotype and were not collected in situ. All that can be confidently said about these talonid fragments is that they are from a uruguaytheriine astrapotheriid. Similarly, DGM 611-M, a perfectly preserved upper premolar identified as a P^3 and also included in the hypodigm was collected at a different locality, not in situ, and the specimen has never been illustrated. Frailey (1986, fig. 26D) refers an upper canine (LACM 117531) from the Río Acre local fauna to *Synastrapotherium amazonense.* This specimen lacks any diagnostic features and can be referred only questionably to Astrapotheria.

Granastrapotherium gen. nov.

DIFFERENTIAL DIAGNOSIS

Granastrapotherium may be distinguished from all other astrapotheres by lacking lower incisors, by possessing large, ever-growing, sexually dimorphic, horizontally implanted lower canine tusks that are round to oval in cross section; by having proportionally longer M$_2$ and M$_3$ crowns compared with M$_1$; in lacking P^3; and by having a P^4 without parastyle. *Granastrapotherium* may be further distinguished from *Uruguaytherium* by possessing traces of the hypoflexid groove on the buccal aspect of the lower molars and by having a more distally oriented and basally rounded lower molar metalophid. *Granastrapotherium* differs from *Xena-*

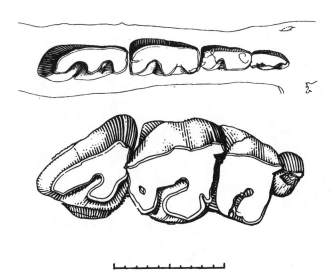

Figure 22.3. *Granastrapotherium snorki,* upper and lower tooth rows showing the last premolar and the molars. *Top,* UCMP 40017, P$_4$, M$_{1-3}$; *bottom,* UCMP 40188, P^4, M$^{1-3}$. Scale bar 10 cm.

strapotherium in having only very weakly developed upper molar lingual cingula and in lacking any connection between the protoloph and ectoloph on the M^3.

Granastrapotherium snorki sp. nov.
(figs. 22.3–22.5)

ETYMOLOGY

The genus-group name is a combination of the Spanish *gran,* meaning great, and *Astrapotherium,* meaning lightning

Figure 22.4. *Granastrapotherium snorki,* UCMP 40358, basicranium, ventral view. Scale bar 10 cm.

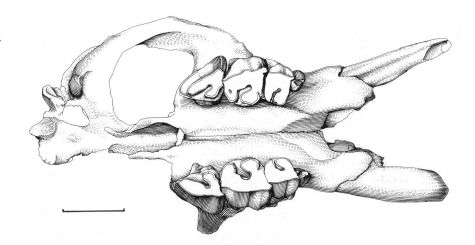

beast. The species-group name *snorki* refers to the apparently well-developed proboscis of this taxon as argued here.

HOLOTYPE

UCMP 40358, a partial skull including the palate and basicranium preserving the complete upper dentition, including right and left upper canine, P^4, and M$^{1-3}$. Preserved also are the internal nares, inferior margin of the right orbit, and an incomplete occipital condyle. The top of the face, neurocranium, and occipital region are missing.

TYPE LOCALITY

UCMP Locality V4525, south of Qda. La Venta, about 6 km east-northeast of Villavieja, Huila Department, Colombia (for more details, see Fields 1959). Stratigraphically, the specimen was collected in the "Unit between the Ferruginous Sands and lower Red Bed" of Fields (1959), in the Baraya Member, Villavieja Formation, Honda Group.

HYPODIGM

UCMP 38397, palate with left P^4, M$^{1-3}$, portions of right M$^{2-3}$, from UCMP Locality V4415, Coyaima. UCMP 37914, right upper canine; and UCMP 39821, right lower canine; both from UCMP Locality V4518. UCMP 39635, M$^{1\ or\ 2}$, from UCMP Locality V4519. UCMP 38007, left dP^4, left M^3; UCMP 38005, upper canine; and UCMP 39819, left lower canine; all from UCMP Locality V4522. UCMP 40017, a mandible with right and left C, P$_4$, M$_{1-3}$,

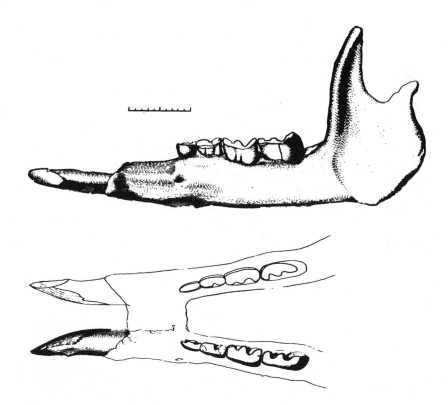

Figure 22.5. *Granastrapotherium snorki,* UCMP 40017, mandible and symphysis: *top,* lateral view; *bottom,* superior view. Scale bar 10 cm.

from UCMP Locality V4525, lying close to the holotype but probably representing another individual. UCMP 40178, left M^{2-3}, M_3; UCMP 40187, left mandible with P_4, M_{1-3}; UCMP 40188, left maxilla with P^4, M^{1-3}; UCMP 40189, mandible symphysis with tusk; UCMP 40190, mandible symphysis with right tusk; UCMP 40196, mandible fragments and condyle; and UCMP 40408, mandible with left and right P_4, M_{1-3}; all from UCMP Locality V5046. In addition, the following postcranial material: UCMP 40191, left radius; UCMP 40192, left humerus; UCMP 40194, left astragalus; UCMP 40409, left tibia (all from UCMP Locality V5046); and UCMP 39211, femur (from UCMP Locality V4517). IGM 183162, isolated right M^2 and M^3, partial right and left lower canines, from Duke Locality 31. IGM 251110, partial right M^2, from Duke Locality 114. IGM 182778, mandible with right C, P_4, M_{1-3}, and left M_{1-3}, from Duke Locality 19. IGM 182647, proximal right tibia, from Duke Locality 5. IGM 183076, right astragalus, from Duke Locality 28. IGM 252946, partial left astragalus, from Duke Locality 146. YPM-PU 21937, associated left M^2 and M^3, right M^2 and M^3, canine tusks; and YPM-PU 21938, unassociated teeth and tooth fragments, including left M_{2-3}, right M_{2-3}, left M^3 (see

later for locality information). MPV-2, an isolated M^2; MPV-5, right M^3; MPV-3, right maxilla with P^4, M^1 (broken), and M^{2-3}; from unknown localities in the La Venta area.

LOCALITIES AND GEOLOGIC HORIZONS

UCMP localities in the Honda Group in Tolima Department include V4415 Coyaima 5, V4420 Carmen de Apicalá 3, and V4422 Carmen de Apicalá 5. UCMP localities in the Villavieja Formation, Baraya Member, in Huila Department include V4518 and V4519, from the Monkey Beds; V4522 from between the Fish Bed and Monkey Beds; V4523 Fish Bed 2; V4524 Fish Bed 3, V4525 Red Bed 1 from between the La Venta red beds and Ferruginous Sandstone Beds; V4527 Red Bed 3 from between the El Cardón and La Venta Red Beds; V5045 in the Ferruginous Sandstone Beds; and V4528 Red Bed 4 in the El Cardón Red Beds, Villavieja Formation, Cerro Colorado Member. UCMP localities in the La Victoria Formation in Huila Department include V4533 "Hacienda San José" and V5046 "Astrapothere Locality," stratigraphically just below the Chunchullo Sandstone Beds. For more details about the geographic and stratigraphic position of the UCMP localities see Fields (1959) and archived files at the UCMP.

Duke University localities include Duke Locality 114, from below the Tatacoa Sandstone, La Victoria Formation, and Duke Localities 19 and 31, from the Monkey Beds, Villavieja Formation. For more complete geographic and stratigraphic information about Duke localities, see Guerrero (chap. 2).

YPM-PU 21937 and YPM-PU 21938 were collected by F. B. Van Houten about 6–8 km northeast of Villavieja, probably in the Villavieja Formation. USNM 10876 was collected by Brother Ariste Joseph from somewhere near Neiva, in Huila Department, Colombia.

Granastrapotherium snorki occurs throughout the Honda Group Section in the La Venta area and from the Monkey Beds equivalent of the Villavieja Formation west of Guacirco (Duke Locality 19) in Huila Department. *Granastrapotherium snorki* is also known from Honda Group deposits in the middle Magdalena River valley near Carmen de Apicalá in Tolima Department. The lowest stratigraphic occurrence of *G. snorki* is in the La Victoria Formation near Coyaima in Tolima Department (UCMP Locality V4415). The highest stratigraphic occurrence is at Duke Locality 126 in the Polonia Red Beds.

DIAGNOSIS

As for the genus.

DESCRIPTION

The upper canines of *G. snorki* are large, ever-growing tusks. Near the tip, the upper canine is triangular in cross section, becoming oval to nearly round at the base. Upper canine tusks at various stages of growth occur in the available collections, the largest measure nearly 1 m long with a basal diameter of 8–9 cm. In unworn specimens the enamel covers nearly the entire circumference of the crown, lacking only a thin longitudinal band along the dorsal surface. The enamel is smooth and lacks the fluting and longitudinal grooves seen on the upper tusks of *Xenastrapotherium kraglievichi*. With continued growth and wear the enamel gradually retreats from the crown circumference until eventually dentine is exposed continuously around the base. An

oblique wear facet for contact with the lower canine is directed medially and attains nearly 10 cm in maximum diameter in larger, older specimens (table 22.4).

Granastrapotherium differs from all other known astrapotheres in lacking P³. P⁴ is bicuspid, two-rooted, and about half the size of M¹ (fig. 22.3). Its two cusps are separated by a deep central fissure and are joined by a crest distally. There is no labial cingulum, but a weak lingual cingulum is variably present. Two slight vertical depressions on the buccal surface indicate a weak parastyle and distostyle. Although it erupts early, the P⁴ crown does not come into occlusion until the upper molar series has become quite worn.

M¹ is a three-rooted, lophodont tooth with a straight ectoloph and slight parastylar bulge. The transverse protoloph is broad and extends distally then lingually around the prominent mesiolingual pocket. The metaloph extends lingually perpendicular to the ectoloph. There is no lingual cingulum or trace of a crista or crochet on even moderately worn specimens. As in M¹, M² is a three-rooted, lophodont tooth, elongate mesiodistally when little worn, but becoming roughly quadrate in occlusal outline with wear. The ectoloph displays a well-defined parastyle in the less-worn condition that becomes obliterated with moderate wear. The protoloph is narrow when little worn but becomes broader with wear. A broad mesiolingual cingulum encloses a deep depression or pocket so that the enamel of the mesial aspect of the protoloph becomes infolded in advanced wear stages. The crown displays an ephemeral medifossette, but the crista becomes confluent with the ectoloph with moderate wear. A postcingulum encloses a posterofossette, which persists into more advanced wear stages. There is no buccal cingulum, and, unlike *Xenastrapotherium*, there is no lingual cingulum in *Granastrapotherium*.

M³ is a three-rooted, roughly triangular tooth. The buccal aspect of the crown displays a bluntly rounded mesial extremity and pointed distal extremity. There is no parastyle. The protoloph gradually broadens with wear. On the anterolingual aspect a broad and indistinct mesiolingual cingulum encloses a deep pocket, and the enamel on the mesial aspect of the protoloph is deeply infolded around this pocket. The crista does not enclose a medifossette and with wear, rapidly becomes confluent with the widening ectoloph. There is no postcingulum or posterofossette, no lin-

Table 22.4. Tooth measurements (cm) for *Granastrapotherium snorki*

Tooth	Measurement[a]	N	Mean	SD	SE	Observed Range	CV
		LOWER DENTITION					
C	Max. diameter	6	5.03	1.12	0.46	3.80–6.70	—
	Trans. diameter	6	3.50	0.89	0.36	2.90–4.65	—
P_4	Length	5	2.95	0.93	0.41	1.98–4.04	—
	Breadth	5	1.84	0.28	0.12	1.56–2.18	—
M_1	Length	7	4.04	0.43	0.16	3.30–4.42	—
	Breadth	7	2.86	0.42	0.16	2.40–3.45	—
M_2	Length	9	6.65	0.37	0.12	5.93–7.08	5.57
	Breadth	9	3.49	0.19	0.06	3.30–3.79	5.44
M_3	Length	9	7.87	0.53	0.18	6.76–8.43	6.76
	Breadth	9	3.48	0.21	0.07	3.22–3.82	6.10
		UPPER DENTITION					
C	Max. diameter	12	6.17	1.60	0.46	4.40–8.60	25.96
	Trans. diameter	12	4.69	1.43	0.41	2.20–6.80	30.45
P^4	Length	6	2.20	0.18	0.07	1.92–2.40	—
	Breadth	5	3.32	0.47	0.21	2.67–3.84	—
M^1	Length	5	5.38	0.64	0.29	4.50–6.05	—
	Breadth	4	5.77	0.58	0.29	5.16–6.51	—
M^2	Length	9	7.05	0.58	0.19	6.01–7.71	8.24
	Breadth	8	6.45	0.61	0.22	5.58–7.32	9.53
M^3	Length	9	8.00	0.44	0.15	7.38–8.65	5.48
	Breadth	8	6.05	0.21	0.07	5.76–6.50	3.40

[a]Maximum diameter = major axis of cross section. Transverse diameter = minor axis of cross section.

gual or buccal cingulum, and the median valley is deeply incised and open lingually on the M³. Unlike in *Xenastrapotherium kraglievichi*, where M² and M³ ectolophs are more nearly equal in length, in *Granastrapotherium* the ectoloph in M³ is consistently longer than that of M².

Granastrapotherium differs from all other known astrapotheres in lacking lower incisors. Undamaged mandibular symphyses (UCMP 40408, IGM 182778) show no evidence of incisor alveoli. The lower canines are large to very large, ever-growing tusks. They are implanted horizontally and in the males project directly toward the front; in females the tusks are somewhat curved laterally. The lower tusks are round in cross section and have a characteristic, dorsolaterally facing, beveled wear surface for contact with the upper tusk. The enamel is nearly circumferential in the unworn condition with two thin enamel-less

bands on the dorsal and ventral surfaces. With growth and wear the enamel retreats from the circumference until it is completely obliterated in very worn specimens.

The lower cheek teeth are separated from the canine tusks by a long diastema. The single lower premolar (P_4) is lophodont and bicrescentic and in size and structure is indistinguishable from that of *Xenastrapotherium kraglievichi*. The three lower molars of *Granastrapotherium* are simple, bicrescentic, and lophodont. The progressive increase in crown length from M_1 to M_3 is greater than in *Xenastrapotherium kraglievichi*. The M_1 in both species is roughly similar in mean length and breadth dimensions, but the M_2 and M_3 of *Granastrapotherium* are 60 and 55% larger than those respective teeth in *X. kraglievichi*. In relatively unworn specimens the narrow paralophid and metalophid are directed somewhat distally. With wear the paralophid and metalophid become broader and more lingually oriented. A distinct raised ridge descends from the distal aspect of the paralophid toward the base of the metalophid. This ridge serves to shallow the base of the paraflexid so that with wear the paraflexid becomes obliterated and the paralophid and metalophid merge. The entoflexid is deep and remains well-defined in even the later wear stages, a unique feature of the lower molars of *Granastrapotherium*. The hypoflexid is indicated as a faint indentation opposite the metalophid. There is no lingual or buccal cingulum.

Several partial and complete mandibles are known for *Granastrapotherium* (UCMP 40187, 40408, and 40017). This material includes a complete and relatively undamaged symphysis (UCMP 40408; see fig. 22.5) and other partial symphyses (UCMP 40190 and 40189). The symphysis is narrow and long anteroposteriorly, extending posteriorly to behind P_4. The dorsal aspect of the symphysis is concave and troughlike. The symphyseal suture is completely fused and obscured in adult specimens. The ventral surface of the symphysis is rugose and pitted, the number and location of foramina appears to be variable although one or more are always positioned ventral to the anterior root of P_4. The lower canine tusks are closely approximated and no incisor alveoli can be seen in the narrow space between them. With no incisors and anteriorly projecting canine tusks, the mandibular symphysis is distinctive in *Granastrapotherium*. The mandible of *Granastrapotherium* otherwise resembles that of *Astrapotherium magnum*, differing in being larger, having a less regularly rounded gonial angle, a higher coronoid, and a more pronounced sigmoid notch.

Granastrapotherium is known from a palate and basicranium (fig. 22.4; table 22.5), including the upper dentition, internal nares, a partial occipital condyle, and the lower margin of the right orbit (UCMP 40358), and a partial palate (UCMP 38397). Neither the tympanic nor auditory meatus are preserved in either specimen. These structures are not known for any Astrapotheriidae and are assumed to have been loosely attached to the skull (Scott 1928; Simpson 1967).

In lateral view the skull has a short facial region. With relatively and absolutely larger upper canine tusks than *Astrapotherium magnum*, the bony alveolar sheaths of the upper tusks of *Granastrapotherium* are thicker and project more forward. As in *Astrapotherium*, the orbit in *Granastrapo-*

Table 22.5. Cranial measurements for *Granastrapotherium snorki* (UCMP 40358)

Variable	Measure (cm)
Basilar length:	
Basion to medial alveolar margin of left C)	52.5
Skull length:	
Condylion to apex of right C	64.8
Condylion to medial alveolar margin of right C	56.5
Bizygomatic breadth (right zygomatic breadth × 2)	43.0
Palatal width:	
Between posteriormost alveolar margin of M^3	17.0
Between anteriormost alveolar margin of P^4	15.0
Neurocranium length:	
Basion to staphylion	22.0
Basion to back of M^3	28.4
Facial length:	
Staphylion to medial alveolar margin of left C	30.5

therium is low, positioned above the posterior root of M^2. The zygomatic arch is long and horizontal, and the posterior root is positioned low on the skull. The zygomatic arch extends posteriorly as a horizontal shelf above the postglenoid and posttympanic processes. The free portion is flat and slender and extends anteriorly nearly to the orbit. There is a large ovoid infraorbital foramen with its long axis oriented vertically.

In basal view the *Granastrapotherium* skull has a short and constricted face with massive, anteriorly extended and projecting alveolar sheaths for the upper canine tusks. The cheek tooth rows are parallel, unlike *Astrapotherium magnum* in which the cheek tooth rows converge posteriorly. The roof of the palate is shallow, not as high or concave as in *A. magnum*. There is a V-shaped, premaxillary fissure and a V-shaped depression extending as far back as the anterior end of M^1 for the loosely attached premaxilla. The internal nares appear to have been situated somewhat more posteriorly in *G. snorki* compared with *A. magnum*, and the zygomatic arch flares more widely. The maxillary root is positioned above M^{2-3} in *G. snorki*, rather than above M^{1-2} as in *A. magnum*. The glenoid cavity is elongate transversely and generally concave. The postglenoid process describes a transverse keel behind the glenoid. The paroccipital process is narrow, thin, anterolaterally concave, and more obliquely oriented than in *A. magnum*.

The greatest zygomatic breadth occurs at the distal end of M³. The glenoid is positioned more posteriorly, and the articular surface for the mandibular condyle shows less relief than in *A. magnum*. Greatest zygomatic breadth in *A. magnum* occurs at a point even with the glenoid, which, in turn, is positioned more anteriorly, level with the basioccipital-basisphenoid suture. The palate in *Granastrapotherium* is broader posteriorly than in *A. magnum*, and there is no lateral palatine notch. Possible functional implications for these differences are discussed later.

Postcranial bones from UCMP Locality V5046 (Savage's "Astrapothere Locality") were found in "a regular bone yard" of astrapothere material. According to Savage's field notes, some of the dental and postcranial material (DES field numbers 643, 644, 645, and 646) was "intermeshed" and may represent the same individual. All dental material collected at UCMP Locality V5046 (representing at least three different individuals) can be referred to *Granastrapotherium*. Based upon this association with the dental remains, these postcranial bones from UCMP Locality V5046 are herein referred to *G. snorki*. These elements include UCMP 40191 left radius, UCMP 40192 left humerus, UCMP 40194 left astragalus, and UCMP 40409 left tibia. Four other postcranial specimens, UCMP 39211 (a femur), IGM 182647 (proximal tibial fragment), IGM 183076 (right astragalus), and IGM 252946 (partial left astragalus), are morphologically and metrically similar to equivalent elements from UCMP Locality V5046 and are also referred to *Granastrapotherium*. Measurements of postcranial bones are given in table 22.6. With the exceptions noted here, these postcranial bones generally resemble those of *A. magnum* described by Scott (1928) but are 20–30% larger in most dimensions.

The humerus of *G. snorki* (UCMP 40192) is broader and more robust with the head oriented more directly upward than in *A. magnum*. The radius (UCMP 40191) has a more robust distal extremity, and the distal articular facet is broader and more ovoid than in *A. magnum*. Proximally, the radius is broad, covering most of the width of the humeral trochlea. The femur (UCMP 39211) is larger and more massive than the same element in *A. magnum*, with a very short and broad neck, and the head oriented perpendicular

to the shaft. The tibia (UCMP 40409) is badly crushed and distorted. The proximal tibial fragment (IGM 182647) is morphologically similar to *A. magnum* but shows a more massive tuberosity for the quadruceps tendon and there is a more well-developed tibial spine, which forms a prominent crest between the two articular surfaces. The astragalus (UCMP 40194, IGM 183076, and IGM 252946) is, in general, like that of *Astrapotherium* but is 35% larger and has a more concave external calcaneal facet (ectal facet).

METRIC AND MORPHOLOGIC VARIATION

Although the Honda Group sample of astrapotheres is relatively numerous compared with samples of astrapotheres from elsewhere in South America, there are no complete skulls yet known for either of the Honda Group astrapotheres, and the number of complete or nearly complete mandibles preserving the incisors, symphysial region, tusks, and cheek teeth together is very limited. Our present understanding of patterns of variation in other astrapotheres is not impressive. The best available quarry or single locality sample of a single species is the FMNH sample of *Parastrapotherium* sp. that was collected by E. S. Riggs and party at the Deseadan type locality at La Flecha, in Santa Cruz Province, Argentina. It is with this collection that dental variation in the Honda Group sample has been compared. Bivariate scatterplots of lower M$_2$ dimensions (fig. 22.6) reveal generally comparable dispersion for the samples of *Xenastrapotherium kraglievichi* and *Granastrapotherium snorki* from the Honda Group.

Canine measurements for Honda Group astrapotheres reveal pronounced canine (sexual) dimorphism in *Granastrapotherium* and moderate canine dimorphism for *Xenastrapotherium kraglievichi* (table 22.7; fig. 22.7). Canine dimorphism in *Granastrapotherium* extends to canine shape, the larger (male?) lower canines are longer, greater in diameter, and only very slightly curved, whereas the smaller (female?) lower canines are shorter, have smaller diameter, and are more curved.

ADAPTATIONS

Gregory (1920) and Scott (1928, 1937) argued that the extremely short or retracted nasals, sinus-filled and domed

Table 22.6. Measurements (cm) of postcranial elements in Miocene Astrapotheriidae

Variable Measured	*Astrapotherium magnum* FMNH 14251	*?Parastrapotherium* sp. FMNH 13437	*Granastrapotherium snorki* UCMP[a]
Humerus:			
Length from head	49.0	—	63.3
Length from external tuberosity	52.3	—	65.5 est.
Greatest thickness of shaft	9.1	—	14.0 est.
Anteroposterior diameter of head	9.9	—	15.5
Transverse diameter of head	9.1	—	13.5
Width of trochlea	8.4	—	12.5
Radius:			
Length	34.8	—	43.5
Width at proximal end	8.1	—	11.5
Thickness at proximal end	5.8	—	8.5
Width at distal end	6.5	—	9.3
Thickness at distal end	5.9	—	6.1
Median width of shaft	4.9	—	5.5
Median thickness of shaft	3.1	—	4.5
Femur:			
Length from head	57.0	—	75.5
Length from greater trochanter	53.0	—	71.5
Proximal width	14.2	—	17.5
Thickness of greater trochanter	5.1	—	5.2
Median width of shaft	6.6	—	12.8
Median thickness of shaft	3.9	—	5.5
Width of distal end above condyles	11.8	—	15.0
Thickness of distal end	10.4	—	8.5
Width of rotular groove	5.2	—	6.5
Tibia:			
Internal length	39.9	40.3	49.0
External length	38.3	37.8	47.8
Width of proximal end	9.0	10.3	13.8
Width of distal end	7.8	8.3	12.8
Thickness of distal end	4.3	4.8	8.1
Astragalus:			
Proximo-distal length	7.3	—	10.0 est.
Width	6.6	—	11.5
Thickness at external calcaneal facet	5.6	—	6.2
Thickness at sustentacular facet	4.2	—	5.3

Source: *Astrapotherium magnum* and *?Parastrapotherium* sp. from Scott (1937).

Note: Est. = estimated measure.

[a]Specimen used for measurements of the humerus, UCMP 40192; radius. UCMP 40191; femur, 39211; tibia, UCMP 40409; astragalus. UCMP 40194.

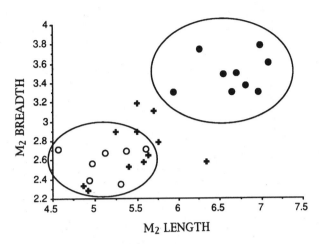

Figure 22.6. Bivariate plots of lower M_2 crown length and breadth (cm) for Honda Group astrapotheres and for *Parastrapotherium* sp. from La Flecha, Chubut, Argentina. *Open circles, Xenastrapotherium kraglievichi; filled circles, Granastrapotherium snorki; plus signs, Parastrapotherium* sp. (a single-species quarry sample).

Table 22.7. Canine measurements (mm) for Honda Group Astrapotheriidae

Specimen	Side[a]	Maximum Diameter	Transverse Diameter
LOWER CANINES			
Xenastrapotherium, sex indet.			
UCMP 39818	R	49.0	32.0
UCMP 40014	R	41.0	26.5
UCMP 38847	R	43.0	23.0
All (N = 3)		44.3	27.2
		(4.16)	(4.54)
Granastrapotherium, females			
UCMP 39821	L	54.7	26.5 est.
UCMP 38106	R	39.0 est.	31.0
UCMP 38114	?	46.8	31.0
UCMP 40189	R	38.0	29.0
All (N = 4)		44.6	29.4
		(7.78)	(2.14)
Granastrapotherium, males			
UCMP 39819	L	56.4	46.5
UCMP 40190	R	67.0	46.0
All (N = 2)		61.7	46.3
		(7.50)	(0.35)

frontal region, anteriorly positioned and heavily worn lower incisors, deep preorbital fossae, and bony rim around the orbits indicated that *A. magnum* probably had an enlarged or even prehensile proboscis. The skulls of Miocene *Parastrapotherium, Astrapotherium,* and *Astrapothericulus* all show these morphological features. Wall (1980) correlated two additional anatomical features with the presence of a prehensile proboscis in *Tapirus*: reduced preorbital portion of the skull and reduced ascending process of the premaxilla with loss of premaxillary-nasal contact. The preserved basicranium of *Granastrapotherium* displays a well-developed maxillary concavity anteromedial to the orbit, a shortened facial region, and a reduced premaxillary region, suggesting that this genus likewise possessed at least a moderately enlarged (tapirlike) prehensile proboscis.

In *Astrapotherium* the upper canine tusks show substantial abrasion and transverse wear striations on the anterior (dorsal) edge, and the lower canine tusks typically show two or more transverse grooves on their anterior edges. Scott (1928, 1937) reasoned that because these grooves occur at points where no other tooth could have occluded with the tusks, they were more likely made by rubbing against some external object. In *Xenastrapotherium kraglievichi* (which had tusks similar in size and configuration to *Astrapotherium*) the anterior face of the upper tusks above the contact facet lacks enamel, and the dentine is heavily polished and abraded.

The enamel adjacent to these areas shows fine wear striations transverse to the long axis of the tooth. The lower canines show a series of transverse grooves on the leading (anterior) edge, and there are numerous transversely oriented wear striations on the superior and inferior enamel surfaces. These wear patterns in *Xenastrapotherium kraglievichi* are consistent with those produced by abrasion against hard external objects while the head was moved sideways.

The skull of *Granastrapotherium* displays features suggesting that it may have possessed a longer proboscis than *Tapirus*. In *Granastrapotherium* the palate is broad and the enlarged upper tusks project forward on each side of a broad nasal aperture. In some specimens the anteriorly directed tusks measure close to 1 m in length and have characteristic worn areas with grooves and striations on the dorsal, medial, and distal (lateral) surfaces near their tips. Equivalent wear on the dorsal surfaces of the tusks of modern proboscideans is produced when the prehensile trunk is used in conjunction with the tusks during leaf and branch stripping

Table 22.7. Continued

Specimen	Side[a]	Maximum Diameter	Transverse Diameter
	UPPER CANINES		
Xenastrapotherium, sex indet.			
UCMP 38922	L	35.0	23.0
UCMP 39820	R	39.0	26.5
UCMP 39818	R	50.0	29.0
All (N = 3)		41.3	26.2
		(7.77)	(3.01)
Granastrapotherium, females			
UCMP 40358	L	45.3	37.5
UCMP 39202	R	55.8	34.9
UCMP 39217	L	44.0	41.5
UCMP 39844	?	45.0	22.0 est.
UCMP 40358	?	45.0	37.5 est.
UCMP 38005	?	47.5	38.2
All (N = 6)		47.1	35.3
		(4.42)	(6.83)
Granastrapotherium, males			
UCMP 37914	R	72.4	54.3
UCMP 40017	L	86.0	56.0
UCMP 38937	?	69.1	44.0
UCMP 39230	?	74.0	63.5
UCMP 38086	?	77.0	65.0
UCMP 38114	?	79.0	68.0
All (N = 6)		76.3	58.5
		(5.90)	(8.85)

Notes: est. = estimated measure; specimen distorted or fragmentary; value not used in computation of means in table 22.4. Where N > 1, values are means (with standard deviations below, in parentheses).

[a]L = left; R = right.

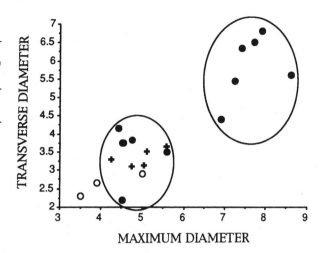

Figure 22.7. Bivariate plots of upper canine dimensions (cm) for Honda Group astrapotheres, showing dimorphism in *Granastrapotherium snorki*, and for *Parastrapotherium* sp. from La Flecha, Chubut, Argentina. *Open circles, Xenastrapotherium kraglievichi; filled circles, Granastrapotherium snorki; plus signs, Parastrapotherium* sp. (a single-species quarry sample).

activity (Kingdon 1979). The large, anteriorly directed tusks of *Granastrapotherium* are reminiscent of some fossil proboscideans, especially the "serridentines" and advanced gomphotheres (Osborn 1936). Modern elephants use tusks in conjunction with the trunk in various sorts of feeding behaviors, such as to strip leaves from branches, to remove bark from trunks, to break tree limbs, and to push over or uproot slender-stemmed or shallow-rooted trees (Kingdon 1979). These activities, which have considerable impact on woodland and forest vegetation (Laws 1970; Owen–Smith 1988; Guy 1989; Campbell 1991), also produce noteworthy wear and abrasion on tusks. Although the wear features

observed on the tusks of *Granastrapotherium* could have been produced without an elongate trunk, to function in opposition with the tusks, the proboscis must have extended at least to the tip of the tusks.

Nonselenodont ungulate regression equations are used to obtain body mass estimates for the Honda Group astrapotheres from dental dimensions (Damuth 1990). The morphological convergence between astrapothere lower molars and those of living rhinoceroses suggest that at least this component of the nonselenodont sample is appropriate. Additional body mass estimates for Honda Group astrapotheres were also obtained from cranial and mandible dimensions. For these dimensions regression equations based upon nineteen species of perissodactyls and hyracoids, including both browsers and grazers, were used (Janis 1990). Body weight estimates for individual measures vary considerably (table 22.8). *Granastrapotherium* and *Xenastrapotherium* display similar M_1 crown lengths but differ appreciably in M_{1-3} length. Differences in tooth crown dimensions between these two species range from 4% for M_1, to 42% for M_2, up to 92% for M_3, with an average size difference of 46%. Dental dimensions are the only directly comparable estimates for the two La Venta taxa, and the body weight estimates based upon lower tooth row length

Table 22.8. Estimates of body mass for astrapotheres in the La Venta fauna

Variable Measured	Observed Value	Regression Formula[a]	Body Mass Estimate (kg)
		GRANASTRAPOTHERIUM	
Dental:			
M_1 length	40.0 mm	$3.170 (\log_x) + 1.040$	1,313.8
M_1 area	1,155.4 mm²	$1.510 (\log_x) + 1.440$	1,160.7
M_{1-3} length	187.5 mm	$3.030 (\log_x) - 0.390$	3,141.9
All 3 variables			1,872.1 ± 636.4[b]
Cranial:[c]			
TSL	64.8 cm	$2.751 (\log_x) - 1.931$	1,128.9
PSL	28.4 cm	$2.763 (\log_x) - 0.988$	1,065.4
PAW	10.0 cm[d]	$3.243 (\log_x) - 0.027$	1,644.4
MFL	31.2 cm[d]	$2.815 (\log_x) - 1.106$	1,259.1
Mandible:[c]			
PJL	18.45 cm	$2.202 (\log_x) + 0.283$	1,176.9
DMA	19.80 cm	$2.824 (\log_x) - 0.979$	481.7
Dental, cranial, mandible:			
All 9 variables			1,374.7 ± 243.0[b]
		XENASTRAPOTHERIUM	
Dental:			
M_1 length	39.5 mm	$3.170 (\log_x) + 1.040$	1,262.4
M_1 area	920.4 mm²	$1.510 (\log_x) + 1.440$	823.3
M_{1-3} length	151.0 mm	$3.030 (\log_x) - 0.390$	1,630.4
All 3 variables			1,238.7 ± 233.3[b]

[a]Regression equations for dental variates are from the nonselenodont ungulate sample of Damuth (1990, 245). Regression equations for cranial and mandibular variates are from the living Perissodactyla plus hyracoid sample of Janis (1990, table 16.8).

[b]Mean ± SE.

[c]Cranial measurements for UCMP 40358 (female): PAW = palate width, PSL = posterior skull length, TSL = total skull length, and MFL = length of the masseteric fossa. Mandible measurements for UCMP 40017: PJL = posterior jaw length and DMA = depth of the mandible angle. For a definition of these cranial and mandible dimensions, see Janis (1990, fig. 13.1).

[d]Estimated.

provide an indication of the difference in body weight between these two species. The body weight estimates for both species based upon M_1 dimensions are more similar to each other than is the estimate based upon tooth row length. Mean estimates based exclusively on dental dimensions suggest that *Xenastrapotherium kraglievichi* was about two-thirds the size of *Granastrapotherium snorki*.

For *Granastrapotherium* body weight estimates based upon the cranial measurements for UCMP 40358 are much smaller than the estimate based upon lower tooth row length and uniformly smaller than the mean estimate derived from dental dimensions. Judging by canine tusk size, the cranium UCMP 40358 is probably a female, and the mandible (UCMP 40017) is probably a male. The low body weight estimates derived from posterior jaw length and depth of the mandible angle are not easily explained.

Honda Group astrapotheres compare in body mass to some living megaherbivores. *Xenastrapotherium kraglievichi* may have had a body mass equivalent to the African black rhinoceros *Diceros bicornis* (900–1,500 kg). *Granastrapotherium snorki* had a body mass comparable to living rhinos and may have attained a mass equivalent to *Hippopotamus amphibius* (2,500–3,500 kg), up to about half the mass of the African elephant (*Loxodonta africana*). In addition to canine dimorphism, body size dimorphism may also characterize *Granastrapotherium snorki*.

Limb proportions and morphology of the postcranial elements of *Granastrapotherium* are comparable with other graviportal ungulates (Hildebrand 1982). Humeral and femoral heads are oriented almost directly upward and proximal limb segments are longer than are distal segments by a 3:2 ratio. In addition, the carpal and tarsal elements are short and "blocklike," and the metapodials are short.

Subfamily Uruguaytheriinae
cf. *Granastrapotherium*

USNM 205350, a lower M_2, and USNM 205383, a left maxilla with P^4, M^{1-3} broken at the base of the crowns, were collected by Harvey Bassler at Locality VF56, Playa Mapuya, along the Río Inuya, Loreto Department, eastern Peru (Willard 1966). In lower molar crown dimensions and morphology this material is similar to *Granastrapotherium*. In the cobble that was once an astrapothere maxilla the upper molar crowns are all broken at the alveolar margin. It is possible, however, to determine that there was no P^3 in this animal. Thus, in preserved features, this material cannot be distinguished from *Granastrapotherium*. With this record the known geographic distribution of this genus is extended into eastern lowland Peru.

Uruguaytheriinae, gen. et sp. incertae sedis

Hoffstetter (1977) referred isolated upper and lower third molars from the middle Miocene deposits at Quebrada Honda (about 22° S) in the Cordillera Oriental of Bolivia to the "*Uruguaytherium-Xenastrapotherium* group" and later, Frailey (1981, 1987) described a partial neurocranium and associated portions of a palate and upper teeth (UF26679) from the same locality. Based upon the bizygomatic breadth measurement given by Frailey, this cranium appears to be larger than *Astrapotherium magnum*. Among the Astrapotheriidae for which this portion of the cranium is known, the general shape and configuration of the base of the neurocranium in UF 26679 is most similar to *Granastrapotherium snorki*. Unfortunately, the upper and lower molars from Quebrada Honda have never been adequately described nor have crown dimensions been published. The Quebrada Honda fauna has been correlated with the Friasian SALMA (Hoffstetter 1977) and the Santacrucian SALMA (Frailey 1981). Based on K-Ar ages and magnetostratigraphy, the mammals from Qda. Honda have an extrapolated age between 13.0 and 12.7 Ma (MacFadden et al. 1990), essentially the same as that of the La Venta astrapotheres.

Phylogenetics

To establish the phylogenetic position of *Xenastrapotherium* and *Granastrapotherium* within Astrapotheriidae, we scored thirty-nine morphological characters of the cranium, postcranium, and dentition in fifteen astrapothere taxa. The taxa used in the phylogenetic analysis include the following: (1) *Eoastrapostylops* Soria 1982 (Trigonostylopidae); (2) *Tetragonostylops* (Trigonostylopidae); (3) *Trigonostylops* Ameghino 1897 (Trigonostylopidae), as described by Simpson (1967); (4) *Albertogaudrya* Ameghino 1901 (Trigonostylopidae); (5) *Scaglia* Simpson 1957; (6) *Astraponotus* Ameghino 1901; (7) *Parastrapotherium* Ameghino 1895 as described by Scott (1937), with additional observations based upon an undescribed sample from the Deseadan type fauna at La Flecha, including FMNH 13329, 13341, 13343, 13347, 13349, 13354, 13357, 13361, 13364, 13365, 13366, 13369, 13456, 13462, 13473, 13492, 13504, 13556, 13561, 13569, and 13572; (8) *Astrapotherium* Burmeister 1879 as monographed by Scott (1928), with additional observations on undescribed material, including FMNH 13168, 13170, 13173, and 14259; (9) *Astrapothericulus* Ameghino 1901 essentially as discussed by Scott (1937), with additional observations on an undescribed cranium (FMNH 13429) and mandible (FMNH 13426) and undescribed dental material from the Pinturas Formation (YPM-PU

15855, 15845, and 15931); (10) *Xenastrapotherium* Kraglie-vich 1928a, including *X. kraglievichi* Cabrera 1929 and *X. christi* Stehlin 1928; and (11) *Granastrapotherium* Johnson and Madden (this chap.). In addition, the following mate-rial was examined but not included in the analysis because they were too poorly known: *Uruguaytherium* Kraglievich 1928b; *Xenastrapotherium amazonense* (Paula Couto 1976); *Xenastrapotherium chaparralensis* Johnson and Madden (this chap.); and UF26679 from Qda. Honda, Bolivia, described by Frailey (1987), herein referred to the Uruguaytheriinae, gen. et sp. incertae sedis.

The characters and character states are described in ap-pendix 22.1 and the complete character/taxon matrix is presented in appendix 22.2. Phylogenetic analyses were performed using PAUP Version 3.0s (Swofford 1991) and subsequent branch manipulation used MacClade Version 3.0 (Maddison and Maddison 1992). In the analysis, charac-ters were treated as unordered (Fitch parsimony). To deter-mine the phylogenetic position of *Granastrapotherium* and *Xenastrapotherium* within Astrapotheriidae, four genera of relatively primitive trigonostylopoid astrapotheres (*Eoastrapostylops, Tetragonostylops, Albertogaudrya,* and *Tri-gonostylops*) were designated as an outgroup set following arguments put forth by Simpson (1967). The PAUP heuris-tics search with random addition sequence (otherwise using factory settings, including mulpars, collapse zero-length branches, etc.) yielded four equal length trees of seventy steps. The strict consensus tree of this tree set has a con-sistency index of 0.94 (fig. 22.8).

The Uruguaytheriinae share five unambiguous syn-apomorphies in the dentition: (1) well-developed M¹ ante-rolingual pocket, (2) loss of the lower molar hypoflexid, (3) reduced upper molar parastyle, (4) lower molar metaflexid enclosed by mesiolingual cingulum, and (5) M³ lingual valley enclosed by protoloph-metaloph ridge. Honda Group *Xenastrapotherium* and *Granastrapotherium* are the best-known genera of the monophyletic subfamily Uru-guaytheriinae and are geochronologically the last occurring genera of the family Astrapotheriidae. The oldest known Uruguaytheriinae, *Xenastrapotherium chaparralensis* of the Chaparral faunule of Colombia and *Uruguaytherium beau-lieui* from the Fray Bentos Formation of Uruguay (Kraglie-vich 1928a; Mones and Ubilla 1978; Mones 1979), all share the loss of the lower molar hypoflexid as first recog-

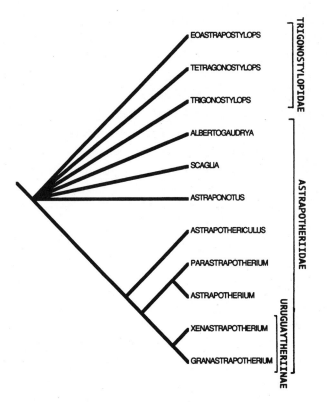

Figure 22.8. Strict consensus maximum parsimony tree for the Astrapotheria.

nized by Kraglievich (1928c). In the FMNH sample of lower molars of *Parastrapotherium* from the type Deseadan at La Flecha some specimens show a reduced hypoflexid. Given this variability and Oligocene to early Miocene (Deseadan) age of the oldest known uruguaytheriines, the evolutionary divergence of the Uruguaytheriinae probably occurred sometime in or shortly before the Deseadan.

The phylogenetic relationship of the Uruguaytheriinae with respect to any particular genus of Oligocene-Miocene Astrapotheriidae from Patagonia (*Parastrapotherium, As-trapotherium,* or *Astrapothericulus*) cannot be resolved on available evidence. The clade comprising *Parastrapotherium* and *Astrapotherium* is supported by only one unambiguous synapomorphy (M³ larger than M²) as is a hypothetical clade combining Uruguaytheriinae with *Parastrapotherium* and *Astrapotherium* (more forward-directed upper canine sheath). By branch rearrangement, *Astrapothericulus* can be placed anywhere with respect to *Parastrapotherium, As-trapotherium,* and the Uruguaytheriinae with only a one-step increase in tree length. An equivalent one-step reduction in parsimony results when all three of the Patagonian Miocene genera are construed as a monophyletic group.

All Oligocene and younger Astrapotheriidae together comprise a monophyletic group sharing five unambiguous synapomorphies: (1) enlarged and ever-growing canine tusks, (2) upper molar metaloph and postcingulum enclose a postfossette, (3) M^1 anterolingual pocket, (4) lower molar hypoconid with anterolingually directed crest, and (5) well-developed lower molar paralophid. These Patagonian Deseadan and later genera plus Uruguaytheriinae demonstrate significant morphologic change over Casamayoran and Mustersan *Scaglia* and *Astraponotus* in these features.

Summary

Two genera of uruguaytheriine astrapothere occur in the Honda Group of Colombia. The smaller species, *Xenastrapotherium kraglievichi* Cabrera 1929, formerly known only by the type specimen, is now known from associated upper and lower dentitions. Based upon this material, an emended diagnosis for the genus *Xenastrapotherium* is provided. *Xenastrapotherium kraglievichi* occurs only in the upper part of the fossiliferous Honda Group at levels corresponding to the interval between 13.0 and 11.9 Ma. A new species, *Xenastrapotherium aequatorialis,* is described from the Miocene Biblián Formation of southern Ecuador in deposits older than 19.0 Ma. A partial maxilla from deposits of uncertain age along the Río Juruá in Acre Province, Brazil, originally described by Paula Couto (1976), is herein referred to *Xenastrapotherium*. Astrapothere material from the Oligocene to early Miocene Chaparral fauna in Colombia should be referred to *Xenastrapotherium* and is recognized as a new species, *X. chaparralensis.*

Granastrapotherium snorki, a morphologically distinct and larger taxon from the Honda Group in Colombia, is a graviportal megaherbivore with tusks and a proboscis. A lower molar and partial maxilla collected in deposits of unknown age in eastern Peru is herein referred to *Granastrapotherium,* extending the known geographic distribution of this genus. The contemporaneous astrapothere material from Qda. Honda in southern Bolivia could also pertain to this genus, but until it is more fully described, it is best referred to the subfamily Uruguaytheriinae, gen. et sp. incertae sedis.

The Uruguaytheriinae share five synapomorphies: well-developed M^1 anterolingual pocket, loss of the lower molar hypoflexid, reduced upper molar parastyle, lower molar metaflexid enclosed by mesiolingual cingulum, and M^3 lingual valley enclosed by a cingular ridge connecting protoloph-metaloph. *Xenastrapotherium* and *Granastrapotherium* are the geochronologically last occurring genera of the family Astrapotheriidae. The Deseadan age of the oldest known Uruguaytheriinae suggests that their evolutionary divergence probably occurred sometime in the Oligocene. The phylogenetic relationship of the Uruguaytheriinae with respect to any particular genus of Oligocene-Miocene Astrapotheriidae from Patagonia (*Parastrapotherium, Astrapotherium,* or *Astrapothericulus*) cannot be established.

All Oligocene and younger Astrapotheriidae together comprise a holophyletic group sharing enlarged and ever-growing canine tusks, upper molar metaloph and postcingulum enclosing a postfossette, M^1 anterolingual pocket, lower molar hypoconid with anterolingually directed crest, and well-developed lower molar paralophid. These taxa demonstrate significant morphologic change over Casamayoran and Mustersan astrapotheres.

Acknowledgments

The heart of this chapter is derived from an unpublished Ph.D. dissertation by Johnson (1984). For this dissertation work, S. Johnson gratefully acknowledges the assistance and support of D. E. Savage, whose friendship and knowledge enriched Steve's years at Berkeley. D. White and W. A. Clemens, Jr., University of California, provided much helpful criticism and encouragement. A. Lucas provided many of the illustrations for the dissertation that are included in this work.

In addition, we thank the authorities of the Museum of Paleontology, University of California at Berkeley, including J. H. Hutchison, M. Goodwin, and M. Greenwald, for many courtesies. The authorities of the Museo de Geologia of the Instituto Nacional de Investigaciones en Geociencias, Minería y Química, Bogotá, Colombia, C. Villarroel of the Departamento de Geociencias of the Universidad Nacional de Colombia in Bogotá and L. Albuja and G. Herrera of the Department of Biology of the Escuela Politecnica Nacional in Quito, Ecuador, permitted us to study the Miocene Astrapotheriidae from Colombia and Ecuador in their collections. We also thank C. Ray and R. Purdy at the United States National Museum in Washington, D.C., J. Ostrom of the Yale Peabody Museum, J. J. Flynn of the Field Museum of Natural History, and M. C. McKenna of the American Museum of Natural History, for access to their important collections of Oligocene and Miocene astrapotheres. J. Guerrero and S. D. Andrade kindly made the collection from the Instituto Huilense de Cultura available for study. M. Bond, Museo de Ciencias Naturales de La Plata, kindly provided a cast of the type of *Xenastrapotherium kraglievichi*. G. H. Jurado Balladares of Quito drew figure 22.2. We thank R. Kay for helpful comments on the manuscript.

Appendix 22.1. Characters, character states, and exemplary taxa used in the phylogenetic analysis

1. Lower incisor crown shape: (0) simple (*Trigonostylops*); (1) weakly bilobed (*Astraponotus*); (2) bilobed (*Astrapotherium*).

2. Canines and canine-premolar diastema: (0) canines not enlarged, diastema absent (*Eoastrapostylops*); (1) canines enlarged and rooted, diastema present (*Trigonostylops*); (2) canines enlarged and ever-growing, diastema pronounced (*Astrapotherium*); (3) canines very large, diastema pronounced (*Granastrapotherium*).

3. Frontal: (0) uninflated, narrow (*Trigonostylops*); (1) broad, slightly domed (*Scaglia*); (2) very broad, domed, and inflated (*Astrapotherium*).

4. Upper/lower permanent premolar number: (0) four/four (*Trigonostylops*); (1) three/two (*Parastrapotherium*); (2) two/one (*Astrapotherium*); (3) one/one (*Granastrapotherium*).

5. Lower canine angle of implantation: (0) vertical (*Trigonostylops*); (1) oblique (*Astrapotherium*); (2) procumbent, directed anteriorly (*Granastrapotherium*).

6. Bulla: (0) attached (*Trigonostylops*); (1) unattached (*Astrapotherium*).

7. Astragalar neck: (0) short neck (*Tetragonostylops*); (1) no neck (*Astrapotherium*).

8. Secondary ectal facet on the astragalus: (0) absent (*Astrapotherium*); (1) present (*Tetragonostylops*).

9. Calcaneum/fibula contact: (0) present (*Tetragonostylops*); (1) absent (*Astrapotherium*).

10. Medial collateral ligament facet on the astragalus: (0) present (*Tetragonostylops*); (1) absent (*Astrapotherium*).

11. Nasals: (0) elongate (*Trigonostylops*); (1) retracted (*Astrapotherium*).

12. M^1 postprotocrista: (0) connects with metacone/

metaconule (*Eoastrapostylops*); (1) connects or points toward hypocone (*Tetragonostylops*); (2) no postprotocrista (*Albertogaudrya*).

13. M^1 hypocone: (0) absent (*Eoastrapostylops*); (1) present (*Trigonostylops*).

14. M^1 metaloph: (0) free lingual terminus (*Astraponotus*); (1) metaloph and postcingulum join to enclose ephemeral postfossette (*Parastrapotherium*); (2) metaloph extended lingually (*Xenastrapotherium*).

15. M^1 lingual cingulum: (0) continuous around the protoloph, extending across the lingual fold, and metaloph (*Astrapotherium*); (1) surrounds protoloph (*Scaglia*); (2) anterolingual cingulum (*Trigonostylops*); (3) no lingual cingulum (*Granastrapotherium*).

16. M^1 anterolingual pocket: (0) absent (*Trigonostylops*); (1) weakly developed (*Parastrapotherium*); (2) well developed (*Xenastrapotherium*).

17. M^1 buccal cingulum: (0) present, well developed (*Astraponotus*); (1) present, weakly developed (*Parastrapotherium*); (2) very weakly developed (*Astrapotherium*); (3) absent (*Granastrapotherium*).

18. Upper molar secondary crest: (0) absent (*Eoastrapostylops*); (1) present, short (*Astraponotus*); (2) long, joins ectoloph to enclose medifossette (*Astrapotherium*).

19. M^2–M^3 relative size: (0) $M^2 = M^3$ (*Eoastrapostylops*); (1) $M^2 > M^3$ (*Trigonostylops*); (2) $M^2 < M^3$ (*Astrapotherium*).

20. M_2 talonid: (0) without accessory structures (*Eoastrapostylops*); (1) hypoconid with anterolingually directed crest which may terminate in an accessory pillar (*Astrapotherium*); (2) hypoflexid lacking (*Xenastrapotherium*).

21. Lower molar paracristids: (0) short (*Eoastrapostylops*); (1) well-developed paralophid (*Parastrapotherium*).

22. Upper molar parastyles: (0) well developed (*Parastrapotherium*); (1) weakly developed (*Granastrapotherium*).

23. M_{1-2} metaflexid: (0) deep (*Astrapotherium*); (1) enclosed by mesiolingual cingulum (*Xenastrapotherium*).

24. M^3 median valley lingual opening: (0) median valley deep and open lingually (*Astrapotherium*); (1) open lingually, beaded cingulum (*Parastrapotherium*); (2) protoloph-metaloph ridge enclosing lingual valley (*Xenastrapotherium*).

25. dP^4 crown morphology: (0) well-developed parastyle, mesiolingual cingulum (*Parastrapotherium*); (2) reduced parastyle, mesiolingual cingulum and pocket (*Xenastrapotherium*); (3) reduced parastyle, mesiolingual pocket (*Granastrapotherium*).

26. Upper canine tusk alveolar "sheath": (0) downward directed (*Trigonostylops*); (1) project forward (*Granastrapotherium*).

27. Choanae: (0) divided by median septum of the palatine (*Trigonostylops*); (1) undivided (*Astrapotherium*).

28. Pterygoid or palatine processes: (0) alate processes anterior to chonae (*Trigonostylops*); (1) slight tubercles lateral to

chonae (*Scaglia*); (2) alate processes lateral to chonae, thickened ventrally (*Granastrapotherium*).

29. Palatal portion of the palatines: (0) broad, without lateral palatine notch (*Trigonostylops*); (1) broad, with lateral palatine notch (*Astrapotherium*); (2) narrow, elongate, without lateral palatine notch (*Granastrapotherium*).

30. Lacrimal and lacrimal foramen: (0) on the orbital rim (*Trigonostylops*); (1) within orbit, orbital rim projecting (*Astrapotherium*).

31. Zygomatic arches: (0) not flaring, dorsoventrally shallow, maxillary root above M$^{1-2}$ (*Trigonostylops*); (1) not flaring, dorsally curved, dorsoventrally thick, maxillary root above M$^{1-2}$ (*Astrapotherium*); (2) flare widely, horizontal, dorsoventrally shallow, maxillary root above M^2 (*Granastrapotherium*).

32. Sagittal and temporal crests: (0) sagittal long, temporal crests weak (*Trigonostylops*); (1) sagittal short, temporal crests strongly developed (*Astrapotherium*); (2) sagittal crest long, temporal crests well developed (Qda. Honda material).

33. Infraorbital or preorbital foramen: (0) multiple foramina, positioned very anterior to orbit (*Trigonostylops*); (1) single, moderate, positioned low and anterior to orbit (*Scaglia*); (2) single, small, elevated in position (*Astrapotherium*); (3) single, large, ovoid, near the orbital, long axis oriented vertically (*Granastrapotherium*).

34. Supraoccipital: (0) not constricted (*Trigonostylops*); (1) constricted (*Astrapotherium*).

35. Occipito-squamosal contact: (0) occipital exposure of mastoid (*Trigonostylops*); (1) extensive, no occipital exposure of mastoid (*Astrapotherium*).

36. Glenoid fossa: (0) at the level of or anterior to the basioccipital-basisphenoid suture (*Trigonostylops*); (1) posteriorly shifted, posterior to suture (*Granastrapotherium*).

37. Postympanic process of the squamosal: (0) very weak, far removed from paroccipital process (*Trigonostylops*); (1) weak, closely applied to paroccipital process (*Granastrapotherium*); (2) strong, closely applied to paroccipital process (*Astrapotherium*).

38. "Lacrimal" fossa anterior to the orbits: (0) absent (*Trigonostylops*); (1) deep (*Astrapotherium*).

39. dP^4 mesiolingual features: (0) mesiolingual cingulum (*Parastrapotherium*); (1) mesiolingual cingulum and pocket (*Xenastrapotherium*); (2) mesiolingual pocket only (*Granastrapotherium*).

Appendix 22.2. Character state–taxon matrix used in the phylogenetic analysis

Taxon	1	2	3	4	5	6	7	8	9	10	11	12	13	14	15	16	17	18	19	20
Eoastrapostylops	?	0	0	0	0	?	?	?	?	?	?	?	0	0	?	0	0	3	0	0
Tetragonostylops	0	1	?	0	0	?	0	1	0	0	0	1	1	0	2	0	0	0	1	0
Trigonostylops	0	1	0	0	1	0	?	?	?	?	?	0	0	1	0	2	0	1	0	1
Albertogaudrya	0	1	?	?	?	?	?	?	?	?	?	2	1	0	0	0	0	0	?	0
Scaglia	?	?	1	?	?	?	?	?	?	?	?	2	0	0	1	0	1	0	?	?
Astraponotus	1	1	?	?	?	?	1	1	0	0	?	2	0	0	1	0	0	1	?	0
Parastrapotherium	2	2	?	1	1	1	1	?	1	1	?	2	0	1	1	1	1	2	2	1
Astrapotherium	2	2	2	2	1	1	1	0	1	1	1	2	0	1	0	1	2	2	2	1
Astrapothericulus	2	2	2	2	1	1	?	?	?	?	?	1	2	0	1	2	1	?	2	1
Uruguaytherium	?	?	?	?	?	?	?	?	?	?	?	?	?	?	?	?	?	?	?	2
Xenastrapotherium	2	2	?	2	1	?	?	?	?	?	?	2	0	1	2	2	3	?	0	2
Granastrapotherium	2	3	?	3	2	1	1	0	?	1	?	2	0	1	3	2	3	2	1	2

Taxon	21	22	23	24	25	26	27	28	29	30	31	32	33	34	35	36	37	38	39
Eoastrapostylops	0	?	?	?	?	?	?	?	?	?	?	?	?	1	?	?	?	?	?
Tetragonostylops	0	?	?	?	?	?	?	?	?	?	?	?	?	?	?	?	?	?	?
Trigonostylops	0	?	?	?	?	?	0	0	0	0	0	0	0	0	0	0	0	0	?
Albertogaudrya	0	?	?	?	?	?	?	?	?	?	?	?	?	?	?	?	?	?	?
Scaglia	?	?	?	?	?	?	1	1	?	1	?	?	1	?	?	?	?	?	?
Astraponotus	0	?	?	?	?	?	?	?	?	?	?	?	?	?	?	?	?	?	?
Parastrapotherium	1	0	0	1	0	?	?	?	?	?	?	?	?	?	?	?	?	?	0
Astrapotherium	1	0	0	0	?	1	1	2	1	1	1	1	2	1	1	0	2	1	?
Astrapothericulus	1	0	0	0	?	0	1	2	1	1	1	1	?	?	?	0	?	1	?
Uruguaytherium	1	?	?	?	?	?	?	?	?	?	?	?	?	?	?	?	?	?	?
Xenastrapotherium	1	1	1	2	2	?	?	?	?	?	?	?	?	?	?	?	?	?	1
Granastrapotherium	1	1	1	2	2	1	1	2	2	?	2	1	3	?	?	1	1	?	2

Note: The characters and character states are defined in appendix 22.1.

23
Sirenia

Daryl P. Domning

RESUMEN

Dos taxa de Sirenia poco conocidos (*Potamosiren magdalenensis* y *Metaxytherium ortegense*) han sido descritos del Grupo Honda y asignados a las familias Trichechidae y Dugongidae, respectivamente. En base a dientes aislados recientemente colectados y descritos aquí, se propone que *Metaxytherium ortegense* es un sinónimo de *Potamosiren magdalenensis*. La asignación de *Potamosiren magdalenensis* a la familia Trichechidae necesita mejores evidencias morfológicas. Suponiendo que *Potamosiren magdalenensis* sea un manatí, es posible proponer como hipótesis un escenario evolutivo que explica tanto la reducción del tamaño de los molares y la gruesa capa de esmalte que rodea las coronas, como la evolución de molares supernumérios de reemplazo interminable, como adaptaciones a una dieta de gramíneas acuáticas silicificadas.

Fossil marine mammals of the order Sirenia (manatees and dugongs) were first reported from the Colombian Miocene by Reinhart (1951), who described *Potamosiren magdalenensis* on the basis of a mandible and tooth collected in the "El Libano sands and clays" near Villavieja. Later Kellogg (1966) described *Metaxytherium ortegense* on the basis of a maxilla with three molars from an uncertain horizon near Ortega. Domning (1982) referred an isolated tooth and an atlas vertebra from the Villavieja Formation and the "Ferruginous Sands," respectively, to *Potamosiren*

sp. Only four additional sirenian specimens have subsequently come to light; hence, it is obvious that the picture of the nature or diversity of the Miocene sirenian fauna of Colombia is still far from clear. Given the fact, however, that the two nominal taxa described are thought to represent different families (Trichechidae and Dugongidae, respectively), it is surprising that the remains so far recovered, scanty as they are, do not unequivocally point to any significant morphological diversity. This has led me to suspect that only one sirenian taxon is actually present in the Honda Group—the view that is defended here.

Systematic Paleontology

Order Sirenia Illiger 1811
Family Trichechidae Gill 1872 (1821)
Subfamily Trichechinae Gill 1872 (1821)
Potamosiren Reinhart 1951, p. 204
Potamosiren magdalenensis Reinhart 1951, p. 204

SYNONYMY

Metaxytherium ortegense Kellogg 1966, p. 93; new synonymy. *Felsinotherium ortegense* Kellogg 1966, pl. 36, figs. 1,

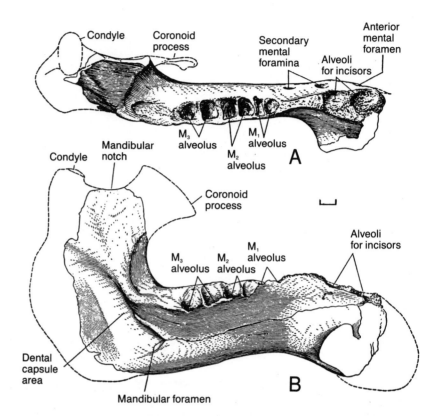

Figure 23.1. *Potamosiren magdalenensis,* holotype, UCMP 39471, left mandible: *A,* occlusal view; *B,* lingual view. Scale bar 1 cm. (Adapted from Reinhart 1951.)

2 (lapsus for *Metaxytherium ortegense*). *Ribodon magdalenensis* (Reinhart) Marshall et al. 1983, p. 52.

HOLOTYPE

UCMP 39471, incomplete left mandible lacking teeth, and a probably associated right M_3 (figs. 23.1, 23.2, 23.3F).

TYPE LOCALITY AND HORIZON

UCMP Locality V4533, approximately 1.75 km N66° W of the high point of Cerro Gordo, north of Villavieja, Huila Department; in the "El Libano sands and clays" (=lower

part of La Victoria Formation, equivalent in level to Duke University Localities 116 and 118), about 80 m below the "Cerbatana gravels and clays" (=upper part of La Victoria Formation); middle Miocene.

PREVIOUSLY REFERRED SPECIMENS

The following two specimens were referred to *Potamosiren* sp. by Domning (1982, 601–2; figs. 1 and 2 therein): (1) YPM-PU 18820, left M_3, collected 6–8 km northeast of Villavieja, probably in the Monkey Beds near the base of the Villavieja Formation. (2) UCMP 104979, atlas vertebra, collected at UCMP Locality V5045, in the Ferruginous Sandstone Beds.

DISCUSSION

The monotypic genus *Potamosiren*, known so far only from Colombia, represents the earliest and most primitive stage of evolution yet recorded for the manatees (Family Trichechidae). Its morphology, however, is not adequately

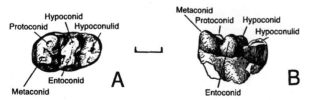

Figure 23.2. *Potamosiren magdalenensis,* holotype, UCMP 39471, right M_3: *A,* occlusal view; *B,* lingual view. Scale bar 1 cm. (Adapted from Reinhart 1951.)

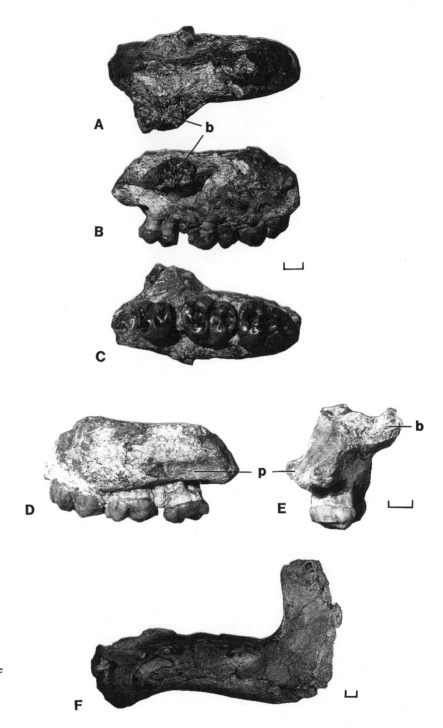

Figure 23.3. *Potamosiren magdalenensis.*
A–E. Holotype of *Metaxytherium ortegense*
USNM 10870, left maxilla with M$^{1-3}$:
A, dorsal view; *B,* lateral view; *C,* ventral
view; *D,* medial view; *E,* anterior
view. *b,* zygomatic-orbital bridge; *p,* palatine
process. *F.* Holotype of *P. magdalenensis,*
UCMP 39471, left mandible, lateral view.
Scale bars 1 cm.

known. The holotype mandible (figs. 23.1, 23.2, 23.3F) clearly shows that *Potamosiren* lacks the horizontally replaced supernumerary cheek teeth that characterize all later trichechids, but it can, nevertheless, be assigned to the Tri-chechidae on at least four grounds of varying persuasiveness: (1) The shape and arrangement of its molar cusps closely resemble those of the Mio-Pliocene manatee *Ribodon* Ameghino 1883; the significance of this is doubt-

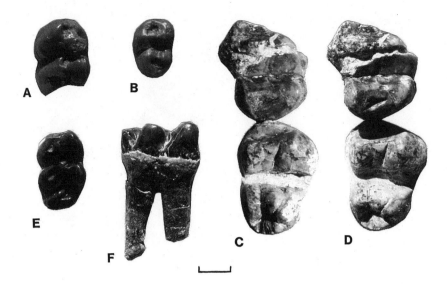

Figure 23.4. Cheek teeth from the Honda Group, Colombia, newly referred to *Potamosiren.*
A. Potamosiren cf. *P. magdalenensis,* cast of MNHN VIV 66, left M_3, occlusal view (anterior at top).
B–F. Potamosiren magdalenensis. B. Cast of IGM 182639, left upper molariform tooth, occlusal view
(anterior at top). *C–D.* IGM 250927, right M^{2-3}, two occlusal views illuminated from different
directions (anterior at top); the anterolabial corner of M^2 is damaged. *E–F.* IGM 251093, left $M_{2?}$:
E, occlusal view (anterior at top); *F,* lingual view. Scale bar 1 cm (approximate).

ful, however, because the arrangement of the cusps is essentially that of primitive sirenians. (2) Its slender mandibular ramus, although also a primitive condition, is too primitive to be assigned to any known non-trichechid taxon of Miocene sirenians (Domning 1982). (3) The moderate and even ventral curvature of its mandibular ramus appears to be a derived trichechid condition (although independently evolved by certain Eocene and Oligocene dugongids and by the hydrodamaline dugongids of the Neogene). (4) The absence of all premolars is also derived in trichechids although independently evolved in Oligocene dugongids.

A more satisfactory phylogenetic analysis must await more complete cranial material. The referred atlas is likewise not certainly diagnostic at the family level.

Metaxytherium ortegense, here synonymized with *P. magdalenensis,* was founded by Kellogg (1966) on part of a left maxilla with M^{1-3} (USNM 10870; fig. 23.3 A–E). It was collected near Ortega, north of the mouth of the Río Saldaña (Tolima Department), in gray to green black-speckled sandstone assumed to be part of the Honda Group. Discussion of its affinities and the reasons for this synonymy are given here.

NEWLY REFERRED SPECIMENS (FIG. 23.4)

The following four specimens are reported here for the first time: (1) MNHN VIV 66, left M_3, collected at Los Mangos in the Cerbatana Conglomerate Beds (=upper part of La Victoria Formation) by R. J. Hoffstetter in 1966. (2) IGM 182639 (a cast of this specimen is cataloged as USNM 411891), left upper molariform tooth, collected near Coyaima, Tolima Department, at UCMP Locality V4411, by R. F. Kay in 1985. (3) IGM 250927, right M^{2-3}, collected at Duke Locality 113, in the Tatacoa Sandstone Beds. (4) IGM 251093, left ?M_2, collected at Duke Locality 80, near the base of the Tatacoa Sandstone Beds.

DESCRIPTION OF NEW SPECIMENS

MNHN VIV 66 (fig. 23.4A) is an M_3 lacking its large hypoconulid lophule; it is identifiable as a third molar because the posterior root was much larger than the anteroposteriorly compressed anterior root. Its anterior crown width (18.5 mm) and posterior width (18.2 mm) are similar to those of YPM-PU 18820 rather than to the smaller and stratigraphically lower UCMP 39471. It is slightly less worn than YPM-PU 18820 and displays an anterior interdental wear facet and a broad, flat facet on the gentle posterior slope of the protoconid as well. The tip of the metaconid is missing. The heights of the crown at the very slightly worn protoconid and entoconid are 12.6 and 13.4 mm, respectively. As in YPM-PU 18820, the cusps are bulbous, with very thick enamel (about 3.4 mm), and on both teeth the hypoconid has a small anterolingual spur that is separated from the entoconid by a tiny cleft. This spur is more pronounced on MNHN VIV 66. It corresponds to the crista obliqua, which is more distinctly developed in UCMP 39471 (fig. 23.2).

IGM 182639 (fig. 23.4B) is a small upper tooth (crown length, 18.6 mm; anterior width, 15.4 mm; posterior width, 12.8 mm) having the same basic crown pattern as sirenian upper molars in general. Its chief peculiarity is that the individual cusps are not distinct despite the lack of significant wear. There appears to be some wear on the gentle anterior slope formed by the protoloph and precingulum and on the metaconule and hypocone as well, and there is an anterior interdental wear facet, but nowhere is the enamel breached. The precingulum is very prominent and massive with only a faint suggestion of individual cuspules; it completely surrounds a deep anterior basin, or pit. The protocone, protoconule, and paracone form a straight transverse line but are almost completely confluent. The protocone and paracone, however, are larger than the protoconule and project backward at either end of the protoloph, constricting the transverse valley both labially and (especially) lingually where the protocone contacts the off-center metaconule. The hypocone lies posterolingual to the latter and is continuous with it, whereas the much smaller metacone is connected to the metaconule by a saddle. The hypocone is also continuous with a large cingular cuspule, which, together with the posterior slope of the metacone, almost encloses the posterior cingular basin.

IGM 250927 (fig. 23.4C, D) comprises two upper molars, probably (given the narrow metaloph of the posterior tooth) the second and third. Both are divided transversely by fissures filled with matrix, so their exact crown lengths are uncertain. The posterior width of M^2 is 19.4 mm; anterior and posterior widths of M^3 are 21.4 and 17.6 mm,

respectively. The protoloph of M^2 is damaged by one of the aforementioned fissures. The precingulum, however, is visible anterior to it and is of peculiar form, lying entirely labial to the tooth's midline and protruding farthest forward at its labial end. Its enamel is slightly breached by wear. The anterior cingular basin, or what remains of it, is very small and is located far labially. The transverse valley is broad, straight, and open, the metaloph is nearly straight, the metaconule is centrally located, and the enamel of the metaconule and hypocone is breached by wear at their tips. A straight, nearly transverse, noncuspidate postcingulum descends gradually from the hypocone and encloses a posterior basin, which (because of wear on the postcingulum) is now very narrow and slitlike. M^3 bears a noncuspidate precingulum that extends across most of the tooth's front end but is very low, possibly in part as a result of wear; the basin behind it is likewise very narrow and slitlike. The straight protoloph towers high above this precingulum and is only very slightly worn. A fissure divides the tooth through the straight transverse valley; the enamel here is about 2 mm thick. In contrast to M^2, the metacone is distinctly lower than and separate from the other posterior cusps, and the hypocone lies slightly more posterior relative to the metaconule; the unworn postcingulum descends more steeply from the hypocone, and the posterior basin is broader between the surrounding crests (although with a narrow, slitlike floor).

IGM 251093 (fig. 23.4E, F) is a lower molar with two long, straight, anteroposteriorly compressed roots, a relatively small hypoconulid lophule, and both anterior and posterior interdental wear facets, so it is not an M_3. As it is comparable in width, however, to the holotype M_3 of *P. magdalenensis*, it is probably an M_2. Its crown length is 23.0 mm; anterior and posterior widths, 15.8 and 16.0 mm, respectively; length of unbroken posterior root from base of crown, 26 mm. It is moderately worn, and its enamel (roughly 3 mm thick) is breached on both the protoconid and hypoconid. The crista obliqua is simple in form like that of UCMP 39471 and does not resemble the more hooklike spur seen in MNHN VIV 66 and YPM-PU 18820. The hypoconulid lophule has a transverse ridge from whose labial end a spur extends anterolingually. Otherwise, the cusp pattern corresponds exactly to that of the holotype of *P. magdalenensis* (fig. 23.2).

DISCUSSION OF REFERRED SPECIMENS

YPM-PU 18820 and MNHN VIV 66 have a general resemblance to *Potamosiren* in the arrangement and bulbous shape of their molar cusps and the thickness of their enamel. Because, however, the first and possibly both of these teeth are from higher horizons than the holotype of *P. magdalenensis* and IGM 251093, because both are larger than the latter two, and because both have relatively smaller crista obliquas, they could represent a distinct species of that genus. If so, this later species would share with *P. magdalenensis* the lack of supernumerary molars. This is demonstrated by the large hypoconulid of YPM-PU 18820: manatees with unlimited horizontal tooth replacement never develop a tooth with this sort of "M_3 morphology." But the same two characteristics that suggest specific distinction for the later form also seem to rule it out of the ancestry of *Ribodon*, which has teeth the size of those in *P. magdalenensis* and possibly more distinct crista obliquas (and thinner enamel as well) than the larger Honda Group teeth. Therefore, this later species, if real, may have been a side branch rather than a direct ancestor of modern manatees. Formal recognition, however, of more than one species of *Potamosiren* would be premature at this time, so for the present I refer YPM-PU 18820 and MNHN VIV 66 to *P.* cf. *P. magdalenensis* and the other new specimens to *P. magdalenensis*.

Whereas the lower teeth discussed earlier are readily referable to *Potamosiren*, the upper teeth from the Honda Group are as easily referable to *Metaxytherium ortegense*. The isolated upper tooth IGM 182639 has thick, smooth, bulbous cusps and a massive precingulum that are very reminiscent of the supposed Colombian dugongid. Its smaller size suggests that it may be a deciduous premolar, possibly DP5. In comparison with a sirenian of about the same dental dimensions as *M. ortegense* (*M. floridanum* Hay 1922; see table 5 in Domning 1988), IGM 182639 does seem too small to be the DP5 of an individual as large as the holotype of *M. ortegense* (whose teeth are larger than the mean for *M. floridanum*); the new upper molars from Colombia (IGM 250927), however, are smaller than the mean for *M. floridanum*, and the Honda species could well have included individuals with a DP5 as small as IGM 182639. In the latter specimen, the location of the hypocone more posterior to

the metaconule is another minor difference from the holotype molars of *M. ortegense*. The development of the cingula and anterior and posterior basins, however, is also closely matched in the Mio-Pliocene trichechid *Ribodon limbatus* Ameghino 1883 from the "Mesopotamiense" beds (Huayquerian-Montehermosan) of Argentina (pl. 2 and fig. 6 in Pascual 1953), indicating that this tooth could just as well be referred to the Honda trichechid, *Potamosiren*.

IGM 250927 likewise has no characters that would necessarily bar its referral to *Metaxytherium ortegense*. The upper teeth of this species (and only upper teeth are known), however, are suspiciously reminiscent of the M_3s (YPM-PU 18820 and MNHN VIV 66) that were discussed previously. Although these lower molars are nearly identical to the holotype of *Potamosiren magdalenensis* in cusp pattern and enamel thickness, they are considerably larger and, indeed, of just the size to be expected in M_3 of *M. ortegense*. Their cusp pattern also resembles that of some *M. floridanum* although their enamel is thicker and their cusps lower and more bulbous. In other words, they differ from North American Miocene dugongids in just the same ways and to the same extent as does *M. ortegense*. This observation demanded serious reconsideration of the affinities of the latter.

SYNONYMY OF *METAXYTHERIUM ORTEGENSE*

Metaxytherium de Christol 1840 is one of the most common and best-known sirenian genera and has a cosmopolitan distribution in deposits of Miocene and Pliocene age. In size and cusp arrangement of the teeth the holotype of *M. ortegense* (fig. 23.3C) can scarcely be distinguished from those of other species of the genus, such as *M. calvertense* Kellogg 1966 (which is known from the Miocene of Peru and North America as well; de Muizon and Domning 1985) or *M. floridanum* (cf. Domning 1988). It stands apart from these and all other *Metaxytherium* species, however, in two unusual features. First, there is a very slight gradient of wear from M1 to M3; the former is only somewhat more worn that the latter, whereas in most *Metaxytherium* M1 is essentially worn out by the time M3 is fully in wear. This could indicate that *M. ortegense* had a less-abrasive diet, but it may simply be the result of the second difference: the enamel of USNM 10870 appears to be significantly thicker than in other *Metaxytherium*, even allowing for its relatively low cusps, its slight degree of wear, and the low angle to the enamel of some of the worn surfaces. Numerous *Metaxytherium* molars of all wear stages are available from Bone Valley, Florida (*M. floridanum*), and none resembles the Colombian specimen in this regard.

M. ortegense might conceivably represent some dugongid other than *Metaxytherium*; specifically, one of the diverse members of the subfamily Dugonginae (includes Rytiodontinae) now known to have inhabited the Caribbean throughout the late Tertiary. None of these, however, appears to have possessed especially thick molar enamel or other derived features that might associate the Colombian species with this group (cf. Domning 1990).

A further possibility is that *M. ortegense* is not a dugongid at all. Apart from the general similarity of its molar cusp arrangement to that of "typical" dugongids (which is essentially a shared primitive character), I was long willing to accept it as a member of that family because of two characters of its maxilla: (1) In the holotype of *M. ortegense*, the zygomatic-orbital bridge of the maxilla, although poorly preserved, lies well (about 1.5 cm) above the level of the bony palate (fig. 23.3B, E). This is a derived state seen in *Metaxytherium* and some other sirenians. I now find, however, that this state is also variably present in all species of *Trichechus* and that it can likewise be interpreted as present in a maxilla referred to *Ribodon* sp. (USNM 167655; Domning 1982, 604). (2) The bony palate in *M. ortegense* is only about 0.9 cm thick at the level of M^1 (fig. 23.3D, E). This condition is seen in other *Metaxytherium* and is primitive for sirenians in general. In contrast, in the oldest known (?early Pliocene) trichechid maxilla (USNM 167655, *Ribodon* sp., cited earlier), the palate is much thicker (2.5–3.0 cm), and this derived condition is also seen in modern *Trichechus*.

Further preparation of the holotype of *M. ortegense* has revealed that its maxilla differs from that of all or most *Metaxytherium* in at least two additional respects (characters 3 and 4): (3) The posterior edge of the zygomatic-orbital bridge is relatively thin and ascends posterodorsal across the side of the maxilla (fig. 23.3B). In *Metaxytherium* the posterior edge is very thick and frequently even forms a nearly flat, vertical surface facing posteriad or posterodorsad; but in *Ribodon* and frequently in *Trichechus* the posterior edge is thinner and more nearly resembles that of *M. ortegense*. The

polarity of this character is uncertain. (4) The posterior edge of the palatine process of the maxilla in *M. ortegense* lies at the level of the front of M^2, hence more than 1 cm posterior to the most anterior point on the after edge of the zygomatic-orbital bridge (fig. 23.3A, C, D). In other words, the palatines apparently did not extend as far forward as the bridge. This is a synapomorphy of trichechines (Domning 1994), but it also evolved independently in *Dugong* and in hydrodamaline dugongids and occurs in at least one Old World species of *Metaxytherium*, *M. serresii* (Gervais 1847). It, however, has not been observed in any New World *Metaxytherium*.

Therefore, characters 1 and 4 might support either a dugongid or a trichechid assignment of USNM 10870; character 2 is not phylogenetically informative; and character 3 is of uncertain polarity and taxonomic value but is a point of phenetic resemblance between USNM 10870 and trichechids. Hence, *M. ortegense* could just as easily be a trichechid as a dugongid. Furthermore, thickening of the palate in *Ribodon* and *Trichechus* (character 2) may be connected with the presence of supernumerary molars (and possibly greater masticatory forces) in these genera, in which case it should not be expected in a more primitive trichechid (such as *Potamosiren*) that lacks the extra teeth.

If *M. ortegense* is indeed such a trichechid, then Kellogg's original interpretation was not only reasonable but inescapable given the state of knowledge thirty years ago. At first glance, the overall proportions of the maxilla are certainly more similar to those of *Metaxytherium* than to those of *Trichechus*. The gross differences in form and proportions of the maxillary alveolar process between *Metaxytherium* and *Trichechus*, however, are (as suggested in the case of thickness of the palate) the result mainly of the derived presence of supernumerary molars in the latter. A primitive trichechid lacking supernumerary molars might be expected to have a maxilla, and a dentition as well, much more like that of a dugongid because trichechids, in fact, seem to be derived from Eocene or early Oligocene dugongids (Domning 1994).

In this light, it seems more suspicious than ever that *M. ortegense* and *Potamosiren*, represented by upper and lower teeth, respectively, just happen to share bulbous molar cusps, unusually thickened enamel, and an exclusively Colombian and middle Miocene distribution. All the Honda

Group sirenian specimens seem to have come from the "fluvial" portion of the section, which is characterized by sandstone channels formed in large, meandering lowland rivers with slow rates of flow. This is typical manatee habitat in South America today, and there is, therefore, a certain attraction in supposing, a priori, that all these fossils represent a single species or genus of trichechid that was endemic to fluviatile habitats in South America. (*Metaxytherium* and other dugongids, however, have never been otherwise reported from continental deposits.) This would explain why no teeth like those described earlier have been found elsewhere in the Caribbean. Although knowledge of Caribbean sirenians is admittedly very incomplete, it has grown rapidly in the last few years, and if *Metaxytherium ortegense* were truly a dugongid and, hence, presumably a marine animal, it should have been distributed throughout much or all of the Caribbean.

Taken together, these lines of evidence make it highly likely that only one form is present in the Honda Group—a trichechid, *Potamosiren*. Pending clarification of whether one or more species of this genus is represented in the available material, I synonymize *M. ortegense* with the single named species, *P. magdalenensis*. The assignment of this species to the Trichechidae is supported by a cladistic analysis of the entire Order Sirenia (using the Hennig86 computer program; © James S. Farris, 1988), in which *P. magdalenensis* (including *M. ortegense*) consistently and with stability forms the plesiomorphic sister group to *Ribodon* and *Trichechus* and is well removed from dugongids such as *Metaxytherium* (Domning 1994).

Parenthetically, this situation neatly exemplifies what might be called an axiom of paleosirenology: sirenian teeth are only good for raising taxonomic questions, not for answering them. Answers require skull bones as in the present case where the familial assignments of *M. ortegense* and *P. magdalenensis* inevitably hinge mainly on the holotype maxilla and mandible, respectively.

THE RÍO ACRE TRICHECHID

An additional manatee tooth (LACM 117532) from the Miocene of South America deserves comment. This was collected from late Miocene (Huayquerian) deposits on the Río Acre (Brazil-Peru border) and described by Frailey (1986) as a left lower molar of *?Ribodon*. Frailey suggested that it might represent a separate evolutionary lineage from that of the Colombian and Argentinian trichechids, apparently basing this opinion largely on the presence in LACM 117532 of an extra anterior cusp (paraconid) not seen in the other trichechid fossils. Such a cusp, however, is often present on the tooth I have identified as DP_4 in *Trichechus* and some other sirenians (Domning 1982), and this fact, together with the relatively small size of LACM 117532, renders it likely that the Río Acre tooth is a DP_4 of *Ribodon* rather than a representative of some previously unknown lineage.

FUNCTIONAL ANATOMY AND PALEOECOLOGY

A remaining point of interest is the functional significance of the thick enamel seen in these Colombian teeth. Among terrestrial mammals, bunodont teeth with thickened enamel are often viewed as being adapted to eating hard seeds, nuts, or fruits (e.g., Kay 1981). Although there is one report of modern Florida manatees feeding on acorns (O'Shea 1986), and fallen fruit can be important in the diets of fish in South American rivers (Goulding 1980), it is novel and somewhat difficult to envisage hard-shelled nuts or fruits as a major component of a sirenian's diet.

A more plausible alternative is offered by Janis and Fortelius (1988), who point out that thick enamel is useful not only for cracking hard food items but also for crushing soft ones without cutting them. The latter function is far more applicable to a diet of aquatic vegetation, especially in view of my earlier suggestions (Domning 1982) that manatees shifted from a diet of relatively soft and mostly nonabrasive water plants (in the case of *Potamosiren*) to a diet of siliceous grasses (in the more derived forms *Ribodon* and *Trichechus*). According to this view, *Potamosiren* would fit the profile of a bunodont herbivore whose (predominantly orthal) mastication of soft food required little or no exact occlusion, but whose diet was imposing some selective pressure for increased wear resistance of the dentition. According to Janis and Fortelius (1988, 212), this could be accomplished by increasing the resistance (e.g., thickness) of the enamel as long as wear could be prevented from penetrating the enamel and destroying the preformed occlusal morphology.

Beyond that point, however, an adaptive threshold would be crossed.

Once the rate of wear of the teeth is fast enough to destroy the preformed occlusal morphology at an early state in the life of an animal, the *amount* of dental tissue, rather than its wear resistance, must be increased in order to produce teeth of sufficient functional durability. The wear resistance of the tissues themselves is probably frequently reduced in this process. (Janis and Fortelius 1988, 213; emphasis in original).

This can be read as a straightforward description of subsequent dental evolution in manatees. Addition of siliceous grasses to the diet of *Potamosiren* would have pushed the animal over the adaptive threshold and begun selecting for a greater amount of less-resistant dental tissue. This took the form of cheaply made, continuously replaced, "throw-away" supernumerary teeth, smaller in size and with thinner enamel than those of *Potamosiren*.

This adaptive scenario appears promising and capable of accommodating all the presently available data, but until one has a better idea of what the Colombian sirenians actually were, both taxonomically and morphologically, such speculations inevitably have an air of the premature about them.

Acknowledgments

I thank C. de Muizon for providing me with casts of MNHN VIV 66. The manuscript was reviewed by M. A. Cozzuol and R. H. Madden. This research was supported by National Science Foundation grants DEB-8020265 and BSR-8416540.

24
Rodents

Anne H. Walton

RESUMEN

Entre los ordenes de mamíferos más diversos en la fauna de La Venta, los roedores están dominados por los Dinomyidae grandes y medianos. Aunque el zarandeo de cinco niveles estratigráficos diferentes ha producido una diversidad importante de pequeños roedores caviomorfos, los Cricetinae de origen norteamericano aún no han sido descubiertos en el Grupo Honda.

Los Erethizontidae están representados por tres taxa, uno algo más grande que el actual *Coendou mexicanus*, otro comparable en tamaño a *Myoprocta acouchy*, y un tercer género y especie, *Microsteiromys jacobsi* Walton, el Erethizontidae más pequeño que se conoce, semejante a *Neotoma cinerea* en tamaño.

Los Echimyidae están representados por tres y posiblemente cuatro taxa. Una especie de *Acarechimys* cercano a *A. minutissimus,* casi del tamaño de *Mesocricetus auratus* y así uno de los más pequeños caviomorfos conocidos, es el único género de roedor compartido entre La Venta y las faunas de Patagonia. *Ricardomys longidens* Walton (Adelphomyinae) era del tamaño del actual *Ctenomys haigi*. Un tercer Echimyidae casi el tamaño del actual *Proechimys* ha sido descrito anteriormente en base a las colecciones viejas. Un cuarto taxon, referido con duda a los Echimyidae, sería el más pequeño caviomorfo conocido.

Entre los Dasyproctidae, *"Neoreomys" huilensis* es distinto morfológicamente de *Neoreomys* de Patagonia. Las afinidades filéticas de los Dasyproctidae fósiles con los géneros actuales (*Dasyprocta* y *Myoprocta*) siguen siendo os-curas. Las dos especies de *Microscleromys* Walton encontradas en la fauna de La Venta tienen afinidades inciertas a nivel de familia.

Los Dinomyidae *"Scleromys"* abarca hasta cuatro especies diferentes. *"Scleromys" schurmanni,* el dinómido más pequeño y la especie de roedor más abundante en las colecciones, era comparable al actual *Myoprocta acouchy* en tamaño. Material de esta especie proveniente de los niveles inferiores del Grupo Honda se distingue en tamaño y morfología del material proveniente de los niveles superiores y así merece reconocimiento como especie distinta de un linaje evolutivo continuo. *"Scleromys" colombianus* es más grande y de coronas más altas, y se asemeja al actual *Dasyprocta punctata* en tamaño. Así como en *"Scleromys" schurmanni,* especímenes de *"S." colombianus* demuestran cambio evolutivo a medida que se sube en la columna estratigráfica.

El Dinomyidae grande, *"Olenopsis,"* está representado por al menos dos especies, la pequeña con una masa corporal estimada cerca de la de *Cuniculus paca*. Esta especie se distingue de la especie grande en el tamaño de los molares e incisivos, y la altura de las coronas. Una forma aun más grande de Dinomyidae está indicada por una mandíbula y unos dientes sueltos.

Se registran dos géneros de Dolichotinae (Caviidae) en la fauna: *Prodolichotis pridiana,* algo más grande que el actual *Dolichotis salinicola*, y dos especies de un segundo género distinguido por presentar el prisma anterior del cuarto premolar débilmente desarrollado.

Los roedores de La Venta se distinguen en composición de los de Patagonia debido a claras diferencias ambientales. Taxa con las coronas de los dientes jugales de crecimiento continuo son escasos en el Grupo Honda, y las familias Chinchillidae y Octodontidae, abundantes en las faunas australes, no se registran en el Grupo Honda. Por el contrario, taxa con dientes de corona baja que se interpretan por analogía como habitantes de zonas boscosas, son comunes en el Grupo Honda, pero ausentes o raros en Patagonia.

Comparándolos en analogía con los caviomorfos vivientes, los roedores de La Venta indican la presencia de bosques. Por ejemplo, los Erethizontidae actuales son folívoros y de hábitos arborícolas o semi-arborícolas. Los Echimyidae también tienen hábitos arborícolas y predominan en bosques tropicales. El actual Dinomyidae *Dinomys* habita bosques, y los Dasyproctidae, aunque de hábitos terrestres, son frugívoro-granívoros y prefieren zonas boscosas. Sin embargo, la presencia de Caviidae en la fauna sugiere que vegetación abierta o mixta pudiera haber estado presente. Los Caviidae actuales están adaptados en dentadura a dietas abrasivas, como suele ocurrir más comunmente en hábitat abierta.

The Miocene La Venta fauna of Colombia is the best-known assemblage of terrestrial vertebrates from tropical South America and the youngest known northern fauna before the arrival of North American immigrants (the Great American Faunal Interchange; Webb 1985b). The fauna has little in common with the faunas of the more temperate and marginal areas of southern South America (Patagonia) where South American mammalian biochronology was established. The fossils are recovered from sediments recording the Neogene history of a tectonically active and geologically complex region and represent a time and place about which little is known. The purpose of this chapter is to compare the La Venta rodent fauna with other South American rodent paleofaunas, incorporating data offered by new discoveries, and to evaluate the paleoenvironmental evidence offered by the rodents.

The La Venta fauna is one of the most diverse fossil vertebrate assemblages in South America, and rodents are one of the most diverse mammalian orders present. They were first reviewed in the ground-breaking monograph by Fields (1957), who described several taxa and first ascribed the most abundant rodent species to the family Dinomyidae. In recent years renewed field collecting by re-

searchers from several institutions, under the auspices of Duke University in cooperation with the Instituto Nacional de Investigaciones en Geociencias, Minería y Química, Museo Geológico (INGEOMINAS) of Colombia, has resulted in the recovery of important new fossils in a sound stratigraphic and geochronologic context (chaps. 2, 3).

Descriptions of the new small rodents from La Venta are to be published elsewhere. Fields (1957) described mainly larger rodents and referred them to the Patagonian taxa *Neoreomys, Scleromys,* and *Olenopsis.* The La Venta rodents are not congeneric with these taxa (for reasons outlined here). Until the redescriptions of the La Venta genera are published, the names erected by Fields (1957) are set in quotes, for example, *"Scleromys" schurmanni. Prodolichotis pridiana,* the most common caviid rodent, was also described by Fields (1957); a new genus of caviid has since been recognized (Walton 1990) and is here referred to as Dolichotinae gen. 2, pending publication of its description.

The greatest range of material has been found by surface collecting. Screenwashing has also been employed, and the La Venta section is presently the only Tertiary section in South America that has been systematically screenwashed. In addition to rodents, screenwashing recovers primates, marsupials, bats, reptiles, amphibians, and fish (described by other researchers). Five stratigraphic levels are presently known to produce microfossils by screenwashing: the Fish Bed, the Monkey Bed, the El Cardón Red Beds, and two localities in the La Victoria Formation.

Overview of La Venta Rodents

Figures 24.1–24.3 and 24.7 show mandibles and tooth rows for some representative rodent taxa. Skulls of *"Olenopsis," "Scleromys,"* and *Prodolichotis pridiana* are elegantly illustrated in Fields (1957). Reconstructions of *Neoreomys australis* and *Eocardia excavata,* taxa related to the La Venta genera here referred to as *"Neoreomys"* and Dolichotinae gen. 2, can be found in Scott (1905).

ERETHIZONTIDAE (PORCUPINES)

An isolated upper molar (fig. 24.2A) and edentulous jaw (IGM 183417) of ?*Steiromys* from the Monkey Beds are close in size and morphology to species of *Steiromys* known

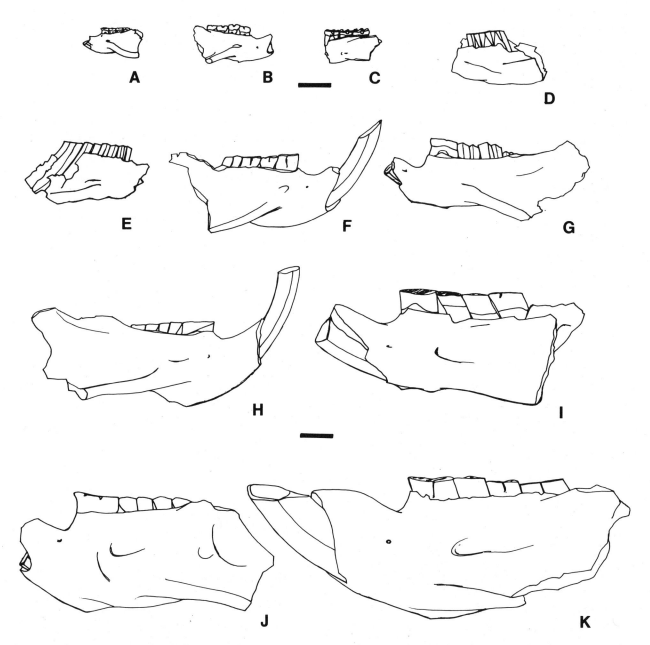

Figure 24.1. Mandibles of representative La Venta rodents in lateral view. *A*. IGM 184229, *Acarechimys* cf. *A. minutissimus,* adult. *B.* FMNH PM 54672, *Microsteiromys jacobsi* Walton, juvenile. *C.* IGM 183847, *Ricardomys longidens* Walton, juvenile. *D.* IGM 183212, Dolichotinae gen. 2, small sp., adult. *E.* IGM 183425, Dolichotinae gen. 2, large sp., adult. *F.* IGM 183061, *"Scleromys"* cf. *"S."* schurmanni, adult. *G.* FMNH PM 54708, *"Neoreomys" huilensis,* juvenile. *H.* IGM 126788, *"Scleromys"* cf. *"S."* colombianus, juvenile. *I.* IGM 250490, *"Olenopsis"* sp., small, adult. *J.* IGM 183682, *"Olenopsis"* sp., large, juvenile. *K.* UCMP 40055, *"Olenopsis"* sp., large, adult. Note that adults have a greater degree of development of the masseteric fossa than do juveniles, in which the lateral surface of the mandible is relatively flat. Juveniles commonly have a swelling in the region of erupting or newly erupted molars. Scale bars 5 mm.

from the Santacrucian of Patagonia. Erethizontids are morphologically conservative, so the generic identification is uncertain. This species is slightly larger than the living prehensile-tailed porcupine *Coendou mexicanus* (1.4–2.6

kg; Emmons and Feer 1990). Rough visual comparisons suggest a possible body weight of 2.5–4 kg.

The teeth of *?Steiromys* sp. described by Fields (1957) are smaller than the IGM material and similar to an un-

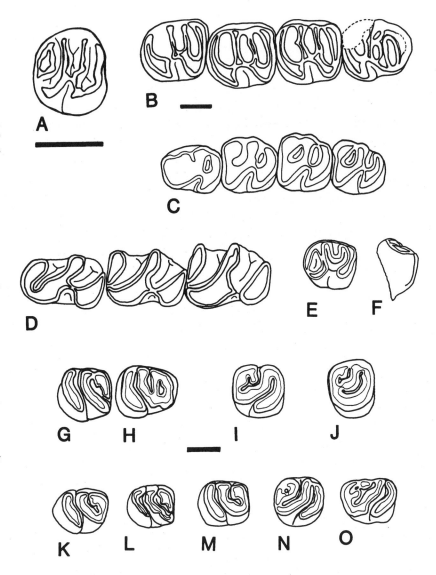

Figure 24.2. Tooth rows of representative La Venta rodents with bunodont and mesodont cheek teeth. All oriented with anterior to right in right teeth, to left in left teeth; buccal side upward in upper teeth, downward in lower teeth. Specimens represent adults unless otherwise indicated. *A, B.* Erethizontidae. *A.* IGM 250611, *?Steiromys* sp. (large), right P4, El Cardón Red Beds. *B.* FMNH PM 54672, *Microsteiromys jacobsi* Walton, juvenile, right m3–dp4, Monkey Beds. *C–F.* Echimyidae: *C.* IGM 184299, *Acarechimys* cf. *A. minutissimus*, left dp4–m3, Monkey Beds. *D.* IGM 183847, *Ricardomys longidens* Walton, juvenile, left dp4–m2, El Cardón Red Beds. *E.* IGM 250295, gen. et sp. indet., ?juvenile, right ?dP4 in occlusal view, Fish Bed. *F.* Same specimen as in *E,* but in posterior view, showing lingual hypsodonty. *G–O.* ?Dasyprocts. *G.* IGM 250319, *Microscleromys paradoxalis* Walton, right m1 or 2, Fish Bed. *H.* IGM 250308, *Microscleromys paradoxalis*, right p4, Fish Bed. *I.* IGM 250321, *Microscleromys paradoxalis*, right M1 or 2, Fish Bed. *J.* IGM 251040, *Microscleromys paradoxalis*, right P4, Fish Bed. *K.* IGM 251020, *Microscleromys cribriphilus* Walton, right m3, Fish Bed. *L.* IGM 250303, *Microscleromys cribriphilus*, unworn right m1 or 2, Fish Bed. *M.* IGM 251018, *Microscleromys cribiphilus*, right p4, Fish Bed. *N.* IGM 250283, *Microscleromys cribriphilus*, left M1 or 2 (reversed), Fish Bed. *O.* IGM 250320, *Microscleromys cribriphilus*, juvenile, right dP4, Fish Bed. Scale bar for *A,* 5 mm. Scale bars for *B–O,* 1 mm.

described specimen (UCMP 38374) from UCMP Locality V4502, on Río Peneya in Colombian Amazonia. The material is comparable in size to the living acouchy, *Myoprocta acouchy* (0.9–1.5 kg; Eisenberg 1989; Emmons and Feer 1990). The UCMP and IGM specimens probably represent two distinct species separable on the basis of size.

Microsteiromys jacobsi Walton is the smallest known erethizontid, living or fossil. Its body size is estimated at 200–300 g, larger than the living echimyid *Mesomys hispidus* (130–220 g; Emmons and Feer 1990) and similar to the packrat *Neotoma cinerea* (200–450 g; Nowak 1991). The two mandibles known for this species, including a juvenile (figs. 24.1B, 24.2B) and an aged adult, have uniquely deep, slitlike genioglossal fossae but are otherwise generalized.

ECHIMYIDAE (SPINY RATS) (FIG. 24.2)

Acarechimys is abundant in the Santa Cruz and Río Frias formations of Patagonia and is the only rodent genus shared in common by these reference faunas and the La Venta fauna of Colombia. Its true range is probably more extensive than now recognized because it is one of the smallest known caviomorph rodents, living or fossil, and little effort has been invested yet in the recovery of small mammals from most South American paleofaunas. It is larger than most cricetids common in the area today, such as *Zygodontomys brevicauda* (54–70 g; Eisenberg 1989), but smaller than large cricetids, such as the marsh rat *Holochilus sciureus* (144–177 g; Eisenberg 1989). The body weight of *Acarechimys* cf.

A. minutissimus is estimated at 90–150 g, based on visual comparisons with the golden hamster, *Mesocricetus auratus* (97–113 g; Nowak 1991). The La Venta material of *A.* cf. *A. minutissimus* includes a complete, well-worn mandible (figs. 24.1A, 24.2C), mandible fragments, and isolated teeth, similar in morphology but marginally larger than the Santacrucian species *Acarechimys minutissimus.*

Ricardomys longidens Walton is represented by a juvenile mandible fragment with three teeth (figs 24.1C, 24.2D) and isolated tooth fragments. It is characterized by long lower molars with three oblique lophids and wide, flat-bottomed reentrants. It is unique to the La Venta fauna but believed to be related to the extinct echimyid subfamily Adelphomyinae, known from the Colhuehuapian through Friasian of Patagonia. Body size is estimated at 150–250 g, comparable to the southern octodontoids *Ctenomys haigi* (164 g; Redford and Eisenberg 1992) or *Octodon degus* (170–260 g; Redford and Eisenberg 1992).

An edentulous mandible of an unknown echimyid, larger than *Acarechimys* or *Ricardomys* but smaller than most living echimyids, was described and illustrated by Fields (1957). Body size is estimated at 200–300 g, based on comparison with *Octodon degus* (170–260 g; Redford and Eisenberg 1992) and specimens of the spiny rat *Proechimys.*

Figure 24.2E and F show two views of the isolated upper deciduous premolar of possibly the smallest known caviomorph rodent. Although its small size and incomplete lophs would seem to exclude it from *Acarechimys* cf. *A. minutissimus,* it possibly represents a variant of this species and is assigned, with query, to the Echimyidae.

DASYPROCTIDAE (AGOUTIS, ACOUCHYS, PACAS, AND THEIR EXTINCT RELATIVES)

"Neoreomys" huilensis was described by Fields (1957) based on two specimens from the Monkey Beds. Fields (1957) recognized its resemblance to the abundant and geologically long-lived rodent *Neoreomys* known from the Colhuehuapian through the Friasian LMA of Patagonia. The La Venta material differs from the Patagonian genus in two characters: (1) the incisor in *"N." huilensis* is slender and ovate, whereas *Neoreomys* has wide, massive incisors with flat buccal surfaces; and (2) *"N." huilensis* has a pen-

talophodont lower deciduous premolar (fig. 24.3A), but *Neoreomys australis* (the most common Santacrucian species) has a tetralophodont dP4. The recovery of more material has revealed a greater size range for *"N." huilensis* than recognized by Fields (figs. 24.1G; 24.3 A–C). The La Venta material is similar in size to the living agouti *Dasyprocta punctata* (3.2–4.2 kg; Emmons and Feer 1990).

The familial assignment of *Neoreomys* to the Capromyidae was accepted at the time of Fields's work, but most workers now refer *Neoreomys* to the Dasyproctidae. The relations of early dasyprocts such as *Neoreomys, Incamys,* and *Alloiomys* to modern Dasyproctidae (*Dasyprocta* and *Myoprocta*), however, remain unclear (Walton 1990).

Screenwashing produced most of the material known of *Microscleromys* Walton, mainly represented by isolated teeth. So far it is unique to the La Venta fauna. Two species may be recognized, separable on the basis of size: *Microscleromys paradoxalis* Walton (fig. 24.2 G–J) is larger and higher-crowned, whereas *Microscleromys cribriphilus* Walton (fig. 24.2 K–O) is slightly smaller. Body size is estimated at 100–200 g although this is especially uncertain given the incomplete material. The genus is assigned, with query, to the Dasyproctidae based on its lophate tooth structure and relatively heavy enamel, similar to the much larger Santacrucian genus *Scleromys.* The recovery of reliable jaw material may place *Microscleromys* in the Echimyidae or in a new family.

DINOMYIDAE (PACARANAS AND THEIR EXTINCT RELATIVES)

The holotype of *"Scleromys" schurmanni* (Basel Museum of Natural History, no number) comes from near Carmen de Apicalá in the Melgar Basin, middle Magdalena River valley, of Colombia, and was referred by Stehlin (1940) to the rare Santacrucian genus *Scleromys.* It is not cogeneric with *Scleromys* based on two characters: (1) The incisors of *Scleromys* are massive, often wider in cross section than they are long, and triangular with flat buccal surfaces; the incisors of La Venta *"Scleromys"* are relatively long anteroposteriorly and gracile (fig. 24.3D), and upper incisors are slightly rounded. (2) *Scleromys* molars are trilophate during most of the animal's lifetime; the molars of La Venta *"Scleromys"* are

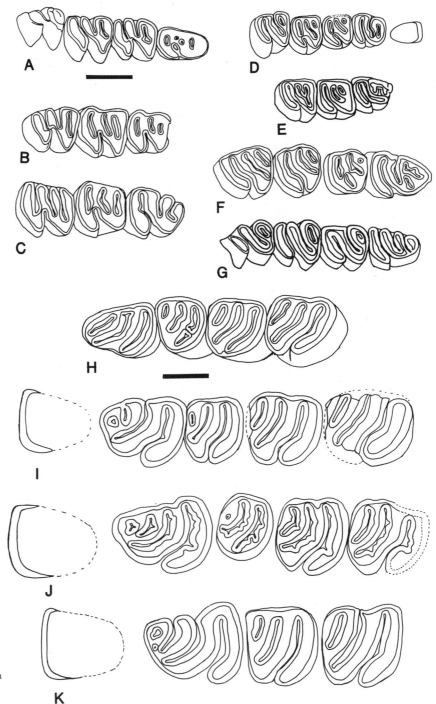

Figure 24.3. Tooth rows of representative La Venta rodents with rooted hypsodont cheek teeth. Orientation as in figure 24.2. All specimens except *A* represent adults. *A–C.* Dasyprocts. *A.* FMNH PM 54708, *"Neoreomys" huilensis,* juvenile, left dp4–m3 (reversed), Monkey Beds. *B.* IGM 183327, *"Neoreomys" huilensis,* right m3–m1, La Victoria Formation. *C.* IGM 183683, *"Neoreomys" huilensis,* right m3–m1, La Victoria Formation. *D–K.* Dinomyidae. *D.* UCMP 38967, *"Scleromys" schurmanni,* left p4–m3, cross section of left lower incisor (reversed), unit between Monkey Beds and Fish Bed. *E.* IGM 184233, *"Scleromys"* cf. *"S." schurmanni,* right m2–p4, La Victoria Formation. *F.* UCMP 37994, *"Scleromys" colombianus,* right m3–p4, unit between Monkey Beds and Fish Bed. *G.* IGM 184464, *"Scleromys"* cf. *"S." colombianus,* right m3–p4, La Victoria Formation. *H.* IGM 250490, *"Olenopsis"* sp. (small), left p4–m3, La Victoria Formation below Tatacoa Sandstone. *I.* TMM 41659-1, *"Olenopsis"* sp. (large), left p4–m3, cross section of left lower incisor, La Venta Red Beds. *J.* IGM 250513, *"Olenopsis"* sp. (large), right m3–p4, cross section of right lower incisor (reversed), La Victoria Formation. *K.* IGM 250820, *"Olenopsis"* sp. (large), partly reconstructed left p4–m2, cross section of left lower incisor, Polonia Red Beds. Scale bars 5 mm.

higher crowned, with greater development of the anterolophid, to the extent that molars are tetralophate during most of the animal's lifetime.

"Scleromys" schurmanni Stehlin is the smallest known dinomyid and the most common rodent species in the La Venta fauna. Size is estimated at 1–2 kg, comparable to that of the living acouchy *Myoprocta acouchy* (1.05–1.45 kg; Emmons and Feer 1990). Specimens from above the Monkey

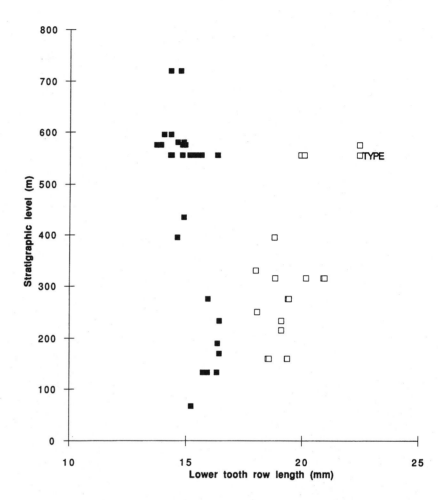

Figure 24.4. Adult tooth row length for *"Scleromys"* spp. by stratigraphic level. *Filled squares, "Scleromys" schurmanni and "S." cf. "S." schurmanni; open squares, "Scleromys" colombianus and "S." cf. "S." colombianus.*

Beds are smaller and slightly higher crowned than those from low in the section (fig. 24.4, filled symbols). A t-test on mean tooth row length shows the size difference is statistically significant at a 95% confidence level. A new species was not erected because there is considerable size overlap between populations from the extremes of the section to the extent that specimens cannot be confidently identified based on size. *"S." schurmanni* here refers to specimens from the upper part of the section and *"S." cf. "S." schurmanni* to those from low in the section with an arbitrary separation below the Monkey Unit. When specimens are recovered from a longer stratigraphic sequence it may be possible to define two species. In many specimens of *"S." cf. "S." schurmanni* from low in the section, the metalophid remains unconnected to the anterolophid until moderate wear so that the anterior fossetids of lower molars describe the letter *Y* (fig. 24.3E); specimens from higher in the section acquire complete lophids with little wear (fig. 24.3D).

"S." colombianus Fields is distinguished from *"S." schurmanni* by larger size and greater crown height. Size is estimated at 2.5–3.5 kg, similar to *Dasyprocta punctata* or *D. fuliginosa* (3.2–4.2 kg, 3.5 kg; Emmons and Feer 1990). In *"S." colombianus* the metalophid is connected to the anterolophid upon tooth eruption and molars are smoothly tetralophate until advanced wear (fig. 24.3F). Figure 24.4 (open symbols) gives a strong impression of an evolving lineage becoming larger and higher crowned upward through the section, in part because no specimens have been recovered from between the Tatacoa Sandstone and the Monkey Beds. As with *"S." schurmanni, "S." colombianus* is distinct from *"S." cf. "S." colombianus* below the Monkey Beds.

Fields (1957) identified the larger La Venta dinomyid *"Olenopsis"* with *Drytomomys aequatorialis* Anthony, known from a fragmentary skull of unknown age from the Nabón region of Ecuador (Anthony 1922). Fields synonymized

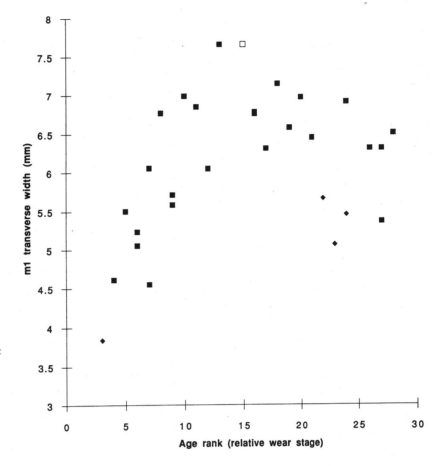

Figure 24.5. Transverse width of lower first molars of *"Olenopsis"* spp. (juveniles and adults), ranked by wear stage. Emplacement of permanent p4 and m3 occurs between wear stages 8 and 12. *Filled squares,* *"Olenopsis"* sp., large; *filled diamonds,* *"Olenopsis"* sp., small; *open square,* Dinomyidae, large.

Drytomomys with *Olenopsis* Ameghino but retained Anthony's specific name, *aequatorialis,* as valid. Patterson and Wood (1982) pointed out that Ameghino's type specimen for *Olenopsis* consists of an unworn upper deciduous premolar, probably of *Neoreomys australis,* and a lower cheek tooth too worn to be recognizable, preferring to refer the La Venta dinomyids to the Friasian genus *Simplimus.* Walton (1990) confirmed the observation by Patterson and Wood (1982) and questioned the synonymy of *Olenopsis* with either *Drytomomys* or the La Venta dinomyids. *Drytomomys* has wide, massive incisors and differs from all of the La Venta material in having six lophids on p4 (in a pattern suggesting the p4 may be a retained deciduous premolar), whereas the La Venta dinomyids have at most five. The resemblance is stronger between the La Venta material and *Simplimus,* known from isolated teeth recovered from the Laguna Blanca Locality in Argentina, but most of the La Venta teeth are smaller, lower crowned, with a more tapering tooth form and a different arrangement of lateral striae.

"Olenopsis" from the La Venta fauna will be renamed as a new genus.

The species of *"Scleromys"* and *"Olenopsis"* describe a near continuum in size and crown height. *"Scleromys"* cf. *"S."* schurmanni can have five lophids on deciduous lower premolars, but the higher-crowned *"Scleromys"* schurmanni and *"S."* colombianus and *"Olenopsis"* spp. all have six. It would not be unreasonable to synonymize the two genera. They are retained with separate names mainly for ease of field identification: even when the species cannot be recognized in a poorly preserved specimen, it is easy to distinguish the "big genus" from the "little genus" (fig. 24.1: F, H versus I–K).

Two species of *"Olenopsis"* are recognized. The less-abundant and more clearly defined is a small species typified by IGM 250490 (figs. 24.1I, 24.3H), characterized by long, oblique reentrants that remain smooth (do not develop enamel infolding) with advanced wear. Estimated size is 8–15 kg, close to that of the modern paca *Cuniculus paca*

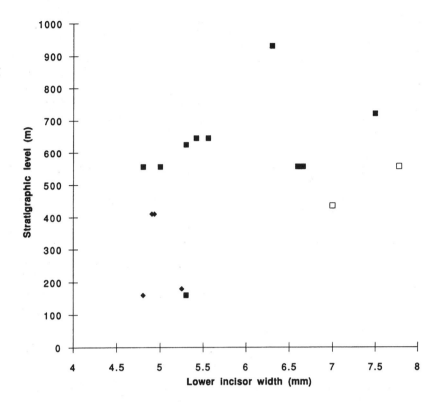

Figure 24.6. Transverse width of lower incisor of *"Olenopsis"* spp. by stratigraphic level. Juveniles and adults are undifferentiated. *Filled squares, "Olenopsis"* sp., large; *filled diamonds, "Olenopsis"* sp., small; *open squares,* Dinomyidae, large.

(5–13 kg; Emmons and Feer 1990) or pacarana *Dinomys branickii* (10–15 kg; Nowak 1991).

Within the larger recognized species of *"Olenopsis"* there is considerable variation in size, relative size of incisors, crown height and shape (columnar or tapering), and enamel infolding at different stages of wear (figs. 24.1J, K; 24.3 I–K). UCMP 40055, the only mature adult illustrated by Fields (1957; fig. 16C, D), typifies the massive mandible and wide incisor attained by some specimens (fig. 24.1K). Biometric definitions of species are complicated by changes in tooth size and shape with wear (fig. 24.5). Size is estimated at 13–20 kg, larger than most *Dinomys branickii* and close to that of small beavers (*Castor canadiensis,* 12–25 kg; Nowak 1991). There is a general increase in size and crown height upward through the section (fig. 24.6).

One poorly preserved jaw (UCMP 136479, from Locality V4932 in the Monkey Beds) and a number of isolated teeth are easily distinguishable from *"Olenopsis"* by the large size, columnar form of the molars, and persistent pentalophodonty, especially at equivalent stages of wear. There appears to be little tendency to develop infolded enamel

with advanced wear. This specimen is probably a larger, higher-crowned member of the *"Olenopsis"* lineage but also invites comparison with the ?Friasian genus *Simplimus* and more advanced dinomyids such as *Eumegamys.*

CAVIIDAE (CAVIES, MARAS) (FIG. 24.7)

Fields (1957) described *Prodolichotis pridiana,* the most common caviid in the La Venta fauna. It shows considerable variation in size, the development of prisms on p4 and M3, width of dentine tracts, and the development of sulci and offset prisms (fig. 24.7 D–I). Size is estimated at 2–3.5 kg, slightly larger than the living dwarf mara *Dolichotis salinicola* (1.0–2.7 kg; Redford and Eisenberg 1992). Ever-growing, rootless molars present special problems for recognizing relationships among taxa so poorly known because it is difficult to evaluate the influence of changes with ontogeny. The range of development of the anteriormost prism on p4 relative to p4 length is shown in figure 24.8.

A second genus of Dolichotinae is distinguished from *Prodolichotis* by the minimal development of the anterior-

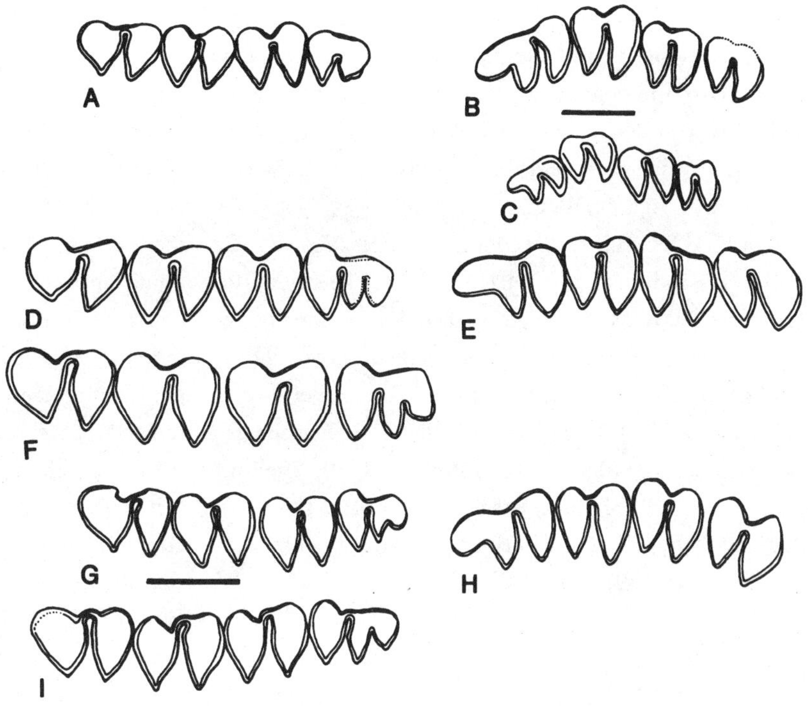

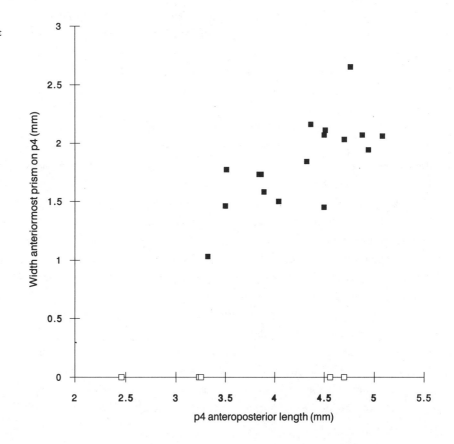

Figure 24.7. Tooth rows of representative La Venta rodents with hypselodont cheek teeth. Orientation as in figure 24.2. Caviidae: *A.* IGM 183425, Dolichotinae gen. 2, large, right m3–4, La Victoria Formation. *B.* IGM 250556, Dolichotinae gen. 2, large, left M3–P4 (reversed), La Victoria Formation. *C.* ING-KU T.W. 20-5, Dolichotinae gen. 2, small, right M3–P4, Monkey Bed. *D.* UCMP 36761, *Prodolichotis pridiana* (Fields 1957), right m3–p4, unit between Monkey Beds and Fish Bed. *E.* Same, right M3–P4. *F.* UCMP 39930, *Prodolichotis pridiana*, right m3–p4, Monkey Beds. *G.* UCMP 41671, *Prodolichotis pridiana*, left m3–p4 (reversed), Monkey Beds. *H.* IGM 183881, *Prodolichotis pridiana*, right M3–P4, La Victoria Formation. *I.* IGM 250459, *Prodolichotis pridiana*, right m3–p4, El Cardón Red Beds. Scale bars 5 mm.

Figure 24.8. Width (mm) of anteriormost prism on p4 versus anteroposterior length (mm) of p4 in Dolichotinae. Juveniles and adults are undifferentiated. *Filled squares, Prodolichotis pridiana; open squares, Dolichotinae gen. 2, large.*

most prism on p4, a primitive character (figs. 24.7A, 24.8). It is too advanced to be assigned to the primitive family Eocardiidae (the only cavioids present in the Santacrucian and Friasian of Patagonia), in that fossette/ids are absent from little-worn teeth and there are two fully developed prisms on P4 (fig. 24.7B, C).

Two species may be recognized: large (figs. 24.1E; 24.7A, B), and small (figs. 24.1D, 24.7C). The size of the larger species (IGM 183425) is estimated at 1–2 kg, slightly smaller than the living dwarf mara *Dolichotis salinicola* (1.0–2.7 kg; Redford and Eisenberg 1992). The smaller species estimated at 0.8–1.3 kg, slightly smaller than the acouchy *Myoprocta acouchy* (0.9–1.5 kg; Eisenberg 1989, Emmons and Feer 1990). The best example is a palate, ING-KU T.W. 20-5 (fig. 24.7C), described but not named by Setoguchi et al. (1979).

Change within the Stratigraphic Section

Figure 24.9 shows the stratigraphic ranges of rodent species in the La Venta section. Several taxa are characteristic of either the La Victoria Formation or the Villavieja Formation with a zone of overlap at the densely sampled Monkey Beds. *"Neoreomys" huilensis,* the small species of *"Olenopsis,"* and Dolichotinae gen. 2 are confined to the lower part of the section and the Monkey Beds; the latter two taxa are primitive relatives of the species of *"Olenopsis"* and *Prodolichotis* found throughout the section. The small rodents *Microsteiromys jacobsi* Walton, *Acarechimys* cf. *A. minutissimus, Ricardomys longidens* Walton, and *Microscleromys* spp. Walton are so far recovered only from the upper part of the section, but this distribution may reflect preservational bias more than a real absence.

Although these differences in the distribution of species and the biometric changes shown for *"Scleromys"* spp. make it possible to distinguish localities from the upper and lower extremes of the section, there is little indication of faunal turnover at the generic level. Both temporal and environmental change (as indicated by changes in style of sedimentation) probably affect the composition of the fauna, but the changes are not extreme. Discussion of a single La Venta rodent fauna seems justified for purposes of intracontinental correlation.

Comparisons with Patagonian Faunas

Figure 24.10 summarizes the present state of knowledge of South American fossil rodents. There are few pre-Pleistocene vertebrate fossil localities known outside of the rich fossil beds of Argentina and neighboring countries. Of the few faunas known from tropical South America the La Venta fauna is the most diverse and best studied.

Quantification of the degree of similarity between the La Venta rodents and the Miocene reference faunas of Patagonia, using Simpson's coefficients on the generic and familial level, is shown in table 24.1. Although this traditional view of Patagonian biostratigraphy is admittedly simplistic, it is evident that the La Venta rodent fauna shares little in common with any of the Miocene rodent faunas of Argentina. On the familial level the La Venta rodents resemble the fauna of the Río Frías and Collón Curá formations slightly more than other Patagonian reference faunas.

The La Venta rodents contrast strongly with those of Patagonia for fairly clear environmental reasons. Forms with extremely high-crowned or ever-growing cheek teeth, characteristic of mesic climates with coarse vegetation, are diverse in the Friasian faunas but notably scarce at La Venta. Vucetich (1984), in a revision of the Friasian rodents of Patagonia, interpreted the southern environments to be mosaics of open grasslands and forests. This reconstruction is corroborated by sedimentological and paleobotanical evidence and by the evidence from other fossil vertebrates (Bondesio et al. 1980). Friasian temperatures were probably lower relative to the preceding Santacrucian Land Mammal Age, corresponding to the worldwide trend toward lower temperatures at high latitudes throughout the Neogene.

Chinchillidae have open-rooted teeth and are diverse and abundant throughout the Miocene in southern faunas (*Prolagostomus* and *Pliolagostomus*) but are completely absent from La Venta. Chinchillids today are generally temperate Andean rodents and are never known to have occurred in the tropics. Octodontids, which similarly have open-rooted teeth and today inhabit temperate upland areas (Contreras et al. 1987), occur in the Santacrucian and Friasian of Patagonia but are absent from La Venta.

Vucetich (1984) listed three genera of caviids for the Collón Curá Formation and an additional genus for the

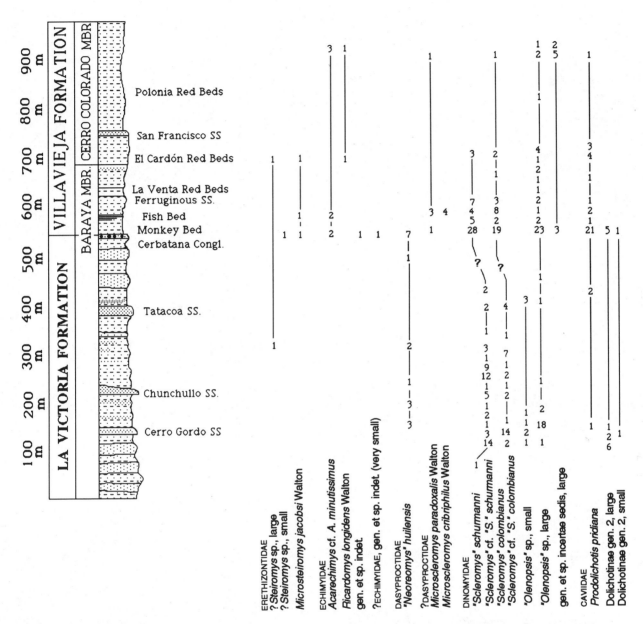

Figure 24.9. Preliminary faunal list and species ranges of rodents in the La Venta section. Numbers represent minimum number of individuals per stratigraphic level. Abbreviations: *SS.*, sandstone beds; *congl.*, conglomerate beds; *mbr.*, member.

Río Frías Formation in Chubut. The La Venta fauna has only two caviid genera, the more primitive of them structurally intermediate between the Eocardiidae (known from the Deseadan to Friasian of Patagonia) and the Caviidae (known from the Chasicoan to Recent of Argentina). Both La Venta genera are allocated to the caviid subfamily Dolichotinae. Modern dolichotines only live in the far

south, from the puna and chaco (arid and semiarid grassland and scrub) of southern Bolivia and Paraguay, to the arid steppes of Patagonia.

A high-crowned relative of *Neoreomys* (*Alloiomys* Vucetich) and a dasyproct with extremely high-crowned, rootless molars occur in the Collón Curá Formation of Patagonia. No high-crowned equivalent occurs in the La

Group	Taxon	Recent, 15°N–10°S	Recent, 10°S–25°S	Recent, >25°S	Pleistocene, Ecuador–Colombia–N. Brazil	Pleistocene, S. Peru–Bolivia–S. Brazil	Uquian–Ensanadan-Lujanian	Pliocene, Colombia–Venezuela	Montehermosan, >25°S	Huayquerian, Colombia–Venezuela	Río Acre, Brazil–Peru	Huayquerian, >25°S	Girón, Ecuador	Chasicoan, >25°S	La Venta–Carmen de Apicalá, Colombia	Collón Curá Fm., Río Negro	Río Frías Fm., Chile, Chubut	Santacrucian, >25°S	Colhuehuapian, >25°S	Salla-Luribay, Bolivia	Deseadan, >25°S
Cavioidea, non-"dasyprocts"	Heteromyidae	*																			
	Geomyidae	*																			
	Sciuridae, Sciurinae	*	*	+	*	*															
	Cricetidae:																				
	Sigmodontinae, large (>250 g)	*	*	+		*	+														
	Sigmodontinae, small (<250 g)	*	*	+	*	*	+		+												
	Hydrochoeridae	*	*	+	*	*	+		+	*	*	+		+							
	Caviinae	*	*	+		*	+		+	*	*	+		+							
	Cardiomyinae								+	*	*	+		*		*	?	?			
	Dolichotinae			+			+		+			+		+		*	?	?			
	Eocardiidae															+	+	+	+	?	+
Cavioidea, "dasyprocts"[e]	Cuniculinae	*	*	+		*															
	Dasyproctinae	*	*	+		*															
	Dinomyidae, large (>1.5 kg)	*	*		?	*		*	+	*	*	+		*		*	+				
	Dinomyidae, small (<1.5 kg)										*										
	Tetralophodont, sigmoidal[g]										*					+	+	+		*	+
	Pentalophodont[f]										*					+	+	+	+	*	
Chinchilloidea	Abrocomidae[d]			+		+			+		+										
	Lagostomiinae			+		+			+			+		+		+	+	+			
	Chinchillinae			+		*	+														
	Neoepiblemidae, large (>1 kg)				*				?	*	*	+		*							
	Neoepiblemidae, small (<1 kg)																	+	+		+
	Cephalomyidae																		+	*	+
Octodontoidea	Myocastoridae		*	+		*	+		+												
	Echimyidae, small[c]		*	+		*									*	+	+	+	?		
	Echimyidae, large[c]	*	*		*				?		*	+		*	*	+	+		?		
	Ctenomyidae		*	+		*	+		+												
	Octodontidae		*	+		*	+					+		+		*					
	"Acaremyidae": Octodontidae[b]																+	+		*	+
	"Acaremyidae": Echimyidae[b]															+		+		*	+
Erethizontoidea[a]	Erethizontidea, small										*				*						
	Erethizontidea, large	*	*			*						+			*	+	+		+		+

Figure 24.10. Stratigraphic ranges of South American rodents in the tropics *(rows with stars)* and the temperate zone south of 25°S *(rows with plus signs)*.

Sources: Data on modern ranges from Mares and Ojeda (1989), Eisenberg (1984), and Nowak (1991). Data for southern South America principally from Bondesio (1986), supplemented by Vucetich (1984) and personal observation. The Friasian of the Collón Curá Formation and Río Frías Formation (in Chile and Chubut) are separated following the recommendations of Marshall and Salinas (1990). Data for tropical faunas from Bondesio and Bocquentin Villanueva (1988), Frailey (1986), Hartenberger et al. (1984), Marshall et al. (1984), Pascual and Díaz de Gamero (1969), Linares (1990), Patterson and Wood (1982), Paula Couto (1975), Paula Couto (1978), Stirton (1953b), and personal observation.

Notes: Question marks indicate uncertain occurrences. *Fm.,* formation.

[a]Subdivision into "small" and "large" was determined by visually evaluating size relative to *Coendou*, the smallest living erethizontid. The Friasian rodent *Disteiromys graciloides* (Ameghino), referred with query by Vucetich (1984) to the Erethizontidae, is here considered an echimyid. The rare genus *Chaetomys*, considered by Patterson and Wood (1982, 394–95) a monotypic subfamily of the Echimyidae, is here ascribed to the Erethizontidae as suggested by Landry (1990).

[b]The term "Acaremyidae" is here used informally in the sense of Spencer (1987) for octodontoids that retain the primitive features of normal placement of p4/P4 and possession of a mental foramen. This group is further broken down into "Echimyidae" and "Octodontidae" in keeping with the identifications of Wood and Patterson (1959) and Patterson and Wood (1982).

[c]Subdivision into "small" and "large" echimyids was determined visually and is not precise (dividing point about 250 g). "Large" echimyids are rat-sized; "small" are between the size of a rat and a hamster.

[d]The Abrocomidae are here considered primitive chinchilloids, having derived characters in common with Lagostomiinae (Glanz and ANderson 1990).

[e]"Dasyproct" is an informal term here used to subdivide the Cavioidea. Included are living dasyproctids, the paca *Cuniculus,* and fossil forms traditionally referred to the Dasyproctidae and Dinomyidae.

[f]The informal designation "pentalophodont dasyprocts" arbitrarily includes all forms with five transverse crests on upper teeth; *Branisamys* (assigned to the Dinomyidae by Patterson and Wood 1982), Dasyproctidae gen. et sp. indet. A and B of Vucetich (1984), and the *Neoreomys* group (*Neoreomys, Alloiomys, Megastus, "Neoreomys" huilensis*).

[g]"Dasyprocts" with "tetralophodont, sigmoidal" teeth include *Incamys, Scleromys,* and *Microscleromys.*

←——————

Table 24.1. Simpson's coefficient of faunal similarity on the generic and family level for rodents from the La Venta fauna and the Miocene reference faunas of Argentina

	Friasian, Patagonia	La Venta, Colombia	Chasicoan, Pampas	Huayquerian, Andes
Santacrucian, Patagonia (N = 18, 8)	46.2/75.0	15.4/50.0	0/67.1	0/71.4
Friasian, Patagonia (N = 13, 8)		15.4/62.5	0/83.3	0/85.7
La Venta, Colombia (N = 13, 5)			0/60.0	0/60.0
Chasicoan, Pampas (N = 17, 6)				47.1/100.0
Huayquerian, Andes (N = 26, 7)				

Sources: Range data for Santacrucian and Friasian rodents from Vucetich (1984) and personal observation; range data for Chasicoan and Huayquerian faunas from Bondesio (1986).

Notes: Simpson's coefficient of faunal similarity = 100 × C/N, where C = the number of taxa (on any taxonomic level) shared in common between two faunas, and N = the total number of taxa at the same level in the smaller sample. "Friasian" here refers to the composite "Friasian" faunas of Chile and Argentina, collectively. Total numbers of genera and families in each fauna are indicated in parentheses in the first column.

Venta fauna. The Patagonian ?dinomyid *Eusigmomys,* with evergrowing molars, is not present at La Venta. In contrast, the dinomyid *Simplimus* from the Río Frías Formation of Chubut has high-crowned but rooted teeth and closely resembles the largest dinomyids from the La Venta fauna.

In contrast, mesodont taxa interpreted as forest-dwelling (figs. 24.2, 24.3) are common in the La Venta fauna but are relatively rare in the Friasian of Patagonia. Vucetich (1984) lists one genus (*Disteiromys*) tentatively referred to the Erethizontidae from the Río Frías Formation of Chubut. A *Steiromys*-size erethizontid also occurs in the Río Frías Formation of Chile (M. G. Vucetich and P. Salinas, pers. comm.). At La Venta there are at least three erethizontid species, one of them very small. Erethizontids are not known in southern faunas after the Huayquerian. Echimyids presumed to be forest forms disappear from the southern faunas at the same time. Where the Friasian has one or possibly two genera of Echimyidae, the La Venta fauna has at least three. Low-crowned forms in general are rare in the Friasian. The La Venta genus *Microscleromys* Walton (fig. 24.2) is smaller and more primitive (in terms of cusps and crown height) than any lophate rodent known from Patagonia, including the Deseadan genus *Incamys,* one of the oldest known caviomorph rodents.

Paleoenvironmental Implications

The latitudinal distribution of Recent South American rodents is represented in the upper lines of figure 24.10.

The closest living relatives of the La Venta rodents are generally forest dwellers.

There are at least three species of Erethizontidae. Of the four genera of living erethizontids, *Sphiggurus* and *Coendou* are arboreal, predominantly folivorous, with a preference for multistratal tropical evergreen forest (Eisenberg 1989); *Echinoprocta* is arboreal; and *Erethizon* is widespread and inhabits a wide variety of environments, including temperate North America, but is semiarboreal and a capable climber (Woods 1984; Nowak 1991).

There are at least three genera with three species of Echimyidae. Echimyids are small, so their true diversity as fossils probably has been underestimated. Modern echimyids are diverse, predominantly tropical, forest dwelling, and many are strongly arboreal (Eisenberg 1989).

There are at least two genera with a minimum of five species of Dinomyidae at La Venta. Although it is difficult to imagine the extinct giant dinomyids climbing trees, the living pacarana *Dinomys branickii* is a forest dweller and a capable climber with a preference for sleeping in high places (Merritt 1984; Nowak 1991). Modern *Dinomys* is mainly confined to the tropical eastern slopes of the Andes and in Colombia to the eastern slope of the Cordillera Oriental (Eisenberg 1989). The postcrania of "*Scleromys*" and "*Olenopsis*" from La Venta show a more cursorial condition than in modern *Dinomys* (e.g., longer limbs) but less so than in the living forest-dwelling dasyproctids *Dasyprocta* and *Myoprocta* (Fields 1957).

Microscleromys Walton is only tentatively assigned to the

Dasyproctidae. *"Neoreomys" huilensis* Fields has much in common with *Neoreomys australis,* one of the most abundant mammals in the Santacrucian fauna of Patagonia, and both are referred to the Dasyproctidae sensu lato. Living Dasyproctidae are terrestrial; the acouchy *Myoprocta* is confined to Amazonian multistratal tropical forest, and the agouti *Dasyprocta* is a less-restricted tropical and subtropical forest dweller (Smythe 1978; Eisenberg 1989; Nowak 1991). *Dasyprocta* is an important element in tropical forest ecosystems because it scatter-hoards fruits, providing an essential means of seed dispersal for some neotropical trees (Smythe 1970). The paca *Cuniculus,* perhaps more directly related to the extinct dasyprocts than *Dasyprocta* or *Myoprocta,* is primarily a frugivore and occupies a variety of forest habitats (Redford and Eisenberg 1992). *Cuniculus* inhabits riparian gallery forests on the llanos of Colombia and Venezuela today (Collett 1981).

The only rodents at La Venta with ever-growing cheek teeth, suggestive of grazing habits and savanna habitats, are the Caviidae (fig. 24.7). The two genera present in the La Venta fauna are more common than the Echimyidae or the Erethizontidae, in part because of sampling bias, but do not approach the ubiquitous Dinomyidae in number of individuals or stratigraphic distribution (fig. 24.9).

The living Dolichotinae occur exclusively in the Southern Cone. *Dolichotis patagonum* occurs in the scrub and grassland habitats of southern Bolivia, Paraguay, and Argentina (Redford and Eisenberg 1992). *Dolichotis (Pediolagus) salinicola* inhabits the puna (arid Andean plateau) and more vegetated chaco scrubland (Nowak 1991). The living Dolichotinae are cursorial open-country forms with long forelimbs and highly specialized locomotion. The body form of the La Venta caviid *Prodolichotis pridiana,* however, has more in common with *Cavia,* the domestic guinea pig, and the generalized Santacrucian caviid *Eocardia* than with cursorial *Dolichotis* (Fields 1957). *Cavia aperea* is the only caviid rodent now inhabiting northern South America (Eisenberg 1989). *Cavia* is a domestic animal, so the scattered and isolated northern colonies may have been imported by humans or may be the result of interbreeding by wild and feral populations (Ojasti 1964). Ojasti (1964) reports *Cavia* occupying tall grass and bushes on hillsides in Venezuela.

The La Venta caviids are more advanced than the Eocar-

diidae known from the Friasian of Patagonia (as discussed earlier). In addition, some specimens of *Prodolichotis pridiana* have molars with deep, asymmetrical sulci that suggest a relationship with *Cardiomys,* not known from the La Venta fauna but present in the late Miocene deposits of the Nabón basin of Ecuador and Chasicoan to Chapadmalalan of Argentina.

The La Venta caviids contrast strongly with the rest of the rodent fauna: (1) they are the rodents most probably adapted to open conditions; (2) they are more advanced than Friasian caviids from Patagonia; (3) they are the only La Venta rodents whose closest modern relatives no longer inhabit the tropical region. These observations suggest that the La Venta paleoenvironment may have been, at least in part, an ecological mosaic more extreme than suspected to the extent that grasslands or scrub (the result of high altitude, soil type, or some other unknown local effect) were in sufficiently close proximity to tropical rain forest that fossiliferous sediments from both areas could sometimes be deposited at the same locality. An alternative explanation is that the La Venta caviids are not directly comparable to their living relatives and that the middle Miocene forms had more diverse or more general ecological requirements. Their present absence from the tropical region would then be ascribed to factors other than environmental change alone, such as competitive exclusion (dolichotines, for instance, could have been effectively displaced from open country in tropical South America by immigrant lagomorphs during the Great American Interchange) or an unknown degree of specialization in combination with the temporary or permanent dissappearance of crucial habitat.

Paleoenvironmental Change and Rodent Paleobiogeography

The paleoenvironmental history of the Cenozoic of Argentina evidenced by fossil vertebrates is generally one of increasing aridity and decreasing temperatures. The rainshadow effect caused by the uplifting Andes was in evidence as early as the Friasian (Vucetich 1984) with a reduction in mammalian diversity and the initiation of desertification (Pascual and Odreman Rivas 1971). The transition from the Friasian to the Chasicoan is accompanied by spatial changes in deposition (from Patagonia to the pampas), style of deposition (from volcaniclastics to loess and silt), climate (cooler and drier), and fauna (an increased diversity of caviids and hypselodont dinomyids). Forests existed in the subtropical Andean foothills during the Huayquerian, evidenced by fossil logs and the last record of an erethizontid in southern faunas (Pascual and Odreman Rivas 1971), while grasslands occupied the pampas. The Huayquerian and Montehermosan (late Miocene and Pliocene) were the heyday of giant rodents in South America; large size and cursorial adaptations evolved in the families Dinomyidae, Neoepiblemidae, and Hydrochoeridae.

The scarcity of fossils from the middle and equatorial latitudes of South America does not permit a secure assessment of latitudinal habitat variation in the Miocene, but the few fossils recovered suggest intriguing biogeographical possibilities. Today, only four caviomorph genera occur in both tropical South America and temperate Argentina (*Cavia, Hydrochoerus, Dasyprocta,* and *Cuniculus*); none of these are shared by both the equatorial tropics and Patagonia. Although Miocene climatic differences between the north and the south of the continent were probably not as severe as they are now, the La Venta fauna documents strong middle Miocene endemism with only one rodent genus (*Acarechimys*) shared in common by the La Venta fauna and the Santacrucian and Friasian faunas of then-subtropical Patagonia. During the Chasicoan and Huayquerian, however, four genera appear in scattered localities in northern South America that also occur in Argentina. The high level of coincidence, especially given the poor material, suggests lower endemism during the late Miocene than during the time of La Venta deposition and possibly lower than exists today.

Rodent evidence for late Miocene faunal exchange between temperate and tropical South America is as follows: (1) a new, relatively small (*D. salinicola*-size) species of *Cardiomys* occurs in the Nabón basin of Ecuador bracketed by dates indicating correlation with temperate faunas of Chasicoan age (Madden et al. 1989). *Cardiomys* occurs in the Chasicoan to Chapadmalalan of Argentina (Bondesio 1986). (2) The dinomyid *Gyriabrus royoi* was recovered from the Caribbean coastal plain of Colombia (Stirton 1947a) from sediments of probable late Miocene age (Porta 1961; Hoffstetter 1971). *Gyriabrus* is known from the Chasicoan to Montehermosan of Argentina (Bondesio 1986). (3) The

better-known Urumaco Formation of Venezuela produces several fossil rodents, including the giant neoepiblemid *Phoberomys,* representing a Huayquerian age (Bondesio and Bocquentin Villanueva 1988). *Phoberomys* (=*Dabbenea*) in Argentina comes from Mesopotamian sediments, including the Huayquerian and Montehermosan ages (Scillato-Yané 1980). *Phoberomys* also occurs in the Río Acre area of Brasil (Paula Couto 1978). (4) *Eumegamys,* a large dinomyid, similarly occurs in both the Urumaco Formation and the Huayquerian and Montehermosan of Argentina (Pascual and Díaz de Gamero 1969). (5) Linares (1990) reports the presence of Hydrochoeridae in the Urumaco Formation, not known from Argentina before the Chasicoan. All the rodents shared by these isolated fossil occurrences in the tropics and Argentina are grazing forms with extremely high-crowned or open-rooted teeth. With the exception of *Cardiomys* sp., all are large (*Dasyprocta*-size or larger) or giant.

Pascual and Odreman Rivas (1971) and Pascual and Ortiz Jaureguizar (1990) report the faunas in subtropical and temperate zones in Argentina intergraded relatively smoothly during the Huayquerian. The tropical Río Acre fauna of Brazil and Peru, with its many similarities to the subtropical Mesopotamian faunas of Argentina and Uruguay, corroborates the conclusion of low endemism during the late Miocene relative to now.

It has been proposed that during the glacial maxima of the Pleistocene an open-country corridor existed along the eastern foothills of the Andes, permitting the disperal of savanna forms (large edentates and large ungulates) from the llanos of the Orinoco basin south to the Llanos de Mojos, chaco, and pampas of Argentina (Webb 1978; Marshall 1979a; Rancy 1990). A similar ecological corridor possibly existed for large grazing rodents during the late Miocene (in contrast with the middle Miocene endemism indicated by the La Venta fauna). Subsequently, climatic fluctuation during the Quaternary intensified, dry periods alternated frequently with warm periods, and the first desert vegetation appears in northern South America during the Pliocene (Solbrig 1976). The environmental barriers of rain forest (to savanna forms) and low temperatures and aridity (to tropical forms) prevent extensive north-south intermigration today and possibly did so during the middle Miocene.

Summary

The La Venta rodent fauna is dominated by medium to large Dinomyidae although screenwashing and intensified surface collecting have revealed an important diversity of small rodents. In contrast with the dinomyids, the caviids of the La Venta section are hypselodont grazing forms and are less common and diverse. The only living relatives of the La Venta Dolichotinae now inhabit arid regions of the Southern Cone, suggesting dolichotines were previously widespread but have been displaced from the tropics, perhaps by habitat loss or competitive exclusion.

Even though small caviomorph rodents are fairly common, sigmodontines (Cricetidae) have not been recovered from the Honda Group. No elements of North American origin have yet been found from the La Venta fauna. Given the extent of collecting and screenwashing by Duke-INGEOMINAS expeditions, the absence is probably real, not artifactual.

The La Venta fauna has little in common with the faunas of Patagonia. The La Venta rodent fauna has marginally greater similarity to Friasian faunas than to those of any other South American land mammal age. The La Venta caviids are more advanced than those of the Friasian although other rodents are not.

There was little interchange of terrestrial fauna between northern and southern South America during the time of La Venta deposition as is the case now. During the late Miocene, however, there is some evidence for the dispersal of large grazing rodents between northern Argentina and the equatorial tropics. The late Miocene was generally a period of cool global climates and savanna expansion on other continents.

Acknowledgments

Field work in Colombia was partially supported by grants to the author from Institute for the Study of Earth and Man, the Geological Society of America, and the Society of Sigma Xi. Partial support was also provided by the National Science Foundation (NSF BSR 8614533/8614133) to R. F. Kay and J. J. Flynn and BSR 89-18657 to R. F. Kay. A year of museum and field work in Argentina, Chile, and Colombia was supported by a Fulbright Fellowship and a supplementary fellowship from the Organization of American States. A collections study grant was provided by the Field Museum of Natural History. Much of the matrix from the La Venta section was washed and picked at the Oklahoma Museum of Natural History, University of Oklahoma, under the supervision of R. Cifelli and N. Czaplewski. The Hocol Petroleum Company graciously allowed us to use their grounds for screenwashing.

I thank L. Jacobs, R. F. Kay, and R. H. Madden for their help with the initiation of this project. I am especially grateful to M. G. Vucetich of the Museo de La Plata who gave freely of her time, expertise, and insight. C. Villarroel of the Universidad Nacional de Colombia also offered invaluable advice and discussion, as did J. Guerrero, B. Jacobs, A. Winkler, and many others. Thoughtful reviews of the manuscript were provided by N. Czaplewski, L. Flynn, E. Lindsay, J. Flynn, and R. Madden.

25
Chiroptera

Nicholas J. Czaplewski

RESUMEN

En contraste con la diversidad extraordinaria de los mur-
ciélagos (Chiroptera) actuales en América del Sur, se con-
oce prácticamente nada de su historía en el archivo fósil
antes del Pleistoceno. Con el desarrollo de las técnicas del
zarandeo o tamizado de sedimentos y la prospección in-
tensiva de la superficie en afloramientos de sedimentos
fluviales del Grupo Honda, se han descubierto nuevos re-
stos fósiles de murciélagos que representan once taxa de
cinco familias distintas. Estos incluyen un Emballo-
nuridae; un Noctilionidae; los Phyllostomidae *Notonycteris*
magdalenensis, Tonatia **sp., Phyllostominae indet., y Glos-**
sophaginae indet.; dos Thyropteridae, *Thyroptera robusta*
sp. nov. y *T.* **cf.** *T. tricolor;* **y tres Molossidae,** *Mormopterus*
(Neomops) colombiensis **sp. nov.,** *Eumops* **sp., y** *Potamops mas-*
cahehenes **gen. et sp. nov. Los registros en el Mioceno medio**
de un Thyropteridae y un Noctilionidae asignados tenta-
tivamente a especies actuales indican un período largo de
evolución incambiable para sus familias que concuerda
con sus falta de diversidad actual.

A excepción de los Phyllostomidae de hábitos carnívoro,
insectívoro-omnívoro, y nectívoro, todos los murciélagos
del Grupo Honda eran insectívoros en sentido estricto. Es
notable la falta de los frugívoros, sanguínivoros, y
piscívoros actuales tan comúnes en comunidades Neo-
tropicales actuales, pero esta ausencia es probablemente
una consecuencia de la poca abundancia de especímenes
en estos depósitos. En base a las exigencias de la ecología y
comportamiento de taxa estrechamente relacionados o la
analogía morfológica, los murciélagos del Grupo Honda
en general sugieren y corroboran otras evidencias de com-
unidades de vegetación boscosa. Sin embargo, los mur-
ciélagos no sirven para refutar la existencia de sabanas o
zonas abiertas dentro del bosque tal como ha sido sugerido
por otras evidencias.

The fossil record of bats is generally recognized as poor, especially for the Tertiary period. In no continent except Antarctica is this record poorer than in South America. At this writing, Tertiary bats were known there from just two species represented by a total of four specimens. The oldest of these is a molossid, *Mormopterus faustoi* (Paula Couto 1956a), from the Oligocene of southeastern Brazil (Paula Couto 1983; Legendre 1984c). The other species is a large phyllostomid, *Notonycteris magdalenensis,* from Miocene sediments of the Honda Group in the Neiva Basin, Huila Department, Colombia (Savage 1951a).

Recently, another bat has been formally recognized in the La Venta fauna; I believe, however, the specimens involved do not represent a chiropteran. Takai et al. (1988) described two molars screenwashed from sediments in the Villavieja Formation, Honda Group of Colombia, as those of an unidentified chiropteran. More recently Takai et al. (1991) named these specimens *Kiotomops lopezi* and tentatively referred them to the Molossidae. In my view, the teeth probably represent an unusual ?didelphoid marsupial or other mammal whose relationships are enigmatic. The molars bear well-developed stylar cusps in the A (parastyle), B (stylocone), and D (no homologue in Chiroptera) positions (Clemens and Lillegraven 1986; Cifelli 1993b) as in many Marsupialia. Unlike the upper molars of molossids, *Kiotomops* teeth lack a hooked parastyle strongly connected to the preparacrista, bear a swelling extending labially from the protocone into the trigon basin, and have protocones that are posteriorly recumbent rather than anteriorly recumbent, possibly implying a different kind of jaw action from that of molossids.

The poor fossil record of bats in South America contrasts strongly with the present-day richness of these mammals on that continent, especially in tropical areas. Bat diversity ordinarily dominates modern communities of vertebrates in lowland tropical forests in South America, suggesting a long and complex evolutionary history for bats there. The purpose of this chapter is to provide descriptions of fossil bats, including new taxa, based on newly collected specimens originating from Honda Group sediments in the La Venta area, Tatacoa Desert, upper Magdalena River valley northeast of the pueblo of Villavieja, Huila Department, Colombia. Preliminary notes concerning the relationships of the La Venta taxa are given, necessarily based mostly on dental features, with little possibility for understanding even dental variation because of small sample sizes. Nevertheless, these notes provide a first approximation that may serve as an aid for better future work as more specimens and more complete knowledge and analysis of all South American bats, living and extinct, become available. In addition, preliminary conclusions are drawn about the contribution of bats toward the reconstruction of the paleoenvironment of the middle Miocene lowland region that is now the valley between eastern and central cordilleras of the northern Andes Mountains. Although the fossil record is still far from complete, it is now clear that a number of families of Chiroptera were present in South America before the middle Miocene.

Diverse vertebrate fossils originating from Honda Group sediments in the Tatacoa Desert constitute the La Venta fauna. The La Venta fauna is significant as one of the few Miocene faunas from the tropical portion of South America. It has usually been assigned to the Friasian Land Mammal Age (Fields 1957, 1959; Marshall, Hoffstetter, et al. 1983), although Kay et al. (1987) suggested that it possibly is more similar to Santacrucian than Friasian assemblages of Chile and Argentina. Recently, new radiometric and fission-track dates have been provided for the Honda

Group in this area. These dates suggest an age range of approximately 11.0–13.8 Ma for these sediments (Guerrero 1990; also chaps. 2, 3). Several of the richest mammal-producing localities, including some yielding bat remains, occur in the Monkey Beds of the lower part of the Villavieja Formation, where they date from approximately 12.9 Ma (chap. 2).

We have used primarily the screenwashing method to recover small vertebrate fossils from the Honda Group. All La Ventan chiropteran specimens (except University of California Museum of Paleontology specimens; Savage 1951a) recovered to date originated from three localities (Duke Localities 22, 90, and the Fish Bed) in the badlands of the Tatacoa Desert, Huila Department, Colombia. Because of the fragmentary or isolated nature of most specimens recovered by screening, it is not yet possible to identify all chiropteran teeth to family. Therefore, a few of the isolated teeth that have been found, including small premolars, canines, and lower molars, are not described. Many of the bat teeth from Locality 22 exhibit erosion and loss of enamel similar to that of teeth that have undergone digestion within the gut of a predator (Andrews 1990). Specimens discussed in this chapter are in the collections of the Museo Geológico, Instituto Nacional de Investigaciones en Geosciencias, Minería y Química (IGM), Bogotá, Colombia, and the University of California Museum of Paleontology, Berkeley (UCMP). Casts have been retained at the Oklahoma Museum of Natural History and Duke University. Dental terminology used in this chapter is primarily taken from Szalay (1969) and Legendre (1984b).

Systematic Paleontology

Order Chiroptera
Suborder Microchiroptera
Family Emballonuridae
Diclidurus Wied 1820
Diclidurus, sp. indet.
(fig. 25.1; table 25.1)

REFERRED SPECIMEN

IGM 251048, associated right M^2 (broken) and M^3.

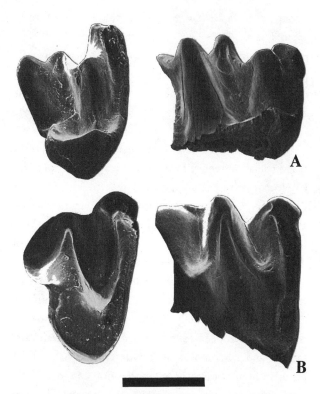

Figure 25.1. Teeth of Emballonuridae from the Villavieja Formation, Colombia. *Diclidurus,* sp. indet., IGM 251048, associated right M²–M³: *A,* lingual views (inverted); *B,* occlusal views. Scale bar 1 mm.

LOCALITY

San Nicolás, Duke Locality 22.

STRATIGRAPHIC POSITION

Honda Group, Villavieja Formation, Monkey Beds.

DESCRIPTION

The posterolingual portion of the M² is broken away, but otherwise both teeth are intact. The teeth represent a fairly large emballonurid approximately the size of *Diclidurus albus.*

The protocone of M² is far forward, level with the position of the paracone and very near to that cusp. The short preprotocrista is continuous with a broad precingulum. A lingual cingulum is absent (at least anterior to the protocone; the tooth is broken immediately posterior to the protocone). The parastyle is large and hooklike. The meso-

style is very short in its anteroposterior dimension. The postmetacrista is straight with no development of a cusp at the position of the metastyle. The paraconule crista is very weak and short, existing only as a sharp keel on the lingual side of the base of the paracone. The metaconule crista also is indistinct, occurring as a very low rounded ridge extending from the mesial side of the base of the metacone to the protocone.

The preprotocrista of M³ is continuous with a broad precingulum. No lingual cingulum is present. The parastyle is large and hooklike. The metacone is taller than the paracone and leans posteriad. A postmetacrista is absent. A weak paraconule crista is developed as in M², but a metaconule crista is absent.

DISCUSSION

The two molars are clearly assignable to Emballonuridae. They differ from the Old World genera *Taphozous* and *Saccolaimus* in having less-reduced M³ and broad precingula on upper molars. They share with all other extant genera of emballonurids the possession of M³ with backward-leaning metacone that is equal in height or taller than the paracone. In addition, the fossil molars match those of many New World species of Emballonuridae in the position and weakness of the conular cristae. The paraconule crista usually forms a short keel on the lingual base of the paracone that barely extends onto the floor of the trigon basin. The metaconule crista is usually absent on M³ (except in *Rhynchonycteris, Cormura, Balantiopteryx,* and some *Diclidurus*) but is present on M² (except in *Embablonura,* some *Taphozous, Coleura, Saccolaimus,* and *Cormura*), in which it normally extends from the mesial side of the base of the metacone to the protocone.

The size of these teeth is consequential. Among extant New World Emballonuridae, only *Diclidurus ingens* exceeds them in size. The fossils equal the size of M² and M³ of *D. albus,* and they exceed *D. scutatus, D. (Depanycteris) isabellus,* and all other Neotropical genera. Because of their large size and morphology, they are assigned to *Diclidurus.*

In the western hemisphere Emballonuridae are known in the Tertiary only in the Oligocene and Miocene of North America (Florida, USA; Morgan 1989). This is the first documented Tertiary occurrence of the family in South America and the first fossil record for *Diclidurus.*

Table 25.1. Tooth measurements (mm) of Miocene bats from Honda Group sediments, Río Magdalena valley, Colombia

Taxon	IGM Number	Tooth	Upper Teeth Labial Length	Lingual Length	Transverse Width
EMBALLONURIDAE					
Diclidurus sp.	251048	Right M^2	1.80	—	(2.25)
	251048	Right M^3	1.43	0.93	2.17
NOCTILIONIDAE					
Noctilio albiventris	252961	Left P^4	1.37	—	2.22
	253006	Right M$^{1 \text{ or } 2}$	(2.10)	—	2.60
PHYLLOSTOMIDAE					
Tonatia sp.	250326	Right M^1	1.95	1.80	2.32
Glossophaginae, gen. et sp. indet.	252863	Left M$^{1 \text{ or } 2}$	—	1.22	1.30
THYROPTERIDAE					
Thyroptera robusta	252868	Right C^1	0.98		0.71
	252876	Right M^1	1.38	1.09	1.71
	252877	Right M^2	(1.48)	1.04	1.94
Thyroptera cf. *T. tricolor*	250327	Right C^1	0.77	—	0.60
	250327	Right P^4	0.94	—	0.94
	250327	Left P^4	0.92	—	0.87
	250327	Right M^1	1.17	0.92	1.45
	250327	Left M^1	1.22	1.05	1.35
	250327	Right M^2	1.23	0.80	1.62
	250327	Left M^2	1.21	—	—
	250327	Right M^3	0.95	0.57	1.41
MOLOSSIDAE					
Mormopterus colombiensis	250997	Left M^1	1.61	1.13	1.70
	250290	Right M^2	1.45	0.97	1.96
	250333	Right M^2	(1.45)	1.05	2.06
	250999	Left M^2	1.48	1.02	1.95
	251000	Right M^2	—	1.05	—
	250277	Left M$^{1 \text{ or } 2}$	—	1.05	—
	250334	Right M^3	1.12	0.68	1.85
Eumops sp.	184794	Left M$^{1 \text{ or } 2}$	(2.2)	1.74	2.91
Potamops mascahehenes	184349	Left P^4	2.00	1.47	3.12
	184349	Left M^1	3.27	2.73	4.04
	184349	Left M^2	3.13	2.47	4.12
	184349	Left M^3	1.92	1.33	3.39

Continued on next page

Note: Parentheses indicate estimated measurements of broken specimens.

Family Noctilionidae
Noctilio Linnaeus 1766
Noctilio albiventris Desmarest 1818
(fig. 25.2; table 25.1)

REFERRED SPECIMENS

IGM 252961, left P^4 missing enamel from labial side of main cusp; 253006, right M^1 or M^2; 252965, right M$_1$; 252964, right M$_1$ or M$_2$ fragment missing most of its enamel; 252967, left M$_1$ or M$_2$ talonid fragment (lingual half); 252968, left M$_1$ or M$_2$ talonid fragment, mostly enamelless; 253005, left M$_2$ in fragment of dentary; 252960, left M$_3$; 252962, right M$_3$, enamelless; 252963, right M$_3$ lacking anterior two-thirds of trigonid.

LOCALITY

San Nicolás, Duke Locality 22.

Table 25.1. Continued

Taxon	IGM Number	Tooth	Lower Teeth		
			Anteroposterior Length	Trigonid Width	Talonid Width
NOCTILIONIDAE					
Noctilio albiventris	252965	Right M$_1$	2.30	1.35	1.87
	253005	Left M$_2$	2.43	1.50	1.87
	252960	Left M$_3$	2.15	1.38	1.18
	252963	Right M$_3$	—	—	1.23
PHYLLOSTOMIDAE					
Notonycteris magdalenensis	252865	Left M$_2$	3.75	2.35	2.32
Phyllostominae incertae sedis	252869	Right M$_2$	2.83	1.72	1.78
THYROPTERIDAE					
Thyroptera robusta	252878	Left M$_{1 \text{ or } 2}$	—	—	0.85
Thyroptera cf. *T. tricolor*	250327	Right P$_{3 \text{ or } 4}$	0.91	0.51	—

STRATIGRAPHIC POSITION

Honda Group, Villavieja Formation, Monkey Beds.

DESCRIPTION

All these specimens are identical in qualitative morphological features with those of modern noctilionids. The lower molars exhibit myotodonty (Menu and Sigé 1971). To the exclusion of all other families of bats, they share with modern Noctilionidae a unique, presumably derived connection of the cristid obliqua with the metaconid (in most Microchiroptera and many other kinds of mammals with tribosphenic teeth, the cristid obliqua extends from the hypoconid to the middle of the posterior wall of the protocristid). As in extant noctilionids, the talonids of M$_1$ and M$_2$ are much wider than the trigonids. In modern *N. albiventris* the ratio of talonid width to trigonid width in M$_1$ averages about 1.30:1.00, in *N. leporinus,* 1.24:1.00. In the M$_1$ from the La Venta fauna, this ratio is 1.38:1.00. In size, all the fossil noctilionid teeth fit within the range of variation observed in modern *N. albiventris* from scattered population samples preserved in museum collections.

Figure 25.2. Teeth of *Noctilio albiventris* from the Villavieja Formation, Colombia. *A.* IGM 252961, left P^4, view occlusal and slightly posterior. *B.* IGM 253006, right M^1 or M^2, occlusal and lingual views. *C.* IGM 252960, left M$_3$, occlusal and lingual views. *D.* IGM 253005, left M$_2$, occlusal and lingual views. Scale bar 1 mm.

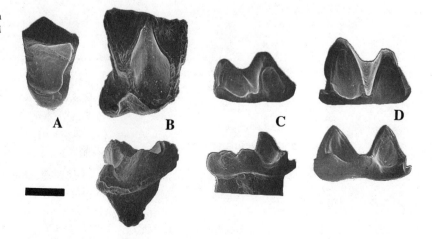

DISCUSSION

Only two species are known in the Noctilionidae, *N. leporinus* and *N. albiventris*. The derived character of the lower molars shared by these two species apparently has been overlooked by previous investigators. The occurrence of this unique character in the fossils clearly unites them with the Noctilionidae. Moreover, the fossils can be referred to *N. albiventris* because of their similar overall dental morphology and size. It follows that *N. albiventris* has been evolutionarily static since 12.5 Ma. Fossil occurrences of the "bulldog bats," Noctilionidae, previously were limited to the islands of Cuba and Puerto Rico where remains of *N. leporinus* have been found in cave deposits of Holocene (or less likely, late Pleistocene) age (Reynolds et al. 1953; Choate and Birney 1968; Silva Taboada 1979). The specimens from the La Venta fauna comprise the first certain pre-Holocene record for the Noctilionidae and the first fossil occurrence for *N. albiventris*.

Family **Phyllostomidae**
Subfamily **Phyllostominae** Gray 1838
Tonatia Gray 1827
Tonatia, sp. indet.
(fig. 25.3E; table 25.1)

REFERRED SPECIMEN

IGM 250326, right M^1.

LOCALITY

Duke Locality CVP 2–4, La Venta area.

STRATIGRAPHIC POSITION

Honda Group, Villavieja Formation, Fish Bed.

DESCRIPTION

Three main cusps (protocone, paracone, and metacone) are present plus a small hypocone. A large, rounded talon is present. The crests extending labially from paracone and metacone form a W-shaped ectoloph in which the anterior vee is smaller than the posterior one. The length of the preparacrista is about 60% of the length of the postparacrista (measured to the terminus of the postparacrista, not to the mesostyle) and about 50% of the length of the postmetacrista. The preparacrista curves posteriorly (rather than anteriorly as in most insectivorous bats) from the stylocone; there it forms a short labial cingulum or stylar ridge that partly closes the "valley" labial to the paracone. This stylar ridge and the stylocone are not confluent with the parastyle. The parastyle is represented by a small swelling on an exceedingly short precingulum anterior and dorsal to the preparacrista. The postparacrista is separated by a minute notch from the mesostyle; the mesostyle is positioned slightly posteriad, at the labial terminus of the premetacrista rather than at the junction of the postparacrista and premetacrista, and is slightly elongated posteriorly, forming a short stylar ridge. The mesostyle is a moderately prominent cusp. The apex of the protocone is closer to the base of the paracone than to that of the metacone. The short, curved preprotocrista connects the protocone with the base of the paracone and bears a swelling that might represent a paraconule. No paraconule cristae are present. The hypocone is situated directly on the nearly straight postprotocrista. The postcingulum along the posterior base of the tooth is barely discernible and appears to be continuous with the cingulum of the talon. The cingulum of the talon, in turn, is continuous with a crest extending posteriorly from the hypocone. The lingual walls of the hypocone and protocone are separated by a shallow depression.

DISCUSSION

In the characteristics listed this M^1 is most similar to certain species of the Phyllostominae. The morphology of the ectoloph is closest to that of *Tonatia*. The ectoloph of the M^1 of *Macrotus* has an even shorter preparacrista relative to the length of the postparacrista than in the fossil; also, the hypocone is absent. *Trachops,* more specialized for carnivory, shows greater discrepancy between the size of the anterior and posterior "vees" of the ectoloph, with a very short preparacrista that is about one-quarter the length of the postmetacrista. In *Trachops* the cingulum of the talon does not connect with the apex of the hypocone and the postprotocrista does not extend to the hypocone in the M^1.

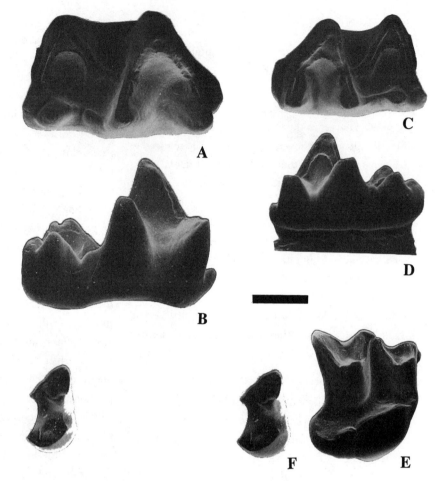

Figure 25.3. Teeth of Phyllostomidae from the Villavieja Formation, Colombia. *A–B. Notonycteris magdalenensis,* IGM 252865, left M_2: *A,* occlusal view; *B,* lingual view. *C–D.* Phyllostominae, unnamed new taxon, IGM 252869, right M_2: *C,* occlusal view; *D,* lingual view. *E. Tonatia,* sp. indet., IGM 250326, right M^1, occlusal view. *F.* Glossophaginae, gen. et sp. indet., IGM 252863, left M^1 or M^2, occlusal views, stereo pair. Scale bar 1 mm.

In *Micronycteris* M^1 is transversely narrowed (especially its ectoloph); there is no notch at the mesostyle and no posteriorly extended mesostylar ridge. In *Phyllostomus* the postcingulum is intermediate in its development. The M^1 of *Phyllostomus discolor* has a less-trenchant ectoloph, lacks a hypocone, and bears a premetaconule crista, unlike the fossil. In *Phyllostomus hastatus* and *P. elongatus* the parastyle is not strongly curved posteriorly (although it does join with a short labial cingulum), the postprotocrista is sigmoid-shaped and does not reach the hypocone, and there is a deep notch in the lingual margin of the tooth between hypocone and protocone. In *Phylloderma* the parastyle is large and continuous with the short stylar ridge extending posteriorly from the stylocone. In *Macrophyllum* the postparacrista and premetacrista are not separated by a notch and the mesostyle is not extended posteriorly. In general, the morphology of the fossil is closest to the M^1 of *Mimon* and *Tonatia;* in *Mimon,* however, the hypocone (absent in *M.*

crenulatum) is more lingually situated than in the fossil and the postprotocrista is strongly sigmoid rather than straight. Although direct comparison with *Notonycteris* (for which only lower teeth are known) is presently impossible, this M^1 is much smaller than would be expected for *Notonycteris*. Therefore, among relevant genera of Phyllostomidae, the fossil is most similar to *Tonatia* in all morphological characteristics and in its proportions. It is slightly smaller in size than *T. silvicola*.

Although *Tonatia* species are known as fossils in the late Quaternary (Paula Couto 1946; Koopman and Williams 1951; Czaplewski 1990), this is the first record of the genus in the Tertiary and only the third certain record of a phyllostomid from the Tertiary of the New World. Other than the forms in the La Venta fauna, the only record is that of the vampire *Desmodus archaeodaptes* in the late Pliocene (very early Irvingtonian) in Florida, USA (Morgan et al. 1988).

Notonycteris magdalenensis Savage 1951
(fig. 25.3A, B; table 25.1)

REFERRED SPECIMENS

IGM 251047, lower canine fragment; 252865, left M_2.

LOCALITY

San Nicolás, Duke Locality 22; UCMP Localities 4517 and 4536.

STRATIGRAPHIC POSITION

Honda Group, Villavieja Formation, Monkey Beds.

DESCRIPTION

The first specimen is a fragment of the crown of a right lower canine lacking the cingulum except for a tiny smooth area representing the mesial area of the cingulum dorsal to its point of contact with an outer incisor. The crown shape is that of a large and curved hemicone with a flat posterior face. A rounded ridge emphasizes the posterolabial corner of the crown. A second ridge ascends the mesial portion of the rounded anterior face above the postulated point of contact with a lower incisor. The height of the crown, measured from its apex to the mesial portion of the cingulum dorsal to its point of contact with a lower distal incisor is 3.33 mm.

The left M_2 is almost identical in size and shape to the M_2 in the left lower jaw fragment of UCMP 39963, the paratype of Notonycteris magdalenensis. The enamel is broken away along the labial base of the tooth, and the apex of the hypoconid is slightly damaged. The only noteworthy difference between the two specimens is that in UCMP 39963 the entocristid forms a continuous sharp ridge reaching the posterior base of the metaconid, whereas in IGM 252865 the same ridge is interrupted by a notch. This difference probably represents minor individual variation.

DISCUSSION

As in the other known teeth, the C_1 is characteristic of that of other large phyllostomines. Notonycteris magdalenensis must have been a relatively common member of the La Venta fauna; these teeth represent the fourth and fifth known individuals from San Nicolás (parts of three individuals were described by Savage [1951a], and the new specimens described here came from screened matrix samples collected one year apart, which presumably represent different individuals) in the Monkey Beds.

Phyllostominae, gen. et sp. indet.
(fig. 25.3C, D; table 25.1)

SPECIMEN

IGM 252869, right M_2 in a fragment of the dentary bone.

LOCALITY

San Nicolás, Duke Locality 22.

STRATIGRAPHIC POSITION

Honda Group, Villavieja Formation, Monkey Beds.

DESCRIPTION

This molar represents a large phyllostomid. It is about three-fourths the size of the M_2 in Notonycteris magdalenensis and slightly smaller than in Chrotopterus auritus; its size is approximately equal to M_2 in Phyllostomus elongatus. The tooth is in a moderate stage of wear. Enamel has been lost all along the labial base. The paraconid is well developed and nearly as tall as the metaconid; the paracristid is relatively high. The talonid is slightly wider than the trigonid, and the talonid basin is deep. The cristid obliqua meets the trigonid below the notch in the protocristid. The postcristid is continuous with the hypoconulid, which, in turn, occurs as a curved ridge separated from the entoconid by a groove. A small interdental contact facet is present on the posterior of the tooth beneath the hypoconulid. The entoconid is large and distinct and is connected with the metaconid by a relatively high entocristid. The talonid crests (cristid obliqua and postcristid) lack carnassial-like notches.

DISCUSSION

Except for its smaller size, the tooth is nearly identical in occlusal view to the M_2 of *N. magdalenensis* and *Chrotopterus auritus*. The interdental contact facet on the rear of the tooth and the presence of an unreduced talonid indicate that its locus is M_1 or M_2, eliminating the possibility that it is the M_3 of *N. magdalenensis* or some other phyllostomid. Height of the paracristid and paraconid relative to that of the metaconid further indicates the tooth is M_2, because M_1 of phyllostomines typically has a very low paraconid and paracristid.

Relative to M_2s of generally comparable phyllostomid genera, the fossil M_2 has the following morphological differences: (1) The trigonid is more open, that is, the angle formed by the paracristid and postcristid is less acute than in *Mimon* and is more acute than in *Lonchorhina, Micronycteris, Phylloderma,* and *Tonatia*. (2) The paraconid is nearly equal in height to the metaconid, as in most genera, but relatively larger and higher than in *Trachops*. (3) The cristid obliqua meets nearer the trigonid below the notch in the protocristid, much more labially than in *Macrotus* (in which it meets nearer the metaconid) and slightly more lingually than in *Phyllostomus discolor* and *Tonatia bidens* (but not other *Phyllostomus* or *Tonatia* spp.). (4) The cristid obliqua and postcristid lack carnassial-like notches, unlike the talonid crests in *Phylloderma, Tonatia, Lonchorhina,* and *Micronycteris*. (5) The entoconid is much lower relative to the hypoconulid than in *Macrophyllum*. (6) The talonid is not reduced in size relative to the trigonid as in *Vampyrum*. (7) The postcristid's connection to a large hypoconulid widely separated from the entoconid is in contrast with the configurations in certain other genera: in *Macrotus* and *Micronycteris* the postcristid connects instead to the entoconid, whereas in *P. hastatus, P. elongatus,* and *Tonatia* (all species except *T. bidens*) the postcristid connects with the hypoconulid but the hypoconulid is weak and closely appressed to the entoconid. (8) Although cusp deployment and crest morphology are quite similar to *Chrotopterus auritus,* the height of the crown is relatively lower than in that species. The fossil is 87% of the length of the M_2 of *Chrotopterus*, but the height of its protoconid is only 73% of that of an equivalently worn M_2 of *Chrotopterus*. Relative height of the crown is similar to that of *Notonycteris*; length of IGM 252869 is approxi-

mately 76% and protoconid height is 74%, respectively, of equivalently worn teeth of *Notonycteris*. Cusp deployment and crest morphology are almost identical to those of *Notonycteris magdalenensis*.

Thus IGM 252869 is most like *Notonycteris* and (somewhat less similar to) *Chrotopterus*, yet it can be shown to be distinct from other known genera of Phyllostomidae; perhaps it represents a new species of *Notonycteris*. Based on its phenetic resemblance, it seems most likely to be related to *Notonycteris*. Despite its distinctness, a single M_2 is insufficient to serve as the basis for describing a new taxon, so it is left unnamed until more and better material is discovered.

Subfamily Glossophaginae Gray 1838
Gen. et sp. indet.
(fig. 25.3F; table 25.1)

REFERRED SPECIMEN

IGM 252863, left M^1 or M^2 lacking paracone.

LOCALITY

San Nicolás, Duke Locality 22.

STRATIGRAPHIC POSITION

Honda Group, Villavieja Formation, Monkey Beds.

DESCRIPTION

This molar represents a small phyllostomid probably about the size of one of the larger extant glossophagines. Although the anterolabial portion of the molar crown including the paracone is broken away, the paracone was probably well formed based on the preserved anterior wall of the trigon basin. Therefore, the tooth probably had a W-shaped ectoloph. The tooth is wider transversely than in most glossophagines. The crown height, protocone, and metacone all are reduced relative to phyllostomids such as *Macrotus, Micronycteris,* and *Tonatia,* but these features are developed equivalently to those of modern glossophagines (e.g., as in *Glossophaga*). A tiny angle occurs in the postprotocrista at the posterolingual corner of the tooth, but a true hypocone and talon are absent. The mesostyle is much reduced; its position is shifted posteriorly onto the end of the premetacrista where it occurs as a slight swelling.

DISCUSSION

This tooth seems clearly to represent a member of the Glossophaginae as specialized dentally as several of the more conservative living genera. The combination of characteristics listed earlier are similar to those in *Lonchophylla, Glossophaga, Anoura,* and *Monophyllus,* but some differences in the details of morphology exist. In the fossil the cusps and stylar shelf are more reduced than in *Lonchophylla,* and the hypoconal basin seen in *Lonchophylla* is lacking. Compared to *Anoura,* the fossil lacks a shallow hypoconal basin, the mesostyle is not a prominent cone, and the protocone is not a ridgelike cusp. In contrast to the molars of *Glossophaga* and *Monophyllus,* the mesostyle is not formed by equal contributions from the anterior and posterior elements of the ectoloph (postparacrista and premetacrista), and the protocone is more than a low ridgelike cusp. In addition, the fossil lacks the unique hypoconal wing or "buttress" of *Monophyllus* upper molars.

Despite these differences it should be emphasized that intraspecific variation in the dentition of glossophagines is great and, moreover, there appears to be extensive convergent evolution in the dental characteristics among the different genera (Phillips 1971). Thus, it would be premature to name a new taxon or to base phylogenetic conclusions on one incomplete tooth. Nevertheless, the tooth remains a significant discovery in that it represents the first Tertiary record for the subfamily (or tribe Glossophagini of Baker et al. 1989) and documents the existence of a relatively specialized "nectar bat" as early as 12.5 Ma.

Family Thyropteridae
Thyroptera Spix 1823
Thyroptera robusta sp. nov.
(fig. 25.4; table 25.1)

TYPE

IGM 252876, right M^1.

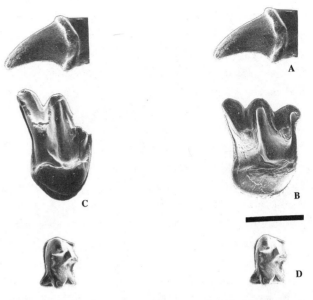

Figure 25.4. Teeth of *Thyroptera robusta* sp. nov. from the Villavieja Formation, Colombia. *A.* IGM 252868, right C^1, lingual view, stereo pair. *B.* IGM 252876, holotype, right M^1, occlusal view. *C.* IGM 252877, right M^2, occlusal view. *D.* IGM 252878, left M_1 or M_2 talonid, occlusal view, stereo pair. Scale bar 1 mm.

HYPODIGM

Type specimen; IGM 252868, right C^1; 252877, right M^2; and 252878, talonid of left M_1 or M_2.

LOCALITY

San Nicolás, Duke Locality 22.

STRATIGRAPHIC POSITION

Honda Group, Villavieja Formation, Monkey Beds.

DIAGNOSIS

Thyroptera with large teeth, exceeding the teeth of *T. tricolor* (and the La Venta fossils herein referred to *T.* cf. *T. tricolor*) in its measurements by about 4–32%. The teeth of *T. robusta* exceed those of *T. discifera* in various dimensions by 21–50%.

DESCRIPTION

The upper canine has a broad lingual cingulum that bears weak anterior and medial cuspules, both of which are slightly damaged (missing enamel at their apices). No posterolingual cuspule is developed. In these characteristics the canine is intermediate between those of *T. discifera* and *T. tricolor*. *Thyroptera discifera* bears three (anterior, medial, and posterior) cuspules on the lingual cingulum, all of which are moderately well developed. *Thyroptera tricolor* has upper canines with no lingual cingular cusps.

Enamel seems to be missing from parts of the M^1 specimen. The tooth is similar in essential aspects of its morphology to the same element in extant *Thyroptera* spp., but it is distinctly larger. The mesostyle appears particularly robust. The postparaconule and premetaconule cristae are relatively prominent.

The available specimen of M^2 is broken and, as a result, lacks its anterolabial corner. In addition, the enamel is damaged on the apices of the metacone and hypoconal ridge. Except for its larger size, this molar is nearly identical to the M^2 of *T. tricolor* and *T. discifera*.

A lower molar, either M_1 or M_2, is represented by a talonid fragment that shows the morphology unique to the Thyropteridae, Furipteridae, and Natalidae. The cristid obliqua and postcristid both are notched near the hypoconid, and the entocristid is notched near the entoconid. The postcristid extends to the entoconid leaving the hypoconulid isolated and situated well below the level of the postcristid (i.e., the tooth exhibits "myotodonty"; Menu and Sigé 1971). As in upper teeth included in the hypodigm, this specimen appears relatively larger and more robust than in *T. tricolor* and *T. discifera*.

DISCUSSION

It is difficult to interpret the phylogenetic significance of *Thyroptera robusta* when the species is based on such meager material. The intermediate morphology of its upper canine suggests that it might be ancestral to one or both of the two living species of Thyropteridae. The close correspondence in dental morphology among the three species of *Thyroptera* indicates conservatism or stasis in their dental evolution. The existence of Thyropteridae in the middle Miocene indicates they had already differentiated from related vespertilionoids (especially Furipteridae and Natalidae) by then. This is the first fossil record for the Thyropteridae.

Thyroptera cf. *T. tricolor* Spix 1823
(fig. 25.5; table 25.1)

REFERRED MATERIAL

IGM 250327, series of isolated but associated teeth, probably from one skull.

LOCALITY

San Nicolás, Duke Locality 22.

STRATIGRAPHIC POSITION

Honda Group, Villavieja Formation, Monkey Beds.

DESCRIPTION

This bat is represented by nine teeth, eight uppers and one lower. These teeth were dissociated by screenwashing but are similar in preservation and degree of wear and probably represent parts of the maxillary toothrows and a lower jaw of an individual skull. They include the right C^1, both P^4, both M^1, both M^2 (left M^2 broken), right M^3, and right P_3 or P_4.

The morphology of the upper canine (C^1) is simple and differs little from the condition seen in *Stehlinia* or in modern *T. tricolor*. The outline of the cingulum, seen in occlusal view, is ovoid. It forms a wide shelf lingually but is very narrow and even discontinuous labially. No cuspules or other projections occur on the cingulum. The cusp is tall with smooth sides and is straight except for a very slight posteriad curve at the apex. In cross section the cusp is teardrop-shaped, rounded anteriorly and having flattened lingual and posterolabial faces that meet along a posterior vertical ridge.

In occlusal view, the cingulum of P^4 is concave anteriorly, convex lingually, and nearly straight along posterior and labial sides. The cingulum slopes dorsally from the anterolabial corner to the posterolabial corner. Anterolingually, the cingulum is developed into a cuspule; from the apex of this cuspule the cingulum sweeps upward posteriad around a rounded talon. The main cone is tall with a convex anterior face. No anterolabial crest or lingual crest is

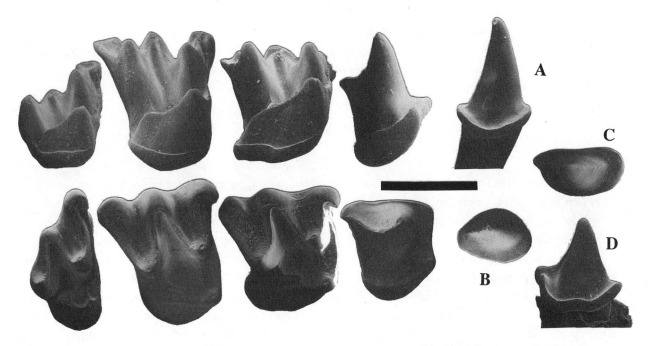

Figure 25.5. *Thyroptera* cf. *T. tricolor* associated teeth, IGM 250327, from the Villavieja Formation, Colombia. *A.* Reconstructed partial right upper tooth row including C^1 and P^4–M^3, inverted lingual views. *B.* Same, except that M^1 shown is reversed image of left M^1, occlusal views. *C–D.* Right P_3 or P_4: *C,* occlusal view; *D,* labial view. Scale bar 1 mm.

present. A well-developed, recurved crest extends posteriad from the main cone to a "metastyle" at the posterolabial corner of the cingulum. The face of the tooth labial to this crest is concave; the lingual face is flat.

The first two upper molars are basically similar to each other except that the talon is better developed in M^1, and M^2 is wider than M^1. Both teeth bear a distinct, sharp, sigmoid premetaconule crista extending from the base of the metacone to the postprotocrista. A weak postparaconule crista is present on M^1 only at the base of the paracone, but a stronger one is present in M^2 that does not quite reach the protocone. Conules and hypocones are lacking in both teeth. A lingual cingulum reaches around the base of the protocone and talon to the metastyle as a postcingulum. Precingula are absent. The lingual border of M^2 is evenly rounded; that of M^1 is slightly indented along the cingulum between the base of the protocone and the talon. The preprotocrista extends to the parastyle. The postprotocrista extends distally as a simple ridge swollen at its base, but it lacks identifiable cusps and is larger in M^1 than in M^2.

The other upper molar specimen (M^3) is unreduced rela-

tive to that of the most primitive known bats, that is, the ectoloph consists of three commissures, the preparacrista, postparacrista, and premetacrista; the metacone is well developed. The trigon basin is open posteriorly through a deep notch between the metacone and the postprotocrista. A weak, very short premetaconule crista branches off the postprotocrista. A distinct postparaconule crista extends from the protocone to the paracone. Conules and hypocone are absent. A lingual cingulum is present between the metacone and the base of the protocone, but it disappears on the anterior flank of the protocone. The preprotocrista extends from protocone to reach the parastyle.

A right lower premolar (P_3 or P_4) is preserved in a fragment of the dentary. The outline of the cingulum in occlusal view is rather D-shaped with a flat lingual side and a posterior projection. The cingulum completely encircles the tooth and forms a slightly broader shelf anteriorly and posteriorly than labially and lingually. The cingulum varies in elevation around the tooth with its lowest point along the posterolabial margin anterior to an interdental contact facet. The main cusp is a little higher than long and is hemiconical with the nearly flat lingual wall separated from the convex labial wall by anterior and posterior cristids. No strong posterior vertical concavity is present along the posterior cristid as in lower premolars of living thyropterids. The posterior cristid extends down onto the posterior projection of the cingulum. The tip of the main cusp tilts backward slightly.

DISCUSSION

These teeth almost certainly belonged to a single skull. They were recovered from a very small amount of matrix at a single locality, show identical preservation and wear, and there are no redundant elements.

The fossils were compared with teeth of the two extant species of the Thyropteridae, *T. tricolor* and *T. discifera* (and several other families as well). In size the fossils are similar to teeth of *T. tricolor,* smaller than those of *T. robusta,* and larger than those of *T. discifera.* In details of dental morphology the modern species are very similar to one another and exhibit a fair amount of geographic variability. The main distinguishing feature other than size between the extant species is in the shape of the upper canine. In *T. tricolor* C¹ lacks cingular cusps, whereas in *T. discifera* C¹ bears antero-, medio-, and posterolingual cusps on the cingulum. The upper canine of *T. robusta* is intermediate. The upper canine of the fossils at hand clearly resembles *T. tricolor* and differs from *T. discifera* and *T. robusta* in this feature.

These fossils have P⁴ with the anterolingual cingular cusp relatively smaller and less labially positioned than in *T. discifera.* Furthermore, in the fossils the molars bear slightly stronger premetaconule cristae and slightly weaker postparaconule cristae than in *T. tricolor* and *T. discifera.* Talons on M¹ and M² are less well developed than in *T. tricolor* but about the same as in *T. discifera.*

These differences seem to be relatively minor and may not exceed the expected limits of individual variation. Based on present evidence, the teeth correspond to *T. tricolor.* Although with more and better specimens it may eventually be determined that these teeth represent an undescribed species of *Thyroptera,* I cannot find sufficient reason to name a new species based on the fossils, despite the fact that they are separated by some 12.5 million years from their living relatives. These specimens indicate a significant period of evolutionary stasis for disk-winged bats consistent with their present lack of diversity. Moreover, these Miocene teeth are little derived with respect to those of their putative ancestor *Stehlinia* of the upper Eocene and Oligocene of Europe (Sigé 1974) and possibly North America (Ostrander 1985).

Family Molossidae

The first Tertiary molossid to be discovered in South America, named *Tadarida faustoi* by Paula Couto (1956a) was based on a partial skeleton from ?lacustrine sediments of the Tremembé-Taubaté basin, São Paulo state, Brazil. Although they were originally considered to be Pleistocene in age, these beds are now thought to reflect the Oligocene epoch (Paula Couto and Mezzalira 1971; Paula Couto 1983), and the associated fauna, still poorly known, might represent the Deseadan Land Mammal Age (Soria and Alvarenga 1987; Madden 1990). Legendre (1984c) transferred *Tadarida faustoi,* known only from the holotype (Museo Nacional de Rio de Janeiro 3000-V), to the genus *Mormopterus* Peters 1865 and placed it in a new subgenus, *Neomops,* defined strictly on dental characteristics and including only the single extinct species (Legendre 1984b, 1984c, 1985; Legendre et al. 1988).

Mormopterus Peters 1865
Mormopterus colombiensis sp. nov.
(fig. 25.6 A–F; table 25.1)

TYPE

IGM 250997, right M¹.

HYPODIGM

Type specimen and IGM 250290, 250333, 250999, 251000, M²s; 250334, M³ (mesostyle broken); 250277, lingual half of M¹ or M².

LOCALITY

All specimens originated from Duke Localities CVP 5, 9, 10, and 13C in the Fish Bed, northeast of Villavieja, Magdalena River valley, Huila Department, Colombia.

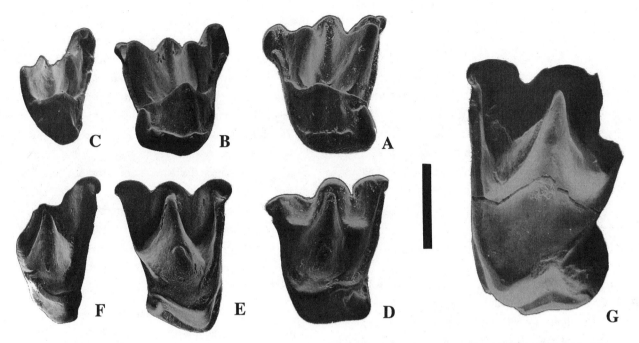

Figure 25.6. Teeth of Molossidae from the Honda Group, Colombia. *A–F. Mormopterus colombiensis* sp. nov. *A.* IGM 250997, holotype, right M¹, lingual view (inverted). *B.* IGM 250290, right M², lingual view (inverted). *C.* IGM 250334, right M³, lingual view (inverted). *D.* Same as *A*, occlusal view. *E.* Same as *B*, occlusal view. *F.* Same as *C*, occlusal view. *G. Eumops* sp. indet., IGM 184794, left M¹ or M², occlusal view. Scale bar 1 mm.

STRATIGRAPHIC POSITION

Honda Group, Villavieja Formation, Fish Bed.

DIAGNOSIS

M¹ and M² with hypocone tall, conical, and isolated from the postprotocrista; with trigon basin (=protofossa) that is closed posteriorly by the postprotocrista; with strong lingual cingulum; and with prominent paraconule and metaconule cristae that connect like keels to the bases of the paracone and metacone, respectively. M³ with strong paraconule crista and rudimentary hypocone.

DESCRIPTION

The first two upper molars are basically similar to one another except in their proportions. In occlusal view M¹ is narrower transversely than M². In M¹ the pre- and postprotocrista meet and form a right (or slightly obtuse) angle; in M² the same angle is acute. In both these teeth the hypocone is tall (although its apex does not reach the level of the postprotocrista), vertically oriented, and conical to subcylindrical. The distinctness of the hypocone is variable; in the M¹ and two specimens of M² it is isolated and in the other two M²s it is partly fused to the posterolingual wall of the tooth above (dorsal to) the postprotocrista. Even in the last two specimens the hypocone has a distinct apex, but after slight wear it would probably appear to be indistinct and completely fused with the preprotocrista. Talons are absent. The preprotocrista is continuous with the precingulum. The postprotocrista is continuous with the postcingulum. The paraconule and metaconule cristae are always present as sharp ridges at the bases of the paracone and metacone, respectively, that curve toward but do not reach the protocone or protocristae. There is no tendency for the metaconule crista to pass toward the trigon basin side of the metacone as in the Old World forms of *Mormopterus* (Legendre 1985). Instead, it descends the lingual base of the metacone like a keel. The trigon basin is closed posteriorly by the postprotocrista and metaconule crista. The labial margins of M¹ and M² are deeply notched between the parastyle and mesostyle and between the mesostyle and metastyle. Prominent lingual cingula surround the bases of protocones and extend to but not around the bases of the hypocones.

In M^3, the metacone is large, nearly as tall as the paracone. The premetacrista is equal in length to the postparacrista; the postmetacrista is lacking. A strong paraconule crista occurs, but a metaconule is represented only by a slight swelling on the postprotocrista. A short cingulum is present on the anterolingual side of the protocone. A tiny rudimentary hypocone occurs distal to the protocone.

DISCUSSION AND COMPARISONS

Legendre (1985) characterized the genus *Mormopterus* as having lower molars with the condition of myotodonty or submyotodonty, a tall coronoid process, and the general presence of a paraloph (paraconule crista) and metaloph (metaconule crista) enclosing the protofossa (trigon basin). Inasmuch as the first two of these features are as yet unknown in *M. colombiensis,* the species' membership in *Mormopterus* is tentative and is based on the presence of the third feature and on the similarity of its upper molars to those of other species, especially *M. faustoi.*

Mormopterus colombiensis differs from the Miocene Eurasian species *M. stehlini, M. helveticus,* and *M. nonghenensis* (all placed in the subgenus *Hydromops* by Legendre [1984c; Legendre et al. 1988]), and from all extant species of *Mormopterus* (including *Sauromys* and *Platymops*) in having M^1 and M^2 with vertically oriented, subcylindrical hypocone. In this feature it is unlike all other Molossidae except *M. faustoi.*

Legendre (1984c) erected a new subgenus *Neomops* for *M. faustoi,* diagnosed by its (1) lower molars showing myotodonty; (2) M^1 and M^2 with hypocone tall, conical, and isolated; and (3) M^1 and M^2 lacking a metaloph but having a closed protofossa. *Mormopterus colombiensis* clearly shares with *M. faustoi* its distinctively shaped hypocone, but it differs from *M. faustoi* (and resembles other *Mormopterus* spp.) in having M^1 and M^2 with paraconule crista, metaconule crista, and prominent lingual cingula. Molars of *M. colombiensis* are about three-fourths the size of those of *M. faustoi.*

The strong paraconule crista on M^3 in *M. colombiensis*

occurs also in *M. stehlini, M. helveticus,* and *M. nonghenensis,* but the lingual portion of the M^3 is unknown for *M. faustoi.*

In at least one specimen of *M. colombiensis* (IGM 250277) the conule cristae are better developed and longer than in *M. faustoi* or any other genus of Molossidae except *Nyctinomops.* The unusual configuration of the conule cristae approaches but does not equal the unique derived configuration seen in *Nyctinomops* in which the long, prominent conule cristae actually meet one another before continuing as a single ridge reaching the protocone. Thus, the fragment might reflect a convergent morphology or an intermediate stage in dental evolution between *Mormopterus* and *Nyctinomops,* or it may indicate the presence of an unrecognized, additional taxon in the La Venta fauna.

Eumops Miller 1906
Eumops, sp. indet.
(fig. 25.6G; table 25.1)

REFERRED SPECIMEN

IGM 184794, broken left M^1 or M^2.

LOCALITY

Duke Locality CVP 9, Fish Bed.

STRATIGRAPHIC POSITION

Honda Group, Villavieja Formation, Fish Bed.

DESCRIPTION

This molar is damaged. The metastyle and postmetacrista are broken away, and a crack runs anteroposteriorly across the entire tooth. Enamel is broken away from small areas of the preparacrista, parastyle, mesostyle, hypocone, and adjacent postprotocrista, and the lingual cingulum directly above (dorsal to) the protocone.

In size this molar is similar to upper first and second molars of *Eumops glaucinus.* In occlusal view the protocone is situated anteriorly, level with the paracone. The preprotocrista is continuous with the precingulum and the postprotocrista is continuous with the postcingulum. Conules and conule cristae are absent. The hypocone is hemiconical, occurs on a medium-sized (for a molossid) rounded talon, and is fused to the flat posterior wall of the trigon where its summit is adjacent to the postprotocrista. A broad shelflike lingual cingulum extends completely around the lingual base of the tooth except where it is interrupted shortly at the base of the hypocone.

DISCUSSION

This tooth is referred to *Eumops* based on its relatively large size and the combination of characteristics mentioned above, especially in the absence of conule cristae and the configuration of the talon and hypocone.

Legendre (1984b) placed *Eumops* in a new subfamily, Molossinae, partially diagnosed as having upper molars with weak or absent hypocone with a talon very little developed. Although these characteristics occur in some genera he placed in the subfamily, they do not describe *Eumops*. In most species of *Eumops* examined by me (including *E. auripendulus, E. bonariensis, E. dabbenei, E. glaucinus, E. hansae, E. perotis, E. trumbulli,* and *E. underwoodi*) the hypocone is strong, in others moderate. In no species is it absent in both M¹ and M² (see also Miller 1907) but it may be merged to the postprotocrista in M² in *E. auripendulus* and *E. hansae*. The talon is moderate to large by comparison with all extant and extinct genera of molossids. This is the first Tertiary record for the genus in South America.

Potamops gen. nov.

TYPE SPECIES

Potamops mascahehenes sp. nov.

ETYMOLOGY

Potamos, Greek, river; *mops,* a commonly used root word for molossid genera; *mascar,* Spanish, to chew; *jejenes,* Spanish, insects (more specifically, biting gnats).

DIAGNOSIS

A large molossid, similar in the size of its teeth to *Eumops perotis* (smaller than in *E. dabbenei;* larger than in *Cheiromeles*

parvidens but smaller than in *C. torquatus,* the largest living molossid bat), and unique in that its upper first and second molars lack a hypocone and lingual cingulum but have a postprotocrista that is continuous with the cingulum of a small talon, which, in turn, is continuous with the postcingulum.

Potamops mascahehenes sp. nov.
(figs. 25.7, 25.8; table 25.1)

TYPE

IGM 184349, skull fragment including left maxilla with P⁴–M³, anterior root of zygomatic arch, anterior margin of orbit, and portions of the palate, plus associated radius fragment.

HYPODIGM

Known only from the type specimen.

LOCALITY

Duke Locality 90, northeast of Villavieja, Magdalena River valley, Huila Department, Colombia.

STRATIGRAPHIC POSITION

Honda Group, La Victoria Formation, about 60 m above the Chunchullo Sạndstone Beds.

DIAGNOSIS

As for the genus.

MEASUREMENTS

Length of the cheek-tooth row from P⁴ to M³ is 9.3 mm.; individual tooth measurements are provided in table 25.1. Transverse diameter of the infraorbital foramen is 1.04 mm. Transverse diameter of the posterior orifice of the infraorbital canal is 1.32 mm. Diameters of the shaft of the radius bone, measured at the level of the groove for the abductor pollicis longus muscle, are 2.4 mm (dorsoventral) and 2.9 mm (anteroposterior).

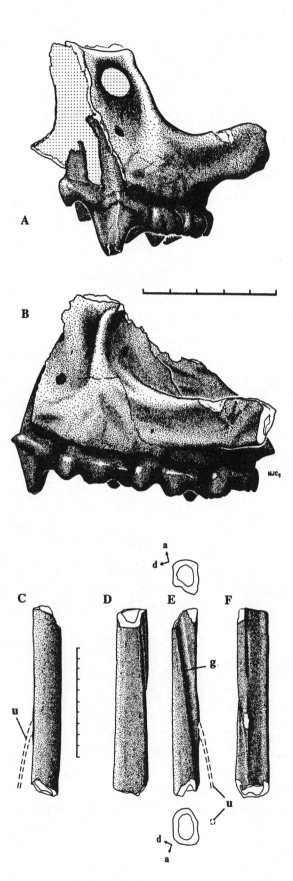

A

B

C **D** **E** **F**

u

g

u

a
d

d
a

Figure 25.7. *Potamops mascahehenes* gen. et sp. nov., IGM 184349, holotype, fossils from La Victoria Formation of Colombia. *A–B.* Left maxilla: *A,* anterior view; *B,* lateral view. *C–F.* Fragment of right radius (proximal end at bottom): *C,* anteroventral view; *D,* anterodorsal view; *E,* posterodorsal view and cross sections; *F,* posteroventral view. *u,* restored ulna; *g,* groove for abductor pollicis longus muscle; *a,* anterior; *d,* dorsal. Scale bar for *A–B,* 5 mm. Scale bar for *C–F,* 10 mm.

←

DESCRIPTION

The skull fragment (fig. 25.7) includes the left maxilla with P⁴–M³, the anterior portion of the zygomatic arch, and the anterior margin of the orbit. The lacrimal ridge is strongly developed. The infraorbital foramen opens anteriorly within a groove formed between the side of the rostrum and the lacrimal ridge so that, in lateral view, the infraorbi-

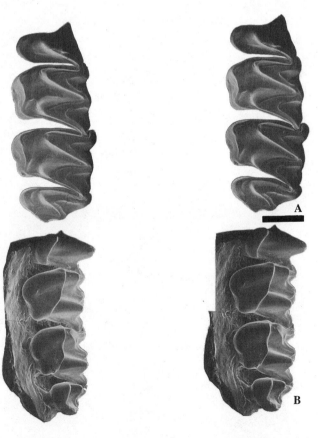

A

B

Figure 25.8. *Potamops mascahehenes* gen. et sp. nov., IGM 184349, composite stereo pairs of partial left maxillary tooth row: *A,* occlusal views; *B,* lingual views. Scale bar 2 mm.

tal foramen is barely visible. The posterior opening of the infraorbital canal is broken but appears to have been large (low but wide). The lacrimal foramen opens just inside the orbit at the dorsal end of the lacrimal ridge. The anterior edge of the orbit is even with the mesostyle of M^1. The zygomatic arch arises above the posterolabial corner of the toothrow, dorsal to the point of contact between the metastyle of M^2 and parastyle of M^3. From the remnant of the palatal branch of the maxilla it appears that the palate was moderately arched (from side to side) but not domed (from front to back).

The teeth (fig. 25.8) are in a moderate stage of wear. The vees formed between shearing crests of the maxillary teeth (interlophs and intralophs in the dental-functional terminology of Freeman 1984) are more acute than in other large molossids. Interlophs and intralophs are approximately equal in size.

In occlusal view, the posterior upper premolar is wider than long, and its margins are straight posteriorly, nearly straight (slightly wavy) labially, concave anteriorly, and convex (semicircular) lingually. A cingulum encircles the base of the entire tooth except at the posterolabial corner. The anterolabial corner of the tooth is pointed in occlusal view and lacks a cingular cusp. A moderate cingulum cusp is present at the base of the main cusp on the anterior edge of the lingual heel.

The first and second upper molars are wider than long and are basically similar to one another in size and shape, except that M^2 is wider than M^1. The protocone is situated far forward, near the paracone. A metaconule is absent on M^2 but on M^1 the metaconule appears to be represented as a low rounded ridge on the postprotocrista, extending toward but not reaching the metacone. Hypocones and lingual cingula are absent. On each of M^1 and M^2 a small rounded talon is present that is separated from the base of the protocone by a broad, shallow lingual groove. The postprotocrista, cingulum of the talon (or postmetaconule crista?), and postcingulum form a continuous sinuous ridge from protocone to metastyle; therefore, the postprotocrista does not separate the trigon basin from the talon. A precingulum also is present and is continuous with the preprotocrista. Labial cingula are exceedingly weak and appear on the stylar shelf of M^1 and M^2 only as a short ridge anterior to the metastyle. On M^2 the paracone of this indi-

vidual might have been broken in life; it is worn much lower than all other cusps.

The crown area of M^3 is slightly less than one-half that of M^1 or M^2. Protocone and paracone are large; metacone is small. The premetacrista is one-half the length of the preparacrista. Postmetacrista, hypocone, lingual cingulum, and conules are absent. Small wear facets on the postprotocrista and the lingual face of the metacone indicate the probable occurrences, respectively, of a strong postcristid and a hypoconulid on M_3.

The fragment of radius represents the middle portion of the shaft of that bone. The shaft is angular in cross section (fig. 25.7). The groove for the tendon of the abductor pollicis longus muscle is relatively deep and narrow. The ulna was very slender but probably ossified as evidenced by the presence of a small swelling on the posteroventral surface of the radius (adjacent to the groove for the *M. abductor pollicis longus*) that represents the point of fusion of these two bones.

DISCUSSION AND COMPARISONS

Potamops mascahehenes differs from most known Vespertilionoidea in the absence of lingual cingula on its upper molars and from all known Vespertilionoidea except certain Molossidae and Philisidae in its great size. Although in its size (and a few other features) it resembles *Philisis sphingis*, *Potamops* differs from the extinct Philisidae (represented by *P. sphingis* and the much smaller *Dizzya exsultans*; Sigé 1985, 1991) in the same ways that *Philisis* differs from many Molossidae (Sigé 1985).

Although no features considered to be diagnostic of the Family Molossidae are preserved in the maxilla and radius fragments, *Potamops mascahehenes* is tentatively referred to the Molossidae because of its large size and because the ulna was presumably "less reduced than in the Vespertilionidae, [its] very slender shaft usually about half as long as [the] radius" (as in Molossidae; Miller 1907). Miller (1907) further noted that the "posterior orifice of [the] antorbital canal [is] not enlarged" in Molossidae; I find this structure to be enlarged, however, in a number of extant species including *Nyctinomops femorosacca*, *N. macrotis*, *Molossus ater*, *Otomops martiensseni*, and *Chaerephon ansorgei*. Thus, the occurrence of an enlarged posterior opening of the infraorbital foramen in *Potamops* is not inconsistent with membership in the Molossidae.

A complete character analysis, taking into account individual variation of the details of tooth morphology and certain osteological features in all Molossidae would be necessary to estimate the phylogenetic relationships of *P. mascahehenes*. Such an analysis is beyond the scope of this chapter. Some preliminary comparisons, however, were made between *Potamops* and one or more species from all genera of Molossidae.

The only hints about postcranial anatomy of *Potamops* are provided by the fragment of radius. The shaft of this bone is angular in cross section as in molossids that are adapted for high-speed flight (Vaughan and Bateman 1980). The groove for traverse of the abductor pollicis longus muscle is deep and narrow as in *Molossus ater* and *Nyctinomops macrotis*, rather than shallow and wide as in *Eumops perotis* and *Otomops martiensseni*. The condition of the ulna, not actually preserved but assumably ossified and relatively stout in *Potamops*, is shared with *Molossus*, *Tadarida*, and *Nyctinomops*, but not with *Eumops* and *Otomops*.

Few osteological features of the cranium are available for assessment in the skull fragment of *Potamops*, and those that are probably are correlated. The lacrimal ridge in *Potamops* is strong, as it also is in *Eumops*, *Myopterus*, *Tadarida*, *Molossops*, some *Mops*, some *Mormopterus*, and some *Chaerephon*, but not in *Cheiromeles*, *Nyctinomops*, *Otomops*, *Promops*, *Molossus*, some *Mops*, some *Mormopterus*, and some *Chaerephon*. In *Potamops* the infraorbital foramen opens forward within a groove formed between the lacrimal ridge and the side of the rostrum. This condition is shared with *Myopterus*, *Molossops*, and *Mormopterus* but differs from that in *Eumops*, *Cheiromeles*, *Nyctinomops*, *Tadarida*, *Otomops*, *Chaerephon*, *Promops*, and *Molossus*, in which the infraorbital foramen opens more or less laterad.

Preliminary comparisons of dental traits between *P. mascahehenes* and extant molossid genera have not allowed me to discern its relationship within the family. It has not been possible even to discern whether the species fits into any of the subfamilies (Tadarinae, Molossinae, Cheiromelinae) established on dental morphology by Legendre (1984b). The species, however, does not seem to be closely related to *Nyctinomops* and *Tadarida* based on dental features.

Potamops mascahehenes differs from the extinct *Cuvier-*

imops parisiensis (Legendre and Sigé 1982) and *Wallia scalopidens* (named by Storer [1984] as a member of the Proscalopidae and transferred to the Molossidae by Legendre [1985]) in its much greater size and lack of a hypocone on upper first and second molars.

Potamops mascahehenes differs from *Kiotomops lopezi* (recently named by Takai et al. [1991]) in the same manner by which all molossids differ from *K. lopezi*. That is, unlike the upper molars of *Kiotomops*, the upper molars of *P. mascahehenes* and other molossids lack stylar cusps, bear a hooked parastyle strongly connected to the preparacrista, lack a swelling extending labially from the protocone into the trigon basin, and have protocones that are canted anteriorly from base to apex rather than posteriorly.

The development of acutely angled, equal-length interloph and intraloph shearing crests among the maxillary teeth indicates that *Potamops mascahehenes* was insectivorous (Freeman 1984). The stratigraphic position of this large bat is 200 m lower than the lowest occurrence of *Notonycteris magdalenensis* (and other bats of the La Venta fauna described herein), making it the oldest bat yet known from Colombia.

Contribution of Bats to Paleoenvironmental Reconstruction

A great diversity of living bats occurs in tropical South America (Koopman 1982); 152 species are known in the modern fauna of Colombia, and another 20 species are likely to occur there (Cuervo et al. 1986). They include the members of all nine Neotropical families, Emballonuridae, Mormoopidae, Noctilionidae, Phyllostomidae, Thyropteridae, Natalidae, Furipteridae, Vespertilionidae, and Molossidae. Their ecological roles are diverse and their contributions to community structure substantial (Tamsitt 1967; McNab 1971; Wilson 1973; LaVal and Fitch 1977; Bonaccorso 1979; Bonaccorso and Humphrey 1984; Willig 1986; Fleming 1988; Muñoz 1990). What do the fossil bats of the La Venta fauna imply about the paleoenvironment of this part of tropical South America? Currently available data on modern bat assemblages from different plant communities are inadequate (Fleming 1986) although they generally suggest that bats are not life zone specialists but instead are eurytopic, tolerant of a range of habitats (Willig and Mares 1989). Moreover, the assemblage of bats is simply too small yet to allow meaningful detailed conclusions about paleocommunity structure in the La Venta fauna, let alone tropical South America in the middle Miocene (see also Mares and Willig 1994). Interpretations, however, of foraging flight style, trophic role, and foraging and roosting habitat selection of the fossil bats, based on the ecology of their closest living relatives or modern analogs, can provide a crude insight into the mid-Miocene paleo-environment.

The La Ventan emballonurid *Diclidurus* probably was an aerial pursuit insectivore, and like extant white bats it probably foraged above the forest canopy. Modern white bats (Diclidurinae) usually select moist habitats in multistratal evergreen forests for roosting and foraging, but they are tolerant of human-made clearings. They tend to roost in high places including the foliage of palms. Other Neotropical Emballonuridae vary widely in their use of natural day roosts, which occur on tree trunks and buttress cavities, in tree hollows, in and under fallen logs, and in limestone and coral caves (Bradbury and Vehrencamp 1976; Handley 1976; Linares 1986; Eisenberg 1989).

The noctilionid *Noctilio albiventris* is primarily insectivorous, feeding largely on aquatic insects swimming near the surface in rivers, streams, marshes, and lakes. It, however, sometimes feeds on fish and occasionally even on fruits of *Brosimum* and *Morus* (both in Moraceae) that might also be taken from the surface of the water where the fruits are floating. This bat is strongly associated with streamside vegetation in lowland mesic tropical forests but tolerates a wide variety of habitats. Lesser bulldog bats roost in hollow trees and sometimes in small caves (Howell and Burch 1974; Hood and Pitocchelli 1983; Linares 1986; Eisenberg 1989). Interestingly, one of the plant genera (*Brosimum*) whose fruits *N. albiventris* infrequently has been known to eat also is known as a fossil from the Miocene of Colombia.

Phyllostomidae are the most structurally varied bats. This structural variability is partly the result of an adaptive radiation into diverse feeding niches and foraging styles. The La Venta *Tonatia* was nearly the size of *T. silvicola* and *T. bidens,* which are large species of the genus. *Tonatia* are aerial foliage-gleaning insectivores that sometimes eat fruit (the larger species may rarely kill small vertebrates). A surface-gleaning foraging style in bats demands highly maneuverable flight. *Tonatia* species usually occupy mature multistratal evergreen forest but occasionally use deciduous forests or "matas" in the llanos. When they occur sympatrically in forests, different species forage at different levels from groundstory to top of canopy. *Tonatia* usually roost in tree hollows and sometimes in abandoned termite nests (Humphrey et al. 1983; Linares 1986; Belwood and Morris 1987; Eisenberg 1989).

Notonycteris was a phyllostomid slightly larger than *Chrotopterus auritus*. Postcranially, it was most similar to *Vampyrum spectrum*. Dentally, it bore specializations for carnivory similar to *Vampyrum* and *Chrotopterus* (Savage 1951a). These bats are aerial gleaning carnivores that occasionally eat insects and fruit, and they take prey animals from vertical surfaces or horizontal roosts. Like carnivores of many types, *Vampyrum* and *Chrotopterus* select a great variety of habitats ranging from deciduous and gallery forests to lowland humid forest and premontane forest. *Vampyrum spectrum* forages by flying low and slowly over deciduous and secondary growth vegetation, in broken woodland, and forest edge situations. Although *Chrotopterus* sometimes roosts in caves, both it and *Vampyrum* often roost in hollow tree trunks (Wilson 1973; Gardner 1977; Vehrencamp et al. 1977). Presumably, *Notonycteris* acted similarly. The relative abundance of *Notonycteris magdalenensis* in the La Venta fauna suggests the existence of a broad prey base of small vertebrates sufficient to support fair numbers of this tertiary consumer (see also Savage 1951a). The unnamed new phyllostomine described earlier also might have been similar to *Chrotopterus auritus*. Unlike *Chrotopterus,* however, its teeth were much lower crowned and, therefore, less specialized for carnivory. It probably was insectivorous / omnivorous.

The glossophagine bat was as specialized dentally as some other genera (*Lonchophylla, Glossophaga, Anoura,* and *Monophyllus*) of its subfamily and, perhaps, filled a similar ecological role. These modern Glossophaginae are best known for feeding on nectar, pollen, fruit, and insects. All can hover in flight to drink nectar, and some can also glean insects from foliage. The various species in these genera occupy a variety of habitats from tropical dry deciduous forest to multistratal tropical evergreen forest, and they also roost in a variety of sites in caves, rock crevices, tree hollows, and under dry tree roots (Linares 1986; Eisenberg 1989).

The family Thyropteridae includes just two living species restricted to Neotropica. They are aerial pursuit insectivores probably using slow, maneuverable flight and able to remain continuously on the wing when foraging. Thyropterids, commonly known as "disk-winged" bats, have remarkable roosting specializations including suction disks on their wrists and feet that allow them to cling to clean smooth surfaces. *Thyroptera tricolor* is apparently a species of lowland disturbed forest areas (edges, treefall gaps, successional areas). It roosts in the young, unfurling leaves of subcanopy forest plants including *Heliconia* (Musaceae) and occasionally *Calathea* (Marantaceae) that are partially or completely shaded by taller canopy trees (Findley and Wilson 1974; Wilson and Findley 1977; Linares 1986).

Paleomacrofloras from Colombia comprise part of the evidence for architecturally complex forest plant communities in the Miocene of tropical South America (Englehardt 1895; Berry 1935, 1936; Schönfeld 1947; Pons 1969). The platanillos *Heliconia* used by *Thyroptera tricolor* are included among fossil plants described by Englehardt (1895) from late Miocene–Pliocene (Pons 1969) tuffaceous beds near Santa Ana (Tolima Department) in the upper Magdalena River valley. A substantial period (nearly 13 Ma) of evolutionary stasis for *T. tricolor* is implied by its occurrence in the La Venta fauna. This stasis also implies the continuous availability, since at least the middle Miocene, of suitable roosting and foraging habitat for the species (and for *Noctilio albiventris* as well) in northwestern South America. The edges and successional sites often used by *T. tricolor* could have been provided by natural disturbances resulting from seasonal and longer-term dynamics of river wander and flood of the ancestral Río Magdalena like those described for the upper Amazon (e.g., Salo et al. 1986) and to which much of the biota of the Amazon flood plain and river itself has become adapted or is dependent.

Three kinds of Molossidae, *Mormopterus colombiensis, Eumops* sp., and *Potamops mascahehenes,* all represent aerial pursuit insectivores even though they varied in body mass from approximately 5–10 g in *Mormopterus* to approximately 50–60 g in *Potamops* (body mass of *Potamops* is based on the large extant *Eumops* species; if body build, however, was like that of *Cheiromeles,* body weight could have been 140–

175 g). Little is known about the ecology of living *Mormopterus* species and other tropical molossids. Most molossids, however, are rapid, direct, and enduring fliers less maneuverable than other bats but with the ability to travel long distances during foraging periods. Thus, most fly in open air, regularly foraging above the forest canopy (sometimes at high elevations) or over open water (ponds, lakes, streams). Tropical molossids have been known to roost in rock crevices, in tree cavities, and beneath palm leaves (Wilson 1973; Vaughan and Bateman 1980; Janzen and Wilson 1983; Chase et al. 1991).

Most of the bats of the La Venta fauna (at least in the Monkey Beds and Fish Bed) suggest the occurrence of forests (and in the Fish Bed, open water) through their foraging and roosting site selection, but they do not indicate the type and extent of these forests. The presence of a forest community is corroborated by the presence of fossil monkeys, smaller sloths, and opossums (Fields 1957), most of the rodents (Walton 1990; also chap. 24), and possibly also the hoatzin (*Hoazinoides magdalenae*) in the Monkey Beds (Miller 1953; also chap. 10). The bats, however, do not refute the postulated existence of savannas or open areas within the forest that are indicated by some hypselodont rodents and notoungulates (Fields 1957; Walton 1990). The distinctive, extensive, 2- to 6-m-thick Fish Bed is thought to reflect a temporary change in depositional environment to a shallow lake or marsh (*ciénaga*) possibly created by meandering rivers (Guerrero 1990). The relative abundance of open-air- and over-water-foraging molossids (*Mormopterus colombiensis, Eumops* sp.) is concordant with the presence of some type of open water during deposition of the Fish Bed as are the abundant freshwater fishes, crocodilians, and turtles. The occurrence in the Fish Bed of *Tonatia* suggests the forest was not far away. Interestingly, the depositional setting of the Fish Bed and some of its vertebrate fossils are similar to those of the Oligocene Tremembé-Taubaté basin that include serranid and characinid fishes, turtles, crocodilians, and the bat, *Mormopterus faustoi* (Paula Couto 1956). More specific information about local community structure cannot be inferred without knowing more precisely the identities of those bats that are presently identifiable only to subfamily or genus.

The chiropteran frugivores, sanguinivores, and piscivores (unless *Noctilio albiventris* can be considered a fish eater) of extant Neotropical communities are conspicuous by their absence. Frugivores are especially so in light of the probable availability of plants (based on related fossil families and genera) whose parts are used by extant Phyllostomidae (Fleming 1988). Such plant fossils known in the Miocene of Colombia include members of the families Anacardiaceae, Moraceae (*Ficus* spp.), and Sapotaceae but not necessarily those belonging to genera and species known to be used by extant bats. The absence of bats filling frugivorous and other trophic roles is puzzling but may simply be an artifact of the small number of specimens yet known from the Miocene.

The families Mormoopidae, Natalidae, Furipteridae (sometimes included as a subfamily of the Natalidae; Van Valen 1979), and Vespertilionidae are not yet represented in the late Tertiary fossil record of South America although all of these except Furipteridae have recently been discovered in the late Oligocene and early Miocene of Florida, USA (Morgan 1989), on the periphery of the Neotropical region. The Floridian fossils originate in karst deposits and represent cavernicolous bats, whereas all the presently known South American Tertiary bats originate from "open" sites in fluvio-lacustrine sedimentary rocks and represent species that roosted in tree cavities or foliage. Thus, the South American record of bats during the Tertiary presently is biased away from the cave-roosting types that have contributed so significantly to the Tertiary fossil record in Europe (Sigé and Legendre 1983) and recently in Australia (Archer et al. 1991). This bias might soon be lessened if chiropteran fossils are found in recently revealed fissure fillings of Oligocene age in Peruvian paleokarst (Hartenberger et al. 1984). Bat fossils are completely absent in the taxonomically diverse mammalian fauna of the fossiliferous karstic fillings of middle Paleocene age at São José de Itaboraí, Brazil (Marshall, Hoffstetter, et al. 1983).

Including the new records of bats in the Miocene of Colombia, the chiropteran fauna of the Tertiary period in South America now consists of *Diclidurus* sp. (Emballonuridae); *Noctilio albiventris* (Noctilionidae); *Notonycteris magdalenensis*, *Tonatia* sp., Phyllostominae, gen. and sp. indet., and Glossophaginae, gen. and sp. indet. (Phyllostomidae); *Thyroptera robusta* and *T*. cf. *T. tricolor* (Thyropteridae); and *Mormopterus faustoi*, *M. colombiensis*, *Potamops mascahehenes*, and *Eumops* sp. (Molossidae). These records indicate the arrival in South America and/or autochthonous radiations of each of these families or subfamilies there before, maybe much before, the middle Miocene. Because all the genera except *Notonycteris* and *Potamops* are extant in South America, a high degree of evolutionary conservatism is indicated among the chiropteran fauna of the late Tertiary. Moreover, the middle Miocene existence of fossils referable to the extant species *N. albiventris* and *T*. cf. *T. tricolor* indicates a significant period of morphological evolutionary stasis for these species. Such an interpretation based on the fossil record is concordant with that predicted by previous workers on the basis of the Recent fauna and the two previously known Tertiary bats (*N. magdalenensis* and *M. faustoi*) from South America (Koopman 1976; Reig 1981; Legendre 1984a, 1984b, 1985; Fenton 1992) as well as on other continents (Hall 1984; Sigé 1991; Beard et al. 1992).

Acknowledgments

This research was supported by grants from the National Geographic Society and the National Science Foundation (BSR 8614533, BSR 8918657) to R. F. Kay. Field work was aided by the Colombian Instituto Nacional de Investigaciones Geociencias, Minería y Química (INGEOMINAS), Bogotá, and by R. L. Cifelli, J. Guerrero Díaz, R. F. Kay, R. H. Madden, J. M. Plavcan, C. Ross, A. Serna Gonzalez, and A. H. Walton. Lab assistance was provided by C. E. Alexander and E. Miller. Scanning electron micrographs were produced by C. E. Alexander, E. M. Larson, E. Miller, E. Sherburn, and R. Shivers. For constructive comments and helpful discussion, I thank A. Cadena F., P. W. Freeman, O. J. Linares, M. A. Mares, G. S. Morgan, and the editors of this volume. Access to collections of Recent specimens was provided by J. K. Braun and M. A. Mares, Oklahoma Museum of Natural History, Norman, Oklahoma; J. S. Findley, Museum of Southwestern Biology, Albuquerque, New Mexico; and L. K. Gordon, D. Schmidt, and D. E. Wilson, National Museum of Natural History, Washington, D.C. Comparative casts and loans of fossils were provided by B. Sigé, Institut des Sciences de l'Evolution, Montpellier, France; and by J. H. Hutchison, University of California Museum of Paleontology, Berkeley. To all I give my sincerest thanks.

Part Six
Primates

26
A New Small Platyrrhine and the Phyletic Position of Callitrichinae

Richard F. Kay and D. Jeffrey Meldrum

RESUMEN

El descubrimiento y estudio de un nuevo y pequeño mono, *Patasola magdalenae,* provee una nueva comprensión de la posición filética de los Callitrichinae, *Callimico,* y *Saimiri.* Un análisis de los caracteres dentales conduce a la siguiente interpretación filogenética. La mezcla de caracteres compartidos entre *Patasola,* los Callitrichinae, y *Callimico* por un lado, y con *Neosaimiri* y *Saimiri* por el otro, sugiere que el nuevo taxon fósil podía ocupar una posición filética intermedia entre estos dos grupos. Además, las diferencias numerosas entre *Saimiri* y *Cebus* que se evidencian en el análisis están en contra de una relación estrecha entre ellos como había sido propuesto por otros investigadores. *Saimiri* difiere de *Cebus* y se asemeja más a los Callitrichinae en características de la hilera premolar decídua como, por ejemplo, una mayor compresión bucolingual, trigónidos más inclinados, talónidos abiertos, y metacónidos relativamente menores. Esta tendencia se manifiesta también en la hilera premolar permanente, aunque en grado menor. Por ejemplo, el segundo premolar de *Saimiri* carece de un metacónido, que está presente en *Cebus.* El área oclusal del cuarto premolar y primer molar inferior es menor en *Saimiri* que en *Cebus.* En los molares, el cíngulo lingual está más desarrollado en *Saimiri,* mientras que *Cebus* retiene un metacónulo bien desarrollado que se pierde en *Saimiri.* El análisis de estos y otros caracteres indica que *Saimiri* es el grupo hermano de los Callitrichinae y *Callimico,* omitiendo a *Cebus.*

El reconocimiento de especímenes aislados de dientes de pequeños platirrinos en el Grupo Honda como taxa no es recomendado dado el estado actual de conocimiento. Por ejemplo, un ejemplar proveniente de las Capas Rojas de Polonia sería el platirrino más reciente del Grupo Honda. La estructura de los incisivos superiores es diferente a la de *Saimiri* y el molar superior es diferente al de las especies de *Neosaimiri.* La asignación de este material a *Patasola* sería posible si una mejora documentación de los dientes superiores estuviera disponible. Algunos molares descubiertos recientemente en el zarandeo de los sedimentos de la Capa de Peces son referidos, con reservas, al género *Neosaimiri.*

La preservación del holotipo de *Micodon* de la Capa de Los Monos (coetáneo con *Neosaimiri fieldsi*) es muy deficiente. En su morfología y en poseer un hipocono más grande y una prehipocrista, este espécimen se distingue tanto de los especímenes de las Capas Rojas de Polonia como de los Callitrichinae. Excepto en el desarrollo de una prehipocrista y un tamaño ligeramente más pequeño, no se lo puede distinguir de *Neosaimiri annectens* y fácilmente podría pertenecer a *Neosaimiri fieldsi* del mismo nivel estratigráfico. Los dos otros dientes asignados a *Micodon* son aun más enigmáticos. El incisivo superior pertenece a un Callitrichinae mientras el premolar inferior, aunque más pequeño, se asemeja más a *Soriacebus,* del Santacrucense.

La presencia en el Grupo Honda de *Patasola,* un primate pequeño, insectívoro, y frugívoro, sugiere un hábitat boscoso. La diversidad a nivel genérico de los primates de La Venta, junto con las características postcraneales de al-

gunas especies, sugiere también la presencia de bosques húmedos en el área durante el Mioceno medio.

Two phenetic clusters of platyrrhine genera, often described taxonomically as families, are recognized traditionally: the Callitrichidae is used for marmosets and tamarins and the Cebidae for the rest (Pocock 1925; Simpson 1945; Hill 1957, 1960, 1962; Cabrera 1958; Napier and Napier 1967). Of late, most workers have come to accept that some or all the morphological traits of the marmoset/tamarin group—small body size, clawed digits, tricuspid upper molars, and the loss of the third molar—are synapomorphies of a holophyletic cluster, whereas the remaining platyrrhines are possibly a paraphyletic assemblage, some members of which are closer to marmosets and tamarins than others (but see Hershkovitz 1977).

One living platyrrhine, *Callimico,* displays a mosaic of features, combining the small body size and clawed digits of the marmosets and tamarins with hypocones and the three molars of the cebids. The phylogenetic implication is that *Callimico,* marmosets, and tamarins are a holophyletic group because the features *Callimico* shares with marmosets and tamarins are synapomorphies, whereas those it shares with other cebids are symplesiomorphies (e.g., Rosenberger 1980). It is debated how this should be reflected in a taxonomic arrangement at the family level. Thorington and Anderson (1984) allocate marmosets and tamarins to subfamily status, as the Callitrichinae, and place *Callimico* in its own subfamily, Callimiconinae, but offer no overarching familial name for the two groups together as a holophyletic group even while recognizing that there is strong evidence for such an association. In contrast, Rosenberger et al. (1990) recognize marmosets and tamarins as one tribe and *Callimico* as another tribe within a holophyletic subfamily Callitrichinae. We provisionally follow the scheme of Thorington and Anderson (1984). Thus, when we refer to Callitrichidae, we mean marmosets, tamarins, and *Callimico;* by Callitrichinae, we mean marmosets and tamarins only; by Callimiconinae we mean *Callimico* only.

Another genus of living platyrrhine, *Saimiri,* has been mentioned as more distantly related to callitrichids (Rosenberger 1981b, 1988b). Immunological-distance studies, however, although admittedly phenetic, support no special link between *Saimiri* and callitrichids (Cronin and Sarich 1975; Baba et al. 1979) and Ford (1986) in her study of platyrrhine postcranial anatomy finds a sister-grouping of *Saimiri* with *Aotus* and *Callicebus* rather than callitrichids.

Rosenberger, Setoguchi, et al. (1991) have described a new genus of primate *Laventiana* from the Miocene of Colombia, which they consider to support a close association of *Saimiri* and *Cebus* with callitrichids.

Small fossil primates from the Miocene of Colombia have a bearing on the relationship between callitrichids and *Saimiri.* Table 26.1 summarizes their occurrence. *Neosaimiri fieldsi* was the first small primate described from La Venta (Stirton 1951). The type specimen, from the Monkey Beds, is a lower jaw preserving I_2 and $C-M_2$. Stirton noted the close resemblence to living *Saimiri* and Rosenberger, Hartwig, Takai, et al. (1991) have argued that the two should be considered congeneric (but see here). In 1985 Setoguchi and Rosenberger described several isolated teeth from the same geologic horizon, the Monkey Beds. They made one specimen, a worn and chemically eroded upper molar, the type of another genus *Micodon kiotensis.* Although the specimen has a fairly large hypocone, the apparent small size of the tooth led them to suggest *Micodon* as a possible relative of tamarins and marmosets. In 1991 Rosenberger, Hartwig, et al. (1991) described another small platyrrhine, *Laventiana annectens,* based on a mandible with the canines and cheek teeth from the El Cardón Red Beds about 100 m stratigraphically above the Monkey Beds. They argued that the geologically younger *Laventiana* is generally more primitive than *Neosaimiri fieldsi* (=*Saimiri fieldsi*) and, although it has a few specializations of its own, closely approximates the ancestral morphotype of callitrichids.

Takai (1994) has described a large sample of primate teeth from the *Laventiana annectens* quarry. Takai notes that the variation within this quarry sample encompasses that of both *Neosaimiri fieldsi* and *Laventiana annectens* and concludes that the two should be considered conspecific. It is our position, based on our cladistic analysis, that this material should be treated as two species, *Neosaimiri fieldsi* and *N. annectens.* Furthermore, the upper incisors, first upper premolars, and molars of this taxon convincingly show *Neosaimiri* to be quite distinct morphologically from *Saimiri* and not congeneric with it as hitherto claimed (Rosenberger, Hartwig, Takai, et al. 1991). Finally, the *Neosaimiri* upper molars described by Takai also bear a striking resemblance

Table 26.1. Stratigraphic occurrence of specimens described in the text

Specimen Number	Stratigraphic Position	Locality Information[a]	Description	Comments[b]
IGM 184332 (Field no. 88-275)	La Victoria Formation, below Chunchullo Sandstone	Duke Locality 40 (13.4 Ma)	Mandible with dP_2–M_3, P_{3-4}	Type specimen of *Patasola magdalanae*
IGM 250829 (Field nos. 89-483, 89-640, 90-100, and three screenwash specimens without field nos.)	Villavieja Formation, Cerro Colorado Member, Polonia Red Beds	Duke Locality 126 (11.8 Ma)	I^{1-2}, C^1, right and left C_1, P_4 (broken), $M^{1 \text{ or } 2}$	cf. *Patasola*
IGM 250346 (Field no. 88-846)	Villavieja Formation, Baraya Member, Fish Bed	Duke Locality CVP 13 (12.9 Ma, but younger than UCMP Locality V4517)	Broken left $M^{1 \text{ or } 2}$	*Neosaimiri* cf. *N. fieldsi*
IGM 250613 (Field no. 89-267)	Villavieja Formation, Baraya Member, Fish Bed	Duke Locality CVP 13c (12.9 Ma)	Left $M^{1 \text{ or } 2}$	*Neosaimiri* cf. *N. fieldsi*
UCMP 39205	Villavieja Formation, Baraya Member, Monkey Bed	UCMP Locality V4517 (12.9 Ma)	Mandible with left I_1–M_2, right P_3–M_2	Type of *Neosaimiri fieldsi* Stirton 1951
IGM-KU 8401	Villavieja Formation, Baraya Member, Monkey Bed	Kyōto Site (Setoguchi and Rosenberger 1985a) (12.9 Ma)	$M^{1 \text{ or } 2}$	cf. *Neosaimiri fieldsi* [type of *Micodon kiotensis* Setoguchi & Rosenberger 1985a]
IGM-KU 8402	Villavieja Formation, Baraya Member, Monkey Bed	Kyōto Site (12.9 Ma)	Right I^1	Callitrichidae, gen. et sp. indet. (Setoguchi & Rosenberger 1985a)
IGM-KU 8403	Villavieja Formation, Baraya Member, Monkey Bed	Kyōto Site (12.9 Ma)	Left P_4	cf. *Soriacebus* sp. [Callitrichidae, gen. et sp. indet. (Setoguchi & Rosenberger 1985a)]
IGM-KU 8801a	Villavieja Formation, Baraya Member, El Cardón Red Beds	Masato Site equivalent to Duke Locality 32 (12.6 Ma)	Mandible with left C_1–M_2, right C_1–M_2	*Neosaimiri annectens* [type of *Laventiana annectens* Rosenberger, Setoguchi, & Hartwig 1991]
IGM-KU specimens enumerated by Takai (1994)	Villavieja Formation, Baraya Member, El Cardón Red Beds	Masato Site equivalent to Duke Locality 32 (12.6 Ma)	Many isolated teeth, several upper and lower jaws; includes deciduous teeth	*Neosaimiri annectens* [referred to *N. fieldsi* (Takai 1994)]

Notes: Field no., Duke-INGEOMINAS field number; IGM, INGEOMINAS Museo Geológico, Bogota; UCMP, University of California Museum of Paleontology, Berkeley; KU, Kyōto University.

[a] The estimated age of each locality is from Flynn et al. (chap. 3) and Guerrero (chap. 2).

[b] Our allocation of each specimen and, in square brackets, allocations by other workers.

to the type of *Micodon* and raise the possibility that the latter specimen should also be referred to *Neosaimiri*.

Between 1988 and 1990, expeditions by Duke University and the Instituto Nacional de Investigaciones en Geociencias, Minería y Química, Museo Geológico, IN-GEOMINAS, to the La Venta region recovered several additional specimens of small primates described here. One of these specimens from much lower in the stratigraphic section is made the type of a new genus and species that we consider to be a callitrichine. Our analysis, based upon all available material of small fossil primates from La Venta, lends further support to Rosenberger's contention that *Saimiri* is the sister group of callitrichids.

Systematics

Infraorder Platyrrhini
Family Callitrichidae
Subfamily incertae sedis
Patasola gen. nov.

TYPE SPECIES

Patasola magdalenae sp. nov.

INCLUDED SPECIES, DISTRIBUTION, AND DIAGNOSIS

As for the type and only species.

ETYMOLOGY

Named for patasola, a forest spirit of the Grand Tolima region, Magdalena Valley, Colombia.

Patasola magdalenae sp. nov.
(figs. 26.1–26.4)

TYPE SPECIMEN

IGM 184332, a lower jaw with dP_{2-4}, M_{1-2}, P_{2-4} crowns fully developed but in crypt, M_3 partially erupted, and roots of dI_{1-2}, and dC_1.

HYPODIGM

The type only. Dimensions of the type are given in table 26.2.

HORIZON AND LOCALITY

The type is from Duke Locality 40 in the stratigraphic interval between the Cerro Gordo Sandstone Beds and the Chunchullo Sandstone Beds, La Victoria Formation (middle Miocene), Colombia (Guerrero 1990; also chap. 2). This stratigraphic interval is approximately 560 m below the Monkey Beds, from which *Neosaimiri fieldsi* is recovered, and a further 720 m below the stratigraphic level of *Neosaimiri annectens* in the El Cardón Red Beds.

DIAGNOSIS

Patasola is a small platyrrhine, smaller than *Saimiri* or *Neosaimiri* spp., dentally about the size of *Leontopithecus rosalia*—this small size is a synapomorphy with callitrichids. Other synapomorphies shared by *Patasola* and some or all callitrichids include, in the deciduous dentition, procumbent incisor roots with a mesiodistally compressed dI_2 root, premolar trigonid basins steeply sloping mesiolingually, absence of a metaconid on dP_{2-3}, and buccolingual compression of the deciduous premolar row. The permanent M_{1-2} protoconids are much taller than the metaconids as in callitrichines. *Patasola* differs from, and is more primitive than, living callitrichines by having third molars, that, although small, are not as reduced as in *Callimico,* the only living callitrichid with three molars.

Differences from *Neosaimiri annectens, Neosaimiri fieldsi,* and *Saimiri* include the following: the P_3 of *Patasola* is buccolingually narrower, and P_{3-4} have smaller metaconids and more sloping trigonids. The molars are significantly more buccolingually compressed, and the protoconid is taller than the metaconid as compared with *Saimiri* or *Neosaimiri*.

Patasola, like *N. annectens,* lacks buccal cingula on the P_{3-4}, whereas a strong but broken buccal cingulum is present in *N. fieldsi*. The premolars of *Patasola* and *N. annectens* are distinctly smaller proportionate to the molars than those of *N. fieldsi*. In *Patasola* and *N. annectens* the M_1 hypoconulids are more distinct than in *N. fieldsi* and the M_1 cristid obliqua reaches only part way up the distal wall of the

Table 26.2. Selected dental measurements (mm) of La Venta specimens and *Saimiri sciureus*

Variable Measured		IGM 250829	IGM 184332	UCMP 29205	IGM-KU 88-001[a]	IGM-KU 88-001[b]	El Cardón Sample[c]		*Saimiri sciureus*[d]	
							Mean	SD	Mean	SD
I_2	length	—	—	1.54	—	—	1.52	0.07	1.91	0.14
	breadth	—	—	1.70	—	—	1.94	0.09	2.41	0.13
C_1	length	3.09	—	3.20	2.56	—	—	—	3.24	0.24
	breadth	2.06	—	2.23	1.81	—	—	—	2.39	0.33
	height	—	—	4.82	4.37	—	—	—	5.12	1.41
P_2	length	—	—	2.59	2.37	2.44	2.41	0.09	2.99	0.26
	breadth	—	—	2.43	1.88	2.44	2.42	0.07	2.22	0.27
P_3	length	—	1.94	2.37	2.19	2.28	2.29	0.08	2.42	0.10
	breadth	—	2.02	2.53	2.25	2.44	2.39	0.09	2.67	0.13
P_4	length	—	2.11	2.35	2.19	2.40	2.29	0.07	2.38	0.08
	breadth	—	2.02	2.79	2.31	2.60	2.50	0.14	2.65	0.18
M_1	length	—	2.73	3.02	3.25	3.52	3.29	0.14	2.67	0.12
	trigonid breadth	—	2.21	2.87	2.50	2.56	2.52	0.15	2.74	0.14
	talonid breadth	—	2.34	2.67	2.75	2.76	2.65	0.09	2.70	0.13
M_2	length	—	2.80	2.93	3.13	2.93	3.06	0.13	2.49	0.10
	trigonid breadth	—	2.08	2.63	2.38	2.63	2.52	0.11	2.45	0.13
	talonid breadth	—	2.02	2.61	2.13	2.61	2.44	0.10	2.36	0.13
M_3	length	2.44	—	—	—	—	2.48	0.05	2.13	0.11
	trigonid breadth	2.13	—	—	—	—	2.01	0.03	2.00	0.10
	talonid breadth	2.00	—	—	—	—	—	—	1.67	0.09
DP_2	length	—	1.88	—	—	—	—	—	—	—
	breadth	—	1.50	—	—	—	—	—	—	—
DP_3	length	—	2.06	—	—	—	—	—	—	—
	breadth	—	1.63	—	—	—	—	—	—	—
DP_4	length	—	2.50	—	—	—	—	—	—	—
	trigonid breadth	—	1.50	—	—	—	—	—	—	—
	talonid breadth	—	1.82	—	—	—	—	—	—	—
I^1	length	2.70	—	—	—	—	2.43	—	—	—
	breadth	1.65	—	—	—	—	1.71	—	—	—
I^2	length	2.33	—	—	—	—	—	—	—	—
	breadth	1.86	—	—	—	—	—	—	—	—
M^1	length	2.68	—	—	—	—	—	—	—	—
	breadth	4.44	—	—	—	—	—	—	—	—
M^2	length	—	—	—	—	—	—	—	—	—
	breadth	—	—	—	—	—	—	—	—	—

Notes: IGM 250829, Polonia Red Beds specimen; IGM 184332, type specimen of *Patasola magdalenae*; UCMP 29205, type specimen of *Neosaimiri fieldsi*; IGM-KU 88-001, type specimen of *Laventiana annectens*.

[a] Measurements made by Kay and Meldrum.

[b] Measurements made by Takai (1994).

[c] Sample statistics of specimens reported by Takai (1994) from type site of *Laventiana annectens* (right and left teeth of the same specimen were averaged; type specimen is excluded).

[d] Measurements made by R. F. Kay and J. M. Plavcan.

trigonid, not to the protoconid as in *Neosaimiri fieldsi*. *Patasola* and *N. fieldsi* lower molars lack posterointernal basins, whereas they are commonly present in *N. annectens*. Many other differences from living small-bodied platyrrhines are detailed here and in table 26.6.

ETYMOLOGY

Named for the Magdalena River, which is near the localities where these specimens were collected.

DESCRIPTION

The type mandible of *Patasola magdalenae* (IGM 184332) is a juvenile specimen. This specimen is compared to specimens of contemporary La Venta small primates and extant species *Saimiri sciureus, Aotus trivirgatus, Callicebus moloch, Saguinus oedipus, Leontopithecus rosalia, Callithrix jacchus,* and *Callimico goeldii* from the Department of Vertebrate Zoology collections of the Smithsonian Institution, Washington, D.C., and the Department of Mammalogy, Field Museum of Natural History, Chicago.

Mandible

The right mandibular corpus is preserved from a point just to the left of the symphysis to the root of the ascending ramus (fig. 26.1). The corpus is shallow, measuring approximately 6.0 mm in depth at the level of dP_2. It deepens very slightly distally. The symphysis has a rather low angle of inclination, 40° relative to the basal mandibular plane. This falls within the range for adult *Callithrix*, 19–45°, and *Leontopithecus*, 36–51° (Hershkovitz 1977). *Saguinus* and *Callimico* have slightly greater symphyseal angles, ranging from 41 to 63° (Hershkovitz 1977). *Saimiri* has an even greater angle, ranging from 61 to 74° (Hershkovitz 1984b). *Neosaimiri* and *Laventiana* appear to fall within the range of *Saimiri*. Preliminary observations were made of a limited sample of juvenile mandibles to determine whether the symphyseal angle changes with age and tooth eruption. Juvenile values fell within adult ranges.

Deciduous Incisors and Canine

The dI_1 root of *Patasola* is a rounded oval in cross section. The dI_2 root is larger and mesiodistally compressed with the long axis of the cross section oriented buccolingually (fig. 26.2). In orientation and cross-sectional shape *Patasola*'s deciduous incisor roots are very similar to those of *Callimico*; *Callithrix* also has a compressed dI_2 root, but the orientation of the long axis of the root cross section is more mesiodistal. In contrast, in the other taxa examined the deciduous incisor roots are more rounded. No deciduous lower incisors have been described for *Neosaimiri* spp.

The dC_1 root is moderately compressed buccolingually, most closely resembling *Leontopithecus, Saimiri,* and *Callicebus* but unlike *Callithrix* where it is very buccolingually compressed.

Deciduous Premolars

All three deciduous premolars are preserved in the type (fig. 26.3). The trigonid of dP_2 is much longer mesiodistally than the talonid and dominates the occlusal surface. The trigonid basin slopes very steeply mesiolingually and is partly bounded by a very weak mesiolingual marginal crest. This crest is incomplete distally, leaving the trigonid open distolingually. A metaconid and premetacristid are absent. These conditions are most similar to those seen in callitrichid dP_2s. *Neosaimiri annectens* (fig. 12 in Takai 1994) has a small dP_2 metaconid. *Saimiri* has a lingually closed trigonid. The dP_2 talonid of *Patasola* is small, lacks cusps, and is open lingually. This is most similar to *Callithrix, Saguinus, Callimico,* and *Saimiri*. In *N. annectens* a hypoconid is present and the talonid is partially enclosed lingually.

The trigonid of dP_3 is slightly longer mesiodistally than the talonid. The trigonid basin slopes steeply mesiolingually as in dP_2, but it is fully enclosed by a continuous weak marginal crest. A swelling is present on the protocristid, in the metaconid position, but no distinct cusp is present. The latter condition is most similar to the condition observed in *Callithrix, Saguinus,* and *Leontopithecus*. In *N. annectens* the metaconid is smaller than the protoconid (Takai 1994). In *Saimiri, Callimico, Aotus,* and *Callicebus* the metaconid is well developed and nearly as large as the protoconid. A dP_3 entoconid is lacking, as in the callitrichines, in contrast to *N. annectens* and *Saimiri*, which have a distinct cusp. The occlusal outline of the crown of the dP_3 is compressed buccolingually to a similar degree as in the dP_2. The condition is intermediate between that observed in

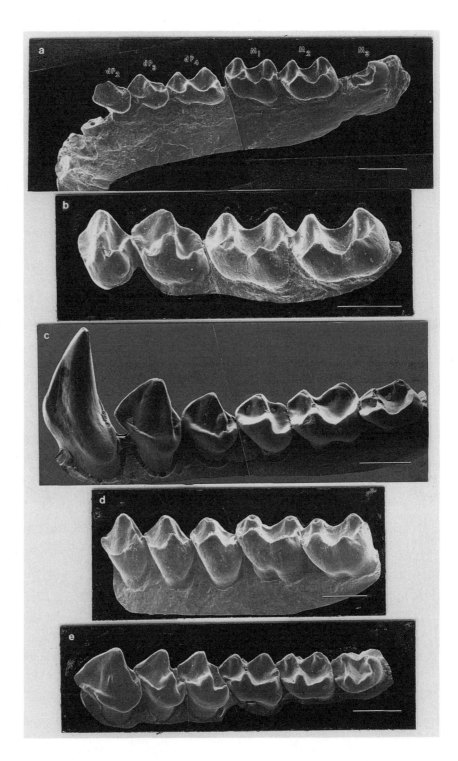

Figure 26.1. Lingual views of mandibular dentition. *a. Patasola magdalenae,* type, IGM 184332; note that the premolars in the mandible of *Patasola* are deciduous. *b.* Photomontage of *P. magdalenae* mandible with permanent P_3–P_4 in place. *c. Saguinus oedipus,* USNM 301653. *d. Neosaimiri fieldsi,* type, UCMP 32905. *e. Saimiri sciureus.* Scale bars 2 mm.

Callithrix, in which the deciduous premolars are greatly compressed buccolingually, and *N. annectens* (Takai 1994) and *Saimiri,* in which the premolars are more rounded.

The cusp relief of dP_4 is moderate. The trigonid is much narrower than the talonid buccolingually and is dominated by the protoconid. A much smaller metaconid lies distolingually and very close to the protoconid. The proportions of the protoconid and metaconid are comparable to those of *Callithrix* and *Saguinus.* In *N. annectens* the protoconid and metaconid are of similar height (Takai 1994).

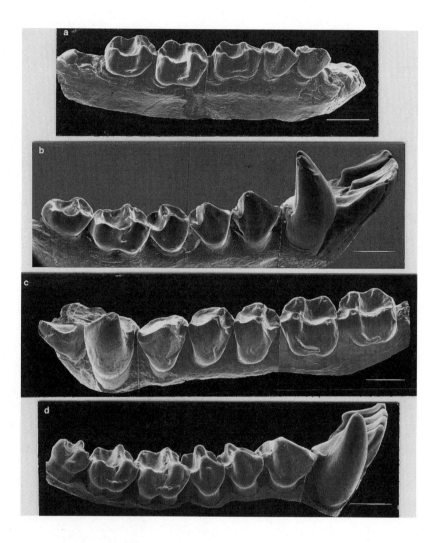

Figure 26.2. Buccal views of mandibular dentition. *a. Patasola magdalenae,* IGM 184332, with deciduous premolars. *b. Saguinus oedipus,* USNM 301653. *c. Saimiri sciureus. d. Neosaimiri fieldsi,* type, UCMP 39205. Scale bars 2 mm.

In *Saimiri* and *Callimico* the metaconid projects higher than the protoconid. A paraconid is absent, and the trigonid basin is closed lingually by a very weak marginal crest. The entoconid is distinct as in *N. annectens, Callimico,* and *Saimiri* but unlike callitrichines in which it is cristiform. A small hypoconulid is present, twinned with the entoconid. A trace of a postentoconid sulcus is present. *N. annectens* has a small, variably present, lingually placed hypoconulid and postentoconid notch (Takai 1994). In *Leontopithecus, Callimico,* and some *Saguinus* the hypoconulid and entoconid are also twinned. In *Saimiri* the hypoconulid is larger and situated in the midline. The cristid obliqua reaches the trigonid wall at a point distolingual to the protoconid and extends part way up the distal trigonid wall. The hypoflexid

is rather deep as in *N. annectens* (Takai 1994), and, as in *N. annectens,* is further emphasized by a strong buccal cingulum. The occlusal outline of the tooth crown is buccolingually compressed, most resembling *Callithrix; N. annectens* (Takai 1994) and *Saimiri* have broader dP$_4$s.

Eruption Pattern of the Teeth

The deciduous incisor and canine roots are present in the type mandible (IGM 184332) as are the crowns of the deciduous premolars (fig. 26.4). Fully formed rootless crowns of P$_3$ and P$_4$ were excavated and removed from their crypts beneath dP$_{3-4}$. A rootless P$_2$ crown is also present but was not removed for study to avoid excessive damage to the

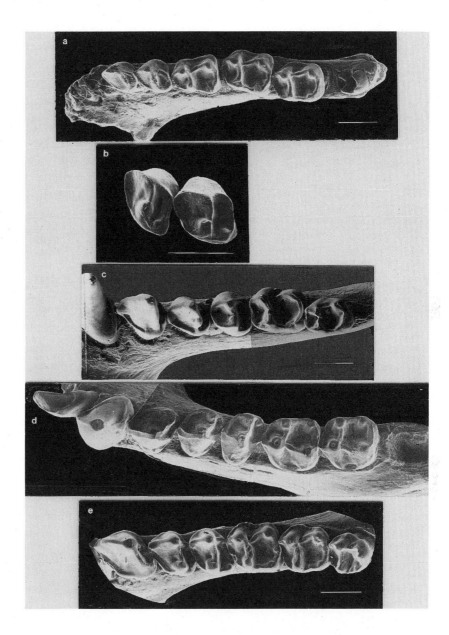

Figure 26.3. Occlusal views of mandibular dentition. *a. Patasola magdalenae,* type, IGM 184332. *b.* Permanent P$_3$–P$_4$ of *Patasola*. *c. Saguinus oedipus,* USNM 301653. *d. Neosaimiri fieldsi,* type, UCMP 39205. *e. Saimiri sciureus.* Scale bars 2 mm.

mandibular corpus. Based on their proximity to the occlusal margin of the jaw, P$_2$ and P$_4$ would have erupted before P$_3$. *Patasola* has M$_{1-2}$ fully erupted; M$_3$ was in the process of eruption but is still partially enclosed in the crypt. Thus, all three lower molars most likely would have erupted before any permanent incisor or premolar had erupted. Takai (1994) reports that in *N. annectens* the M$^{1-2}$ are fully erupted, whereas dP$^{2-4}$ are still in occlusion; thus, the eruption pattern of the two appears similar.

Lower Premolars

The trigonids of P$_{3-4}$ are very large (fig. 26.3); in each they are closed lingually by a raised marginal ridge running from the mesial apex of the paracristid to the base of the metaconid. On P$_3$ the trigonid basin dips steeply mesiolingually, whereas on P$_4$ the dip is much less. On P$_3$ the occlusal outline of the trigonid is triangular, whereas on P$_4$ it is more rectangular because the mesiolingual aspect of the trigonid

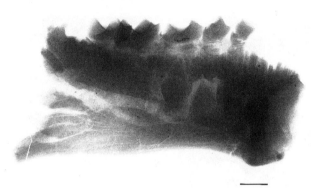

Figure 26.4. Radiograph of the mandible of *Patasola magdalenae,* IGM 184332, lateral view. Note the fully formed crowns of the unerupted permanent premolars in their crypts, the partially formed crown of the permanent canine, and the partially erupted M_3. Scale bar 2 mm.

is more filled out. Similar steeply sloping triangular trigonids are seen in callitrichines and *Saimiri;* in *Aotus* and *Callimico* the P_3 trigonid is more nearly rectangular. *Neosaimiri* spp. resemble the latter taxa and are distinct from *Patasola* in this respect, having a more filled out mesiolingual trigonid profile on both P_{3-4}.

P_{3-4} of *Patasola* have well-developed metaconids although this cusp is distinctly smaller than the protoconid. The relative size of these cusps in P_3 and P_4 most closely resembles the condition seen in *Neosaimiri* spp., whereas among callitrichines the P_3 metaconid is very small or absent. By contrast, *Saimiri, Aotus,* and *Callimico* have much larger P_{3-4} metaconids. Like callitrichines, metaconids in *Patasola* and *Neosaimiri* spp. are in close proximity to the protoconids, whereas in *Callimico, Saimiri,* and *Aotus* they are much more widely separated. In *Patasola* there is a continuous raised crest connecting the protoconid and metaconid, walling off the raised trigonid from the lower and smaller talonid. This is the same in all other small-bodied extant platyrrhines except *Callimico* in which the crest is deeply notched.

Patasola has a small, discrete entoconid standing out on the lingual talonid marginal crest of P_{3-4} as in *Callicebus, Neosaimiri annectens,* and *Saimiri.* In *Callithrix* the entoconid is absent on P_{3-4}. In other callitrichids and *Aotus* the entoconid is cristiform. As in all extant small platyrrhines, *Laventiana,* and *Neosaimiri, Patasola* has no discernible hypoconid on the buccal margin of the talonid. Only a slight

trace of a buccal cingulum is visible on P_{3-4} of *Patasola.* The buccal cingulum is also weak or absent in callitrichines, *Callimico, Saimiri,* and *Aotus.* Buccal cingula are better developed in *Neosaimiri.*

The premolars of *Patasola* are proportionally small compared with the molars. A ratio of P_4 to M_1 areas is 0.66. This most closely approximates the proportions of *Aotus* (0.65) and *Callicebus* (0.61). *N. annectens* also has small premolars with a ratio of 0.64 (table 26.3). *Saimiri* and callitrichids fall at the opposite extreme with comparatively large premolars ranging from 0.76 to 1.00 (table 26.3). *Neosaimiri fieldsi* falls with the latter, having a value of 0.79.

Lower Molars

The crown relief of the molars of *Patasola* is high, resembling callitrichids, *Neosaimiri* spp., and *Saimiri;* other small-bodied extant platyrrhines (*Aotus, Callicebus*) have lower crowns with less cusp relief. Molar breadth across the trigonid and talonid are approximately equal as in the other taxa compared; *Callithrix* has a uniquely narrow trigonid. As in all other taxa examined, the M_1 trigonid is closed lingually by a ridge running from the mesial end of the paracristid distolingually to the tip of the metaconid and there is no trace of a paraconid on M_{1-2}. The protoconid of M_{1-2} is taller than the metaconid as it is in *Callithrix* but unlike other platyrrhines, including *Neosaimiri* spp., in which the two cusps are of similar height. The distal edge of the trigonid is trenchant with a continuous raised crest joining protoconid and metaconid. This crest is oriented transversely on M_{1-2} with the metaconid lingual and slightly distal to the protoconid.

In *Patasola,* as in callitrichids, *Saimiri,* and *Neosaimiri* spp., the talonid basin of M_{1-2} is closed lingually by strong crests running between the metaconid and entoconid. *Patasola's* entoconid is a comparatively large cusp as in *Saimiri, Neosaimiri* spp., and *Aotus,* whereas in callitrichines and *Callimico* the entoconid is cristiform. The M_1 of *Patasola* has a small hypoconulid situated slightly lingual to the midline of the talonid basin although not twinned with the entoconid. A similar situation is found in *Saimiri, Neosaimiri fieldsi,* and *N. annectens.* Callitrichids, *Callimico,* and *Aotus* lack hypoconulids on M_{1-2}. A possible exception is *Saguinus* in which a hypoconulid is occasionally present on

Table 26.3. Dental proportions of living small-bodied platyrrhines compared with those of *Neosaimiri* spp. and *Patasola magdalenae*

Taxon or Sample	n	P_4 Area $\div M_1$ Area	M_1 Trigonid Br $\div$ Talonid Br	md $\div$ bl	M_2 Area $\div M_3$ Area	I^1 md $\div I_1$ bl	M^1 md $\div M^1$ bl
Aotus trivirgatus	29	0.65 (0.06)	0.95 (0.04)	1.09 (0.05)	1.11 (0.12)	1.39 (0.08)	0.82 (0.04)
Callicebus moloch	19	0.61 (0.05)	0.95 (0.03)	1.03 (0.03)	1.16 (0.11)	1.13 (0.07)	0.76 (0.04)
Saimiri sciureus	46	0.88 (0.06)	1.01 (0.03)	0.99 (0.04)	1.38 (0.11)	1.15 (0.07)	0.62 (0.02)
Callimico goeldii	2	0.76 (—)	1.06 (—)	1.16 (—)	1.70 (—)	1.37 (—)	0.72 (—)
Saguinus oedipus	42	0.84 (0.07)	1.01 (0.05)	1.29 (0.05)	—	1.26 (0.09)	0.81 (0.05)
Leontopithecus rosalia	17	0.96 (0.09)	1.04 (0.05)	1.16 (0.05)	—	1.16 (0.09)	0.79 (0.03)
Callithrix jacchus	16	1.00 (0.28)	1.12 (0.06)	1.15 (0.10)	—	1.19 (0.20)	0.77 (0.06)
UCMP 39205	1	0.81 (—)	1.07 (—)	1.05 (—)	—	—	—
IGM-KU 88-001	1	0.64 (—)	0.93 (—)	1.27 (—)	—	—	—
El Cardón sample	1–10	0.63 (—) (n = 2)	0.95 (0.01) (n = 10)	1.24 (0.06) (n = 10)	1.55 (—) (n = 1)	1.42 (0.06) (n = 8)	0.73 (0.01) (n = 9)
IGM 184332	1	0.67 (—)	0.94 (—)	1.17 (—)	—	1.63 (—)	0.60 (—)

Notes: Values are means (with SDs in parentheses). See table 26.2 for definition of samples and specimen numbers. Br: breadth; bl: buccolingual length; md: mesiodistal length.

the distolingual side of the talonid separated by a narrow sulcus from an indistinct entoconid.

M_{1-2} cristid obliquas are very strong. They reach the base of the distal wall of the trigonid at a point distolingual to the protoconid before turning buccally toward the protoconid. On M_1 this crest reaches only part way up the distal wall of the trigonid; on M_2 the crest reaches to the protocristid. There are well-developed hypocristids on M_{1-2}. The buccal cingula of M_{1-2} are very strongly developed. On M_1 the buccal cingulum is broken at the protoconid but wraps around the base of the hypoconid. On M_2 the cingulum runs around both cusps without a break. The greatest development of the buccal cingulum is in the hypoflexid. *Saimiri, Saguinus,* and *Neosaimiri* spp. also have well-developed buccal cingula, whereas the cingulum is poorly developed or absent in other callitrichids, *Aotus,* and *Callicebus.* The hypoflexids of M_{1-2} are deeply incised as in *Saimiri* and *Neosaimiri* spp.; callitrichids, *Aotus,* and *Callicebus* have shallower hypoflexids.

On M_{1-2} of *Patasola*, crests join hypoconulid and entoconid and there is no trace of posterointernal basins on either tooth. The same configuration is found in *Callimico, Aotus, Saimiri,* and *Neosaimiri fieldsi. N. annectens, Cebus,* and *Callicebus* are distinct in having a posterointernal basin. In

Cebus and *Callicebus,* however, the posterointernal basin is linked with the talonid by a sulcus, whereas in *N. annectens* the sulcus is absent.

The size of M_3 in *Patasola* was only estimated because it is partially encrypted, but this was apparently similar to *Neosaimiri.* Extant callitrichids have lost M_3. *Callimico, Neosaimiri annectens, Saimiri, Callicebus,* and *Aotus* may be arrayed in order of increasing relative M_3 size. The buccolingual breadth of M_3 in *Patasola* is less than that of M_2 and the proportions appear to be similar to *Neosaimiri annectens* and *Saimiri.*

The M_1 trigonid breadth of *Patasola* is narrow compared to the talonid with a ratio of 0.94. This is comparable to the proportions of M_1 in *Neosaimiri annectens* at 0.95. These proportions fall close to the means of *Aotus* and *Callicebus* but below the observed ranges of large samples of *Saimiri* and callitrichids, which, like *Neosaimiri fieldsi,* have relatively broadened trigonids (table 26.3).

M_1 of *Patasola* is relatively long and narrow with a ratio of mesiodistal length to maximum breadth of 1.16. This is a resemblance to living callitrichids, which range from 1.16 to 1.28 (table 26.3). *Neosaimiri annectens,* with a ratio of 1.24, also closely resembles *Patasola. Neosaimiri fieldsi* has broader molars than *Patasola* and *Neosaimiri annectens,* with a

Table 26.4. Definitions of dental character states used in cladistic analysis

Character	State 0	State 1	State 2 (and 3, if applicable)
1. dI_1 root cross section shape	Small and round	Larger and oval	—
2. dI_2 root cross section shape	Round	Oval, aligned with tooth row	(2) Compressed buccolingually, aligned with tooth row; (3) compressed buccolingually, perpendicular to tooth row
3. dC_1 root cross section shape	Round	Slightly compressed buccolingually	Very compressed buccolingually
4. dP_2 metaconid	Absent	Present, widely spaced	Present, narrowly spaced
5. dP_2 trigonid slope	Steep	Moderate	—
6. dP_2 trigonid orientation	Open	Bounded by a weak marginal crest	Closed
7. dP_2 talonid	Open lingually	Closed lingually	—
8. dP_2 entoconid	Absent	Present, but does not stand out on lingual talonid margin	Present as a small discrete cusp
9. dP_3 metaconid	Absent	Present, small	Present, large
10. dP_3 trigonid and talonid, relative lengths (md)	Trigonid >> talonid	Trigonid > talonid	Trigonid = talonid
11. dP_3 entoconid	Absent	Present, but does not stand out on lingual talonid margin	Present as a discrete cusp
12. dP_3 shape	Very compressed buccolingually	Moderately compressed buccolingually	Rounded oval
13. dP_4 protoconid and metaconid, relative heights	Protoconid > metaconid	Protoconid = metaconid	Protoconid < metaconid
14. dP_4 entoconid	Absent	Present, cristiform	(2) Present as a small discrete cusp; (3) present, large
15. dP_4 hypoconulid size	Large	Medium	(2) Small; (3) absent
16. dP_4 hypoconulid locus	Twinned to entoconid	Slightly lingual to midline	In midline
17. dP_4 buccal cingulum	Absent	Present, partial	Present, complete
18. dP_4 hypoflexid depth	Very shallow	Shallow	Deep
19. dP_4 shape	Very compressed buccolingually	Moderately compressed buccolingually	Broad
20. P_3 trigonid slope	Steeply sloping lingually	Moderately sloping lingually	Nearly horizontal to occlusal plane
21. P_3 trigonid shape	Triangular	Rectangular	—
22. P_3 metaconid	Absent or trace	Present, small	Present, large
23. P_3 metaconid and protoconid, relative position	Close together	Widely spaced	—
24. P_3 entoconid	Absent	Present, cristiform	Present as a discrete cusp
25. P_4 trigonid and talonid, relative lengths (md)	Trigonid >> talonid	Trigonid = talonid	Trigonid < talonid

Table 26.4. Continued

Character	State 0	State 1	State 2 (and 3, if applicable)
26. P_4 metaconid and protoconid, relative position	Close together	Widely spaced	—
27. P_4 metaconid	Absent or trace	Present, small	Present, large
28. P_4 entoconid	Absent	Present, cristiform	Present as a small discrete cusp
29. M_{1-2} crown relief	Moderate to high	Low	—
30. M_1 protoconid and metaconid, relative heights	Protoconid > metaconid	Protoconid = metaconid	Protoconid < metaconid
31. M_2 protoconid and metaconid, relative heights	Protoconid > metaconid	Protoconid = metaconid	Protoconid < metaconid
32. M_{1-2} entoconid	Absent	Present, cristiform	Present as a small discrete cusp
33. M_1 hypoconulid size	Large	Medium	(2) Small; (3) absent
34. M_1 hypoconulid locus	Twinned to entoconid	Lingual to midline	In midline
35. M_2 hypoconulid size	Large	Medium	(2) Small; (3) absent
36. M_{1-2} point at which oblique cristid reaches trigonid wall	Distal to protoconid	Distolingual to protoconid	Distal to metaconid
37. M_1 end point of oblique cristid	Base of trigonid	Part way up the distal trigonid wall	Protoconid or protocristid
38. M_2 end point of oblique cristid	Base of trigonid	Part way up the distal trigonid wall	Protoconid or protocristid
39. M_{1-2} buccal cingulum	Absent or trace	Present, partial (broken at protoconid and hypoconid)	Present, complete
40. M_3	Present	Absent	—
41. M_3 and M_2, relative lengths (md)	$M_3 \ll M_2$; M_2 md/M_3 md $\geq$ 1.10	$M_3 < M_2$; M_2 md/M_3 md < 1.10 but > 1.00	$M_3 > M_2$; M_2 md/M_3 md < 1.00
42. M_3, timing of eruption	Erupts before P_{2-4}	Erupts after P_{2-4}	—
43. P_4 and M_1, relative areas (P_4 area/M_1 area)	Ratio < 0.70	Ratio > 0.70 but < 0.80	Ratio > 0.80
44. M_1 shape (length/breadth)	Ratio < 1.10	Ratio > 1.10	—
45. I^1 occlusal shape	Rounded oval (md/bl < 1.05)	Compressed buccolingually (md/bl > 1.05 but < 1.30)	Extremely compressed buccolingually (md/bl > 1.30)
46. I^{1-2} lingual cingulum	Discontinuous	Moderate	Strong
47. I^1 lingual fovea	Simple	Dual with midcrown pillar	—
48. I^1 and I^2, relative areas	$I^1 = I^2$	$I^1 > I^2$	$I^1 \gg I^2$
49. M^{1-2} hypocone	Large	Small	Absent
50. M^{1-2} prehypocrista	Absent	Present, weak	Present, strong
51. M^{1-2} lingual cingulum	Absent or weak	Present, strong	—
52. M^{1-2} pericone	Absent	Present, small	Present, large
53. M^{1-2} paraconule	Attached	Unattached	Absent
54. M^{1-2} metaconule	Absent	Present, small	Present, large
55. P_4 hypocone	Absent	Present	—

Notes: bl: buccolingual length; md: mesiodistal length.

Table 26.5. Character states for selected extant and fossil platyrrhine taxa

Taxon	Character																										
	1	2	3	4	5	6	7	8	9	10	11	12	13	14	15	16	17	18	19	20	21	22	23	24	25	26	27
Callithrix	0	2	2	0	0	0	0	0	0	0	0	0	0	0	0	1	3	?	0	1	0	0	0	0	0	0	2
Saguinus	0	1	0	0	0	0	0	0	0	0	0	0	0	0	1	2	0	0	2	1	0	0	0	1	0	0	2
Leontopithecus	0	0	1	0	0	1	1	0	0	2	0	1	1	1	2	0	1	1	1	0	0	1	1	1	1	1	2
Callimico	1	3	1	0	0	1	0	0	2	2	0	1	2	2	2	0	0	2	2	2	1	2	1	1	1	1	2
Saimiri	0	0	1	0	0	2	0	0	2	0	1	2	2	3	1	2	1	2	1	1	0	2	1	2	0	1	2
Cebus	0	0	1	1	1	2	1	2	2	2	2	2	1	2	1	2	0	1	2	2	1	2	1	2	1	1	2
Aotus	1	1	1	0	1	2	1	0	2	1	0	2	1	2	3	?	0	1	2	1	1	2	1	1	1	1	2
Callicebus	0	0	0	0	1	2	1	0	2	3	0	2	1	2	1	2	0	1	2	2	1	2	1	1	1	1	2
Patasola	1	3	1	0	0	1	0	0	0	1	0	1	0	2	2	0	2	2	0	0	0	2	0	2	0	0	1
Neosaimiri fieldsi	?	?	?	?	?	?	?	?	?	?	?	?	?	?	?	?	?	?	?	1	1	1	1	0	1	1	2
Neosaimiri annectens	?	?	?	2	?	1	1	?	1	?	2	2	1	2	2	0	2	2	1	1	1	1	1	0	1	1	2

Taxon	Character																											
	28	29	30	31	32	33	34	35	36	37	38	39	40	41	42	43	44	45	46	47	48	49	50	51	52	53	54	55
Callithrix	0	0	0	0	1	3	?	3	0	0	0	0	1	?	?	2	1	2	0	1	2	2	?	0	0	2	0	0
Saguinus	1	0	0	2	1	1	1	1	0	1	1	1	1	?	?	2	1	1	1	0	2	2	?	1	0	2	0	0
Leontopithecus	1	0	0	2	2	2	?	2	0	1	1	0	1	?	?	2	1	1	1	0	1	2	?	1	1	2	0	0
Callimico	1	0	2	2	1	3	?	3	0	0	0	0	0	0	0	1	1	1	1	0	1	1	?	1	0	0	1	0
Saimiri	2	0	2	2	3	2	1	2	1	0	0	1	0	0	1	2	0	0	2	0	2	0	1	1	1	2	0	0
Cebus	2	1	2	2	3	1	2	1	0	2	2	0	0	0	1	1	0	0	2	0	1	0	2	0	0	0	2	1
Aotus	1	0	2	2	3	3	3	3	0	2	2	0	0	1	0	0	0	0	2	0	2	0	2	0	0	2	0	?
Callicebus	2	1	1	2	3	1	2	1	0	2	2	0	0	1	1	0	0	0	2	0	1	0	2	1	0	2	2	1
Patasola	2	0	0	0	3	2	1	3	1	1	1	1	0	0	0	0	1	?	?	0	?	?	?	?	?	?	?	?
Neosaimiri fieldsi	2	0	1	2	3	2	1	2	0	2	2	1	0	?	?	1	0	?	?	0	?	?	?	?	?	?	?	?
Neosaimiri annectens	2	0	1	3	3	2	1	2	0	2	2	1	0	0	0	0	0	1	1	0	1	0	0	1	0	2	0	1

Note: Characters 1–55 and their states are as defined in table 26.4.

ratio of 1.05, close to the condition in *Aotus* and *Callicebus* (1.03 and 1.08). *Saimiri* has molars with almost the same length as breadth (0.99).

PHYLETIC POSITION OF *PATASOLA*

A phylogenetic analysis was undertaken of *Patasola magdalenae*, *Neosaimiri fieldsi*, and *N. annectens*, and eight extant platyrrhines (Callitrichinae: *Callithrix*, *Saguinus*, *Leontopithecus*; Callimiconinae: *Callimico*; *Saimiri*, *Aotus*, *Cebus*, *Callicebus*). We used the phylogenetic analysis program PAUP (Version 3.0s; Swofford 1991) for the analyses. The fifty-five dental characters used were limited to those that can be observed in *Patasola* and/or *Neosaimiri* spp. and the extant taxa. Character definitions and states are given in table 26.4; the character/taxon matrix is presented in table 26.5. No character weighting was imposed on the data, nor were the multistate characters ordered.

A parsimony analysis using a PAUP heuristic search of eleven taxa, with random addition sequence, and ten repetitions yields three Wagner networks of equivalent parsimony (fig. 26.5 a–c). These networks have 156 steps with a consistency index of 0.544 when uninformative characters are excluded. The retention index is 0.550 and the homoplasy index 0.456. These networks were folded so as to make *Aotus*, *Cebus*, and *Callicebus* the outgroups.

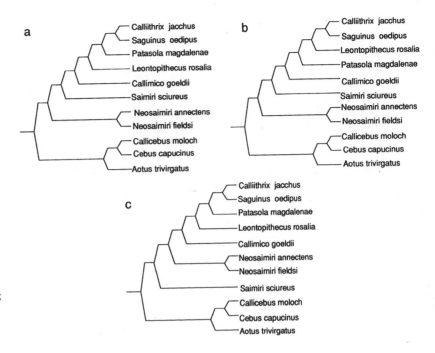

Figure 26.5. Three maximum parsimony trees based on fifty-five characters, illustrating the proposed phyletic position of *Patasola magdalanae*. Each cladogram requires 1,256 steps.

All three maximum-parsimony trees in figure 26.5 share a number of common features: (1) a holophyletic cluster is found corresponding to living Callitrichinae (*Callithrix, Saguinus,* and *Leontopithecus*) with *Patasola* embedded within it. (2) *Callimico,* with callitrichines, form a holophyletic clade Callitrichidae. (3) Two species of *Neosaimiri* cluster as sister taxa. (4) *Saimiri* and *Neosaimiri* are the sister taxa to Callitrichidae. (5) *Cebus* and *Callicebus* cluster as a clade with *Aotus* as their sister taxon. The three maximum-parsimony trees vary in the ordering of *Patasola* within callitrichines. In two trees *Patasola* is the sister group of *Callithrix* and *Saguinus* with *Leontopithecus* as the next outgroup; in the third it is the sister group of all three two-molared callitrichids. The trees also differ in the placement of *Neosaimiri* and *Saimiri*. In two, *Saimiri* is the sister group of callitrichids with *Neosaimiri* as the next outgroup; in the other, the positions of *Neosaimiri* and *Saimiri* are reversed.

Character Transformations

Character state distributions are shown in table 26.6 for the maximum-parsimony tree depicted in figure 26.6, which retains all two-molared callitrichines as a holophyletic group; the other two equally parsimonious cladograms would entail independent loss of the third molars in *Leontopithecus* compared with other callitrichines and imply that *Patasola* has lost its hypocone, an unlikely scenario because a wear facet is found on the trigonid of M_1 that suggests the presence of that cusp.

Characters of the last common ancestor (LCA, fig. 26.6) of *Neosaimiri, Saimiri,* and callitrichids, as we reconstruct them from the cladistic analysis, are as follows.

The lower deciduous incisor roots are small and rounded; the lower deciduous canine root is slightly compressed buccolingually (characters 1–3). DP_2 lacks a metaconid; its trigonid is steeply sloping but is bounded lingually by a weak marginal crest; its talonid is lingually open and lacks an entoconid (4–8). DP_3 is rounded with a fairly large trigonid sporting a well-developed metaconid; its talonid is simple and lacks an entoconid (9–13). DP_4 is molarized, moderately compressed bucculingually, and has an incomplete buccal cingulum; its metaconid and protoconid are of similar height; it has a deeply incised hypoflexid and a talonid with a well-defined entoconid, and a small hypoconulid twinned to entoconid (14–19).

P_3 has a mesiodistally short, broad trigonid, roughly

equal in length to the talonid; it has a large metaconid widely spaced from protoconid; its trigonid basin is moderately sloping and entoconid cristiform (characters 20–24). P_4 is between 70 and 80% of the area of M_1. It has a trigonid roughly equal in length to talonid, a large metaconid widely spaced from the protoconid, and the entoconid is a distinct cusp (25–28). M_{1-2} is buccolingually more or less squared with moderate to high cuspal relief; these teeth have incomplete buccal cingula, metaconids larger than protoconids, small hypoconulids situated lingual to the midline of the crowns, and cristid obliquas that run to the protoconids (29–39, 43, 44). M_3 is small and erupts before P_{2-4} (40–42).

I^1 is slightly larger than I^2; the tooth is buccolingually compressed with a low, complete lingual cingulum and simple, undivided lingual fovea (characters 45–47). P^4 has a cingulum-developed hypocone (55). M^{1-2} has a well-developed hypocone on a strong lingual cingulum lacking a prehypocrista; a pericone is absent as are the paraconule and metaconule (48–54).

Cladogenesis leading from the morphotype described here is as follows (refer to numbered "nodes," i.e., the morphotypes, in fig. 26.6).

Neosaimiri spp. (node 6) departs from this morphotype by adding a dP_3 metaconid close to the protoconid (4). Additionally, dP_4 is altered by reducing the size of metaconid (9),

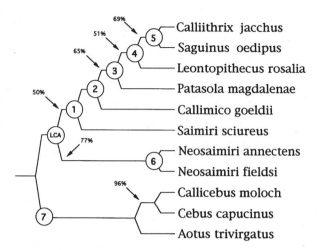

Figure 26.6. Favored hypothesis of the cladogenesis (=maximum parsimony tree *b* in fig. 26.5). The derived character states that support each lineage segment are given in table 26.6. Percentages refer to the frequency of occurrence of the groups in a bootstrap analysis. Branches that are not labeled with a percentage occur less than 50% of the time. *LCA*, last common ancestor.

adding a discrete entoconid (11), and developing a stronger buccal cingulum (17). P_3 has the metaconid reduced in size (22) and has lost the entoconid (24). On M_1 the protoconid and metaconid are more nearly the same height. Further changes from *N. fieldsi* to *N. annectens* include reduction in the size of P_4 (43) and addition of a variable present postentoconid fossa on the molars. We can identify no characters

Table 26.6. Details of the maximum parsimony cladogram in figure 26.6

A. Reconstructions for the internal nodes

Node	Character (1–55)
	1 2 3 4 5 6 7 8 9 0(10) 1 2 3 4 5 6 7 8 9 0(20) 1 2 3 4 5 6 7 8 9 0(30) 1 2 3 4 5 6 7 8 9 0(40) 1 2 3 4 5 6 7 8 9 0(50) 1 2 3 4 5
LCA	0 0 1 0 0 1 1 0 2 1 0 2 1 2 2 0 1 2 1 1 1 2 1 1 1 1 2 2 0 2 2 3 2 1 2 0 2 2 1 0 0 0 1 0 1 1 0 1 0 0 1 0 2 0 1
1	0 0 1 0 0 1 0 0 2 1 0 2 2 2 2 0 1 2 1 1 0 2 1 1 0 1 2 2 0 2 2 3 2 1 2 0 0 0 1 0 0 0 1 0 1 1 0 1 0 1 1 0 2 0 0
2	1 3 1 0 0 1 0 0 2 1 0 1 2 2 2 0 1 2 1 1 0 2 1 1 0 1 2 1 0 2 2 1 2 1 3 0 0 0 0 0 0 1 1 1 1 0 1 1 1 1 0 2 0 0
3	1 3 1 0 0 1 0 0 0 1 0 1 0 2 2 0 1 2 1 0 0 2 0 1 0 0 2 1 0 0 2 1 2 1 3 0 1 1 0 0 0 0 1 1 1 1 0 1 2 1 1 0 2 0 0
4	0 0 1 0 0 1 0 0 0 1 0 1 0 1 2 0 1 1 1 0 0 0 1 0 0 2 1 0 0 2 1 2 1 2 0 1 1 0 1 0 0 2 1 1 1 0 1 2 1 1 0 2 0 0
5	0 1 0 0 0 0 0 0 0 0 0 0 1 2 0 0 1 1 0 0 0 0 1 0 0 2 1 0 0 2 1 2 1 2 0 1 1 0 1 0 0 2 1 1 1 0 2 2 1 1 0 2 0 0
6	0 0 1 2 0 1 1 0 1 1 2 2 1 2 2 0 2 2 1 1 1 1 1 0 1 1 2 2 0 1 2 3 2 1 2 0 2 2 1 0 0 0 1 0 1 1 0 1 0 0 1 0 2 0 1
7	0 0 1 0 1 2 1 0 2 1 0 2 1 2 2 0 0 1 2 1 1 2 1 1 1 1 2 2 0 2 2 3 2 1 2 0 2 2 0 0 0 0 0 0 0 2 0 1 0 2 0 0 2 0 1

Notes: LCA, last common ancestor. Characters 1–55 and their states are as defined in table 26.4.

Table 26.6. Continued

B. Character state transformations along each of the eighteen lineage segments

Lineage Segment	Character	Change	Lineage Segment	Character	Change
(1) LCA to *Saimiri* + Callitrichidae (between LCA and node 1)	7	1 → 0	(6) Autapomorphies of *Callithrix* (*continued*)	28	1 → 0
	13	1 → 2		31	2 → 0
	21	1 → 0		33	2 → 3
	25	1 → 0		35	2 → 3
	37	2 → 0		37	1 → 0
	38	2 → 0		38	1 → 0
	50	0 → 1		45	1 → 2
	55	1 → 0		46	1 → 0
				47	0 → 1
(2) *Saimiri* branch-point to Callitrichidae (between nodes 1 and 2)	1	0 → 1		51	1 → 0
	2	0 → 3			
	12	2 → 1	(7) Autapomorphies of *Saguinus*	18	1 → 2
	28	2 → 1		33	2 → 1
	32	3 → 1		35	2 → 1
	35	2 → 3		39	0 → 1
	39	1 → 0	(8) Autapomorphies of *Leontopithecus*	7	0 → 1
	44	0 → 1		10	1 → 2
	49	0 → 1		13	0 → 1
				22	0 → 1
(3) *Callimico* branch-point to Callitrichinae, including *Patasola* (between nodes 2 and 3)	9	2 → 0		23	0 → 1
	13	2 → 0		25	0 → 1
	20	1 → 0		26	0 → 1
	23	1 → 0		32	1 → 2
	26	1 → 0		52	0 → 1
	30	2 → 0	(9) Autapomorphies of *Patasola*	17	1 → 2
	37	0 → 1		19	1 → 0
	38	0 → 1		24	1 → 2
(4) *Patasola* branch-point to living Callitrichinae (between nodes 3 and 4)	1	1 → 0		27	2 → 1
	2	3 → 0		28	1 → 2
	14	2 → 1		31	2 → 0
	18	2 → 1		32	1 → 3
	22	2 → 0		36	0 → 1
	35	3 → 2		39	0 → 1
	40	0 → 1		43	1 → 0
	43	1 → 2	(10) Autapomorphies of *Callimico*	10	1 → 2
	49	1 → 2		17	1 → 0
(5) *Leontopithecus* branch-point to *Callithrix, Saguinus* (between nodes 4 and 5)	2	0 → 1		19	1 → 2
	3	1 → 0		20	1 → 2
	6	1 → 0		21	0 → 1
	10	1 → 0		25	0 → 1
	12	1 → 0		33	2 → 3
	17	1 → 0		53	2 → 0
	48	1 → 2		54	0 → 1
(6) Autapomorphies of *Callithrix*	2	1 → 2	(11) Autapomorphies of *Saimiri*	6	1 → 2
	3	0 → 2		10	1 → 0
	15	2 → 3		11	0 → 1
	19	1 → 0			
	24	1 → 0			

Continued on next page

Table 26.6 Continued

Lineage Segment	Character	Change	Lineage Segment	Character	Change
(11) Autapomorphies of *Saimiri*	14	$2 \rightarrow 3$	(17) Autapomorphies of *Callicebus*	30	$2 \rightarrow 1$
(*continued*)	15	$2 \rightarrow 1$	(*continued*)	41	$0 \rightarrow 1$
	16	$0 \rightarrow 2$		51	$0 \rightarrow 1$
	24	$1 \rightarrow 2$	(18) Autapomorphies of *Aotus*	1	$0 \rightarrow 1$
	36	$0 \rightarrow 1$		2	$0 \rightarrow 1$
	42	$0 \rightarrow 1$		15	$2 \rightarrow 3$
	43	$1 \rightarrow 2$		28	$2 \rightarrow 1$
	45	$1 \rightarrow 0$		33	$2 \rightarrow 3$
	46	$1 \rightarrow 2$		34	$1 \rightarrow 3$
	48	$1 \rightarrow 2$		35	$2 \rightarrow 3$
	52	$0 \rightarrow 1$		41	$0 \rightarrow 1$
(12) Autapomorphies of *Neosaimiri*	4	$0 \rightarrow 2$		48	$1 \rightarrow 2$
(=*Neosaimiri fieldsi*) (between LCA and	9	$2 \rightarrow 1$			
node 6)	11	$0 \rightarrow 2$			
	17	$1 \rightarrow 2$			
	22	$2 \rightarrow 1$			
	24	$1 \rightarrow 0$			
	30	$2 \rightarrow 1$			
(13) Autapomorphies of *Neosaimiri*	43	$1 \rightarrow 0$			
annectens					
(14) Between LCA and node for	5	$1 \rightarrow 0$			
Cebus, Callicebus, Aotus	6	$2 \rightarrow 1$			
	17	$0 \rightarrow 1$			
	18	$1 \rightarrow 2$			
	19	$2 \rightarrow 1$			
	39	$0 \rightarrow 1$			
	43	$0 \rightarrow 1$			
	45	$0 \rightarrow 1$			
	46	$2 \rightarrow 1$			
	50	$2 \rightarrow 0$			
	51	$0 \rightarrow 1$			
(15) Autapomorphies of	10	$1 \rightarrow 2$			
Callicebus/Cebus branch segment	15	$2 \rightarrow 1$			
	16	$0 \rightarrow 2$			
	20	$1 \rightarrow 2$			
	29	$0 \rightarrow 1$			
	33	$2 \rightarrow 1$			
	34	$1 \rightarrow 2$			
	35	$2 \rightarrow 1$			
	42	$0 \rightarrow 1$			
	54	$0 \rightarrow 2$			
(16) Autapomorphies of *Cebus*	4	$0 \rightarrow 1$			
	8	$0 \rightarrow 2$			
	11	$0 \rightarrow 2$			
	24	$1 \rightarrow 2$			
	43	$0 \rightarrow 1$			
	53	$2 \rightarrow 0$			
(17) Autapomorphies of *Callicebus*	3	$1 \rightarrow 0$			
	10	$2 \rightarrow 3$			

to exclude the possibility that *N. fieldsi* is the ancestor of *N. annectens,* and stratigraphic position allows this also.

The hypothetical ancestor of *Saimiri* and callitrichids (node 1) departs from LCA in a somewhat different direction. The dP$_2$ trigonid is opened lingually (7), dP$_4$ metaconid is reduced (13); the P$_3$ trigonid becomes more triangular by reduction of the mesiolingual corner (21), and the trigonid of P$_4$ is mesiodistally more elongate (25). On M$_{1-2}$ the cristid obliquas are abreviated to reach only to the base of the trigonids (37, 38). The hypocone is lost on the upper premolars (55) and on the molars, a weak prehypocrista develops between the hypocone and postprotocrista (50).

From node 1 the cladogram branches to *Saimiri* through changes in fourteen characters enumerated in table 26.6. In another direction the following changes lead from node 1 to the morphotype of callitrichids (node 2): the deciduous lower incisor roots become more compressed mesiodistally (1, 2) and the crown of dP$_3$ becomes more buccolingually compressed. P$_4$–M$_2$ entoconids become cristiform (28, 32). M$_{1-2}$ are buccolingually compressed (44), their hypoconulids are lost (35), and their buccal cingulua are greatly reduced (39). The M$^{1-2}$ hypocones are reduced in size but still retained.

From node 2 the cladogram branches to *Callimico* through nine changes enumerated in table 26.6. Along the other branch eight other changes lead to the morphotype of *Patasola* and living Callitrichinae (node 3). The dP$_3$ metaconid is lost, and the dP$_4$ metaconid is further reduced (9, 13). On P$_3$ the trigonid basin becomes even more steeply sloping mesiolingually (23). The P$_3$ metaconid is reduced, and that of P$_4$ is approximated to the protoconid (26, 30). In a homoplasy from node 16, M$_{1-2}$ cristid obliquas again reach part way up the back of the trigonids (37, 38).

From node 3 the cladogram again branches to *Patasola* in one direction through the following, not necessarily unique, autapomorphies: dP$_4$ buccal cingulum strong and complete (17); dP4 very compressed buccolingually (19); dP$_4$, P$_4$, and M$_1$ entoconids are homoplasiously reacquired as small discrete cusps (24, 28, 32); the P$_4$ metaconid is reduced in size (27); the M$_2$ protoconid is taller than the metaconid (31), and its cristid obliqua is oriented more mesiodistally toward the protoconid (36). Finally, M$_{1-2}$ buccal cingula are again broken (39), a homoplasy from node 1.

Node 3 also branches to the morphotype of living callitrichines by a series of nine changes listed in table 26.6, the most obvious of which are the loss of M$_3$ and the upper molar hypocones.

Decay and Bootstrap Analyses

To assess the strength of the most parsimonious cladograms, following the recommendations of Mishler (1994), we examine how these proposed clusters "decay" as network topologies of one, two, three, and four additional steps are included. Intuitively, we consider a proposed association to be stronger when it holds together for more steps. We also employ another measure of the robusticity of a proposed relationship, the "bootstrap" approach of Felsenstein (1985), which assesses the strength of the phylogenetic signal across the data set as a whole.

Some of the clusters of taxa are much more robust than others and should be accepted with a higher degree of confidence. Decay analysis shows that the clustering of *Patasola* with living callitrichines holds together in the ten most parsimonious networks discovered that are equal to or one step less parsimonious than the maximum parsimonious network. In this data set only the association between *Cebus* and *Callicebus* is more robust, holding for the first 107 cladograms equal to or three steps less parsimonious than

the maximum parsimonious network. In a bootstrap analysis of randomly selected character sets (fig. 26.6), *Cebus* plus *Callicebus* was again the most robust group with *Neosaimiri* species, *Saguinus* plus *Callithrix*, and *Patasola* plus living callitrichids being, in sequence, less well supported.

The maximum-parsimony trees in figure 26.5 may be used as a yardstick to compare other proposed phylogenies drawn from the literature.

In the most fully resolved phylogenetic hypotheses, Rosenberger, Setoguchi, et al. (1991) suggest that *Laventiana* (=*Neosaimiri*) *annectens* may be a sister group to *Saimiri* (including *N. fieldsi* as a primitive member of the latter) and *Cebus*. *Cebus, Saimiri, Neosaimiri,* and *Laventiana* were depicted as the sister group to callitrichids (*Saguinus* and *Callimico*) (fig. 2 in Rosenberger, Setoguchi, et al. 1991). Using PAUP, we rearranged our taxa to conform with this proposal and found that 172 steps were required, 16 steps more than the maximum parsimony tree. When we clustered *Saimiri* with *N. fieldsi* but left *Cebus* where it fits best in our data, that is, with *Callicebus* 162 steps were required, 7 more than the minimum. A clustering of *Cebus* and *Saimiri* with all other parts of the tree left alone requires 166 steps, 10 steps over the minimum. Thus, this limited data set lends no support to the proposal that *Saimiri* and *Cebus* are a monophyletic group. Nor is the clustering of *Neosaimiri* with *Saimiri* supported.

The possibility must also be entertained that *Patasola* is simply a more primitive species of *Neosaimiri*. We find such a linkage of *Patasola* with *Neosaimiri* to be implausible. Such a reorganization of taxa requires 161 steps, 5 more than the minimum. Other possible arrangements of *Patasola* are also less parsimonious: linking *Patasola* with *Saimiri* requires 160 steps (+4), with *Callimico* 161 steps (+5).

Other Small Platyrrhine Material from La Venta

A number of other isolated finds have been made at La Venta (refer to table 26.1 for a summary of the stratigraphic occurrence of this material). These are arranged stratigraphically from youngest to oldest in the following discussions.

POLONIA RED BEDS, CERRO COLORADO MEMBER, VILLAVIEJA FORMATION

The material from this stratigraphic interval is approximately 11.8 Ma, 0.8 million years younger than the next oldest primate material in the El Cardón Red Beds (chaps. 2, 3). A single individual, IGM 250829 (fig. 26.7), has been recovered from this stratigraphic horizon. The specimen consists of a right P_4, a right C_1, a left C_1, and a right M_3, a left upper molar, and left I^{1-2}, C^1. Dimensions of these teeth are given in table 26.2. The specimen comes from Duke Locality 126.

Both right and left lower canines were unerupted or partially erupted. The roots are unformed and the bases of the enamel of the crowns is incomplete at the cervical margin. These were projecting teeth slightly larger than the types of *Neosaimiri annectens* and *N. fieldsi,* which they closely resemble. The lingual cingulum is complete and runs from the mesial terminus of the paracrista to the distal edge of the tooth where it ends in a small tubercle (fig. 26.7).

P_4 is broken with the mesial part of the trigonid gone. Distally, the tooth is complete. There is no wear on the crown, and the root is incompletely formed, suggesting that it may not have been erupted. The tooth resembles *Neosaimiri* in size and in being more buccally inflated; both these features are distinctions from *Patasola,* which has a smaller and less-inflated P_4. M_3 is also similar to *Patasola, Neosaimiri annectens,* and *Saimiri* with a broader trigonid than talonid and in lacking a heel.

The upper incisors and canine of the Polonia specimen (fig. 26.7) are heteromorphic as in other platyrrhines: I^1 is spatulate, and I^2 is lanceolate. In absolute and relative size these teeth most closely resemble *Neosaimiri annectens, Saimiri,* and *Callimico.* The incisors, however, are very compressed labiolingually as in *Neosaimiri annectens* and callitrichids and distinctly unlike *Saimiri.* I^{1-2} have low, well-defined, and complete lingual cingula and no basal lingual tubercles, a close resemblance to *Neosaimiri annectens, Callimico, Saguinus,* and *Leontopithecus.* The lingual cingulum of *Saimiri* is somewhat more inflated. *Callithrix* and *Cebuella* have weak incomplete lingual cingula. The upper canine has the beginnings of a mesial groove, but at the stage of crown formation it is impossible to say if this

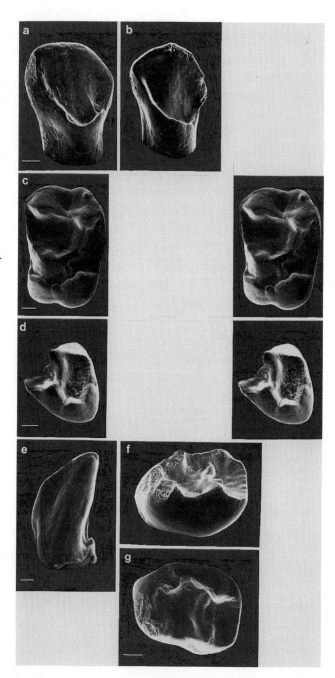

Figure 26.7. Isolated material from Polonia Red Beds representing a single individual, IGM 250829: *a.* I^1, lingual view. *b.* I^2, lingual view. *c.* $M^{1 \text{ or } 2}$, stereo pair in occlusal view. *d.* P_4, stereo pair in occlusal view. *e.* C_1, lingual view *f, g.* M_3, occlusolingual and occlusal views. Scale bars 1 mm.

groove would have extended onto the root or if there was a lingual cingulum.

M^1 or M^2 (fig. 26.7) is very broad buccolingually, most closely resembling *Saimiri* and *Callimico* and very unlike the more squared upper molars of *Neosaimiri annectens*. It has four principal cusps with the small hypocone situated distolingual to the protocone on a prominent lingual cingulum that stretches lingually around the base of the protocone. The hypocone is separated from the post-protocrista by a sulcus; the prehypocrista is absent. The preprotocrista supports a weak paraconule; no metaconule is visible on the strong postprotocrista that reaches the base of the metacone. Remnants of a buccal cingulum are present mesially and distally. A small cusp is found buccal to the metacone. A pericone is present lingual to the protocone on the strong lingual cingulum. Overall, this tooth strikingly resembles the molars of *Saimiri* and *Callimico*. It differs from the former in having a paraconule and slightly better-developed hypocone. In addition to being broader the tooth differs from *Neosaimiri annectens* in having a pericone and paraconule.

The Polonia primate presents some uncertainties as to allocation. Reference to *Neosaimiri* is doubtful because of the shape and structure of the upper molar. Reference to *Saimiri* is likewise unlikely because of the differing shape of the upper incisors, which we interpret as being more primitive than those of *Saimiri*. Allocation to *Patasola magdalenae* is unlikely owing to size differences. Pending recovery of more material of *Patasola*, especially upper molars, positive identification of the Polonia specimen is doubtful.

FISH BED, BARAYA MEMBER, VILLAVIEJA FORMATION

Material from the Fish Bed is considered to be approximately 12.9 Ma, the same age as the Monkey Beds, but younger by stratigraphic position (chaps. 2, 3). Two small isolated teeth were collected from screenwash concentrate from Duke Locality CVP-13C in this interval. These specimens probably do not pertain to *Patasola*. They may ultimately be assignable to *Neosaimiri*.

IGM 250613 (fig. 26.8) is a left M¹ or M². The mesiodistal length is 2.8 mm (this is a conservative measure because enamel is eroded from the mesial surface of the paracone). The buccolingual breadth is 3.4 mm (this definitely underestimates the actual breadth, which, allowing for the missing enamel on the buccal side of the tooth, was likely nearer 3.5–3.6 mm). Therefore, the occlusal area (length ×

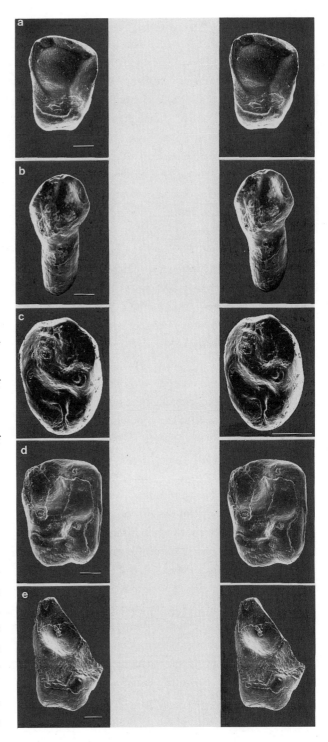

Figure 26.8. Stereo pairs of other small platyrrhine dental material discussed in the text. *a.* IGM-KU 8401, type, *"Micodon kiotensis." b.* IGM-KU 8402. *c.* IGM-KU 8403. *d.* IGM 250613. *e.* IGM 250346. Note that *b* and *c* were not referred to *"Micodon"* in the original description. Scale bars 0.5 mm.

breadth) for this tooth is between 9.8 and 10.4 mm². These values fall within the range for *Neosaimiri annectens* M²s (mean = 8.74 mm²; SD = 0.52 mm²; N = 10; measurements from Takai 1994). The tooth is nearly square in outline, being less mesiodistally compressed than the M¹ or M² of *Saimiri*. It is most similar in proportion to *Neosaimiri annectens* or IGM-KU 8401, the type of *Micodon kyotensis*. The crown was considerably worn in life, with dentine exposed on all cusps; therefore, the original crown relief cannot be inferred with certainty although it seems likely to have been relatively low, similar to *Neosaimiri annectens*. The trigon is quadrangular and resembles that taxon in that the postprotocrista runs distally before turning sharply buccally toward the metacone. In *Saimiri*, the callitrichids, and the Polonia specimen the trigon is triangular. The hypocone is raised to the level of the protocone, and the prehypocrista, although quite worn, must have been strong and continuous with the postprotocista as in *Aotus/Callicebus*. This differs from *Neosaimiri annectens, Saimiri, Callimico,* and the Polonia specimen in which the hypocone is lower relative to the protocone and is separated from the postprotocrista by a sulcus. The lingual cingulum is strong at the base of the protocone, very similar to the condition in *Neosaimiri annectens, Saimiri,* and the Polonia specimen. Overall this tooth shows great similarity to *Neosaimiri annectens*.

IGM 250346 (fig. 26.8) is an incomplete, damaged left upper M¹ or M². Enamel is missing over the paracone and from the lingual border of the crown, and the distobuccal corner of the tooth is missing. It is slightly broader buccolingually (~3.95 mm, allowing for missing enamel) and slightly shorter mesiodistally (~2.7–2.8 mm) than IGM 250613. The lingual border is narrow, giving a "pinched" appearance. The trigon is quadrangular with the postprotocrista running distally before curving buccally toward the metacone. The protocone and paracone are rather tall, suggesting that the cusp relief was moderate. The hypocone is displaced lingual to the protocone and is much shorter than the latter. The hypocone is separated from the postprotocrista by a sulcus. This tooth differs from the Polonia specimen in the narrowness of the lingual margin and the quadrangular shape of the trigon. It resembles *Neosaimiri annectens* even more closely than IGM 250613, but its incompleteness and eroded condition preclude definite allocation.

MONKEY BEDS, BARAYA MEMBER, VILLAVIEJA FORMATION

The age of the Monkey Beds is approximately 12.9 Ma (chaps. 2, 3). Setoguchi and Rosenberger (1985a) named a small platyrrhine from La Venta *Micodon kiotensis*. The type and only specimen, IGM-KU 8401 (fig. 26.8), was identified as a left M¹ but may be an M² because of the small size and lingual displacement of the metacone by analogy with M² of *Aotus trivirgatus* and *Neosaimiri annectens*. Two other specimens were described by these authors in the same paper but were not included in the proposed hypodigm of the species. These are discussed separately here. Through postmortem erosion, the crown of IGM-KU 8401 is nearly devoid of enamel except for a small area between the protocone and hypocone. This obscures most of its occlusal details. There are four principal cusps, including a large hypocone. The inferred size of the hypocone would be similar to that of *Neosaimiri annectens, Aotus,* or *Callicebus*. *Saimiri* and the Polonia specimen have smaller hypocones. Hypocones are greatly reduced in *Callimico* and absent in callitrichids. In occlusal outline it is more squared than *Saimiri, Patasola,* or *Callimico* and close to *Neosaimiri annectens, Aotus,* or *Callicebus*. Compensating for the missing enamel in this specimen, assuming enamel thickness was comparable to that in similar-sized extant platyrrhines, the buccolingual breadth of this tooth was between 3.45 and 3.55 mm and the mesiodistal length was between 2.65 and 2.75 mm. This gives an occlusal area of between 9.14 and 9.76 mm². These values fall within the lower limit of M² area in *Aotus trivirgatus* but are smaller than the sample of teeth of *Neosaimiri annectens* described by Takai (1994) (table 26.1). The trigon is quadrangular, resulting from the postprotocrista running distally before turning buccally toward the metacone. The shape of the postprotocristid closely approximates *Neosaimiri annectens* and *Aotus* and is distinct from the Polonia specimen, *Saimiri,* and callitrichids in which the crest takes a straighter course toward the metacone. A prehypocrista appears to have been present in *"Micodon"* as in *Aotus* and *Callicebus* but is reduced and often absent in *Neosaimiri annectens, Saimiri,* and the Polonia specimen. The lingual cingulum is well developed, extending from the hypocone mesiolingually around the protocone as in *Neosaimiri annectens,* the Polonia specimen, *Callicebus,* or *Saimiri,* but not in *Aotus.*

Allocation of this tooth presents many difficulties, not the least of which is that most of the enamel is eroded away. The large size of the hypocone precludes allocation of this tooth to any callitrichid. The more squared shape of IGM-KU 8401 together with its larger hypocone sets it apart from the Polonia specimen. The tooth appears to be too small to belong to *Neosaimiri annectens* with which it is otherwise similar. *Neosaimiri fieldsi,* however, from the same stratigraphic level as IGM-KU 8401 is smaller than *N. annectens.* When upper teeth of *N. fieldsi* become known, it may be possible to allocate this specimen to it.

Two other teeth from the Monkey Beds were described with *Micodon kiotensis* by Setoguchi and Rosenberger (1985a) but not placed by them in the hypodigm of that taxon.

IGM-KU 8402 (fig. 26.8) is a very small right upper incisor closely resembling I^1 of some callitrichids. Therefore, we prefer to identify this tooth as an I^1. (Setoguchi and Rosenberger [1985a] described this tooth as I^2, but later Rosenberger et al. [1990] changed the identification to an I^1.) The crown occlusal outline is labiolingually compressed, resembling the callitrichids, *Neosaimiri annectens,* and the Polonia species. The mesial border is prominent and angles sharply lingually, whereas the distal border slopes steeply distally. This shape resembles the I^1 of *Callithrix* and *Cebuella* and is an important distinction from the other callitrichids, *Neosaimiri annectens,* and the Polonia species in which the mesial border has no lingual inflection. The lingual surface of the crown is marked by a central pillar, creating a bifoveate lingual surface, and a lingual basal crest is absent as is also characteristic of *Callithrix* and *Cebuella* and unlike the aforementioned taxa. The distinctive morphology of this incisor is very suggestive of the presence of a small *Callithrix*-like primate, but we do not feel naming a taxon based on a single isolated tooth is warranted at this time.

IGM-KU 8403 (fig. 26.8), a left P$_4$, is a very small tooth, measuring 2.2 × 1.9 mm (mesiodistal × buccolingual). The occlusal area is 4.18 mm^2, well within the range for a sample of *Callithrix jacchus* and *Saguinus fuscicollis* (table 26.2). The protoconid is slightly damaged by erosion but would have been larger and more projecting than the metaconid. The trigonid is nearly square, shallow, and enclosed lingually. A small, indistinct paraconid is present. The mesiobuccal border of the crown is inflated. The talonid is very narrow relative to the trigonid. The entoconid is large and rounded and separated from a large hypoconulid by a deep sulcus. The hypoconulid sits directly on the midline. In contrast, the entoconid of the P$_4$ of *Patasola* is small and discrete and no hypoconulid is present.

Rosenberger et al. (1990) note unspecified similarities between IGM-KU 8403 and *Leontopithecus.* The configuration of the trigonid, however, is not similar to that of the callitrichids. The trigonid basin is horizontal, more similar to the pitheciines and *Callicebus* and unlike *Leontopithecus* in which it is steeply sloping. The large entoconid resembles *Saimiri* and *Cebus* (and atelines). The presence of a hypoconulid is unusual on a premolar. IGM-KU 8403 shares many similarities with *Soriacebus* of the Miocene of Patagonia (Fleagle 1990). These include a raised nearly square trigonid; the presence of a paraconid; a bowed distal wall of the trigonid; an entoconid separated from a centrally placed hypoconulid by a sulcus; similar proportions of talonid and trigonid and a mesiobuccal inflated crown. IGM-KU 8403, with an area of 4.18 mm^2 is below the P$_4$ area for the smallest *S. adrianae,* which is 4.8 mm^2 (Fleagle 1990). The principal difference, other than smaller size, is that IGM-KU 8403 is compressed buccolingually in contrast to *Soriacebus* in which the buccolingual breadth is greater than the mesiodistal length.

Adaptations of *Patasola*

Determination of body weight for *Patasola* may be based on its molar occlusal area. From body weights of females of fifteen platyrrhine species found in the literature and molar areas of the same species, we derive the following least-squares regression with an r^2 of 0.935:

$$\ln \text{female body weight} = \ln \text{M}_1 \text{ area } (1.565) + 3.272.$$

Based on this equation the estimated weight of IGM 184332, if a female, would be approximately 480 g, that is, about the same size as *Leontopithecus rosalia.* It is impossible to determine if the available specimens belong to males or females or if there was any sexual dimorphism.

DIET

Fleagle et al. (chap. 28) have assessed the development of shearing on the molars of various living and fossil platyrrhines. They report that small-bodied living platyrrhines with well-developed molar shearing like *Saimiri,* and especially *Callimico,* eat substantial quantities of insects. In contrast, small species that eat more plant gum (e.g., *Saguinus, Callithrix*) or fruit (*Aotus*) tend to have less-well-developed shearing. The shearing index reported for *Patasola* falls within the range of gum- or fruit-eaters, suggesting that this species had a similar diet. The possibility that *Patasola* ate vegetable gum, as do marmosets and tamarins, is further hinted by some anterior dental resemblances between *Patasola* and callitrichids, especially *Saguinus.*

Summary and Conclusions

The discovery and analysis of the small platyrrhine monkey, *Patasola magdalenae,* provides new insights into the phyletic position of the callitrichids, *Callimico,* and *Saimiri.* An analysis of dental characters leads us to the phylogenetic interpretation summarized in figure 26.6.

The mosaic of characters shared by *Patasola* with both the callitrichids and *Callimico,* on the one hand, *Neosaimiri* and *Saimiri,* on the other hand, suggest that this fossil taxon may occupy a phyletic position intermediate to these two groups. Furthermore, the numerous distinctions between *Saimiri* and *Cebus* that are evident from this analysis argue against their sister relationship as proposed by Rosenberger (1979b, 1981b). *Saimiri* differs from *Cebus,* being more similar to the callitrichids, in features of the deciduous premolar row involving greater buccolingual compression, steeper trigonids and open talonids, and relatively smaller metaconids. This trend is also apparent, to a lesser degree, in the permanent premolar row. For example, the P_2 in *Saimiri* lacks a metaconid, which is present in *Cebus.* The area of P_4/M_1 is smaller in *Saimiri* than *Cebus.* Turning to the molars, the lingual cingulum is well developed in *Saimiri* and not *Cebus. Cebus* retains a well-developed metaconule that is lost in *Saimiri.* Analysis of these and other characters

indicates that *Saimiri* is the sister taxon to the callitrichids and *Callimico* to the exclusion of *Cebus.*

It will have become clear from the review here of isolated dental specimens of small platyrrhines from La Venta that recognition of any as a distinct and diagnosible taxon is unwarranted at this time. An enigmatic specimen from the Polonia Red Beds is the youngest known platyrrhine from La Venta. The structure of its upper incisors is unlike *Saimiri,* and the upper molar is unlike *Neosaimiri* spp. Allocation to *Patasola* is possible but must await fuller documentation of the upper teeth of the latter. Several molars recently recovered by screenwashing in the Fish Bed are poorly preserved but are tentatively assignable to *Neosaimiri* spp. From the Monkey Beds, contemporaneous with *Neosaimiri fieldsi,* comes the isolated type of *Micodon.* This specimen is in very poor condition. By its shape and larger hypocone and prehypocrista, it differs from the Polonia specimen and callitrichids. Except by the apparently better-developed prehypocrista and slightly smaller size it cannot be distinguished from *Neosaimiri annectens.* Possibly, it belongs to *Neosaimiri fieldsi* from the same stratigraphic level. Two other teeth originally described by Setoguchi and Rosenberger (1985a) are even more enigmatic. An upper incisor apparently belongs to a callitrichid, whereas a lower premolar most closely resembles *Soriacebus* of the early Santacrucian of Patagonia although it is much smaller.

The presence of *Patasola,* a small insectivorous/frugivorous primate, suggests forested habitats in the Miocene of La Venta. The overall generic diversity of the primates at La Venta, along with postcranial features of some, further reinforces this possibility and suggests the presence of rain forest.

Acknowledgments

This research was supported by NSF grant BSR 8614533 and a Smithsonian Institution Fellowship to R. F. Kay, and NSF grant BNS 8719448 to D. J. Meldrum. Measurements of extant specimens were made with the help of J. M. Plavcan. We thank J. J. Flynn, A. L. Rosenberger, J. G. Fleagle, and R. H. Madden for their helpful comments and criticisms.

27
Postcranial Skeleton of
Laventan Platyrrhines

D. Jeffrey Meldrum and Richard F. Kay

RESUMEN

El archivo fósil para elementos del esqueleto postcraneal para primates en el Mioceno de Colombia es relativamente pobre, por lo que los elementos conocidos se puede atribuir solamente a cuatro taxa. De esta manera, no es posible reconstruir la diversidad total de comportamientos de postura en la comunidad de primates en la fauna de La Venta.

En este capítulo se resume la información disponible sobre los elementos postcraneales fósiles atribuidos al orden provenientes del Grupo Honda y se describe algunos especímenes nuevos asignados a *Cebupithecia* (Pitheciinae) y *Neosaimiri* (=*Laventiana,* Saimiriinae). Luego, se discute el origen de los distintos morfotipos de ciertas subfamilias de platirrinos actuales.

La comparación de elementos del esqueleto de *Cebupithecia* con *Neosaimiri* permite inferir algunos aspectos de la diversidad adaptativa de estos dos taxa. Dada las semejanzas marcadas entre los dos taxa fósiles y sus contrapartes actuales, parece razonable inferir que *Cebupithecia* y *Neosaimiri* habían logrado subdividir la parte baja de la copa arbórea en una manera análoga a *Pithecia* y *Saimiri* en la actualidad. Sin embargo, numerosos rasgos primitivos en el esqueleto de *Cebupithecia* y en la dentadura de *Neosaimiri* demuestran que las características modernas de estas dos subfamilias no pudieron haber evolucionado en sincronía. Inferencias acerca de las adaptaciones locomotoras en otros taxa fósiles reconocidos solamente en base a dentaduras serán posibles con el descubrimiento de los elementos postcraneales respectivos.

The living platyrrhine monkeys of South America span a considerable range of body size, from the 100-g pygmy marmoset, *Cebuella pygmaea,* to the 10-kg woolly spider monkey, *Brachyteles arachnoides.* The evolution of body size as an adaptive character related to food resource exploitation is recognized as a principal factor in the diversification of locomotor and postural adaptations. Consequently, much of the variation in craniodental and postcranial skeletal features of platyrrhines is correlated with increasing and decreasing body size in an arboreal habitat. The radiation of the platyrrhine primates can also be defined in terms of locomotor diversification associated with ecological adaptation and niche partitioning. Novel locomotor strategies have played a prominent role in the adaptive radiation of primate taxa (e.g., Dagosto 1988). Reconstructing the evolutionary history of this diversification is one of the principal aims of paleoprimatology. Unfortunately, the postcranial fossil record for Miocene Colombian primates is relatively sparse (see table 27.1), somewhat limiting the potential for reconstructing positional behavioral diversity in extinct Colombian primate

Table 27.1. Platyrrhine primate postcranial fossils from the Miocene of Colombia

Specimen	Stratigraphic Position	Locality	Description	Taxon Assignment
UCMP 38762	Villavieja Formation, Baraya Member, Monkey Beds	UC Locality V4517	Partial skeleton ~70% complete	Holotype of *Cebupithecia sarmientoi* (Stirton & Savage 1951)
IGM 184667	La Victoria Formation, below Chunchullo Sandstone Beds	Duke Locality 79	Partial pelvis and pelvic limbs	*Cebupithecia sarmientoi* (Meldrum, Fleagle, & Kay 1990a)
IGM 183420	La Victoria Formation, below Chunchullo Sandstone Beds	Duke Locality 21	Left distal humerus	*Cebupithecia sarmientoi* (Meldrum et al. 1990a)
IGM 182858	La Victoria Formation, below Chunchullo Sandstone Beds	Duke Locality 21	Left proximal humerus	*Cebupithecia sarmientoi* (Meldrum & Kay, chap. 27)
IGM 184074	Villavieja Formation, Cerro Colorado Member, El Cardón Red Beds	Duke Locality 43	Left talus	Gen. et sp. indet. (Ford et al. 1991)
IGM 183512	Villavieja Formation, Baraya Member, Monkey Beds	Duke Locality 54	Left distal humerus	*Neosaimiri* cf. *N. fieldsi* (Meldrum et al. 1990a)
IGM 250436	Villavieja Formation, Cerro Colorado Member, El Cardón Red Beds	Duke Locality 32	Left distal tibia	*Neosaimiri annectens* (Meldrum & Kay, chap. 27)
IGM-KU 8803	Villavieja Formation, Cerro Colorado Member, El Cardón Red Beds	Kyōto Locality Masato site	Right talus	*Laventiana* (=*Neosaimiri*) *annectens* (Gebo et al. 1990; Rosenberger, Setoguchi, & Hartwig 1991)
IGM-KU 8802	Villavieja Formation, Monkey Beds	Kyōto Locality 9-86A	Right talus	*Aotus* cf. *A. dindensis* (Gebo et al. 1990)

Notes: IGM, INGEOMINAS Bogotá; UCMP, University of California Museum of Paleontology, Berkeley; KU, Kyōto University.

communities. Fieldwork, however, continues to produce new finds, adding further elements to the picture of fossil platyrrhine postcranial morphology and diversity. In this chapter we review the platyrrhine postcranial fossils from the Miocene of Colombia, describe several new specimens, and discuss the establishment of the distinctive morphotypes of selected extant platyrrhine subfamilies.

Materials

CEBUPITHECIA SARMIENTOI

The first and, by far, the most significant discovery of a platyrrhine postcranial fossil in Colombia was made during R. A. Stirton's 1944–45 joint paleontological expedition of the University of California and the Servicio Geológico de Colombia to the La Venta badlands (Stirton 1951; Stirton and Savage 1951). The skeleton of a small monkey (UCMP 38762) (museum abbreviations are given in table 27.1) was recovered, with an estimated body size of 2–3 kg (Fleagle 1988). The holotype of *Cebupithecia sarmientoi* includes the postcranial skeleton that is approximately 70% complete (fig. 27.1). In the past it had been erroneously suggested that the craniodental and the postcranial elements of this specimen were incompatible, that is, the skull of an immature individual associated with the adult postcranium of a second individual or possibly a different taxon (Rosenberger 1979a; Ford 1980a, 1986, but see Ford 1990a for

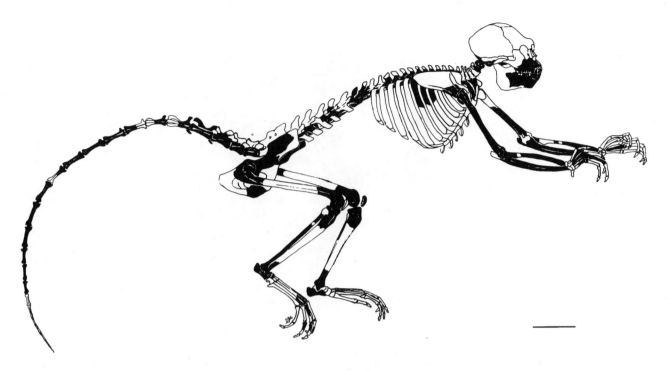

Figure 27.1. Reconstruction of the skeletal anatomy of *Cebupithecia sarmientoi*. Darkened areas indicate preserved portions of the skeleton. Scale bar 5 cm.

revised position). In fact, more recent studies have convincingly demonstrated the integrity of the entire skeleton based on dental eruption and wear, basicranial morphology, and postcranial affinities (Davis 1987; Meldrum and Fleagle 1988; Kay 1990a; Meldrum and Lemelin 1991). The discovery of the complementary fragments of the basicranium, including the occipital condyles, together with the atlas, which articulate with one another, clearly established the unity of this specimen (Meldrum and Lemelin 1991).

Stirton observed that *Cebupithecia* bears a close resemblance to the Pitheciinae, and to *Pithecia* in particular, while noting some distinctions of the postcranium, such as similarity of the femur to callitrichines, as elaborated by Ciochon and Corruccini (1975) and Davis (1987). More extensive analysis of the postcranial anatomy confirms Stirton's overall characterization of *Cebupithecia* (e.g., Fleagle and Meldrum 1988; Meldrum and Fleagle 1988; Ford 1990a; Meldrum and Lemelin 1991). These postcranial similarities suggest that *Cebupithecia* was basically an arboreal quadruped that frequently adopted vertical clinging postures and was an adept leaper. Many aspects of the

postcranium, however, are rather primitive in character and do not show the derived conditions that characterize the extant pitheciines. These are discussed in more detail here.

NEW POSTCRANIAL REMAINS OF *CEBUPITHECIA SARMIENTOI*

Humerus

Ongoing paleontological fieldwork by Duke University, in cooperation with INGEOMINAS, has produced additional postcranial fossils of *Cebupithecia* (table 27.1). IGM 183420 is a left distal humerus fragment allocated to *Cebupithecia* (Meldrum et al. 1990a). This fossil is less than one million years older than the holotype of *Cebupithecia* and is extremely similar in size and morphology. It presents a number of features associated with clinging postures similar to those present in the humerus of the extant pitheciine *Pithecia pithecia,* including a medial epicondyle with very little dorsal angulation, a cylindrical trochlea, and a contact facet for the coronoid process of the ulna (fig. 27.2).

A fragmentary left proximal humerus (fig. 27.3) and por-

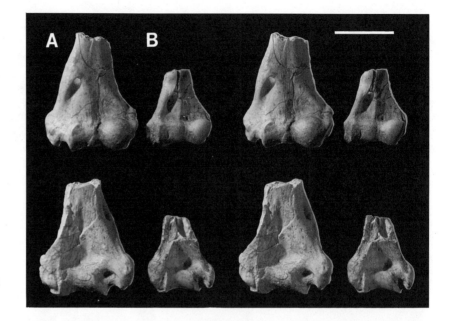

Figure 27.2. Stereophotographs of the left distal humeri of two platyrrhine species: *A, Cebupithecia sarmientoi,* IGM 183420; and *B, Neosaimiri fieldsi,* IGM 183512. Scale bar 1 cm.

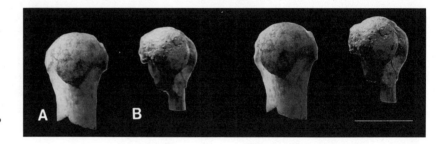

Figure 27.3. Stereophotographs of the proximal humerus of *Cebupithecia: A,* UCMP 39762; *B,* IGM 182858. Scale bar 1 cm.

tion of the diaphysis were recovered in 1985 and are described here for the first time. IGM 182858 was recovered at Duke Locality 21, in the La Victoria Formation, Perico Member, below the Chunchullo Sandstone Beds. The humeral head is oval in outline and less beaked than that of UCMP 38762. The lesser tubercle is very pronounced and demarcated from the articular surface by a distinct groove. A prominent crest extends distally from the lesser tubercle, providing attachment for the teres major muscle. The greater tubercle is considerably eroded and shattered but appears to have been very similar to UCMP 38762. The portion of the diaphysis articulates loosely with the distal humeral fragment (IGM 183420) noted earlier from the same locality. It seems likely that the fragments represent the humerus of a single individual. The shaft fragment preserves the brachialis flange, forming a distinct crest along the lateral border.

Pelvis

In 1988 a partial pelvis and pelvic limbs were recovered at Duke Locality 79 in the La Victoria Formation, Perico Member, immediately below the Chunchullo Sandstone Beds (fig. 27.4). The specimen (IGM 184667) is also referred to *Cebupithecia sarmientoi* (Meldrum et al. 1990a).

The pelvic fragments of IGM 184667 consist of the caudal half of the left ilium, the cranial two-thirds of the right ischium, and the caudal one-third of the left ischium. The iliac fragment (fig. 27.4A) preserves the cranial half of the acetabulum, which measures 12.0 mm in diameter. Just medial to the ventral lip of the acetabulum is a small crest. Dorsal to the acetabulum, the beginning of a ridge is evident that leads to the ischial spine. Cranial to the acetabulum, the anterior inferior iliac spine is prominent and robust. The caudal portion of the gluteal fossa is quite

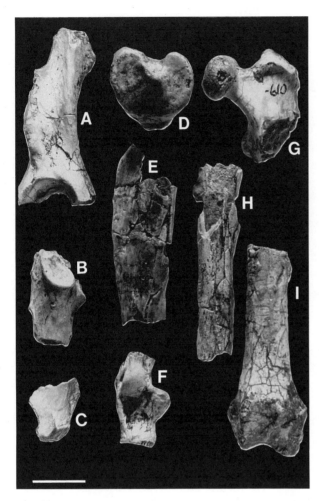

Figure 27.4. Partial pelvis and pelvic limbs of *Cebupithecia sarmientoi*, IGM 184667: *A*, left ilium, lateral view; *B*, right ischium, lateral view; *C*, left ischial tuberosity, lateral view; *D*, right tibial condyles, proximal view; *E*, right tibia, medial view; *F*, right calcaneus, dorsal view; *G*, left proximal femur, dorsal view; *H*, right femoral diaphysis, dorsal view; *I*, right distal femur, dorsal view. Scale bar 1 cm.

acetabulum. The ischial ramus is broken away. Its point of origin, however, is evident on the ventral border.

The left ischial fragment (fig. 27.4C) preserves the caudal end of the ischium, including the ischial tuberosity and the initial part of the ischial ramus. The mediolateral convexity of the ischium continues to the tuberosity. The medial surface is slightly concave owing to a thickening of the dorsal border. The dorsal border is also very rugous along its lateral surface. On the ventral border the cranial edge of the ischial ramus is preserved as indicated by the presence of a small area of cortical bone. The distance from the cranial border of the ischial ramus to the tuberosity measures approximately 10.4 mm.

These pelvic fragments fill in many of the regions missing in the holotype. For example, the overlapping fragments of the ischium permit a reasonably accurate estimate of the length of the ischium, which can be expressed as a ratio of the lower iliac length (the distance between the center of the acetabulum and the auricular surface). With a value 0.96, *Cebupithecia* has a relatively long ischium, comparable to means for *Pithecia monachus* (0.95; range, 0.94–1.01; N = 4) or *Pithecia pithecia* (0.92; range, 0.87–0.99; N = 5; Meldrum, unpublished data). By combining the characteristics of the two sets of fragments a reconstruction of the pelvis was undertaken. The result bears a number of similarities to the pelvis of *Pithecia monachus*, as distinct from *P. pithecia* (fig. 27.5). These are especially evident in the superoinferiorly broad and relatively long pubis. The ratio of pubis length to acetabulum width is 1.64, compared to *P. monachus* with a mean of 1.66 (range, 1.45–1.93; N = 8) and *P. pithecia* with a mean of 1.20 (range, 1.12–1.26; N = 5). The ilium is, however, rather short in comparison to *Pithecia* and more similar to *Cebus, Callicebus,* or the callitrichines. More extensive analysis of the functional implications of this combination of pelvic traits is in preparation.

Femur

The proximal end of the left femur is very well preserved (fig. 27.4G). The head is spherical, and the articular surface extends smoothly onto the posterior surface of the neck. This feature is characteristic of leaping primates and other mammals in which the hind limbs move in an adducted

concave. This is accentuated by the thickening of the dorsal margin of the ilium along the border of the sciatic notch, extending lateral to the auricular surface. The distance between the cranial border of the acetabulum and the caudal border of the auricular surface measures 20.6 mm.

The fragment of the right ischium (fig. 27.4B) preserves a caudal portion of the acetabulum and just over 11.0 mm of the ischium caudal to its border. The ischium is mediolaterally convex, and the break at the caudal end reveals a thickening of the cortical bone on the raised dorsal surface. The dorsal border is marked by a ridge that supports a small ischial spine situated just caudal to the level of the

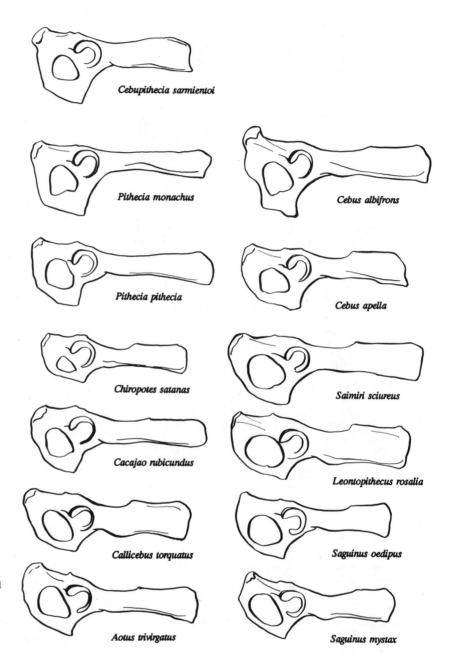

Figure 27.5. Composite reconstruction of the os coxae of *Cebupithecia sarmientoi,* based on UCMP 38762 and IGM 184667, compared with the os coxae of selected platyrrhine species. All specimens scaled to same acetabular width.

parasagittal plane (Walker 1974; Jenkins and Camazine 1977; Fleagle and Meldrum 1988). The femoral neck is rather short and set nearly perpendicular to the long axis of the femoral diaphysis. On the dorsal surface of the neck the intertrochanteric line is evident, which marks the point of attachment of the iliofemoral and ischiofemoral ligaments. These ligaments function to check extreme extension of the hip and the intertrochanteric line is commonly well developed in leaping primates (Fleagle and Meldrum

1988). On the ventral surface of the neck the crista para-trochanterica, for the insertion of the capsular ligament, is well developed. The greater trochanter is dorsoventrally deep and overhangs the femoral shaft dorsally. The greater trochanter does not project above the level of the femoral head. The lesser trochanter is rather robust. The beginning of a ridge on the lateral border suggests the presence of a third trochanter.

The right femur is represented by a shattered portion of

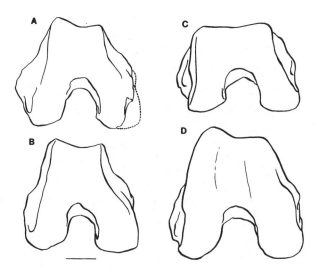

Figure 27.6. Reconstruction of the distal femur of *Cebupithecia sarmientoi: A,* IGM 184667; *B,* UCMP 38762; *C, Pithecia pithecia,* USNM 339660; and *D, Lemur macaco,* ANSP 1422. Scale bar 1 cm.

the midshaft approximately 27 mm long (fig. 27.4H). It is circular in cross section, but no other morphology is discernible. The distal segment of the right femoral shaft is also shattered, but is in a much better state of preservation (fig. 27.4I). It is likewise circular in cross section and flares only very slightly as it meets the condyles. The supracondylar fossae are very faint and shallow. The surfaces of the condyles are eroded, revealing the underlying cancelous bone. The condyles are dorsoventrally deep. The biepicondylar width measures 15.3 mm. The patellar groove is moderately broad, measuring approximately 8.6 mm just dorsal to the intercondylar fossa. The medial and lateral lips of the patellar groove are very prominent. The lateral lip is slightly more pronounced and continues farther proximally than the medial lip. Working from existing landmarks, a conservative reconstruction of the distal outline of the femur was made (fig. 27.6). The result indicates a high lateral lip of the patellar groove and a relatively narrow biepicondylar width as compared to the dorsoventral depth of the condyles. *Cebupithecia* has a value for this ratio of 0.90, which is just outside the range for a sample of platyrrhines (0.65–0.89; N = 109), and is most closely approached by *Aotus* (mean 0.82; range, 0.77–0.89; N = 15) and *Saguinus* (mean, 0.83; range, 0.79–0.86; N = 17). Comparable values are common among prosimian leapers.

Tibia

The proximal portion of the right tibia is preserved in a shattered condition. The tibial condyles are broad (16.0 mm) and are set at a low angle to the long axis of the shaft (fig. 27.4D). The tibial tuberosity is located very proximally. The tibial shaft is very deep dorsoventrally and very mediolaterally compressed (fig. 27.4E). This configuration resists bending moments incurred in the anteroposterior plane as in forceful leaping. The estimated ratio of proximal shaft width to depth is less than 0.48, below the range for a sample of platyrrhines (0.51–0.82; N = 109), but is most closely approximated by *Saguinus* with a mean of 0.56 (range, 0.51–0.65; N = 17).

Calcaneus

The calcaneus is well preserved and lacks only the calcaneal tuberosity (fig. 27.4F). It measures 17.0 mm from the posterior margin of the posterior talar facet to the cuboid facet. The long axis of the posterior talar facet is only slightly oblique to that of the calcaneus. The distal surface of the facet is quite perpendicular to the long axis of the calcaneus. The sustentaculum is triangular. The distal margin of the middle facet is indistinct, and the anterior facet is very narrow and poorly defined. The area of the sinus tarsi, which gives attachment to the cervical ligament, is very rugous. The cuboid facet is dorsoplantarly compressed, and the medial aspect of the facet is much narrower than the lateral surface.

In all features this new specimen reaffirms the interpretation that *Cebupithecia sarmientoi* was an accomplished leaper.

NEOSAIMIRI FIELDSI

It has been proposed to synonymize *Neosaimiri* with the genus *Saimiri* (Rosenberger, Hartwig, et al. 1991). Given the significant differences in dental proportions and mandibular morphology between the two taxa, however, we do not feel synonymy is warranted and will continue to recognize the generic distinction of *Neosaimiri* (chap. 26). This is a reasonable position, in view of the rather minor differences used by Rosenberger, Setoguchi, and Hartwig

(1991) to justify generic distinction of *Laventiana*, as discussed in chapter 26.

A single postcranial fossil has been attributed to *Neosaimiri* (Meldrum et al. 1990a). IGM 183512 is a small left distal humerus (fig. 27.2B). This specimen was recovered from the same stratigraphic horizon as the type specimen of *Neosaimiri*. It is very similar in size and morphology to the extant squirrel monkey, *Saimiri*. The medial epicondyle is more dorsally angled, the medial lip of the trochlea is more prominent, and the capitulum is less spherical by comparison to *Cebupithecia* (fig. 27.2A,B). These anatomical details suggest that the forelimb of *Neosaimiri* was employed in arboreal quadrupedalism much like that seen in the extant genus *Saimiri*.

NEOSAIMIRI (LAVENTIANA) ANNECTENS

This fossil saimiriine was described on the basis of a nearly complete mandible and an associated talus (Setoguchi et al. 1990; Rosenberger, Setoguchi, et al. 1991). The dental morphology of *Laventiana annectens* is extremely similar to that of *Neosaimiri fieldsi*, differing only in having relatively wider and longer M_{1-2}, with differentiated hypoconulids and prominent postentoconid sulci. Takai (1994) suggests synonomy of *Laventiana* with *Neosaimiri*, a view we adopt here.

During the 1989 Duke/INGEOMINAS field season, the distal fragment of an isolated small primate tibia was recovered (IGM 250436). It was recovered at Duke Locality 32, in the El Cardón Red Beds, Cerro Colorado Member, Villavieja Formation. The fossil is the distal end of a right tibia, approximately 16.5 mm long. The shaft is mediolaterally broad and the lateral border is marked by a pronounced crest (fig. 27.7A), indicating a well-developed syndesmosis between the tibia and fibula that extended almost the entire lateral length of the fragment. There is no indication of synostosis of the inferior tibiofibular joint, but there was apparently a tight appression of the distal tibia and fibula. The fibular notch is shallow, and there are distinct fossae for the inferior tibiofibular ligaments at the anterior and posterior margins of the fibular notch (fig. 27.7B). A smooth articular facet for the fibula is not discernible. Instead, the fibular incisure is very rugous.

The posterior surface of the shaft is flat. At the disto-

medial border of the posterior surface is a small ridge (fig. 27.7C) bounding the posterior malleolar groove. The groove is relatively shallow. The medial malleolus is moderately long and oriented perpendicular to the trochlear surface. The fossa for the attachment of the deltoid ligament is deep and very distinct on the posterolateral edge of the malleolus (fig. 27.7D).

The trochlear surface is nearly equilateral in distal view. The lateral border is slightly longer than the medial. The trochlear surface is concave with a very slight median ridge. The ridge angles slightly from anterolateral to posteromedial edge of the trochlea. The posterior edge of the facet is sharp. The anterior edge is less sharp, but there is no indication of an extension of the articular surface onto the anterior aspect of the shaft.

The fossil tibia is very similar in both size and morphology to the extant squirrel monkey. Five measurements of the distal tibia were taken from the fossil and a sample of twelve adult *Saimiri sciureus* (fig. 27.8). The results are summarized in table 27.2 and reveal that IGM 250436 falls consistently just below the mean values of each variable for *Saimiri*. Nonmetric features described above for the fossil specimen also agree very closely with the tibial morphology of *Saimiri*. Of particular note is the frequent occurrence of a well-developed syndesmosis of the distal tibiofibular joint in *Saimiri*. The posterior surface of the distal tibia of *Saimiri* is quite flat, and the lateral border is marked by a pronounced ridge along which the tibia is appressed. Such an appression of the distal tibia and fibula occurs in several platyrrhine genera in addition to *Saimiri*, including *Aotus*, *Callithrix*, *Cebuella*, and *Pithecia* (fig. 27.8; see also Fleagle and Simons 1983; Fleagle and Meldrum 1988). Only *Saimiri*, however, commonly displays the distinctive mediolateral widening and posterior flattening of the distal tibial shaft also evident in the fossil tibia. In contrast, the tibial shaft of *Callithrix* and *Cebuella* remains narrow, and the fibula is simply appressed to the tibial shaft. In Aotus the syndesmosis occurs less frequently and is quite restricted in length along the lateral border of the distal tibial shaft.

The extensive syndesmosis of the distal tibiofibular joint serves to limit lateral and rotational movement of the tibiotalar joint by restricting movement of the talus to flexion-extension in the sagittal plane (fig. 27.9). This morphology is correlated with frequent leaping. Its presence in

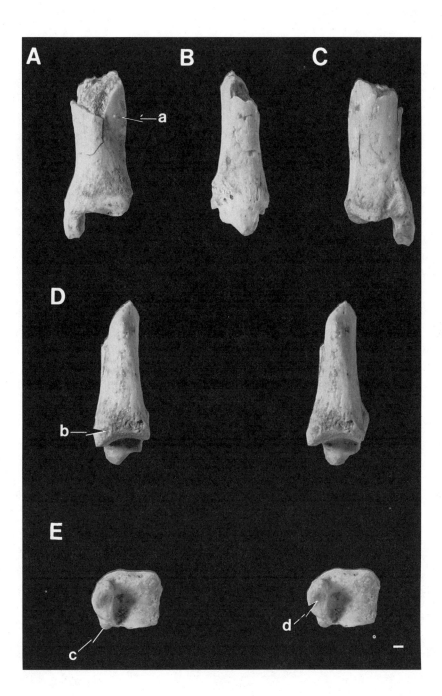

Figure 27.7. Distal fragment of a left tibia, IGM 250436: *A,* anterior view; *B,* medial view; *C,* posterior view; *D,* lateral view (stereo pair); *E,* distal view (stereo pair). Features indicated by lowercase letters are discussed in text. Scale bar 1 mm.

Saimiri agrees with reports of frequent leaping in *Saimiri sciureus* (Fleagle et al. 1981) although some species of *Saimiri,* for example, *S. oerstedii,* reportedly leap less frequently (Boinski 1989). The presence of a well-developed syndesmosis in IGM 250436 indicates that this small fossil saimiriine also employed frequent leaps during travel and foraging.

The fossil tibia differs from *Saimiri* and from most callitrichines and small-bodied platyrrhines in that the articular surface of the tibial trochlea does not extend onto the anterior surface of the tibial shaft (fig. 27.10A). This feature in extant platyrrhines is matched by a small depression at the anterior end of the dorsal surface of the talar trochlea, referred to as the "dorsal tibial stop." The tibial stop receives the anterior edge of the distal tibial epiphysis during extreme dorsiflexion of the talocrural joint.

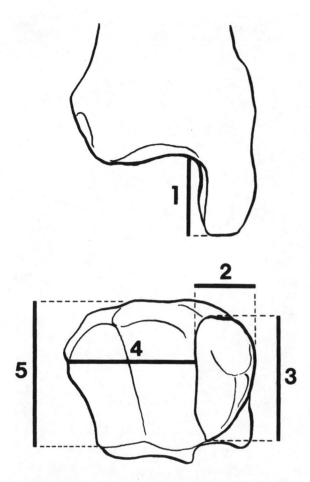

Figure 27.8. Anterior and distal views of a schematic tibia, illustrating the measurements reported in table 27.2: *1,* projection of the medial malleolus below the trochlear surface; *2,* maximum mediolateral width of the medial malleolus; *3,* maximum anteroposterior length of the medial malleolus; *4,* maximum mediolateral length of the trochlear surface; *5,* maximum anteroposterior length of the trochlear surface.

Table 27.2. Five selected measurements (mm) of the fossil tibia IGM 250436 versus those for a sample of tibia from the extant cebid *Saimiri sciureus*

Specimen(s)	1	2	3	4	5
IGM 250436	3.05	2.18	4.13	5.45	4.92
Saimiri sciureus					
Mean	3.14	2.24	4.30	5.53	5.07
Minimum	2.56	2.06	3.88	4.23	4.30
Maximum	3.73	2.38	4.67	6.00	5.90
SD	0.34	0.11	0.23	0.46	0.46

Note: Measurements are described and illustrated in figure 27.8. For *S. sciureus*, n = 12.

The lack of this articular surface in the fossil tibia is particularly interesting in connection with the talus (IGM-KU 8803) associated with the type mandible of *Neosaimiri annectens* from the same locality (=Masato site), which was recovered by the Japanese/American team in cooperation with INGEOMINAS (Gebo et al. 1990). The talus is also very similar to *Saimiri* but lacks a dorsal tibial stop (fig. 27.10B). The overall concordance in size and articular morphology between the tibia and talus suggests that these fossils likely represent the same taxon. The talus (IGM-KU 8803) falls well within the range of twelve linear dimensions for a sample of thirty adult *Saimiri* spp. reported in Meldrum (1990). It is most similar to *Saimiri* among the extant platyrrhines in having a relatively long, narrow talar neck; narrow, rounded talar head; a narrow talar trochlea with a sharp, curved lateral rim; and a small nonfaceted medial protuberance.

"AOTUS DINDENSIS" (MOHANAMICO HERSHKOVITZI)

A small isolated talus (IGM-KU 8802) was recovered within the Monkey Beds (Gebo et al. 1990). It falls within the size range of extant *Aotus* and was referred to cf. *Aotus dindensis* on the basis of a suite of characters consisting of a moderately short talar neck, wide talar head, moderately high talar body, long and narrow talar body, and medium to large medial protuberance, sometimes faceted, which characterize the Aotini, that is *Aotus* and *Callicebus,* as recognized by Gebo et al. (1990). Particular emphasis is placed by Gebo et al. (1990) on the similar development of the medial protuberance in IGM-KU 8802 and *Aotus,* which provides attachment for the posterior talotibial ligament. The condition of this feature is quite variable, and a moderate tubercle is also present in *Pithecia* and some of the callitrichines. The presence of the facet on the medial tubercle is also quite variable, present in only one-third of the specimens of Aotinae and also present in a few specimens of *Cebus* and the fossil talus of *Cebupithecia* (UCMP 38762) (table 2 in Gebo et al. 1990).

Controversy surrounds the validity of this taxon, which affects the interpretation of the affinities of the talus. Kay (1990b) demonstrated that the type of *Aotus dindensis* does

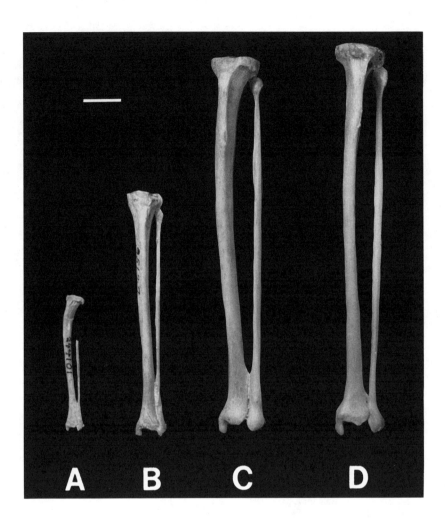

Figure 27.9. Left tibia and fibula of *(A) Cebuella pygmaea, (B) Callithrix jacchus, (C) Saimiri sciureus,* and *(D) Aotus trivirgatus,* illustrating the occurrence of a syndesmosis of the distal tibiofibular joint. Note the distinctive widening of the distal tibia of *Saimiri* and its similarity to IGM 250436 *(S. sciureus)*. Scale bar 1 cm.

not differ significantly from the type specimen of *Mohanamico hershkovitzi* (Luchterhand et al. 1986), and the former should be considered a junior synonym of *Mohanamico hershkovitzi.* There is no compelling evidence to suggest a close affinity of *Mohanamico* to the Aotinae; in fact, Rosenberger et al. (1990) have suggested that *Mohanamico* shares many similarities with the callitrichines and *Callimico,* whereas others have noted affinities to the pitheciines (Luchterhand et al. 1986; Kay 1990b; Meldrum and Kay 1992). This raises the question of the soundness of referring the talus (IGM-KU 8802) to *"Aotus dindensis"* (*Mohanamico hershkovitzi*) on the basis of perceived similarities to *Aotus* or *Callicebus.* Gebo et al. (1990), however, note that IGM-KU 8802 clearly differs from extant *Aotus* in having a more square (relatively wide and short) talar body.

The proportions and configuration of the talar head and trochlea are very similar to those reported for callitrichines (Meldrum 1990). A satisfactory resolution of this issue must await the discovery of associated cranial and postcranial materials of *Mohanamico.*

SPECIMENS OF UNCERTAIN ALLOCATION

Another talus (IGM 184074) was recovered from higher in the section (Ford et al. 1991). It is a left specimen, reasonably well preserved with superficial erosion of the head, posterior tubercles and calcaneal facets. It measures approximately 14.0 mm long, suggesting a monkey intermediate in size to *Callicebus* and *Pithecia* (Meldrum 1990). It shares a number of similarities with the talus of *Cebupithecia* but has

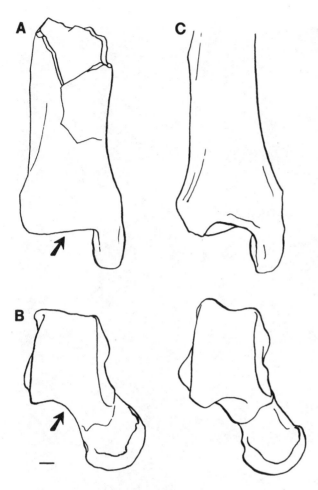

Figure 27.10. Distal tibia and talus of *Saimiri sciureus: A,* USNM 397810; *B,* IGM 250436; *C,* IGM-KU 8803. Scale bar 1 mm.

a somewhat narrower head and neck, and a slightly more rounded lateral trochlear crest. The mandible of a new species of pitheciine (IGM 251074) was recovered from almost the same locality as the talus (Meldrum and Kay 1992). The correlation of talar length and M_1 length suggests that the talus potentially may pertain to this new species.

Adaptive Diversity and the Establishment of Extant Subfamilial Morphotypes

By the middle Miocene several of the modern platyrrhine subfamilies (sensu Thorington and Anderson 1984) are represented in the fossil record: the Alouattinae represented by *Stirtonia tatacoensis* and *S. victoriae;* the Pitheciinae by *Cebupithecia sarmientoi* and IGM 251074 gen. et sp. nov.; the Saimiriinae by *Neosaimiri fieldsi* and *N. annectens;* and the Callitrichidae by *Lagonimico conclucatus* (Kay 1994), *Patasola magdalenae,* and, perhaps, IGM-KU 8402 (an isolated premolar, gen. et sp. indet.). Additionally, taxa of uncertain subfamilial allocation occur, such as *Mohanamico hershkovitzi* (=*Aotus dindensis*). This assessment of taxonomic diversity is based almost solely upon craniodental anatomy. Had these Miocene fossil taxa also achieved the level of diversity in postcranial adaptation that characterizes the extant platyrrhine subfamilies? The postcranial fossils described and discussed in the preceding section provide a limited answer to this question. Only two subfamilies are well represented by fossil postcrania, the Pitheciinae and the Saimiriinae. To what extent do these fossils resemble the postcrania of extant pitheciines and saimiriines?

Ford (1988, 1990a) reconstructs a hypothetical ancestral morphotype for platyrrhines that postcranially resembles *Aotus* (~1.0 kg) with a locomotor repertoire dominated by arboreal quadrupedalism and limited leaping and suspensory behavior. If this hypothetical ancestor is accurate, to what extent had the modern subfamilial postcranial morphotypes differentiated from it by the middle Miocene?

The extant pitheciines are larger than the hypothetical ancestor, weighing on average between 2.0 and 3.5 kg (Fleagle 1988). *Cebupithecia* was within the lower limits of this range, with an estimated body weight of 2.2 kg (Fleagle 1988). The extant pitheciines display more climbing and suspensory behaviors than the hypothetical ancestor, as reflected in their proportionately longer limbs. The relative length of the forelimb, (humerus + radius)/trunk, ranges between 70 and 96 for species of *Chiropotes* and *Pithecia* (Hershkovitz 1985, 1987). Using an estimate of trunk length (cervical to sacral vertebrae inclusive) of 250 mm (Meldrum and Lemelin 1990), *Cebupithecia* had forelimbs relatively shorter than the extant pitheciines, with an estimated value of 65. This falls within the range of values for *Aotus, Callicebus,* and *Saimiri* (52–73; Hershkovitz 1990). The proclivity for suspensory behaviors in the extant pitheciines is also reflected in the configuration of their limb joints. For example, the distal femur of the extant pitheciines is relatively broad and shallow with a wide shallow patellar groove, very similar to the condition in the extant atelines. Likewise, the talar trochlea is low with a rounded lateral margin approaching the condition found in the ateline talus (Meldrum 1990). These and other features of the pitheciine limbs that resemble the ateline condition, which are best explained as skeletal correlates of climbing and suspensory behaviors, are generally lacking in the postcranium of *Cebupithecia*. The distal femur of *Cebupithecia* is very narrow and deep and the talar trochlea is moderately high with a sharply crested lateral margin. It more closely resembles such forms as *Aotus, Callicebus,* or some callitrichines (Ciochon and Corruccini 1975; Davis 1988). Therefore, although *Cebupithecia* had evolved the body size and dental morphology characteristic of modern pitheciines, it still retained many aspects of the hypothetical ancestral postcranial skeleton.

By comparison, much less is known concerning the postcranial anatomy of *Neosaimiri*. The distal humerus, distal tibia, and talus that are known are nearly identical in size and morphology to those of the extant squirrel monkey, *Saimiri*. Analysis of dental characteristics suggests that *Saimiri* and *Neosaimiri* form the sister group to the Callitrichinae and Callimiconinae to the exclusion of *Cebus* (chap. 26). Therefore, the saimiriines may represent part of a radiation of platyrrhines, including the Callitrichinae and Callimiconinae, which underwent reduction in body size to exploit a primarily insectivorous/frugivorous diet. *Saimiri* differs from the hypothetical platyrrhine ancestral morphotype in several ways. Squirrel monkeys are moderately smaller with mean body weights of 960 g for males and 750 g for females (Fleagle 1988). *Neosaimiri fieldsi* had an estimated body weight of 840 g (Fleagle 1988). *Saimiri* has relatively longer hindlimbs than the hypothetical ancestor and leaps with greater frequency. Although nothing can be said about the limb proportions of *Neosaimiri,* the morphology of the distal tibia and talus of *N. annectans* indicates adaptations for frequent leaping. This is particularly evident in the extensive syndesmosis of the distal tibiofibular joint.

By comparing corresponding elements of the postcranium of *Cebupithecia* and *Neosaimiri,* some aspects of the ecological adaptive diversity present in the primate community of the Colombian Miocene can be inferred. Their modern counterparts, *Pithecia* and *Saimiri,* occur sympatrically over much of their respective ranges. *Pithecia* has a rather broad range of forest habitat tolerance. It frequents the understory and lower canopy strata and is often encountered on vertical supports. Locomotion is primarily by leaping and bounding. Few observations are available for postures during feeding. It is almost exclusively a frugivore/seed predator. The morphology of the distal humerus of *Cebupithecia* suggests that it also adopted frequent clinging postures on vertical supports (Meldrum et al. 1990b). The morphology of the hindlimb also suggests frequent leaping although after a fashion somewhat different than *Pithecia* (Fleagle and Meldrum 1988). The dental and mandibular structure of *Cebupithecia* strongly implies that it had progressed toward a seed-eating diet as well. This is especially indicated by the broad, flat molars, large, chisellike canines (Kay 1990b), and procumbent, mesiodistally compressed lower incisors of a recently recovered specimen (IGM 251074).

Saimiri is also environmentally tolerant within the context of forest habitats and also occupies the understory and lower canopy strata. It moves primarily by quadrupedal walking and running, interspersed with frequent leaps. Its diet differs from *Pithecia* in that it consists largely of insects and fruit. By comparison, the distal humerus of *Neosaimiri* also suggests arboreal quadrupedalism, and the extensive syndesmosis of the distal tibiofibular joint suggests frequent leaping. The dentition of both *Neosaimiri* species also indicates a primarily frugivorous/insectivorous diet as especially illustrated by the well-developed but not fully trenchant molar shearing crests (chap. 28). Given these sometimes striking similarities between the fossil taxa and their modern "counterparts," it seems reasonable to infer that the sympatric *Cebupithecia* and *Neosaimiri* had partitioned the forest understory in a manner somewhat comparable to that of living *Pithecia* and *Saimiri.*

The numerous primitive features of the postcranial skeleton of *Cebupithecia* demonstrate that the distinctive modern features of the dentition and postcranium, characteristic of the extant members of at least two platyrrhine subfamilies, may not have evolved in synchrony. Therefore, it is impossible to predict confidently the postcranial adaptations of fossil platyrrhines known only from dental remains. For example, the platyrrhine subfamily Alouattinae is represented in the Miocene of Colombia by *Stirtonia*, which displays many similarities to the modern genus *Alouatta* and had a comparably large body size (Kay et al. 1987). But only the discovery of postcranial fossils of *Stirtonia* will determine whether the distinctive climbing and suspensory adaptations associated with the foraging strategy of the large-bodied alouattines had emerged at this point in their evolutionary history.

Acknowledgments

We thank J. G. Fleagle, D. Gebo, and S. Ford for comments; D. Gebo kindly provided casts of IGM-KU 8802 and 8803. This research was supported in part by NSF grant BNS 8719448 and a Smithsonian Institution short-term fellowship to D. J. Meldrum, and NSF grant BSR 8614533 to R. F. Kay.

28
Fossil New World Monkeys

John G. Fleagle, Richard F. Kay, and
Mark R. L. Anthony

RESUMEN

Con el notable mejoramiento del archivo fósil de los platirrinos en los últimos diez años, nuestro conocimiento de su
história evolutiva es más refinado. Fósiles de platirrinos
han sido hallados en depósitos en cinco áreas distintas
en geografía y cronología. *Branisella* y *Szalatavus* del
Deseadense (Oligoceno tardío–Mioceno temprano) de
Bolivia constituyen los más antiguos registros de platirrinos en el Hemisfério Occidental. En términos generales,
se considera a *Branisella* un platirrino de afinidades pa-
trísticas inciertas.

Cinco géneros de platirrinos fósiles han sido descubiertos en el Mioceno de Patagonia. Cuatro géneros
provienen del Colhuehuapense. Algunos investigadores
consideran a *Dolichocebus* como el antecesor de *Saimiri*,
pero los autores prefieren considerar que las semejanzas
morfológicas numerosas entre estos dos taxa son caracteres primitivos retenidos. *Tremacebus*, un mono con ór-
bitas moderadamente grandes, resulta semejante a *Aotus*
en convergencia. *Soriacebus* posee una mezcla de rasgos
derivados de Pitheciinae y caracteres extremadamente
primitivos que sugieren a su vez una convergencia en hábitos alimenticios. *Carlocebus* es semejante a *Callicebus*, pero
estas semejanzas son también primitivas y no indicadores
de afinidad filética. La anatomía dental de *Homunculus*, el
primer platirrino fósil descubierto y de Edad Santacrucense, no es bien conocida y falta aún un estudio
detallado. En general, es difícil establecer para los monos

fósiles de Argentina una relación filogenética estrecha con
las subfamilias de platirrinos actuales.

Por el contrario, los platirrinos fósiles del Grupo Honda
son de fácil asignación a subfamilias de platirrinos vivos.
Por ejemplo, *Neosaimiri* pertenece a la subfamilia
Saimiriinae, mientras que *Patasola* pertenece al grupo
monofilético que abarca a *Saimiri*, *Callithrix*, *Saguinus*,
Cebuella y *Callimico*, sugiriendo que estos géneros están estrechamente relacionados. *Lagonimico* parece relacionarse
a los Callitrichinae y *Callimico* también, aunque más
grande que *Cebuella*, *Callithrix*, y *Saguinus*. *Cebupithecia* es
un Pitheciinae sin lugar a dudas, y *Stirtonia* está cercana-
mente relacionada a *Alouatta*. Las afinidades filéticas de
Mohanamico son algo menos ciertas, aunque podría repre-
sentar un Pitheciinae.

Material fragmentario proveniente del Mioceno tardío
de Brasil sugiere una ampliación en la distribución
geográfica y cronológica del género *Stirtonia* y también el
registro más antiguo del grupo monofilético de *Cebus*.

Los platirrinos fósiles más recientes se encuentran en el
Pleistoceno/Reciente de Brasil y algunas islas del Caribe.
Protopithecus brasiliensis de las cavernas de Lagoa Santa en
Brasil era un Atelinae muy grande que pesaba, por lo
menos, dos veces más que cualquier platirrino vivo. De las
islas mayores de las Antillas, provienen *Ateles anthropomorphus* y *Paralouatta*, este último parecido a *Alouatta*.
"Saimiri" bernensis de la Hispaniola se asemeja tanto a *Cebus*
como a *Saimiri* en dentadura, pero no en el esqueleto

postcraneal. *Xenothrix* de Jamaica ha perdido el tercer molar como los Callitrichinae, pero algunos investigadores lo consideran suficientemente distinto para merecer una familia distinta y monotípica. El orígen biogeográfico de los monos del Caribe es todavía oscuro pero al juzgar por el grado de su divergencia morfológica de los demás monos de América continental, su radiación evolutiva pudiera haber sido muy antigua.

El nuevo material fósil nos presenta una serie de desafíos para los esquemas previos de la evolución de los platirrinos. Por ejemplo, la mezcla de carácteres primitivos y derivados en *Cebupithecia* sugiere que los caracteres que vinculan a Pitheciinae con Atelinae, o alternativamente, a Pitheciinae con *Callicebus*, han evolucionado en paralelo. Asimismo, la simplificación de los molares en Callitrichinae ha sido considerada una simple consecuencia de la evolución hacia tamaños corporales pequeños. Sin embargo, el descubrimiento de *Lagonimico*, un mono fósil más grande con estas mismas simplificaciones, sirve para descartar esta asociación. Además, nuevos hallazgos y el reestudio del material ya existente del Mioceno y Pleistoceno de Brasil demuestra claramente que platirrinos de otras épocas alcanzaban pesos más grandes que todas las especies vivas.

La comparación de la morfología de los molares e incisivos en los platirrinos fósiles nos permite la siguiente reconstrucción de las adaptaciones para dieta, aunque con menos confianza en el caso de los monos más antiguos. En general, los platirrinos fósiles muestran una diversidad de dietas, entre ellas folivoría (*Stirtonia*), frugivoría (*Neosaimiri*), nectarivoría (*Lagonimico*), y granivoría (*Cebupithecia*).

No aspect of higher primate evolution has been so slowly documented and so poorly understood as the history of the platyrrhine monkeys of Central and South America. As recently as the 1950s, all fossil platyrrhines (remains of fewer than ten individuals) were generally placed in a single genus, *Homunculus*. In the 1970s only fifteen specimens of fossils of New World monkeys had been placed in six genera. Moreover, in many cases, different species were known from different anatomical parts, precluding any meaningful comparison.

Since the late 1980s, the fossil record of platyrrhines has increased considerably in both the quantity and quality of material (fig. 28.1). Several new taxa have been described, and new material has been referred to every "old" taxon. In addition, a number of taxa are now known from dental, cranial, and limb elements, permitting a wider range of both functional and phylogenetic analyses than was previously possible (e.g., Ford 1988, 1990a, 1990b; Kay 1990b; Rosenberger et al. 1990). Although we now know much more about the diversity and biogeography of platyrrhines during past epochs, this record still consists of only a limited number of species, sampling the last 26 million years (figs. 28.2, 28.3). Not surprisingly, there are many unresolved questions and debates concerning the systematics within individual taxa, their relationships to extant groups, and the origin of the entire platyrrhine radiation.

Fossil platyrrhines are known from only a few, widely separated regions of South America and the Caribbean (fig. 28.2). There is almost no overlap between the distribution of extant platyrrhines and the sites that have yielded fossil monkeys. Only the middle Miocene Honda Group in Colombia and a few sites of various ages in Brazil (Lund 1838; Kay and Frailey 1993) are near areas where monkeys currently live. The rocks yielding fossil platyrrhines can be divided into five groups based on geographical and temporal grounds. These are, from oldest to youngest: (1) late Oligocene/early Miocene Salla beds in Bolivia; (2) early to middle Miocene rocks in southern Argentina; (3) middle Miocene Honda Group rocks in Colombia; (4) late Miocene rocks from Río Acre, Brazil; and (5) Pleistocene-Recent cave deposits in Brazil and the Caribbean. In this chapter we first review the platyrrhine fossil record in chronological order, placing the greatest emphasis upon the fossil primates from Colombia. We examine evidence concerning the phylogenetic relationships among living platyrrhines and divergence of the various platyrrhine clades as well. We then discuss the ecological adaptations of fossil platyrrhines, again with special emphasis upon the Miocene Colombian species.

Late Oligocene / Early Miocene Primates from Bolivia

The earliest record of fossil primates in South America comes from the Deseadan Land Mammal Age (LMA)(late Oligocene/early Miocene) of Bolivia. Two species, *Branisella boliviana* (Hoffstetter 1969) and *Szalatavus attricuspis* (Rosenberger, Hartwig, and Wolff 1991) have been described from a single horizon, dated at 25–26 Ma (Mac-

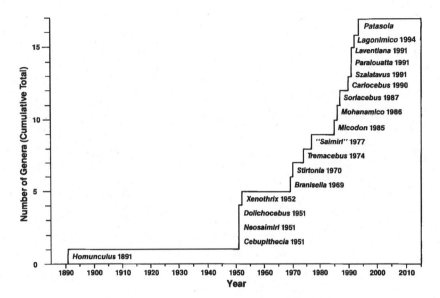

Figure 28.1. Graph showing the increasing rate at which fossil platyrrhines have been described in the past forty years. (Adapted from Fleagle and Rosenberger 1980, with updated information. *Patasola* is described in chapter 26 of this volume.)

Fadden 1990). Paleoclimatic data indicate that temperatures during the Deseadan were cooler than those characteristic of later Santacrucian times (Pascual and Ortiz Jaureguizar 1990), and MacFadden (1990) has recently suggested that the Salla primates may have lived in dryer environments than any extant platyrrhine.

These two early platyrrhines (fig. 28.4) are similar in size (approximately 1,000 g) but differ in the shape of upper molars and the robustness of the mandible. This suggests more taxonomic and dietary diversity among Oligocene platyrrhines than was previously realized, but there is little evidence of particular phyletic relationships between these early taxa and any single group of more recent New World monkeys.

Miocene Platyrrhines from Argentina

There are numerous fossil primates from Miocene deposits in southern Argentina (Patagonia) that are between 3 and 10 million years younger than the Bolivian monkeys and are associated with faunas placed in the successively younger Colhuehuapian, Santacrucian, and Friasian LMAs. At present, there are nine species and six or more genera of these early to middle Miocene platyrrhines, demonstrating a continuous low level of diversity of monkeys at very high latitudes throughout this time.

Several fossil monkeys are known from the Colhuehua-

pian LMA which is, thus far, only recognized for certain in Chubut Province. The precise age of the Colhuehuapian is uncertain and debated (e.g., MacFadden 1990). Some regard it as a primitive stage of the Santacrucian LMA and, thus, presumably, of similar absolute age; preliminary radiometric dates from several sites, however, suggest that Colhuehuapian deposits may be substantially older than the best known Santacrucian rocks (Swisher, pers. comm.). Two Colhuehuapian monkeys, *Dolichocebus* and *Tremacebus*, have been known for about fifty years from damaged skulls and more recently from additional, not yet fully published, dental and postcranial remains (e.g., Fleagle and Bown 1983; Fleagle and Kay 1989).

Dolichocebus gaimanensis (fig. 28.5) is known from a crushed skull, approximately twenty isolated teeth, a partial mandible, and an isolated talus, all from the locality of Gaiman, Chubut Province, Argentina. *Dolichocebus* was a medium-sized monkey similar to a small capuchin. Functional studies of the talus attributed to this species suggest quadrupedal leaping habits (Reeser 1984; Gebo and Simons 1987; Meldrum 1990).

The phyletic affinities of *Dolichocebus gaimanensis* are unresolved and the subject of considerable debate. Rosenberger (1979b) has argued that this species is ancestral to the extant squirrel monkey, *Saimiri,* on the basis of the shared possession of an interorbital fenestra and other cranial similarities. Hershkovitz (1982b), in contrast, challenged the

Figure 28.2. Map of South America and the Caribbean, showing sites that have yielded fossil platyrrhines.

identification of the interorbital fenestra and gracile zygomata and noted striking differences between *Dolichocebus* and *Saimiri* in several aspects of cranial anatomy, particularly in the relative size and arrangement of the dental arcade. He argued that *Dolichocebus* is a divergent, very primitive platyrrhine unlikely to be closely related to any single extant lineage. Analyses of recently recovered dental and postcranial remains attributed to this same taxon have failed to provide any unambiguous evidence of affinities with any one modern genus, but instead show that *Dolichocebus* may be placed in a variety of equally parsimonious phyletic positions (Gebo and Simons 1987; Fleagle and Kay 1989; Meldrum 1990). Analyses concern-

ing its phylogenetic position are further hindered by the difficulty of identifying derived features of the squirrel monkey lineage as distinct from primitive platyrrhine features.

Tremacebus (Hershkovitz 1974) is known from a single skull found at Sacanana, Chubut Province, Argentina. This was a slightly smaller monkey than *Dolichocebus,* more comparable in size to the living saki, *Pithecia.* The most striking aspect of this skull (fig. 28.6) is the moderately large size of the orbits, which most authors have interpreted as evidence of a close phyletic relationship with the extant owl monkey, *Aotus* (e.g., Fleagle and Rosenberger 1983; Setoguchi and Rosenberger 1987; but see Hershkovitz 1974). Unfortu-

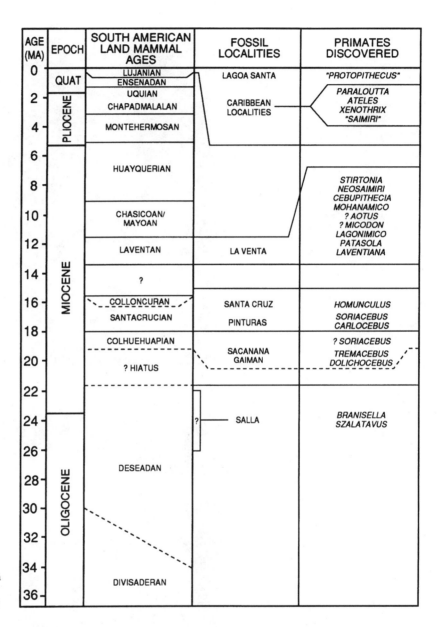

Figure 28.3. Geological time scale for South American land mammal ages and primate-bearing fossil sites. (Redrawn from chap. 29.)

nately, little additional information about the morphology of this taxon is available. A mandible from the same locality was originally allocated to *Tremacebus* (Fleagle and Bown 1983) but may instead belong to *Soriacebus*.

The Pinturas Formation of western Santz Cruz Province (fig. 28.7) has yielded nearly three hundred primate specimens from several sites, more primate fossils than any other formation in South America. These have been placed in two very distinct genera, *Soriacebus* and *Carlocebus,* each with two species (Fleagle et al. 1987; Fleagle 1990). The Pinturas Formation dates from approximately 17.5 to 16.5

Ma, and the mammalian fauna from Pinturas has been considered intermediate in age between Santacrucian and Colhuehuapian faunas (e.g., Barrio et al. 1984). The Pinturas Formation was deposited primarily as volcanic ash in valleys adjacent to the rising Andes, a process that began in earnest in the Miocene and continues today. The strata of the formation yielding primates were formed under warmer, more tropical climatic conditions although the formation also records episodes of more arid conditions (Bown and Larriestra 1990). The Pinturas Formation differs from the better-known east-coastal Santa Cruz For-

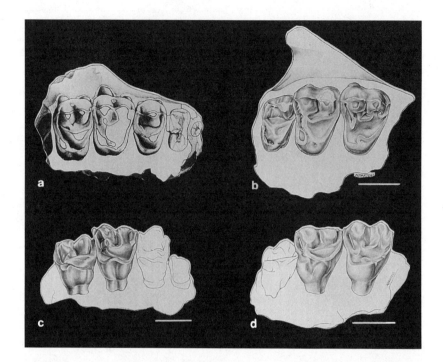

Figure 28.4. Maxillary remains of (*a, c*) *Branisella boliviana* and (*b, d*) *Szalatavus attricuspis* from Salla, Bolivia. Scale bars 2 mm. (Courtesy of A. L. Rosenberger.)

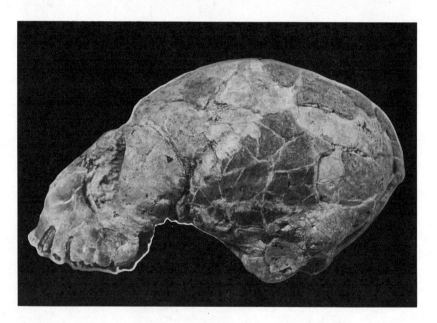

Figure 28.5. Skull of *Dolichocebus gaimanensis* from Sacanana, Chubut Province, Argentina: left lateral view, about 60 mm in length.

mation in the composition of its mammalian faunas in general, and in its primate fauna as well, but it is not yet clear whether these faunal discrepancies reflect primarily temporal or ecological differences, or both.

Soriacebus is a very unusual platyrrhine characterized by a deep, V-shaped mandible, relatively large, laterally compressed lower incisors, stout canines, and small low-cusped molars (fig. 28.8). There are two species. *S. ameghinorum* is a larger, saki-sized species that weighed around 2 kg from the lower part of the Pinturas Formation. This species is known from several mandibles, maxillae, dozens of isolated teeth, and several postcranial remains (Fleagle et al. 1987; Fleagle 1990; Ford 1990a; Meldrum 1990). Functional studies of the dentition suggest *Soriacebus* had a frugivorous diet, pos-

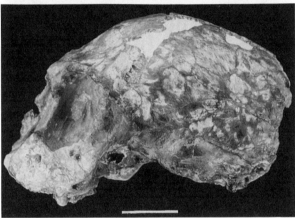

Figure 28.6. Cranium of *Tremacebus harringtoni* from Gaiman, Chubut Province, Argentina: (*top*) frontal view and (*bottom*) left lateral view. Scale bar (both views) 10 mm.

sibly with emphasis on "seed predation." Studies of the isolated limb elements attributed to this species suggest an arboreal species with some clinging or suspensory habits. *Soriacebus adrianae* is a smaller species (800–900 g) from higher levels in the Pinturas Formation (Fleagle 1990). It is known from several mandibular and maxillary remains and about one dozen isolated teeth. It differs from *S. ameghinorum* primarily in size and in reduction of the upper molar hypocones. *Soriacebus* possesses a mosaic of traits resembling various modern platyrrhine groups. *Soriacebus* has been identified as a primitive pitheciine on the basis of its deep mandible, large styliform incisors, and large P_2 (Rosenberger et al. 1990), or, alternatively, as the sister taxon of all extant platyrrhines because it lacks a number of shared-derived molar features of extant platyrrhines (Kay 1990b). Regardless of the phyletic position in which this genus is

placed, *Soriacebus* forces a revision of earlier notions of primitive platyrrhine morphotypes (e.g., Fleagle et al. 1987; Fleagle 1990; Kay 1990b; Rosenberger et al. 1990). For example, most authorities have argued previously that the enlarged incisors and robust, splayed canines of pitheciines evolved as a functionally integrated unit that was the basal adaptation of the group. *Soriacebus,* however, has large, mesiodistally compressed incisors with canines adjacent to them, not separated by a diastema. Thus, if *Soriacebus* is an early pitheciine, the incisor/canine complex characteristic of extant pitheciines evolved late in the history of the group with changes in incisor shape preceding canine enlargement and broadening of the anterior part of the mandible. If *Soriacebus* is not a pitheciine, then some pitheciine dental specializations, including large incisors and anterior premolars and a distinctive mandible shape as well, either evolved several times in parallel among New World monkeys or are primitive for a wider range of platyrrhines.

Carlocebus is a more generalized primate from the Pinturas Formation (Fleagle 1990) (fig. 28.8). Compared with *Soriacebus, Carlocebus* has relatively smaller incisors and larger molars and more well-developed hypocones and lingual cingula on the upper molars. Two species have been described. *C. intermedius,* a relatively small species comparable in size to *Soriacebus ameghinorum* or a saki monkey, is known from a single site in the lower part of the formation. *C. carmenensis* is a larger species present throughout the formation and is known from numerous mandibles, several facial remains, dozens of isolated teeth, and several isolated limb elements. The dentition of *Carlocebus* resembles that of living *Callicebus,* which is primarily frugivorous but which also eats leaves and insects. The forelimb and foot bones that have been attributed to *Carlocebus* suggest a quadrupedal monkey with the possibility of some suspensory abilities in the larger taxa (Anapol and Fleagle 1988; Meldrum 1990).

Carlocebus shows greatest dental similarities to the younger Santacrucian *Homunculus* among fossil taxa and to *Callicebus* among extant platyrrhines. If, however, *Callicebus* is dentally more primitive than other living platyrrhines (Kay 1990b), this would suggest that many of the features shared by *Carlocebus* and extant titi monkeys (including prominent hypocones on upper premolars and prominent lingual cingula on upper cheek teeth) may be primitive retentions.

Figure 28.7. Two formations in Santa Cruz Province, Argentina, that have yielded different fossil primate fauna: *A,* the Pinturas Formation south of Perito Moreno; *B,* Monte Observación, a rich Santacrucian locality on the Atlantic coast.

Figure 28.8. Fossil primates from Argentina. *a. Soriacebus ameghinorum,* Pinturas Formation. *b.* cf. *Soriacebus* sp. from Gran Barranca. *c, d. Carlocebus carmenensis,* Pinturas Formation. Scale bar 10 mm.

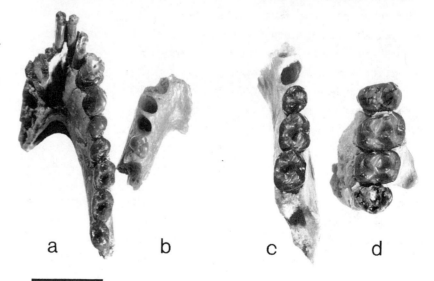

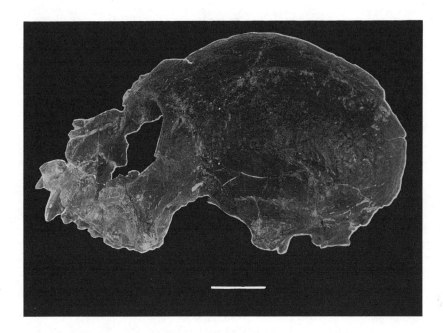

Figure 28.9. A new partial cranium of *Homunculus patagonicus* from the Santacrucian of Argentina: left lateral view. Scale bar 10 mm. (Courtesy of A. Tauber.)

Fossil primates from the Santacrucian LMA are less diverse than those of the Colhuehuapian. *Homunculus,* from the Santa Cruz Formation (middle Miocene, Santacrucian LMA; see fig. 9 in Marshall et al. 1986) was the second fossil platyrrhine ever discovered. Dates for the Santa Cruz Formation deposits yielding *Homunculus* are approximately 16 million years old (Marshall, Drake, et al. 1986; MacFadden 1990). The Santa Cruz Formation contains some of the richest fossil mammal localities in the world (e.g., Simpson 1984). In contrast with the dry, temperate climate found in southern Argentina today, Santacrucian times have been reconstructed on the basis of the rodent fauna as having "warm and wet climates with forested habitats" (Pascual and Ortiz Jaureguizar 1990). This climatic reconstruction is supported by data from paleobotanical studies and oxygen isotope analyses of marine foraminifera (reviewed by the same authors).

Homunculus (fig. 28.9) is known from several mandibular fragments, two partial skulls, and parts of an associated skeleton (Ameghino 1891; Bluntschli 1931; Tauber 1991). The dental anatomy of this taxon is poorly known, and, for most of this century, fossil primates from Argentina and other parts of South America were placed in this genus simply on geographic or temporal criteria. Many of the non-Santacrucian fossils (e.g., those representing species of *Dolichocebus, Tremacebus,* and *Stirtonia*) subsequently have

been removed, but the fossils attributed to *Homunculus* from the Santa Cruz Formation in Argentina are still in great need of proper documentation and systematic analysis (e.g., Fleagle et al. 1987). At present, all primate fossils from this formation are placed in a single species, *Homunculus patagonicus.*

The phylogenetic relationships of *Homunculus* are not clear. Its nineteenth-century describer, Florentino Ameghino, thought it was in the ancestry of humans, hence his choice of nomenclature. It is not. *Homunculus* has been considered, by various authorities, to be close to *Callicebus,* the titi monkey (Szalay and Delson 1979; Fleagle and Rosenberger 1983; Rosenberger 1988b; Rosenberger et al. 1990), *Aotus,* the owl monkey (Bluntschli 1931), *Alouatta,* the howling monkey (Hershkovitz 1970; but see Hershkovitz 1984a), and the Pitheciinae, sakis, and uakaries (Tauber 1991) based on different types of analyses of different anatomical regions. A comprehensive review of *Homunculus* that includes recently recovered material is long overdue.

From Cañadon del Tordillo, Neuquén Province, within the Collón Curá Formation, comes a collection of primate teeth that represents the youngest fossil primates yet known from Argentina. These teeth were recovered from a soil horizon within an ignimbrite dated at around 15.5 Ma (chap. 29) and possibly represent two taxa. Several lower incisors are mesiodistally compressed and styliform, greatly resembling those of *Soriacebus,* whereas several molars are distinct from those of *Soriacebus, Carlocebus,* or *Homunculus* from the Pinturas and Santa Cruz formations. Given the poor quality of the Cañadon del Tordillo fossil sample, little more can be said concerning their affinities.

Miocene Argentine fossil platyrrhines cannot be readily allied with the commonly recognized clades of living platyrrhines. Rather, they are generally more primitive, each possessing a mosaic of similarities to extant platyrrhine taxa that is incongruent with many current phyletic reconstructions based solely on the anatomy of living New World monkeys. Another striking feature of the Argentine faunas is that primate diversity—for any locality or in any formation, at either the specific or generic level—is low compared to that of Recent platyrrhines living in closed-canopy tropical rain forests. Possibly, this reflects the presence of more marginal, possibly dryer or more seasonal, habitats for the Argentine primates, a setting comparable to modern platyrrhines at the edges of their present ranges in places such as Salta Province, Argentina (Ojeda and Mares 1989).

Miocene Primates from Colombia

The middle Miocene Honda Group from the La Venta area, Huila, Colombia, has yielded an unparalleled diversity of fossil platyrrhines (figs. 28.10–28.13). Many of these are clearly related to extant subfamilies, whereas the affinities of others are the topic of considerable debate.

The Honda Group is of middle Miocene age, deposited in a relatively short time interval, between 13.5 and 11.8 Ma (Guerrero 1990; also chaps. 2, 3). Honda Group rocks and their contained faunas are younger than those of the type Friasian LMA in Chile and the Collón Curá Formation in Argentina nearby, which are approximately 15–16 million years old (chaps. 2, 3, 29). Associated vertebrate faunas from the La Venta fauna indicate a humid tropical environment in the Miocene (chap. 30). Separated from the Patagonian faunas by 10,000 km, this Colombian fauna bears little resemblance to either Friasian or Santacrucian faunas already mentioned and has almost no diagnostic faunal elements in common with them.

Several of the Honda Group primates are readily attrib-

utable to modern subfamilies of platyrrhines. *Neosaimiri fieldsi* (chap. 26, figs. 26.1–26.3), described by Stirton (1951), is very similar to the extant squirrel monkey. Known dental material of *Neosaimiri fieldsi* comes mainly from the Monkey Beds of the Villavieja Formation although some isolated teeth from the Fish Bed may pertain to this species (chap. 26). Some authors have suggested that the type dental material of *N. fieldsi* cannot be distinguished from that of the modern genus *Saimiri* (Szalay and Delson 1979; Hartwig et al. 1990; Rosenberger, Hartwig, et al. 1991). Some significant differences remain, however, as pointed out by Stirton (1951), such as proportionately smaller incisors in *N. fieldsi,* as noted in table 26.1 by Kay and Meldrum. A ratio of the lower lateral incisor area to M_1 area is 0.33 in *Neosaimiri,* whereas in both species of *Saimiri* it is nearly twice as large, suggesting that the incisors of *N. fieldsi* are a little more than one-half the size of those of *Saimiri.*

Fragmentary limb bones attributed to *Neosaimiri fieldsi* have proved indistinguishable from those of extant squirrel monkeys (Meldrum et al. 1990b; also chap 26). Clearly, *Neosaimiri* is phylogenetically very close to *Saimiri* and may even be its direct ancestor (Rosenberger, Setoguchi, et al. 1991). Whether this animal should be accorded separate generic status from *Saimiri* seems to be a matter of taste. Our own preference is to recognize its distinctness at the generic level (see also Takai 1994).

Expeditions to the La Venta area by Setoguchi has recovered a new animal, *Laventiana annectens,* that is grouped with *Neosaimiri* and *Saimiri* (Rosenberger, Hartwig, et al. 1991). Dental material of this taxon comes from the El Cardón Red Beds of the Villavieja Formation. The described lower dentition of this monkey is very similar to that of *Neosaimiri* but is distinct in having a well-developed postentoconid notch on the molars. Takai (1994) has described a larger sample of this species from the type locality. He concludes that *Laventiana annectens* belongs to the species *Neosaimiri fieldsi.* Kay and Meldrum (chap. 26) sustain the specific distinctness of the two within the genus *Neosaimiri. Patasola magdalenae,* found by the Duke expedition in 1988 (Kay 1989; also chap. 26), is found in both the La Victoria and Villavieja formations. Based on molar dimensions its body size is inferred to have been intermediate between *Saimiri* and living callitrichines. This animal

a

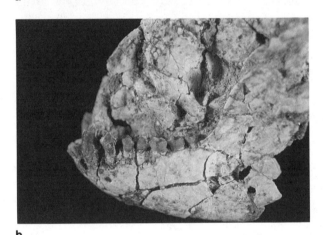

b

Figure 28.10. *Lagonimico conclucatus,* La Victoria Formation, Huila Department, Colombia: *a,* maxilla; *b,* mandible.

shares many derived features with *Neosaimiri* and *Saimiri,* on the one hand, and living callitrichines and *Callimico,* on the other, suggesting that the subfamilies Saimiriinae, Callitrichinae, and Callimiconinae are closely related (chap. 26; see also Rosenberger et al. 1990).

Another distinct Colombian Miocene platyrrhine found in 1988, *Lagonimico conclucatus,* has recently been described (Kay 1990a, 1994; also fig. 28.10). The only specimen known is a crushed skull, preserving most of the dentition, from the La Victoria Formation. It has greatly reduced third molars, simplified premolars, and lacks the molar hypocones, indicating closer affinity to callitrichines than to *Callimico* or squirrel monkeys. Consequently, a surprising feature of this animal is its relatively large size. The mandible suggests an animal the size of *Callicebus,* approximately

1,000 g, larger than *Callimico* or any living marmoset or tamarin.

Micodon kiotensis, described by Setoguchi and Rosenberger (1985a), was allocated by them to the Callitrichinae. The type specimen of *Micodon* is a single upper molar collected in 1984 from the Monkey Bed Villavieja Formation. This isolated tooth, however, lacks any morphological similarities to extant callitrichines, and *Micodon* has been placed in this subfamily entirely on the basis of its small size (Rosenberger et al. 1990). Indeed, there seems to be insufficient morphology on which to base a generic diagnosis (chap. 26). Another isolated premolar described by Rosenberger and Setoguchi (1985) from the same stratigraphic level, but not allocated to any taxon, certainly belongs to another small platyrrhine.

Like *Neosaimiri, Cebupithecia sarmientoi* (fig. 28.11), the first Colombian fossil primate discovered (in 1945) can also be linked with an extant platyrrhine subfamily, sharing several derived dental features with the extant pitheciines, *Pithecia, Chiropotes,* and *Cacajao* (Stirton 1951; Rosenberger 1979a, 1984; Setoguchi et al. 1987; Kay 1990b). These characteristics include the enlarged, chisel-like canines and low-crowned, "inflated" molars characteristic of extant pitheciines. The dental morphology of this species, however, is more primitive than that of the living pitheciines in having unmolarized premolars and smooth molar enamel; therefore, it is most likely a sister-group to all the living taxa (Kay 1990b). In its postcranial anatomy, *C sarmientoi* lacks many distinctive features shared by all extant pitheciines while it possesses striking similarities to particular species of *Pithecia* in a few anatomical details (Davis 1987; Meldrum and Fleagle 1987; Meldrum and Lemelin 1990; Meldrum et al. 1990b; also chap. 27). This reinforces the view that *Cebupithecia* is the sister taxon of all three extant pitheciine genera (Rosenberger 1979a, 1984; Kay 1990b). In 1989 yet another pitheciine was discovered. This as yet unnamed taxon from the El Cardón Red Beds, Villavieja Formation, shows many similarities to *Cebupithecia* (Meldrum and Kay 1992). Preserved for the first time are the lower incisors of an undoubted Miocene pitheciine, showing that this group had already evolved the mesiodistally compressed styliform incisors so characteristic of the extant genera.

Stirtonia tatacoensis (including fossils previously attributed to *Kondous laventicus* [Setoguchi 1985]) from the Villavieja

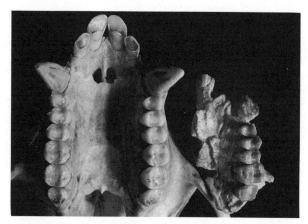

a

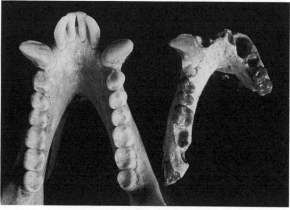

b

Figure 28.11. Dental comparisons of Miocene and extant Pitheciinae. *a.* Maxillae of *Cebupithecia sarmientoi,* Villavieja Formation, Huila Department, Colombia (*right*), and of *Pithecia monachus* (*left*). *b.* Mandibles of *C. sarmientoi* (*right*) and of *P. monachus* (*left*).

Formation and its sister taxon *S. victoriae* from the La Victoria Formation are very large fossil monkeys comparable in size to living howling monkeys and woolly spider monkeys (Stirton 1951; Hershkovitz 1970; Kay et al. 1987) (fig. 28.12). Although the relationship between *Stirtonia* and *Alouatta* has been questioned by some (Hershkovitz 1984a),

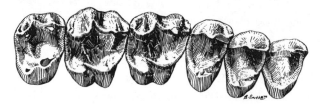

Figure 28.12. Reconstructed maxillary postcanine dentition of *Stirtonia tatacoensis,* Monkey Beds, showing P$^{2-4}$, M$^{1-3}$. (Drawing by Bradley Smith.)

others have demonstrated that this large, folivorous monkey is most closely related to the living howling monkey (e.g., Rosenberger 1979a, 1984; Setoguchi et al. 1981; Kay et al. 1987).

The nature and relationships of the other Miocene primates from La Venta area are more controversial. *Mohanamico hershkovitzi,* known by a single *Aotus*-sized mandible from the Monkey Bed, Villavieja Formation (fig. 28.13), was described by Luchterhand et al. (1986) as a possible primitive pitheciine. *"Aotus dindensis"* from the same locality and stratigraphic level was described by Setoguchi and Rosenberger (1987) as a Miocene owl monkey on the basis of dental and mandibular similarities to the extant genus and on facial remains that could suggest a very large orbit. More recently, Kay (1990b) has argued that *Mohanamico hershkovitzi* and *"A." dindensis* are the same species, that the orbit size is indeterminate, and that this taxon lacks any derived similarities with owl monkeys to the exclusion of other platyrrhines. Following Luchterhand et al. (1986), he notes that the combined *"Aotus dindensis"*–*Mohanamico* sample (N = 2) shows a few features that suggest it is a very primitive pitheciine (Kay 1990b). In contrast, Rosenberger et al. (1990) maintain that *Mohanamico* shares derived features with *Callimico* and maintain that *"Aotus" dindensis* is indeed an owl monkey. At the heart of this disagreement is the simple issue of comparing the variability between two fossil specimens with the intraspecific variability of an extant taxon. Kay maintains that the few features distinguishing these two proposed species, each represented only by its type specimen collected from the same fossil locality, are of the sort commonly found in a single species of extant platyrrhine. To the contrary, Rosenberger and colleagues maintain not only that the morphological distinctness of the two specimens is outside the range found in a single modern species but that each actually represents a widely disparate clade separate for as long as 10 million years. In addition, there is a clear disagreement concerning the identification of shared-derived traits that can be used to determine the phylogenetic relationships of *Aotus, Callicebus, Callimico,* callitrichines, and pitheciines among themselves.

The Honda Group documents a diverse middle Miocene primate fauna with at least seven and, possibly, as many as nine penecontemporaneous platyrrhine genera. This diver-

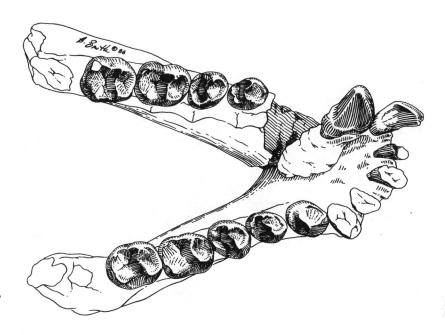

Figure 28.13. *Mohanamico hershkovitzi,* type, from La Venta Monkey Beds. (After Luchterhand et al. 1986.)

sity approaches that found in platyrrhine communities that live in humid tropical forests today. As detailed here, it appears that generic diversity of Colombian primates is also mirrored by the diversity in their ecological roles (Rosenberger 1992). Another striking feature of this assemblage is that many of the taxa are closely related to living platyrrhine subfamilies. Both in terms of diversity and phyletic relationships, this presents a sharp contrast to slightly earlier Argentine Miocene primates; the latter are less diverse and show no undisputed relationship with any extant platyrrhine groups.

Late Miocene Platyrrhines

Kay and Frailey (1993) reported the occurrence of two teeth, belonging to two extinct primates, at a fossil vertebrate locality along the Río Acre, Brazil. The fauna represented at this locality has been assigned to the late Miocene Huayquerian LMA (approximately 9–6 Ma) based on the rodents (Frailey 1986). These specimens are the first fossil primates recovered from the interval of time between approximately 11.8 Ma (Colombian Miocene) and the Pleistocene/Recent.

One of the two teeth, a lower molar, compares well with *Stirtonia* from the Colombian middle Miocene but less

closely resembles *Alouatta,* the living howler monkey. All three taxa have high-crowned M_1s of similar size with trenchant crests, a narrow trigonid, and an oblique distal trigonid wall. In both *Stirtonia* and the new specimen, however, the cristid obliqua is directed relatively more buccally, the hypoconulid is larger, and the postentoconid notch is more deeply incised than in *Alouatta.* On this basis the tooth is tentatively assigned to *Stirtonia,* extending both the temporal and geographic range of this genus.

The second tooth, an upper molar discovered at the same locality, represents a new kind of monkey. From the size of the molar, it appears that this animal was distinctly larger than any living platyrrhine. This specimen resembles *Cebus* in the derived feature of having a very strong metaconule connected to the postprotocrista, but differs from this genus by having a well-developed lingual cingulum and a more squared crown. It is also much larger; mesiodistal length is approximately twice the mean for *Cebus apella.* This specimen may document the oldest occurence of a primate specially related to the *Cebus* lineage.

Pleistocene/Recent Platyrrhines

The first platyrrhine fossil and one of the first fossil primates discovered and recognized was a large femur named *Pro-*

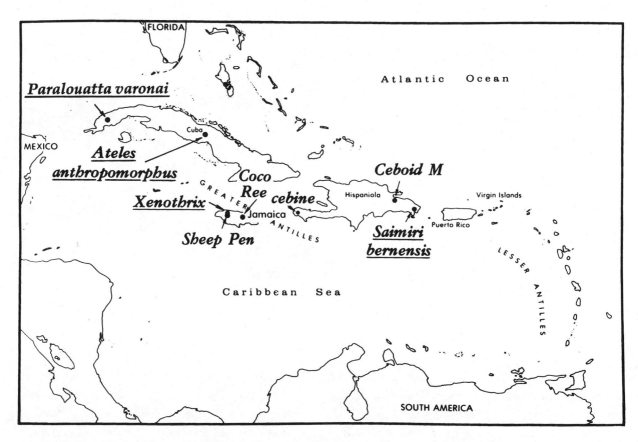

Figure 28.14. Map of the Caribbean showing locations of fossil primates from the Greater Antilles. (Courtesy of S. Ford.)

topithecus brasiliensis from Pleistocene cave deposits of Lagoa Santa in Brazil (Lund 1838). For many years this specimen was regarded as indistinguishable from the living muriqui, *Brachyteles arachnoides,* from nearby parts of the same country. Recent studies, however, of the original specimen and a humerus from the same caves demonstrate that *Protopithecus* was more than twice as large is the living muriqui in linear dimensions and probably had a body size of more than 20 kg, making it the largest known platyrrhine and suggesting that it may well deserve separate taxonomic status (Sandman and Hartwig 1993).

The youngest and northernmost fossil platyrrhines are found in Pleistocene and Recent cave deposits on several islands of the Greater Antilles (fig. 28.14). Although no platyrrhines currently live there, cave faunas on these islands have revealed an unexpected diversity of primate—at least five endemic species—living in the Caribbean until quite recently (Ford 1990b). There are two species from Cuba, *Ateles* (=*Montaneia*) *anthropomorphus* (Ameghino 1910) and

Paralouatta varonai (fig. 1 in Rivero and Arredondo 1991). Similar genera are found today nearby on the Yucatán peninsula, but the Cuban fossils, especially *Paralouatta,* are very distinctive from living species.

The island of Hispaniola southwest of Cuba has yielded several isolated fossil remains dating from the last 10,000 years. These were originally placed in the taxon *"Saimiri" bernensis* (Rimoli 1977), which some authors (Rosenberger 1978; MacPhee and Woods 1982) thought showed greater similarities to *Cebus.* More recently, a nearly complete dentition and several new limb elements of this same primate have been recovered, confirming that it is a distinct form with a dentition that somewhat resembles both *Saimiri* and *Cebus* and a postcranial skeleton that resembles neither (Ford and Woods 1990).

Although numerous primate fossils are known from Jamaica, there is little consensus on either the number of taxa they represent or their phyletic affinities. *Xenothrix mcgregori* (Williams and Koopman 1952; Rosenberger 1975) is known from a mandible and several recently attributed limb bones that show striking morphological distinctions compared to all other platyrrhines (MacPhee and Fleagle 1991). Although Rosenberger (e.g., Rosenberger et al. 1990) has argued for phyletic affinities to the extant *Callicebus*, most other authors have found this taxon distinct enough to deserve placement in a separate family within platyrrhines (e.g., Hershkovitz 1970; MacPhee and Fleagle 1990). There are two other primate femora from separate localities on Jamaica, one dated between 100,000 and 200,000 years old and the other between 30,000 and 50,000 years old (Ford 1990b). One of these bones seems attributable to *Xenothrix,* and the other appears to be of another taxon.

The place of origin of the island faunas in the Caribbean and the length of their isolation from mainland faunas are impossible to reconstruct from current evidence (e.g., MacPhee and Woods 1982; Ford 1990b; MacPhee and Fleagle 1991). Most, however, are morphologically distinct from mainland taxa, and several have radiometric dates that precede the arrival of humans in the New World, indicating that their distribution must be accounted for by factors other than human transportation. Moreover, it has been noted that many of the Pleistocene/Recent mammals from the Caribbean show greater similarities to the Miocene faunas of southern South America than to more recent faunas from central and northern South America, suggesting a very ancient colonization (e.g., Rosenberger 1978; Ford 1990b; Pascual and Ortiz Jaureguizar 1990).

Fossils and Our View of Platyrrhine Evolution

The relevance of fossils to interpreting phylogenetic patterns has been regarded from two extreme points of view: those who claim that fossils give us almost no information about the phylogeny of any vertebrate group that was not already evident from the study of living taxa (Patterson 1981) and those who argue that fossils—even incomplete ones—contribute considerably to our understanding of vertebrate phylogeny (Gautier et al. 1988; Donoghue et al. 1989). Indeed, it is probable that our understanding of the phylogeny and evolution of any one group will, in the end, involve an appreciation of both its extant and extinct representatives. Unfortunately, current views of platyrrhine phy-

logeny have been based entirely on extant taxa, with fossils forms "mapped" on only after the fact (fig. 28.15) (see Rosenberger 1981a; Thorington and Anderson 1984; Ford 1986; Kay 1990b). Yet the data available from extant taxa have proven insufficient for resolving many aspects of platyrrhine phylogeny; at the same time, recently recovered fossil material provides a number of direct challenges to present interpretations.

Although there is general agreement among current New World monkey phylogenies (fig. 28.15) that marmosets, tamarins, and Goeldi's monkey (*Cebuella, Callithrix, Saguinus, Leontopithecus, Callimico*) form one distinct clade, pitheciines (*Pithecia, Chiropotes, Cacajao*) another, and atelines and alouattines (*Ateles, Brachyteles, Lagothrix, Alouatta*) a third, there is no agreement concerning the relationships among these clades or concerning the phyletic position of *Saimiri, Cebus, Callicebus,* or *Aotus* (see Rosenberger 1981a; Ford 1986; Fleagle 1988; Kay 1990b). Nor do we at present understand the character transformations involved in the evolution of the modern platyrrhine clades. The increasing platyrrhine fossil record, however, provides both a challenge and an opportunity for testing and refining these conflicting "extant" phylogenies.

Opinions concerning the relationship between fossil and extant platyrrhines have generally been quite polarized. Some authorities (Hershkovitz 1970, 1977, 1982b; see also Simons 1972) have argued that the platyrrhine fossil record provides little evidence concerning the history of extant taxa. Others, especially A. L. Rosenberger, have argued that the platyrrhine fossil record provides evidence that most modern clades (or even genera) appeared very early in the evolutionary history of the group (Rosenberger 1979a, 1984; also Szalay and Delson 1979; Fleagle and Rosenberger 1983; Delson and Rosenberger 1984) and that most fossil taxa can be confidently referred to extant lineages.

We fall somewhere between these extremes. New, diverse, and more complete fossil remains have demonstrated that the evolutionary history of New World monkeys is more complex and more interesting than either of these simple alternatives would allow. Our assessment is that, by the middle to late Miocene, it is possible to identify lineages leading uniquely to extant pitheciines (*Cebupithecia*), *Alouatta* (*Stirtonia*), *Saimiri* (*Neosaimiri*), possibly *Aotus* ("*Aotus*" *dindensis*; but see Kay 1990b), *Cebus*

Figure 28.15. Alternative views of platyrrhine relationships: *a.* Rosenberger (1984); *b.* Ford (1986); *c.* Sarich and Cronin (1980).

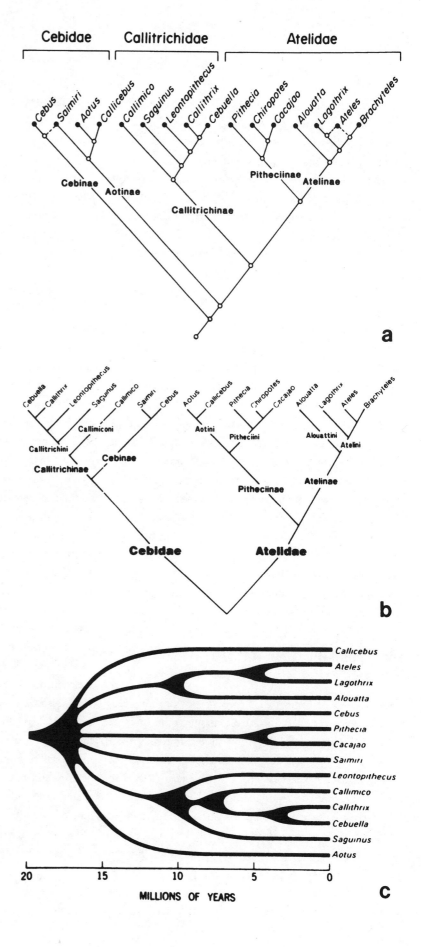

(new taxon from Río Acre) and, perhaps, the marmoset–tamarin–Goeldi's monkey clade (*Patasola, Lagonimico*) (chap. 26; also Setoguchi and Rosenberger 1987; Kay 1990a, 1994; Rosenberger et al. 1990). Evidence relating any of the Santacrucian, Colhuehuapian, or Deseadan taxa to extant genera or clades is more problematic. Moreover, the scattered temporal and geographic ranges of the known platyrrhine fossils further add to the problem of reconstructing platyrrhine phylogeny.

The fossil record holds clues to the nature and timing of many morphological transformations that took place during platyrrhine evolution—events that could not be reconstructed from a knowledge of extant taxa alone. An instance of how the fossil record has altered our view of character transformations in platyrrhine evolution is to be seen in the Miocene pitheciine *Cebupithecia*. Extant pitheciines (*Pithecia, Chiropotes,* and *Cacajao*) are a relatively uniform clade that share numerous derived features of the mandible and teeth, skull, and limb skeleton that distinguish them from other platyrrhines (Rosenberger 1979a; Ford 1986; Fleagle and Meldrum 1988; Kay 1990b). Many, however, of the apparent synapomorphies that these extant taxa share with certain other platyrrhine groups, such as *Alouatta* and *Ateles,* are not present in the more primitive pitheciine *Cebupithecia* (see Kay 1990b; also chap. 27). For example, Ford (1986) hypothesized that *Alouatta* and the atelines are the sister group of extant pitheciines, based on a number of shared-derived features, such as the rounded deltopectoral crest and widened distal femoral epiphysis. These particular features are not present in *Cebupithecia,* however, suggesting that they may have evolved independently in pitheciines and atelines and thereby weakening the evidence linking these two subfamilies. Similarly, Rosenberger (e.g., 1979a) has argued that *Callicebus* is the sister taxon of pitheciines based upon hypothesized shared-derived dental features of extant pitheciines and *Callicebus.* One such character is molarization of the upper and lower premolars. The premolars of *Cebupithecia,* however, especially the lowers, are less molarized than those of either *Callicebus* or extant sakis and uakaries, again suggesting that this similarity was independently acquired.

Platyrrhine evolution is ready for incorporation of fossil data into scenarios proposed for the evolution of extant taxa. For example, callitrichines, marmosets, and tamarins are distinctive among platyrrhines because of their small size, possession of digital claws, reduced dental formula, simplified upper molars, and twin births with extensive parental investment. Ford (1980c) and Leutenegger (1980), among others, argue that these features are all derived features of callitrichines and are linked evolutionarily with phyletic dwarfing. That is, it is argued that they have evolved in conjunction with, and as a consequence of, the small size of marmosets and tamarins. Among extant platyrrhines, these features are uniquely characteristic of this single group and, to a lesser degree, *Callimico.* The fossil record, however, provides additional taxa to test these associations and suggested mechanisms.

For example, the Jamaican *Xenothrix* also lacks third molars but is more than twice as large in estimated body weight as any callitrichine (Rosenberger et al. 1990). Ford has argued that this island species may be secondarily enlarged from a callitrichine ancestry, whereas Rosenberger (1975) has provided alternative arguments. A new genus from the Miocene of Colombia, *Lagonimico,* is substantially larger than extant callitrichines or *Callimico* but lacks a hypocone and has a very reduced third molar (Kay 1990a, 1994). This primitive sister taxon of extant callitrichines shows that many of the dental features found among the small callitrichines evolved before the size reduction characteristic of the extant members of that subfamily. Thus, although small size and dental reduction may well be adaptively associated in the extant marmosets and tamarins, they are not necessarily linked causally because the dental features evolved at a larger size and dwarfing was subsequent (see also Rosenberger 1984; Sussman and Kinzey 1984). Although the present fossil evidence is certainly insufficient to resolve the origin of callitrichines (Rosenberger et al. 1990), it certainly provides the best opportunity to test evolutionary hypotheses concerning their origin.

Ecological Adaptations in Platyrrhine Evolution

Although studies of catarrhine evolution have long emphasized ecological adaptations (e.g., Andrews and Van Couvering 1975; Andrews et al. 1979; Bown et al. 1982; Pickford 1983; Fleagle and Kay 1985), until recently not much attention has been paid to reconstructing the adaptations of extinct platyrrhines or the paleoenvironments that they occupied (Rosenberger 1992). This omission is probably the result, in part, of the overall paucity of the platyrrhine fossil record. It is an omission that must be corrected if we wish to understand the evolutionary history of platyrrhines over the past 25 million years, and the fossil record of platyrrhines is becoming substantial enough that there are numerous ecological questions about platyrrhine evolution that can be addressed.

Compared with the primates of the Old World, platyrrhines are unusual in their relatively small size, low number of folivorous taxa, large number of taxa with prehensile tails, and the absence of terrestrial species. Some have argued that this is because ecological niches occupied by primates in other parts of the world have been preempted by other mammals in South America, such as the folivorous sloths and insectivorous and frugivorous marsupials, whereas others suggest continental differences in resource availability to explain these differences (e.g., Eisenberg et al. 1972; Terborgh and Van Schaik 1987). For example, Ortiz and Pascual (1988; also Pascual and Ortiz Jaureguizar 1990) have noted that the evolution of very small platyrrhines seems to be coincident with a simultaneous decline in the number of small caenolestid marsupials. More detailed analysis of the adaptive characteristics and relative abundance and diversity of both the marsupials and the small platyrrhines could provide evidence about whether this was likely to be an example of either competition, conicidence, or replacement (e.g., Maas et al. 1988).

Interpreting the adaptive evolution of platyrrhines is linked with our understanding of phyletic evolution in platyrrhines. For example, the evolution of Old World anthropoids shows a pattern of successive radiations occupying similar spans of adaptive diversity during the last 20 million years (Fleagle and Kay 1985; Fleagle 1988). It has been argued, however, that platyrrhine evolution shows a different pattern, in which many extant lineages have been present since the initial radiation of the group (Delson and Rosenberger 1984) and have occupied essentially the same niches as platyrrhines do today. Indeed, Rosenberger (1980) has argued that the phyletic radiation of the group is intimately related to an initial trophic polychotomy. An alternative view is that morphological similarities between the Miocene platyrrhines of southern South America and those from northern South America reflect adaptive similarities in separate radiations (as in the Old World) rather than phyletic lineages. As it becomes better known, the fossil material is beginning to lend support to the latter interpretation. The actual adaptive pattern, however, was related to both the evolution of other mammalian groups and to climatic and biogeographic changes in South America.

The adaptive evolution of platyrrhines can also provide a perspective for our understanding of the uniqueness of anthropoid evolution in the Old World. In reviewing the ecological history of Old World anthropoids Fleagle and Kay (1985; see also Kay and Simons 1983) noted that the earliest Old World anthropoids, the parapithecids and propliopithecids of Egypt, are adaptively more similar to platyrrhines than to later anthropoids in their relatively small size, paucity of folivorous taxa, and absence of terrestrial forms. As a group, platyrrhines seem to have gone a different adaptive way than later Old World anthropoids, which subsequently evolved large size, folivory, and terrestriality. As suggested earlier, the history of platyrrhine evolution has almost certainly been influenced by the nature of their mammalian competitors in South America, including the lack of prosimians. Primate evolution in South America, however, may also have been strongly influenced and constrained by the characteristics of the ancestral platyrrhine population.

As extinct primate faunas become better represented in the fossil record of South America, we are able to examine morphological parameters of these assemblages that can tell us about the evolution of platyrrhine adaptations and, reciprocally, that allow us to infer about the paleoenvironments from which these faunas come. The parameters that can be compared between living and extinct primate assemblages include taxonomic diversity, body size, and dietary patterns.

Living platyrrhines are represented by sixteen genera and more than seventy species. Diversity in any given area is the result of a complex interplay of factors of which rainfall seems to be dominant. The largest number of sympatric species occur in tropical humid forests with rainfall in excess of 2,000 mm/year, as in Cocha Cashu, Peru, where there are nine sympatric genera (Terborgh 1983), and in Raleigh-Voltzberg, Surinam, where there are eight (Mittermeier and Van Roosmalen 1981). In drier environments, like the Caatinga and Cerrado of Brazil, with prolonged seasonal dry periods of up to ten months, diversity declines to two or three genera (Mares et al. 1981, 1989). The same is true in the dry forests of Venezuela in which just three genera are represented (Handley 1976). In Salta Province, Argentina, at the southern limit of platyrrhine distribution, there are just two genera. Here, as elsewhere, low rainfall is implicated in the low diversity although it has been suggested that the southern limit of primates is tied to the southern limits of fig trees upon which many primate species depend for food (Ojeda and Mares 1989). This relationship of rainfall to primate species diversity is dramatically illustrated in figure 28.16.

The best-sampled Miocene fossil platyrrhine assemblages reveals interesting distinctions between the faunas of southern Argentina and those of Colombia, differences that may be the result of seasonality, rainfall, or other climatic differences. In the Pinturas and Santa Cruz Formations of Patagonia primate diversity is low with just two genera and three species and a single genus and species represented, respectively. Especially in the Santa Cruz Formation, this low diversity is unlikely to be an artifact of poor sampling. Fossil mammals are extremely abundant here with many thousands of specimens of primate-sized mammals in museum collections. Yet just a handful of fossil primate specimens are known, and all fit comfortably within a single genus, *Homunculus*. Although wet, equable environments have often been inferred for the Santa Cruz Formation (e.g., Pascual and Ortiz Jaureguizar 1990), the low diversity of primates would be more consistent with drier forests or gallery forests. It is important to remember that the Santa Cruz Formation was deposited at extreme southern latitudes—greater than 50° south—indicating that there must have been considerable seasonal variation in day length and, presumably, attendant seasonality in plant growth and reproduction.

The diversity of Colombian Miocene primates presents a striking contrast with that of the Argentine Miocene. In the Honda Group at La Venta at least seven, and as many as nine, genera of platyrrhines are represented (*Stirtonia, Cebupithecia, Lagonimico, Laventiana, Neosaimiri, Patasola, Mohanamico,* an unnamed small platyrrhine, and an unnamed pitheciine). In the Monkey Bed, a minimum of five

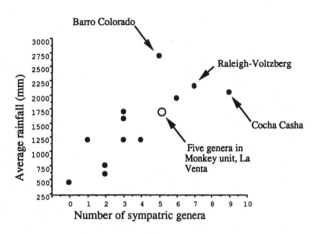

Figure 28.16. Bivariate plot of generic abundance (number of sympatric genera) versus annual rainfall for selected localities of modern South American mammals. Species abundance was known for primates in the Monkey Beds.

genera are sympatric and synchronic. This diversity rivals that of Neotropical rain forests today and would appear to indicate between 1,500 and 2,000 mm of rainfall per annum (fig. 28.16). It is not consistent, however, with reconstructions that imply a patchwork environmental setting where forests were restricted to river courses and the areas between these forest ribbons were filled in by grassland and scrub (Stirton 1953b; and chap. 2). As noted earlier, under those conditions today, primate diversity is much lower.

It is also possible to examine extant platyrrhine diversity in terms of body size. Living platyrrhines range in size through two orders of magnitude from *Cebuella pygmaea* (50–100 g) to *Brachyteles arachnoides* (~10 kg). Using tooth size to infer body size, it is clear that the present-day body-size range of platyrrhines was already established by middle Miocene times in Colombia (compare tooth sizes of living and extinct species in table 28.1). Recent studies, however, of the *Protopithecus* femur from the Pleistocene (Sandman and Hartwig 1993) and the isolated teeth from the late Miocene of Brazil (Kay and Frailey 1993) demonstrate that the present size range of platyrrhines in restricted compared with that of earlier epochs.

Neither presently nor, as far as we know, in the past did platyrrhines exploit terrestrial settings. Living and fossil primates representing other independent adaptive radiations—strepsirrhines in Madagascar or Old World

Table 28.1. The relative development of shearing crests on M_1 in platyrrhines

Taxon	N	Measured Values (mm)		Expected M_1 Shear (mm)	Shearing Quotient	Major Dietary Feature
		M_1 Length	M_1 Shear			
EXTANT TAXA						
Callimico goeldii	3	2.60	5.48	4.72	16.14	Insects
Brachyteles arachnoides	9	7.22	15.19	13.10	15.93	Leaves
Alouatta palliata	10	6.92	13.91	12.56	10.76	Leaves
Alouatta caraya	6	6.72	13.09	12.20	7.33	Leaves
Alouatta fusca	6	6.70	12.94	12.16	6.42	Leaves
Aotus trivirgatus	10	3.06	6.16	5.55	10.92	Fruit/Leaves
Saimiri sciureus	5	2.87	5.54	5.21	6.36	Insects/Fruit
Lagothrix lagotricha	8	5.47	10.12	9.93	1.94	Fruit/Leaves
Leontopithecus rosalia	5	3.09	5.62	5.61	0.22	Fruit/Insects
Ateles geoffroyi	10	5.26	9.31	9.55	−2.47	Fruit
Callicebus moloch	10	3.18	5.50	5.77	−4.70	Fruit
Saguinas mystax	5	2.52	4.03	4.57	−11.88	Fruit/Insects
Callithrix argentata	4	2.22	4.08	4.03	1.27	Fruit/Gum
Cebuella pygmaea	4	1.78	3.26	3.23	0.92	Gum/Fruit
Pithecia monachus	4	4.00	6.78	7.26	−6.60	Fruit/Seeds
Cebus apella	5	4.79	7.71	8.69	−11.31	Fruit/Seeds
Chiropotes satanas	5	3.64	5.50	6.61	−15.53	Seeds/Fruit
Cacajao melanocephalus	2	3.97	5.90	7.20	−17.70	Seeds/Fruit
PINTURAS FORMATION						
Soriacebus ameghinorum	1	3.60	5.87	6.53	−10.15	Fruit/Seeds
Carlocebus carmanensis	1	4.67	7.33	8.48	−13.51	Fruit
Soriacebus adrianae	1	3.23	4.67	5.86	−20.33	Seeds/Fruit
SANTA CRUZ FORMATION						
Homunculus patagonicus	5	4.20	7.57	7.62	−0.69	Fruit
HONDA GROUP						
"Aotus dindensis"	1	3.23	6.14	5.86	4.74	Insects/Fruit
Stirtonia victoriae	1	7.67	14.53	13.92	4.38	Leaves
Stirtonia tatacoensis	2	6.24	11.50	11.32	1.55	Fruit/Leaves
Patasola magdelenae	1	2.50	4.22	4.54	−6.99	Fruit
Lagonimico conclucatus	1	3.01	5.06	5.46	−7.37	Fruit/Gum
Neosaimiri annectens	1	3.26	5.41	5.92	−8.56	Fruit
Neosaimiri fieldsi	1	2.91	4.74	5.28	−10.25	Fruit/Seeds
Mohanamico hershkovitzi	1	3.20	4.96	5.81	−14.59	Fruit/Seeds
Cebupithecia sarmientoi	1	3.52	5.15	6.39	−19.38	Seeds/Fruit
Pitheciinae, gen. et sp. nov.	1	4.00	5.47	7.26	−24.65	Seeds/Fruit
CARIBBEAN						
Xenothrix mcgregori	1	5.80	8.53	10.53	−18.96	Seeds/Fruit

Notes: The estimate of shearing development is based on measurements of six lower molar crests (see Kay 1975 for anatomical details). A line was assigned to a bivariate cluster of the natural log of M_1 length (lnML) versus the natural log of the sum of the measured shearing crests (lnSH). The line was assigned a slope of 1.0 (slope of isometry) and passed through the mean lnML and mean lnSH for extant taxa. The equation expressing this line is lnSH = 1.0(lnML) + 0.596. For each taxon, the expected lnSH was calculated from this equation. The observed (measured) lnSH for each species was compared with the expected and expressed as a residual (shearing quotient, or SQ) SQ = 100 · (observed − expected)/(expected). Extant and extinct taxa are listed separately according to dietary categories (Fleagle 1988). Diet is inferred for extinct taxa by comparison to a modern analog with a similar SQ. (Features of the anterior dentition were also used in making some inferences; see text for explanation.)

anthropoids—did produce terrestrial species, a move that was generally accompanied by an increase in body size. For example, terrestrial cercopithecids are larger than their arboreal close relatives, and the same is generally true of hominoids. Likewise, some of the extinct malagasy strepsirrhines attained large body size in association with terrestrial adaptations. It is not clear why some platyrrhines did not follow the path of other primates toward terrestriality. One possible explanation may be interaction with terrestrial mammalian competitors in South America. As relative latecomers on that continent, primates would have encountered a considerable diversity of sloths already occupying terrestrial niches.

Another way to look at extant platyrrhine diversity is in terms of dietary adaptations (Rosenberger 1992). Living platyrrhines demonstrate a considerable diversity of dietary habits with species specializing on insects, small vertebrates, gums, nuts and seeds, fruit, and leaves. Fortunately, we can get at the problem of what platyrrhines were eating in the past by inference from the dental patterns of living species.

The functional design of the cheek teeth of primates, especially the molars, is selectively modified to best deal with the physical properties of the foods the species eats (Kay 1975; Kay and Covert 1984; Strait 1990; Kay and Anthony 1991; Anthony and Kay 1993). These differences can be quantified by examining the relative development of the shearing crests on the molars after molar size is taken into account. Table 28.1 quantifies the development of molar shearing in living platyrrhines as a "shearing quotient." As calculated, the better developed the shearing crests, the larger the shearing quotient. In table 28.1 the shearing quotient for living species is broken down by dietary preference. Species that eat considerable amounts of fibrous foods, such as leaves high in cellulose (*Alouatta, Brachyteles*) or chitinous insects (*Aotus, Callimico*), have well-developed shearing edges on the molars. In contrast, species that feed on less-fibrous, soft fruits (*Ateles, Callicebus*), and tree gum (*Callithrix, Cebuella*) tend to have relatively flatter teeth with shorter, more rounded shearing crests. The teeth of species that specialize in eating hard seeds or in splitting open tough, hard fruits (pitheciines and *Cebus*) tend to resemble those of the soft-fruit eaters but may have thicker enamel (Kay 1981) or specialized enamel structure (Maas 1986).

The differences in dietary patterns also extend to the anterior dentition—incisors and canines. These teeth are highly modified for gouging bark in gum-feeding species such as *Callithrix* (Coimbra-Filho and Mittermeier 1978) or for splitting fruit husks in seed-eating species such as *Chiropotes* and other pitheciines (Kay 1990b; Kinzey and Norconk 1990). Folivorous species, such as *Alouatta*, have proportionately smaller incisors than their frugivorous relatives, such as *Ateles* (Eaglen 1984; Anthony and Kay 1993), perhaps because the frugivores use these teeth for husking fruits.

In table 28.1 the shearing quotients of various extinct platyrrhines are listed and compared with those of living species. One caveat that must be considered while interpreting the data for the fossils is that most of the samples for these extinct species are often single specimens. This introduces a considerable degree of uncertainty in our inferences given the variation that occurs in samples of living species.

Soriacebus species have very poorly developed shearing crests on their molars. This fact, and the gnathic and anterior dental adaptations outlined here, makes it probable that these animals were specialized seed eaters resembling living pitheciines. *Carlocebus* also has poorly developed shearing on its teeth consistent with fruit, seed, or gum feeding. Among these possibilities, fruit eating is the more likely given the apparent lack of anterior dental specializations found in seed and gum eaters. Likewise, Santacrucian *Homunculus* appears to have been a frugivore comparable in dental structure to *Callicebus*; shearing crest development in *Homunculus*, however, is substantially greater than in *Carlocebus*. In addition, many *Homunculus* dental specimens are extremely worn, suggesting an appreciable amount of fibrous food in this animal's diet.

Considerably more diversity of feeding patterns is demonstrated by the Colombian Miocene taxa. *Stirtonia tatacoensis* closely resembles *Lagothrix*, a mixed fruit and leaf feeder, in its molar shearing development. *Stirtonia victoriae*, a larger species, has even better-developed shearing on its molars and is inferred to have been even more folivorous, similar to *Alouatta*. *Cebupithecia* was certainly a specialized seed eater based on the many dental similarities with living pitheciines. Despite its apparently close phyletic affinities with *Saimiri*, however, *Neosaimiri* may have been more frugivorous and less committed to insect eating than its living relative. Both *Lagonimico* and *Patasola* have poorly developed shearing. *Lagonimico* has accompanying incisal elongation, suggesting a tendency toward mixed gum and fruit eating, whereas *Patasola,* with its unspecialized anterior teeth, seems to have been a fruit and insect eater.

Mohanamico has very poorly developed molar shearing and anterior dental modifications incipiently resembling pitheciines. Fruit eating, possibly with some seed eating, is inferred. Interestingly, *"Aotus dindensis"* seems to depart markedly from *Mohanamico* with which it has been synonymized (Kay 1990b). More study of variation in living species is needed to assess whether the difference in these two specimens is compatible with variation within a single species. If this distinction were upheld by more specimens, it would be consistent with an insectivorous/herbivorous adaptive pattern for *"A." dindensis* much like *Saimiri* or *Aotus* today.

It is clear that dietary differentiation had already occurred in platyrrhines by the early Miocene, and the primate fauna of the Honda Group approached the dietary diversity seen today in humid tropical forests.

Although there are isolated postcranial bones attributable to many fossil platyrrhine taxa, much of this material is undescribed, and many of the described remains have yielded only very general information about locomotor adaptations of these extinct monkeys (see review by Meldrum 1993; also chap. 27). There are no postcranial remains of the Oligocene primates from Bolivia. For the early to middle Miocene platyrrhines from southern Argentina there are numerous pedal remains attributable to several taxa, isolated elements of *Soriacebus* and *Carlocebus,* and a partial skeleton of *Homunculus.* All of these remains suggest quadrupeal animals with possible tendencies toward leaping or suspensory behavior in some taxa. There is no evidence in the described remains of specialized locomotor adaptations as found, for example, in extant *Ateles.*

Among the middle Miocene primates from Colombia, there is good associated postcranial material for *Cebupithecia* and isolated bones attributed to *Neosaimiri* and possibly several other taxa as well (Meldrum et al. 1990b; Meldrum and Lemelim 1991a, 1991b; Meldrum 1993; also chap. 27). Compared with the striking taxonomic and dental diversity among the Colombian Miocene platyrrhines, the available material is inadequate to document a comparable locomo-tor diversity. The remains of *Neosaimiri* suggest an animal very similar to the extant squirrel monkey, but the skeleton of *Cebupithecia* lacks many presumably derived skeletal features (e.g., tail reduction) shared by all extant pitheciines. Unfortunately, there are no indications of locomotor abilities of either the very small or the very large species from La Venta.

Perhaps the most suggestive evidence of striking diversity in locomotor and ecological adaptations is among the Pleistocene-Recent platyrrhines of the Caribbean. In the Recent the islands of the Greater Antilles were home to a considerable diversity of primate species, most of which were clearly different from any primates alive today (Ford 1990a; MacPhee and Fleagle 1991). *Xenothrix* appears to have been a robust, quadrupedal climbing or suspensory loris-like creature, and other Jamaican primates show skeletal features unlike any extant platyrrhine. The increasingly well-known primates from Cuba include limb bones very different from any extant platyrrhines.

The available postcranial elements of fossil platyrrhines are generally too meager to permit any detailed analysis of locomotor diversity. This is especially true for the Patagonian and Colombian remains. The continued recovery of primate fossils from these and other areas can only add to our understanding of this aspect of platyrrhine evolution.

Summary

The history of platyrrhine primates in Central and South America has long been poorly understood. With a platyrrhine fossil record much improved over the last decade, however, previous views of New World monkey evolution can be both refined and challenged.

Fossil platyrrhines have been found in deposits separable into five groups based on geographical and temporal grounds. *Branisella* and *Szalatavus* are known from the Deseadan LMA (late Oligocene/early Miocene) of Bolivia, marking the first occurrence of platyrrhines in the New World. *Branisella* is generally regarded as a primitive platyrrhine with uncertain phyletic affinities.

A number of fossil platyrrhines are known from the Miocene of Argentina. Four genera come from the Colhuehuapian LMA. *Dolichocebus* is thought by some workers to be ancestral to *Saimiri,* but we prefer the view that many morphological similarities between the two are primitive retentions. *Tremacebus* is a monkey with moderately large orbits, convergently similar to extant *Aotus*. *Soriacebus* possesses a mosaic of derived, very pitheciine-like features and extremely primitive features, suggesting convergence on a seed-eating life style. *Carlocebus* is similar to extant *Callicebus* although again the resemblance may be the result of primitive retentions. *Homunculus,* the first fossil platyrrhine discovered, is Santacrucian in age. Its dental anatomy is poorly known and in need of comprehensive review. In general, the Argentinean fossils are difficult to link with extant platyrrhine subfamilies.

In contrast, many fossil platyrrhines from the middle Miocene deposits of the Honda Group, Colombia, are readily attributable to extant subfamilies. *Neosaimiri* can be grouped with extant squirrel monkeys, whereas *Patasola* is similar to living squirrel monkeys, marmosets, tamarins, and Goeldi's monkey, suggesting these groups are closely related. *Lagonimico* appears related to callitrichines and *Callimico* as well, although it is larger than any living marmoset or tamarin. *Cebupithecia* is an undoubted Miocene pitheciine, and *Stirtonia,* a large fossil monkey, is closely related to extant *Alouatta*. The phyletic affinities of *Mohanamico* are unclear, although this taxon may represent an early pitheciine.

Fragmentary late Miocene material from Brazil suggests a geographic and temporal range extension of the Colombian Miocene genus *Stirtonia* and the oldest occurrence of a *Cebus* clade in the platyrrhine fossil record for the first time.

The youngest fossil platyrrhines are found in Pleistocene/Recent deposits from Brazil and from several Caribbean islands. *Protopithecus brasiliensis* from Lagoa Santa in Brazil was a very large ateline that probably weighed at least twice as much as any extant platyrrhine. Known from the Greater Antilles are *Ateles anthropomorphus* and an *Alouatta*-like monkey, *Paralouatta*. From Hispaniola comes *"Saimiri" bernensis,* a form that resembles both extant *Cebus* and *Saimiri* dentally and neither postcranially. Known from Jamaica, *Xenothrix* has lost its third molar as have callitrichines, but some workers have considered this fossil distinct enough to deserve its own family. The biogeographical origins of these Caribbean monkeys are unclear but may be quite ancient.

New fossil material has provided a number of challenges to previous views of platyrrhine evolution. For example, the mosaic pattern of primitive and derived traits in the Miocene pitheciine *Cebupithecia* suggests that characters used to link pitheciines and atelines or, alternatively, pitheciines and *Callicebus,* evolved in parallel. Similarly, molar simplification in callitrichines has traditionally been regarded as an effect of phyletic dwarfing. The presence, however, of a much larger fossil monkey, *Lagonimico,* possessing these same traits weakens this assumption. Moreover, new finds and reanalysis of old material from the Miocene and Pleistocene of Brazil have demonstrated quite clearly that some platyrrhines of earlier epochs were distinctly larger than any extant species.

Comparing the molar and incisor morphology of fossil platyrrhines to that in extant forms allows the reconstruction of fossil platyrrhine dietary adaptations. In general, fossil platyrrhines show a diversity of dietary habits, including leaf eating (*Stirtonia*), fruit eating (*Neosaimiri*), gum eating (*Lagonimico*), and seed eating (*Cebupithecia*), although this is less true for the earlier Argentinean monkeys than for the later Colombian monkeys.

Acknowledgments

We thank S. Ford, A. L. Rosenberger, and R. H. Madden for providing many helpful comments on the manuscript. This work was supported in part by grants from the National Science Foundation (BNS 8606796 and BRS 8614533, BSR 8896178 and BSR 8918657), the L. S. B. Leakey Foundation, and the National Geographic Society.

Part Seven
Summary

29
The Laventan Stage and Age

Richard H. Madden, Javier Guerrero,
Richard F. Kay, John J. Flynn, Carl C. Swisher III,
and Anne H. Walton

RESUMEN

La correlación original de Stirton, ubicando la fauna de La Venta en la Edad Mamífero "Friasense," se hizo en base al estado general de la evolución de ciertos taxa. La fauna de mamíferos de La Venta demuestra bajos niveles de semejanza con las faunas de las Edades Mamíferas Colhuehuapense y Santacrucense y con faunas del Colloncurense y Friasense. Semejantes niveles de similitud no permiten una correlación precisa. Asimismo, taxa "guías" conducen a correlaciones ambíguas, ya que tanto fósiles guías para la Edad Santacrucense como para el "Friasense" estan registradas en la fauna de La Venta. Las distribuciones bioestratigráficas de los géneros registrados en el Mioceno de la Argentina y en el Grupo Honda indican una mejora ubicación para la fauna de La Venta entre las faunas del Colloncurense y Mayoense, es decir, dentro de la Edad Mamífero "Friasense."

Una pronunciada discontinuidad o cambio faunístico ocurre en la secuencia Argentina durante el Mioceno medio, entre la fauna del Colloncurense y las faunas de la Edad Mamífero Chasiquense. En Patagonia, este cambio faunístico esta registrado dentro de la Edad Mamífero "Friasense," es decir, entre las faunas del Colloncurense y las del Mayoense, donde se puede documentar un hiato cronológico entre 15,7 y 11,8 Ma para las rocas portadoras de estas faunas. En base a la evidencia disponible ahora, éste hiato temporal en la secuencia para el Mioceno medio no está representada en Argentina por rocas portadoras de mamíferos fósiles.

Se propone una nueva unidad cronoestratigráfica, el Piso Laventense, caracterizado por la Zona de Conjunto *Miocochilius,* correspondiente a la parte fosilífera del Grupo Honda portador de la fauna de La Venta. El Piso Laventense se extiende desde 11,8 hasta 13,5 Ma, en base a una serie de fechados ^{40}Ar/^{39}Ar y correlación magnetoestratigráfica con la Escala Temporal Global de Polaridad Magnética. La Edad Laventense, una nueva unidad geocronológica equivalente a y basada en el Piso Laventense, se ubica dentro del hiato temporal reconocido en la secuencia mamífera en Patagonia. Se propone que la Edad Laventense sea ubicada en la secuencia de Edades Mamíferas Sudaméricanas entre el Colloncurense y el Chasiquense. El límite superior de la Edad Mamífera Colloncurense está modificada hasta conformarse con el limite inferior de la Edad Laventense, o sea, en aproximadamente 13.5 Ma, y que el limite inferior de la Edad Mamífera Chasiquense corresponde al limite superior de la Edad Laventense, aproximadamente 11,8 Ma.

La fauna de Quebrada Honda del Departamento de Tarija en Bolivia es de Edad Laventense, pero su asignación al Piso Laventense requiere la determinación en la fauna de Quebrada Honda de taxa diagnósticos para el Piso Laventense. Hasta la fecha, la fauna de Quebrada Honda es la única conocida de Edad Laventense que se ubica geográficamente en el límite austral de la zona tropical. Será preciso averiguar la utilidad práctica de la Zona de Conjunto *Miocochilius* y del Piso Laventense a través de estudios detallados en las secuencias miocénicas del Ecuador y Bolivia.

The La Venta fauna of Colombia is now among the best-known and most thoroughly studied terrestrial vertebrate paleofaunas of South America. Many of the fossil mammals from La Venta were originally described by R. A. Stirton, his students, and others at the University of California Museum of Paleontology (chap. 1). Based upon the stage of evolution of some of these mammals, Stirton concluded that the La Venta fauna is Miocene in age, postdating the well-known Santacrucian fauna of Argentine Patagonia. Of the faunal units of the Argentine Miocene sequence, Stirton (1953b) proposed that the La Venta fauna is "nearer in age to the Friasian of Argentina," a "stage" based primarily upon the fossils from Río Cisnes as discussed by Simpson (1940) and described by Kraglievich (1930a). (Ambiguities in the use of the name Friasian are discussed further here.)

Stirton's original Friasian correlation principally was based on the "advanced" stage of evolution of the interatheriid notoungulate *Miocochilius,* which was distinguished from Santacrucian interatheriids by relatively derived postcranial, cranial, and dental characters (Stirton 1953a). Supporting evidence for a post-Santacrucian, or Friasian, correlation was provided by the relatively "advanced" stage of evolution of La Venta rodent species compared with congeners from the Santacrucian (Fields 1957). Similarly supporting evidence for this correlation was reported among the marsupials and sloths (Marshall 1977, 1978; Hirschfeld 1985).

In the mid-1980s attempts were made to correlate the La Venta fauna with the sequence of Miocene faunas in Argentina by overall taxonomic similarity of the mammals as measured by faunal resemblance indices. For example, based upon fauna lists published more than ten years ago (Hirschfeld and Marshall 1976; Marshall et al. 1983), Kay et al. (1987), using the Simpson coefficient of faunal similarity, proposed a Santacrucian (pre-Friasian) correlation for the La Venta fauna. More recently, applying the Jaccard coefficient and more complete fauna lists for the Patagonian middle Miocene, Ortiz Jaureguizar et al. (1993) found no evidence to favor a particular correlation between La Venta and any of the Miocene land mammal ages (LMA) of Patagonia. This lack of precision is also reflected in the fact that guide fossils (sensu Marshall et al. 1983) for three different land mammal ages of Patagonia (*Pachybiotherium* and *Pro-*

thoatherium for the Colhuehuapian LMA, *Asterostemma* for the Santacrucian LMA, and *Glossotheriopsis* and *Neonematherium* for the "Friasian" LMA) all occur in the La Venta fauna.

Here, we discuss some of the problems encountered in the use of mammalian fossils for continentwide correlation in South America. Although of great utility in comparing faunas within Patagonia, significant obstacles are encountered when attempting continentwide correlation in the Miocene using the evidence of fossil mammals alone. First, mammal-based correlation is made problematical by the great latitudinal span of South America. Land mammal age correlation over great latitudinal spans is normally not a problem in Europe (Lindsay et al. 1989) nor generally in North America (Woodburne 1987). North American land mammal ages have been established on the basis of faunas distributed between 30° and 60°N latitude, and most of the area of the North American continent lies within these middle and temperate latitudes with only a narrow peninsula extending southward beyond the Tropic of Cancer. Even late Miocene faunas from tropical latitudes in Honduras (10°N) are readily correlated with type Clarendonian and Hemphillian faunas from the Texas panhandle (35°N) and with faunas from as far north as Montana and Washington states (49°N) (Woodburne 1987; Savage and Russell 1983).

By contrast, South America spans 65° of latitude, from 10°N to 55°S, and 70% of the area of South America occurs in the equatorial tropics north of the Tropic of Capricorn. Only the narrow southern peninsula of Patagonia extends into temperate latitudes. Miocene fossil mammal localities in the vast South American lowland tropics are few, and equatorial localities are generally limited to the western, tectonically active, Pacific-Andean margin. Although the record for the vast lowlands of the stable Brazilian and Guyana shields remains largely unknown, most of the Miocene fossil record is known from the narrow southern peninsula of Patagonia at south temperate latitudes (between 40° and 55°S) where the standard sequence of Miocene South American land mammal ages (SALMA) has been established.

A further difficulty for faunal correlation in South America is that direct stratigraphic superposition of Miocene faunas is rare in Patagonia. With few exceptions the record is discontinuous, characterized by relatively brief intervals of sediment deposition with fossils and long temporal hiatuses with no fossil record. The incompleteness of the stratigraphic record generally has made it difficult to establish phylogenies for most mammal groups, and without stable phylogenies, determining temporal relationships by stage of evolution is difficult. Finally, during the early and middle Miocene, there were no dispersal events between adjacent continents (nor documented dispersal events within the continent such as, for example, the diachronous *Hipparion* datum across Eurasia [Sen 1989]), so that datum planes based upon the evolutionary first appearance or time of arrival of immigrant taxa are of only limited utility.

As a contribution toward the establishment of the age of the La Venta fauna, we summarize new information about the stratigraphic distributions of the nonvolant mammal species in the Honda Group of the La Venta area. On the basis of the stratigraphic ranges of fossil mammals in the Honda Group we propose a new mammalian biozone. This biozone provides the basis for naming a new chronostratigraphic unit, the Laventan Stage, and to propose a new geochronologic unit, the Laventan Age, equivalent in magnitude to land mammal ages as they have been understood in South America. The boundaries of the new stage are precisely defined at the proposed unit stratotype. The Laventan Stage/Age falls within a temporal hiatus in the mammalian record of Patagonia. It is noted that other South American Miocene land mammal ages, the Santacrucian LMA, Friasian LMA, and Chasicoan LMA, are informal in the sense that they are not based upon chronostratigraphic units with boundaries defined by observable paleontological features in a single designated stratotype. For example, the Friasian LMA is based upon several distinct faunas from different geographic localities and rock units, and, because it lacks a unique stratotype (stage), its temporal boundaries are ambiguous.

For the naming of new stratigraphic units, this chapter follows the 1983 North American Stratigraphic Code (NACSN 1983) and the recommendations of the International Stratigraphic Guide (ISSC 1976). We generally follow current practice in Argentina as exemplified by Cione and Tonni (1995) and many (but not all) of the recommendations of Woodburne (1977, 1987) and Lindsay et al. (1989). Our discussion of South American land mammal ages generally follows the treatment of Marshall et al. (1983) with some significant geochronologic and faunal refinements relating to the middle Miocene (Marshall, Drake, et al. 1986; Marshall and Salinas 1990; Bown and Fleagle 1993).

In this chapter we do not propose to completely resolve the fate of the problematic "Friasian" LMA because that is the topic of work still in progress (although see further). In the present discussion "Friasian" LMA, in quotes, refers to the composite LMA of Bondesio, Laza, et al. (1980) and Marshall et al. (1983). The Friasian, without quotes, refers to the fossil-bearing exposures of the Río Frías Formation along the Río Cisnes in Chile. The Colloncuran refers to the fossiliferous Collón Curá Formation along the Río Collón Curá in Neuquén Province and in the Pilcaniyeu region of Río Negro Province. The Mayoan (not "Mayan" as in Simpson [1940]) refers to the fauna from the fossil-bearing exposures of the "Río Mayo Formation" on the Pampa de Chalia and Meseta de Guenguel in southwestern Chubut Province and the adjacent area of northwestern Santa Cruz Province.

Problems of Continentwide Correlation Based on Regional Mammalian Fossil Evidence

South American land mammal ages typically have been defined by a list of "guide" fossils (genera) and characterized by lists of known taxa (Marshall et al. 1983). Most SALMAs of the early and middle Miocene are based upon faunas from Patagonia and have proven useful for regional correlation within that part of the continent. But the middle Miocene La Venta fauna is separated from similar age faunas in Patagonia by up to 45° latitude. The utility of these Patagonian faunas and SALMAs for temporal correlations in the equatorial part of the continent during this time interval is tested in what follows. The problem might best be framed as a question: Can we, using faunal evidence alone, correctly place the La Venta fauna within the sequence of Patagonian faunas? To answer this question we look at faunal resemblance at the generic level, guide fossils, and biostratigraphic range information. As shown here, the evidence from faunal resemblance and guide fossils yields imprecise temporal correlations and generally tends to overestimate the age of the La Venta fauna.

CORRELATION BY FAUNAL RESEMBLANCE

Measures of faunal similarity (faunal resemblance indices or FRIs) were computed using the Simpson coefficient, C/N_s, where C = the number of genera in common, and N_s = the number of genera in the smaller fauna. Similar studies have used the Jaccard coefficient, $C/N_1 + N_2 - C$, where N_1 = number of genera in the first fauna and N_2 = number of genera in the second fauna (see Ortiz Jaureguizar et al. 1993). The Simpson coefficient is used here because its behavior in zoogeography and especially in temporal correlation is better known (Flynn 1986; Bernor and Pavlakis 1987) and it has been shown to be preferable when the number of taxa in the smaller fauna is unusually low, which is often the case in paleontology (Simpson 1960; Shuey et al. 1978).

FRIs have been computed to compare the La Venta fauna to middle to late Miocene LMAs and their component faunas elsewhere in South America (table 29.1). The taxonomic identifications for the La Venta fauna follow revisions in this volume. Fauna lists for Patagonian faunas of Colhuehuapian, Santacrucian, and Huayquerian LMAs are taken from Marshall et al. (1983). Faunal lists for the Mesopotamian fauna of the Ituzaingo Formation follow Pascual and Odreman (1971), Scillato-Yané (1980, 1981a, 1981b), and Bianchini and Bianchini (1971). Those for the Vivero and Las Barrancas members of the Arroyo Chásico Formation follow Scillato-Yané (1977a, 1977b, 1978), Marshall (1978, 1980), and Bondesio, Laza, et al. (1980). For the Río Acre local fauna we follow Frailey (1986). The fauna list for the Urumaco Formation in Falcón State, Venezuela, follows Linares (1990), Pascual and Díaz (1970), Díaz and Linares (1989). Fauna lists for the Friasian, Colloncuran, and Mayoan components of the "Friasian" LMA follow preliminary identifications from work in progress as reported by Ortiz Jaureguizar et al. (1993).

Of all the Miocene faunas of South America, the La Venta fauna shares greatest similarity with the Friasian local fauna (FRI = 19). Of these, there is greater similarity between the La Venta fauna and the type Friasian. Thus, FRIs

confirm the original Friasian correlation established by Stirton (1953b).

The La Venta fauna shows similar low levels of faunal resemblance with Colhuehuapian, Santacrucian, Friasian, and Colloncuran faunas (see table 29.1). If the Santacrucian LMA and the Friasian local fauna are temporally sequential, the greater similarity between the La Venta fauna and the Friasian is consistent with the known geochronologic age of the fossiliferous Honda Group and the La Venta fauna (chap. 3). If, however, the Santacrucian and Friasian faunas (previously considered to be temporally sequential) turn out to overlap in time, FRI values with La Venta are more variable among time-equivalent faunas in Patagonia than they are between earlier (Colhuehuapian) or later (Colloncuran) faunas.

The La Venta fauna is more similar to early and middle Miocene faunas of Patagonia (Colhuehuapian through Colloncuran) than with late Miocene faunas there (the Mayoan). More significantly, the La Venta fauna is arguably less similar to late Miocene faunas (whether these are from tropical or temperate latitudes) than it is to the early and middle Miocene faunas of Patagonia. The La Venta fauna has FRIs ranging from 13 to 19 with early and middle Miocene faunas (Colhuehuapian–Colloncuran), whereas FRIs with late Miocene faunas (Mayoan–Huayquerian) range between 0 and 10 (see table 29.1).

CORRELATION BY "GUIDE" FOSSILS

Some mammal genera have been proposed as "guide" fossils for SALMA on the basis of their conspicuous presence in austral local faunas and presumed restricted temporal range (Ameghino 1906; Marshall et al. 1983). Although never made explicit, guide fossils (sensu Marshall et al. 1983) appear to be neither index fossils nor characteristic taxa, and their utility for continental correlation has never been tested. Of the seventy-four guide fossils for LMAs in the middle Miocene sequence proposed by Marshall et al. (1983) (twenty-five for the Colhuehuapian, thirty for the Santacrucian, and nineteen for the "Friasian"), only five have been found in the La Venta fauna (*Pachybiotherium, Prothoatherium, Asterostemma, Glossotheriopsis,* and *Neonematherium*). These five include guide fossils for the Col-

Table 29.1. Faunal resemblance indices (Simpson coefficients) for Miocene faunas and South American land mammal ages

	Colhuehuapian	Santacrucian	Friasian	Colloncuran	Laventan
Santacrucian	45				
Friasian	59	81			
Colloncuran	46	76	69		
Laventan	15	13	19	14	
Mayoan	7	21	21	36	7
Chasicoan (Vivero Member)	6	11	11	11	5
Chasicoan (Las Barrancas Member)	0	7	7	7	0
Urumaco Formation	15	15	15	8	8
Acre Local	0	0	0	0	10
"Mesopotamian" (Ituzaingo Formation)	3	5	6	3	0
Huayquerian	0	0	0	0	2

huehuapian (two), Santacrucian (one), and "Friasian" (two) LMAs. The fact that only five of the seventy-four proposed guide fossils occur at all in the La Venta fauna indicates that the value of the proposed guide fossils for continental correlation in South America is indeed limited. The temporal ambiguity of the correlation suggested by these five austral guide fossils indicates that even when found in distant faunas their value for correlation is limited because their temporal ranges are not adequately constrained. Without benefit of an independently dated succession of Miocene faunas in the tropics, the limits of the geographic and temporal ranges of these taxa cannot be constrained nor can their relative value for local, provincial, or continental correlation be determined.

CORRELATION BY BIOSTRATIGRAPHIC RANGES

The biostratigraphic ranges of the fifteen mammal genera that occur both in the La Venta fauna and in the Patagonian sequence (and elsewhere) are depicted in figure 29.1. In the figure the La Venta fauna is placed in biostratigraphic position by maximizing the number of first and last occurrences and minimizing the number of taxon ranges that extend both below and above this horizon. As depicted, of the fifteen genera, ten represent last occurrences, four represent first occurrences, and only one genus, the macraucheniid *Theosodon,* has a known range in Argentina both older and younger than this estimated "age" of the La Venta fauna.

The ten last occurrences include three marsupials. Of these the rarest, *Pitheculites chenche* and *Pachybiotherium minor,* are only known in the Honda Group from single specimens. *Pitheculites* (Abderitidae) is known from the Colhuehuapian LMA and the Friasian (chap. 12; also L. G. Marshall 1990), *Pachybiotherium* (Microbiotheriidae) only from the Colhuehuapian LMA, and the prothylacynine *Lycopsis* only from the Santacrucian LMA (Marshall 1977, 1978). The proterotheriid litopterns *Prolicaphrium* and *Prothoatherium* are known only from the Colhuehuapian LMA of Patagonia (chap. 18). Among the rodents the rare erethizontid *Steiromys* in the Honda Group is known in Patagonia from the Colhuehuapian LMA, Santacrucian LMA (including the Cerro Boleadoras Formation of the Río Zeballos Group), the Río Frías Formation at Río Cisnes, in Chile, and in the Collón Curá Formation. The small echimyid *Acarechimys* has been registered in faunas from the Santa Cruz, Río Frías, and Collón Curá formations of Patagonia. The records for two sloth genera at La Venta represent geochronologic last occurrences: *Pseudoprepotherium confusum* and *Glossotheriopsis pascuali* are otherwise known only from the Friasian at Río Cisnes (chap. 15).

In contrast to these ten last occurrences, only four first occurrences are recorded among the fifteen genera common to Colombia and Patagonia. These include the first appearance of the extant marmosin didelphids *Thylamys* and *Micoureus* (chap. 11), and the well-known genus *Prodolichotis* (Dolichotinae, Caviidae). The La Venta species,

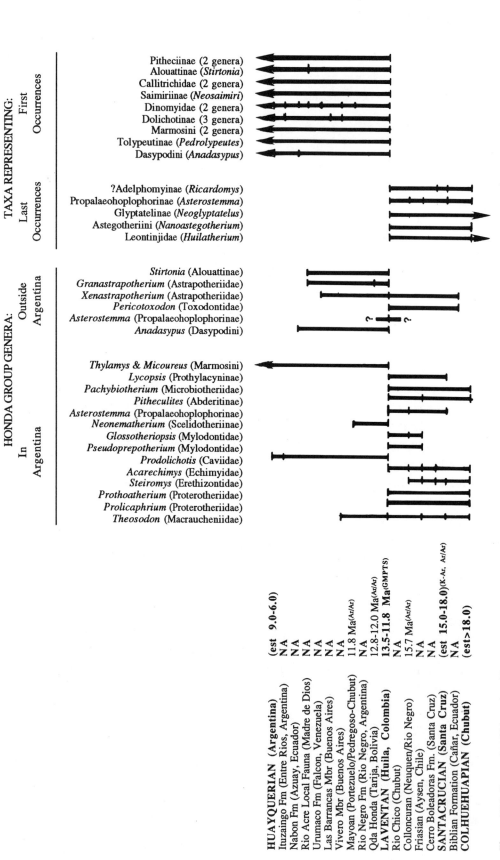

Figure 29.1. Temporal and biostratigraphic ranges of mammal genera and higher categories occurring in the La Venta fauna and known from dated and undated faunas in Argentina and elsewhere in South America. South American land mammal ages and chronostratigraphic stages are indicated in boldface type. Provinces or departments and countries of named local faunas and the sources for the indicated age estimates are given in parentheses. Abbreviations: *est.*, estimated age; *fm.*, formation; *mbr.*, member; *NA*, precise age unknown; *qda.*, quebrada.

Prodolichotis pridiana, is the oldest known representative of the genus, otherwise known from deposits of Huayquerian age in Argentina (Kraglievich 1934; Fields 1957; Walton 1990). The presence of the sloth *Neonematherium flabellatum* in the La Venta fauna has been confirmed (Hirschfeld 1985; also chap. 15). This genus was first described by Ameghino (1904) from Río Fénix, a Mayoan faunule (Kraglievich 1934; Bondesio, Rabassa, et al. 1980).

Three genera (*Pachybiotherium, Prothoatherium,* and *Prolicaphrium*) represent late surviving genera that are not registered in the immediately preceeding LMA but occur in the older Colhuehuapian LMA. Three other genera (*Prodolichotis, Thylamys,* and *Micoureus*) make precocious first appearances in the South American fossil record at La Venta; that is, these three genera are not registered in the immediately younger Chasicoan LMA but are known elsewhere in South America from younger intervals. The phenomenon of "jumping" taxa was first pointed out by McKenna (1956), with respect to *Megadolodus* and Leontiniidae, and Fields (1957) with respect to *Prodolichotis.* This phenomenon now appears to be even more frequent among mammals in the La Venta fauna, involving six genera. Although jumping taxa may serve only to point out possible sampling deficiencies in the fossil record, the fact that jumping taxa occur in many SALMAs, even within Patagonia (e.g., the absence of Homalodotheriidae in the Colhuehuapian LMA and of Mesotheriidae in the Colhuehuapian-Santacrucian interval), suggests that in the future it may be possible to establish geographic isolation, climatic stability, or heterochrony from among the possible controls underlying these occurrences.

By maximizing the number of first and last occurrences among the genera in common, the position of the La Venta fauna can be proposed relative to Patagonian Miocene faunas and its geochronologic age estimated by reference to the dated faunas between which it best fits. Using the criterion of maximizing first and last occurrences, the La Venta fauna falls in relative biostratigraphic position between the Colloncuran and Mayoan faunas. Based upon the prevalence of last occurrences in the biostratigraphic ranges of these fifteen taxa, the La Venta fauna would seem to have a preferred correlation closer to middle Miocene faunas, rather than to late Miocene faunas.

COMMENTS

Of the three approaches to faunal correlation between the La Venta fauna and the Patagonian sequence used here, two yield ambiguous results. Faunal similarity suggests a broad correlation with early to middle Miocene faunas in Patagonia, but not with late Miocene faunas. Faunal similarity, however, does not provide convincing evidence to support referring the La Venta fauna to any one particular Miocene LMA or fauna. Guide fossils appear to support this broad correlation but, like faunal resemblance, do not provide an unambiguous guide to any one particular LMA. With the third approach, based on the biostratigraphic ranges of taxa in common, the La Venta fauna appears to fit best between Colloncuran and Mayoan faunas, that is, between two of the components of the "Friasian" LMA of Patagonia.

Under ideal circumstances a taxonomically diverse and species rich fauna such as La Venta should be quickly referrable to one of the LMAs of the South American standard sequence. That this has not been satisfactorily achieved cannot be attributed to the incomplete state of our knowledge about the La Venta fauna. Instead, these ambiguous results lead us to question whether middle Miocene SALMAs based on southern faunas have continental scope. As presently defined and characterized, the Miocene SALMAs do not appear to be very powerful tools for refined temporal correlation across the latitudinal extremes of the continent.

The La Venta Mammal Fauna and Honda Group Biostratigraphy

Seventy-one mammal species (excluding Chiroptera) have been described from the Honda Group section in the upper Magdalena valley. These species belong to fifty-three genera in twenty-nine families. A full fauna list is presented in appendix 30.1. These fossil mammals constitute the La Venta fauna.

Fifty-two stratigraphically distinct fossil levels have been identified in the measured Honda Group section (chap. 2; fig. 29.2). Generically identifiable fossil mammals have been recovered from forty-six of these levels. As used here, a

Stratigraphic Unit	Fossil Level	Duke Locality	Screenwash Locality	UCMP Locality
Villavieja Formation				
Polonia Red Beds	52	101		
	51	131,144		
	50	78		
	49	140		
	48	126	126:1	
	47	141		
	46	125		
San Francisco Sandstone Beds				4529
El Cardón Red Beds	45	32	32:1–6	4528
Fossil levels between named units	44	25, 43		
	43	60,71		4527, 4938
La Venta Red Beds	42	14, 44		4526, 4538
Fossil levels between named units	41	24		
	40	66		4525
Ferruginous Beds	39			5045
Fossil level between named units	38	45, 52, 72		4524
Fish Bed	37	23, 46, 51, 128	CVP:1–14	4523
Fossil level between named units	36	59, 73		4522
Monkey Beds	35	5, 6, 12, 16, 17, 19, 22, 31, 50, 54, 57, 64, 65, 68, 69, 74, 82, 83, 93, 98, 99, 130, 142	6:1–4; 22:1	4517–21, 4536, 4541–42, 4932, 4934, 4936–37, 5043–44
La Victoria Formation				
Cerbatana Conglomerate Beds	34	7		
Fossil levels between named units	33	100,129		
	32	58		
	31	53		4933
	30	30		
	29	109		
	28	8, 108		
	27	97, 106, 107		
	26	9, 11A		
Tatacoa Sandstone Beds	25	11B		
Fossil levels between named units	24	29, 75, 86	75:1,2,B,M	4534
	23	87, 114, 115, 136		
	22	96, 137		
	21	26, 133, 138		
	20	41, 90, 95		4935
	19	116, 118		4533
	18	39, 84		
	17	117, 121		
	16	70, 113		
	15	112, 139		
	14	38, 63, 80, 103, 132, 134		4531
	13	36, 62, 81, 88, 102		4532
Chunchullo Sandstone Beds	12	127		
	11	79, 135		5046, 5047
	10	21, 27, 42, 61, 67, 91		4530
	9	76, 77, 145		
Fossil levels between named units	8	34, 104, 105	34:A,B,C	
	7	33, 85, 123	33:1–3	4535
	6	40		
	5	89, 92, 119, 122		
Cerro Gordo Sandstone Beds				
	4	111	111:1	
Fossil levels between named units	3	28, 120		
	2	48		
	1	49, 110, 146		

Figure 29.2. La Venta fossil locality correlation table.

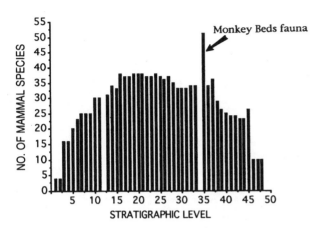

Figure 29.3. Number of mammal species by stratigraphic level in the Honda Group of the La Venta area. Gaps at stratigraphic levels 12 and 33 indicate levels where only nonmammalian vertebrates have been collected. Stratigraphic levels as in figure 29.2.

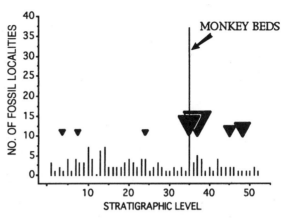

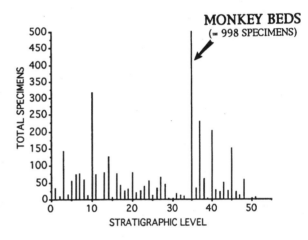

Figure 29.4. Number of fossil localities and specimens collected by surface prospecting at each stratigraphic level in the Honda Group (see figure 29.2). *Top:* Total number of fossil localities at each stratigraphic level. *Triangles,* stratigraphic levels where screenwashing has been undertaken; the size of the triangle indicates the relative intensity of the screenwashing effort. *Bottom:* Total number of fossil vertebrate specimens collected by surface prospecting at each stratigraphic level in the Honda Group.

fossil level refers to a set of collecting localities that cannot be subdivided stratigraphically. The relative stratigraphic position of each level has been established by lithologic correlation and tied to measured sections (chap. 2). Change in total mammalian species diversity through the Honda Group section is manifested principally as decreasing diversity toward the base of the section and near the top of the section (fig. 29.3) and is related in part to the reduced area of exposure of the lowest levels of the La Victoria Formation near Cerro Chacarón and of the highest levels of the Villavieja Formation in the El Cardón and Polonia red beds. Additionally, the low diversity at the top of the section may be related to a slowing in the rate of net sediment accumulation, increased paleosol maturity (laterization), and the consequent destruction and poor preservation of bones (chaps. 2, 3). With only one exception (chap. 11), fossil mammals are not known from levels above Duke Locality 126 (Level 48) in the Polonia Red Beds.

Frequency histograms of the number of collecting localities and specimens collected by stratigraphic level reveal some variation in the intensity of collecting effort at each level through the section (fig. 29.4, top). Screenwashing has been undertaken at seven levels in the Honda Group, and the relative intensity of this sediment washing effort is also indicated. In terms of the number of collecting localities, number of specimens collected, and diversity of collecting methods the greatest collecting effort has been made in the

Monkey Beds. Fifty-one nonvolant mammal species are known from the Monkey Beds. These species are referred to collectively as the Monkey Beds fauna. (For a list of the mammalian taxa of the Monkey Beds fauna, see table 30.2.)

The University of California collection of fossil vertebrates from the La Venta area was recovered by surface prospecting principally from the Villavieja Formation as that unit is now understood (chap. 2). Duke University/INGEOMINAS surface prospecting has been concentrated in the La Victoria Formation. Relatively few localities are known from stratigraphic levels below the

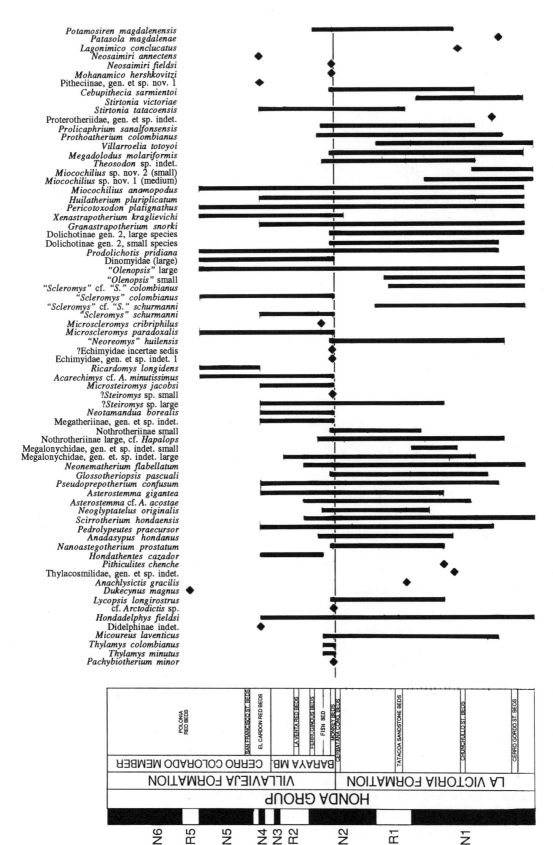

Figure 29.5. Biostratigraphic ranges of all nonvolant mammal species in the Honda Group. Stratigraphic nomenclature follows Guerrero (chap. 2); magnetostratigraphy after Flynn et al. (chap. 3). The base of the Monkey Beds is indicated by the horizontal line. Abbreviations: *St.*, sandstone; *cong.*, conglomerate. Unique occurrences are indicated by filled diamonds.

level of Duke Locality 28 in the area just south of Cerro Chacarón. Three levels in the La Victoria Formation were briefly sampled by wet screening and were found to be unproductive. Within the Villavieja Formation, the Monkey Beds, Fish Bed, El Cardón Red Beds, and Polonia Red Beds have been intensively screenwashed with considerable success in producing small mammals and vertebrates. Fossil vertebrate localities become less numerous in the red beds, and fossils are not abundant in the Polonia Red Beds.

The stratigraphic ranges of all mammal species described from the Honda Group in the La Venta area are indicated in figure 29.5. These stratigraphic ranges are compiled primarily from occurrences of type and hypodigm material in publications and museum records at the UCMP and IN-GEOMINAS (for all mammal taxa except *Neoglyptatelus* and the Dasypodoidea [Cingulata], which are based only on INGEOMINAS material), and represent total inferred or filled ranges between lowermost and uppermost occurrences. Based on these taxic ranges, we propose the following biostratigraphic unit within the Honda Group in the La Venta area.

The *Miocochilius* Assemblage Zone

DEFINITION

The "*Miocochilius* Assemblage Zone" is named after *Miocochilius*, a small quadrupedal, herbivorous typotherian notoungulate of the family Interatheriidae, the most common fossil mammal of the Honda Group in the La Venta area (Stirton 1953a). The base of the *Miocochilius* Assemblage Zone is defined by the coincident stratigraphic first occurrences of *Miocochilius* sp. nov. 1 (unnamed), "*Scleromys*" *colombianus, Granastrapotherium snorki,* and *Scirrotherium hondaensis,* and the top of the *Miocochilius* Assemblage Zone is defined by the coincident stratigraphic last occurrences of *Miocochilius anamopodus,* "*Scleromys*" *schurmanni, Prodolichotis pridiana,* "*Olenopsis*" large species, *Xenastrapotherium kraglievichi,* and *Pericotoxodon platignathus.* Other characteristic taxa of the *Miocochilius* Assemblage Zone include *Pedrolypeutes praecursor,* "*Neoreomys*" *huilensis, Stirtonia* spp., *Huilatherium pluriplicatum,* and *Glossotheriopsis pascuali.*

STRATOTYPE

The stratotype of the *Miocochilius* Assemblage Zone extends stratigraphically from the level of Duke Localities 49, 110, and 146 in the La Victoria Formation (Level 1) below the Cerro Gordo Sandstone Beds to the level of Duke Locality 126 in the Polonia Red Beds of the Villavieja Formation (Level 48) (fig. 29.6). The stratotype of the *Miocochilius* Assemblage Zone is found in the La Victoria and Villavieja formations exposed in the Cerro Chacarón–Cerro Alto section of the Honda Group in the La Venta area (figs. 2.8, 2.9). This area corresponds to that mapped in six 1:25,000 sheets of the Carta General de Colombia (Instituto Geográfico "Agustin Codazzi"), 302-II-D, 302-IV-B, 302-IV-D, 303-I-C, 303-III-A, 303-III-C, a quadrangle delimited by local coordinates X = 870.000 to the north, X = 840.000 to the south, Y = 895.000 to the east, and Y = 865.000 to the west.

The Honda Group exposures in the vicinity of Coyaima (Tolima Department, at approximately 3°45′ N and 75°10′ W, corresponding to the vicinity of X = 877.000 and Y = 912.000 on the 1983 Carta General, Plancha 282, 1:25,000 of the Instituto Geográfico "Agustin Codazzi") and in the Dina Oil Field (Neiva Department, southwest of Aipe, 4 km WNW of Guacirco, at 3°06′ N and 75°18′ W, corresponding to X = 862.000 and Y = 835.000 on the 1972 Carta General, Plancha 323, 1:100,000 of the Instituto Geográfico "Agustin Codazzi") are designated as reference sections (fig. 29.7).

DISCUSSION

As formulated, the *Miocochilius* Assemblage Zone conforms to the definition of the "Oppel zone" of the North American Stratigraphic Code (NACSN 1983) and of the "Oppel-Zone" of the International Stratigraphic Guide (ISSC 1976). To conform with the recommendation of Woodburne (1987) to provide a single-taxon definition, an informal taxon range zone, equivalent in stratigraphic and temporal scope to the *Miocochilius* Assemblage Zone, can be formulated on the basis of the stratigraphic first and last occurrences of the genus *Miocochilius.* Our rationale for preferring to propose formally a multiple taxon assemblage zone (rather than a single-taxon range zone) relates to our belief that it will eventually prove to be more stable and more useful for biocorrelation.

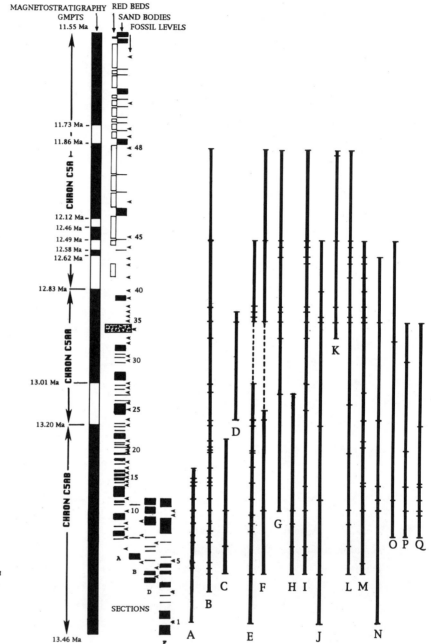

Figure 29.6. Biostratigraphic ranges of the defining and characteristic taxa of the *Miocochilius* Assemblage Zone. Geochronology follows Berggren et al. (1985). Honda Group magnetic polarity stratigraphy is correlated to GMPTS following Flynn et al. (chap. 3). *A, Miocochilius* sp. nov. 1 (medium); *B, Miocochilius anamopodus; C, Stirtonia victoriae; D, Stirtonia tatacoensis; E, "Scleromys" colombianus* species lineage; *F, "Scleromys" schurmanni* species lineage; *G, Prodolichotis pridiana; H, "Olenopsis"* small species; *I, "Olenopsis"* large species; *J, Granastrapotherium snorki; K, Xenastrapotherium kraglievichi; L, Pericotoxodon platignathus; M, Huilatherium pluriplicatum; N, Scirrotherium hondaensis; O, Pedrolypeutes praecursor; P, "Neoreomys" huilensis; Q, Glossotheriopsis pascuali.*

The defining and characterizing taxa of the *Miocochilius* Assemblage Zone are among the best-known fossil taxa and have the most densely sampled and complete stratigraphic ranges of all mammal taxa in the Honda Group (table 29.2). *Miocochilius* Stirton is the most abundant and characteristic fossil mammal of the Honda Group and is the only mammal genus that has stratigraphic occurrences and an inferred stratigraphic range extending from the lowest to highest stratigraphic levels of the *Miocochilius* Assemblage Zone. Three species of *Miocochilius* with overlapping stratigraphic ranges occur in the stratotype section of the Honda Group in the La Venta area (descriptions of two of these species have yet to be published). *Miocochilius anamopodus* Stirton occurs from Level 3 up to the upper limit of the *Miocochilius* Assemblage Zone at Level 48 (fig. 29.6B). *Miocochilius* sp. nov. 1 (a medium-sized species) occurs from the lowest

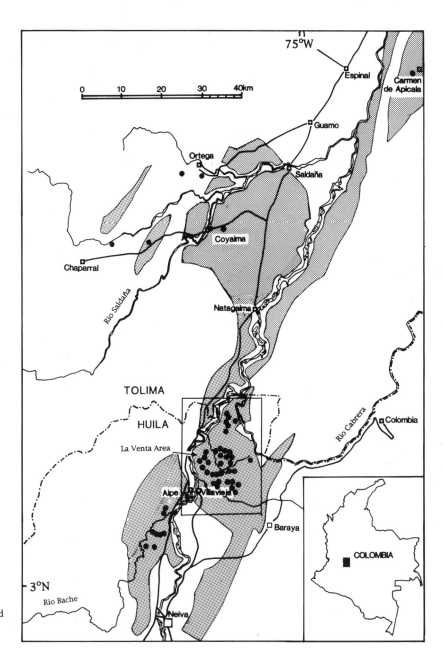

Figure 29.7. Distribution map of fossiliferous Honda Group sediments in the Magdalena River valley, detailing the locations of the stratotype area at La Venta (see also chap. 2) and the areas of the referred sections near Coyaima and Carmen de Apicalá.

level of the *Miocochilius* Assemblage Zone up to Level 20 (fig. 29.6A). *Miocochilius* sp. nov. 2 (a small-sized species) has a more restricted stratigraphic range and is not included among the defining or characteristic taxa of the *Miocochilius* Assemblage Zone.

The abundant small dinomyid material from the Honda Group described by Stehlin (1940) and Fields (1957) was referred to *Scleromys* Ameghino. This generic allocation is incorrect, and pending published revision of this taxon, it is herein referred to as *"Scleromys."* This taxon is among the

most abundant and longest ranging of mammals from the Honda Group. Two species have been described, *"Scleromys" schurmanni* Stehlin and *"Scleromys" colombianus* Fields. Each species exhibits morphological change upsection interpreted as anagenesis. Morphologically primitive species of each lineage occur in the lower levels of the La Victoria Formation; *"Scleromys"* cf. *"S." schurmanni* between Levels 1 and 29 and *"Scleromys"* cf. *"S." colombianus* between Levels 4 and 25 (fig. 29.6E, F). More advanced species, indistinguishable from the holotypes, occur in the Villavieja Formation:

Table 29.2. Sampling density and stratigraphic ranges of taxa in the *Miocochilius* Assemblage Zone

Taxon	Family	Occurrences[a]
Miocochilius anamopodus	Interatheriidae	28/41
Stirtonia spp.	Cebidae	7/30
Pedrolypeutes praecursor	Dasypodidae	5/33
"Olenopsis" sp. (large)	Dinomyidae	16/41
"Scleromys" colombianus	Dinomyidae	18/41
Pericotoxodon platignathus	Toxodontidae	15/41
Scirrotherium hondaensis	Pampatheriidae	7/37
Huilatherium pluriplicatum	Leontiniidae	17/37
Granastrapotherium snorki	Astrapotheriidae	8/41
Miocochilius sp. nov. 1 (medium)	Interatheriidae	14/20
Xenastrapotherium kraglievichi	Astrapotheriidae	6/13
Prodolichotis pridiana	Caviidae	10/34
"Neoreomys" huilensis	Dasyproctidae	6/28
Glossotheriopsis pascuali	Mylodontidae	8/28

[a]Occurrences represent stratigraphic levels actually recorded (numerator) within the total inferred or filled stratigraphic range of the taxon as defined by its lowermost and uppermost occurrences (denominator).

"Scleromys" schurmanni between Levels 35 and 45 and *"Scleromys" colombianus* between Levels 35 and 48. Neither species has been recovered in the stratigraphic interval between the Tatacoa Sandstone Beds and the Cerbatana Conglomerate Beds (between Levels 29 and 35). Apparent anagenetic change in these lineages may eventually serve as the basis for more refined, hierarchically subordinate, taxon range zones.

Among the astrapotheres the stratigraphic range of *Granastrapotherium snorki* Johnson and Madden (fig. 29.6J) extends from the base of the *Miocochilius* Assemblage Zone (Level 1) up to the level of the El Cardón Red Beds (Level 45). *Xenastrapotherium kraglievichi* (fig. 29.6K) is known from a stratigraphic level below the Cerbatana Conglomerate Beds (Level 33) up to the top of the *Miocochilius* Range Zone (Level 48).

The stratigraphic range of *Scirrotherium hondaensis* Edmund and Theodor, the distinctive pampathere from the Honda Group, is based upon type and hypodigm material from both the UCMP and INGEOMINAS collections but does not include all the known specimens from the Honda Group that may eventually be referred to this taxon (chap. 14; also fig. 29.6N). The stratigraphic range of *Scirrotherium hondaensis* extends from the base of the *Miocochilius* Assemblage Zone at Level 1 upward to Level 43.

Prodolichotis pridiana Fields (fig. 29.6G) extends from Level 10 upward to the upper limit of the *Miocochilius* Assemblage Zone (Level 48). Some large dinomyid rodent specimens from the Honda Group were referred to *Olenopsis* Ameghino by Fields (1957). This allocation is incorrect, and moreover, the type specimen of *"Olenopsis" aequatorialis* (Anthony) from the vicinity of Nabón in southern Ecuador is morphologically distinct from all species of large dinomyid that occur in the Honda Group (chap. 24). Pending revision of this genus and the description of two new species, Honda Group material is herein designated *"Olenopsis"* spp. Despite this taxonomic instability, this important and distinctive taxon, generally well described by Fields (1957), is included in the *Miocochilius* Assemblage Zone. The combined stratigraphic ranges of two species of *"Olenopsis"* (fig. 29.6H, I) extend from Level 3 to the upper limit of the *Miocochilius* Assemblage Zone in the Polonia Red Beds at Level 48.

Pericotoxodon platignathus Madden, a toxodontid notoungulate, occurs at many fossil levels throughout the Honda Group in the stratotype area (fig. 29.6L). The known stratigraphic range extends from Level 3 in the La Victoria Formation upward to the top of the *Miocochilius* Assemblage Zone in the Polonia Red Beds at Level 48. A complete list of all referred material, with locality and stratigraphic information, is provided in Madden (1990).

The known stratigraphic range of *Huilatherium pluriplicatum* Villarroel and Guerrero, the advanced leontiniid notoungulate described by Villarroel and Danis (chap. 19), in the stratotype section extends upward from Level 3 in the La Victoria Formation to Level 45 in the El Cardón Red Beds (fig. 29.6M).

The stratigraphic range of the armored xenarthran, *Pedrolypeutes praecursor* Carlini, Vizcaíno, and Scillato-Yané, which extends from Level 7 to Level 45, is based only upon holotype and hypodigm material from the INGEOMINAS collection. The stratigraphic occurrences of specimens in the UCMP collections that may be referred to this taxon are not yet known (fig. 29.6O).

The dasyproctid *"Neoreomys" huilensis* Fields was originally allocated to *Neoreomys* Ameghino by Fields (1957). This allocation is incorrect, and the La Venta species will be made the type of a new, closely related genus. Pending the published description of this new genus, this material is herein referred to as *"Neoreomys" huilensis.* The stratigraphic range of *"Neoreomys" huilensis* extends from Level 7 to Level 35 in the Monkey Beds (fig. 29.6P).

Stirtonia tatacoensis (Stirton and Savage) Hershkovitz is recorded in the upper part of the Honda Group between Levels 24 and 37 (fig. 29.6C). *Stirtonia victoriae* comes from the La Victoria Formation in the lower part of the Honda Group (Kay et al. 1987; Madden et al. 1989) between Levels 4 and 22 (fig. 29.6D). *Stirtonia* is the most common fossil primate in the Honda Group (chap. 28).

Finally, new material of the mylodontine *Glossotheriopsis pascuali* Scillato-Yané from the Honda Group in the La Venta area has been described by McDonald (chap. 15). This species has a stratigraphic range in the stratotype section from Level 7 to Level 35 (fig. 29.6Q).

Hierarchically Subordinate Biozones within the *Miocochilius* Assemblage Zone

Numerous stratigraphic first occurrences are recorded for mammal species in the Honda Group on the basis of any one of which it would be possible to subdivide the *Miocochilius* Assemblage Zone into hierarchically subordinate biozones (fig. 29.5). The stratigraphic first occurrences of the best-known species, the defining and characterizing

species of the *Miocochilius* Assemblage Zone (fig. 29.6), also suggest that subdivision may be warranted. Pending formal taxonomic revision of *Miocochilius, "Scleromys," "Olenopsis,"* and *"Neoreomys,"* knowledge of the complete stratigraphic ranges of all taxa of Cingulata, and a clear demonstration of the utility of such subdivision for more refined correlation outside the La Venta area, we refrain from formalizing any subdivision of the *Miocochilius* Assemblage Zone at this time.

Faunal Turnover

Ten species of mammals have their stratigraphic first occurrences in the Monkey Beds (table 29.3). (Species known by unique occurrences in the Monkey Beds are not considered.) Of these ten, five are very small species (*Thylamys minutus, Thylamys colombianus, Microsteiromys jacobsi* Walton, *Acarechimys* cf. *A. minutissimus,* and *Microscleromys paradoxalis* Walton) known only through screenwashing. Given the relatively meager return on screen-washing efforts at levels below the Monkey Beds (because of the generally less-suitable sediments of the La Victoria Formation), the five biostratigraphic first occurrences of these small species may reflect sampling bias. Three other first occurrences are known only by single specimens recovered at two or three stratigraphic levels (Dinomyidae [large], Megatheriiinae, and *Neotamandua borealis*) so that confidence intervals (following C. R. Marshall 1990) on their stratigraphic ranges are very large. Thus, eight of the ten first occurrences may be attributable to sampling deficiencies.

The two best established stratigraphic first occurrences in the Monkey Beds are *"Scleromys" schurmanni* and *"Scleromys" colombianus.* As noted earlier, material from lower levels in the La Victoria Formation referred to the closely related species *"S."* cf. *"S." schurmanni* and *"S."* cf. *"S." colombianus* may represent the direct ancestors of *"Scleromys" schurmanni* and *"Scleromys" colombianus.* These are the only two examples of stratigraphically successive species lineages in the Honda Group section in the La Venta area, and the first appearances of the two successor *"Scleromys"* species appear to be coincident. There is, however, an important gap in the stratigraphic record of both lineages below the Monkey Beds, in the interval between the Tatacoa Sandstone Beds and the Cerbatana Conglomerate, where the Honda Group section is not well exposed.

Table 29.3. Mammal species with first occurrences and those with last occurrences in the Monkey Beds

Species	Family	Collection Method Used	Occurrences[a]
		FIRST OCCURRENCES	
Thylamys minutus	Didelphidae	Screenwash	2/3
Thylamys colombianus	Didelphidae	Screenwash	2/3
Microsteiromys jacobsi	Erethizontidae	Screenwash	3/10
Acarechimys cf. *A. minutissimus*	Echimyidae	Screenwash	3/13
Microscleromys paradoxalis	?Dasyproctidae	Screenwash	3/12
Megatheriinae, gen. et sp. indet.	Megatheriidae	Surface	3/10
Neotamandua borealis	Myrmecophagidae	Surface	2/10
"Scleromys" schurmanni	Dinomyidae	Surface and screenwash	5/10
"Scleromys" colombianus	Dinomyidae	Surface and screenwash	7/13
Dinomyidae, large	Dinomyidae	Surface	3/14
		LAST OCCURRENCES	
Lycopsis longirostrus	Borhyaenidae	Surface	2/19
Nanoastegotherium prostatum	Dasypodidae	Surface	2/19
Glossotheriopsis pascuali	Mylodontidae	Surface	8/26
Nothrotheriinae, small	Megatheriidae	Surface	2/12
"Neoreomys" huilensis	Dasyproctidae	Surface	6/29
Dolichotinae gen. 2 (large)	Caviidae	Surface	4/31
Dolichotinae gen. 2 (small)	Caviidae	Surface	2/31
Megadolodus molariformis	Proterotheriidae	Surface	4/31

[a]Occurrences represent stratigraphic levels actually recorded (numerator) within the total inferred or filled stratigraphic range of the taxon as defined by its lowermost and uppermost occurrences (denominator).

Thus, in our judgment the apparent faunal turnover suggested by the biostratigraphic range chart (fig. 29.5) is most probably an artifact of sampling bias, reflecting both the stratigraphically uneven sampling of small mammals by screenwashing and the uniquely and extensively well-exposed sediments of the Monkey Beds. Given the stratigraphic gap in the ranges of *"Scleromys"* species, it is not possible to establish a precise level for the first occurrences of *"Scleromys" colombianus* and *"S." schurmanni*.

The Laventan Stage

NAME AND DEFINITION

The Laventan Stage, a chronostratigraphic unit, is named after Quebrada La Venta and the surrounding semidesertic region, well known (Stirton 1953b; Fields 1959; also chap.

2) for the abundance of fossil mammals occuring in the excellent exposures of the Honda Group. The base of the Laventan Stage is defined at the stratigraphic level of Duke Localities 49, 110, and 146 in the La Victoria Formation in Cerro Chacarón Section F of the Honda Group in the La Venta area (chap. 2). The base of the Laventan Stage coincides with the first appearance of *Miocochilius* sp. nov. 1 and the first appearances of the *"S." colombianus* species lineage, *Granastrapotherium snorki* and *Scirrotherium hondaensis*, as well. The base of the Laventan Stage is not defined by either a single or multiple taxon first occurrence but by the stratigraphic horizon of Duke Localities 49, 110, and 146 at stratigraphic level 1 (fig. 29.2).

STRATOTYPE

The stratotype of the Laventan Stage is the Cerro Chacarón–Cerro Alto section of the Honda Group in the La

Venta area (figs. 2.8, 2.9). This is also the stratotype of the *Miocochilius* Assemblage Zone.

BOUNDARIES

The lower boundary of the Laventan Stage coincides with the lower boundary of the *Miocochilius* Assemblage Zone in the La Victoria Formation, which in the type area is placed at about 13.5 Ma, based upon radioisotopic dates and the inferred age of the lower boundary of the lowest polarity interval (N1) sampled in the stratotype section (chap. 3). Tentatively, the upper boundary of the Laventan Stage coincides with the upper boundary of the *Miocochilius* Assemblage Zone, which in its type locality is placed at about 11.8 Ma (fig. 29.6), based upon the radioisotopic dates and the inferred age of the upper boundary of polarity interval N5 in the stratotype section. Thus, the Laventan Stage, as represented in the type section, spans a time interval between approximately 13.5 and 11.8 Ma.

REFERRED AND CORRELATIVE FAUNAS

The fossiliferous Honda Group near Carmen de Apicalá, Cundinamarca Province, Colombia (fig. 29.7), is referred to the Laventan Stage based upon the occurrences there of *"Scleromys"* and *Granastrapotherium* (Stehlin 1940; and chap. 22). The Honda Group outcropping in the vicinity of Coyaima, Tolima Department, Colombia, also is referred to the Laventan Stage based on the occurrence there of *"Scleromys," Granastrapotherium, Huilatherium,* and *Pedrolypeutes* (chaps. 13, 19, 22). The Ayancay Group outcropping south of Girón, in Azuay Province, Ecuador, may eventually be correlated with the Laventan Stage on the basis of the presence there of *"Olenopsis"* (Madden et al. 1991).

The Laventan Age

A Laventan Age is herein proposed as a geochronologic subdivision of the Miocene of South America corresponding to the interval of time represented by the Laventan Stage. The upper and lower boundaries of the Laventan Age are equivalent to the upper and lower boundaries of the Laventan Stage as defined at the stratotype area (chaps. 2, 3).

The Laventan Age fills a temporal hiatus in the sequence of Miocene biochronologic units in South America between the Colloncuran and Chasicoan land mammal ages. The former boundary between the Colloncuran and the Chasicoan LMAs was arbitrarily set at 12 Ma (Marshall 1985). Accomodating the Laventan Age within the Miocene sequence of South America requires a downward adjustment of the upper boundary of the Colloncuran LMA to conform with the lower boundary of the Laventan Age at approximately 13.5 Ma and an upward adjustment of the lower boundary of the Chasicoan LMA from about 12 Ma to about 11.8 Ma.

DISCUSSION

The following discussion about the temporal relationships between the Laventan Stage and previously recognized South American land mammal ages takes into account the common use by South American paleontologists of two different systems of stratigraphic classification and nomenclature: (1) units such as the Laventan and Marplatan (Cione and Tonni 1995), which follow the recommendations of the International Stratigraphic Guide incorporated, for example, in the North American Stratigraphic Code with its explicit procedures for distinguishing and naming biostratigraphic and chronostratigraphic units and upon which formal ages (geochronologic units) should be based; and (2) an informal system of land mammal ages/subages, erected directly as biochrons without reference to, or association with, a formally defined stage and without explicit rules for their definition or for the naming of new units.

We begin by considering the temporal relationships between the Laventan Stage and the LMAs of the early and middle Miocene of Argentina and Chile. The relevant ones in this time interval, in order of successively younger age, are the Santacrucian LMA, "Friasian" LMA, and the Chasicoan LMA. In South America these LMAs are named for aggregates of mammalian faunas (or assemblages) sometimes, but not usually, confined to particular rock units that are thought to be roughly contemporaneous or restricted to an interval of geologic time presumed to be equivalent to a stage. Typically, the temporal boundaries of South American LMAs are often more arbitrarily, and always less precisely, defined than chronostratigraphic stages in that they are not tied to biostratigraphic zones with precise upper and lower boundaries as defined at a unique unit stratotype.

Santacrucian Land Mammal Age

The Santacrucian LMA, characterized by faunas preserved in the Santa Cruz Formation, appears to be too old to overlap with the Laventan Stage. Several radioisotopic ages have been reported for the Santa Cruz Formation. Marshall, Drake, et al. (1986) gave K-Ar dates for the Santa Cruz Formation ranging from 17.3 ± 0.3 and 16.5 ± 0.4 Ma (at Monte León), to 16.7 ± 0.2 Ma (at Karaikén), and 16.0 ± 0.8 Ma (at Rincón del Buque). The accompanying magnetostratigraphic section included a long interval of reversed polarity correlated with either Chron C5CR (17.6–17.0 Ma) or with Chron C5BR (16.2–15.3 Ma) (Marshall, Drake, et al. 1986). Bown and Fleagle (1993) reported $^{40}Ar/^{39}Ar$ dates for the Santa Cruz Formation at Monte León and Monte Observación ranging between about 16.57 and 16.06 Ma. Thus, the radioisotopic and magnetic polarity chronology of the composite Santa Cruz Formation indicate an age of between about 17.3 and 15.4 Ma for classical coastal Santacrucian LMA faunas.

"Friasian" Land Mammal Age

The next younger SALMA is the "Friasian" LMA. As mentioned in the introduction, the "Friasian" LMA has generally been understood to be composed of three parts (hierarchically equivalent to land mammal subages in North America); the Friasian (sensu stricto), the Colloncuran, and the Mayoan as formulated by Kraglievich (1930a).

FRIASIAN Recent debate has centered on whether the Friasian fauna from the Río Frías Formation along the Río Cisnes in Aysén Province, Chile, should be referred to the Santacrucian LMA (following Marshall and Salinas 1990) or retained in a composite "Friasian" LMA along with the Colloncuran and Mayoan (Vucetich et al. 1993). A preliminary date of about 17 Ma reported for strata containing the Friasian fauna at Río Cisnes suggests that the Friasian fauna is temporally equivalent to part of the Santacrucian LMA (Marshall and Salinas 1990). Although some notoungulate

and rodent genera are distinct, Friasian mammals in general (and especially other rodents and marsupials) closely resemble those of the Santacrucian (Madden 1990; L. G. Marshall 1990; Ortiz Jaureguizar et al. 1993; Vucetich et al. 1993).

COLLONCURAN The radioisotopic age for the Collón Curá Formation and its contained fauna is derived from the Pilcaniyeu Ignimbrite. This ignimbrite occurs within the fossiliferous part of that formation and can be taken to date those faunas. Published K-Ar dates for the Pilcaniyeu Ignimbrite include four determinations ranging between 15.4 and 14.0 Ma reported by Marshall et al. (1977) and one of 15.0 Ma reported by Bondesio, Rabassa, et al. (1980). Additional whole-rock K-Ar determinations reported by Mazzoni and Benvenuto (1990) are consistent with these dates. The mammalian fauna from the Collón Curá Formation along the Río Collón Curá has its greatest overall resemblance with faunas of the Santacrucian LMA (Ortiz Jaureguizar et al. 1993; Vucetich et al. 1993). As in the case of the Friasian fauna, some rodents and notoungulates in the Colloncuran fauna are generically distinct from those of the Santacrucian LMA (Madden 1990; Vucetich et al. 1993). Interestingly, these are the same relatively derived taxa that occur in the Friasian fauna. The faunal differences between the Santacrucian LMA and the Colloncuran might seem to be consistent with the generally somewhat younger K-Ar age of the Collón Curá Formation. Our C40Ar/^{39}Ar determinations for the Pilcaniyeu Ignimbrite of 15.7 Ma, however, suggest instead that the Colloncuran fauna is contemporaneous with at least part of the Santacrucian LMA. This opens the possibility that the relatively minor differences between the faunas of the Santacrucian LMA, on the one hand, and the Friasian and Colloncuran faunas, on the other hand, may relate to geographic differences between southern and northern Patagonia during the middle Miocene (the two faunas are separated by 2° to as much as 10° latitude). Whether the Colloncuran fauna is eventually referred to the Santacrucian LMA or retained along with the Friasian as a subage of the Santacrucian LMA or as a composite "Friasian" LMA, all the available evidence indicates that the Colloncuran and Friasian faunas are probably at least 1.5 million years older than the base of the Laventan Stage.

MAYOAN   Our preliminary $^{40}Ar/^{39}Ar$ dates of about 11.8 Ma for a level just below the Mayoan assemblage at El Portezuelo on Cerro Guenguel are substantially younger than those for the Colloncuran. The Mayoan fauna appears to be distinct from the Colloncuran fauna, sharing very few genera (Ortiz Jaureguizar et al. 1993). Thus, in Patagonia, there is both a significant faunal turnover between the Colloncuran and Mayoan and a temporal hiatus, representing some part of the interval between approximately 15.7 and 11.8 Ma.

Chasicoan Land Mammal Age

The radioisotopic age of the Arroyo Chásico Formation and its contained faunas is not known. Less well-known local faunas from Catamarca, San Juan, and San Luis Provinces in Argentina have been referred to the Chasicoan LMA (Bondesio, Rabassa, et al. 1980; Marshall et al. 1983) as were the fossil mammals described by Rovereto (1914) from the "Rionegrense" in northern Patagonia. With this one exception the Chasicoan is known almost exclusively from outside Patagonia. Marshall and others (Marshall et al. 1983; Marshall 1985) suggest an arbitrary date of around 12.0 Ma for the boundary between the "Friasian" LMA and the Chasicoan LMA.

A significant temporal hiatus between faunas of the Santacrucian LMA and late Miocene faunas in Argentina was first inferred by Ameghino (1902a). Ameghino continued to recognize this hiatus after the discovery and first descriptions of the Friasian, Colloncuran, and Mayoan faunas of western Chubut (Ameghino 1906).

The fauna from Arroyo Chásico in southern Buenos Aires Province provides further documentation of the nearly complete replacement of "pan-Santacrucian" taxa (taxa ranging through the Santacrucian LMA, Colloncuran, and Friasian) by uniformly more advanced late Miocene taxa of the Chasicoan and Huayquerian LMAs, referred to collectively as the "pan-Araucanian" fauna (Bondesio, Laza, et al. 1980; Pascual et al. 1985; Pascual and Ortiz Jaureguizar 1990).

A "Friasian"-Chasicoan hiatus is often depicted in tables of Miocene mammal chronology for South America. The temporal hiatus appears in the chronologies of Marshall et al. (1977), Marshall and Pascual (1978), Pascual et al.

(1984), and MacFadden (1990), but not in Patterson and Pascual (1972), Marshall (1985), Marshall, Drake, et al. (1986), Pascual and Ortiz Jaureguizar (1990), or Marshall and Salinas (1990). This "Friasian"-Chasicoan hiatus has been variably estimated to fall between 12.0 and 10.0 Ma or between 12.0 and 11.5 Ma (Marshall et al. 1977; Marshall and Pascual 1978; Marshall et al. 1983). Most recently, MacFadden (1990) depicts the "Friasian" LMA–Chasicoan LMA hiatus at between 12.0 and 11.0 Ma. It now appears possible that the timing of the faunal turnover and hiatus in the Miocene sequence of Argentina may fall between the Colloncuran and Mayoan in Patagonia, that is, within what have been considered component faunas of the "Friasian" LMA (Ortiz Jaureguizar et al. 1983).

The Boundaries of the Laventan Age

The Laventan Age represents the time interval between 13.5 and 11.8 Ma. This interval falls between the Colloncuran LMA and Chasicoan LMA as depicted by Marshall and Salinas (1990). The Laventan fills at least part of the important temporal hiatus between middle and late Miocene faunas in Argentina (Pascual et al. 1985) (fig. 29.8). The lower boundary of the Laventan Age, in the stratotype area of the Laventan Stage, is placed at about 13.5 Ma, although it may be as much as 13.8–14.0 Ma, depending on the relative weight given to discrete radioisotopic dates, magnetostratigraphic correlations, and the specific GMPTS used (chap. 3). The temporal range of the Laventan does not overlap the published K-Ar age or the $^{40}Ar/^{39}Ar$ age of the Pilcaniyeu Ignimbrite of the Collón Curá Formation or the inferred age of the Colloncuran fauna.

The upper boundary of the Laventan Age is placed at about 11.8 Ma. This upper boundary overlaps the arbitrarily determined 12 Ma lower boundary of the Chasicoan LMA (Marshall 1985). Thus, inserting the Laventan into the sequence of South American Miocene biochrons requires a slight upward modification of the arbitrary lower boundary of the Chasicoan LMA.

For the present, the Laventan Age seems to be without representation in Argentina where geochronologically contemporaneous mammal faunas are as yet unknown. For the present, the immediately younger and older portions of

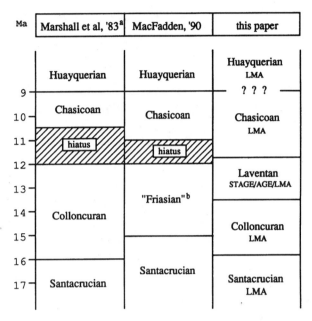

Figure 29.8. Middle Miocene geochronology and land mammal biochronology in South America. *LMA*, land mammal age.

[a]Marshall et al. (1983) as modified by Marshall and Salinas (1990).
[b]Composite "Friasian" LMA as used in text.

the middle Miocene (Colloncuran LMA and Chasicoan LMA) are also without known representation in the tropics.

Referred Faunas

The fossiliferous sediments yielding the Quebrada Honda local fauna in southern Bolivia are well dated, but the contained fossils have not been fully described (Hoffstetter 1977; MacFadden and Wolff 1981; Takai et al. 1984; Frailey 1987, 1988). Based on ^{40}K-^{40}Ar ages and magnetostratigraphy, the mammals from Quebrada Honda have been reported to have an extrapolated age of between 13.0 and 12.7 million years (MacFadden et al. 1990). Thus, the Quebrada Honda fauna is of Laventan Age. The Quebrada Honda fauna can only be referred to the Laventan Stage if the defining and characteristic taxa of the Laventan Stage are shown to be present. The geographic position of this faunule at a latitude intermediate between Colombia and Patagonia makes its study of compelling interest.

Occurrences of defining or characteristic taxa of the *Miocochilius* Assemblage Zone elsewhere in South America, such as *Stirtonia* in the Río Acre local fauna (Kay and Frailey 1993) and *Granastrapotherium* at Playa Mapuya on the Río

Inuya, Loreto Province, Peru (chap. 22), suggest that the associated "faunas" and fossil-bearing rocks may be correlatives of, or assignable to, the Laventan Stage. Should the weight of all the available evidence from the associated fossil mammals or independent isotopic age determinations demonstrate a different temporal correlation, either older or younger than the Laventan, it will extend the known geochronological range of these two genera. Such an eventuality does not change the stratigraphic range of these taxa in the *Miocochilius* Assemblage Zone in the stratotype section of the Laventan Stage.

A "Laventan Land Mammal Age"

A "Laventan Land Mammal Age" would be the informal, biochronologic equivalent of the Laventan Age defined by the mammalian taxa of the Laventan Stage and the *Miocochilius* Assemblage Zone and characterized by the mammalian taxa of the La Venta fauna. A "Laventan LMA," equivalent in temporal scope to the Laventan Stage and in content to the *Miocochilius* Assemblage Zone and the La Venta fauna, would be the first and, to date, only SALMA based upon a well-defined stage and biostratigraphic unit. As such, a "Laventan LMA" would be fully justified given its (1) clear faunal characterization, (2) distinct chronostratigraphic distribution, (3) demonstrated applicability within the tropics, and (4) well-constrained temporal span comparable to land mammal ages elsewhere. The relative uniqueness of the La Venta fauna in terms of latitudinal position, faunal diversity, and tropical lowland environment suggests that it may be difficult to demonstrate the practical utility of a "Laventan LMA" as a standard of reference for intracontinental faunal correlation across the full latitudinal extent of South America. The Laventan Stage, however, is a chronostratigraphic unit of demonstrated regional scope, and the Laventan Age is of continental scope by definition. The argument that a "Laventan LMA" should not be proposed because the Laventan Stage is of only regional scope is an argument that can be made about nearly all SALMAs. A high, but not unusual, degree of endemism is evident in the La Venta fauna. Although this endemism may be attributed to sampling deficiencies of the South American fossil record and to the fact that contemporaneous faunas are not yet known from Argentina (although fossil mammal faunas are known from the immediately preceeding and succeeding intervals), this level of endemism is typical for living south temperate and equatorial tropical mammalian faunas separated by as much as 45° of latitude (chap. 30).

We are equivocal about the proposition of a formal "Laventan LMA." Our uncertainty has to do with the fact that there are no guidelines for such propositions. More importantly, we are unwilling to propose new guidelines for what we consider to be a superfluous unit of stratigraphic classification given that explicit rules for the proposition of biostratigraphic, chronostratigraphic, and geochronologic units are available, which together conceptually encompass LMAs.

Summary and Conclusions

Stirton's original "Friasian" LMA correlation for the La Venta mammal fauna based upon general stage of evolution is not supported by either overall faunal resemblance or "guide" fossils. The mammals of the La Venta fauna shows similar, low levels of faunal resemblance with faunas of the Colhuehuapian LMA, Santacrucian LMA, and with Colloncuran and Friasian faunas, indicating that temporal correlation by faunal resemblance is not precise. Guide (or index) fossils for Colhuehuapian LMA, Santacrucian LMA, and "Friasian" LMA occur in the La Venta fauna. These index fossils make ambiguous correlators because their temporal ranges are poorly constrained. The biostratigraphic ranges of mammalian genera occurring both in the Miocene of Argentina and La Venta indicate that the La Venta fauna best fits between the Colloncuran and Mayoan faunas, that is, within the "Friasian" LMA.

A pronounced faunal discontinuity or turnover occurs in the middle Miocene fossil record of Argentina between the Colloncuran and Chasicoan LMAs. In Patatonia this faunal turnover occurs within the "Friasian" LMA, between the Colloncuran and the Mayoan, where it can be calibrated to between approximately 15.7 and 11.8 Ma. This temporal gap in the middle Miocene sequence apparently does not yet have representation in the mammal-bearing rock record anywhere in Argentina.

The Laventan Stage, characterized by the *Miocochilius* Assemblage Zone within the fossilferous Honda Group of Colombia (containing the La Venta fauna), is herein proposed as a new chronostratigraphic unit. The Laventan

Stage spans at least the interval of geologic time between about 13.5 and 11.8 Ma as documented radioisotopically and by magnetostratigraphic correlation with the GMPTS (chap. 3). The Laventan Age, a new geochronologic unit equivalent to and based upon the Laventan Stage, fits into the temporal hiatus recognized in the Patagonian mammal record. We propose that the Laventan Age be inserted in the SALMA sequence between the Colloncuran and Chasicoan LMAs. This requires that the arbitrary 12 Ma boundary between the Colloncuran and Chasicoan LMAs be changed. The upper boundary of the Colloncuran LMA is changed to conform with the lower boundary of the Laventan Age and LMA at about 13.5 Ma, and the lower boundary of the Chasicoan LMA is changed to conform with the upper boundary of the Laventan Age and LMA at approximately 11.8 Ma.

The Quebrada Honda fauna from Tarija Department in southern Bolivia is geochronologically contemporaneous with the Laventan Stage and is of Laventan Age, but its assignment to the Laventan Stage will require identification of diagnostic Laventan taxa in Bolivia. To date, this is the only other fauna outside the equatorial tropics of Colombia, Brazil, and Ecuador that is of Laventan Age. It will be possible to test the utility of the *Miocochilius* Assemblage Zone and the Laventan Stage through more detailed study of the middle Miocene sequences of Ecuador and Bolivia. Likewise, it will be possible to test the utility of the Laventan Stage by the discovery of contemporaneous mammalian faunas in southern South America.

Acknowledgments

The research on which this chapter was based was made possible by research grants from the National Geographic Society 2964-84 and 3292-86 and National Science Foundation BSR 86-14533 and 89-18657 to R. F. Kay, and NSF BSR 88-96178 and 86-14133 to J. J. Flynn. We thank all those who contributed to the study of the fossil mammals from La Venta and whose chapters appear in this volume. These contributions made this chapter possible. We have benefited from discussions with A. Cione, G. Vucetich, R. Pascual, L. G. Marshall, R. Cifelli, and D. Savage. We thank B. MacFadden for numerous helpful comments on an earlier version.

30
Paleogeography and Paleoecology

Richard F. Kay and Richard H. Madden

RESUMEN

El resumen de todas las evidencias disponibles acerca del entorno geográfico regional y ambiental durante el Mioceno medio en el área de La Venta conduce a las siguientes conclusiones generales. El área de La Venta se ubicó durante el Mioceno medio sobre una península aislada del resto del continente por un brazo del mar que se extendía desde la cuenca del Maracaibo hasta la Amazonía occidental. Comparaciones entre faunas mamíferas en base a índices de semejanza sugieren que el aislamiento geográfico parcial, distancia, y/o diferencias climáticas entre la zona ecuatorial y Patagonia podían haber sido responsables del alto grado de endemismo registrado para la fauna de La Venta.

El área de La Venta correspondía a un medio ecuatorial de bajas alturas, con poco relieve local. Fuentes de sedimento ubicadas al oeste contribuyeron sedimentos que fueron transportados por ríos que bajaban hacia el sureste. Inundaciones periódicas están indicadas por los sedimentos de llanura de inundación del Grupo Honda y sugeridas por los diversos peces adaptados a aguas temporales y que se alimentan dentro de selvas inundadas. Estas inundaciones pudieron haber sido provocadas por fluctuaciones en lluvias. Sin embargo, estas mismas evidencias no permiten concluir que el ambiente sufrió un déficit de agua estacional prolongado. La actual distribución de peces pulmonados, por ejemplo, no está restringida a ambientes con estaciones marcadas de sequía.

Los ambientes de deposición en el Grupo Honda no incluyen características generalmente asociadas a déficits estacionales prolongados de lluvias.

Los vertebrados fósiles demuestran evidencias más fehacientes de presencia de bosques, y son uniformes en indicar la presencia de biotopos terrestres de bosque dentro del área de La Venta durante el intervalo de deposición del Grupo Honda. La diversidad de adaptaciones presentada por la fauna de mamíferos de la Capa de Los Monos demuestra una alta proporción de especies de hábitos arborícolas, semejante a faunas Neotropicales actuales, como, por ejemplo, las de Cocha Cashu en el Parque Nacional del Manú y del alto Río Marañón, Perú. El alto porcentaje de especies ramoneadoras en comparación con especies pastoreadores es semejante a faunas tropicales actuales de ambientes con precipitaciones de más de 2.000 mm por año.

La única especie de planta fósil del Grupo Honda es afin al género *Goupia,* taxon que se encuentra en la actualidad restringida a ambientes ribereños en zonas de más de 2.000 mm de precipitacíon anual. La formación vegetal sugerida para el Mioceno medio quizás experimentó una estación seca, pero ésta hubiera sido menos de tres meses de duración.

La reconstrucción del ambiente de La Venta apoyada por estas evidencias se distingue del panorama de llanos extendidos entre bosques angostos de galería presentado por Stirton.

The preceding chapters of this book have presented detailed information and interpretation about the paleobiology of La Venta animals. In this chapter we summarize information about the paleogeography and sedimentary environments at La Venta and the evidence used by many contributors to infer the paleobiology of La Venta animals. Finally, the composition of the fauna as a whole is compared with selected tropical faunas in South America today in an effort to portray the paleoenvironment of La Venta in the middle Miocene.

First efforts to understand the paleoecology of La Venta were made more than thirty years ago by R. A. Stirton in his 1953 monograph on *Miocochilius* (Interatheriidae, Notoungulata). Stirton was impressed by this animal's cementum-covered, ever-growing cheek teeth and possible cursorial locomotion, features he took to be signs of open-country grazing habits. At the same time he sought to balance this interpretation against the apparently conflicting evidence of monkey fossils that implied forests at La Venta:

Our evidence will not permit, at this time, an accurate restoration of the environment throughout the area, since it probably varied in different places or even in the same place at different times. Nevertheless, there must have been wide stretches of savanna in the La Venta, Río Tatacoa and Cerro Gordo areas with heavy forests near the streams. Oxbows, developed and isolated by meandering streams, became mud traps for terrestrial animals as water evaporated during the dry seasons. . . . The only obviously arboreal animals were monkeys, though some of the smaller sloths may have climbed the trees. (Stirton 1953a, 268)

Fields (1957, 1959) reached similar conclusions, noting that La Venta sediment deposition had occurred on a broad flood plain through which large rivers meandered. He suggested that the climate was seasonal and that stream borders were heavily forested while there were broad expanses of open lands on the slightly higher ground stretching back from the mud flats. He identified primates, opossums, and small sloths as denizens of the forest and interatheres, toxodonts, and a rodent *Prodolichotis* (whose closest living relative is *Dolichotis,* the "Patagonian mara") as living in nearby savannas.

No one has questioned, nor do we, that the La Venta fauna was situated in a tropical region of low relief near sea level. Nor is there any question that broad flood plains and oxbow lakes existed within a meander belt characterized by periods of flooding. The important unresolved questions revolve around the severity and length of the dry season, total annual rainfall, and whether both forested and more open-country environments were present and, if so, in what proportions. Upon review of all the available evidence and comparisons with modern Neotropical faunas, we are led to the conclusion that the vertebrate fauna does not present a picture of extensive, interfluvial savanna grasslands with narrow, semideciduous gallery forests as Stirton and Fields claimed. We contend that moist evergreen forests were far more extensive and the open areas probably much less so.

Support for our conclusion is seen in the La Venta vertebrate fauna, especially (1) the large number of nonvolant mammalian species; (2) the high proportion of browsing species and relatively few grazing species; (3) the large number of frugivorous taxa; (4) the large number of arboreal species, especially primates; and (5) the presence of still-extant species of forest-dwelling marsupials, bats, lizards, and birds. All these indicate that moist forests were extensive and open areas less so. From comparisons with Recent Neotropical faunas and their environments, it appears that rainfall must have been 1,500–2,000 mm/year. If the area experienced a seasonal water deficit, the dry season would not have exceeded about three to four months. We propose a vegetational setting similar to the riparian succession described at Cocha Cashu, eastern Peru (Terborgh 1983; Foster 1986; Salo et al. 1986; Foster 1990; Janson and Emmons 1990). This would have been a vegetation mosaic characterized by successional stages initiated by changes in the courses of the rivers, and possibly modified by large herbivores.

Paleogeography and Geology

COLOMBIA IN THE MIDDLE MIOCENE

In the middle Miocene the area of northern Ecuador, central and western Colombia, and western Venezuela formed a peninsula (fig. 30.1). To the north and west the peninsula was bordered by the Pacific and Caribbean oceans. No land connection existed with Central America,

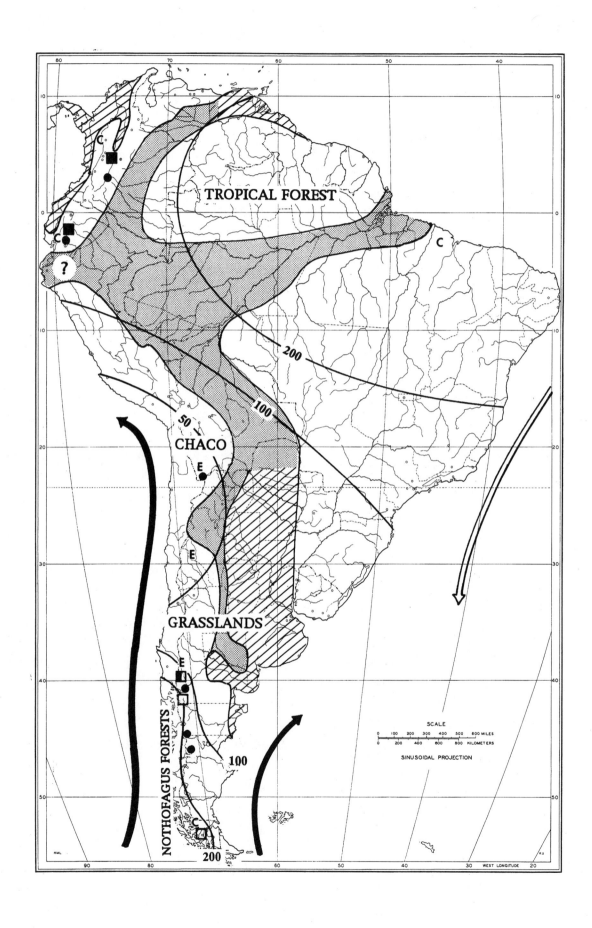

TROPICAL FOREST

CHACO

GRASSLANDS

NOTHOFAGUS FORESTS

200

100

50

100

200

SCALE

0 100 200 300 400 500 600 MILES

0   200   400   600   800 KILOMETERS

SINUSOIDAL PROJECTION

and these two oceans were joined across what is now the Isthmus of Panama (Whitmore and Stewart 1965; Duque-Caro 1979, 1980). A seaway also extended southward from the Caribbean through the Maracaibo basin into the upper Amazon basin, separating the peninsula from the Guyana shield (Nutall 1990). Domning (1982) and others have speculated that Colombia may also have been cut off to the south by the Marañon or Guayaquil Portal, a marine or continental lowland connection between the upper Amazon and the Pacific.

In the middle Miocene the South American continent occupied a latitudinal position roughly similar to or a few degrees south of where it is today. Thus, the La Venta fauna, located today at approximately 3°N, during the middle Miocene fell squarely within the equatorial tropics, within 5° north or south of the geographic equator.

The middle Miocene peninsula had a mountainous backbone, and the La Venta area was situated on the eastern lowland region. Sediments shed by the Central Cordillera of the Andes (to the west) were deposited continuously across the area (Campbell and Bürgl 1965; Lundberg et al. 1986). The freshwater fish and decapods described by Lundberg (chap. 5) and Rodríguez (chap. 4) include many species congeneric or conspecific with taxa found today in the Amazon, Maracaibo, and Orinoco basins. Between 13.8 and 12.0 Ma these lowland rivers flowed eastward. After about 12 Ma drainage in the La Venta area shifted to flow westward, signaling the formation of the eastern Cordillera (chap. 2). Today, the eastern Cordillera channels the rainfall of the La Venta area into the Magdalena River and thence northward into the Caribbean Sea.

The La Venta area was several hundred kilometers from the Maracaibo/upper Amazon seaway. There is no evidence from the fossil vertebrates or invertebrates to suggest marine or coastal influences. None of the La Venta freshwater fish have affinities with brackish or marine taxa (chap. 5), and the sirenian is most closely related to freshwater forms rather than marine taxa (chap. 23). Likewise, the fossil crab in the La Venta fauna, *Silviocarcinus piriformis,* belongs to the living freshwater family Trichodactylidae (chap. 4).

LOCAL TOPOGRAPHY

We infer that this area was lowland because of the presence of animals with narrow elevation or temperature tolerances and limited dispersal abilities. Among freshwater fish there are many very large species and species that today inhabit slow-moving, lowland rivers. Included among the large turtles is *Podocnemis,* found today only below 100 m elevation (Hoogmoed 1979; Lynch 1979; Rivero-Blanco and Dixon 1979). Likewise, limbless amphibians of the family Typhlonectidae are found today only between sea level and 40 m. Geologic structures in the area surrounding the Tatacoa Desert, however, suggest some local relief. Hills in the La Venta area, such as Cerro Chacarón and Cerro Gordo, composed of the Jurassic Payandé Group were exposed in Miocene times as evidenced by the presence of angular clasts of the Payandé Group in sediments of the La Victoria Formation (chap. 2).

REGIONAL AND LOCAL GEOLOGY

Fossils of Miocene age, described in this volume, come principally from the region informally called the Tatacoa

Figure 30.1. Paleogeography of South America in the middle Miocene. *Filled circles,* middle Miocene fossil vertebrate localities, from north to south: La Venta area of Colombia, Girón basin of southern Ecuador, Quebrada Honda in southern Bolivia, Río Collón-Curá in Río Negro and Neuquén Provinces, Argentina, Río Cisnes in southern Chile, Río Mayo vicinity of Chubut Province, Argentina. *Squares,* middle to late Miocene paleofloras: *filled squares,* tropical floras; *half-filled squares,* mixed-temperate and warm-climate floras; *open squares,* temperate broad-leaved evergreen floras (Berry 1925a, 1925b, 1934, 1936, 1938, 1945; Menendez 1971). *Question mark,* Marañon Portal. Climate indicators: *C,* coal deposits; *E,* evaporite deposits. Sediment facies: *diagonal hachure,* areas covered by marine transgression; *stippling,* areas covered by brackish water facies (Weeks 1947; Harrington 1962; Camacho 1967; Beurlen 1970; Bigarrella 1973; Bigarrella and Ferreira 1985; Duque-Caro 1990). Ocean surface currents: *open arrow,* warm current; *filled arrows,* cold currents (Frakes and Kemp 1972; Berggren and Hollister 1974; Hodell and Kenett 1985). Vegetation zones follow Solbrig (1976). Relative precipitation isohyets after Parrish et al. (1982). Base map after H. M. Leppard 1961, Goode Base Map Series, Department of Geography, University of Chicago.

Desert, or La Venta area, adjacent to the towns of Polonia, Villavieja, and La Victoria, south of the Río Cabrera, along the eastern side of the Río Magdalena (see fig. 2.9). A few fossils were collected along the eastern side of the river south of the town of Aipe. Today, the area is sparsely vegetated by woody cactus, sparse grasses, and thorny acacia and shrubs. Few trees and thickets occur except along river and stream courses. The Honda Group in this region was deposited in a structural depression. The base of the section is exposed at the northern end of the area where the group laps unconformably onto Jurassic andesites of the Saldaña Formation at the foot of Cerro Gordo and Cerro Chacarón. Except for some local faulting and minor folding, the Honda Group beds dip gently to the south so that successively younger rocks are exposed as one proceeds southward. The Neiva Formation disconformably overlies the Honda Group (chap. 2).

In the La Venta area the Honda Group is subdivided into two constituent formations, the La Victoria Formation and the conformably overlying Villavieja Formation. As noted earlier, the La Victoria Formation begins with a disconformity on Mesozoic rocks. In the type area it is composed of 460–650 m of repeating "fining-upward" sequences of sandstones and mudstones. The sandstones are fine grained but contain some pebbly layers. They alternate vertically and laterally with reddish-brown to greenish-gray mudstones. At several levels the sandstones are stacked vertically in thicknesses up to 30 m, representing persistent river channels. These thick sandstone units can be traced for tens of kilometers in the direction of stream flow and are used in the La Venta area as stratigraphic marker horizons. At the top of the La Victoria Formation is a 10-m-thick coarse-pebble conglomerate called the Cerbatana Conglomerate Beds.

The Villavieja Formation rests conformably on the Cerbatana Gravel Beds of the La Victoria Formation. The Villavieja Formation is composed of 580 m of variegated mudstones with minor layers of sandstone. The sandstones are more abundant in the lower part of the Formation (the Baraya Member) and far less so in the overlying Cerro Colorado Member. It is much less common for these sandstones to be stacked vertically. Only four stacked sandstones were encountered that are greater than 10 m in thickness; by comparison eleven such units are encountered in the La

Victoria Formation. Red beds increase in number and thickness up-section in the Villavieja Formation. At the top of the formation occur the Polonia Red Beds where sandstone is uncommon (chap. 2). For the purposes of faunal analysis in this chapter, we pay special attention to one unit, the Monkey Beds. This unit, of approximately 14 m, is the lowest part of the Villavieja Formation. Other units mentioned in the text may be identified by referring to Guerrero (chap. 2).

The geologic time represented by the Honda Group is tightly constrained by a combination of radiometric and paleomagnetic evidence to a 2.1 million year interval (chaps. 2, 3). According to this evidence, deposition of the La Victoria Formation began about 13.5 Ma and ended at about 12.9 Ma. The Villavieja Formation spans the time interval from 12.9 Ma to 11.4 Ma. A temporal hiatus of approximately 1.3 million years occurs between the Honda Group and the overlying Neiva Formation.

At present, the Eastern Cordillera of the Andes intervenes between the Magdalena River and the Amazon Basin in the vicinity of La Venta. Guerrero (chap. 2) notes that the time of uplift of this mountain range is reflected in a shift in the direction of river flow at about 11.8 Ma. Before this shift, sediment shed from the Central Cordillera was transported generally southeastward into what is now the Amazon Basin. The Pebas Formation in eastern Peru and Colombia represents the temporal equivalent of the Honda Group to the east. The Pebas Formation contains benthic foraminifera and brackish-water molluscs with Caribbean affinities.

The La Venta Fauna

In appendix 30.1 we summarize the vertebrate fauna in the Honda Group in the La Venta area. Madden et al. (chap. 29) provide added information about the temporal distribution of these taxa within the stratigraphic section. In the following accounts emphasis is placed upon what the species at La Venta tell us about the ecological setting at that time.

FISH

The diverse freshwater fish fauna of thirteen families and twenty-three species is a lowland fluvial assemblage (chap.

5). Among the described taxa are no elements indicative of upland environments. The fish indicate that heterogeneous aquatic environments were present, including both large, open river and in-shore habitats, marginal shallow waters, and relatively still, even anoxic, temporary waters. This diversity of aquatic biotopes is typical of modern lowland meandering stream systems. The fish fauna presents wide dietary diversity (algivores, detritivores, carnivores, etc.) and includes the frugivorous/granivorous *Colossoma macropomum* (Lundberg et al. 1986), suggesting the presence of fruiting plants along the river margins or in the flood plain.

AMPHIBIANS, SNAKES, AND LIZARDS

A toad, *Bufo marinus,* was reported by Estes and Wassersug (1963) from the Monkey Beds. This living species inhabits Amazonian flood plain forests but also has a broader geographic and altitudinal range; Rivero-Blanco and Dixon (1979) recorded it in dry forests and very dry forests. A limbless aquatic caecilian of the family Typhlonectidae is also represented at La Venta (chap. 6). Living typhlonectids are distributed discontinuously in the Caribbean and Amazonian lowlands and in the Paraná Basin. The three extant genera that live in northern South America occur in marshes and rivers of lowland forests. The group has been hypothesized to have originated in lowland rain forests (Lynch 1979).

Two genera of aniliid snakes are represented at La Venta, *Colombophis* and another unnamed genus. Found also are boids, including *Eunectes* (an anaconda) and representatives of the Colubroidea and Scolecophidia. Of the La Venta snakes three taxa are fossorial or forest-floor leaf-litter inhabitants (chap. 6).

Lizards reported by Sullivan and Estes (chap. 7) substantially amplify what was known hitherto about this group at La Venta. Poorly preserved and indeterminate iguanid material has been reported. In the entire collection just one specimen is still referable to *Tupinambis*. Most of the material previously allocated to *Tupinambis* and also to *Dracaena* is now allocated to a new genus *Paradracaena,* which is morphologically intermediate between extant *Dracaena* and *Crocodilurus*. The reallocation of this material is important for understanding the paleoenvironment at La Venta because *Tupinambis* occurs today only in dry and very dry

forests. In contrast, *Dracaena* and *Crocodilurus* are found only at elevations below 90 m and both inhabit rain forests although *Dracaena* may also be found along forest edges (Hoogmoed 1979; Rivero-Blanco and Dixon 1979).

CROCODILES

In terms of number of taxa, body size range, and presumed feeding ecology, crocodilian diversity in the La Venta fauna is unparalleled in the Cenozoic record of tropical South America (chap. 8). Long-snouted, presumably piscivorous crocodiles are especially diverse in the La Venta fauna, being represented by gavials (*Gryposuchus*), the unique crocodilid *Charactosuchus,* and the broad, flat-snouted *Mourasuchus*. Among the caimans, specimens of *Purusaurus neivensis* had a body length estimated to be between 7 and 8 m and an estimated body mass of more than 1,000 kg (Langston 1965).

The ziphodont *Sebecus* is reconstructed to have been a powerful terrestrial carnivore (Colbert 1946; Buffetaut and Hoffstetter 1977; Buffetaut 1980). Sebecosuchia had a wide distribution in Late Cretaceous and Cenozoic South America, last occurring in the middle and late Miocene tropical lowlands of Colombia and eastern Peru (Gasparini 1984). Buffetaut (1980) speculated that their extinction may have been linked to the arrival of Carnivora (Felidae, Canidae, Procyonidae) in South America during the Great American Faunal Interchange.

TURTLES

Three families of turtles are represented at La Venta. The aquatic Chelidae (represented by one species of *Chelus*) and Pelomedusidae (represented by three species of *Podocnemis*) are quite common. These aquatic genera have living representatives in the Venezuelan llanos today. Specimens of Testudinidae (land tortoises) are more scarce at La Venta. There appear to have been two species, an extremely large species of *Geochelone* and a second smaller one. *Geochelone* is found today only in the northern fringes of the Venezuelan llanos where it prefers forested areas rather than the extensive open savannas (Rivero-Blanco and Dixon 1979; also chap. 9).

BIRDS

Fossil birds often provide an unambiguous picture of paleoenvironments because their closely related living analogues have well-known and rather specific habitat preferences (Olson and Rasmussen 1986; also chap. 10). Based upon the habits of their closest living relatives, most of the fossil birds in the La Venta fauna inhabited riparian environments. The extant *Anhinga anhinga* (Anhingidae), a riparian piscivore, inhabits freshwater marshes in both the Amazon and Orinoco river systems. The living limpkin (*Aramus guarauna*) prefers heavily vegetated freshwater marshes, wooded swamps, and other similar fluvial wetlands. The jabiru (*Jabiru mycteria*) occurs both in wetlands and more open grasslands.

Two of the fossil birds suggest the presence of nearby forests. Jacamars (*Galbula*) occupy the canopy of tall primary forests, mature secondary forests, or forest-edge habitats. Finally, the living hoatzin (*Opisthocomus hoazin*), similar to the extinct *Hoazinoides,* is an obligate folivore and not a proficient flier. This species occurs today along the banks of forested streams in both the Orinoco and Amazon river systems where it clambers in low trees and shrubs near the water's edge.

MARSUPIALS

Eleven species and ten genera of marsupials are now known for the La Venta fauna (Bown and Fleagle 1993; also chaps. 11, 12). Several of these marsupials have recognizable adaptive counterparts in the Neotropical marsupial fauna of today. For example, new species of the extant insectivorous/frugivorous arboreal opossums *Thylamys* and *Micoureus* (Didelphinae) are described (chap. 11).

La Venta marsupial diversity includes a hyenalike borhyaenid, *Arctodictis* (Marshall 1976a, 1978), doglike prothylacynines, *Lycopsis longirostrus* and *Dukecynus magnus,* and a sabre-toothed thylacosmilid, *Anachlysictis* (Marshall 1976a, 1978; also chap. 11). Caenolestoids *Pithiculites* and *Hondathentes,* the first ever described from tropical lowlands, have modern adaptive analogs among frugivorous or nectarivorous Australian phalangeroid marsupials (Pascual and Ortiz Jaureguizar 1990; Strait et al. 1990; also chap. 12) and may have been replaced by immigrant arboreal squirrels. Finally, the La Venta marsupials include the shrewlike

microbiothere *Pachybiotherium.* The living microbiothere, *Dromiciops,* is a forest-dwelling animal.

XENARTHRANS

The xenarthrans of La Venta are abundant and diverse. In this volume the sloths, armadillos, and glyptodonts are described.

Compared with living sloths, the sloths of the Miocene La Venta fauna are remarkable for their diversity (Hirschfeld 1985; also chap. 15). Seven sloth species are reported in the La Venta fauna (chap. 15), including representatives of three families—mylodontids, megalonychids, and megatheriids. All but one of these sloth species show postcranial morphology associated with arboreal habits although none shows specializations for suspensory behavior like those of the living tree sloths *Bradypus* and *Choloepus* (chap. 16). The body size range of these species is great. Judging from distal femur bicondylar widths, all were larger than the living *Myrmecophaga.*

The mylodontine mylodontid *Pseudoprepotherium* is the largest, with a skull length of about 0.3 m and an estimated shoulder height in quadrupedal posture of 1.3 m, suggesting it was roughly half the weight of *Mylodon,* or about 500 kg. *Pseudoprepotherium* exhibits a suite of features indicative of terrestrial habits, including walking on abducted hindlimbs and inverted hind feet (chap. 16). The cheektooth morphology of this animal suggests a folivorous diet.

The scelidotheriine mylodontid *Neonematherium* was a smaller sloth with a skull length of approximately 0.15–0.2 m. Postcranially, it greatly resembled the living anteater *Tamandua,* suggesting a mix of arboreal and terrestrial habits. *Neonematherium* displays a single transverse loph on its cheek teeth, suggesting that food was being sliced, not ground and, thus, that leaves or other fibrous foods may have comprised its diet (chap. 15). Postcranial bones are not known for the somewhat smaller mylodontine *Glossotheriopsis.*

All the remaining sloths were smaller than *Neonematherium.* Among megatheriids, *Hapalops* and a smaller nothrothere exhibit primarily arboreal or climbing adaptations (chap. 16). The megalonychids are represented by two species, the smaller of which has distal phalanges suggesting arboreality, whereas the larger one has a distal femur more nearly resembling that of terrestrial sloths.

One genus of anteater, *Neotamandua*, occurred in the La Venta area (Hirschfeld 1976). The similar-sized living Neotropical anteater *Tamandua* is an accomplished arborealist.

There is a great diversity of armored xenarthrans in the Honda Group (chaps. 13, 14), including two tribes of dasypodid armadillos, Dasypodini (*Anadasypus*) and Astegotheriini (*Nanoastegotherium*), a tolypeutine (*Pedrolypeutes*), a pampathere (*Scirrotherium*), and two glyptodonts (*Neoglyptatelus* and *Asterostemma*). This taxonomic representation is strikingly different from that observed among armored edentates from the middle Miocene of Patagonia. In the Santacrucian of Patagonia Euphractinae are dominant and occur along with Peltephilinae and Stegotheriini. Neither Astegotheriini, Dasypodini, nor Tolypeutinae occur in the middle Miocene of Patagonia. No hypothesis has yet been advanced to explain this difference.

Given the lack of published descriptions of dental or skeletal material, the behavior and ecology of these xenarthrans must be inferred from generalizations about the living taxa to which they are most closely related and from the pattern of diversity itself. All of these taxa were almost certainly terrestrial and possibly semifossorial. Judged by comparison with body lengths of living armadillos, body weights for La Venta armadillos and glyptodonts are comparable to those of living armadillos and do not include giant taxa of a size comparable to those of the Pleistocene pampas. The largest armored xenarthrans in the La Venta fauna (*Asterostemma gigantea* and *Scirrotherium hondaensis*) are comparable in carapace size to the living giant armadillo *Priodontes*, which weighs around 30 kg. Interestingly, *Priodontes* is one of the most broadly tolerant of modern tropical mammals, occurring in both wet and dry habitats (see tables 30.1, 30.2).

Carlini et al. (chap. 13) suggest that the diversity of armored forms could indicate environmental heterogeneity. They state that two tribes of armadillos with living representatives, Dasypodini and Astegotheriini, suggest the presence of humid forests, and that the glyptodontids and pampathere may have inhabited grasslands. The presence of a tolypeutine suggests ecotonal settings such as where *Tolypeutes* occurs in the Federal District, Brazil, where rainfall exceeds 1,500 mm/year (tables 30.1, 30.2). It is often

stated that the diets of the larger cingulates (glyptodonts and pampatheres) were probably dominated by coarse vegetable matter, perhaps including grasses. We view this interpretation with caution. These taxa have neither living close relatives nor plausible modern analogs on which dietary interpretations can be based. In the absence of more convincing evidence from tooth function in living armadillos and microwear or stable isotope studies it is plausible that either or both these taxa could have been low-vegetation browsers or frugivores.

RODENTS

Substantially increased rodent diversity is now known for the La Venta fauna; Walton (chap. 24) recognizes five families, thirteen genera, and twenty-eight species, including some that are within the size range of Neotropical murid rodents. All of the rodent taxa found at La Venta are caviomorphs. Thus, murids, notwithstanding their substantial diversity in South America today, probably had not arrived in South America by the middle Miocene.

As with the cingulates, most of what is now known about the adaptive roles of the La Venta rodents is based on taxonomic affinity rather than detailed morphological assessment. For example, there are at least three species of porcupines (Erethizontidae) (two of ?*Steiromys* and one *Microsteiromys* Walton). Because living erethizontids are capable climbers and several are restricted to tropical evergreen forests, the presence of erithizontids suggests forested environments. Such an approach leads to valid conclusions only if a fossilized species is identical to its living descendant (in which case it is not a distinct species), or differs morphologically only in ways that are unrelated to habitat preference. Walton (chap. 24) points up an example of the pitfalls of such an approach. She reports two genera of dolichotine caviids, *Prodolichotis* and another unnamed genus. Living dolichotines occur exclusively outside the tropics. They are cursorial open-country forms inhabiting the arid Andean/Patagonian steppes. She notes, however, that the postcraniums of *Prodolichotis* more closely resemble that of another caviid, *Cavia*, than they do *Dolichotis*.

With the above caveats it is possibly significant that the remainder of the rodent fauna of La Venta has taxonomic affinities with forest-dwelling living species. Three genera

Table 30.1. Characteristics of the modern Neotropical mammal fauna sampling localities discussed in the text

Locality	Area Sampled	State or Province, Country	Latitude	Longitude	Altitude (m)	Annual Rainfall (mm)	Vegetation and Length of Dry Season	Source(s)
Guatopo	926 km²	Miranda, Venezuela	10°N	66°W	250–1,430	1,500	Semideciduous, submontane to montane forest; 6 months	Eisenberg et al. 1979
Masguaral	30 km²	Guarico, Venezuela	8°34'N	67°35'W	75	1,250	Subtropical vegetational mosaic high savanna; 7 months	Eisenberg et al. 1979
Federal District	16 km²	Brasilia, Brazil	15°57'S	47°54'W	1,100	1,586	Seasonal xerophyllous savanna grasslands and gallery forests; 4 months	Mares et al. 1989
Cayapas	450 km²	Esmeraldas, Ecuador	0°N	78°W	300–600	>5,000	Evergreen pluvial forest; 0 months	Albuja and Orces, manuscript; Madden and Albuja 1989
Low Montane	Regional fauna	Salta, Argentina	22°–24°S	64°–66° W	500–1,500	800–1,200	Lower montane moist forest	Ojeda and Mares 1989; Mares et al. 1989
Chaco	Regional fauna	Salta, Argentina	22°–24°S	63°W	200–500	700	Subtropical, drought-resistant, thorn forest; 6 months	Ojeda and Mares 1989; Mares et al. 1989
Manaus	5,000 km²	Amazonas, Brazil	2°30'S	60°W	10	2,200	Primary upland terra firme forest; 3 months	Malcolm 1990
Cocha Cashu	10 km²	Madre de Dios, Peru	12°S	70°W	400	2,080	Lowland floodplain rain forest; 4 months	Janson and Emmons 1990
Patagonia	Regional fauna	Chubut, Argentina	42°S	69°W	200	200	Temperate tussock grass steppe / scrubland	Redford and Eisenberg 1992
Pampas	Regional fauna	Buenos Aires, Argentina	34°S	58°W	50	800–1,000	Temperate grasslands	Redford and Eisenberg 1992
Alto Marañon	3 localities of restricted area	Amazonas, Peru	4°47'S	78°17'W	210	2,880	Abandoned fields, secondary regrowth riparian forest, undisturbed humid forest	Patton et al. 1982

Table 30.2. Distributions and adaptations of tropical South American mammals. *(See notes on page 533.)*

Taxon	Gu	Ms	FD	Ca	Ch	Mn	CC	AM	Diet	Substrate	Body Weight Category
			Locality								
METATHERIA											
DIDELPHIMORPHIA											
Caluromys derbianus	•	•	•	+	•	•	•	•	F_I	A	II
Caluromys lanatus	•	•	•	•	•	+	+	+	F_I	A	II
Caluromys philander	+	•	•	•	•	+	•	•	F_I	A	II
Caluromysiops irrupta	•	•	•	•	•	•	+	•	F_I	A	II
Chironectes minimus	+	•	+	+	•	•	•	+	Ve	Q_T	II
Didelphis albiventris	•	•	+	•	+	•	•	•	I_F	A_T	III
Didelphis marsupialis	+	+	•	+	•	+	+	+	I_F	A_T	III
Glironia venusta	•	•	•	•	•	•	+	•	•	A	II
Gracilinanus agilis	•	•	+	•	•	•	•	•	•	•	I
Marmosa murina	+	•	+	•	•	+	•	+	I_F	A	I
Marmosa robinsoni	•	+	•	+	•	•	•	•	I_F	A_T	I
Marmosa rubra	•	•	•	•	•	•	•	+	I_F	A	I
?Marmosa sp.	•	•	+	•	•	•	•	•	I_F	A	•
Marmosops fuscatus	+	•	•	•	•	•	•	•	I_F	A_T	I
Marmosops impavidus	•	•	•	+	•	•	•	•	I_F	A_T	•
Marmosops noctivagus	•	•	•	•	•	•	+	•	I_F	A_T	I
Marmosops parvidens	•	•	•	•	•	+	•	•	I_F	A_T	I
Metachirus nudicaudatus	•	•	•	+	•	+	+	+	Myr	T	II
Micoureus demerarae	+	•	•	•	•	+	+	•	I_F	A	II
Micoureus regina	•	•	•	+	•	•	•	+	I_F	A	•
Monodelphis adusta	•	•	•	•	•	•	•	+	I_F	T	I
Monodelphis americana	•	•	+	•	•	•	•	•	I_F	T	I
Monodelphis brevicaudata	+	•	•	•	•	+	•	•	I_F	T	I
Monodelphis domestica	•	•	+	•	•	•	•	•	I_F	T	I
Monodelphis kunsi	•	•	+	•	•	•	•	•	I_F	T	I
Philander opossum	+	•	•	+	•	+	+	+	I_F	A_T	II
Thylamys elegans	•	•	•	•	+	•	•	•	I_F	A	I
Thylamys pusilla	•	•	•	•	+	•	•	•	•	•	I
EUTHERIA											
PRIMATES											
Alouatta caraya	•	•	+	•	•	•	•	•	L	A	III
Alouatta palliata	•	•	•	+	•	•	•	•	L	A	III
Alouatta seniculus	+	+	•	•	•	+	+	+	L	A	III
Aotus trivirgatus	•	•	•	•	•	•	+	+	F_L	A	II
Ateles belzebuth	+	•	•	•	•	•	•	•	F_L	A	III
Ateles fusciceps	•	•	•	+	•	•	•	•	F_L	A	III
Ateles paniscus	•	•	•	•	•	+	+	•	F_L	A	III
Callicebus moloch	•	•	•	•	•	•	+	+	F_L	A	II
Callimico goeldii	•	•	•	•	•	•	+	•	I_F	A	II
Callithrix jacchus	•	•	+	•	•	•	•	•	F_I	A	II
Cebuella pygmaea	•	•	•	•	•	•	+	•	F_I	A	II
Cebus albifrons	•	•	•	+	•	•	+	+	F_I	A	III
Cebus apella	•	•	+	•	•	+	+	•	F_I	A	III
Cebus capucinus	•	•	•	+	•	•	•	•	F_I	A	III

Continued on next page

Table 30.2. Continued

Taxon	Locality								Diet	Substrate	Body Weight Category
	Gu	Ms	FD	Ca	Ch	Mn	CC	AM			
Cebus nigrivittatus	+	+	•	•	•	•	•	•	F_I	A	III
Chiropotes satanas	•	•	•	•	•	+	•	•	F_L	A	III
Lagothrix lagothricha	•	•	•	•	•	•	+	•	F_L	A	III
Pithecia monachus	•	•	•	•	•	+	+	•	F_I	A	III
Saguinus fuscicollis	•	•	•	•	•	•	+	•	F_I	A	II
Saguinus imperator	•	•	•	•	•	•	+	•	F_I	A	II
Saguinus midas	•	•	•	•	•	+	•	•	F_I	A	II
Saimiri sciureus	•	•	•	•	•	•	+	+	I_F	A	II
CARNIVORA											
Atelocynus microtis	•	•	•	•	•	•	•	+	•	T	III
Bassaricyon gabbii	•	•	•	+	•	•	+	+	F_I	A	III
Cerdocyon thous	•	+	+	•	+	•	•	•	Ve	T	III
Chrysocyon brachyurus	•	•	+	•	•	•	•	•	F_I	T	IV
Conepatus chinga	•	•	•	•	+	•	•	•	I_F	T	III
Conepatus semistriatus	+	+	•	+	•	•	•	•	I_F	T	III
Eira barbara	+	+	+	+	•	+	+	+	Ve	A_T	III
Galictis cuja	•	•	•	•	+	•	•	•	Ve	T	III
Galictis vittata	•	+	•	+	•	•	+	+	Ve	T	III
Herpailurus yaguaroundi	+	+	+	+	+	•	+	+	Ve	T	III
Leopardus pardalis	+	+	+	+	•	•	+	+	Ve	T	IV
Leopardus wiedii	•	•	•	+	•	•	•	+	Ve	A_T	III
Lontra longicaudus	•	•	•	+	•	+	+	+	Ve	Q_T	IV
Nasua nasua	•	•	+	+	•	+	+	+	F_I	A_T	III
Oncifelis geoffroyi	•	•	•	•	+	•	•	•	Ve	T	III
Panthera onca	+	•	•	+	+	+	+	+	Ve	T	V
Potos flavus	+	•	•	+	•	+	+	+	F_I	A	III
Procyon cancrivorus	+	+	+	+	•	•	+	+	I_F	A_T	III
Pseudalopex griseus	•	•	•	•	+	•	•	•	Ve	T	III
Pseudalopex gymnocercus	•	•	•	•	+	•	•	•	Ve	T	III
Pteronura brasiliensis	•	•	•	•	•	•	+	•	Ve	Q_T	IV
Puma concolor	+	+	+	•	+	+	+	•	Ve	T	IV
Speothos venaticus	•	•	+	•	•	+	•	+	Ve	T	III
Tremarctos ornatus	•	•	•	+	•	•	•	+	F_I	T	V
RODENTIA											
Agouti paca	+	+	•	+	•	+	+	+	F_L	T	IV
Akodon cursor	•	•	+	•	•	•	•	•	S	T	I
Akodon lindberghi	•	•	+	•	•	•	•	•	•	•	•
Akodon mollis	•	•	•	+	•	•	•	•	S	T	I
Akodon reinhardti	•	•	+	•	•	•	•	•	S	T	I
Akodon sp. 1	•	•	+	•	•	•	•	•	S	T	I
Akodon sp. 2	•	•	+	•	•	•	•	•	S	T	I
Akodon urichi	+	•	•	•	•	•	•	•	S	T	I
Akodon varius	•	•	•	•	+	•	•	•	S	T	I
Bolomys lasiurus	•	•	+	•	•	•	•	•	S	T	I
Calomys callosus	•	•	+	•	•	•	•	•	F_I	A_T	I
Calomys laucha	•	•	•	•	+	•	•	•	F_I	A_T	I
Calomys tener	•	•	+	•	•	•	•	•	F_I	A_T	I
Carterodon sulcidens	•	•	+	•	•	•	•	•	L	T	II
Cavia aperea	•	•	+	•	•	•	•	•	G	T	II

Table 30.2. Continued

Taxon	Locality								Diet	Substrate	Body Weight Category
	Gu	Ms	FD	Ca	Ch	Mn	CC	AM			
Clyomys laticeps	•	•	+	•	•	•	•	•	•	T	II
Coendou bicolor	•	•	•	+	•	•	•	+	F_L	A	III
Coendou prehensilis	•	+	+	•	•	+	+	•	F_L	A	III
Ctenomys mendocinus	•	•	•	•	+	•	•	•	L	T	II
Dactylomys dactylinus	•	•	•	•	•	•	+	•	L	A	II
Dasyprocta fuliginosa	•	•	•	•	•	•	•	+	F_L	T	III
Dasyprocta leporina	+	+	•	•	•	+	•	•	F_L	T	III
Dasyprocta punctata	•	•	•	+	•	•	+	•	F_L	T	III
Dasyprocta sp.	•	•	+	•	•	•	•	•	F_L	T	III
Dinomys branickii	•	•	•	•	•	•	+	+	F_L	A	IV
Diplomys caniceps	•	•	•	+	•	•	•	•	F_L	A	II
Dolichotis salinicola	•	•	•	•	+	•	•	•	G	T	III
Echimys braziliensis	•	•	•	+	•	•	•	+	F_L	A	II
Echimys semivillosus	+	+	•	•	•	•	•	•	F_L	A	II
Echimys sp.	•	•	•	•	•	•	+	•	F_L	A	II
Galea musteloides	•	•	•	•	+	•	•	•	L	T	II
Galea spixii	•	•	+	•	•	•	•	•	L	T	II
Graomys griseoflavus	•	•	•	•	+	•	•	•	F_I	A_T	I
Heteromys anomalus	+	+	•	•	•	•	•	•	S	T	I
Heteromys australis	•	•	•	+	•	•	•	•	S	T	I
Holochilus brasiliensis	•	•	+	•	+	•	•	•	L	A_T	II
Hoplomys gymnurus	•	•	•	+	•	•	•	•	F_L	T	II
Hydrochaeris hydrochaeris	•	+	+	•	•	•	+	+	G	Q_T	IV
Ichthyomys hydrobates	•	•	•	+	•	•	•	•	Ve	Q_T	II
Isothrix pagurus	•	•	•	•	•	+	•	•	•	A	II
Juscelinomys candango	•	•	+	•	•	•	•	•	•	T	•
Kunsia fronto	•	•	+	•	•	•	•	•	•	T	II
Lagostomus maximus	•	•	•	•	+	•	•	•	G	T	III
Melanomys caliginosus	•	•	•	+	•	•	•	•	S	A	I
Mesomys hispidus	•	•	•	•	•	+	+	•	F_L	A	II
Microcavia australis	•	•	•	•	+	•	•	•	G	A_T	II
Microsciurus flaviventer	•	•	•	+	•	•	•	+	I_F	A	I
Myocastor coypus	•	•	•	•	+	•	•	•	L	Q_T	III
Myoprocta acouchy	•	•	•	•	•	+	+	+	S	T	III
Neacomys guianae	•	•	•	•	•	+	•	•	I_F	T	I
Neacomys spinosus	•	•	•	•	•	•	•	+	I_F	T	I
Neacomys tenuipes	+	•	•	•	•	•	•	•	I_F	T	I
Nectomys squamipes	•	•	+	•	•	•	+	+	I_F	T	I
Oecomys bicolor	+	+	+	•	•	+	+	+	S	A	I
Oecomys concolor	+	•	+	•	•	•	•	+	S	A	I
Oecomys paricola	•	•	•	•	•	+	•	•	S	A	I
Oecomys superans	•	•	•	•	•	•	+	+	S	A	I
Oligoryzomys longicaudatus	•	•	•	•	+	•	•	•	S	A_T	I
Oligoryzomys microtis	•	•	+	•	•	•	+	•	S	T	I
Oligoryzomys nigripes	•	•	+	•	•	•	•	•	S	A	I

Continued on next page

Table 30.2. Continued

Taxon	Locality								Diet	Substrate	Body Weight Category
	Gu	Ms	FD	Ca	Ch	Mn	CC	AM			
Oligoryzomys spodiurus	•	•	•	+	•	•	•	•	S	A	I
Oligoryzomys utiaritensis	•	•	+	•	•	•	•	•	S	A	I
Oryzomys albigularis	+	•	•	•	•	•	•	•	S	A	I
Oryzomys alfaroi	•	•	•	+	•	•	•	•	S	A	I
Oryzomys capito	+	•	+	+	•	+	+	+	S	T	I
Oryzomys hammondi	•	•	•	+	•	•	•	•	S	A	I
Oryzomys lamia	•	•	+	•	•	•	•	•	S	A	I
Oryzomys macconnelli	•	•	•	•	•	+	+	+	S	A	I
Oryzomys nitidus	•	•	•	•	•	•	+	•	S	A	I
Oryzomys subflavus	•	•	+	•	•	•	•	•	S	A	I
Oxymycterus roberti	•	•	+	•	•	•	•	•	I_F	T	I
Oxymycterus sp.	•	•	•	•	•	•	+	•	I_F	T	I
Proechimys brevicauda	•	•	•	•	•	•	+	•	S	T	II
Proechimys cayennensis	•	•	•	+	•	+	•	•	S	T	II
Proechimys cuvieri	•	•	•	•	•	+	•	•	S	T	II
Proechimys longicaudatus	•	•	+	•	•	•	•	+	S	T	II
Proechimys semispinosus	+	•	•	+	•	•	•	•	S	T	II
Proechimys simonsi	•	•	•	•	•	•	+	+	S	T	II
Proechimys sp.	•	•	+	•	•	•	•	•	S	T	II
Proechimys steerei	•	•	•	•	•	•	+	•	S	T	II
Pseudoryzomys simplex	•	•	+	•	•	•	•	•	•	Q_T	•
Rhipidomys couesi	•	•	•	•	•	•	+	•	F_L	A	I
Rhipidomys latimanus	•	•	•	+	•	•	•	•	F_L	A	I
Rhipidomys leucodactylus	•	•	•	+	•	•	•	•	F_L	A	I
Rhipidomys mastacalis	•	•	+	•	•	+	•	•	F_L	A	I
Rhipidomys sp. 1	•	+	•	•	•	•	•	•	F_L	A	I
Rhipidomys sp. 2	•	•	+	•	•	•	•	•	F_L	A	I
Rhipidomys venezuelae	+	•	•	•	•	•	•	•	F_L	A	I
Sciurus gilvigularis	•	•	•	•	•	+	•	•	F_L	A_T	II
Sciurus granatensis	+	+	•	+	•	•	•	•	F_L	A_T	II
Sciurus ignitus	•	•	•	•	•	•	+	•	F_L	A_T	II
Sciurus igniventris	•	•	•	•	•	•	•	+	F_L	A_T	II
Sciurus sanborni	•	•	•	•	•	•	+	•	F_L	A_T	II
Sciurus spadiceus	•	•	•	•	•	•	+	+	F_L	A_T	II
Sigmodon alstoni	•	+	•	•	•	•	•	•	L	•	•
Sigmodon hispidus	•	•	•	+	•	•	•	•	F_I	•	II
Sigmodontomys alfari	•	•	•	+	•	•	•	•	I_F	T	I
Thrichomys apereoides	•	•	+	•	•	•	•	•	F_L	•	II
Tylomys mirae	•	•	•	+	•	•	•	•	L	A_T	II
Zygodontomys brevicauda	•	+	•	•	•	•	•	•	S	T	I
LAGOMORPHA											
Sylvilagus brasiliensis	+	•	+	+	+	•	+	+	G	T	II
Sylvilagus floridanus	•	+	•	•	•	•	•	•	G	T	III
PERISSODACTYLA											
Tapirus bairdii	•	•	•	+	•	•	•	•	L	T	V
Tapirus terrestris	+	•	+	•	•	+	+	+	L	T	V

Table 30.2. Continued

Taxon	Gu	Ms	FD	Ca	Ch	Mn	CC	AM	Diet	Substrate	Body Weight Category
				Locality							
ARTIODACTYLA											
Blastocerus dichotomus	•	•	•	•	+	•	•	•	G	T	V
Catagonus wagneri	•	•	•	•	+	•	•	•	L	T	IV
Mazama americana	+	•	+	+	•	+	+	+	L	T	IV
Mazama gouazoupira	•	•	•	•	+	+	+	•	L	T	IV
Odocoileus virginianus	•	+	•	•	•	•	•	•	L	T	IV
Ozotoceros bezoarticus	•	•	+	•	+	•	•	•	L	T	IV
Pecari tajacu	+	+	+	+	+	+	+	+	L	T	IV
Tayassu pecari	•	•	+	+	•	+	+	+	F_L	T	IV
XENARTHRA											
Bradypus tridactylus	•	•	•	•	•	+	•	•	L	A	III
Bradypus variegatus	+	•	•	•	•	•	+	+	L	A	III
Cabassous unicinctus	+	•	+	•	•	•	•	+	Myr	T	III
Chaetophractus vellerosus	•	•	•	•	+	•	•	•	I_F	T	III
Chlamyphorus retusus	•	•	•	•	+	•	•	•	I_F	T	II
Choloepus didactylus	•	•	•	•	•	+	•	•	L	A	III
Choloepus hoffmanni	•	•	•	+	•	•	+	+	L	A	III
Cyclopes didactylus	•	•	•	+	•	+	+	+	Myr	A	II
Dasypus kappleri	•	•	•	•	•	+	•	•	Myr	T	III
Dasypus novemcinctus	+	+	+	+	•	+	+	+	I_F	T	III
Dasypus septemcinctus	•	•	+	•	•	•	•	•	I_F	T	III
Euphractus sexcinctus	•	•	+	•	+	•	•	•	Myr	T	III
Myrmecophaga tridactyla	•	+	+	+	+	+	+	+	Myr	T	IV
Priodontes maximus	+	•	+	•	+	+	+	+	Myr	T	IV
Tamandua tetradactyla	+	+	+	+	+	+	+	+	Myr	A_T	III
Tolypeutes matacus	•	•	•	•	+	•	•	•	Myr	T	III
Tolypeutes tricinctus	•	•	+	•	•	•	•	•	Myr	T	III

Sources: Species lists referenced in table 30.1. Behavior and size data from Eisenberg (1989), Mares et al. (1989), Ojeda and Mares (1989), Emmons and Feer (1990), and Redford and Eisenberg (1992). If behavior or size has not been reported for a particular species, it is extrapolated from a congener.

Notes: Localities: GU, Guatopo, Venezuela; Ms, Masaguaral, Venezuela; FD, Federal District, Brazil; CA, Cayapas, Ecuador; Ch, Chaco, Argentina; Mn, Manaus, Brazil; CC, Cocha Cashu, Peru; AM, Alto Marañon, Peru. Dietary categories: Ve, vertebrate prey; Myr, termites and ants; I_F, insects and fruit or nectar; F_I, fruit with invertebrates; S, small seeds of grasses and other plants; F_L, fruit with leaves; L, leaves (browse); G, grass stems and leaves (graze). Locomotor or substrate preferences: A, arboreal; A_T arboreal and terrestrial (scansorial); T, terrestrial and fossorial; Q_T, semiaquatic. Body weight: I, 10 g to 100 g; II, 100 g to 1 kg; III, 1 kg to 10 kg; IV, 10 kg to 100 kg; V, 100 kg to 500 kg; VI, >500 kg. Symbols: •, unknown or not present; +, known presence.

of echimyids are present (*Acarechimys, Ricardomys* Walton, and an unnamed larger species). *Acarechimys,* at an estimated weight of 90–150 g, falls within the size range of modern murids; the other two were larger, between 150 and 300 g. Modern echimyids are mainly forest dwellers and many are arboreal.

There are at least two genera of dinomyids, each represented by two species, *"Scleromys"* species range from 1 to 3.5 kg, and *"Olenopsis"* species from 8 to 20 kg. The 2- to 3-kg dasyproctid *"Neoreomys"* is abundant at La Venta. Additionally, there are two species of a new genus, *Microscleromys* Walton, small rodents (100–200 g) with dynomyid or dasyproctid affinities. Walton suggests a phylogenetic relationship between the La Venta dasyproctids and extant *Cuniculus,* the paca, a forest-dwelling frugivore that is found in Colombia and Venezuela.

LITOPTERNS

Litopterns were especially diverse at La Venta. Four species of proterotheriids and one macraucheniid are described (chaps. 17, 18).

Megadolodus, once thought to be a condylarth, is a specialized proterotheriid litoptern (chap 17). *Megadolodus* was a robustly built, tridactyl animal with short limbs. Its canines were modified into tusks, and the polycuspate cheek teeth with thick enamel suggest frugivory. In its locomotion and diet it may have resembled living tayassuid artiodactyls.

Three other proterotheriid genera are reported from La Venta: *Prothoatherium*, *Prolicaphrium*, and *Villarroelia*. Postcranially, *Prothoatherium* (Cifelli and Guerrero 1989) was more cursorial than *Megadolodus*. Cifelli and Guerrero (1989) draw analogies with small (5–10 kg) forest-dwelling artiodactyls of Africa. Its cheek teeth are bunodont, although perhaps less so than those of *Megadolodus*, and suggest a frugivorous/herbivorous diet. Like *Prothoatherium*, *Prolicaphrium* has extremely low-crowned, bunodont or subbunodont teeth and may have had an omnivorous, frugivorous, or browsing diet. No living ungulate occurring today in open environments has cheek teeth as low crowned as these litoptern genera (chap. 18). *Villarroelia*, a larger and somewhat more high-crowned proterotheriid, is inferred to have been a "mixed" feeder, i.e., a frugivore and browser (chapter 18).

Theosodon is the only macraucheniid represented at La Venta. Santacrucian material of this genus suggests it had long limbs, an elongate neck, and a proboscis that would have equipped it to browse leaves in taller vegetation (Scott 1913).

HIGH-CROWNED UNGULATES

Ungulates with high-crowned cheek teeth are relatively abundant but low in diversity at La Venta with three or possibly four species in two genera. A single species of *Miocochilius* (Interatheriidae) was described by Stirton (Stirton 1953a). New larger samples (still under study) make it clear that two or possibly three sympatric species occurred in the lower part of the stratigraphic section. *Miocochilius* is the most abundant fossil mammal at La Venta and is known from a number of complete and partial skeletons. Com-

pared with other interatheres, *Miocochilius* was a more cursorial animal, paraxonic, with reduced lateral digits and a compact carpus/tarsus. The rostrum narrows anteriorly, and the incisors form a selective food-cropping mechanism. The cheek teeth were ever-growing and encased by thick external cementum. The enamel microstructure shows adaptations for resisting heavy wear (chap. 20). Based upon dental dimensions, these animals ranged in body size from 2 to 11 kg, similar to that of living African forest tragulids. A possible ecological analog for *Miocochilius* is the lagomorph *Sylvilagus*, a Neotropical grazer with a broad habitat range.

Pericotoxodon was a large (~800 kg), slightly dimorphic toxodontid with ever-growing cheek teeth and extremely high-crowned incisors, suggesting an abrasive diet (chap. 21). As with *Miocochilius*, *Pericotoxodon* remains are abundant in the Honda Group.

Grazing animals such as *Miocochilius* and *Pericotoxodon* are a conspicuous component of the mammal fauna at La Venta. This led Stirton (1953a) to depict the La Venta area in the Miocene as a fairly open grassland with trees restricted to a narrow band along the river courses. Alternative interpretations of the evidence are plausible. In the lowland tropics of today, grasses occur not only in open savannas but also in riparian and terra firme forest. Aquatic grasses occur in early successional communities along meandering rivers. Herbaceous bamboos and grasses are important pioneer species in tree-fall gaps. Thus, in a riparian mosaic, aquatic, river-margin, or successional grasses may have provided sufficient forage for the abundant but not species-rich grazing herbivore fauna.

"MEGAHERBIVORES"

Very large terrestrial herbivores (i.e., >500 kg) do not occur in the modern lowland tropics of South America where the largest living herbivores are the swamp deer (*Blastoceros*, 100–300 kg) and the tapir (*Tapirus*, ~300 kg). In the Honda Group very large terrestrial herbivores were common. A large terrestrial sloth, *Pseudoprepotherium* (~500 kg), and the large toxodont *Pericotoxodon* (~800 kg) have already been mentioned. In addition, *Huilatherium* (Leontiniidae) was a large notoungulate with relatively low-crowned, thick-enameled cheek teeth (chap. 19), and two large astrapotheres (1,200–3,000 kg), *Xenastrapotherium* and *Gra-*

nastrapotherium, also occur in the fauna (chap. 22). The leontiniid and astrapotheres evidently were adapted for feeding upon less-abrasive vegetation, given that their cheek teeth are relatively low-crowned.

In Africa today, open clearings are maintained within forests by the destructive habits of megaherbivores, particularly the tusked species (Laws 1970; Owen-Smith 1988; Campbell 1991). Four of the large mammals in the La Venta mammal community, *Pseudoprepotherium*, *Huilatherium*, *Granastrapotherium*, and *Xenastrapotherium*, had tusks, and the tusks of astrapotheres were used in food acquisition and branch stripping as described by Johnson and Madden (chap. 22). Megaherbivores may have helped create and maintain edge habitat within the forests at La Venta. This, in turn, could explain the diverse and abundant browsing and grazing herbivores.

SIRENIANS

All sirenian material in the Honda Group is referred to *Potamosiren magdalenensis* (chap. 23). These trichechid remains, although rare, occur throughout most of the fossiliferous Honda Group section. Domning (chap. 23) calls attention to the bunodonty and remarkably thick enamel covering the cheek teeth of *Potamosiren*. He considers this to be an adaptation for munching soft, mostly nonabrasive aquatic vegetation, a dietary preference unlike that of the living *Trichechus* and *Ribodon* that have diets of siliceous grasses. The latter are dentally more derived with batteries of continuously erupting cheek teeth.

BATS

Czaplewski (chap. 25) describes a diverse chiropteran fauna (five families, nine genera, and eleven species) from La Venta, including a number of representatives of living genera and even some living species. Except where indicated, these species are found only in the Monkey Beds.

Diclidurus, the white bat (Emballonuridae), is an extant aerial pursuit insectivore that usually roosts and forages in multistory evergreen forests. The extant phyllostomine, *Tonatia,* from the Fish Bed, is an aerial foliage-gleaning insectivore. This genus usually forages and roosts in mature multistory evergreen forests but is occasionally found in deciduous forests. Another phyllostomine, *Notonycteris,* resembles the living *Vampyrum* postcranially and dentally. The latter is an aerial gleaning vertebrate eater that occasionally also eats insects and fruit. *Vampyrum* lives in a great variety of forest habitats, including deciduous and gallery forests, lowland moist forests, and premontane forests. Another unnamed phyllostomine was apparently a less-specialized carnivore and probably was more insectivorous/frugivorous.

Glossophagines are also represented by an unnamed species. Living glossophagines are tolerant of a wide variety of habitats. All living species can hover in flight to drink nectar.

A noctilionid, *Noctilio,* from the Monkey Beds belongs to a living species that gaffs aquatic insects and sometimes tiny fish from water surfaces by dragging its hind feet. This species is found in streamside habitats and roosts in trees.

Two species of disk-wing bats, *Thyroptera* (Thyropteridae), one assigned to a living species, are found in the La Venta fauna. These are aerial pursuit insectivores that use slow, maneuverable flight. They tend to forage and roost in lowland forest edges, in tree gaps and successional areas.

Collectively, the bat fauna of La Venta strongly indicates the presence of forest habitat. Living species of *Diclidurus* and *Tonatia* usually forage and roost in mature multistory evergreen forests, and, except for Thyroptera, all the La Venta bats (or their extant close relatives) roost in trees.

PRIMATES

Honda Group rocks have yielded a diversity of fossil platyrrhines unparalleled by any other fossil assemblage on the continent. Many of these are clearly related to extant subfamilies, whereas the affinities of others are debated (chaps. 26–28).

Callitrichidae (encompassing marmosets, tamarins, and *Callimico,* with Callitrichinae here used only for marmosets and tamarins, sensu Rosenberger et al. 1990) is represented by three species. *Lagonimico conclucatus,* from the La Victoria Formation, is the sister group of living marmosets and tamarins. Consequently, a surprising feature of this animal is its relatively large size, approximately 1 kg, which is larger than *Callimico* or any living marmoset or tamarin. *Lagonimico* had poorly developed molar crests and elongate incisors, suggesting a diet of gum and fruit eating as among many living marmosets and tamarins (Kay 1994).

Patasola magdalenae, from quite low in the La Victoria Formation, a more primitive species of callitrichine than *Lagonimico,* also is intermediate in body size between *Saimiri* and living callitrichines. *Patasola* had poorly developed molar shearing capabilities. This, coupled with its unspecialized premolar teeth, suggests it was a mixed fruit and insect eater. A third callitrichine, known only from a single tooth in the Monkey Beds, was very small, perhaps comparable to living *Callithrix* or *Saguinus.*

Neosaimiri, a close relative of the living squirrel monkeys (*Saimiri,* Saimiriinae), is represented by time-successive species at La Venta, *N. fieldsi* from the Monkey Beds and *N.* (=*Laventiana*) *annectens* from the El Cardón Red Beds. These species were very similar in size to the extant squirrel monkeys, ranging between 600 and 900 g. *Neosaimiri* species may have been more frugivorous and less committed to insect eating than their living relative. Fragmentary limb elements have proved indistinguishable from those of extant squirrel monkeys, implying arboreal quadrupedalism (chap. 27).

Cebupithecia sarmientoi is one of three taxa representing the Pitheciinae (sakis and uakaris). This animal weighed about 1.5 kg. It is dentally very similar to the extant pitheciines *Pithecia, Chiropotes,* and *Cacajao.* Like them, its enlarged, chisel-like canines and low-crowned, "inflated" molars indicate seed-eating habits. There is very good postcranial material for *Cebupithecia.* This animal was an accomplished arboreal quadruped that frequently adopted vertical clinging postures and was an adept leaper (Meldrum and Kay 1992). *Cebupithecia* comes from the Monkey Beds with some material from lower in the La Victoria Formation. *Mohanamico hershkovitzi,* also from the Monkey Beds, was somewhat smaller, more primitive pitheciine with more gracile jaws than living pitheciines. Nevertheless, it had poorly developed molar shearing and procumbent incisors that may have allowed it to husk and pry open hard fruits. Fruit eating, possibly with some seed eating, is inferred.

A third, as yet unnamed, pitheciine from the El Cardón Red Beds, shows many similarities to *Cebupithecia* (Meldrum and Kay 1992). This taxon had already evolved the mesiodistally compressed styliform incisors so char-

acteristic of the extant species and indicative of fruit husking.

Stirtonia victoriae and its time-successive congener *S. tatacoensis* were very large fossil monkeys, up to 8 kg, comparable in size to living howling monkeys. *Stirtonia victoriae* was a folivorous monkey, dentally closely resembling the living howling monkey. *Stirtonia tatacoensis* more closely resembles *Lagothrix* in its dentition, suggesting that it was a mixed leaf and fruit feeder. Unfortunately neither is known from postcranial material.

Where it is best known from fossil specimens in the Monkey Beds, the diversity of La Venta platyrrhines resembles that found in platyrrhine communities that live in modern humid tropical forests today with 1,500–2,000 mm rainfall per annum. At least five genera of platyrrhines occur in the Monkey Beds: *Mohanamico hershkovitzi, Stirtonia tatacoensis, Cebupithecia sarmientoi, Neosaimiri fieldsi,* and an unnamed small callitrichine. The presence of several pitheciines suggests evergreen forest conditions because living pitheciines are not found in deciduous or gallery forests today.

Paleoenvironment at La Venta

SEDIMENTOLOGICAL EVIDENCE

Guerrero (chap. 2) interprets mudstones in the Honda Group to be paleosols in which occur dispersed calcite nodules, calcite-replaced root casts, desiccation cracks, drab-haloed root casts, and other bioturbation. This suite of features could indicate seasonally dry intervals. He interprets the sequence of colors in the mudstones—from almost unaltered greenish gray to deeply altered red and purple—to represent paleosols at increasing stages of maturity. When river-flow geometry is controlled for, the soils toward the top of the Villavieja Formation are more mature than those throughout the La Victoria Formation and in the lower parts of the Villavieja Formation. This appears to be related to changes in geologic rather than climatic factors because paleomagnetic evidence (chap. 3) shows that the rate of basin subsidence and corresponding net sediment accumulation slows dramatically up-section. Under these circumstances, attributing the differences in soil maturity to climatic change is not warranted.

PALEOBOTANICAL EVIDENCE

Fragments of fossilized trunks assigned to *Goupioxylon stutzeri* (Celastraceae) were first described from the La Venta area by Schönfeld (1947). Since that time, additional specimens have been collected from numerous Miocene localities throughout Colombia, including both the Honda and Mesa Groups (Pons 1969). Living species of *Goupia* (Celastraceae) are large trees (25–40 m) that occur along rivers in lowland forest between sea level and 1,000 m elevation in the equatorial tropics. Species of this family today occur in tropical life zones with at least 2,000 mm rainfall per year and mean annual temperatures between 23 and 30°C.

MAMMALIAN COMMUNITY STRUCTURE

Approach and Methods

As we observed earlier, many aspects of the anatomy of extinct species at La Venta yield some information about body size, diet, and locomotion or substrate preference. In this section we examine the overall structure of the nonvolant mammalian fauna of La Venta in comparison with that of eight modern South American tropical faunas listed in table 30.1.

The modern faunas can be arrayed along an axis of total rainfall and length of the dry season. At one extreme, at Cayapas, Ecuador, rainfall exceeds 5,000 mm/year and no appreciable dry season exists; at the other extreme is the Chaco in Salta Province, Argentina, where rainfall is as low as 300 mm and the dry season extends for half the year. Total rainfall and seasonality have a direct bearing on the type of vegetation that prevails in any given area. In the wetter environments, with rainfall exceeding 2,000 mm/year and with a dry season of four or fewer months (Cayapas, Alto Marañon, Manaus, and Cocha Cashu), evergreen rain forests predominate. In drier regions, with less than 1,000 mm rain and dry intervals longer than six months (Chaco), the dominant vegetation is drought-resistant, subtropical, deciduous thorn forest. Areas of intermediate rainfall (1,000–2,000 mm/year, with four to six months of drought) tend to exhibit semideciduous gallery forests with intervening savanna grasslands (Federal District). To examine the relationship between niche structure and habitat, as

a first approximation, total annual rainfall is used as a surrogate for vegetation type.

A list of the nonvolant mammalian species that live in each of the eight modern faunas is given in table 30.2. From the literature, the physical attributes and adaptations of each of these species have been compiled. These attributes are

(1) BODY WEIGHT. We recognize six categories of body size: 10–100 g, 100–1,000 g, 1–10 kg, 10–100 kg, 100–500 kg, and ≥500 kg.

(2) LOCOMOTION. The categories used here are essentially those of Fleming (1973) and Andrews et al. (1979), wherein animals are divided into zones corresponding to the physical layers of the environment. We recognize semi-aquatic, terrestrial (including fossorial), scansorial, and arboreal categories.

(3) DIET. Because we are not confident that it is possible to subdivide dietary categories as finely for extinct species as can be done for living ones, we have not subdivided the diets of living species as finely as have other workers (e.g., Eisenberg 1981; Eisenberg and Redford 1982; Robinson and Redford 1986, 1989). We employ seven dietary categories: fruit with insects, fruit with leaves, small seeds of grasses and other plants, ants and termites, vertebrate prey; leaves (browse), and stems and leaves of grasses (graze).

The same sort of ecological data were also assembled for the nonvolant La Venta mammals of the Monkey Beds fauna (see table 30.3). This list was compiled from the taxonomic revisions reported in this book. The La Venta fauna represents an interval of about 2.1 million years, and faunal sampling is not continuous or uniform throughout the section. Despite the fact that there is no signficant faunal turnover in the section, only in the Monkey Beds do we feel our sampling approaches the actual diversity present in any restricted time interval. For this reason only those species found in the Monkey Beds are used in our formal analyses.

Body weights of the mammalian species from the Monkey Beds are estimated from linear dimensions of the teeth and bones based on least-squares regression models from their living close relatives. Weights for marsupials, rodents, and primates were estimated from published equations for lower first molar crown area versus body weight

Table 30.3. Hypothesized specializations for Honda Group nonvolant mammals

Taxon	Stratigraphic Occurrence	Diet	Substrate Preference	Body Weight Category
METATHERIA				
Pithiculites chenche	+	F_I	A_T	I
Hondathentes cazador	+	F_I	A_T	I
Pachybiotherium minor	P	I_F	A_T	•
Micoureus laventicus	P	I_F	A	I
Thylamys colombianus	P	I_F	A	I
Thylamys minutus	P	I_F	A	I
Hondadelphys fieldsi	P	I_F	T	III
Anachlysictis gracilis	+	Ve	T	IV
Thylacosmilidae, gen. et sp. indet.	+	Ve	T	IV
Dukecynus magnus	+	Ve	T	IV
cf. *Arctodictis* sp.	P	Ve	T	•
Lycopsis longirostrus	P	Ve	T	IV
EUTHERIA				
PRIMATES				
Neosaimiri annectens	+	F_I	A	II
Neosaimiri fieldsi	P	F_I	A	II
Patasola magdalenae	+	I_F	A	II
Cebupithecia sarmientoi	P	F_L	A	III
Mohanamico hershkovitzi	P	F_L	A	II
Pitheciinae, gen. et sp. nov. 1	+	F_L	A	III
Lagonimico conclucatus	+	F_L	A	III
?Callitrichidae, gen. et sp. nov. 1	P	F_I	A	II
Stirtonia tatacoensis	P	L	A	III
Stirtonia victoriae	+	L	A	III
RODENTIA				
Echimyidae, gen. et sp. indet. 1	P	F_L	A_T	II
?Echimyidae, gen. et sp. incertae sedis	P	S	A_T	I
Ricardomys longidens Walton	+	S	A_T	II
Acarechimys cf. *A. minutissimus*	P	S	A_T	II
"Olenopsis" sp. 1 (small)	P	F_L	T	IV
"Olenopsis" sp. 2 (large)		F_L	T	IV
"Scleromys" colombianus	P	F_L	T	III
"Scleromys" schurmanni	P	F_L	T	III
"Scleromys" cf. "S." colombianus	P	F_L	T	III
"Scleromys" cf. "S." schurmanni	P	F_L	T	II
Microscleromys paradoxalis	P	S	T	II
Microscleromys cribriphilus	+	S	T	II
Dinomyidae, gen. et sp. incertae sedis (cf. *Simplimus*)	P	F_L	A	IV
"Neoreomys" huilensis	P	S	T	III
Microsteiromys jacobsi Walton	P	F_L	A	II
?*Steiromys* sp. (large)	P	F_L	A	III
?*Steiromys* sp. (small)	P	F_L	A	III
Prodolichotis pridiana	P	G	T	III
Dolichotinae, gen. 2 (large)	P	G	T	III

Table 30.3. Continued

Taxon	Stratigraphic Occurrence	Diet	Substrate Preference	Body Weight Category
Dolichotinae, gen. 2 (small)	P	G	T	III
Caviomorpha, gen. et sp. indet. (very small)	+	●	●	I
LITOPTERNA				
Prolicaphrium sanalfonsensis	P	L	T	IV
Prothoatherium colombianus	P	F_L	T	III
Villarroelia totoyoi	+	L	T	IV
Megadolodus molariformis	P	F_L	T	V
Theosodon sp.	P	L	T	V
NOTOUNGULATA				
Miocochilius anamopodus	P	G	T	III
Miocochilius sp. nov. 1 (medium)	+	G	T	III
Miocochilius sp. nov. 2 (small)	+	G	T	III
Huilatherium pluriplicatum	P	L	Q_T	VI
Pericotoxodon platignathus	P	G	T	VI
ASTRAPOTHERIA				
Xenastrapotherium kraglievichi	P	L	Q_T	VI
Granastrapotherium snorki	P	L	Q_T	VI
XENARTHRA				
Neotamandua borealis	P	Myr	A_T	IV
cf. *Hapalops*	P	L	A_T	IV
Neonematherium flabellatum	P	L	A	IV
Nothrotheriinae, gen. et sp. nov. (small)	P	L	T	●
Glossotheriopsis pascuali	P	L	A	III
Megalonychidae, gen. et sp. indet. (large)	+	L	T	●
Megalonychidae, gen. et sp. indet. (small)	P	L	T	●
Megatheriinae, gen. et sp. indet.	P	L	T	●
Anadasypus hondanus	P	I_F	T	III
Pseudoprepotherium confusum	P	L	T	IV
Pedrolypeutes praecursor	P	Myr	T	III
Scirrotherium hondaensis	P	I_F	T	IV
Asterostemma gigantea	P	L	T	IV
Asterostemma cf. *A. acostae*	P	L	T	IV
Neoglyptatelus originalis	P	G	T	IV
Nanoastegotherium prostatum	P	Myr	T	IV

Notes: Symbols/localities: +, Present in the Colombian middle Miocene; ●, unknown; P, present in the Monkey Beds, Villavieja Formation. Dietary categories: F_I, fruit with insects; F_L, fruit with leaves; Myr, ants and termites; Ve, vertebrate prey; L, leaves (browse); G, grass stems and leaves (graze); S, small seeds of grasses and other plants. Locomotor or substrate preferences: A, arboreal; A_T arboreal and terrestrial (scansorial); T, terrestrial and fossorial; Q_T, semiaquatic. Body weight: I, 10 g to 100 g; II, 100 g to 1 kg; III, 1 kg to 10 kg; IV, 10 kg to 100 kg; V, 100 kg to 500 kg; VI, >500 kg.

(Conroy 1987; Legendre 1989). Those for armadillos and glyptodonts use regressions of carapace length (or body length) versus body weight from published measurements for living armadillos (Wetzel 1985b). Weights of sloths and anteaters are based on a regression of femur bicondylar width versus body weight for living sloths and anteaters. Weights for litopterns, notoungulates, and astrapotheres were estimated using dental dimensions (Litopterna) and mean estimates from diverse skeletal, cranial, and dental dimensions (Notoungulata and Astrapotheria) (chaps. 1, 17, 18, 21, 22). Using these estimates, individual species were assigned to body weight classes.

The assignment of locomotor and dietary behavior to extinct species is based upon analogy with living species

Table 30.4. Summary data on the niche specializations of tropical faunas of South America

Fauna	Annual Rainfall (mm)[a]	Vegetation	Locomotion			
			A	A$_T$	T	Q$_T$
Cayapas (n = 58)	>5,000 (8)	Evergreen rain forest	20	11	24	3
Alto Marañon (n = 60)	2,880 (7)	Evergreen rain forest	22	9	26	3
Manaus (n = 50)	2,200 (6)	Evergreen rain forest	21	8	21	1
Cocha Cashu (n = 70)	2,080 (5)	Evergreen rain forest	32	10	25	3
Federal District (n = 68)	1,600 (4)	Gallery/semideciduous forest and savanna	15	8	41	3
Guatopo (n = 41)	1,500 (3)	Gallery/semideciduous forest	14	7	20	1
Masagural (n = 29)	1,250 (2)	Savanna mosaic	6	6	15	1
Chaco (n = 36)	500–800 (1)	Thorn forest	1	8	26	1
Caatingas (n = 20)	<500 (0)	Semiarid	2	5	11	0
Monkey Beds (n = 50)	—	—	14	6	26	3

Fauna	Body Weight Category					
	I	II	III	IV	V	VI
Cayapas (n = 58)	12	15	19	7	3	0
Alto Marañon (n = 60)	11	14	21	10	3	0
Manaus (n = 50)	9	12	18	9	2	0
Cocha Cashu (n = 70)	10	24	21	13	2	0
Federal District (n = 68)	24	13	17	10	1	0
Guatopo (n = 41)	11	8	14	6	2	0
Masagural (n = 29)	6	2	14	7	0	0
Chaco (n = 36)	6	6	15	7	2	0
Monkey Beds (n = 50)	4	6	17	13	2	4

Fauna	Diet							
	Ve	Myr	I$_F$	F$_I$	S	F$_L$	L	G
Cayapas (n = 58)	9	4	10	8	9	11	6	1
Alto Marañon (n = 60)	9	6	12	6	8	10	6	2
Manaus (n = 50)	5	6	8	7	7	9	7	0
Cocha Cashu (n = 70)	8	5	10	11	10	15	8	2
Federal District (n = 68)	7	6	11	6	15	6	8	3
Guatopo (n = 41)	6	3	10	3	7	6	5	1
Masagural (n = 29)	6	2	5	1	3	6	4	2
Chaco (n = 36)	8	5	6	2	2	1	7	5
Monkey Beds (n = 50)	2	3	7	2	3	13	15	6

Notes: n = number of species. Dietary categories: F$_I$, fruit with insects; F$_L$, fruit with leaves; Myr, ants and termites; Ve, vertebrate prey; L, leaves (browse); G, grass stems and leaves (graze); S, small seeds of grasses and other plants. Locomotor or substrate preferences: A, arboreal; A$_T$, arboreal and terrestrial (scansorial); T, terrestrial and fossorial; Q$_T$, semiaquatic. Body weight: I, 10 g to 100 g; II, 100 g to 1 kg; III, 1 kg to 10 kg; IV, 10 kg to 100 kg; V, 100 kg to 500 kg; VI, >500 kg.

[a]Rank order is in parentheses.

Table 30.5. Comparison of faunas from selected tropical environments in South America today with fauna of the Monkey Beds, Villavieja Formation, middle Miocene

Locality	Browser/ Grazer Index[a]	Frugivore Index[b]	Arboreality Index[c]
Cayapas (n = 59)	86	80	53
Alto Marañon (n = 59)	75	75	52
Manaus (n = 51)	100	77	58
Cocha Cashu (n = 70)	80	78	60
Federal District (n = 68)	73	70	33
Guatopo (n = 41)	83	73	49
Masaguaral (n = 29)	67	62	43
Chaco (n = 36)	58	29	25
Monkey Beds (n = 50)	71	46	41

Notes: n = number of species. Dietary categories: F_I, fruit with invertebrates; S, small seeds of grasses and other plants; F_L, fruit with leaves; L, leaves (browse); G, grass stems and leaves (graze). Locomotor or substrate preferences: A, arboreal; A_T, arboreal and terrestrial (scansorial); T, terrestrial and fossorial; Q_T, semiaquatic.

[a] $100 \cdot (L)/(L + G)$.

[b] $100 \cdot (F_I + F_L + S)/(F_I + F_L + S + L + G)$.

[c] $100 \cdot (A + A_T)/(A + A_T + Q_T + T)$.

with similar morphology, known locomotion, and diet. For fossil herbivorous mammals with no living relatives (e.g., litopterns, notoungulates, astrapotheres) the hypsodonty index of Janis (1988) was useful for assigning species to fruit-with-leaves, browsing, and grazing categories.

Modern South American Mammal Faunas

A number of significant associations are found between the numbers and percentages of species occupying various dietary and substrate categories. Tables 30.4 and 30.5 summarize the specializations of the mammal species occurring in both the modern Neotropical faunas and the Monkey Beds fauna at La Venta. Statistically significant relationships are enumerated in table 30.6.

DIETARY TRENDS We find a significant positive correlation (Spearman's rho and Kendall's tau) at the p = 0.05 level between the number of nonvolant frugivorous species and rainfall (=habitat). Wetter environments supporting evergreen forests always have more frugivores than drier gallery forest/savanna or thorn forest environments. At Cayapas there are twenty-eight species of frugivores, whereas in the Chaco there are just five. This relationship also is significant for frugivorous species that obtain substantial amounts of added energy and protein from insects or from leaves.

Frugivorous species account for a larger proportion of the total number of nonvolant species in wetter environments. Our "frugivore index" expresses the proportion of frugivorous species to the total number of plant-eating species in a fauna:

$$100 \cdot (F_I + F_L + S)/(F_I + F_L + S + L + G),$$

Table 30.6. Significant ecological distinctions among extant Neotropical faunas

Variables Compared	Spearman's Rank			Kendall Rank		
	rho	Z	p	tau	Z	p
RAINFALL VERSUS:						
Number of primates	0.75	2.00	0.05	0.55	1.89	0.06
Number of F_I	0.77	2.06	0.04	0.61	2.10	0.03
Number of F_L	0.79	2.20	0.03	0.68	2.35	0.01
Number of $F_I + F_L$	0.84	2.22	0.02	0.75	2.60	0.01
Number of arboreal species	0.79	2.08	0.04	0.64	2.23	0.03
Browser/grazer index	0.71	1.89	0.06	0.71	2.47	0.01
Frugivore index	0.88	2.33	0.02	0.71	2.47	0.01
Arboreality index	0.71	1.89	0.06	0.50	1.73	0.08

Notes: Abbreviations and indices are defined in table 30.5. Level of significance, p ≤ 0.05.

where F_I = fruit with invertebrates, S = small seeds of grasses and other plants, F_L = fruit with leaves, L = leaves (browse), and G = grass stems and leaves (graze). It is significantly positively correlated with rainfall at the p = 0.02 and 0.01 for Spearman's rho and Kendall's tau, respectively (table 30.6). At Cayapas, 80% of herbivores are frugivorous, whereas in the Chaco 29% are frugivorous.

We find no significant relationships between rainfall and the number of grazing or browsing species as expressed by a "browsing index":

$$100 \cdot (L)/(L + G),$$

although the largest number of grazing species is found in the drier, more open environments (table 30.4). A weak but significant relationship is found between the proportion of browsers and the proportion of grazers (p = 0.06 and 0.05 for Spearman's rho and Kendall's tau, respectively; table 30.6).

We find no significant relationships between rainfall and the numbers of carnivorous species, ant and termite feeders, or more generalized insect eaters. Nor are the overall proportions of these feeding types significantly affected by variations in rainfall.

BODY SIZE TRENDS   Significant relationships were found between rainfall (=habitat) and the number of species within two body size classes (table 30.4), classes II and III at the level of p = 0.05. This finding accords with Andrews et al. (1979) and Legendre (1989), who emphasize the presence of such a relationship for many communities of living mammals.

SUBSTRATE PREFERENCE   A significant correlation is found (Spearman's rho or Kendall's tau at p = 0.04 and 0.03, respectively, table 30.6) between the number of arboreal species and rainfall (=habitat). Wetter environments supporting evergreen forests always have more arboreal species than semideciduous gallery forest/savanna or thorn forest environments. In the rain-forest environments there are twenty to thirty-two arboreal species, whereas in savanna mosaic and thorn forest areas there are just one to six (table 30.4). As noted by Fleagle et al. (chap. 28) primates, being arboreal species, are more abundant in rain forests than in any other habitat in our sample (table 30.6).

An "arboreality index" was used to express the proportions of arboreal species compared with the total number of nonvolant species (table 30.5):

$$100 \cdot (A + A_T)/(A + A_T + Q_T + T),$$

where A = arboreal, A_T = arboreal and terrestrial (scansorial), T = terrestrial and fossorial, and Q_T = semiaquatic. It is scarcely surprising that a larger proportion of arboreal species are present in forested environments than in gallery forests or thorn forest environments, although our correlations were of marginal significance (p = 0.06 and 0.08, table 30.6).

No significant relationships were found between rainfall and the numbers of scansorial, terrestrial, or semiaquatic species in our sample of tropical faunas.

The Monkey Beds Mammalian Community

Possible pitfalls such as the following should be identified before any effort is made to interpret the faunal composition of the Monkey Beds fauna based upon a comparison with modern South American faunas.

(1) Phylogenetic factors must play a role in the way niches are subdivided. Many of the mammalian groups living today in South America are very different from those of the Miocene. For example, artiodactyls and scuirid and murid rodents are dominant mammalian herbivores today, in terms of species numbers, but were absent at La Venta where notoungulates, sloths, litopterns, and caviomorph rodents prevailed. It is unreasonable to expect that different groups of mammals that have long-independent phylogenetic histories will adapt themselves in identical ways to similar environmental circumstances.

(2) The composition of the floras of Miocene tropical South America is poorly understood, so feeding niches may have been different at that time.

(3) Niche competition and predation may have been different in the Miocene than it is now. For example, Langston (1965) has noted that crocodilian diversity may have been higher at La Venta than today and may have included large, active terrestrial carnivores.

(4) South American faunas of today are not in equilibrium with the major changes in ecosystems caused by the

arrival of *Homo sapiens* on the continent. Some evidence for this is suggested by the extinction of many of the large mammals on the continent over the past 20,000 years; for example, toxodonts, gomphotheres, giant sloths, and glyptodonts.

(5) Beyond the factors mentioned, taphonomic and sampling biases may impair knowledge of what species were actually present at La Venta. Taxa at the same locality could have been derived from several different nearby habitats. We do not attempt to deal with this last source of bias in this chapter. It is a legitimate question to inquire whether after death animal carcasses at La Venta were buried where they lived or whether carcasses, or their parts, were transported before final permanent burial. We cannot arrive at any meaningful conclusions about these taphonomic processes because we remain ignorant of nearly all taphonomic processes in tropical lowland forest environments. Nor do we feel justified in assuming that modern processes mediating these things elsewhere in the lowland tropics would be similar in any meaningful way to Neotropical middle Miocene processes. For example, the predators and scavengers present in Neotropical Miocene terrestrial fauna were phylogenetically unrelated to those of modern Neotropical faunas.

With all these caveats in mind, we offer the following generalizations about the Monkey Beds nonvolant mammalian fauna. In overall diversity (fifty species), the Monkey Beds fauna compares favorably with living forest communities in the modern Neotropics and has 40% or more species diversity than our savanna-mosaic or thorn forest examples at Masaguaral and Chaco (twenty-nine and thirty-six species, respectively) (table 30.4). The estimate of species may be low because the number of small mammals may be inadequately sampled. Only four species of mammals less than 100 g are reported from the Monkey Beds, whereas the total number of these species in the modern faunas ranges from six to twenty-four. This difference, however, may be because of a true absence of very small mammals at La Venta. Most of the small mammals in the modern faunas are murid rodents, a group that evidently had not entered the continent by Miocene times.

The body size diversity spectrum for the Monkey Beds mammal fauna differs from all modern South American

mammal faunas in displaying a higher percentage of large-mammal species (table 30.4). Five species in the Monkey Beds fauna exceeded 400 kg, including a sloth, two notoungulates, and two astrapotheres. This discrepancy may have to do with Recent megafauna extinctions, perhaps, in part, human-induced, in South America. The relatively high percentage of large mammals compares more favorably with tropical African mammal faunas where humans coevolved with the vertebrate fauna.

There is a high percentage of arboreal species in the Monkey Beds fauna (arboreality index = 41; table 30.5). This compares more closely with extant mammal faunas from forested areas although it is not as high as in modern rain forests (table 30.5). The presence of five primate species in the Monkey Beds also suggests that forested environments were prevalent. Again, we suspect that taphonomic and collecting biases may have led us to substantially undersample the diversity of small species, many of which may have been arboreal.

A notable feature of the Monkey Beds fauna is the high percentage of browsing species compared with grazers. Seventy-one percent of the nonfrugivorous herbivores are browsers (table 30.5). This compares favorably with modern environments with annual rainfall exceeding 1,500 mm and suggests that La Venta was a forested environment, but more likely a forest mosaic—for example, a disturbed riparian succession—than a continuous, uninterrupted, multistratal evergreen forest.

Endemicity in the Middle Miocene

We have depicted La Venta to be situated on either a peninsula or an island in the middle Miocene. It is appropriate to ask whether this had a measurable effect on endemism. To examine this question we assembled lists of mammalian genera from the modern Neotropics and compared them with similar lists from the middle Miocene of Colombia and Patagonia. If endemism were found to be similar to, or less, in the middle Miocene than it is today, this might suggest that La Venta and the region around it may not have been isolated from the rest of the continent for a long period. If the differences are greater than today, geographic isolation by a physical barrier would be among the possible hypotheses available to explain the difference. Another pos-

Table 30.7. Faunal resemblance indices for mammal genera in modern Neotropical nonvolant mammal faunas

Fauna Compared	Latitude	FRI
PATAGONIA VERSUS:		
Guatopo	10°N	13
Masaguaral	9°N	8
Cayapas	1°N	17
Manaus	3°S	8
Alto Marañon	5°S	13
Cocha Cashu	12°S	17
Federal District	16°S	17
Low Montane (Salta)	23°S	25
Chaco (Salta)	23°S	50
Pampas	34°S	50
Patagonia	42°S	100

Note: See table 30.1 for a description of these localities.

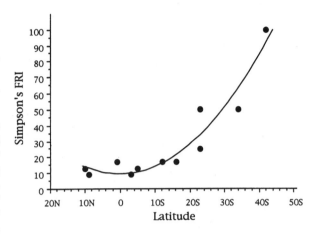

Figure 30.2. Bivariate plot showing the relationship between Simpson's faunal resemblance index (FRI) and latitude. Latitudes north of the equator are assigned negative numbers. Reference fauna is that from Patagonia so that FRI for Patagonia is 100. The faunas used in the analysis are listed in table 30.2. Second order polynomial is fit to the data: FRI $= 14.611 - 0.089 \cdot L + 0.047 \cdot L^2$, where L is latitude. Coefficient of multiple correlation, $R^2 = 0.92$.

sible explanation might be that in the middle Miocene there was a steeper latitudinal climatic gradient between the equator and Patagonia than exists today.

Table 30.1 summarizes the physical and biotic features of eleven living lowland Neotropical mammal communities used in our comparisons. Measures of faunal similarity (faunal resemblance indices, or FRIs) were computed between Patagonia and other faunas using the Simpson coefficient, C/N_s, where C = the number of genera in common and N_s = the number of genera in the smaller fauna (table 30.7). The data clearly show that faunal resemblance today is greatly influenced by latitude. Correlation, R, in a polynomial regression, is 0.95 and latitude explains 92% of the variance in FRI. FRI values decrease with increasing latitudinal separation (fig. 30.2). FRIs comparing Patagonia's fauna to those within 5° of the equator range from 8 (versus Manaus, 3°S) to 17 (versus Cayapas, 1°N). These levels of faunal resemblance between faunas separated by 40° to 45° latitude correspond to between two and five genera in common.

Of the nonvolant mammalian genera in the La Venta fauna fifteen are known elsewhere in Miocene to Recent faunas in Argentina (table 30.8). Twenty-three genera are endemic to the La Venta fauna (table 30.1). The remaining six genera have geographic distributions restricted to the tropical zone of South America. *Miocochilius, Anadasypus,*

Pericotoxodon, and *Xenastrapotherium* occur in the Miocene of southern Ecuador (Carlini, Vizcaíno, et al. 1989; Madden et al. 1989), *Xenastrapotherium* in the Miocene of Venezuela (Stehlin 1928), and *Granastrapotherium* and *Stirtonia* in the Miocene of western Brazil (Kay and Frailey 1993; chap. 22).

The La Venta fauna shares thirteen genera with Miocene faunas of southern South America (table 30.8), representing one-third of the total nonvolant La Venta fauna. This would at first seem to indicate a greater degree of latitudinal faunal continuity in the Miocene than is the case today. This, however, does not reflect the true degree of endemism in the La Venta fauna. As shown in table 30.8, some of the genera in our list are shared between La Venta and the Colhuehuapian of Patagonia alone; others are in common only during the Santacrucian, and so on. To take into account changing levels of endemism with time we calculated FRIs comparing the La Venta fauna with various Miocene faunas and assemblages (table 30.9).

Temporally, the La Venta fauna (13.5–11.8 Ma) falls between the older Colloncuran (15.7 Ma) and younger Mayoan Subages (11.8 Ma), but does not overlap with either (chap. 29). The La Venta fauna shows its greatest faunal resemblance (FRI) with older faunas (Colhuehuapian, Santacrucian, Friasian, and Colloncuran; table 30.9). FRIs

Table 30.8. Genera of nonvolant mammals shared between the La Venta fauna and Miocene to Recent faunas of southern South America

Taxon	Co	Sa	Fr	CC	Ma	Ch	Hu	Me	R
Theosodon	★	★	★	–	–	★	–	–	–
Prolicaphrium	★	–	–	–	–	–	–	–	–
Prothoatherium	★	–	–	–	–	–	–	–	–
?Steiromys	★	★	★	★	–	–	–	–	–
Acarechimys	★	★	★	★	–	–	–	–	–
Prodolichotis	–	–	–	–	–	–	★	★	–
Pseudoprepotherium	–	–	★	–	–	–	–	–	–
Glossotheriopsis	–	–	★	★	–	–	–	–	–
Neonematherium	–	–	–	–	★	–	–	–	–
Asterostemma	–	★	–	★	–	–	–	–	–
Pithiculites	★	–	★	–	–	–	–	–	–
Pachybiotherium	★	–	–	–	–	–	–	–	–
Lycopsis	–	★	–	–	–	–	–	–	–
Micoureus	–	–	–	–	–	–	–	–	★
Thylamys	–	–	–	–	–	–	–	–	★

Sources: Faunal lists for Patagonian faunas of Colhuehuapian, Santacrucian, and Huayquerian land mammal ages (LMA) are taken from Marshall et al. (1983). Faunal lists for the Mesopotamian fauna of the Ituzaingo Formation follow Pascual and Odreman (1971), Scillato-Yané (1980, 1981a, 1981b), and Bianchini and Bianchini (1971). Those for the Vivero and Las Barrancas Members of the Arroyo Chásico Formation follow Scillato-Yané (1977a, 1977b, 1978), Marshall (1978, 1980), and Bondesio, Laza, et al. (1980). For the Río Acre local fauna we follow Frailey (1986). The faunal list for the Urumaco Formation in Falcón State, Venezuela, follows Linares (1990), Pascual and Díaz (1969), and Díaz and Linares (1989). Faunal lists for the Río Cisnes, Colloncuran, and Mayoan components of the "Friasian" follow preliminary identifications from work in progress as reported by Ortiz Jaureguizar et al. (1993).

Notes: Co, Colhuehuapian; Sa, Santacrucian; Fr, Friasian fauna of Río Cisnes, Chile; CC, Colloncuran fauna; Ma, Mayoan fauna; Ch, Chasicoan; Me, Mesopotamian LMA; Hu, Huayquerian LMA; R, Recent. The taxonomic identifications for the La Venta fauna follow revisions in this volume. Symbols: ★, present; –, absent.

Table 30.9. Faunal resemblance indices (Simpson coefficients) for Miocene faunas and South American land mammal ages and assemblages

	Santacrucian	Friasian	Colloncuran	LAVENTAN
Colhuehuapian	45	59	46	15
Santacrucian	100	81	76	13
Friasian		100	69	19
Colloncuran			100	14
LAVENTAN				100
Mayoan	21	21	36	0
Chasicoan (Vivero Member)	11	11	11	0
Chasicoan (Las Barrancas Member)	7	7	7	0
Urumaco Formation	15	15	8	8
Acre Local Fauna	0	0	0	10
"Mesopotamian" (Ituzaingo Formation)	5	6	3	0
Huayquerian	0	0	0	2

Sources: As for table 30.8.

with younger Miocene faunas (Mayoan and Chasicoan) are much lower (table 30.9): Just one genus is shared between La Venta and each of these latter faunas (table 30.8). This suggests that some factor or factors came into play between 15.7 and 11.8 Ma to produce a greatly increased endemicity between La Venta and Patagonia. Whether this was the result of a developing geographic isolation of Colombia by a marine barrier to dispersal of terrestrial mammals or to a greater contrast between the climates of the equatorial zone and Patagonia than today ultimately may become testable when we have more information about Miocene faunas elsewhere on the continent.

Summary and Conclusions

In this chapter we have reviewed the evidence concerning the regional geographic setting and middle Miocene environment of the Honda Group in the La Venta area. This review leads to the following general conclusions.

The La Venta area was located on a northward projecting peninsula, isolated from the rest of the continent to the east by a seaway extending south from the Maracaibo basin into the Amazon basin. Some faunal-resemblance evidence suggests weakly that the isolation of the La Venta fauna may have been responsible for the high endemism after 11.8 Ma. The high levels of endemism could equally have been the result of greater climatic difference between the equatorial zone and Patagonia in the Miocene than today.

The La Venta area was an equatorial lowland with minimal local relief. Sediment sources were from the west. The Eastern Cordillera uplifted in this region at about 12 Ma. So rivers flowed southeastward into the Maracaibo/Amazon seaway in a region occupied today by the llanos until 12 Ma; afterwards, flow was northerly.

Evidence for periodic flooding is indicated by the predominance of overbank sediments within the Honda Group and is suggested by the presence of fishes that can survive in temporary waters (*Lepidosiren*) and species that forage in inundated forest (*Colossoma*). This periodic flooding may have been caused by fluctuations in rainfall in the La Venta area. It has been argued that the presence of the lungfish *Lepidosiren* is prima facie evidence for seasonal water deficit at La Venta (Bondesio and Pascual 1977). The

geographic distribution of extant *Lepidosiren,* however, is not confined to areas having markedly seasonal environments. Thus, the presence of a *Lepidosiren* species in the Monkey Beds fauna cannot be used to infer a seasonal water deficit sufficient to have produced syncroneity of plant phenophases and fluctuations in food resource supply for primary consumers, such as vertebrates. Moreover, the fluvial sediments of the Honda Group do not display features associated with prolonged seasonal water deficit. Such features might include hardened and cracked surfaces, evidence of wind erosion, alteration by fire, evaporation, hardpan, impeded drainage, slow chemical weathering, and low root biomass. If eventually found in the distal facies of Honda Group sediments, these features might provide independent evidence for savanna grasslands.

The fossil vertebrates provide undeniable evidence for forest cover, which can be summarized as follows: (1) freshwater fish (*Colossoma macropomum*) exploited periodically flooded flood-plain forest; (2) snakes inhabited forest leaf litter; (3) the lizard (*Paradracaena*) is closely related to living teiids of lowland forests; (4) the land tortoise *Geochelone* was a forest-dweller; (5) a forest-dwelling jacamar (*Galbula*) and hoatzin (*Opisthocomus*) were present; (6) diverse arboreal marsupials were found; (7) all but one of the seven sloth species show arboreal and/or climbing adaptations; (8) the closest taxonomic affinities of some armored edentates and most rodents are with forest-dwelling extant species; (9) five litopterns display low-crowned teeth associated with frugivory and browsing habits in living ungulates; (10) of the few ungulate species with high-crowned cheek teeth, only one (*Pericotoxodon*) is morphologically consistent across all features of its cranium and dentition with a grazing habitus; (11) none of the diverse tusked "megaherbivores" are grazing animals; (12) the Monkey Beds fauna includes bats that are known to roost and forage in multistratal evergreen forests; and (13) all the monkeys in the Monkey Beds fauna are arboreal, and in particular, pitheciines today are not known to occur in seasonally dry forest. Thus, the evidence from diverse groups of fossil vertebrates consistently demonstrates that terrestrial forest biotopes occurred in the La Venta area during the deposition of the Honda Group.

The substrate diversity spectrum for the Monkey Beds mammal fauna displays a high percentage of arboreal mammal species comparable with modern Neotropical mammal faunas from forested environments.

The high percentage of browsing species compared with grazing species in the Monkey Beds fauna compares with modern Neotropical faunas from forested environments where annual rainfall is 1,500–2,000 mm. Evidence that rainfall may have been at least 2,000 mm/year also is suggested by the one known plant species that has minimum annual rainfall requirements of 2,000 mm. The extensive forest cover suggested by the evidence mentioned usually is associated with seasonal water deficits of less than three months.

Reconstructions of the La Venta forests favored by us are interspersed throughout this book and differ considerably from the expansive savanna grasslands depicted for the Honda Group by Stirton (1953a).

Acknowledgments

We thank R. Cifelli, N. Czaplewski, M. Maas, and B. A. Williams for helpful comments and suggestions. This work was supported by NSF grants for research on the Colombian Miocene.

Appendix 30.1. The vertebrate fauna from the Miocene Honda Group, Colombia

A star (*) indicates that a nonvolant mammal is endemic to the La Venta fauna.

CHONDRICHTHYES

ELASMOBRANCHII

MYLIOBATIFORMES
  cf. Potamotrygonidae
    gen. et sp. incertae sedis

OSTEICHTHYES

DIPNOI

LEPIDOSIRENIFORMES
  Lepidosirenidae
    Lepidosiren paradoxa

ACTINOPTERYGII

OSTEOGLOSSIFORMES
  Arapaimidae
    Arapaima sp.
CHARACIFORMES
  Characidae
    cf. Tetragonopterinae
      gen. et sp. indet.
    Serrasalminae
      Colossoma macropomum
      Serrasalmus, Pygocentrus, or *Pristobrycon* sp.
      cf. *Myletes* sp.
      gen. et sp. incertae sedis
  Erythrinidae
    Hoplias sp.
  Cynodontidae
    Hydrolycus sp.
  Anostomidae
    cf. *Leporinus*
SILURIFORMES
  Ariidae
    gen. et sp. incertae sedis
  Callichthyidae
    cf. *Hoplosternum*
    cf. *Corydoras* sp.
  Doradidae
    gen. et sp. incertae sedis 1
    gen. et sp. incertae sedis 2
    gen. et sp. incertae sedis 3
  Pimelodidae
    Pimelodinae
      Brachyplatystoma cf. *B. vaillanti*
      Brachyplatystoma sp.
      cf. *Pimelodus*
      Phractocephalus hemiliopterus
  Loricariidae
    gen. et sp. incertae sedis 1
    gen. et sp. incertae sedis 2
    Ancistrinae
      cf. *Acanthicus*
    Hypostominae
      cf. *Hypostomus*
PERCIFORMES
  Cichlidae
    gen. et sp. incertae sedis

AMPHIBIA

ANURA

  Bufonidae
    Bufo cf. *B. marinus*
GYMNOPHIONA

Typhlonectidae
gen. et sp. indet.

REPTILIA

SQUAMATA

LACERTILIA
SCINCOMORPHA
Teiidae
Tupinambinae
Tupinambis sp.
Paradracaena colombiana
cf. *Paradracaena colombiana*
Iguanidae
gen. et sp. indet.
OPHIDIA
ALETHINOPHIDIA
Colubroidea
fam. indet.
gen. et sp. indet.
Boidae
gen. et sp. indet.
Eunectes stirtoni
Eunectes sp.
Anilioidea
Aniliidae
Colombophis portai
cf. Aniliidae
gen. et sp. nov.
SCOLECOPHIDIA
fam. indet.
gen. et sp. indet.

CROCODYLIA

SEBECOSUCHIA
Sebecidae
Sebecus huilensis
Sebecus sp.
EUSUCHIA
Gavialidae
Gryposuchus colombianus
Crocodylidae
cf. Tomistominae
Charactosuchus fieldsi
Nettosuchidae
Mourasuchus atopus
Alligatoridae
gen. et sp. incertae sedis
Eocaiman sp.
Caiman cf. *C. lutescens*

Purrusaurus (=Caiman) neivensis
Caiman sp.
Balanerodus logimus

TESTUDINES

PLEURODIRA
Chelidae
Chelus colombianus
Pelomedusidae
Podocnemis pritchardi
Podocnemis medemi
Podocnemis cf. *P. expansa*
Podocnemis, sp. indet.
Testudinidae
Geochelone hesterna
Geochelone, sp. indet.
CRYPTODIRA
Emydidae
gen. et sp. indet.
Testudinidae
Geochelone hesterna
Geochelone, sp. indet.

AVES

CUCULIFORMES
Hoazinoididae
Hoazinoides magdalenae
PELECANIFORMES
Anhingidae
Anhinga cf. *A. grandis*
GRUIFORMES
Aramidae
Aramus paludigrus
Ciconiidae
Jabiru sp. aff. *J. mycteria*
CORACIIFORMES
Galbulidae
Galbula hylochoreutes

MAMMALIA

METATHERIA

SPARASSODONTA
Borhyaenoidea
Hondadelphidae
Hondadelphys fieldsi
Borhyaenidae
Borhyaeninae

cf. *Arctodictis* sp.
  Prothylacyninae
    Lycopsis longirostrus
    ★*Dukecynus magnus*
  Hathlyacyninae
    aff. *Cladosictis*
  Thylacosmilidae
    ★*Anachlysictis gracilis*
  fam. incertae sedis (Thylacosmilidae?)
    gen. et sp. incertae sedis
DIDELPHIMORPHIA
Didelphoidea
  Didelphidae
    Didelphinae
      Micoureus laventicus
      Thylamys minutus
      Thylamys colombianus
      gen. et sp. indet. (marmosine)
      gen. et sp. indet. (non-marmosine?)
PAUCITUBERCULATA
Caenolestoidea
  Abderitidae
    Pithiculites chenche
  Palaeothentidae
    ★*Hondathentes cazador*
MICROBIOTHERIA
Microbiotherioidea
  Microbiotheriidae
    Pachybiotherium minor

EUTHERIA

XENARTHRA
CINGULATA
Dasypodoidea
  Dasypodidae
    Dasypodinae
      Astegotheriini
        ★*Nanoastegotherium prostatum*
      Dasypodini
        Anadasypus hondanus
    Tolypeutinae
      ★*Pedrolypeutes praecursor*
  Pampatheriidae
    ★*Scirrotherium hondaensis*
Glyptodontoidea
  Glyptodontidae
    Propalaeohoplophorinae
      Asterostemma gigantea
      cf. *Asterostemma* (cf. *A. acostae*)
    Glyptatelinae
      ★*Neoglyptatelus originalis*

PILOSA
  Mylodontidae
    Mylodontinae
      Pseudoprepotherium confusum
      Glossotheriopsis pascuali
    Scelidotheriinae
      Neonematherium flabellatum
  Megalonychidae
    Megalonychinae
      gen. et sp. indet. (large)
      gen. et sp. indet. (small)
    Nothrotheriinae
      cf. *Hapalops* (large)
      gen. et sp. indet. (small)
  Megatheriidae
    Megatheriinae
      gen. et sp. indet.
VERMILINGUA
  Myrmecophagidae
    Neotamandua borealis
RODENTIA
CAVIOMORPHA
  fam. incertae sedis
    gen. et sp. indet. (very small)
  Echimyidae
    Echimyinae
      Acarechimys cf. *A. minutissimus*
      ★*Ricardomys longidens* Walton
      gen. et sp. indet.
  Dinomyidae
    ★*"Scleromys" schurmanni*
    ★*"Scleromys"* cf. *"S." schurmanni*
    ★*"Scleromys" colombianus*
    ★*"Scleromys"* cf. *"S." colombianus*
    ★gen. et sp. incertae sedis (large)
    ★*"Olenopsis"* sp. (small)
    ★*"Olenopsis"* sp. (large)
  Dasyproctidae
    ★*"Neoreomys" huilensis*
    ★*"Neoreomys"* sp. nov. (large)
  ?Dasyproctidae
    ★*Microscleromys paradoxalis* Walton
    ★*Microscleromys cribriphilus* Walton
  Caviidae
    Dolichotinae
      Prodolichotis pridiana
    ★gen. 2 (large)
    ★gen. 2 (small)
  Erethizontidae
    ?*Steiromys* sp. (large)
    ?*Steiromys* sp. (medium)
    ★*Microsteiromys jacobsi* Walton

ASTRAPOTHERIA
  Astrapotheriidae
    Uruguaytheriinae
      Granastrapotherium snorki
      Xenastrapotherium kraglievichi
LITOPTERNA
  Macraucheniidae
      Theosodon, sp. indet.
  Proterotheriidae
      ★*Megadolodus molariformis*
      ★*Villarroelia totoyoi*
      Prothoatherium colombianus
      ★*Prolicaphrium sanalfonensis*
      gen. et sp. indet.
NOTOUNGULATA
TOXODONTIA
  Toxodontidae
      Pericotoxodon platignathus
  Leontiniidae
      ★*Huilatherium pluriplicatum*
TYPOTHERIA
  Interatheriidae
      Miocochilius anamopodus
      Miocochilius sp. nov. 1
      Miocochilius sp. nov. 2
PRIMATES
  Callitrichidae
      ★*Lagonimico conclucatus*
      ★*Patasola magdalenae*
      gen. et sp. indet.
  Cebidae
    Pitheciinae

      ★*Mohanamico hershkovitzi*
      ★*Cebupithecia sarmientoi*
      ★gen. et sp. nov.
    Alouattinae
      Stirtonia tatacoensis
      Stirtonia victoriae
    Saimiriinae
      ★*Neosaimiri fieldsi*
      ★*Neosaimiri annectens*
CHIROPTERA
MICROCHIROPTERA
  Phyllostomidae
      Notonycteris magdalenensis
      Tonatia, sp. indet.
    Phyllostominae
      gen. et sp. nov.
    Glossophaginae
      gen. et sp. indet.
  Molossidae
      Mormopterus colombiensis
      Eumops, sp. indet.
      ★*Potamops mascahehenes*
  Thyropteridae
      Thyroptera robusta
      Thyroptera cf. *T. tricolor*
  Noctilionidae
      Noctilio albiventris
  Emballonuridae
      Diclidurus, sp. indet.
SIRENIA
  Trichechidae
      Potamosiren magdalenensis

Literature Cited

Alarcón Pardo, H. 1969. Contribución al conocimiento de la morfología, ecología, comportamiento y distribución geográfica de *Podocnemis volgi,* Testudinata (Pelomedusidae). Revista de la Academia Colombiana de Ciencias Exactas, Físicas y Naturales, Bogotá 13: 303–26.

Albrecht, G. H. 1978. Some comments on the use of ratios. Systematic Zoology 27: 67–71.

Allen, J. R. L. 1963. The classification of cross-stratified units, with notes on their origin. Sedimentology 2: 93–114.

Allen, J. R. L. 1970. Studies in fluviatile sedimentation: A comparison of fining-upwards cyclothems, with special reference to coarse-member composition and interpretation. Journal of Sedimentary Petrology 40: 298–323.

Alvarado, B. 1940. Informe del Jefe del Servicio Geológico. Memorias del Ministerio de Minas y Petróleo, Bogotá 1940: 29–38.

Alvarado, B. 1941. Labores desarrolladas por el Servicio Geológico. Memorias del Ministerio de Minas y Petróleo, Bogotá 1941: 141–73.

Alvarado, B. 1959. José Royo y Gómez. Boletín de Geología, Bogotá 7: i–iv.

Ameghino, F. 1891. Nuevos restos de mamíferos fósiles descubiertos por C. Ameghino en el Eoceno inferior de la Patagonia austral. Especies nuevas, adiciónes y correcciónes. Revista Argentina de Historia Natural 1: 289–328.

Ameghino, F. 1895. Première contribution à la connaissance de la faune mammalogique des couches à *Pyrotherium*. Boletín del Instituto Geográfico de Argentina (Buenos Aires) 15: 60–660.

Ameghino, F. 1902a. Cuadro sinóptico de las formaciones sedimentarias terciarias y cretáceas de la Argentina, en relación con el desarrollo y descendencia de los mamíferos. Anales de Museo Nacional, Buenos Aires 3: 1–12.

Ameghino, F. 1902b. Nuevas especies de mamíferos cretáceos y terciarios de la República Argentina. Anales de la Sociedad Científica Argentina, vols. 56–58.

Ameghino, F. 1902c. Première contribution à la conaissance de la faune mammalogique des couches à *Colpodon*. Boletín de la Académia de Ciencias (Córdoba) 17: 71–141.

Ameghino, F. 1904. Nuevas especies de mamíferos cretáceos y terciarios de la República Argentina. Anales de la Sociedad Científica Argentina 56–58: 1–142.

Ameghino, F. 1906. Les formations sédimentaires du Crétacé supérieur et du Tertiaire de Patagonie, avec un parallèle entre leurs faunes mammalogiques et celles de l'ancien continent. Anales del Museo Nacional de Historia Natural, Buenos Aires, ser. 3: 1–568.

Ameghino, F. 1910. *Montaneia anthropomorpha*. Un género de monos hoy extinguido de la isla de Cuba. Anales del Museo Nacional de Buenos Aires, ser. 3, 13: 317–18.

Anapol, F., and J. G. Fleagle. 1988. Fossil platyrrhine forelimb bones from the early Miocene of Argentina. American Journal of Physical Anthropology 76: 417–28.

Anderson, T. A. 1972. Paleogene nonmarine Gualanday Group, Neiva basin, Colombia, and regional development of the Colombian Andes. Bulletin of the Geological Society of America 83: 2423–38.

Andrews, C. W. 1901. Preliminary note on some recently

discovered extinct vertebrates from Egypt. Geology Magazine 8: 436–44.

Andrews, C. W. 1906. A descriptive catalogue of the Tertiary Vertebrata of the Fayûm, Egypt. London: British Museum of Natural History.

Andrews, P., and J. A. H. Van Couvering. 1975. Palaeoenvironments in the East African Miocene. In F. S. Szalay, ed., Approaches to Primate Paleobiology. Contributions to Primatology 5: 62–103. Basel: S. Karger.

Andrews, P., J. M. Lord, and E. M. Nesbit-Evans. 1979. Patterns of ecological diversity in fossil and modern mammalian faunas. Biological Journal of the Linnean Society 11: 177–205.

Andrews, P. 1990. Small mammal taphonomy. In E. H. Lindsay, ed., European Neogene Mammal Chronology. Pp. 487–94 New York: Plenum Press.

Anthony, H. E. 1916. Preliminary report of fossil mammals from Puerto Rico. Annals of the New York Academy of Science 27: 193–203.

Anthony, H. E. 1922. A new fossil rodent from Ecuador. American Museum Novitates 35: 1–4.

Anthony, M. R. L., and R. F. Kay. 1993. Tooth form and diet in ateline and alouattine primates: Reflections on the comparative method. American Journal of Science 283A: 356–82.

Aoki, R. 1976. On the generic status of *Mecistops* (Crocodylidae), and the origin of *Tomistoma* and *Gavialis* (in Japanese). Bulletin of the Atagawa Institute 6–7: 23–30.

Aoki, R. 1977. Globular teeth in *Osteolaemus tetraspis*. Nippon Herpetological Journal 8: 15–18.

Aoki, R. 1989. The jaw mechanics in heterodont crocodilians. Current Herpetology in East Asia, Herpetological Society of Japan, pp. 17–21.

Aplin, K. P., and M. Archer. 1987. Recent advances in marsupial systematics with a new syncretic classification. In M. Archer, ed., Possums and Opossums: Studies in Evolution. Sydney: Surrey Beatty and Sons and the Royal Zoological Society of New South Wales. Pp. xv–lxxii.

Arambourg, C. 1948. Mission scientifique de l'Omo 1932–1933. Pt. 1, Géologie-Anthropologie. Muséum National d'Histoire Naturelle 3: 75–562.

Archer, M. 1978. The nature of the molar-premolar boundary in marsupials and a reinterpretation of the homologies of marsupial cheek-teeth. Memoirs of the Queensland Museum 18: 157–64.

Archer, M., S. J. Hand, and H. Godthelp. 1991. Riversleigh: The Story of Animals in Ancient Rainforests of Inland Australia. Balgowlah, New South Wales: Reed Books.

Ariste Joseph, H., and H. Nicéforo María. 1923. Estudios científicos de la Excursión La Salle. Boletín Instituto de La Salle, Bogotá, Colombia 1923: 155–77.

Armesto, J. J., R. Oozzi, M. P. Sabag, and C. Sabag. 1987. Plant/frugivore interactions in South American temperate forests. Revista Chilena de Historia Natural 60: 321–36.

Armstrong-Ziegler, J. G. 1980. Amphibia and Reptilia from the Campanian of New Mexico. Fieldiana, Geology 4: 1–39.

Atchley, W. R., C. T. Gaskins, and D. Anderson. 1976. Statistical properties of ratios. Pt. 1, Empirical results. Systematic Zoology 25: 137–48.

Auffenberg, W. 1954. Additional specimens of *Gavialosuchus americanus* (Sellards) from a new locality in Florida. Quarterly Journal of the Florida Academy of Sciences 17: 185–209.

Auffenberg, W. 1971. A new fossil tortoise, with remarks on the origin of South American testudinines. Copeia 1971: 106–17.

Azpelicueta de las Mercedes, M., J. D. Williams, and E. Gudynas. 1987. Osteología y notas miológicas en la Cecilia Neotropical *Chthonerpeton indistinctum* (Reinhardt & Lutkin, 1861), con una diagnosis de la familia Typhlonectidae (Amphibia, Gymnophiona). Iheringia, Série Zoologia 66: 69–81.

Baba, M. L., L. L. Darga, and M. Goodman. 1979. Immunodiffusion systematics of the primates. Folia Primatologica 32: 207–38.

Badgley, C., and L. Tauxe. 1990. Paleomagnetic stratigraphy and time in sediments: Studies in alluvial Siwalik rocks of Pakistan. Journal of Geology 98: 457–77.

Báez, A. M., and Z. Gasparini. 1977. Origines y evolución de los anfíbios y reptiles del Cenozóico de America del Sur. Acta Geológica Lilloiana 14: 149–232.

Baker, R. J., C. S. Hood, and R. L. Honeycutt. 1989. Phylogenetic relationships and classification of the higher categories of the New World bat family Phyllostomidae. Systematic Zoology 38: 228–38.

Bardack, D. 1961. New Tertiary teleosts from South America. American Museum Novitates 2041: 1–27.

Barlow, J. C. 1984. Xenarthrans and pholidotes. In S. Anderson and J. K. Jones, Jr., ed., Orders and Families of Recent Mammals of the World. New York: John Wiley. Pp. 219–39.

Barrio, R. E. de, G. J. Scillato-Yané, and M. Bond. 1984. La Formación Santa Cruz en el borde occidental del Macizo del Deseado (Provincia de Santa Cruz) y su contenido paleontológico. Actas IX Congreso Geológico Argentino (San Carlos de Bariloche) 4: 539–56.

Barron, E. J., C. G. A. Harrison, J. L. Sloan II, and W. A. Hay. 1981. Paleogeography, 180 million years ago to present. Eclogae Geologicae Helvetiae 74: 443–70.

Barron, E. J., and W. H. Peterson. 1991. The Cenozoic ocean circulation based on Ocean General Circulation Model results. Palaeogeography, Palaeoclimatology, Palaeoecology 83: 1–28.

Bartels, W. S. 1984. Osteology and systematic affinities of the horned alligator *Ceratosuchus* (Reptilia, Crocodilia). Journal of Paleontology 58: 1347–53.

Beard, K. C., B. Sigé, and L. Krishtalka. 1992. A primitive vespertilionoid bat from the early Eocene of central Wyoming. Comptes Rendus de l'Académie des Sciences, Paris 314: 735–41.

Becker, J. J. 1987. The fossil birds of the late Miocene and early Pliocene of Florida. I. Geology, correlation, and systematic overview. Documents des Laboratoires de Géologie, Lyon 99: 159–71.

Beer, J. A. 1990. Steady sedimentation and lithologic completeness, Bermejo basin, Argentina. Journal of Geology 98: 501–17.

Belwood, J. J., and G. K. Morris. 1987. Bat predation and its influence on calling behavior in Neotropical katydids. Science 238: 64–67.

Berg, D. 1969. *Charactosuchus kugleri,* eine neue Krokodilart aus dem Eozän von Jamaica. Ecolgae Geologicae Helvetiae 62: 731–35.

Berggren, W. A., and C. D. Hollister. 1974. Paleogeography, paleobiogeography and the history of circulation in the Atlantic Ocean. In C. A. Ross, ed., Paleogeographic Provinces and Provinciality. Society of Economic Paleontology and Mineralogy, Special Paper. Pp. 126–86.

Berggren, W. A., D. V. Kent, J. J. Flynn, and J. A. Van Couvering. 1985. Cenozoic geochronology. Bulletin of the Geological Society of America 96: 1407–18.

Bernor, R. L., and P. P. Pavlakis. 1987. Zoogeographic relationships of the Sahabi large mammal fauna (early Pliocene, Libya). In N. T. Boaz, A. El-Arnauti, A. W. Gaziry, J. de Heinzelin, and D. D. Boaz, eds., Neogene Paleontology and Geology of Sahabi. Virginia Museum of Natural History, Martinsville, Virginia. New York: Alan R. Liss. Pp. 349–83.

Berry, E. W. 1925a. Fossil plants from the Tertiary of Patagonia and their significance. Proceedings of the United States National Academy of Sciences 11: 404–5.

Berry, E. W. 1925b. A Miocene flora from Patagonia. Johns Hopkins University Studies in Geology 6: 183–249.

Berry, E. W. 1929. Miocene plants from Colombia, South America. Proceedings of the United States National Museum 75(24): 1–12.

Berry, E. W. 1934. Miocene Patagonia. Proceedings of the National Academy of Sciences 20: 280–82.

Berry, E. W. 1935. Tertiary fossil plants from Colombia, South America. Proceedings of the United States National Museum 75(24): 1–12.

Berry, E. W. 1936. Miocene plants from Colombia. Bulletin of the Torrey Botanical Club 63: 53–66.

Berry, E. W. 1938. Tertiary flora from the Río Pichileufú, Argentina. Geological Society of America, Special Paper 12: 1–149.

Berry, E. W. 1945. Fossil floras from southern Ecuador. Johns Hopkins University Studies in Geology 14: 93–150.

Beurlen, K. 1970. Geologie von Brasilien. Berlin: Gebruder Borntraeger.

Bianchini, L. H., and J. J. Bianchini. 1971. Revisión de los Proterotheriinae (Mammalia, Litopterna) del "Mesopotamiense." Ameghiniana 8: 1–24.

Bigarrella, J. J. 1973. Geology of the Amazon and Parnaiba basins. In A. E. M. Nairn and F. G. Stehli, eds., The Ocean Basins and Margins. Vol. 1, The South Atlantic. New York: Plenum Press. Pp. 25–86.

Bigarrella, J. J., and A. M. M. Ferreira. 1985. Amazonian geology and the Pleistocene and the Cenozoic environments and paleoclimates. In G. T. Prance and T. E. Lovejoy, eds., Amazonia. New York: Pergamon Press. Pp. 49–71.

Blake, E. R. 1977. Manual of Neotropical Birds. Vol. 1, Spheniscidae (Penguins) to Laridae (Gulls and Allies). Chicago: University of Chicago Press.

Bluntschli, J. H. 1931. *Homunculus patagonicus* und die ihm zugereithen Fossilfunde aus den Santa-Cruz-Schichten Patagoniens. Morphologisches Jahrbuch 67: 811–92.

Bocquentin Villanueva, J. 1984. Un nuevo Nettosuchidae (Crocodylia, Eusuchia) proveniente de la Formación Urumaco (Mioceno superior), Venezuela. Ameghiniana 21: 3–8.

Bocquentin Villanueva, J., and E. Buffetaut. 1981. *Hesperogavialis cruxenti* gen. nov., sp. nov., nouveau gavialide (Crocodylia, Eusuchia) du Miocène supérieur (Huayquerien) d'Urumaco (Venezuela). Géobios 14: 415–19.

Bocquentin Villanueva, J., and J. P. Souza Filho. 1990. O crocodiliano sul-americano *Carandaisuchus* como sinônimía de *Mourasuchus* (Nettosuchidae). Revista Brasileira de Geociências 20: 230–33.

Bocquentin Villanueva, J., J. P. Souza Filho, E. Buffetaut, and F. R. Negri. 1989. Nova interpretacáo do gênero *Purussaurus* (Crocodylia, Alligatoridae). Anais do XI Congresso Brasileiro de Paleontologia (Curitiba) 1: 428–38.

Bodmer, R. E. 1990. Fruit patch size and frugivory in the lowland tapir (*Tapirus terrestris*). Journal of Zoology (London) 222: 121–28.

Böhlke, J. E., S. H. Weitzman, and N. A. Menezes. 1978. Estado atual da sistematica de peixes da agua doce da America do Sul. Acta Amazonica 8: 657–77.

Boinski, S. 1989. The positional behavior and substrate use of squirrel monkeys: Ecological implications. Journal of Human Evolution 18: 659–77.

Bonaccorso, F. J. 1979. Foraging and reproductive ecology in a Panamanian bat community. Bulletin of the Florida State Museum, Biological Sciences 24: 359–408.

Bonaccorso, F. J., and S. R. Humphrey. 1984. Fruit bat niche dynamics: Their role in maintaining tropical forest diversity. In A. C. Chadwick and S. L. Sutton, eds., Tropical Rainforest, The Leeds Symposium. Leeds: Special Publication of the Leeds Philosophical and Literary Society. Pp. 169–83.

Bondesio, P. 1986. Lista sistemática de los vertebrados terrestres del Cenozóico de Argentina. Actas IV Congreso de Paleontología y Bioestratigrafía (Mendoza, 1986) 2: 187–90.

Bondesio, P., and J. Bocquentin Villanueva. 1988. Novedosos restos de Neoepiblemidae (Rodentia, Hystricognathi) del Mioceno tardío de Venezuela: Inferencias paleoambientales. Ameghiniana 25: 31–37.

Bondesio, P., J. H. Laza, G. J. Scillato-Yané, E. P. Tonni, and M. G. Vucetich. 1980. Estado actual del conocimiento de los vertebrados de la Formación Arroyo Chásico (Plioceno temprano) de la Provincia de Buenos Aires. Actas II Congreso Argentino de Paleontología y Bioestratigrafía y 1 Congreso Latinoamericano de Paleontologia (Buenos Aires, 1978) 3: 101–27.

Bondesio, P., and R. Pascual. 1977. Restos de Lepidosirenidae (Osteichthyes, Dipnoi) del Grupo Honda (Mioceno tardío) de Colombia: Sus indicaciónes paleoambientales. Revista de la Asociación Geológica de Argentina 32: 34–43.

Bondesio, P., J. Rabassa, R. Pascual, M. G. Vucetich, and G. J. Scillato-Yané. 1980. La Formación Collón Curá de Pilcaniyeu Viejo y sus alrededores (Río Negro, República Argentina): Su antigüedad y las condiciones ambientales según su distribución, su litogénesis y sus vertebrados. Actas II Congreso Argentino de Paleontología y Bioestratigrafía (Buenos Aires, 1978) 3: 85–99.

Botero, G. 1937. Bosquejo de la paleontología colombiana. Revista de Indias, Bogotá 3: 2–84.

Boussingault, M. 1849. Viajes científicos a los Andes ecuatoriales ó colección de memorias sobre física, química é historia natural de la Neuva Granada, Ecuador y Venezuela. Paris: Libreria Castellana.

Bown, T. M., and J. G. Fleagle. 1993. Systematics, biostratigraphy, and dental evolution of the Palaeothentidae, later Oligocene to early-middle Miocene (Deseadan-Santacrucian) caenolestoid marsupials of South America. Journal of Paleontology 67 (suppl. to no. 2): 1–76.

Bown, T. M., and C. N. Larriestra. 1990. Sedimentary paleoenvironments of fossil platyrrhine localities, Miocene Pinturas Formation, Santa Cruz Province, Argentina. Journal of Human Evolution 19: 87–119.

Bown, T. M., and M. J. Kraus. 1987. Integration of channel and floodplain suites. Pt. 1, Developmental sequence and lateral relations of alluvial paleosols. Journal of Sedimentary Petrology 57: 587–601.

Bown, T. M., E. R. Dumont, J. G. Fleagle, and O. A. Reig. 1990. Abderitine and palaeothentine caenolestid marsupials from Santacrucian deposits in southern Argentina. Journal of Vertebrate Paleontology 10A: 15.

Bown, T. M., M. J. Kraus, S. L. Wing, J. F. Fleagle, and B. Tiffany. 1982. The Fayum primate forest revisited. Journal of Human Evolution 11: 603–32.

Boyde, A. 1964. The structure and development of mammalian enamel. Ph.D. diss., University College, London.

Boyde, A. 1969. Electron microscope observations relating to the nature and development of prism decussation in mammalian dental enamel. Bulletin du Groupement International pour la Recherche Scientifique en Stomatologie 12: 151–207.

Boyde, A. 1971. Comparative histology of mammalian teeth. In A. Dahlberg, ed., Dental Morphology and Evolution. Chicago: University of Chicago Press. Pp. 81–93.

Boyde, A. 1984. Dependence of rate of physical erosion on orientation and density in mineralised tissues. Anatomy and Embryology 170: 57–62.

Bradbury, J. W., and S. L. Vehrencamp. 1976. Social organization and foraging in emballonurid bats. Pt. 1, Field studies. Behavioral Ecology and Sociobiology 1: 337–81.

Braun, H. 1985. The assasination of Gaitán: Public life and urban violence in Colombia. Madison: University of Wisconsin Press.

Bridge, J. S. 1984. Large-scale facies sequences in alluvial overbank environments. Journal of Sedimentary Petrology 54: 583–88.

Bristow, C. R. 1973. Guide to the geology of the Cuenca basin, southern Ecuador. Quito: Ecuadorean Geological and Geophysical Society.

Bristow, C. R., and J. J. Parodiz. 1982. The stratigraphical paleontology of the Tertiary non-marine sediments of Ecuador. Bulletin of the Carnegie Museum of Natural History 19: 1–53.

Buffetaut, E. 1978. Sur l'histoire phylogénétique et biogéographique des Gavialidae (Crocodylia, Eusuchia). Comptes Rendus de l'Académie des Sciences, Paris, sér. D, 287: 911–14.

Buffetaut, E. 1980. Histoire biogéographique des Sebecosuchia (Crocodylia, Mesosuchia): Un essai d'interprétation. Annales de Paléontologie, Vertébrés, Paris 66: 1–18.

Buffetaut, E. 1982. Systématique, origine et évolution des Gavialidae sud-américains. In E. Buffetaut, ed., Phylogénie et paléobiologéographie: Livre jubilaire en l'honneur de Robert Hoffstetter. Mémoire Spécial no. 6. Lyon: Géobios. Pp. 127–40.

Buffetaut, E. 1985. The place of *Gavialis* and *Tomistoma* in eusuchian evolution: A reconciliation of paleontological and biochemical data. Neues Jahrbuch für Geologie und Paläontologie, Mh. 1985: 707–16.

Buffetaut, E., and R. Hoffstetter. 1977. Découverte du crocodilien *Sebecus* dans le Miocène du Pérou oriental. Comptes Rendus de l'Académie des Sciences, Paris, sér. D, 284: 1663–66.

Burmeister, G. 1885. Exámen crítico de los mamíferos y reptiles fósiles denominados por D. Augusto Bravard y mencionados en su obra precedente. Anales del Museo Nacional de Historia Natural, Buenos Aires 3: 95–174.

Burt, W. H., and R. A. Stirton. 1961. The mammals of El Salvador. Miscellaneous Publications, Museum of Zoology, University of Michigan 117: 1–69.

Burton, J. A. 1987. The Collins Guide to the Rare Mammals of the World. Lexington: Stephen Greene Press.

Burton, P. J. K. 1985. Anatomy and evolution of the feeding apparatus in the avian orders Coraciiformes and Piciformes. Bulletin of the British Museum (Natural History), Zoological Series 47: 331–443.

Busbey, A. B., III. 1986a. New material of *Sebecus* cf. *huilensis* (Crocodilia: Sebecosuchidae) from the Miocene La Venta Formation of Colombia. Journal of Vertebrate Paleontology 6: 20–27.

Busbey, A. B., III. 1986b. *Pristichampsus* cf. *P. vorax* (Eusuchia: Pristichampsinae) from the Uintan of west Texas. Journal of Vertebrate Paleontology 6: 101–3.

Bustard, H. R. 1976. A future for the gharial. Cheetal 17: 3–8.

Butler, J. W. 1942. Geology of Honda district. American Association of Petroleum Geologists Bulletin 26: 793–837.

Butler, R. F., L. G. Marshall, R. E. Drake, and G. H. Curtis. 1984. Magnetic polarity stratigraphy and K-Ar dating of late Miocene and early Pliocene continental deposits, Catamarca Province, NW Argentina. Journal of Geology 92: 623–36.

Cabrera, A. 1929. Un astropotérido de Colombia. Physis 9: 436–39.

Cabrera, A. 1958. Catálogo de los mamíferos de América del Sur. Revista del Museo Argentino de Ciencias Naturales "Bernardino Rivadavia" 4: 1–307.

Cabrera, A., and L. Kraglievich. 1931. Diagnosis previas de los ungulados fósiles del Arroyo Chasicó. Notas Preliminares, Museo de La Plata 1: 107–13.

Camacho, H. H. 1967. Las transgresiones del Cretácico superior y Terciario de la Argentina. Revista de la Asociación Geológica Argentina 22(4): 253–80.

Campbell, C. J., and H. Bürgl. 1965. Section through the eastern Cordillera of Colombia, South America. Bulletin of the Geological Society of America 76: 567–90.

Campbell, D. G. 1991. Gap formation in tropical forest canopy by elephants, Oveng, Gabon, Central Africa. Biotrópica 23: 195–96.

Campbell, K. E., Jr. 1976. The late Pleistocene avifauna of La Carolina, southwestern Ecuador. Smithsonian Contributions to Paleobiology 27: 155–68.

Campbell, K. E., Jr. 1979. The non-passerine Pleistocene avifauna of the Talara tar seeps, northwestern Peru. Life Sciences Contributions, Royal Ontario Museum 118: 1–203.

Campbell, K. E., Jr., and E. P. Tonni. 1980. A new genus of teratorn from the Huayquerian of Argentina (Aves: Teratornithidae). Contributions in Science, Natural History Museum of Los Angeles County 330: 59–68.

Cande, S. C., and D. V. Kent. 1992. A new geomagnetic polarity time scale for the late Cretaceous and Cenozoic. Journal of Geophysical Research 97: 13917–51.

Carlini, A. A., G. J. Scillato-Yané, and S. F. Vizcaíno. 1989. Los Cingulata (Mammalia, Edentata) del Mioceno medio de Colombia: Un singular conjunto. Resumenes VII Jornadas Argentinas de Paleontología de Vertebrados (Buenos Aires). Ameghiniana 26: 242.

Carlini, A. A., S. F. Vizcaíno, and G. J. Scillato-Yané. 1989. Novedosos Cingulata (Mammalia, Edentata) del Mioceno de Ecuador. Resumenes VI Jornadas Argentinas de Paleontología de Vertebrados (San Juan). Pp. 23–25.

Carlson, S. J., and D. W. Krause. 1985. Enamel ultrastructure of multituberculate mammals: An investigation of variability. Contributions, Museum of Paleontology, University of Michigan 27: 1–50.

Carroll, R. 1988. Vertebrate Paleontology and Evolution. New York: W. H. Freeman.

Cartelle, C., and J. S. Fonseca. 1983. Contribuçao ao melhor conhecimento da pequena preguica terricola *Nothrotherium maquinense* (Lund) Lydekker, 1889. Lundiana 2: 127–81.

Cartmill, M. 1974. Pads and claws in arboreal locomotion. In F. A. Jenkins, ed., Primate Locomotion. New York: Academic Press. Pp. 45–83.

Castellanos, A. 1946. Una nueva especie de clamitério *Vassallia maxima* sp. nov. Publicaciones del Instituto de Fisiografía y Geología, Universidad Nacional del Litoral, Rosario, Argentina 6: 5–47.

Cebula, G. T., M. J. Kunk, H. H. Mehnert, C. W. Naeser, and J. O. Obradovich. 1986. The Fish Canyon tuff, a potential standard for the $^{40}Ar/^{39}Ar$ and fission-track methods [abstract]. Terra Cognita 6: 139–40.

Chaffee, R. G. 1952. The Deseadan vertebrate fauna of the Scarritt Pocket, Patagonia. Bulletin of the American Museum of Natural History 98: 509–62.

Chang, A., G. Arratia, and G. Alfaro. 1978 *Percichthys lonquimayiensis* sp. nov. from the upper Paleocene of Chile (Pisces, Perciformes, Serranidae). Journal of Paleontology 52: 727–36.

Chase, J., M. Yepes Small, E. A. Weiss, D. Sharma, and S. Sharma. 1991. Crepuscular activity of *Molossus molossus*. Journal of Mammalogy 72: 414–18.

Choate, J. R., and E. C. Birney. 1968. Sub-Recent Insectivora and Chiroptera from Puerto Rico, with the description of a new bat of the genus *Stenoderma*. Journal of Mammalogy 49: 400–412.

Cifelli, R. L. 1983a. Eutherian tarsals from the late Paleocene of Brazil. American Museum Novitates 2761: 1–31.

Cifelli, R. L. 1983b. The origin and affinities of the South American Condylarthra and early Tertiary Litopterna (Mammalia). American Museum Novitates 2772: 1–49.

Cifelli, R. L. 1985. South American ungulate evolution and extinction. In F. G. Stehli and S. D. Webb, eds., The Great American Biotic Interchange. New York: Plenum Press. Pp. 249–66.

Cifelli, R. L. 1993a. The phylogeny of the native South American ungulates. In F. S. Szalay, M. J. Novacek, and M. C. McKenna, eds., Mammalian Phylogeny. New York: Springer-Verlag. Pp. 195–216.

Cifelli, R. L. 1993b. Theria of metatherian-eutherian grade and the origin of marsupials. In F. S. Szalay, M. J. Novacek, and M. C. McKenna, eds., Mammal Phylogeny: Mesozoic Differentiation, Multituberculates, Monotremes, Early Therians, and Marsupials. New York: Springer-Verlag. Pp. 205–15.

Cifelli, R. L., and J. Guerrero. 1989. New remains of *Prothoatherium colombianus* (Litopterna, Mammalia) from the Miocene of Colombia. Journal of Vertebrate Paleontology 9: 222–31.

Ciochon, R. L., and R. S. Corruccini. 1975. Morphometric analysis of platyrrhine femora with taxonomic implications and notes on two fossil forms. Journal of Human Evolution 17: 193–217.

Cione, A. L. 1978. Aportes paleoictiólogicos al conocimiento de la evolución de las paleotemperaturas en el área austral de América del Sur durante el Cenozóico: Aspectos zoogeográficos y ecológicos conexos. Ameghiniana 15: 183–208.

Cione, A. L. 1986. Los peces continentales del Cenozóico de Argentina. Actas IV Congreso Argentino de Paleontología y Bioestratigrafía, Symposio sobre Evolución de los Vertebrados Cenozóicos. Pp. 101–6.

Cione, A. L., and E. P. Tonni. 1995. Chronostratigraphy and "Land-Mammal Ages" in the Cenozoic of southern South America: The "Uquian" problem. Journal of Paleontology 69: 135–59.

Clemens, W. A., and J. A. Lillegraven. 1986. New late Cretaceous North American advanced therian mammals that fit neither the marsupial nor eutherian molds. Contributions to Geology, University of Wyoming, Special Paper 3: 55–85.

Clemens, W. A., and W. Von Koenigswald. 1991. *Purgatorius,* plesiadapiforms, and evolution of Hunter-Schreger bands. Journal of Vertebrate Paleontology 11: 23A.

Coimbra-Filho, A. F., and R. A. Mittermeier. 1978. Tree-gouging, exudate-eating, and the "short-tusked" condition in *Callithrix* and *Cebuella*. In D. G. Kleiman, ed., The Biology and Conservation of the Callitrichidae. Washington D.C.: Smithsonian Institution Press. Pp. 105–17.

Colbert, E. H. 1946. *Sebecus,* representative of a peculiar suborder of fossil Crocodilia from Patagonia. Bulletin of the American Museum of Natural History 87(4): 217–70.

Collett, S. F. 1981. Population characteristics of *Agouti paca* (Rodentia) in Colombia. Publications of the Museum, Michigan State University, Biological Series 5: 485–602.

Collinson, J. D. 1978. Vertical sequence and sand body shape in alluvial sequences. In A. D. Miall, ed., Fluvial Sedimentology. Memoir of the Canadian Society of Petroleum Geology no. 5. Pp. 577–86.

Colwell, J. 1965. A new notoungulate of the family Leontiniidae from the Miocene of Colombia. Master's thesis, University of California, Berkeley.

Contreras, L. C., J. C. Torres-Mura, and J. L. Yáñez. 1987. Biogeography of octodontid rodents: An eco-evolutionary hypothesis. In B. D. Patterson and R. M. Timm, eds., Studies in Neotropical Mammalogy: Essays in Honor of Philip Hershkovitz. Fieldiana, Zoology 39: 401–11.

Conroy, G. 1987. Problems of body-weight estimation in fossil primates. International Journal of Primatology 8: 115–37.

Coombs, M. C. 1983. Large mammalian clawed herbivores: A comparative study. Transactions of the American Philosophical Society (Philadelphia) 73: 1–96.

Costa, E. V. 1981. Revisão gastrópodes fósseis da localidade do Tres Unidos, Formaçao Pebas, Plioceno do Alto Amazonas, Brasil. Anais II Congreso Latinoamericano Paleontologia (Pôrto Alegre), 635–49.

Cracraft, J. 1971. A new family of hoatzin-like birds (order Opisthocomiformes) from the Eocene of South America. Ibis 113: 229–33.

Cracraft, J. 1973. Systematics and evolution of the Gruiformes (Class Aves). Pt. 3, Phylogeny of the suborder Grues. Bulletin of the American Museum of Natural History 151: 1–128.

Crochet, J. Y. 1980. Les marsupiaux du Tertiaire d'Europe. Paris: Editorial Foundation Singer-Polignac.

Cronin, J. E., and V. M. Sarich. 1975. Molecular systematics of the New World monkeys. Journal of Human Evolution 4: 357–75.

Cuervo Díaz, A., J. Hernández-Camacho, and A. Cadena. 1986. Lista actualizada de los mamíferos de Colombia: Anotaciones sobre su distribución. Caldasia 15: 471–501.

Czaplewski, N. J. 1990. Late Pleistocene (Lujanian) occurrence of *Tonatia silvicola* in the Talara Tar Seeps, Peru. Anais Academia Brasileira de Ciências 62: 235–38.

Dagosto, M. 1988. Implications of postcranial evidence for the origin of Euprimates. Journal of Human Evolution 17: 35–56.

Dahl, G. 1971. Los Peces del Norte de Colombia. Bogotá: Inderena.

Dalrymple, G. B. 1979. Critical table for the conversion of K-Ar ages from old to new constants. Geology 7: 558–60.

Dalrymple, G. B., E. C. J. Alexander, M. A. Lanphere, and G. P. Kraker. 1981. Irradiation of samples for ^{40}Ar/^{39}Ar dating using the TRIGA reactor. United States Geological Survey, Professional Paper 1176: 1–55.

Damuth, J. 1990. Problems in estimating body masses of archaic ungulates using dental measurements. In J. Damuth and B. J.

MacFadden, eds., Body Size in Mammalian Paleobiology. Cambridge: Cambridge University Press. Pp. 229–54.

Damuth, J., and B. J. MacFadden, eds. 1990. Body Size in Mammalian Paleobiology. Cambridge: Cambridge University Press.

Darwin, C. 1871. The Descent of Man and Selection in Relation to Sex. (2d ed., 1874.) New York: D. Appleton.

Davis, L. C. 1987. Morphological evidence of positional behavior in the hindlimb of *Cebupithecia sarmientoi* (Primates: Platyrrhini). Master's thesis, Arizona State University, Tempe.

Davis, L. C. 1988. Morphological evidence of locomotor behavior in the hindlimb of *Cebupithecia sarmientoi* (Primates: Platyrrhini). American Journal of Physical Anthropology 75: 202.

Deino, A., and R. Potts. 1990. Single-crystal ^{40}Ar/^{39}Ar dating of the Olorgesailie Formation, southern Kenya Rift. Journal of Geophysical Research 95: 8453–70.

Deino, A., L. Tauxe, M. Monaghan, and R. Drake. 1990. ^{40}Ar/^{39}Ar age calibration of the litho- and paleomagnetic stratigraphies of the Ngorora Formation, Kenya. Journal of Geology 98: 567–87.

Del Rio, A. 1945. Informe del Servicio Geológico. Memorias del Ministerio de Minas y Petróleo, Bogotá 1945: 3–31.

Delson, E., and A. L. Rosenberger. 1984. Are there any anthropoid primate living fossils? In N. Eldredge and S. M. Stanley, eds., Living Fossils. New York: Springer-Verlag. Pp. 50–61.

Desmarest, A. G. 1818. Noctilion ou bec de lièvre. Nouveau dictionnaire d'histoire naturelle, appliquée aux arts, principalement à l'agriculture et à l'économie rurale et domestique; par une société de naturalistes. New ed. 23: 14–16. Paris: Deterville.

Díaz de Gamero, M. L., and O. J. Linares. 1989. Estratigrafía y paleontología de la Formación Urumaco, del Mioceno tardío de Falcón noroccidental. Memorias del VIII Congreso Geológico Venezolano 1: 419–38.

Dixon, J. R. 1979. Origin and distribution of reptiles in lowland tropical rainforests of South America. Museum of Natural History, University of Kansas, Monograph 7: 217–40.

Dodson, P. 1975. Functional and ecological significance of relative growth in *Alligator*. Journal of Zoology (London) 175: 315–55.

Dollman, G. 1933. Primates, 3d ser. London: Trustees of the British Museum.

Domning, D. P. 1982. Evolution of manatees: A speculative history. Journal of Paleontology 56: 599–619.

Domning, D. P. 1988. Fossil Sirenia of the West Atlantic and Caribbean region. I. *Metaxytherium floridanum* Hay, 1922. Journal of Vertebrate Paleontology 8: 395–426.

Domning, D. P. 1990. Fossil Sirenia of the West Atlantic and Caribbean region. IV. *Corystosiren varguezi*, gen. et sp. nov. Journal of Vertebrate Paleontology 10: 361–71.

Domning, D. P. 1994. A phylogenetic analysis of the Sirenia. In A. Berta and T. A. Deméré, eds., Contributions in Marine Mammal Paleontology Honoring Frank C. Whitmore, Jr., Proceedings of the San Diego Society of Natural History 29: 177–89.

Donadio, O. E. 1985. Un nuevo lacertilio (Squamata, Sauria, Teiidae) de la Formación Lumbrera (Eoceno Temprano), Provincia de Salta, Argentina. Ameghiniana 22: 221–28.

Donoghue, M. J., J. A. Doyle, J. Gauthier, A. G. Kluge, and T. Rowe. 1989. The importance of fossils in phylogeny reconstruction. Annual Review of Ecology and Systematics 20: 431–60.

Dubost, G. 1978. Un aperçu sur l'écologie du chevrotain africain *Hyemoschus aquaticus* Ogilby, artiodactyle tragulide. Mammalia 42: 1–62.

Duque-Caro, H. 1979. Major structural elements and evolution of northwestern Colombia. In J. S. Watkins, ed., Geological and Geophysical Investigations of Continental Margins. American Association of Petroleum Geology Memoir 29: 329–51.

Duque-Caro, H. 1980. Geotectónica y evolución de la región noroccidental Colombiana. Boletín Geológico, INGEOMINAS 23(3): 4–37.

Duque-Caro, H. 1990. Neogene stratigraphy, paleoceanography and paleobiogeography in northwest South America and the evolution of the Panama Seaway. Palaeogeography, Palaeoclimatology, Palaeoecology 77: 203–34.

Eaglen, R. H. 1984. Incisor size and diet revisited: The view from a platyrrhine perspective. American Journal of Physical Anthropology 64: 263–75.

Edmund, A. G. 1985. The fossil giant armadillos of North America (Pampatheriinae, Xenarthra = Edentata). In G. G. Montgomery, ed., The Evolution and Ecology of Armadillos, Sloths, and Vermilinguas. Washington, D.C.: Smithsonian Institution Press. Pp. 83–93.

Edmund, G. G. 1985. The armor of fossil giant armadillos (Pampatheriidae, Mammalia). Texas Memorial Museum, University of Texas at Austin, Pierce-Sellards ser. 40: 1–20.

Edmund, G. G. 1987. Evolution of the genus *Holmesina* (Pampatheriidae, Mammalia). Texas Memorial Museum, University of Texas at Austin, Pierce-Sellards ser. 45: 1–20.

Edwards, M. C., K. A. Eriksson, and R. S. Kier. 1983. Paleochannel geometry and flow patterns determined from exhumed Permian point bars in north-central Texas. Journal of Sedimentary Petrology 53: 1261–70.

Eigenmann, C. H. 1920. The Magdalena basin and the horizontal and vertical distribution of its fishes. Indiana University Studies 7: 21–34.

Eigenmann, C. H., and G. S. Myers. 1929. The American Characidae. Memoirs, Museum of Comparative Zoology, Harvard University 43: 429–558.

Eisenberg, J. F. 1981. The Mammalian Radiations. Chicago: University of Chicago Press.

Eisenberg, J. F. 1984. New World rats and mice. In D. W. Mac-Donald, ed., The Encyclopedia of Mammals. New York: Facts on File. Pp. 640–49.

Eisenberg, J. F. 1989. Mammals of the Neotropics. Vol. 1, Panama, Colombia, Venezuela, Guyana, Suriname, French Guiana. Chicago: University of Chicago Press.

Eisenberg, J. F., M. A. O'Connell, and P. V. August. 1979. Density, productivity, and distribution of mammals in two Venezuelan habitats. In J. F. Eisenberg, ed., Vertebrate Ecology in the Northern Neotropics. Washington, D.C.: Smithsonian Institution Press. Pp. 187–207.

Eisenberg, J. F., and K. H. Redford. 1982. Comparative niche structure and evolution of mammals of the Nearctic and southern South America. In M. A. Mares and H. H. Genoways, eds., Mammalian Biology in South America. Special Publications of the Pymatuning Laboratory of Ecology, University of Pittsburg, Pennsylvania. Pp. 77–84.

Elías del Hierro, J. 1949. Servicio Geológico. Memorias del Ministerio de Minas y Petróleo, Bogotá 1949: 42–92.

Emmons, L. H., and F. Feer. 1990. Neotropical Rainforest Mammals: A Field Guide. Chicago: University of Chicago Press.

Engelmann, G. F. 1985. The phylogeny of the Xenarthra. In G. G. Montgomery, ed., The Evolution and Ecology of Armadillos, Sloths, and Vermilinguas. Washington, D.C.: Smithsonian Institution Press. Pp. 51–64.

Englehardt, H. 1895. Über neue Tertiärpflanzen Süd-Amerikas. Abhandlungen Herausgegeben von der Senckenbergischen Naturforschenden Gesellschaft 19: 1–47.

Estes, R. 1961. Miocene lizards from Colombia, South America. Breviora, Museum of Comparative Zoology, Harvard University 143: 1–11.

Estes, R. 1964. Fossil vertebrates from the late Cretaceous Lance Formation, eastern Wyoming. University of California Publications in Geological Sciences 49: 1–180.

Estes, R. 1969. Relationships of two Cretaceous lizards (Sauria, Teiidae). Breviora, Museum of Comparative Zoology, Harvard University 317: 1–8.

Estes, R. 1970. Origin of the Recent North American lower vertebrate fauna: An inquiry into the fossil record. Forma et Functio 3: 139–63.

Estes, R. 1981. Gymnophiona, Caudata. In P. Wellnhofer, ed., Handbuch der Palaeoherpetologie, Teil 2. Stuttgart: Gustav Fischer Verlag. Pp. 1–115.

Estes, R. 1983a. The fossil record and the early distribution of lizards. In A. Rhodin and K. Miyata, eds., Advances in Herpetology and Evolutionary Biology: Essays in Honor of Ernest E. Williams. Cambridge: Museum of Comparative Zoology, Harvard University. Pp. 365–98.

Estes, R. 1983b. Sauria terrestria, Amphisbaenia. Handbuch der Palaeoherpetologie, Teil 10A. Stuttgart: Gustav Fisher Verlag. xxii + 249 pp.

Estes, R., and A. Baez. 1985. Herpetofaunas of North and South America during the late Cretaceous and Cenozoic: Evidence for interchange? In F. G. Stehli and S. D. Webb, eds., The Great American Biotic Interchange. New York: Plenum Press. Pp. 139–97.

Estes, R., and R. Wassersug. 1963. A Miocene toad from Colombia, South America. Breviora, Museum of Comparative Zoology, Harvard University 193: 1–13.

Estes, R., P. Berberian, and C. Meszoely. 1969. Lower vertebrates from the late Cretaceous Hell Creek Formation, McCone County, Montana. Breviora, Museum of Comparative Zoology, Harvard University 337: 1–33.

Estes, R., K. De Queiroz, and J. A. Gauthier. 1988. Phylogenetic relationships within Squamata. In R. Estes and G. Pregill, eds., Phylogenetic Relationships of the Lizard Families. Palo Alto: Stanford University Press. Pp. 119–281.

Etter, A., and P. J. Botero. 1990. La variedad edáfica de hormigas (Atta laevigata) y su relación con la dinámica sabana / bosque en los Llanos Orientales (Colombia). Colombia Amazonica 4: 77–95.

Evans, J. E. 1991. Facies relationships, alluvial architecture, and paleohydrology of a Paleogene, humid-tropical alluvial-fan system, Chumstick Formation, Washington State, U.S.A. Journal of Sedimentary Petrology 61: 732–55.

Felsenstein, J. 1985. Confidence limits on phylogenies: An approach using the bootstrap. Evolution 39: 783–91.

Fenton, M. B. 1992. Wounds and the origin of blood-feeding in bats. Biological Journal of the Linnean Society 47: 161–71.

Fernández, J., R. Pascual, and P. Bondesio. 1973. Restos de Lepidosiren paradoxa (Osteichthyes, Dipnoi) de la Formación Lumbrera (Eogeno) de Jujuy: Consideraciones estratigráficas, paleoecológicas y paleozoogeográficas. Ameghiniana 10: 152–72.

Ferreira, J. M., P. P. Phakey, J. Palamara, W. A. Rachinger, and H. J. Orams. 1989. Electron microscopic investigation relating the occlusal morphology to the underlying enamel structure of molar teeth of the wombat (Vombatus ursinus). Journal of Morphology 200: 141–49.

Fields, R. W. 1957. Hystricomorph rodents from the late Miocene of Colombia, South America. University of California Publications in Geological Sciences 32: 207–403.

Fields, R. W. 1959. Geology of the La Venta badlands, Colombia, South America. University of California Publications in Geological Sciences 32: 405–44.

Findley, J. S., and D. E. Wilson. 1974. Observations on the Neotropical disk-winged bat, Thyroptera tricolor. Journal of Mammalogy 55: 562–71.

Fisher, R. A. 1953. Disperson on a sphere. Proceedings of the Royal Society of London, A217: 295–305.

Flach, P. D., and G. D. Mossop. 1985. Depositional environments of lower Cretaceous McMurray Formation, Athabasca Oil Sands, Alberta. American Association of Petroleum Geologists Bulletin 69: 1195–1207.

Fleagle, J. G. 1988. Primate Adaptation and Evolution. San Diego, Calif.: Academic Press.

Fleagle, J. G. 1990. New fossil platyrrhines from the Pinturas Formation, Southern Argentina. Journal of Human Evolution 19: 61–85.

Fleagle, J. G., and A. L. Rosenberger, eds. 1980. The Platyrrhine Fossil Record. San Diego, Calif.: Academic Press.

Fleagle, J. G., and T. M. Bown. 1983. New primate fossils from late Oligocene (Colhuehuapian) localities of Chubut Province, Argentina. Folia Primatologica 41: 240–66.

Fleagle, J. G., G. A. Buckley, and M. E. Schloeder. 1988. New fossil primates from Monte Observación, Santa Cruz Formation (lower Miocene), Santa Cruz Province, Argentina. Journal of Vertebrate Paleontology 8: 14A.

Fleagle, J. G., and R. F. Kay. 1985. The paleobiology of catarrhines. In E. Delson, ed., Ancestors: The Hard Evidence. New York: Alan R. Liss. Pp. 23–36.

Fleagle, J. G., and R. F. Kay. 1989. The dental morphology of Dolichocebus gaimanensis, a fossil monkey from Argentina. American Journal of Physical Anthropology 78: 221.

Fleagle, J. G., and D. J. Meldrum. 1988. Locomotor behavior and skeletal morphology of two sympatric pitheciine monkeys, Pithecia pithecia and Chiropotes satanas. American Journal of Primatology 16: 227–49.

Fleagle, J. G., R. A. Mittermeier, and A. L. Skopec. 1981. Differential habitat use by Cebus apella and Saimiri sciureus in Central Surinam. Primates 22: 361–67.

Fleagle, J. G., J. D. Obradovich, and E. L. Simons. 1986. Age of the earliest African anthropoids. Science 234: 1247–49.

Fleagle, J. G., D. W. Powers, G. C. Conroy, and J. P. Watters. 1987. New fossil platyrrhines from Santa Cruz Province, Argentina. Folia Primatologica 48: 65–77.

Fleagle, J. G., and A. L. Rosenberger, eds. 1980. The Platyrrhine Fossil Record. San Diego, Calif.: Academic Press.

Fleagle, J. G., and A. L. Rosenberger. 1983. Cranial morphology of the earliest anthropoids. In M. Sakka, ed., Morphologie et évolution morphogénèse du crane et origine de l'homme. Paris: Centre National de Recherche Scientifique. Pp. 141–55.

Fleagle, J. G., and E. L. Simons. 1983. The tibio-fibular articulation in Apidium phiomense, an Oligocene anthropoid. Nature 301: 238–39.

Fleming, T. H. 1973. Numbers of mammalian species in North and Central American forest communities. Ecology 54: 555–63.

Fleming, T. H. 1986. The structure of Neotropical bat communities: A preliminary analysis. Revista Chilena de Historia Natural 59: 135–50.

Fleming, T. H. 1988. The Short-Tailed Fruit Bat: A Study in Plant–Animal Interactions. Chicago: University of Chicago Press.

Flower, W. H. 1882. Abstract of lectures on the anatomy, physiology, and zoology of the Edentata, I–IV. Abstracts of the British Medical Journal 1: 649–50, 694–95, 737–38, 768, 901, 937–38.

Flynn, J. J. 1986. Faunal provinces and the Simpson coefficient. In K. Flanagan and J. L. Lillegraven, eds., Vertebrates, Phylogeny and Philosophy. G. G. Simpson Memorial Volume, University of Wyoming Contributions to Geology, Special Paper 3: 317–38.

Flynn, J. J., and C. C. Swisher III. In press. Cenozoic South American land mammal ages: Correlation to global geochronologies. In W. A. Berggren, D. V. Kent, and J. Hardenbol, eds., Geochronology, Time Scales, and Stratigraphic Correlation. SEPM Special Publication.

Flynn, J. J., L. Marshall, J. Guerrero, and P. Salinas. 1989. Geochronology of middle Miocene ("Friasian" Land Mammal Age) faunas from Chile and Colombia. Abstracts with Programs, Annual Meeting, Geological Society of America 21: A133.

Folk, R. L. 1980. Petrology of Sedimentary Rocks, 2d ed. Austin, Tex.: Hemphill.

Ford, S. M. 1980a. Callitrichids as phyletic dwarfs, and the place of the Callitrichidae in Platyrrhini. Primates 21: 31–43.

Ford, S. M. 1980b. Phylogenetic relationships of the Platyrrhini: The evidence of the femur. In R. L. Ciochon and A. B. Chiarelli, eds., Evolutionary Biology of the New World Monkeys and Continental Drift. New York: Plenum Press. Pp. 317–29.

Ford, S. M. 1980c. A systematic revision of the Platyrrhini based on features of the postcranium. Ph. D. diss., University of Pittsburgh, Pittsburgh, Pa.

Ford, S. M. 1986. Systematics of the New World monkeys. In D. R. Swindler and J. Erwin, eds., Comparative Primate Biology. Vol. 1, Systematics, Evolution, and Anatomy. New York: Alan R. Liss. Pp. 73–135.

Ford, S. M. 1988. Postcranial adaptations of the earliest platyrrhine. Journal of Human Evolution 17: 155–92.

Ford, S. M. 1990a. Locomotor adaptation of fossil platyrrhines. Journal of Human Evolution 19: 141–73.

Ford, S. M. 1990b. Platyrrhine evolution in the West Indies. Journal of Human Evolution 19: 237–54.

Ford, S. M., and C. A. Woods. 1990. New platyrrhine fossil material from Haiti. American Journal of Physical Anthropology 81: 223.

Ford, S. M., L. C. Davis, and R. F. Kay. 1991. New platyrrhine astragalus from the Miocene of Colombia. American Journal of Physical Anthropology (suppl.) 12: 73–74.

Fortelius, M. 1985. Ungulate cheek teeth: Developmental, functional and evolutionary interrelations. Acta Zoologica Fennica 180: 1–76.

Fosse, G. 1968. A quantitative analysis of the numerical density and distributional pattern of prisms and ameloblasts in dental enamel and tooth germs. Pt. 3, The calculation of prism diameters and numbers of prisms per unit area in dental enamel. Acta Odontologica Scandanavica 26: 315–36.

Foster, R. 1986. Dispersal and the sequential plant communities in Amazonian Peruvian floodplains. In A. Estrada and T. H. Fleming, eds., Frugivores and Seed Dispersal. Dordrecht: W. Junk. Pp. 357–70.

Foster, R. 1990. The floristic composition of the Río Manú floodplain forest. In A. H. Gentry, ed., Four Neotropical Rainforests. New Haven, Conn.: Yale University Press. Pp. 99–111.

Fourtau, R. 1920. Supplement à la contribution à l'étude des vertébrés miocènes de l'Egypte. Cairo: Ministry of Finance, Survey Department.

Frailey, C. D. 1981. Studies on the Cenozoic Vertebrata of Bolivia and Peru. Ph.D. diss., University of Kansas, Lawrence.

Frailey, C. D. 1986. Late Miocene and Holocene mammals, exclusive of Notoungulata, of the Río Acre region, western Amazonia. Contributions to Science, Natural History Museum of Los Angeles County 364: 1–46.

Frailey, C. D. 1987. The Miocene vertebrates of Quebrada Honda, Bolivia. Pt. 1, Astrapotheria. Occasional Papers, Museum of Natural History, University of Kansas 122: 1–15.

Frailey, C. D. 1988. The Miocene vertebrates of Quebrada Honda, Bolivia. Pt. 2, Edentata. Occasional Papers, Museum of Natural History, University of Kansas, 123: 1–13.

Frailey, C. D., E. L. Lavina, A. Rancy, and J. P. Souza Filho. 1988. A proposed Pleistocene/Holocene lake in the Amazon basin and its significance to Amazonian geology and biogeography. Acta Amazonica 18: 119–43.

Frakes, L. A., and E. M. Kemp. 1972. Influence of continental positions on early Tertiary climates. Nature 240: 97–100.

Freeman, P. W. 1984. Functional cranial analysis of large animalivorous bats (Microchiroptera). Biological Journal of the Linnean Society 21: 387–408.

Gaffney, E., and R. Zangerl. 1968. A revision of the chelonian genus Bothremys (Pleurodira: Pelomedusidae). Fieldiana, Geology 16: 193–239.

Gao, K., and R. Fox. 1991. New teiid lizards from the upper Cretaceous Oldman Formation (Judithian) of southeastern Alberta, Canada, with a review of the Cretaceous record of teiids. Annals of the Carnegie Museum 60: 145–62.

Gardner, A. L. 1977. Feeding habits. In R. J. Baker, J. K. Jones, Jr., and D. C. Carter, eds., Biology of Bats of the New World Family Phyllostomidae. Special Publications of the Museum 13. Lubbock: Texas Tech Press. Pp. 293–350.

Gardner, A. L., and G. K. Creighton. 1989. A new generic name for Tate's (1933) Microtarsus group of South American mouse opossums (Marsupialia, Didelphidae). Proceedings of the Biological Society of Washington 102: 3–7.

Gardner, T. W. 1983. Paleohydrology and paleomorphology of a Carboniferous, meandering, fluvial sandstone. Journal of Sedimentary Petrology 53: 991–1005.

Gasparini, Z. 1968. Nuevas restos de Rhamphostomopsis neogaeus (Burm.) Rusconi 1933, (Reptilia, Crocodilia) del "Mesopotamiense" (Plioceno medio-superior) de Argentina. Ameghiniana 5: 299–311.

Gasparini, Z. 1972. Los Sebecosuchia (Crocodilia) del Territorio Argentino, consideraciones sobre su "status" taxonomico. Ameghiniana 9: 23–34.

Gasparini, Z. 1981. Los Crocodylia fósiles de la Argentina. Ameghiniana 18: 177–205.

Gasparini, Z. 1984. New Tertiary Sebecosuchia (Crocodylia: Mesosuchia) from Argentina. Journal of Vertebrate Paleontology 4: 85–95.

Gasparini, Z. 1985. Un nuevo cocodrilo (Eusuchia) del Cenozóico de América del Sur. Actas 1er Congreso Argentino de Paleontología y Bioestratigrafía, 2: 51–53.

Gasparini, Z. 1985. Un nuevo cocodrilo (Eusuchia) del Cenozoico de América del Sur. VIII Congresso Brasileiro de Paleontologia (1983), Coletânea de Trabalhos Paleontológicos MME-DNPM, ser. Geologia, Brasília, 27, Paleont./Estrat. 2(1985): 51–53.

Gasparini, Z., and A. M. Báez. 1975. Aportes al conocimiento de la herpetofauna Terciaria de la Argentina. Actas del Primer Congreso Argentino de Paleontología y Bioestratigrafía 2: 377–415.

Gasparini, Z., L. Chiappe, and M. Fernández. 1991. A new Senonian peirosaurid (Crocedylomorpha) from Argentina and a synopsis of the South American Cretaceous crocodilians. Journal of Vertebrate Paleontology 11: 316–33.

Gasparini, Z., M. Fernández, and J. Powell. 1993. New Tertiary sebecosuchians (Crocodylomorpha) from South America: Phylogenetic implications. Historical Biology 7(1): 1–19.

Gaudry, A. 1906. Fossiles de Patagonie: Attitudes de quelques animaux. Annales de Paléontologie 1: 1–28.

Gautier, J., A. F. Kluge, and T. Rowe. 1988. Amniote phylogeny and the importance of fossils. Cladistics 4: 105–209.

Gebo, D. L., and E. L. Simons. 1987. Morphology and locomotor adaptations of the foot in early Oligocene anthropoids. American Journal of Physical Anthropology 74: 83–101.

Gebo, D. L., M. Dagosto, A. L. Rosenberger, and T. Setoguchi. 1990. New platyrrhine tali from La Venta, Colombia. Journal of Human Evolution 19: 737–46.

Gery, J. 1977. Characoids of the World. Neptune, N.J.: TFH Publications.

Gingerich, P. D. 1990. Prediction of body mass in mammalian species from long bone lengths and diameters. Contributions from the Museum of Paleontology, University of Michigan 28: 79–92.

Gingerich, P. D. 1992. Marine mammals (Cetacea and Sirenia) from the Eocene of Gebel Mokattam and Fayum, Egypt: Stratigraphy, age, and paleoenvironments. Papers in Paleontology, University of Michigan 30: i–ix, 1–84.

Gingerich, P. D., B. H. Smith, and K. Rosenberg. 1982. Allometric scaling in the dentition of primates and prediction of body weight from tooth size in fossils. American Journal of Physical Anthropology 58: 81–100.

Glaessner, M. F. 1969. Decapoda. In R. C. Moore, ed., Treatise on Invertebrate Paleontology, pt. R, Arthropoda 4. 2: R401–R533.

Glanz, W. E., and S. Anderson. 1990. Notes on Bolivian Mammals. Pt. 7, A new species of *Abrocoma* (Rodentia) and relationships of the Abrocomidae. American Museum Novitates 2992: 1–32.

Goffart, M. 1971. Function and Form in the Sloth. New York: Pergamon Press.

Goin, F. G. 1988. Imbricación molar en los marsupiales Borhyaenidae: Implicancias biomecánicas y sistemáticas. Resumenes V Jornadas Argentinas de Paleontología de Vertebrados (La Plata) P. 101.

Goin, F. G. 1991. Los Didelphoidea (Mammalia, Marsupialia) del Cenozóico tardío de la Región Pampeana. Ph.D. diss., Universidad Nacional de La Plata, Argentina.

Goin, F. J., and R. Pascual. 1987. News on the biology and taxonomy of the marsupials Thylacosmilidae (late Tertiary of Argentina). Anales de la Académia Argentina de Ciencias Exactas, Físicas y Naturales, Buenos Aires 39: 219–46.

Gorroño, R., R. Pascual, and R. Pombo. 1979. Hallazgo de mamíferos eógenos en el sur de Mendoza: Su implicancia en las dataciones de los "Rodados Lustrosos" y del primer episodio orogénico del Terciario en esa región. Actas VII Congreso Geológico Argentino (Neuquén) 2: 475–87.

Gosline, W. A. 1940. A revision of the Neotropical catfishes of the family Callichthyidae. Stanford Ichthyology Bulletin 2: 1–29.

Goulding, W. M. 1980. The Fishes and the Forest. Berkeley: University of California Press.

Graves, G. R., and R. L. Zusi. 1989. Avian body weights from the lower Río Xingú, Brazil. Bulletin, British Ornithologists' Club 110: 20–25.

Gray, J. E. 1827. Synopsis of the species of the class Mammalia as arranged with references to their organization . . . with specific characters, synonyms, etc. In G. Cuvier, ed., Animal Kingdom . . . with Additional Descriptions by Edward Griffith and Others. Mammalia. London: George B. Whittaker. 5: 1–391.

Gray, J. E. 1838. A revision of the genera of bats (Vespertilionidae), and the description of some new genera and species. Magazine of Zoology and Botany 2: 486–505.

Greenwood, P. H. 1974. Review of the Cenozoic freshwater fish faunas of Africa. Annals of the Geological Survey of Egypt 4: 211–32.

Gregory, W. K. 1920. Studies in comparative myology and osteology. Pt. 5, On the anatomy of the preorbital fossae of Equidae and other ungulates. Bulletin of the American Museum of Natural History 42: 265–83.

Gregory, W. K. 1922. The Origin and Evolution of the Human Dentition. Baltimore: Williams and Wilkins.

Gregory, W. K. 1929. Adaptation to locomotion in the limbs of the fleet (cursorial) and the ponderous (graviportal) types of titanotheres and other hoofed quadrupeds. In H. F. Osborn, ed., The Titanotheres of Ancient Wyoming, Dakota, and Nebraska. Washington, D.C.: United States Geological Survey Monograph. Pp. 727–46.

Grine, F. E., D. W. Krause, G. Fosse, and W. L. Jungers. 1987. Analysis of individual, intraspecific and interspecific variability in quantitative parameters of caprine tooth enamel structure. Acta Odontologica Scandinavica 45: 1–23.

Grine, F. E., G. Fosse, D. W. Krause, and W. L. Jungers. 1986. Analysis of enamel ultrastructure in archaeology: The identification of *Ovis aries* and *Capra hircus* dental remains. Journal of Archaeological Science 13: 579–95.

Guerrero, J. 1985. Geología de un sector al noreste del Municipio de Villavieja y descripción de un mamífero del Orden Litopterna. Tésis de Grado, Universidad Nacional, Bogotá.

Guerrero, J. 1990. Stratigraphy and sedimentary environments of the mammal-bearing middle Miocene Honda Group of Colombia. Master's thesis, Duke University, Durham, N.C.

Guerrero, J. 1993. Magnetostratigraphy of the upper part of the Honda Group and Neiva Formation. Miocene uplift of the Colombian Andes. Ph.D. diss., Duke University, Durham, N.C.

Gürich, G. 1912. *Gryposuchus jessei,* ein neues schmalschnauziges Krokodil aus den jüngeren Ablagerungen des oberen Amazonas-Gebietes. Jahrbuch der Hambergischen Wissenschaftlichen Austalten 29 (1911) (4, Beiheft: Mitteilungen aus den Mineralogisch-Geologischen Institut): 59–71.

Guthrie, R. D. 1984. Mosaics, alleochemics and nutrients. In P. S. Martin and R. G. Klein, eds., Quaternary Extinctions: A Prehistoric Revolution. Tucson: University of Arizona Press. Pp. 259–98.

Guy, P. R. 1989. The influence of elephants and fire on a *Brachystegia-Julbernardia* woodland in Zimbabwe. Journal of Tropical Ecology 5: 215–26.

Haffer, J. A. 1982. General aspects of the refuge theory. In G. T. Prance, ed., Biological Diversification in the Tropics. New York: Columbia University Press. Pp. 2–24.

Hall, L. 1984. And then there were bats. In M. Archer and G. Clayton, eds., Vertebrate Zoogeography and Evolution in Australasia. Victoria Park, W. Australia: Hesperian Press. Pp. 837–52.

Hammen, T. von der. 1979. History of flora, vegetation and climate in the Colombian Cordillera Oriental during the last five million years. In K. Larsen and L. B. der Holm Nielsen, eds., Tropical Botany. New York: Academic Press. Pp. 25–32.

Hammen, T. von der, J. H. Werner, and H. Van Dommelen. 1973. Palynological record of the upheaval of the northern Andes: A study of the Pliocene and lower Quaternary of the Colombian Eastern Cordillera and the early evolution of its high-Andean biota. Review of Palaeobotany and Palynology 16: 1–22.

Handley, C. O., Jr. 1976. Mammals of the Smithsonian Venezuelan project. Brigham Young University Science Bulletin, Biology 20: 1–91.

Hansen, R. M. 1978. Shasta ground sloth food habits, Rampart Cave, Arizona. Paleobiology 4: 302–19.

Haq, B. U. 1981. Paleogene paleooceanography: Early Cenozoic oceans revisited. Oceanologica Acta 1981: 71–82.

Harrington, H. J. 1962. Paleogeographic development of South America. Bulletin of the American Association of Petroleum Geologists 46: 1773–1814.

Hartenberger, J.-L., F. Mégard, and B. Sigé. 1984. Faunules à rongeurs de l'Oligocène inférieure à Lircay (Andes du Pérou central); datation d'un épisode karstique; interêt paléobiogéographique des remplissages tertiaires en Amérique du Sud. Comptes Rendus de l'Académie des Sciences, Paris, ser. 3, 299: 565–68.

Hartwig, W. C., A. L. Rosenberger, and T. Setoguchi. 1990. New fossil platyrrhines from the late Oligocene and middle Miocene. American Journal of Physical Anthropology 81: 237.

Haverschmidt, F. 1968. Birds of Surinam. Wynnewood, Pa.: Livingston Publishing.

Hayashida, A. 1984. Paleomagnetic study of the Miocene continental deposits in the La Venta badlands, Colombia. Kyōto University Overseas Research, Reports of New World Monkeys 4: 77–83.

Hecht, M. K., and T. C. LaDuke. 1988. Bolyerine vertebral variation: A problem for paleoherpetology. Acta Zoologica Cracovienska 31: 605–14.

Hecht, M. K., and B. Malone. 1972. On the early history of the gavialid crocodilians. Herpetologica 28: 281–84.

Hedberg, H. D. 1976. International Stratigraphic Guide. International Subcommission on Stratigraphic Classification of the IUGS Commission on Stratigraphy, New York: John Wiley.

Henao-Londoño, D., and R. W. Fields. 1949. Honda Formation of the upper Magdalena River basin, Colombia, South America. Bulletin of the Geological Society of America, [Abstracts] El Paso, November 10–12, 1949, 60: 18–94.

Hennig, W. 1966. Phylogenetic Systematics. Urbana: University of Illinois Press.

Hershkovitz, P. 1970. Notes on Tertiary platyrrhine monkeys and description of a new genus from the late Miocene of Colombia. Folia Primatologica 12: 1–37.

Hershkovitz, P. 1974. A new genus of late Oligocene monkey (Cebidae, Platyrrhini) with notes on postorbital closure and platyrrhine evolution. Folia Primatologica 21: 1–35.

Hershkovitz, P. 1977. Living New World Monkeys (Platyrrhini). Vol. 1, with an Introduction to Primates. Chicago: University of Chicago Press.

Hershkovitz, P. 1981. Comparative aanatomy of platyrrhine mandibular cheek teeth dpm4, pm4, m1, with particular reference to those of Homunculus (Cebidae), and comments on platyrrhine origins. Folia Primatologica 35: 179–217.

Hershkovitz, P. 1982a. The staggered marsupial lower third incisor (I3). Géobios 6: 191–200.

Hershkovitz, P. 1982b. Supposed squirrel monkey affinities of Dolichocebus gaimenensis. Nature 298: 202.

Hershkovitz, P. 1984a. More on Homunculus dpm4, and m1, and comparisons with Alouatta and Stirtonia (Primates, Platyrrhini, Cebidae). American Journal of Primatology 7: 261–83.

Hershkovitz, P. 1984b. Taxonomy of squirrel monkeys genus Saimiri: A preliminary report with description of a hitherto unnamed form. American Journal of Primatology 7: 155–210.

Hershkovitz, P. 1985. A preliminary taxonomic review of the South American bearded saki monkey genus Chiropotes (Cebidae, Platyrrhini), with the description of a new species. Fieldiana, Zoology, n. ser., vol. 27.

Hershkovitz, P. 1987. The taxonomy of South American sakis, genus Pithecia (Cebidae, Platyrrhini): A preliminary report and critical review with the description of a new species and a new subspecies. American Journal of Primatology 12: 387–468.

Hershkovitz, P. 1990. Titis, New World monkeys of the genus Callicebus (Cebidae, Platyrrhini): A preliminary taxonomic review. Fieldiana, Zoology, n. ser., 55: 1–109.

Hettner, A. 1892. Die Kordillere von Bogotá. Petermanns Geographische Mitteilungen, Ergänzungsheft, Heft. 104 (Bd. 22).

Hildebrand, M. 1960. How animals run. Scientific American 202: 148–57.

Hildebrand, M. 1974. Analysis of Vertebrate Structure. New York: John Wiley.

Hildebrand, M. 1982. Analysis of Vertebrate Structure. 2d ed. New York: John Wiley.

Hildebrand, M. 1985. Digging of quadrupeds. In M. Hildebrand, D. M. Bramble, K. F. Liem, and D. B. Wake, eds., Functional Vertebrate Morphology. Cambridge: Belknap Press. Pp. 89–109.

Hill, W. C. O. 1957. Primates. Comparative Anatomy and Taxonomy. Vol. III, Pithecoidea, Platyrrhini, Hapalidae. Edinburgh: University Press.

Hill, W. C. O. 1960. Primates. Comparative Anatomy and Taxonomy. Vol. IV, Cebidae. Pt. A. Edinburgh: University Press.

Hill, W. C. O. 1962. Primates. Comparative Anatomy and Taxonomy. Vol. V, Cebidae, Pt. B. Edinburgh: University Press.

Hirschfeld, S. E. 1971. Description of the ground sloths and anteaters from the Tertiary of Colombia, South America, and comparison with other Tertiary faunas. Ph.D. diss., University of California, Berkeley.

Hirschfeld, S. E. 1976. A new fossil anteater (Edentata, Mammalia) from Colombia, South America, and evolution of the Vermilingua. Journal of Paleontology 50: 419–432.

Hirschfeld, S. E. 1985. Ground sloths from the Friasian La Venta fauna, with additions to the pre-Friasian Coyaima fauna of Colombia, South America. University of California Publications in Geological Sciences 28: 1–91.

Hirschfeld, S. E., and L. G. Marshall. 1976. Revised faunal list of the La Venta fauna (Friasian-Miocene) of Colombia, South America. Journal of Paleontology 50: 433–36.

Hirschfeld, S. E., and S. D. Webb. 1968. Plio-Pleistocene megalonychid sloths of North America. Bulletin of the Florida State Museum 12: 213–96.

Hodell, D. A., and J. P. Kennett. 1985. Miocene paleoceanography of the South Atlantic Ocean at 22, 16, and 8 Ma. Geological Society of America, Memoir 163: 317–37.

Hoffstetter, R. 1954a. Les gravigrades cuirasses du Déséadien de Patagonie. Mammalia 18: 159–69.

Hoffstetter, R. 1954b. Phylogenie des édentes xenarthres. Bulletin Muséum National d'Histoire Naturelle 2: 433–38.

Hoffstetter, R. 1958. Xenarthra. In J. Piveteau, ed., Traité de paléontologie. Paris: Masson. Tome 6, 2: 535–636.

Hoffstetter, R. 1961a. Description d'un squelette de *Planops* (gravigrade du Miocène de Patagonie). Mammalia 25: 57–96.

Hoffstetter, R. 1961b. Remarques sur la phylogenie et la classification des édentes xenarthres (mammifères) actuels et fossiles. Bulletin Muséum National d'Histoire Naturelle 41: 92–103.

Hoffstetter, R. 1967a. Observations additionelles sur les serpents du Miocène de Colombie et rectification concernant la date d'arrivée des Colubridés en Amérique du Sud. Comptes Rendus Sommaire Séances Société Géologique de France (June 5, 1967): 209–10.

Hoffstetter, R. 1967b. Remarques sur les dates d'implantation des différents groupes de serpents terrestres en Amérique du Sud. Comptes Rendus Sommaire Séances Société Géologique de France (March 6, 1967): 93–94.

Hoffstetter, R. 1969. Un primate de l'Oligocène inferieur sud-americain: *Branisella boliviana,* gen. et sp. nov. Comptes Rendus de l'Académie des Sciences, Paris, sér. D, 269: 434–37.

Hoffstetter, R. 1970a. Les paléomammalogistes français en Amérique Latine. Bulletin Académie Société Lorraines Sciences 9: 233–43.

Hoffstetter, R. 1970b. Vertebrados Cenozóicos de Colombia. Actas IV Congreso Latinoamericano de Zoologia 2: 931–54.

Hoffstetter, R. 1971. Los vertebrados Cenozóicos de Colombia: Yacimientos, faunas, problemas planteados. Geología Colombiana, Universidad Nacional de Colombia, Bogotá 8: 37–62.

Hoffstetter, R. 1976. Investigaciones de paleomammalogía en los países andinos intertropicales. Primer Congreso Colombiano de Geología, Bogotá, 1969. Pp. 135–54.

Hoffstetter, R. 1977. Un gisement de mammifères miocènes à Quebrada Honda (Sud Bolivien). Comptes Rendus de l'Académie des Sciences, Paris, sér. D, 284: 1517–20.

Hoffstetter, R. 1986. High Andean mammalian faunas during the Plio-Pleistocene. In F. Vuilleumier and M. Monasterio, eds., High Altitude Tropical Biogeography. New York: Oxford University Press. Pp. 218–45.

Hoffstetter, R., and J. C. Rage. 1977. Le gisement de vertébrés miocènes de La Venta (Colombie) et sa faune de serpents. Annales Paléontologie, Vertébrés, Paris 63: 161–88.

Hoffstetter, R., and M. F. Soria. 1986. *Neodolodus colombianus* gen. et sp. nov., un nouveau condylarthe (Mammalia) dans le Miocène de Colombie. Comptes Rendus de l'Académie des Sciences, Paris, sér. 2, 303: 1619–22.

Hood, C. S., and J. Pitocchelli. 1983. *Noctilio albiventris.* Mammalian Species, no. 19.

Hoogmoed, M. S. 1979. The herpetofauna of the Guianan region. In W. E. Duellman, ed., The South American Herpetofauna. Lawrence: Museum of Natural History, University of Kansas. Pp. 241–79.

Hopping, C. A. 1967. Palynology and the oil industry. Review of Palaeobotany and Palynology 2: 23–48.

Hopson, J. A. 1979. Paleoneurology. In C. Gans, R. O. Northcutt, and P. Ulinski, eds., Biology of the Reptilia. Vol. 9, Neurology A. New York: Academic Press. Pp. 39–146.

Hopwood, T. 1928. *Gyrinodon quassus,* a new genus and species of toxodont from western Buchivacoa (Venezuela). Quarterly Journal of the Geological Society of London 84: 573–83.

Horn, C. 1990. Evolución de los ambientes sedimentarios durante el Terciario y el Cuaternario en la Amazonia Colombiana. Colombia Amazónica 4: 97–126.

Howe, M. W. 1969. Geologic studies of the Mesa Group (Pliocene?), upper Magdalena valley, Colombia. Ph.D. diss., Princeton University, Princeton, N.J.

Howe, M. W. 1974. Nonmarine Neiva Formation (Pliocene?), upper Magdalena valley, Colombia: Regional tectonism. Bulletin of the Geological Society of America 85: 1031–42.

Howell, A. B. 1944. Speed in Animals: Their Specialization for Running and Leaping. New York: Hafner Publishing.

Howell, D. J., and D. Burch. 1974. Food habits of some Costa Rican bats. Revista de Biología Tropical 21: 281–94.

Hubach, E. 1930. Apreciación de los llanos del Tolima y de sus tierras agrícolas, según puntos de vista geológicos. Boletín Ministerio de Minas y Petróleo, Bogotá 1930: 209–34.

Hudson, G. E., and P. J. Lanzillotti. 1955. Gross anatomy of the wing muscles in the family Corvidae. American Midland Naturalist 53: 1–44.

Humphrey, S. R., F. J. Bonaccorso, and T. L. Zinn. 1983. Guild structure of surface-gleaning bats in Panama. Ecology 64: 284–94.

INGEOMINAS. 1957a. Mapa geológico escala 1:200,000 de la Plancha K9 "ARMERO." Bogotá, INGEOMINAS.

INGEOMINAS. 1957b. Mapa geológico escala 1:200,000 de la Plancha M8 "ATACO." Bogotá, INGEOMINAS.

INGEOMINAS. 1959. Mapa geológico escala 1:200,000 de la Plancha N8 "NEIVA." Bogotá, INGEOMINAS.

INGEOMINAS. 1988. Mapa geológico de Colombia escala 1: 1' 500.000. Bogotá, INGEOMINAS.

International Subcommission on Stratigraphic Classification. 1976. International Stratigraphic Guide: A Guide to Stratigraphic Classification, Terminology, and Procedure. New York: John Wiley.

Iordansky, N. N. 1964. The jaw muscles of the crocodiles and some relating structures of the crocodilian skull. Anatomischer Anzeiger 115: 256–80.

Iordansky, N. N. 1973. The skull of the Crocodilia. In C. Gans and T. S. Parsons, eds., Biology of the Reptilia. Vol. 4, Morphology D. New York: Academic Press. Pp. 201–62.

James, F. A., and C. E. McCulloch. 1990. Multivariate analysis in ecology and systematics: Panacea or Pandora's box? Annual Review of Ecology and Systematics 21: 129–66.

Janis, C. M. 1982. Evolution of horns in ungulates: Ecology and paleoecology. Biological Review 57: 261–318.

Janis, C. M. 1988. An estimation of tooth volume and hypsodonty indices in ungulate mammals, and the correlation of these factors with dietary preference. Mémoires, Muséum National d'Histoire Naturelle, Paris, sér. C 53: 367–87.

Janis, C. M. 1990. Correlation of cranial and dental variables with body size in ungulates and macropodoids. In J. Damuth and B. J. MacFadden, eds., Body Size in Mammalian Paleobiology. Cambridge: Cambridge University Press. Pp. 255–99.

Janis, C. M., and D. Ehrhardt. 1988. Correlation of relative muzzle width and relative incisor width with dietary preference in ungulates. Zoological Journal of the Linnaean Society 92: 267–84.

Janis, C. M., and M. Fortelius. 1988. On the means whereby mammals achieve increased functional durability of their dentitions, with special reference to limiting factors. Biological Review 63: 197–230.

Janson, C. H., and L. H. Emmons. 1990. Ecological structure of the nonflying mammal community at Cocha Cashu Biological Station, Manú National Park, Peru. In A. H. Gentry, ed., Four Neotropical Rainforests. New Haven, Conn.: Yale University Press. Pp. 314–38.

Janzen, C. M. 1981. Digestive seed predation by a Costa Rican Baird's tapir. Biotropica 13: 59–63.

Janzen, D. H. 1983. *Tapirus bairdi* (Danto, Danta, Baird's tapir). In D. H. Janzen, ed., Costa Rican Natural History. Chicago: University of Chicago Press. Pp. 496–97.

Janzen, D. H., and D. E. Wilson. 1983. Mammals. In D. H. Janzen, ed., Costa Rican Natural History. Chicago: University of Chicago Press. Pp. 426–501.

Janzen, D. H., and P. S. Martin. 1982. Neotropical anachronisms: The fruits the gomphotheres ate. Science 215: 19–27.

Jarman, P. 1983. Mating system and sexual dimorphism in large, terrestrial, mammalian herbivores. Biological Review 58: 485–520.

Jenkins, F. A., Jr., and D. McClearn. 1984. Mechanisms of hind foot reversal in climbing mammals. Journal of Morphology 182: 197–219.

Jenkins, F. A., Jr., and D. W. Krause. 1983. Adaptations for climbing in North American multituberculates (Mammalia). Science 220: 712–15.

Jenkins, F. A., Jr., and S. M. Camazine. 1977. Hip structure and locomotion in ambulatory and cursorial carnivores. Journal of Zoology, London 182: 197–219.

Johnson, S. C. 1984. Astrapotheres from the Miocene of Colombia, South America. Ph.D. diss., University of California, Berkeley.

Johnsson, M. J., and R. H. Meade. 1990. Chemical weathering of fluvial sediments during alluvial storage: The Macuapanim Island point bar, Solimoes River, Brazil. Journal of Sedimentary Petrology 60: 827–42.

Jolly, C. J. 1972a. Changing views of hominid origins. Yearbook of Physical Anthropology 16: 1–17.

Jolly, C. J. 1972b. The classification and natural history of *Theropithecus* (*Simopithecus*) (Andrews, 1916): Baboons of the African Plio-Pleistocene. Bulletin of the British Museum of Natural History, Geology 22: 1–123.

Julivert, M. 1970. Cover and basement tectonics in the Cordillera Oriental of Colombia, South America, and a comparison with some other folded chains. Bulletin of the Geological Society of America 81: 3623–46.

Jungers, W. L. 1976. Hindlimb and pelvic adaptations to vertical climbing and clinging in *Megaladapis*, a giant subfossil prosimian from Madagascar. Yearbook of Physical Anthropology 20: 508–24.

Kälin, J. A. 1931. Ueber die Stellung der Gavialiden im System der Crocodilia. Revue Suisse de Zoologie 38: 379–88.

Kälin, J. S. 1933. Beitrage zur vergleichenden Osteologie des Crocodilidenschädels. Zoologische Jahrbüch 57 (Anat.): 535–714.

Kappelman, J. 1988. Morphology and locomotor adaptations of the bovid femur in relation to habitat. Journal of Morphology 198: 119–30.

Kawai, N. 1955. Comparative anatomy of the bands of Schreger. Okajimas Folia Anatomica Japonica 27: 115–31.

Kay, R. F. 1975. The functional adaptations of primate molar teeth. American Journal of Physical Anthropology 43: 195–216.

Kay, R. F. 1980. Platyrrhine origins: A reappraisal of the dental evidence. In R. L. Ciochon and A. B. Chiarelli, eds., Evolutionary Biology of the New World Monkeys and Continental Drift. New York: Plenum Press. Pp. 159–88.

Kay, R. F. 1981. The nut-crackers—a new theory of the adaptations of the Ramapithecinae. American Journal of Physical Anthropology 55: 141–52.

Kay, R. F. 1985. Dental evidence for the diet of *Australopithecus.* Annual Review of Anthropology 14: 315–41.

Kay, R. F. 1989. A new small platyrrhine from the Miocene of Colombia and the phyletic position of the callitrichines. American Journal of Physical Anthropology 78: 251.

Kay, R. F. 1990a. The phyletic relationships of extant and fossil Pitheciinae (Platyrrhini, Anthropoidea). Journal of Human Evolution 19: 175–208.

Kay, R. F. 1990b. A possible "giant" tamarin from the Miocene of Colombia. American Journal of Physical Anthropology 81: 248.

Kay, R. F. 1994. "Giant" tamarin from the Miocene of Colombia. American Journal of Physical Anthropology 96: 34–48.

Kay, R. F., and M. R. L. Anthony. 1991. Dietary evolution in platyrrhine primates and the comparative method. Journal of Vertebrate Paleontology 11: 39.

Kay, R. F., and H. H. Covert. 1984. Anatomy and the behaviour of extinct primates. In D. Chivers, B. A. Wood, and A. Bilsborough, eds., Food Acquisition and Processing in Primates. New York: Plenum Press. Pp. 467–508.

Kay, R. F., and C. D. Frailey. 1993. Fossil platyrrhines from the Río Acre fauna, late Miocene, western Amazonia. Journal of Human Evolution 25: 319–27.

Kay, R. F., R. H. Madden, J. M. Plavcan, R. L. Cifelli, and J. Guerrero Díaz. 1987. *Stirtonia victoriae,* a new species of Miocene Colombian primate. Journal of Human Evolution 16: 173–96.

Kay, R. F., and E. L. Simons. 1983. Dental formulae and dental eruption patterns in Parapithecidae (Primates, Anthropoidea). American Journal of Physical Anthropology 62: 363–75.

Keller, G., and J. A. Barron. 1987. Paleodepth distribution of Neogene deep-sea hiatuses. Paleoceanography 2: 697–713.

Kellner, A. W. P. 1987. Occurrência de um novo crocodiliano do Cretáceo inferior da Bacia do Araripe, nordeste do Brasil. Anais Academia Brasileira de Ciências 59: 219–32.

Kellogg, R. 1966. Fossil marine mammals from the Miocene Calvert Formation of Maryland and Virginia. Pt. 3, New species of extinct Miocene Sirenia. Bulletin of the United States National Museum 247: 65–98.

Kelly, F. 1989. The Paintings of Frederic Edwin Church. Washington, D.C.: National Gallery of Art and Smithsonian Institution Press.

Kennett, J. P., G. Keller, and M. S. Srinivasan. 1985. Miocene planktonic foraminiferal biogeography and paleoceanographic development of the Indo-Pacific region. Memoirs of the Geological Society of America 163: 197–236.

Kiltie, R. A. 1981a. Distribution of palm fruits on a rain forest floor: Why white-lipped peccaries forage near objects. Biotropica 13: 141–45.

Kiltie, R. A. 1981b. Stomach contents of peccaries. Biotropica 13: 234–36.

Kiltie, R. A. 1982. Bite force as a basis for niche differentiation between rain forest peccaries (*Tayassu tajacu* and *T. pecari*). Biotropica 14: 188–95.

Kimmel, P. G. 1975. Fishes of the Mio-Pliocene Deer Butte Formation, southeast Oregon. University of Michigan Museum of Paleontology, Papers in Paleontology 14: 69–87.

Kingdon, J. 1979. East African Mammals. Vol. 3(B), Large Mammals. New York: Academic Press.

Kinzey, W., and M. A. Norconk. 1990. Hardness as a basis of fruit choice in two sympatric primates. American Journal of Physical Anthropology 81: 5–15.

Kirschvink, J. L. 1980. The least-squares line and plane and the analysis of paleomagnetic data. Geophysical Journal of the Royal Astronomical Society 62: 699–718.

Kluge, A. G. 1991. Boine snake phylogeny and research cycles. Miscellaneous Publications, Museum of Zoology, University of Michigan 178: 1–58.

Koopman, K. F. 1976. Zoogeography. In R. J. Baker, J. K. Jones, Jr., and D. C. Carter, eds., Biology of Bats of the New World Family Phyllostomidae. Pt. 1. Lubbock: Special Publications, Museum, Texas Tech University 10: 39–47.

Koopman, K. F. 1982. Biogeography of the bats of South America. In M. A. Mares and H. H. Genoways, eds., Mammalian Biology in South America. Special Publication of the Pymatuning Laboratory of Ecology. Pittsburgh: University of Pittsburgh Press. Pp. 273–302.

Koopman, K. F., and E. E. Williams. 1951. Fossil Chiroptera collected by H. E. Anthony in Jamaica, 1919–1920. American Museum Novitates 1519: 1–29.

Kraglievich, L. 1928a. Apuntes para la geología y paleontología de la República Oriental del Uruguay. Revista, Sociedad Amigos de la Arqueología, Montevideo 2: 5–61.

Kraglievich, L. 1928b. Descripción de los astrágalos de los gravigrados terciarios de la subfamilia "Nothrotheriinae." Anales de la Sociedad Científica Argentina 105: 332–42.

Kraglievich, L. 1928c. Sobre el supuesto *Astrapotherium christi* Stehlin, descubierto en Venezuela (*Xenastrapotherium* gen. nov.) y sus relaciones con *Astrapotherium magnum* y *Uruguaytherium beaulieui.* Buenos Aires: Editorial Franco.

Kraglievich, L. 1930a. La Formación Friaseana del Río Frías, Río Fénix, Laguna Blanca, etc. y su fauna de Mamíferos. Physis 10: 127–61.

Kraglievich, L. 1930b. Hallazgo de un proterotérido en la República del Uruguay: *Proterotherium berroi* sp. nov. Revista, Sociedad Amigos de la Arqueología, Montevideo 4: 197–203.

Kraglievich, L. 1934. La antigüedad pliocena de las faunas de Monte Hermoso y Chapadmalal, deducidas de su comparación con las que le precedieron y sucedieron. Montevideo: Imprenta "El Siglo Ilustrado."

Kraglievich, L., and L. J. Parodi. 1931. *Theosodon pozzii,* sp. nov., el mayor Teosodonte Santacruceano. Physis 10: 326–27.

Kraus, M. J. 1980. Genesis of a fluvial sheet sandstone, Willwood Formation, northwest Wyoming. University of Michigan, Papers in Paleontology 24: 87–94.

Kraus, M. J. 1985. Sedimentology of early Tertiary rocks, northern Bighorn basin. In R. M. Flores, ed., Field Guide to Modern and Ancient Fluvial Systems in the United States. Third International Fluvial Sedimentology Conference, Guidebook. Pp. 26–39.

Krause, D. W., and S. J. Carlson. 1986. The enamel ultrastructure of multituberculate mammals: A review. Scanning Electron Microscopy 1986: 1591–1607.

Krohn, I. 1978. Functional adaptation in Miocene Glyptodontidae (Edentata, Mammalia) from the Honda Group, Colombia, South America. Ph.D. diss., University of California, Berkeley.

LaBrecque, J. L., D. V. Kent, and S. C. Cande. 1977. Revised magnetic polarity time scale for late Cretaceous and Cenozoic time. Geology 5: 330–35.

LaDuke, T. C. 1991. The fossil snakes of Pit 91, Rancho La Brea, California. Contributions in Science, Los Angeles County Museum 424: 1–28.

Landry, S. O. 1990. The systematic position of the genus *Chaetomys.* Abstracts, Joint Meeting of the Argentine and American Mammal Societies (ASM-SAREM), Buenos Aires. P. 52.

Langdon, J. H. 1984. A comparative functional study of the Miocene hominoid foot remains. Ph.D. diss., Yale University, New Haven, Conn.

Langston, W., Jr. 1956. The Sebecosuchia: Cosmopolitan crocodilians? American Journal of Science 254: 605–14.

Langston, W., Jr. 1965. Fossil crocodilians from Colombia and the Cenozoic history of the Crocodilia in South America. University of California Publications in Geological Sciences 52: 1–157.

Langston, W., Jr. 1975. Ziphodont crocodiles: *Pristichampsus vorax* (Troxell) new combination, from the Eocene of North America. Fieldiana, Geology 33: 291–314.

LaVal, R. K., and H. S. Fitch. 1977. Structure, movements and reproduction in three Costa Rican bat communities. Occasional Papers, Museum of Natural History, University of Kansas 69: 1–28.

Laws, R. M. 1970. Elephants as agents of habitat and landscape change in East Africa. Oikos 21: 1–15.

Leeder, M. R. 1973. Sedimentology and paleontology of the upper Old Red Sandstone in the Scottish Border basin. Scottish Journal of Geology 9: 117–44.

Legendre, S. 1984a. Essai de biogéographie phylogénique des molossides (Chiroptera). Myotis 21–22(1983–84): 30–36.

Legendre, S. 1984b. Étude odontologique des représentants actuels du groupe *Tadarida* (Chiroptera, Molossidae): Implications phylogéniques, systématiques et zoogéographiques. Revue Suisse de Zoologie 91: 399–442.

Legendre, S. 1984c. Identification de deux sous-genres fossiles et compréhension phylogénique du genre *Mormopterus* (Molossidae, Chiroptera). Comptes Rendus de l'Académie des Sciences, Paris, sér. 2, 298: 715–20.

Legendre, S. 1985. Molossidés (Mammalia, Chiroptera) cénozoiques de l'Ancien et du Nouveau Monde: Statut systématique; intégration phylogénique de données. Neues Jahrbuch für Geologie und Paläontologie, Abhandlungen 170: 205–27.

Legendre, S. 1986. Analysis of mammalian communities from the late Eocene and Oligocene of southern France. Palaeovertebrata 16: 191–212.

Legendre, S. 1989. Les communautés de mammifères du Paléogène (Eocène supérieur et Oligocène) d'Europe occidentale: Structures, milieux et évolution. Münchner Geowissenschaftliche Abhandlungen, Reihe A, Geologie und Paläontologie 16:1–110.

Legendre, S., and B. Sigé. 1982. La place de "Vespertilion de Montmartre" dans l'histoire des chiroptères molossides. In E. Buffetaut, J. M. Mazin, and E. Salmon, eds., Actes du Symposium Paléontologique G. Cuvier (Montbeliard). Pp. 347–61.

Legendre, S., T. H. V. Rich, P. V. Rich, G. J. Knox, and P. Punyaprasiddhi, et al. 1988. Miocene fossil vertebrates from the Nong Hen-I(A) exploration well of Thai Shell Exploration and Production Company Limited, Phitsanulok Basin, Thailand. Journal of Vertebrate Paleontology 8: 278–89.

Leppard, H. M. 1961. South America. Goode Base Map Series, Department of Geography. Chicago: University of Chicago Press.

Leutenegger, W. 1980. Monogamy in callitrichids: A consequence of phyletic dwarfism? International Journal of Primatology 1: 95–98.

Lewis, O. J. 1962. The comparative morphology of m. flexor accessorius and the associated long flexor tendons. Journal of Anatomy 96: 321–33.

Linares, O. J. 1986. Murciélagos de Venezuela. Caracas: Cuadernos Lagoven.

Linares, O. J. 1988. José Royo y Gómez. Diccionario de Historia de Venezuela. Caracas: Funcación Polar, Editorial Ex Libris. P. 485.

Linares, O. J. 1990. Mamíferos del Mioceno medio-tardío de Urumaco, Venezuela: Implicaciones paleobiogeográficas. Abstracts, Joint Meeting of the Argentine and American Mammal Societies (ASM-SAREM), Buenos Aires. P. 42.

Lindsay, E. H., V. Fahlbusch, and P. Mein. 1989. European Neogene Mammal Chronology. New York: Plenum Press.

Linnaeus, C. 1766. Systema naturae per regna tria naturae, secundum classis, ordines, genera, species, cum characteribus, differentiis synonymis, locis. Vol. 1, Regnum animale, pt. 1. Stockholm: Laurentii Salvii.

List, J. C. 1966. Comparative osteology of the snake families Typhlopidae and Leptotyphlopidae. Illinois Biological Monographs 36: 1–112.

Loomis, F. B. 1914. The Deseado Formation of Patagonia. Concord, N.H.: Rumford Press.

López, H. 1989. Contribucíon a los Lasallistas a las Ciencias Naturales en Colombia. Bogotá: Fondo FEN Colombia.

Lowrie, W., and W. Alvarez. 1981. One hundred million years of geomagnetic polarity history. Geology 9: 392–97.

Luchterhand, K., R. F. Kay, and R. H. Madden. 1986. *Mohanamico hershkovitzi*, gen. et sp. nov., un primate du Miocène moyen d'Amérique du Sud. Comptes Rendus de l'Académie des Sciences, Paris 303: 1753–58.

Lund, P. W. 1838. Coup-d'oeil sur les espèces éteints de mammifères du Brésil. Annales des Sciences Naturelles, Paris, sér. 1, 11: 214–34.

Lundberg, J. G. 1993. African-South American freshwater fish clades and continental drift: Problems with a paradigm. In P. Goldblatt, ed., Biological Relationships between Africa and South America. New Haven, Conn.: Yale University Press. Pp. 156–99.

Lundberg, J. F., and B. Chernoff. 1992. A fossil of the Amazon fish *Arapaima gigas* (Teleostei, Osteoglossidae) from the Miocene La Venta fauna of Colombia, South America. Biotropica 24: 2–14.

Lundberg, J. G., O. Linares, P. Nass, and M. E. Antonio. 1988. *Phractocephalus hemiliopterus* (Pimelodidae, Siluriformes) from the late Miocene Urumaco Formation, Venezuela: A further case of evolutionary stasis and local extinction among South American fishes. Journal of Vertebrate Paleontology 8: 131–38.

Lundberg, J. F., A. Machado-Allison, and R. F. Kay. 1986. Miocene characid fishes from Colombia: Evolutionary stasis and extirpation. Science 234: 208–9.

Lundberg, J. G., F. Mago-Leccia, and P. Nass. 1991. *Exallodontus aguanai*, a new genus and species of Pimelodidae (Teleostei, Siluriformes) from deep river channels of South America and delimitation of the sub-family Pimelodinae. Proceedings of the Biological Society Washington 104(4): 840–69.

Lynch, J. D. 1979. The amphibians of the lowland tropical forest. In W. E. Duellman, ed., The South American Herpeto-

fauna: Its Origin, Evolution, and Dispersal. Lawrence: Museum of Natural History, University of Kansas. Pp. 189–217.

Maas, M. C. 1986. Function and variation of enamel prism decussation in ceboid primates. American Journal of Physical Anthropology 69: 233–34.

Maas, M. C. 1989. Enamel microstructure in Miocene notoungulates: Diversity and function. Journal of Vertebrate Paleontology 9: 31A.

Maas, M. C. 1991. Enamel microstructure and microwear: An experimental study of the response of enamel to shearing force. American Journal of Physical Anthropology 85: 31–49.

Maas, M. C., D. W. Krause, and S. G. Strait. 1988. The decline and extinction of Plesiadapiformes (Mammalia: ?Primates) in North America: Displacement or replacement? Paleobiology 14: 410–31.

MacDonald, W. D. 1980. Anomalous paleomagnetic directions in late Tertiary andesitic intrusions of the Cauca depression, Colombian Andes. Tectonophysics 68: 339–48.

MacFadden, B. J. 1990. Chronology of Cenozoic primate localities in South America. Journal of Human Evolution 19: 7–21.

MacFadden, B. J., and F. Anaya. 1993. Geochronology and paleontology of Miocene mammals from the Bolivian Andes. Resumenes X Jornadas Argentinas de Paleontología de Vertebrados (Buenos Aires). Ameghiana 30(3): 350–51.

MacFadden, B. J., F. Anaya, H. Perez, C. W. Naeser, and P. K. Zeitler. 1990. Late Cenozoic paleomagnetism and chronology of Andean basins of Bolivia: Evidence for possible oroclinal bending. Journal of Geology 98: 541–55.

MacFadden, B. J., and R. G. Wolff. 1981. Geological investigations of late Cenozoic vertebrate-bearing deposits in southern Bolivia. Anais II Congresso Latino-Americano da Paleontología (Pôrto Alegre) 2: 765–78.

Machado-Allison, A. 1983. Estudios sobre la subfamilia Serrasalminae (Teleostei-Characidae). Pt. 1, Discusión sobre la condición monofilética de la subfamilia. Acta Biologica Venezolana 11: 145–96.

MacPhee, R. D. E., and C. A. Woods. 1982. A new fossil cebine from Hispaniola. American Journal of Physical Anthropology 58: 419–36.

MacPhee, R. D. E., and J. G. Fleagle. 1991. Postcranial remains of *Xenothrix macgregori* (Primates, Xenotrichidae) and other late Quaternary mammals from Long Mile Cave, Jamaica. Bulletin of the American Museum of Natural History 206: 287–321.

Madden, R. H. 1990. Miocene Toxodontidae (Notoungulata, Mammalia) from Colombia, Ecuador, and Chile. Ph.D. diss., Duke University, Durham, N.C.

Madden, R. H., and L. Albuja. 1989. Estado actual de *Ateles fusciceps fusciceps* en el Noroccidente Ecuatoriano. Politécnica, Biologia 2, Revista Informativa Técnico-Cientifico, Quito 14(3): 113–57.

Madden, R. H., R. F. Kay, and J. Guerrero. 1989. New *Stirtonia victoriae* material from the Miocene of Colombia. American Journal of Physical Anthropology 78: 265.

Madden, R. H., R. F. Kay, J. G. Lundberg, and G. Scillato-Yané. 1989. Vertebrate paleontology, stratigraphy, and biochronology of the Miocene of southern Ecuador. Journal of Vertebrate Paleontology 9: 31A.

Madden, R. H., R. F. Kay, G. Vucetich, C. C. Swisher III, and M. Franchi. 1991. The Friasian of Patagonia. Journal of Vertebrate Paleontology 11: 135.

Maddison, W. P., and D. R. Maddison. 1992. MacClade: Analysis of Phylogeny and Character Evolution. Sunderland, Mass.: Sinauer Associates.

Malabarba, M. C. d. S. L. 1988. Loricariid dermal plate and pectoral spines from the Tertiary of São Paulo, Brazil (Osteichthys, Siluriformes). Comunicaçoes do Museo da Ciencias, PUCRS, Séries Zoologia, Pôrto Alegre 1: 5–12.

Malcolm, J. R. 1990. Estimation of mammalian densities in continuous forest north of Manaus. In A. H. Gentry, ed., Four Neotropical Rainforests. New Haven, Conn.: Yale University Press. Pp. 339–57.

Manu, H., and B. Sigé. 1971. Nyctalodontie et myotodontie, importants caractères de grades évolutifs chez les chiroptères entomophages. Comptes Rendus des Séances de l'Académie des Sciences 272: 1735–38.

Mares, M. A., and R. A. Ojeda. 1981. Patterns of diversity and adaptation in South American hystricognath rodents. In M. A. Mares and H. H. Genoways, eds., Mammalian Biology in South America. Special Publication Series, Pymatuning Laboratory of Ecology. Pittsburgh, Pa.: University of Pittsburgh Press. 6: 393–432.

Mares, M. A., and R. A. Ojeda. 1989. Guide to the Mammals of Salta Province, Argentina. Norman: University of Oklahoma Press.

Mares, M. A., and M. R. Willig. 1994. Inferring biome associations of Recent mammals from samples of temperate and tropical faunas: Paleoecological considerations. Historical Biology 8: 31–48.

Mares, M. A., J. K. Braun, and D. Gettinger. 1989. Observations of the distribution and ecology of the mammals of the Cerrado grasslands of central Brazil. Annals of the Carnegie Museum 58: 1–60.

Mares, M. A., M. R. Willig, K. E. Streilein, and T. E. Lacher, Jr. 1981. The mammals of northeastern Brazil: A preliminary assessment. Annals of the Carnegie Museum 50(4): 81–137.

Marshall, C. R. 1990. Confidence intervals on stratigraphic ranges. Paleobiology 16(1): 1–10.

Marshall, L. G. 1976a. New didelphine marsupials from the La Venta fauna (Miocene) of Colombia, South America. Journal of Paleontology 50: 402–18.

Marshall, L. G. 1976b. Revision of the South American marsupial subfamily Abderitinae (Mammalia, Caenolestidae).

Museo Municipal de Ciencias Naturales de Mar del Plata "Lorenzo Scaglia" 2: 57–90.

Marshall, L. G. 1977. A new species of *Lycopsis* (Borhyaenidae; Marsupialia) from the La Venta fauna (late Miocene) of Colombia, South America. Journal of Paleontology 51: 633–42.

Marshall, L. G. 1978. Evolution of the Borhyaenidae, extinct South American predaceous marsupials. University of California, Publications in Geological Sciences 117: 1–89.

Marshall, L. G. 1979a. A model for the paleobiogeography of South American cricetine rodents. Paleobiology 5: 126–32.

Marshall, L. G. 1979b. Review of the Prothylacyninae, an extinct subfamily of South American "dog-like" marsupials. Fieldiana, Geology 3: 1–50.

Marshall, L. G. 1980. Systematics of the South American marsupial family Caenolestidae. Fieldiana, Geology, n. ser. 5: 1–145.

Marshall, L. G. 1982a. Systematics of the South American marsupial family Microbiotheriidae. Fieldiana 10: 1–75.

Marshall, L. G. 1982b. Evolution of South American Marsupialia. In M. Mares and G. G. Genoways, eds., Mammalian Biology in South America. Special Publication Series, Pymatuning Laboratory of Ecology. Pittsburgh, Pa.: University of Pittsburgh Press. 6: 251–72.

Marshall, L. G. 1985. Geochronology and land-mammal biochronology of the transamerican faunal interchange. In F. G. Stehli and S. D. Webb, eds., The Great American Biotic Interchange. New York: Plenum Press. Pp. 49–85.

Marshall, L. G. 1987. Systematics of Itaboraian (Middle Paleocene) age "possum-like" marsupials from the limestone quarry at São José de Itaboraí, Brazil. In M. Archer, ed., Possums and Opossums: Studies in Evolution. Sydney: Surrey Beatty and Sons and the Royal Zoological Society of New South Wales. Pp. 91–160.

Marshall, L. G. 1990. Fossil Marsupialia from the type Friasian Land Mammal Age (Miocene), Alto Río Cisnes, Aisén, Chile. Revista Geológica de Chile 17: 19–55.

Marshall, L. G., and R. L. Cifelli. 1990. Analysis of changing diversity patterns in Cenozoic land mammal age faunas, South America. Palaeovertebrata 19: 169–210.

Marshall, L. G., and R. Pascual. 1978. Una escala radiométrica preliminar de las edades-mamífero del Cenozóico medio y tardío Sudamericano. Obra del Centenario del Museo de La Plata 5: 11–28.

Marshall, L. G., and P. Salinas. 1990. Stratigraphy of the Río Frías Formation (Miocene), along the Alto Río Cisnes, Aisén, Chile. Revista Geológica de Chile 17: 57–87.

Marshall, L. G., A. Berta, R. Hoffstetter, R. Pascual, and O. A. Reig. 1984. Mammals and stratigraphy: Geochronology of the continental mammal-bearing Quaternary of South America. Palaeovertebrata, Mémoire Extraordinaire 1984: 1–76.

Marshall, L. G., J. Case, and M. O. Woodburne. 1989. Phylogenetic relationships of the families of marsupials. Current Mammalogy 2: 433–502.

Marshall, L. G., R. L. Cifelli, R. E. Drake, and G. H. Curtis. 1986. Vertebrate paleontology, geology, and geochronology of the Tapera de Lopez and Scarritt Pocket Chubut Province, Argentina. Journal of Paleontology 60: 920–51.

Marshall, L. G., R. E. Drake, G. H. Curtis, R. F. Butler, K. M. Flanagan, and C. W. Naeser. 1986. Geochronology of type Santacrucian (middle Tertiary) Land Mammal Age, Patagonia, Argentina. Journal of Geology 94: 449–57.

Marshall, L. G., R. Hoffstetter, and R. Pascual. 1983. Mammals and stratigraphy: Geochronology of the continental mammal-bearing Tertiary of South America. Palaeovertebrata, Mémoire Extraordinaire 1983: 1–93.

Marshall, L. G., R. Pascual, G. H. Curtis, and R. E. Drake. 1977. South American geochronology: Radiometric time scale for middle to late Tertiary mammal-bearing horizons in Patagonia. Science 195: 1325–28.

Martin, B. G. H., and A. d'A. Bellairs. 1977. The narial excrescence and pterygoid bulla of the gharial, *Gavialis gengeticus* (Crocodilia). Journal of Zoology (London) 182: 541–58.

Martin, L. B., A. Boyde, and F. E. Grine. 1988. Enamel structure in primates: A review of scanning electron microscope studies. Scanning Microscopy 2: 1503–26.

Martin, L. D., and R. M. Mengel. 1975. A new species of anhinga (Anhingidae) from the upper Pliocene of Nebraska. Auk 92: 137–40.

Matthew, W. D., and C. de Paula Couto. 1959. The Cuban edentates. Bulletin of the American Museum of Natural History 117: 1–56.

Maynard Smith, J., and R. J. G. Savage. 1959. Some locomotory adaptations in mammals. Journal of the Linnean Society 42: 603–22.

Mayo, N. 1980. Nueva especie de *Neocnus* (Edentata: Megalonychidae de Cuba) y consideraciones sobre la evolución, edad, y paleoecología de las especies de este genero. Actas II Congreso Argentino de Paleontología y Bioestratigrafía y I Congreso Latinoamericano de Paleontología (Buenos Aires, 1978) 3: 223–36.

Mazzoni, M. M., and A. Benvenuto. 1990. Radiometric ages of Tertiary ignimbrites and the Collón Curá Formation, northwestern Patagonia. Actas del Décimo Primer Congreso Geológico Argentino (San Juan, 1990) 1:87–90.

Mazzotti, F., and W. A. Dunson. 1989. Osmoregulation in crocodilians. American Zoologist 29: 903–20.

McDonald, H. G. 1977. Description of the osteology of the extinct gravigrade edentate *Megalonyx* with observations on its ontogeny, phylogeny, and functional anatomy. Master's thesis, University of Florida, Gainesville.

McDougall, I., L. Kristjansson, and K. Saemundsson. 1984. Magnetostratigraphy and geochronology of northwest Iceland. Journal of Geophysical Research 89: 7029–60.

McDowell, S. B. 1987. Systematics. In R. A. Siegal, J. T. Collins, and S. S. Novak, eds., Snakes: Ecology and Evolutionary Biology. New York: Macmillan. Pp. 3–50.

McHenry, H. M. 1973. Early hominid humerus from East Rudolf, Kenya. Science 180: 739–41.

McKenna, M. C. 1956. Survival of primitive notoungulates and condylarths into the Miocene of Colombia. American Journal of Science 254: 736–43.

McNab, B. K. 1971. The structure of tropical bat faunas. Ecology 52: 352–58.

McRae, L. E. 1990a. Paleomagnetic isochrons, unsteadiness, and non-uniformity of sedimentation in Miocene fluvial strata of the Siwalik Group, northern Pakistan. Journal of Geology 98: 433–56.

McRae, L. E. 1990b. Paleomagnetic isochrons, unsteadiness, and uniformity of sedimentation in Miocene intermontane basin sediments at Salla, Eastern Andean Cordillera, Bolivia. Journal of Geology 98: 479–500.

Medem, F. 1981. Los Crocodylia de Sur America. Vol. 1, Los Crocodylia de Colombia. Ministerio de Educatión Nacional, Colciencias, Bogotá, Colombia: Editorial Carrera Limitada.

Medem, F. 1983. Los Crocodylia de Sur America. Vol. 2, Venezuela, Trinidad, Tobago, Guyana, Suriname, Guayana Francesa, Ecuador, Perú, Bolivia, Brasis, Paraguay, Argentina, Uruguay. Bogotá, Colombia: Instituto de Ciencias Naturales, Museo de Historia Natural, Universidad Nacional de Colombia.

Medem, F., O. V. Castaño, and M. Lugo. 1979. Contribución al conocimiento sobre la reproducción y el crecimiento de los "morrocoyes" (*Geochelone carbonaria* y *G. denticulata;* Testudines, Testudinidae). Caldasia 12: 497–511.

Medina, C. J. 1976. Crocodilians from the late Tertiary of northwestern Venezuela: *Melanosuchus fisheri* sp. nov. Breviora, Museum of Comparative Zoology 438: 1–14.

Megard, F. 1989. The evolution of the Pacific Ocean margin in South America north of the Arica Elbow (18°S). In Z. Ben-Avraham, ed., The Evolution of the Pacific Ocean Margins. Oxford: Oxford University Press. Pp. 208–30.

Meldrum, D. J. 1990. New fossil platyrrhine tali from the early Miocene of Argentina. American Journal of Physical Anthropology 83: 403–18.

Meldrum, D. J. 1991. Kinematics of the cercopithecine foot on arboreal and terrestrial substrates with implications for the interpretation of hominid terrestrial adaptations. American Journal of Physical Anthropology 84: 273–89.

Meldrum, D. J. 1993. Postcranial adaptations and positional behavior in fossil platyrrhines. In D. Gebo, ed., Postcranial Adaptations in Nonhuman Primates. DeKalb: Northern Illinois University Press. Pp. 235–51.

Meldrum, D. J., and J. G. Fleagle. 1988. Morphological affinities of the postcranial skeleton of *Cebupithecia sarmientoi*. American Journal of Physical Anthropology 75: 249.

Meldrum, D. J., J. G. Fleagle, and R. F. Kay. 1990a. A new partial skeleton of *Cebupithecia sarmientoi* from the Miocene of Colombia. American Journal of Physical Anthropology 81: 267.

Meldrum, D. J., J. G. Fleagle, and R. F. Kay. 1990b. Partial humeri of two Miocene Colombian primates. American Journal of Physical Anthropology 81: 413–22.

Meldrum, D. J., and R. F. Kay. 1992. A new specimen of pitheciine primate from the Miocene of Colombia. American Journal of Physical Anthropology, suppl. 14: 121.

Meldrum, D. J., and P. Lemelin. 1991. Axial skeleton of *Cebupithecia sarmientoi* (Pitheciinae, Platyrrhini) from the middle Miocene of La Venta, Colombia. Amerian Journal of Primatology 25: 69–90.

Mendel, F. C. 1979. The wrist joint of two-toed sloths and its relevance to brachiating adaptations in the Hominoidea. Journal of Morphology 162: 413–24.

Mendel, F. C. 1981a. Foot of two-toed sloths: Its anatomy and potential uses relative to size of support. Journal of Morphology 170: 357–72.

Mendel, F. C. 1981b. Use of hands and feet of two-toed sloths (*Choloepus hoffmanni*) during climbing and terrestrial locomotion. Journal of Morphology 62: 413–21.

Mendel, F. C. 1985a. Adaptations for suspensory behavior in the limbs of two-toed sloths. In G. G. Montgomery, ed., The Evolution and Ecology of Armadillos, Sloths, and Vermilinguas. Washington, D.C.: Smithsonian Institution Press. Pp. 151–62.

Mendel, F. C. 1985b. Use of hands and feet of three-toed sloths (*Bradypus variegatus*) during climbing and terrestrial locomotion. Journal of Morphology 66: 359–66.

Menendez, C. A. 1971. Floras terciarias de la Argentina. Ameghiniana 8: 357–68.

Menu, H., and B. Sigé. 1971. Nyctalodontie et myotodontie, importants caractères de grades évolutifs chez les chiroptères entomophages. Comptes Rendus des Séances de l'Académie des Sciences, Paris 272: 1735–38.

Merritt, D. 1984. The pacarana, *Dinomys branickii*. In O. L. Ryder and M. L. Byrd, eds., One Medicine. Berlin: Springer-Verlag. Pp. 154–61.

Miall, L. C. 1878. The skull of the crocodile: A manual for students. Studies in Comparative Anatomy, no. 1. London: Macmillan. Pp. 1–50.

Miller, A. H. 1953. A fossil hoatzin from the Miocene of Colombia. Auk 70: 484–89.

Miller, G. S. 1906. Twelve new genera of bats. Proceedings of the Biological Society of Washington 19: 83–86.

Miller, G. S. 1907. The families and genera of bats. Bulletin of the United States National Museum 57: 1–282.

Miller, G. S. 1924. A second instance of the development of rodent-like incisors in an artiodactyl. Proceedings of the United States National Museum 66: 1–3.

Miller, R. A. 1935. Functional adaptations in the forelimb of the sloths. Journal of Mammalogy 16: 38–51.

Miller, R. A. 1973. Functional and morphological adaptations in the forelimbs of the slow lemurs. American Journal of Anatomy 42: 153–81.

Mishler, B. D. 1994. The cladistic analysis of molecular and morphological data. American Journal of Physical Anthropology 94: 143–156.

Mittermeier, R. A., and M. G. M. Van Roosmalen. 1981. Preliminary observations on habitat utilization and diet in eight Surinam monkeys. Folia Primatologica 36: 1–39.

Mittermeier, R. A., and R. A. Wilson. 1974. Redescription of *Podocnemis erythrocephala* (Spix, 1824), an Amazonian pelomedusid turtle. Papais Avulsos Zoologia, Rio de Janeiro 28: 147–62.

Mones, A. 1979. Terciario del Uruguay, síntesis geopaleontológica. Revista, Facultad de Humanidades y Ciencias, Série Ciencias de la Tierra 1: 1–27.

Mones, A. 1988. Toxodontia and the principle of authority (Mammalia, Notoungulata). Anais X Congresso Brasileiro de Paleontologia (Rio de Janeiro, 1987). Pp. 211–14.

Mones, A. 1989. Nomen dubium vs. nomen vanum. Journal of Vertebrate Paleontology 9: 232–34.

Mones, A., and M. Ubilla. 1978. La edad Deseadense (Oligoceno inferior) de la Formación Fray Bentos y su contenido paleontológico, con especial referencia a la presencia de *Proborhyaena* cf. *gigantea* Ameghino (Marsupialia; Borhyaenidae) en el Uruguay, nota preliminar. Comunicaciones Paleontológicas, Museo de História Natural de Montevideo 1: 151–58.

Moody, R. T. J. 1976. The discovery of a large pelomedusid turtle from the phosphates of Morocco. Tertiary Research 1: 53–58.

Mook, C. C. 1921. The dermo-supraoccipital bone in the Crocodilia. Bulletin of the American Museum of Natural History 44: 101–3.

Mook, C. C. 1941. A new fossil crocodile from Colombia. Proceedings of the United States National Museum 91: 55–58.

Moore, D. M. 1978. Post-glacial vegetation in the South Patagonian territory of the giant ground sloth, *Mylodon*. Botanical Journal of the Linnean Society 77: 177–202.

Morales, L. G., D. J. Podesta, W. C. Hatfield, H. Tanner, S. H. Jones, M. H. S. Barker, D. J. O'Donoghue, C. E. Mohler, E. P. Dubois, A. D. Graves, C. Jacobs, and C. R. Gross. 1958. General geology and oil occurrences of middle Magdalena valley, Colombia. American Association of Petroleum Geologists, Habitat of Oil Symposium. Pp. 641–95.

Morgan, G. S. 1989. The historical zoogeography of Neotropical bats: A paleontological perspective. Fifth International

Theriological Congress (Rome), Abstract of Papers and Posters 1: 275–76.

Morgan, G. S. 1991. Neotropical Chiroptera from the Pliocene and Pleistocene of Florida. Bulletin of the American Museum of Natural History 206: 176–213.

Morgan, G. S., O. J. Linares, and C. E. Ray. 1988. New species of fossil vampire bats (Mammalia; Chiroptera; Desmodontidae) from Florida and Venezuela. Proceedings of the Biological Society of Washington 101: 912–28.

Morton, S. R. 1980. Ecological correlates of caudal fat storage in small mammals. Australian Mammalogy 3: 81–86.

Moskovits, D. K. 1988. Sexual dimorphism and population estimates of the two Amazonian tortoises (Geochelone carbonaria and G. denticulata) in northwestern Brazil. Herpetologica 44: 209–17.

Moskovits, D. K., and A. R. Kiester. 1987. Activity levels and ranging behavior of the two Amazonian tortoises, Geochelone carbonaria and Geochelone denticulata, in northwestern Brazil. Functional Ecology 1: 203–14.

Moskovits, D. K., and K. A. Bjorndahl. 1990. Diet and food preferences of the tortoises Geochelone carbonaria and G. denticulata in northwestern Brazil. Herpetologica 46: 207–18.

Mourer-Chauviré, C. 1980. The Archaeotrogonidae of the Eocene and Oligocene Phosphorites du Quercy (France). Contributions in Science, Natural History Museum of Los Angeles County 330: 17–31.

Mourer-Chauviré, C. 1982. Les oiseaux fossiles de Phosphorites du Quercy (Eocène supérieur à Oligocène supérieur): Implications paléobiogéographiques. Géobios, Mémoire Spécial 6: 413–26.

Muizon, C. de, and D. P. Domning. 1985. The first records of fossil sirenians in the southeastern Pacific Ocean. Bulletin, Muséum National d'Histoire Naturelle (Paris), sér. 4, vol. 7, sec. C, no. 3: 189–213.

Müller, L. 1927. Ergebnisse der Forschungsreisen Prof. E. Stromers in den Wüsten Agyptens, 5. Tertiäre Wirbeltiere 1. Beiträge zur Kenntnis der Krokodolier des Agyptischen Tertiärs. Abhandlungen der Bayerischen Akademie der Wissenschaften Methemetisch-naturwisenschaftliche 21: 1–96.

Muñoz, J. 1990. Diversidad y hábitos alimenticios de murciélagos en transectos altitudinales a través de la Cordillera Central de los Andes en Colombia. Studies on Neotropical Fauna and Environment 25: 1–17.

Napier, J. R., and P. H. Napier. 1967. A Handbook of Living Primates. New York: Academic Press.

Naples, V. L. 1982. Cranial osteology and function in the tree sloths, Bradypus and Choloepus. American Museum Novitates 2739: 1–41.

Naples, V. L. 1990. Morphological changes in the facial region and a model of dental growth and wear pattern development in Nothrotheriops shastensis. Journal of Vertebrate Paleontology 10: 372–89.

Nass, P. 1991. Anatomía comparada del bagre cunagaro Brachyplatystoma juruense (Boulenger, 1898), incluyendo un analisis filogenético de la familia Pimelodidae. Universidad Central de Venezuela.

Nicéforo María, H. 1952. Testudineos del suborden Pleurodira en el Museo de La Salle. Boletín Instituto de La Salle, Bogotá 39: 14–21.

North American Commission on Stratigraphic Nomenclature. 1983. North American Stratigraphic Code. American Association of Petroleum Geologists Bulletin 67: 841–75.

Norell, M. A. 1989. The higher level relationships of the extant Crocodylia. Journal of Herpetology 23: 325–35.

Norell, M. A., and J. M. Clark. 1990. A reanalysis of Bernissartia fagesii, with comments on its phylogenetic position and its bearing on the origin and diagnosis of the Eusuchia. Bulletin de l'Institut Royal des Sciences Naturelles de Belgique, Sciences de la Terre 60: 115–28.

Nowak, R. M. 1991. Walker's Mammals of the World. 5th ed. Baltimore, Md.: Johns Hopkins University Press.

Nowak, R. M., and J. L. Paradiso. 1975. Walker's Mammals of the World. Baltimore, Md.: Johns Hopkins University Press.

Nowak, R. M., and J. L. Paradiso. 1983. Walker's Mammals of the World. 4th ed. Baltimore, Md.: Johns Hopkins University Press.

Nussbaum, R. A., and M. Wilkinson. 1989. On the classification and phylogeny of caecilians (Amphibia: Gymnophiona): A critical review. Herpetological Monographs 3: 1–42.

Nutall, C. P. 1990. A review of the Tertiary non-marine molluscan faunas of the Pebasian and other inland basins of northwestern South America. Bulletin of the British Museum of Natural History, Geology 45: 165–371.

O'Shea, T. J. 1986. Mast foraging by West Indian manatees (Trichechus manatus). Journal of Mammalogy 67: 183–85.

Ojasti, J. 1964. Notas sobre el género Cavia (Rodentia; Caviidae) en Venezuela con descripción de una nueva subespecie. Acta Biológica Venezolana 4: 145–55.

Ojeda, F. A., and M. A. Mares. 1989. A biogeographic analysis of the mammals of Salta Province, Argentina. Special Publications, Museum of Texas Tech University 27: 1–66.

Olson, S. L. 1976. Oligocene fossils bearing on the origins of the Todidae and the Momotidae (Aves: Coraciiformes). Smithsonian Contributions to Paleobiology 27: 111–19.

Olson, S. L. 1983. Evidence for a polyphyletic origin of the Piciformes. Auk 100: 126–33.

Olson, S. L. 1985a. Faunal turnover in South American fossil avifaunas: The insufficiencies of the fossil record. Evolution 39: 1174–77.

Olson, S. L. 1985b. The fossil record of birds. In D. S. Farner, J. R. King, and K. C. Parkes, eds., Avian Biology, vol. 8. New York: Academic Press. Pp. 79–238.

Olson, S. L. 1987. An early Eocene oilbird from the Green River Formation of Wyoming (Caprimulgiformes: Steatornithidae). Documents des Laboratoires de Géologie, Lyon 99: 57–70.

Olson, S. L. 1992. A new family of primitive land birds from the lower Eocene Green River Formation of Wyoming. Science Series, Natural History Museum of Los Angeles County 36: 127–36.

Olson, S. L., and D. T. Rasmussen. 1986. Paleoenvironment of the earliest hominoids: New evidence from the Oligocene avifauna of Egypt. Science 233: 1202–4.

Opdyke, N. D., E. H. Lindsay, N. M. Johnson, and T. Downs. 1977. The paleomagnetism and magnetic polarity stratigraphy of the mammal-bearing section of Anza Borrego State Park, California. Quaternary Research 7: 316–29.

Ortiz Jaureguizar, E., R. H. Madden, M. G. Vucetich, M. Bond, A. A. Carlini, F. Goin, G. J. Scillato-Yané, and S. Vizcaíno. 1993. Un análisis de similitud entre las faunas de la "Edad-Mamífero Friasense." Resumenes, XI Jornadas Argentinas de Paleontología de Vertebrados (La Plata). Ameghiana 30(3): 351–52.

Ortiz Jaureguizar, E., and R. Pascual. 1988. Distribución de tamaños corporales en primates y marsupiales bunodontideos fósiles sudamericanos: Un probable caso de segregación ecológica. Resumenes V Jornadas Argentinas de Paleontologia de Vertebrados (La Plata) 5: 74.

Osborn, H. F. 1929. The Titanotheres of Ancient Wyoming, Dakota, and Nebraska. Washington, D.C.: Department of the Interior, U.S. Geological Survey.

Osborn, H. F. 1936. Proboscidea: A Monograph of the Discovery, Evolution, Migration and Extinction of the Mastodonts and Elephants of the World. New York: American Museum of Natural History.

Ostrander, G. E. 1985. Correlation of early Oligocene (Chadronian) in northwestern Nebraska. Dakoterra 2: 205–31.

Owen, R. 1842. Description of the Skeleton of an Extinct Gigantic Sloth, *Mylodon robustus* Owen, with Observations on the Osteology, Natural Affinities, and Probable Habits of the Megatherioid Quadrupeds in General. London: R. and J. Taylor.

Owen, R. 1860. On the megatherium (*Megatherium americanum,* Cuvier and Blumenb. ch). Pt. 5, Bones of the posterior extremities. Philosophical Transactions of the Royal Society of London 149: 809–29.

Owen-Smith, R. N. 1988. Megaherbivores: The Influence of Very Large Body Size on Ecology. Cambridge: Cambridge University Press.

Parrish, J. T. 1993a. Paleoclimatic history of the opening South Atlantic. In W. George and R. Lavocat, eds., The Africa-South America Connection. Oxford: Clarendon Press. Pp. 28–43.

Parrish, J. T. 1993b. The paleogeography of the opening South Atlantic. In W. George and R. Lavocat, eds., The Africa-

South America Connection. Oxford: Clarendon Press. Pp. 8–27.

Parrish, J. T., A. M. Zeigler, and C. R. Scotese. 1982. Rainfall patterns and the distribution of coals and evaporites in the Mesozoic and Cenozoic. Palaeogeography, Palaeoclimatology, and Palaeoecology 40: 67–101.

Pascual, R. 1953. Sobre nuevos restos de sirénidos del Mesopotamiense. Revista de la Asociación Geológica Argentina 8: 163–81.

Pascual, R. 1965. Los Toxodontidae (Toxodonta, Notoungulata) de la Formación Arroyo Chasicó (Plioceno inferior) de la Provincia de Buenos Aires. Ameghiniana 4: 101–32.

Pascual, R. 1966. Litopterna. In A. V. Borello, ed., Paleontografía Bonaerense. Vol. 4, Vertebrata. Buenos Aires: Provincia de Buenos Aires. Pp. 161–70.

Pascual, R. 1984. La sucesión de las edades-mamífero, de los climas y del diastrofismo sudamericanos durante el Cenozóico: Fenómenos concurrentes. Anales de la Academia Nacional de Ciencias, Buenos Aires 36: 15–37.

Pascual, R., and M. L. Díaz de Gamero. 1969. Sobre la presencia del género *Eumegamys* (Rodentia, Caviomorpha) en la Formación Urumaco del estado Falcón (Venezuela): Su significación cronológica. Asociación Venezolana de Geología, Minas y Petróleo, Boletín Informativo 12: 369–89.

Pascual, R., and O. Odreman Rivas. 1971. Evolución de las comunidades de los vertebrados del terciario Argentina, los aspectos paleozoogeográficos y paleoclimáticos relacionados. Ameghiniana 7: 372–412.

Pascual, R., E. J. Ortega, D. Gondar, and E. Tonni. 1966. Vertebrata. In A. V. Borello, ed., Paleontografía Bonaerense. Vol. 4, Vertebrata. Buenos Aires: Provincia de Buenos Aires.

Pascual, R., and E. Ortiz Jaureguizar. 1990. Evolving climates and mammal faunas in Cenozoic South America. Journal of Human Evolution 19: 23–60.

Pascual, R., M. G. Vucetich, and G. J. Scillato-Yané. 1990. Extinct and Recent South American and Caribbean edentates and rodents: Outstanding examples of isolation. Roma Accademia Nazionale dei Lincei. Pp. 627–40.

Pascual, R., M. G. Vucetich, G. J. Scillato-Yané, and M. Bond. 1985. Main pathways of mammalian diversification in South America. In F. G. Stehli and S. D. Webb, eds., The Great American Biotic Interchange. New York: Plenum Press. Pp. 219–47.

Patterson, B. 1934a. *Trachytherus,* a typotherid from the Deseado beds of Patagonia. Geological Series, Field Museum of Natural History 6(8): 119–39.

Patterson, B. 1934b. Upper premolar-molar structure in the Notoungulata with notes on taxonomy. Geological Series, Field Museum of Natural History 6(6): 91–111.

Patterson, B. 1936. *Caiman latirostris* from the Pleistocene of Argentina and a summary of South American Cenozoic Crocodilia. Herpetologica 1: 43–54.

Patterson, B., and L. G. Marshall. 1976. The Deseadan, early Oligocene, Marsupialia of South America. Fieldiana, Geology 41: 37–100.

Patterson, B., and R. Pascual. 1968. Evolution of mammals on southern continents. Pt. 6, The fossil mammal fauna of South America. Quarterly Review of Biology 43: 409–51.

Patterson, B., and R. Pascual. 1972. The fossil mammal fauna of South America. In A. Keast, F. C. Erk, and B. Glass, eds., Evolution, Mammals, and Southern Continents. Albany: State University of New York Press. Pp. 247–309.

Patterson, B., and A. E. Wood. 1982. Rodents of the Deseadan Oligocene of Bolivia and the relationships of the Caviomorpha. Bulletin of the Museum of Comparative Zoology, Harvard University 149: 371–543.

Patterson, C. 1981. Significance of fossils in determining evolutionary relationships. Annual Review of Ecology and Systematics 12: 195–223.

Patton, J. L., B. Berlin, and E. A. Berlin. 1982. Aboriginal perspectives of a mammal community in Amazonian Peru: Knowledge and utilization patterns among the Aguaruna Jívaro. In M. A. Mares and H. H. Genoways, eds., Mammalian Biology in South America. Linesville, University of Pittsburgh, Pymatuning Laboratory of Ecology, Special Publications series. Pp. 111–28.

Paula Couto, C. de. 1946. Atualização da nomenclatura genérica usada por Herluf Winge, em "E Museo Lundii." Estudos Brasileiros de Geologia 1: 59–80.

Paula Couto, C. de. 1956a. Une chauve-souris fossile des argiles feuilletées pléistocènes de Tremembé, état de São Paulo (Brésil). Actes du IV Congrès International du Quaternaire (Roma-Pisa) 1: 343–47.

Paula Couto, C. de. 1956b. Mamíferos fósseis do Cenozóico da Amazônia. Boletim Conselho Nacional Pesquisas, Rio de Janeiro 3: 1–121.

Paula Couto, C. de. 1971. On two small Pleistocene ground sloths. Anais Academia Brasileira de Ciências, suppl. 43: 499–513.

Paula Couto, C. de. 1975. Mamíferos fósseis do Quaternário do sudeste Brasileiro. Boletim Paranaense de Geociências 33: 89–132.

Paula Couto, C. de. 1976. Fossil mammals from the Cenozoic of Acre, Brazil. Pt. 1, Astrapotheria. Anais XXVIII Congresso Brasileiro de Geologia (Pôrto Alegre, 1974), Sociedade Brasileira de Geologia 2: 237–49.

Paula Couto, C. de. 1978. Fossil mammals from the Cenozoic of Acre, Brazil. Pt. 2, Rodentia, Caviomorpha, Dinomyidae. Iheringia, Ser. Geologia 5: 3–17.

Paula Couto, C. de. 1979. Tratado de Paleomastozoologia. Rio de Janeiro: Académia Brasileira de Ciências.

Paula Couto, C. de. 1981. Fossil mammals from the Cenozoic of Acre, Brazil. Pt. 4, Notoungulata, Notohippidae and Toxo-

dontidae, Nesodontinae. Anais 2 do Congreso Latíno-Americano de Paleontologia (Pôrto Alegre) 1: 461–77.

Paula Couto, C. de. 1982. Fossil mammals from the Cenozoic of Acre, Brazil. Pt. 5, Notoungulata, Nesodontinae (Pt. 2), Toxodontinae, and Haplodontheriinae, Litopterna, Pyrotheria, and Astrapotheria (Pt. 2). Iheringia, ser. Geologia 7: 5–43.

Paula Couto, C. de. 1983. Geochronology and paleontology of the basin of Tremembé-Taubaté, State of São Paulo. Iheringia 8: 5–31.

Paula Couto, C. de, and S. Mezzalira. 1971. Nova conceituação geocronologica de Tremembé, estado de São Paulo, Brasil. Anais Académia Brasileira de Ciências 43: 473–88.

Pérez Eman, J. L. 1990. Aspectos basicos de la biología, ecología y valor socioeconómico del quelonio cabezón, *Peltocephalus dumerilianus* (Schweiger) (Testudines, Pelomedusidae), en el territorio federal Amazonas. Universidad Simón Bolivar, Caracas, Venezuela.

Peters, W. C. H. 1865. Über die zu den *Vampyri* gehörigen Flederthiere und über die natürliche Stellung der Gattung *Antrozous*. Monatsberichte der Königlichen Preussischen Akademie der Wissenschaften zu Berlin. Pp. 503–24.

Pfretzschner, H.-U. 1986. Structural reinforcement and crack propagation in enamel. In D. E. Russell, J.-P. Santoro, and D. Sigogneau-Russell, eds., Teeth Revisited: Proceedings of the 7th International Symposium on Dental Morphology. Mémoire Muséum National d'Histoire Naturelle, Paris, sér. C, 53: 133–43.

Pfretzschner, H.-U. 1993. Enamel microstructure in the phylogeny of the Equidae. Journal of Vertebrate Paleontology 13: 342–49.

Phillips, C. J. 1971. The dentition of glossophagine bats: Development, morphological characteristics, variation, pathology, and evolution. University of Kansas Museum of Natural History, Miscellaneous Publication 54: 1–138.

Pickford, M. 1983. Sequence and environments of the lower and middle Miocene hominids of western Kenya. In R. L. Ciochon and R. S. Corruccini. eds., New Interpretations of Ape and Human Ancestry. New York: Plenum Press. Pp. 421–39.

Pocock, R. I. 1925. Additional notes on the external characters of some platyrrhine monkeys. Proceedings of the Zoological Society of London 1925: 27–47.

Pons, D. 1969. A propos d'une Goupiaceae du Tertiaire de Colombie: *Goupioxylon stutzeri* Schonfeld. Palaeontographica 128: 65–80.

Porta, J. de. 1961. Algunos problemas estratigráfico-faunísticos de los vertebrados en Colombia (con una bibliografía comentada). Boletín de Geología, Universidad Industrial de Santander, Bucaramanga 7: 83–104.

Porta, J. de. 1962. Edentata Xenarthra del Mioceno de La Venta (Colombia). Pt. 1, Dasypodoidea y Glyptodontoidea. Boletín de Geología, Universidad Industrial de Santander, Bucara-manga 10: 5–32.

Porta, J. de. 1965. Estratigrafía del Cretácio superior y Terciario en el extremo sur del valle medio del Magdalena. Boletín de Geología, Universidad Industrial de Santander, Bucaramanga 19: 5–30.

Porta, J. de. 1966. Geología del extremo Sur del Valle Medio del Magdalena entre Honda y Guataqui (Colombia). Boletín de Geología, Universidad Industrial de Santander, Bucaramanga 22–23: 1–318.

Porta, J. de. 1969. Les vertébrés fossiles de Colombie et les problèmes posés par l'isolement du continent sud-américain. Palaeovertebrata 2: 77–94.

Porta, J. de. 1974. Colombie. Pt. 2., Tertiaire et Quaternaire. In R. Hoffstetter, ed., Amerique Latine, Lexique Stratigraphique International. Vol. 5, pt. 4b. Paris: Centre National de la Recherche Scientifique.

Presch, W. 1973. A review of the Tegus, lizard genus *Tupinambis* (Sauria: Teiidae) from South America. Copeia 1973: 740–46.

Presch, W. 1974. Evolutionary relationships and biogeography of the macroteiid lizards (Family Teiidae, Subfamily Teiinae). Bulletin of the Southern California Academy of Sciences 73: 23–32.

Pretzmann, G. 1968. Die Familie Trichodactylidae (Milne Edwards 1853) Smith (vorläufige Mitteilung). Entomologische Nachrichtchtenblatt (Vienna) 15: 70–76.

Pretzmann, G. 1972. Some fossil chelae of river crabs and the distribution of *Potamon* in the eastern Mediterranean and western Asia. Thalassia Jugoslavica 8: 71–73.

Preuschoft, H. 1970. Functional anatomy of the lower extremity. In G. H. Bourne, ed., The Chimpanzee. Basel: Karger. Pp. 221–94.

Price, L. I. 1945. A new reptile from the Cretaceous of Brazil. Ministério da Agricultura, Notas Preliminares e Estudos 25: 1–8.

Price, L. I. 1950. Os crocodilideos da fauna da formação Baurú, Cretáceo do Estado de Minas Gerais. Anais Académia Brasileira de Ciências 22: 473–90.

Price, L. I. 1955. Novos crocodilídeos dos arenitos da Serie Baurú, Cretáceo do Estado de Minas Gerais. Anais Académia Brasileira de Ciências 27: 487–98.

Price, L. I. 1964. Sôbre o crâneo de um grande crocodilídeo extinto do Alto Río Juruá, Estado do Acre. Anais Académia Brasileira de Ciências 36: 59–68.

Pritchard, P. C. H. 1979. Encyclopedia of Turtles. Neptune, N.J.: TFH Publications.

Pritchard, P. C. H., and P. Trebbau. 1984. Turtles of Venezuela. Society for the Study of Amphibians and Reptiles, Contributions to Herpetology 2: 1–403.

Putnam, P. E. 1983. Fluvial deposits and hydrocarbon accumulations: Examples from the Lloydminster area, Canada. Modern and Ancient Fluvial Systems 6: 517–32.

Putnam, P. E., and T. A. Oliver. 1980. Stratigraphic traps in channel sandstones in the upper Mannville (Albian) of east-central Alberta. Bulletin of the Canadian Society of Petroleum Geologists 28: 489–508.

Radinsky, L. B. 1981. Evolution of skull shape in carnivores. Pt. 1, Representative modern carnivores. Biological Journal of the Linnean Society 15: 369–88.

Rage, J. 1984. Serpentes. Pt. 1 of Encyclopedia of Paleoherpetology. Stuttgart: Gustav Fischer Verlag.

Raikow, R. J. 1985. Locomotor system. In A. S. King and J. McClelland, eds., Form and Function in Birds. New York: Academic Press. Pp. 57–147.

Ramo, C. 1982. Biología del galapago (*Podocnemis vogli* Muller, 1935) en el Hato "El Frío" llanos de Apure (Venezuela). Doñana, Acta Vertebrata 9: 1–161.

Rancy, A. 1990. Pleistocene fossil mammals of western Amazonia. Abstracts, Joint Meeting of the Argentine and American Mammal Societies (ASM-SAREM), Buenos Aires 1990: 40.

Rancy, A., and J. Bocquentin Villanueva. 1987. Dos quelonios do Neogeno do Acre, Brasil. Anais X Congresso Brasileiro de Paleontologia. Pp. 182–87.

Rasmussen, D. T., and R. F. Kay. 1992. A Miocene *Anhinga* from Colombia, and comments on the zoogeographic relationships of South America's Tertiary avifauna. In K. E. Campbell, ed., Papers in Avian Paleontology Honoring Pierce Brodkorb. Vol. 36. Science Series. Los Angeles: Natural History Museum of Los Angeles County. Pp. 225–30.

Rathbun, M. J. 1904. Les crabes d'eau douce (Potamonidae). Nouvelles Archives du Muséum d'Histoire Naturelle, Paris 6: 225–312.

Redford, K. H. 1985. Food habits of armadillos (Xenarthra, Dasypodidae). In G. G. Montgomery, ed., The Evolution and Ecology of Armadillos, Sloths, and Vermilinguas. Washington, D.C.: Smithsonian Institution Press. Pp. 429–38.

Redford, K. H., and J. F. Eisenberg. 1992. Mammals of the Neotropics. Vol. 2, The Southern Cone, Chile, Argentina, Uruguay, Paraguay. Chicago: University of Chicago Press.

Redford, K. H., and R. M. Wetzel. 1985. *Euphractus sexcinctus*. American Society of Mammalogists, Mammalian Species 252: 1–14.

Reeser, L. A. 1984. Morphological affinities of new fossil talus of *Dolichocebus gaimanensis*. American Journal of Physical Anthropology 63: 206–7.

Reig, O. A. 1955. Noticia preliminar sobre la presencia de microbiotherinos vivientes en la fauna sudamericana. Investigaciones Zoológicas Chilenas 2: 121–30.

Reig, O. A. 1981. Teoría del origen y desarrollo de la fauna de mamíferos de América del Sur. Monographiae Naturae, Publicación del Museo Municipal de Ciencias Naturales de Mar del Plata 1: 1–162.

Reig, O. A., J. A. W. Kirsch, and L. G. Marshall. 1985. New conclusions on the relationships of the opossum-like marsupials, with an annotated classification of the Didelphimorphia. Ameghiniana 21: 335–43.

Reig, O. A., J. A. W. Kirsch, and L. G. Marshall. 1987. Systematic relationships of the living and Neocenozoic American "possum-like" marsupials, with comments on the classification of these and of the Cretaceous and Paleogene New World and European metatherians. In M. Archer, ed., Possums and Opossums: Studies in Evolution. Sydney: Surrey Beatty and Sons and the Royal Zoological Society of New South Wales. Pp. 1–89.

Reinhart, R. H. 1951. A new genus of sea cow from the Miocene of Colombia. University of California Publications in Geological Sciences 28: 203–13.

Renne, P. R., and A. R. Basu. 1991. Rapid eruption of the Siberian traps flood basalts at the Permo-Triassic boundary. Science 253: 176–79.

Rensberger, J. M. 1973. An occlusion model for mastication and dental wear in herbivorous mammals. Journal of Paleontology 47: 515–28.

Rensberger, J. M., and W. Von Koenigswald. 1980. Functional and phyletic interpretation of enamel microstructure in rhinoceroses. Paleobiology 6: 477–95.

Repetto, F. 1977. Un mamífero nuevo en el Terciario del Ecuador (Azuay-Cañar). Tecnológica (Escuela Politécnica del Litoral, Guayaquil) 1: 33–38.

Restrepo, J. 1983. Compilación de edades radiométricas en Colombia, departamentos andinos hasta 1982. Boletín de Ciencias de la Tierra, Universidad Nacional, Bogotá 7–8: 201–48.

Retallack, G. J. 1990. Soils of the Past: An Introduction to Paleopedology. Boston: Unwin-Hyman.

Reynolds, T. E., K. F. Koopman, and E. E. Williams. 1953. A cave faunule from western Puerto Rico with a discussion of the genus *Isolobodon*. Breviora, Museum of Comparative Zoology 12: 1–8.

Rimoli, R. O. 1977. Una nueva especie de monos (Cebidae: Saimirinae: *Saimiri*) de la Hispaniola. Cuadernos del Cendia 242: 1–14.

Ringuelet, A. 1953. Revisión de los Didélfidos fósiles argentinos. Revista del Museo de La Plata, n. ser. 2: 265–308.

Rivero, M., and O. Arredondo. 1991. *Paralouatta veronai,* a new Quaternary platyrrhine from Cuba. Journal of Human Evolution 21: 1–12.

Rivero-Blanco, C., and J. R. Dixon. 1979. Origin and distribution of the herpetofauna of the dry lowlands regions of northern South America. In W. E. Duellman, ed., The South American Herpetofauna: Its Origin, Evolution, and Dispersal. Lawrence: Museum of Natural History, University of Kansas. Pp. 218–40.

Roberts, T. R. 1972. Ecology of fishes in the Amazon and Congo basins. Bulletin of the Museum of Comparative Zoology, Harvard University 143: 117–42.

Roberts, T. R. 1975. Characoid fish teeth from Miocene deposits in the Cuenca basin, Ecuador. Journal of Zoology (London) 175: 259–71.

Robertson, J. S. 1976. Latest Pliocene mammals from Haile XV A, Alachua County, Florida. Bulletin of the Florida State Museum, Biological Sciences 20: 111–86.

Robinson, J. G., and K. H. Redford. 1986. Body size, diet and population density in Neotropical forest mammals. American Naturalist 128: 665–80.

Robinson, J. G., and K. H. Redford. 1989. Body size, diet and population variation in Neotropical forest mammal species: Predictors of local extinction? In K. H. Redford and J. F. Eisenberg, eds., Advances in Neotropical Mammalogy. Gainesville, Fla.: Sandhill Crane Press. Pp. 567–94.

Rodríguez, G. 1986. Centers of distribution of Neotropical freshwater crabs. In R. H. Gore and K. L. Heck, eds., Biogeography of the Crustacea. Crustacean Issues. Rotterdam: A. A. Balkema. Pp. 51–67.

Rodríguez, G. 1992. The Freshwater Crabs of America. Family Trichodactylidae, and Supplement to the Family Pseudothelphusidae. Faune Tropicale, Paris: Éditions de l'ORSTOM.

Rodríguez, G., and H. Díaz. 1977. Note sur quelques restes de crabes d'eau douce (Pseudothelphusidae) provenant d'un "Kjokken-Moeding" du Venezuela. Crustaceana 33: 107–8.

Rohlf, F. J. 1988. BIOM: A Package of Statistical Programs to Accompany the Text Biometry. Stony Brook, N.Y.: SUNY at Stony Brook.

Rohlf, F. J. 1989. NTSYS-PC: Numerical Taxonomy and Multivariate Analysis for the IBM-PC Microcomputers (and Compatibles). Setauket: Applied Biostatistics.

Romer, A. S. 1966. Vertebrate Paleontology. Chicago: University of Chicago Press.

Rose, K. D., and R. J. Emry. 1983. Extraordinary fossorial adaptations in the Oligocene palaeodonts *Epoicotherium* and *Xenocranium* (Mammalia). Journal of Morphology 175: 33–56.

Rosenberger, A. L. 1975. On the distinctiveness of *Xenothrix*. American Journal of Physical Anthropology 42: 326.

Rosenberger, A. L. 1978. New species of Hispaniolan monkey: A comment. Anuario Científico, Universidad Central del Este 3: 249–51.

Rosenberger, A. L. 1979a. Cranial anatomy and implications of *Dolichocebus,* a late Oligocene ceboid primate. Nature 279: 416–18.

Rosenberger, A. L. 1979b. Phylogeny, classification and evolution of New World monkeys (Platyrrhini, Primates). Ph.D. diss., City University of New York.

Rosenberger, A. L. 1980. Gradistic views and adaptive radiation of platyrrhine primates. Zeitschrift für Morphologie 71: 157–63.

Rosenberger, A. L. 1981a. A mandible of *Branisella boliviana* (Platyrrhini, Primates) from the Oligocene of South America. International Journal of Primatology 2: 1–17.

Rosenberger, A. L. 1981b. Systematics: The higher taxa. In A. F. Coimbra-Filho and R. A. Mittermeier, ed., Ecology and Behavior of Neotropical Primates. Rio de Janeiro: Académia Brasileira de Ciências. Pp. 9–27.

Rosenberger, A. L. 1984. Fossil New World monkeys dispute the molecular clock. Journal of Human Evolution 13: 737–42.

Rosenberger, A. L. 1988a. Pitheciinae. In I. Tattersall, E. Delson, and J. A. Van Couvering, eds., Encyclopedia of Human Evolution and Prehistory. New York: Garland. Pp. 454–55.

Rosenberger, A. L. 1988b. Platyrrhini. In I. Tattersall, E. Delson, and J. A. Van Couvering, eds., Encyclopedia of Human Evolution and Prehistory. New York: Garland. Pp. 456–59.

Rosenberger, A. L. 1992. The evolution of feeding niches in New World monkeys. American Journal of Physical Anthropology 88: 525–62.

Rosenberger, A. L., W. C. Hartwig, M. Takai, T. Setoguchi, and N. Shigehara. 1991. Dental variability in *Saimiri* and the taxonomic status of *Neosaimiri fieldsi,* an early squirrel monkey from La Venta, Colombia. International Journal of Primatology 12: 291–302.

Rosenberger, A. L., W. C. Hartwig, and R. C. Wolff. 1991. *Szalatavus attricuspis,* an early platyrrhine primate. Folia Primatologica 56: 225–33.

Rosenberger, A. L., T. Setoguchi, and W. C. Hartwig. 1991. *Laventiana annectans,* new genus and species: Fossil evidence for the origins of callitrichine New World monkeys. Proceedings of the U.S. National Academy of Sciences 88: 2137–40.

Rosenberger, A. L., T. Setoguchi, and N. Shigehara. 1990. The fossil record of callitrichine primates. Journal of Human Evolution 19: 209–36.

Rovereto, C. 1912. Los crocodrilos fósiles en las capas de Paraná. Anales del Museo Nacional de Historia Natural, Buenos Aires 22: 339–69.

Rovereto, C. 1914. Los estratos araucanos y sus fósiles. Anales del Museo Nacional de Historia Natural, Buenos Aires 25: 1–247.

Royo y Gómez, J. 1941. Datos para la geología económica del Departamento del Huila. Boletín de Minas y Petróleo, Bogotá 15: 121–205.

Royo y Gómez, J. 1942a. Contribución al conocimiento de la geología del valle superior del Magdalena (Departamento del Huila). Compilación de los Estudios Geológicos Oficiales en Colombia 5: 261–324.

Royo y Gómez, J. 1942b. Un nuevo cocodrílido fósil del Huila. Compilaciones de los Estudios Geológicos Oficiales en Colombia, Bogotá 5: 325–26.

Royo y Gómez, J. 1946. Los vertebrados del Terciario continental colombiano. Revista de la Académia Colombiana de Ciencias Exactas, Físicas y Naturales 6: 496–512.

Royo y Gómez, J. 1960. Los vertebrados de la Formación Urumaco, Estado Falcón. Memorias del Tercer Congreso Venezolano de Geología 2: 506–12.

Roze, J. 1966. La Taxonomía y Zoogeográphia de los Ofidios de Venezuela. Caracas: Ediciones de la Biblioteca.

Rubilar, A., and E. Abad. 1990. *Percicthys sylvia* sp. nov. del Terciario de los Andes sur-centrales de Chile (Pisces, Perciformes, Percichthyidae). Revista Geológica de Chile 17: 197–204.

Ruff, C. 1988. Hindlimb articular surface allometry in Hominoidea and *Macaca,* with comparisons to diaphyseal scaling. Journal of Human Evolution 17: 687–714.

Rusconi, C. 1933. Observaciones críticas sobre reptiles terciarios de Paraná (Alligatoridae). Revista de la Universidad de Córdoba 20: 57–106.

Rusconi, C. 1935. Observaciones sobre los gaviales fósiles argentinos. Anales de la Sociedad Científica Argentina 119: 203–14.

Sadler, P. M. 1981. Sediment accumulation rates and the completeness of stratigraphic sections. Journal of Geology 89: 569–84.

Salo, J., R. Kalliola, I. Häkkinen, Y. Mäkinen, P. Niemelä, et al. 1986. River dynamics and the diversity of Amazon lowland forest. Nature 322: 254–58.

Samson, S. D., and E. C. Alexander, Jr. 1987. Calibration of the interlaboratory ^{40}Ar/^{39}Ar standard, MMhb-I. Chemical Geology Isotope Geoscience 66: 27–34.

Sanchez Villagra, M. R. 1992. Morfometría de la tortuga matamata (*Chelus fimbriatus*) y una revisión de la especie fósil *C. lewisi.* Tésis de Licenciatura, Universidad Simón Bolivar, Caracas, Venezuela.

Sandman, J. D., and W. C. Hartwig. 1993. A possible giant muriquí (Platyrrhini) from the Pleistocene deposits of Lagoa Santa, Minas Gerais, Brazil. American Journal of Physical Anthropology 16: 172.

Sarich, V. M., and J. E. Cronin. 1980. South American mammal molecular systematics, evolutionary clocks, and continental drift. In R. L. Ciochon and A. B. Chiarelli, eds., Evolutionary Biology of the New World Monkeys and Continental Drift. New York: Plenum Press. Pp. 399–421.

Savage, D. E. 1951a. A Miocene phyllostomatid bat from Colombia, South America. University of California Publications in Geological Sciences 28: 357–65.

Savage, D. E. 1951b. Report on fossil vertebrates from the upper Magdalena valley, Colombia. Science 114: 186–87.

Savage, D. E. 1952. Colombia is the key to American vertebrates. Pacific Discovery 5: 18–23.

Savage, D. E. 1962. Cenozoic geochronology of the fossil mammals of the Western Hemisphere. Revista del Museo Argentino de Ciencias Naturales "Bernardino Rivadavia," Ciencias Zoológicas 8(4): 54–67.

Savage, D. E., and D. E. Russell. 1983. Mammalian Paleofaunas of the World. Reading, Mass.: Addison-Wesley.

Schaeffer, B. 1947. Cretaceous and Tertiary actinopterygian fishes from Brazil. Bulletin of the American Museum of Natural History 89: 1–40.

Schindel, D. E. 1980. Microstratigraphic sampling and the limits of paleontologic resolution. Paleobiology 6: 408–26.

Schönfeld, G. 1947. Holzer aus dem Tertiar Kolumbien. Abhandlungen Senckenbergische Naturforschung Gesellschaft 475: 1–53.

Schumacher, G.-H. 1973. The head muscles and hyolaryngeal skeleton of turtles and crocodilians. In C. Gans and T. S. Parsons, eds., Biology of the Reptilia. New York: Academic Press. Pp. 201–62.

Scillato-Yané, G. J. 1976a. El más antiguo Mylodontinae (Edentata Tardigrada) conocido: Glossotheriopsis pascuali gen. nov. sp. nov. del "Colloncurense" (Mioceno superior) de la Provincia de Río Negro (Argentina). Ameghiniana 13: 333–34.

Scillato-Yané, G. J. 1976b. Sobre un Dasypodidae (Mammalia, Xenarthra) de edad Riochiquense (Paleoceno superior) de Itaboraí (Brasil). Anais Académia Brasileira de Ciências 48: 527–30.

Scillato-Yané, G. J. 1977a. Nuevo Megalonychidae (Edentata, Tardigrada) de edad Chasiquense (Plioceno temprano) del sur de la Provincia de Buenos Aires (Argentina): Su importancia filogenética, bioestratigráfica y paleobiogeográfica. Revista de la Asociación de Ciencias Naturales del Litoral 8: 45–54.

Scillato-Yané, G. J. 1977b. Sur quelques Glyptodontidae nouveaux (Mammalia, Edentata) du Déséadien (Oligocène inférieur) de Patagonie (Argentine). Bulletin, Muséum National d'Histoire Naturelle, Paris, sér. 3, Sciences de la Terre 64: 249–62.

Scillato-Yané, G. J. 1978. Nuevo Nothrotheriinae (Edentata, Tardigrada) de edad Chisiquense (Plioceno temprano) del sur de la Provincia de Buenos Aires (Argentina): Su importancia bioestratigráfica, filogenética, y paleobiogeográfica. Actas del VII Congreso Geológico Argentino (Neuquén) 2: 449–57.

Scillato-Yané, G. J. 1980. Catálogo de los Dasypodidae fósiles (Mammalia, Edentata) de la República Argentina. Actas III Congreso Argentino de Paleontología y Bioestratigrafía y I Congreso Latinoamericano de Paleontología (Buenos Aires, 1978). Pp. 7–36.

Scillato-Yané, G. J. 1981a. Nuevo Megalonychidae (Edentata, Tardigrada) del "Mesopotamiense" (Mioceno tardio-Plioceno) de la Provincia de Entre Rios. Ameghiniana 17: 193–99.

Scillato-Yané, G. J. 1981b. Nuevo Mylodontidae (Edentata, Tardigrada) del "Mesopotamiense" (Mioceno tardío-Plioceno) de la Provincia de Entre Ríos. Ameghiniana 18(1–2): 29–34.

Scillato-Yané, G. J. 1986. Los Xenarthra fósiles de Argentina (Mammalia, Edentata). Actas IV Congreso Argentino de Paleontología y Bioestratigrafía (Mendoza, 1986) 2: 151–55.

Scillato-Yané, G. J. 1988. Algunos Cingulata (Mammalia, Edentata) del Oligoceno de Quebrada Fiera (Mendoza, Argentina). Resumenes V Jornadas Argentinas de Paleontología de Vertebrados (La Plata, 1988). Pp. 26–27.

Scillato-Yané, G. J., A. A. Carlini, and S. F. Vizcaíno. 1987. Nuevo Nothrotheriinae (Edentata, Tardigrada) de edad Chasiquense (Mioceno tardío) del sur de la Provincia de Buenos Aires (Argentina). Ameghiniana 24: 211–15.

Scillato-Yané, G. J., A. A. Carlini, and S. F. Vizcaíno. 1991. Aspectos evolutivos de los Tolypeutinae (Mammalia, Dasypodidae). Ameghiniana 27: 392.

Scott, K. M. 1985. Allometric trends and locomotor adaptations in the Bovidae. Bulletin of the American Museum of Natural History 179: 197–288.

Scott, K. M. 1990. Postcranial dimensions of ungulates as predictors of body mass. In J. Damuth and B. J. MacFadden, eds., Body Size in Mammalian Paleobiology. Cambridge: Cambridge University Press. Pp. 301–35.

Scott, W. B. 1903–1904. Mammalia of the Santa Cruz Beds. Vol. 5, Paleontology. Pts. 1–3, Edentata. Reports of the Princeton University Expeditions to Patagonia, 1896–1899. Stuttgart: Princeton University, E. Schweizerbart'sche Verlagshandlung (E. Nàgele). Pp. 1–364.

Scott, W. B. 1905. Mammalia of the Santa Cruz Beds. Vol. 5, Paleontology. Pt. 3, Glires. Reports of the Princeton University Expeditions to Patagonia, 1896–1899. Stuttgart: Princeton University, E. Schweizerbart'sche Verlagshandlung (E. Nàgele). Pp. 365–499.

Scott, W. B. 1910. Mammalia of the Santa Cruz Beds. Vol. 7, Paleontology. Pt. 1, Litopterna. Reports of the Princeton University Expeditions to Patagonia, 1896–1899. Stuttgart: Princeton University, E. Schweizerbart'sche Verlagshandlung (E. Nàgele). Pp. 1–156.

Scott, W. B. 1912. Mammalia of the Santa Cruz Beds. Vol. 6, Paleontology. Pt. 2, Toxodonta. Reports of the Princeton University Expeditions to Patagonia, 1896–1899. Stuttgart: Princeton University, E. Schweizerbart'sche Verlagshandlung (E. Nàgele). Pp. 111–238.

Scott, W. B. 1913. A History of Land Mammals in the Western Hemisphere. New York: Macmillan.

Scott, W. B. 1928. Mammalia of the Santa Cruz Beds. Vol. 6, Paleontology. Pt. 4, Astrapotheria. Reports of the Princeton University Expeditions to Patagonia, 1896–1899. Stuttgart: Princeton University, E. Schweizerbart'sche Verlagshandlung (E. Nàgele). Pp. 301–42.

Scott, W. B. 1937. A History of the Land Mammals of the Western Hemisphere. 2d ed. New York: Macmillan.

Scott, W. B. 1937. The Astrapotheria. Proceedings of the American Philosophical Society 77: 309–93.

Sen, S. 1989. *Hipparion* datum and its chronologic evidence in the Mediterranean area. In E. H. Lindsay, V. Fahlbusch, and P. Mein, eds., European Neogene Mammal Chronology. New York: Plenum Press. Pp. 495–506.

Setoguchi, T. 1980. Discovery of a fossil primate from the Miocene of Colombia. Monkey 24: 64–69.

Setoguchi, T. 1983. Discovery of monkey graveyard from Colombia, South America. Monkey 188: 6–10.

Setoguchi, T. 1985. *Kondous:* A new ceboid primate from the Miocene of La Venta, Colombia, South America. Folia Primatologica 44: 96–101.

Setoguchi, T., and A. L. Rosenberger. 1985a. Miocene marmosets: First fossil evidence. International Journal of Primatology 6: 615–25.

Setoguchi, T., and A. L. Rosenberger. 1985b. Some new ceboid monkeys from the Miocene of Colombia. Memorias VI Congreso Latinoamericano de Geología, Bogotá 1: 287–98.

Setoguchi, T., and A. L. Rosenberger. 1987. A fossil owl monkey from La Venta, Colombia. Nature 326: 692–94.

Setoguchi, T., N. Shigehara, A. L. Rosenberger, and A. Cadena. 1986. Primate fauna from the Miocene of La Venta, in the Tatacoa Desert, Department of Huila, Colombia. Caldasia 15: 761–73.

Setoguchi, T., N. Shigehara, and T. Watanabe. 1979. Description of a new Caviomorpha rodent from Miocene of Colombia, South America. Kyōto University Overseas Research, Reports of New World Monkeys 1: 47–50.

Setoguchi, T., M. Takai, and N. Shigehara. 1990. A new ceboid monkey, closely related to *Neosaimiri,* found in the upper red bed in the La Venta badlands, middle Miocene of Colombia, South America. Kyōto University Overseas Research, Reports of New World Monkeys 7: 9–13.

Setoguchi, T., M. Takai, C. Villarroel, N. Shigehara, and A. L. Rosenberger. 1987. New specimen of *Cebupithecia* from La Venta, Miocene of Colombia, South America. Kyōto University Overseas Research, Reports of New World Monkeys 6: 7–9.

Setoguchi, T., T. Watanabe, and T. Mouri. 1981. The upper dentition of *Stirtonia* (Ceboidea, Primates) from the Miocene of Colombia, South America, and the origin of the postero-internal cusp of upper molars of howler monkeys (*Alouatta*). Kyōto University Overseas Research, Reports of New World Monkeys 2: 51–60.

Shackleton, N. J., and J. P. Kennett. 1975. Paleotemperature history of the Cenozoic and the initiation of Antarctic glaciation: Oxygen and carbon isotopic analyses in DSDP Sites 277, 279, and 281. Initial Reports of the Deep Sea Drilling Project 29: 743–55.

Shaw, J. H., T. S. Carter, and J. C. Machado-Neto. 1985. Ecology of the giant anteater *Myrmecophaga tridactyla* in Serra da Canastra, Minas Gerais, Brazil: A pilot study. In G. G. Montgomery, ed., The Evolution and Ecology of Armadillos, Sloths, and Vermilinguas. Washington, D.C.: Smithsonian Institution Press. Pp. 379–84.

Sherman, J. A. 1984. Functional morphology of the forelimb in nothrotheriine sloths (Mammalia; Xenarthra = Edentata). Ph.D. diss., University of Chicago.

Shuey, R. T., F. H. Brown, G. G. Eck, and F. C. Howell. 1978. A statistical approach to temporal biostratigraphy. In W. W. Bishop, ed., Geological Background to Fossil Man. Edinburgh: Scottish Academic Press. Pp. 103–24.

Sigé, B. 1974. Données nouvelles sur le genre *Stehlinia* (Vespertilionoidea, Chiroptera) du Paléogène d'Europe. Palaeovertebrata 6: 253–72.

Sigé, B. 1985. Les chiroptères oligocènes du Fayum, Egypte. Geológica et Palaeontológica 19: 161–89.

Sigé, B. 1991. Rhinolophoidea et Verpertilionoidea (Chiroptera) du Chambi (Eocène inférieur de Tunisie): Aspects biostratigraphique, biogéographique et paléoécologique de l'origine des chiroptères modernes. Neues Jahrbuch für Geologie und Paläontologie, Abhandlungen 182: 355–76.

Sigé, B., and S. Legendre. 1983. L'histoire des peuplements de chiroptères du bassin méditerranéen: L'apport comparé des remplissages karstiques et des dépôts fluvio-lacustres. Mémoires de Biospéologie 10: 209–25.

Sill, W. D. 1968. The zoogeography of the crocodiles. Copeia 1968: 76–88.

Sill, W. D. 1970. Nota preliminar sobre un nuevo gavial del Plioceno de Venezuela y una discusión de los gaviales sudamericanos. Ameghiniana 7: 151–59.

Silva Santos, R. 1987. *Lepidosiren megalos* sp. nov. do Terciário do Estado do Acre, Brasil. Anais Académia Brasilera de Ciências 59: 375–84.

Silva Taboada, G. 1979. Los murciélagos de Cuba. La Habana, Cuba: Editorial Académia.

Simons, E. L. 1972. Primate Evolution: An Introduction to Man's Place in Nature. New York: Macmillan.

Simpson, G. G. 1932. Some new or little known mammals from the *Colpodon* beds of Patagonia. American Museum Novitates 575: 1–12.

Simpson, G. G. 1934. A new notoungulate from the early Tertiary of Patagonia. American Museum Novitates 735: 1–3.

Simpson, G. G. 1935. Early and middle Tertiary geology of the Gaiman region. American Museum Novitates 775: 1–29.

Simpson, G. G. 1937. New reptiles from the Eocene of South America. American Museum Novitates 927: 1–113.

Simpson, G. G. 1940. Review of the mammal-bearing Tertiary of South America. Proceedings of the American Philosophical Society 83: 649–709.

Simpson, G. G. 1945. The principles of classification and a classification of the mammals. Bulletin of the American Museum of Natural History 85: 1–350.

Simpson, G. G. 1947. A Miocene glyptodont from Venezuela. American Museum Novitates 1368: 1–10.

Simpson, G. G. 1948. The beginning of the age of mammals in South America, Pt. 1. Bulletin of the American Museum of Natural History 91: 1–232.

Simpson, G. G. 1950. Horses: The Story of the Horse Family in the Modern World and through Sixty Million Years of History. New York: Oxford University Press.

Simpson, G. G. 1957. A new Casamayor astrapothere. Revista del Museo Municipal de Ciencias Naturales y Tradicionales del Mar del Plata, Argentina 1: 11–18.

Simpson, G. G. 1960. Notes on the measurement of faunal resemblance. American Journal of Science 258A: 300–11.

Simpson, G. G. 1967. The beginning of the age of mammals in South America. Bulletin of the American Museum of Natural History 137: 1–259.

Simpson, G. G. 1980. Splendid Isolation: The Curious History of South American Mammals. New Haven, Conn.: Yale University Press.

Simpson, G. G. 1984. Discoverers of the Lost World: An Account of Some of Those Who Brought Back to Life South American Mammals Long Buried in the Abyss of Time. New Haven, Conn.: Yale University Press.

Slijper, E. J. 1946. Comparative biologic-anatomical investigations on the vertebral column and spinal musculature of mammals. Verhandlung Koninklijke Nederlandse Akademie van Wetenschappen 2: 1–128.

Smith, A. G., A. M. Hurley, and J. C. Briden. 1981. Phanerozoic PaleoContinental World Maps. London: Cambridge University Press.

Smith, D. G., and P. E. Putnam. 1980. Anastomosed river deposits: Modern and ancient examples in Alberta, Canada. Canadian Journal of Earth Sciences 17: 1396–1406.

Smith, G. R. 1975. Fishes of the Plio-Pleistocene Glenns Ferry Formation, southwest Idaho. Papers in Paleontology, Museum of Paleontology, University of Michigan 14: 1–68.

Smith, G. R. 1981. Late Cenozoic freshwater fishes of North America. Annual Review of Ecology and Systematics 12: 163–93.

Smith, R. J. 1981. Interpretation of correlations in intraspecific and interspecific allometry. Growth 45: 291–97.

Smith, R. J. 1984. Allometric scaling in comparative biology: Problems of concept and method. American Journal of Physiology 256: 152–60.

Smith, R. M. H. 1987. Morphology and depositional history of exhumed Permian point bars in the southwestern Karoo, South Africa. Journal of Sedimentary Petrology 57: 19–29.

Smith, R. M. H. 1990. Alluvial paleosols and pedofacies sequences in the Permian Lower Beaufort of the southwestern Karoo basin, South Africa. Journal of Sedimentary Petrology 60: 258–76.

Smythe, N. 1970. Relationships between fruiting seasons and seed dispersal methods in a Neotropical forest. American Naturalist 104: 25–35.

Smythe, N. 1978. The natural history of the Central American agouti (*Dasyprocta punctata*). Smithsonian Contributions to Zoology 257: 1–52.

Sneath, P. H. A., and R. R. Sokal. 1973. Numerical Taxonomy: The Principles and Practice of Numerical Classification. San Francisco: W. H. Freeman.

Sokal, R. R., and F. J. Rohlf. 1981. Biometry. 2d ed. New York: W. H. Freeman.

Solbrig, O. T. 1976. The origin and floristic affinities of the South American temperate desert and semidesert regions. In D. W. Goodall, ed., Evolution of Desert Biota. Austin: University of Texas Press. Pp. 7–49.

Solounias, N., M. Teaford, and A. Walker. 1988. Interpreting the diet of extinct ruminants: The case of a non-browsing giraffid. Paleobiology 14: 287–300.

Soria, M. F. 1981. Los Litopterna del Colhuehuapense (Oligoceno tardío) de la Argentina. Revista del Museo Argentino de Ciencias Naturales "Bernardino Rivadavia," Paleontología 3: 1–54.

Soria, M. F., and H. Alvarenga. 1987. Un Leontiniidae (Notoungulata, Toxodonta) de la Cuenca de Taubaté, Estado de São Paulo, Brasil. Resumenes IV Jornadas Argentinas de Paleontología de Vertebrados (Comodoro Rivadavia, 1987), pp. 23–24.

Soria, M. F., and H. Alvarenga. 1989. Nuevos restos de mamíferos de la Cuenca de Taubaté, Estado de São Paulo, Brasil. Anais Acádémia Brasileira de Ciências 61: 157–75.

Soria, M. F., and M. Bond. 1988. Asignación del género *Colpodon* Burmeister 1885 a la familia Notohippidae Ameghino, 1894 (Notoungulata; Toxodonta). Resumenes V Jornadas Argentinas de Paleontología de Vertebrados (La Plata, 1988), pp. 36–37.

Soria, M. F., and R. Hoffstetter. 1983. Présence d'un condylarthre (*Salladolodus deuterotheriodes* gen. et sp. nov.) dans le Déséadien (Oligocène inférieur) de Salla, Bolivie. Comptes Rendus de l'Académie des Sciences, Paris, sér. 2, 297: 549–52.

Souza Filho, J. P. 1991. *Charactosuchus sansaoi*, uma nova espécie de Crocodilidae (Crocodylia) do Neógeno do Estado do Acre, Brasil. Actas XII Congresso de Paleontologia (São Paulo, 1991), p. 36.

Souza Filho, J. P., and J. C. Bocquentin Villanueva. 1989. *Brasilosuchus mendesi*, n. g., sp. nov., um novo representante da família Gavialidae no Neógene do Acre, Brasil. Anais do XI Congresso Brasileiro de Paleontologia (Curitiba) 1: 457–63.

Souza Filho, J. P., and J. C. Bocquentin Villanueva. 1991. *Caiman niteroiensis* sp. nov., Alligatoridae (Crocodylia) do Neogeno do Estado do Acre, Brasil. Actas XII Congresso de Paleontologia (São Paulo, 1991), p. 36.

Souza Filho, J. P., J. C. Bocquentin Villanueva, and F. R. Negri. 1993. Um crânio de *Hesperogavialis* sp. (Crocodylia, Gavialidae) Mioceno superior-Plioceno do Estado do Acre, Brasil. Resumens X Jornadas Argentinas de Paleontología de Vertebrados (La Plata). Ameghiniana 30(3): 341–42.

Spencer, L. 1987. Fossil Abrocomidae and Octodontidae (Rodentia; Hystricina). Ph.D. diss., Loma Linda University, Loma Linda, Calif.

Spillmann, F. 1949. Contribución a la paleontología del Perú; una mamifauna fósil de la región del Río Ucayali. Publicaciones del Museo de Historia Natural, Universidad Nacional Mayor de San Marcos, Lima, ser. C, Geología, Paleontología 1: 1–40.

Spix, J. von. 1823. Simiarum et vespertilionum brasiliensium species novae: Ou histoire naturelle des especes nouvelles de singes et de chauve-souris observées et recueillies pendant le voyage dans l'intérieur du Brésil. Monaco: Typis Francisci Serephici Hübschmanni.

Steel, R. 1973. Handbuch der Palaeoherpetologie. Vol. 16, Crocodylia. Stuttgart: Gustav Fischer Verlag.

Stehli, F. G., and S. D. Webb, eds. 1985. The Great American Biotic Interchange. New York: Plenum Press.

Stehlin, H. G. 1928. Ein Astrapotheriumfund aus Venezuela. Schweizerische Paläontologische Gesellschaft Versammlung in Basel 21: 227–32.

Stehlin, H. G. 1940. Ein Nager aus dem Miocene von Colombien. Eclogae Geologicae Helvetiae 32: 179–283.

Steiger, R. H., and E. Jager. 1977. Subcommission on Geochronology: Convention on the use of decay constants in geo- and cosmochronology. Earth and Planetary Science Letters 36: 359–62.

Steininger, F. R. 1983. Vom zerfall der Tethys zu Mediterran und Paratethys. Annalen des Naturhistorischen Museums, Wien 85: 135–63.

Stern, D., A. W. Crompton, and Z. Skobe. 1989. Enamel ultrastructure and masticatory function in molars of the American opossum, *Didelphis virginiana*. Zoological Journal of the Linnean Society 95: 311–34.

Stewart, D. J. 1981. A meander-belt sandstone of the lower Cretaceous of Southern England. Sedimentology 28: 1–20.

Stiles, F. G., A. F. Skutch, and D. Gardner. 1989. A Guide to the Birds of Costa Rica. Ithaca, N.Y.: Comstock Publications.

Stirton, R. A. 1947a. The first lower Oligocene vertebrate fauna from Northern South America. Compilaciones de los Estudios Geológicos Oficiales en Colombia, Bogotá 7: 325–41.

Stirton, R. A. 1947b. A rodent and a peccary from the Cenozoic of Colombia. Compilaciones de los Estudios Geológicos Oficiales en Colombia, Bogotá 7: 317–24.

Stirton, R. A. 1951. Ceboid monkeys from the Miocene of Colombia. University of California Publications in Geological Sciences 20: 315–56.

Stirton, R. A. 1953a. A new genus of interatheres from the Miocene of Colombia. University of California Publications in Geological Sciences 29: 265–348.

Stirton, R. A. 1953b. Vertebrate paleontology and continental stratigraphy in Colombia. Bulletin of the Geological Society of America 64: 603–22.

Stirton, R. A., and D. E. Savage. 1951. A new monkey from the La Venta late Miocene of Colombia. Compilaciones de los Estudios Geológicos Oficiales en Colombia, Bogotá 7: 345–46.

Stock, C. 1917. Structure of the pes in *Mylodon harlani*. University of California Publications in Geological Sciences 10: 267–86.

Stock, C. 1925. Cenozoic gravigrade edentates of western North America with special reference to the Pleistocene Megalonychinae and Mylodontidae of Rancho La Brea. Carnegie Institution of Washington Publication 331: 1–206.

Storer, J. E. 1984. Mammals of the Swift Current Creek local fauna (Eocene: Uintan, Saskatchewan). Saskatchewan Culture and Recreation, Museum of Natural History, Natural History Contribution 7: 1–158.

Strait, S. G. 1990. Food consistency and the dental morphology of extant faunivorous mammals: Implications for reconstructing diet in the fossil record. Journal of Vertebrate Paleontology 9: 44.

Strait, S. G., J. G. Fleagle, T. M. Bown, and E. R. Dumont. 1990. Diversity in body size and dietary habits of fossil caenolestid marsupials from the Miocene of Argentina. Journal of Vertebrate Paleontology 10: 44A.

Sullivan, R. M. 1979. Revision of the Paleogene genus *Glyptosaurus* (Reptilia, Anguidae). Bulletin of the American Museum of Natural History 163: 1–72.

Sullivan, R. M. 1981. Fossil lizards from the San Juan Basin, New Mexico. In S. G. Lucas, K. Rigby, Jr., and B. Kues, eds., Advances in San Juan Basin Paleontology. Albuquerque: University of New Mexico Press. Pp. 76–88.

Sullivan, R. M. 1986. The skull of *Glyptosaurus sylvestris* Marsh 1871. Journal of Vertebrate Paleontology 6: 28–37.

Sussman, R. W., and W. G. Kinzey. 1984. The ecological role of the Callitrichidae: A review. American Journal of Physical Anthropology 64: 419–49.

Swisher, C. C., III, J. M. Grajales-Nishimura, A. Montanari, S. V. Margolis, P. Claeys, W. Alvarez, P. Renne, E. Cedillo-Pardo, F. J-M. R. Maurasse, G. H. Curtis, J. Smit, and M. O. McWilliams. 1992. Coeval ^{40}Ar/^{39}Ar ages of 65.0 million years ago from Chicxulib Crater melt rock and the Cretaceous-Tertiary boundary tektites. Science 257: 954–58.

Swisher, C. C., III, and D. R. Prothero. 1990. Single-crystal ^{40}Ar/^{39}Ar dating of the Eocene-Oligocene transition in North America. Science 249: 760–62.

Swofford, D. L. 1985. PAUP: Phylogenetic Analysis Using Parsimony. Version 2.4. Champaign: Illinois Natural History Survey.

Swofford, D. L. 1991. PAUP: Phylogenetic Analysis Using Parsimony. Version 3.0s. Champaign: Illinois Natural History Survey.

Szalay, F. S. 1969. Mixodectidae, Microsyopidae, and the insectivore-primate transition. Bulletin of the American Museum of Natural History 140: 193–330.

Szalay, F. S., and R. L. Decker. 1974. Origins, evolution, and function of the tarsus in late Cretaceous Eutheria and Paleocene primates. In F. A. Jenkins, Jr., ed., Primate Locomotion. New York: Academic Press. Pp. 223–59.

Szalay, F. S., and E. Delson. 1979. Evolutional History of the Primates. New York: Academic Press.

Takai, M. 1994. New specimens of *Neosaimiri fieldsi*, a middle Miocene ancestor of the squirrel monkeys, Colombia. Journal of Human Evolution 27:329–60.

Takai, M., T. Setoguchi, C. Villarroel, A. Cadena, and N. Shigehara. 1991. A new Miocene molossid bat from La Venta, Colombia, South America. Memoirs of the Faculty of Science, Kyōto University, Series Geology and Mineralogy 56: 1–9.

Takai, M., T. Setoguchi, C. Villarroel, N. Shigehara, and A. L. Rosenberger. 1988. Preliminary report of small mammals from the La Venta fauna, South America. Kyōto University Overseas Research, Reports of New World Monkeys 5: 11–14.

Takemura, K. 1983. Geology of the east side hills of the Río Magdalena from Neiva to Villavieja, Colombia. Kyōto University Overseas Research, Reports of New World Monkeys 3: 19–28.

Takemura, K., and T. Danhara. 1983. Fisson-track age of pumices included in the Gigante Formation, north of Neiva, Colombia. Kyōto University Overseas Research, Reports of New World Monkeys 3: 13–16.

Takemura, K., and T. Danhara. 1986. Fission-track dating of the upper part of Miocene Honda Group in La Venta badlands, Colombia. Kyōto University Overseas Research, Reports of New World Monkeys 5: 31–37.

Tamsitt, J. R. 1967. Niche and species diversity in Neotropical bats. Nature 213: 784–86.

Taplin, L. E., and G. C. Grigg. 1989. Historical zoogeography of the Eusuchian crocodilians: A physiological perspective. American Zoologist 29: 885–901.

Taplin, L. E., G. C. Grigg, and L. Beard. 1985. Salt gland function in fresh water crocodiles: Evidence for a marine phase in eusuchian evolution? In G. C. Grigg, R. Shine, and H.

Ehmann, eds., Biology of Australian Frogs and Reptiles. Sydney: Surrey Beatty and Sons. Pp. 403–10.

Taplin, L. E., and J. P. Loveridge. 1988. Nile crocodiles, *Crocodylus niloticus* and estuarine crocodiles, *Crocodylus porosus* show similar osmoregulatory responses on exposure to sea water. Comparative Biochemistry Physiology 89A: 443–48.

Tarsitano, S. F. 1985. Cranial metamorphosis and the origin of the Eusuchia. Neues Jahrbuch für Geologie und Paläontologie 170: 27–44.

Tarsitano, S. F., E. Frey, and J. Riess. 1989. The evolution of the Crocodilia: A conflict between morphological and biochemical data. American Zoologist 29: 843–56.

Tauber, A. 1991. *Homunculus patagonicus* Ameghino, 1891 (Primates, Cebidae), Mioceno medio, de la costa Atlantica austral, Provincia de Santa Cruz, Republica Argentina. Académia Nacional de Ciencias (Córdoba, Argentina), Miscelanea 82: 1–32.

Tauxe, L., M. Monaghan, R. Drake, G. Curtis, and H. Staudigel. 1985. Paleomagnetism of Miocene East African Rift sediments and the calibration of the geomagnetic reversal time scale. Journal of Geophysical Research 90: 4639–46.

Taverne, L. 1979. Ostéologie, phylogénèse et systématique des téléostéens fossiles et actuels du super-ordre des osteoglossomorphes. Pt. 3, Evolution des structures ostéologiques et conclusions générales relatives à la phylogénèse et à la systematique du super-ordre. Addendum. Académie Royale de Belgique, Mémoires Sciences. Classe des Sciences (2), Collection in 8 parts, 43: 1–168.

Taylor, B. K. 1978. The anatomy of the forelimb in the anteater (*Tamandua*) and its functional implications. Journal of Morphology 157: 347–68.

Taylor, B. K. 1985. Functional anatomy of the forelimb in vermilinguas (anteaters). In G. G. Montgomery, ed., The Evolution and Ecology of Armadillos, Sloths, and Vermilinguas. Washington, D.C.: Smithsonian Institution Press. Pp. 163–71.

Taylor, E. H. 1977. Comparative anatomy of caecilian anterior vertebrae. Kansas Science Bulletin 51: 219–31.

Taylor, J. R. 1982. An Introduction to Error Analysis: The Study of Uncertainties in Physical Measurements. Mill Valley, Calif.: University Science Books.

Taylor, M. E. 1976. The functional anatomy of the hindlimb of some African Viverridae (Carnivora). Journal of Morphology 148: 227–53.

Taylor, M. E. 1989. Locomotor adaptations by carnivores. In J. L. Gittleman, ed., Carnivore Behavior, Ecology, and Evolution. Ithaca, N.Y.: Cornell University Press. Pp. 382–409.

Telles Antunes, M. 1961. *Tomistoma lusitanica*, crocodilien du Miocène du Portugal. Revista da Facultade de Ciências de Lisboa 2: 5–88.

Terborgh, J. 1983. Five New World Primates: A Study in Comparative Ecology. Princeton, N.J.: Princeton University Press.

Terborgh, J., and C. P. Van Schaik. 1987. Convergence vs. non-convergence in primate communities. In J. H. R. Gee and P. S. Giller, eds., Organization of Communities Past and Present. Oxford: Blackwell Scientific Publications. Pp. 205–26.

Thomas, R. G., D. G. Smith, J. M. Wood, J. Visser, E. A. Calverley. 1987. Inclined heterolitic stratification: Terminology, description, interpretation and significance. Sedimentary Geology 53: 123–79.

Thorington, R. W., Jr., and S. Anderson. 1984. Primates. In S. Anderson and J. Knox Jones, Jr., eds., Orders and Families of Recent Mammals of the World. New York: John Wiley. Pp. 187–217.

Tonni, E. P. 1980. The present state of knowledge of the Cenozoic birds of Argentina. Contributions in Science, Natural History Museum of Los Angeles County 330: 105–14.

Travassos, H., and R. Da Silva Santos. 1955. Caracídeos fosseis da bacia do Paraiba. Anais Acadêmia Brasilera de Ciências 27: 313–31.

Underwood, G. 1967. A Contribution to the Classification of Snakes. London: Trustees of the British Museum (Natural History).

Van Houten, F. B. 1976. Late Cenozoic volcaniclastic deposits, Andean foredeep, Colombia. Bulletin of the Geological Society of America 87: 481–95.

Van Houten, F. B., and R. B. Travis. 1968. Cenozoic deposits, upper Magdalena valley, Colombia. Bulletin of the American Association of Petroleum Geologists 52: 675–702.

Van Valen, L. 1978. The statistics of variation. Evolutionary Theory 4: 33–43.

Van Valen, L. 1979. The evolution of bats. Evolutionary Theory 4: 103–21.

Van Valkenburgh, B. 1987. Skeletal indicators of locomotor behavior in living and extinct carnivores. Journal of Vertebrate Paleontology 7: 162–82.

Vari, S. J. 1989. Systematics of the Neotropical characiform genus *Curimata* Bosc (Pisces: Ostariophysi). Smithsonian Contributions to Zoology 474: 1–63.

Vaughan, T. A., and M. M. Bateman. 1980. The molossid wing: Some adaptations for rapid flight. In D. E. Wilson and A. L. Gardner, eds., Proceedings of the 5th International Bat Research Conference. Lubbock: Texas Tech Press. Pp. 69–78.

Vehrencamp, S. L., F. G. Stiles, and J. W. Bradbury. 1977. Observations on the foraging behavior and avian prey of the Neotropical carnivorous bat, *Vampyrum spectrum*. Journal of Mammalogy 58: 469–78.

Villarroel, C. 1977. Revalidación y redescripción de *Posnanskytherium desaguaderoi* Liendo, 1943: Toxodontidae (Notoungulata) del Plioceno superior boliviano. Boletín del Servicio Geológico de Bolivia, ser. A, 1: 21–32.

Villarroel, C. 1983. Descripción de *Asterostemma? acostae,* nueva especie de propalaeohoplophorino (Glyptodontidae, Mammalia) del Mioceno de La Venta, Colombia. Geología Norandina 7: 29–34.

Villarroel, C., J. Brieva, and A. Cadena. 1989. Descubrimiento de mamíferos fósiles de edad Lujanense (Pleistoceno tardío) en el "Desierto" de la Tatacoa (Huila, Colombia). Caldasia 16(76): 119–25.

Villarroel, C., and J. Guerrero. 1985. Un nuevo y singular representante de la familia Leontiniidae? (Notoungulata, Mammalia) en el Mioceno de La Venta, Colombia. Geología Norandina, Bogotá 9: 35–40.

Vizcaíno, S. F. 1990. Sistemática y evolución de los Dasypodinae Bonaparte, 1838 (Mammalia, Dasypodidae). Tésis Doctoral, Universidad Nacional de La Plata, Facultad de Ciencias Naturales y Museo.

Vizcaíno, S. F. 1994. Sistemática y anatomía de los Astegotherini Ameghino, 1906 (Nuevo Rango) (Xenarthra, Dasypodidae, Dasypodinae). Ameghiniana 31(1): 3–13.

Vizcaíno, S. F., A. A. Carlini, and G. J. Scillato-Yané. 1990. Los Dasypodidae miocénicos (Mammalia, Xenarthra): Implicancias en la distribución actual de la familia. Reunión Conjunta de la Sociedad Argentina para el Estudio de los Mamíferos (SAREM) y la American Society of Mammalogists (ASM) (Buenos Aires, Argentina, 1990), Resúmenes, p. 60.

Von Hagen, V. W. 1939. The little silky anteater. Nature Magazine 32: 189–90.

Von Koenigswald, W. 1982. Enamel structure in the molars of Arvicolidae (Rodentia, Mammalia), a key to functional morphology and phylogeny. In B. Kurten, ed., Teeth: Form, Function, and Evolution. New York: Columbia University Press. Pp. 109–22.

Von Koenigswald, W. 1982. Zum Verständnis der Morphologie der Wühlmausmolaren (Arvicolidae, Rodentia, Mammalia). Zeitschrift für Geologische Wissenschaften (Berlin) 10: 951–62.

Von Koenigswald, W., and H.-U. Pfretzschner. 1987. Hunter-Schreger-Bänder im Zahnschmelz von Säugetieren (Mammalia). Zoomorphology 106: 329–38.

Von Koenigswald, W., J. M. Rensberger, and H.-U. Pfretzschner. 1987. Changes in the tooth enamel of early Paleocene mammals allowing increased diet diversity. Nature 328: 150–52.

Von Prahl, H. 1982. Notas sobre *Sylviocarcinus piriformis* (Pretzmann, 1968) (Crustacea: Brachyura: Trichodactylidae) con énfasis en su zoogeografía. Actualidades Biológicas 10: 22–25.

Vucetich, M. G. 1984. Los roedores de la edad Friasense (Mioceno medio) de Patagonia. Revista del Museo de La Plata, n. ser. 8, Paleontología 50: 47–126.

Vucetich, M. G., M. M. Mazzoni, and U. F. J. Pardiñas. 1993. Los roedores de la Formación Collón Curá (Mioceno medio) y la Ignimbrita Pilcaniyeu, Cañadón del Tordillo, Neuquén. Ameghiniana 30(4): 361–81.

Vuilleumier, F. 1985. Fossil and Recent avifaunas and the inter-American interchange. In F. G. Stehli and S. D. Webb, eds., The Great American Biotic Interchange. New York: Plenum Press. Pp. 387–424.

Wake, M. H. 1980. Morphometrics of the skeleton of *Dermophis mexicanus* (Amphibia: Gymnophiona). Pt. 1, The vertebrae, with comparisons to other species. Journal of Morphology 165: 117–30.

Walker, A. 1974. Locomotor adaptations in past and present prosimian primates. In F. A. Jenkins, Jr., ed., Primate Loco-motion. New York: Academic Press. Pp. 349–81.

Walker, R. G., and D. J. Cant. 1984. Sandy fluvial systems. In R. G. Walker, ed., Facies Models. Waterloo, Ontario: Geological Association of Canada. Pp. 71–89.

Wall, W. P. 1980. Cranial evidence for a proboscis in *Cadurcodon* and a review of snout structure in the family Amynodontidae (Perissodactyla, Rhinocerotoidea). Journal of Paleontology 54: 968–77.

Walsh, S. J. 1990. A systematic revision of the Neotropical cat-fish family Ageneiosidae (Teleostei: Ostariophysi: Siluri-formes). Ph.D. diss., University of Florida, Gainesville.

Walton, A. H. 1990. Rodents of the La Venta fauna, Miocene, Colombia: Biostratigraphy and paleoenvironmental implica-tions. Ph.D. diss., Southern Methodist University, Dallas, Texas.

Washburn, S. L. 1960. Tools and human evolution. Scientific American 203: 3–15.

Watanabe, T., T. Setoguchi, and N. Shigehara. 1979. An outline of paleontological investigation. Kyōto University Overseas Research, Reports of New World Monkeys 1: 39–46.

Watson, G. S. 1956a. Analysis of dispersion on a sphere. Royal Astronomical Society Monthly Notices, Geophysics (suppl.) 7: 153–59.

Watson, G. S. 1956b. A test for randomness of directions. Royal Astronomical Society Monthly Notices, Geophysics (suppl.) 7: 160–61.

Weaver, C. E. 1989. Clays, muds, and shales. Developments in Sedimentology 44.

Webb, S. D. 1978. A history of savanna vertebrates in the New World. Pt. 2, South America and the great interchange. An-nual Review of Ecology and Systematics 9: 393–426.

Webb, S. D. 1983. The rise and fall of the late Miocene ungulate fauna in North America. In M. H. Nitecki, ed., Coevolution. Chicago: University of Chicago Press. Pp. 267–306.

Webb, S. D. 1985a. The interrelationships of tree sloths and ground sloths. In G. G. Montgomery, ed., The Evolution and Ecology of Armadillos, Sloths, and Vermilinguas. Wash-ington, D.C.: Smithsonian Institution Press. Pp. 105–12.

Webb, S. D. 1985b. Main pathways of mammalian diversification in North America. In F. Stehli and S. D. Webb, eds., The Great American Biotic Interchange. New York: Plenum Press. Pp. 201–18.

Webb, S. D., and R. C. Hulbert, Jr. 1986. Systematics and evo-lution of *Pseudhipparion* (Mammalia, Equidae) from the late Neogene of the Gulf Coastal Plain and the Great Plains. Con-tributions to Geology, University of Wyoming, Special Paper 3: 237–72.

Webb, S. D., and S. Perrigo. 1985. New megalonychid sloths from El Salvador. In G. G. Montgomery, ed., The Evolution and Ecology of Armadillos, Sloths, and Vermilinguas. Wash-ington, D.C.: Smithsonian Institution Press. Pp. 113–20.

Webb, S. D., and N. Tessman. 1967. Vertebrate evidence of a low sea level in the middle Pliocene. Science 156: 379.

Weeks, L. G. 1947. Paleogeography of South America. Bulletin of the American Association of Petroleum Geologists 41(7): 1194–1241.

Weitzman, S. H., and M. Weitzman. 1982. Biogeography and evolutionary diversification in Neotropical freshwater fishes, with comments on the refuge theory. In G. T. Prance, ed., Bi-ological Diversification in the Tropics. New York: Columbia University Press. Pp. 403–22.

Weitzman, S. H., and W. L. Fink. 1983. Relationships of the neon tetras, a group of South American freshwater fishes (Teleostei; Characidae) with comments on the systematics of New World Characiformes. Bulletin of the Museum of Com-parative Zoology, Harvard University 150: 339–65.

Welles, S. P. 1962. A new species of elasmosaur from the Aptian of Colombia and a review of the Cretaceous plesiosaurs. Uni-versity of California Publications in Geological Sciences 44: 1–96.

Wellman, S. S. 1970. Stratigraphy and petrology of the non-marine Honda Group (Miocene), upper Magdalena valley, Colombia. Bulletin of the Geological Society of America 81: 2353–74.

Wentworth, C. K. 1922. A scale of grade and class terms for clastic sediments. Journal of Geology 30: 377–92.

Wermuth, H. 1964. Das Verhältnis zwischen Kopf-, Rum-ph- und Schwanzelänge bei den rezenten Krokodilen. Senkenbergiana Biologica 45: 369–85.

Wetzel, R. M. 1982. Systematics, distribution, ecology, and conservation of South American edentates. In M. A. Mares and H. H. Genoways, eds., Mammalian Biology in South America, Special Publications of the Pymatuning Laboratory of Ecology, University of Pittsburgh 6: 345–75.

Wetzel, R. M. 1985a. The identification and distribution of Re-cent Xenarthra (=Edentata). In G. G. Montgomery, ed., The Evolution and Ecology of Armadillos, Sloths, and Ver-milinguas. Washington, D.C.: Smithsonian Institution Press. Pp. 5–21.

Wetzel, R. M. 1985b. Taxonomy and distribution of armadillos, Dasypodidae. In G. G. Montgomery, ed., The Evolution and Ecology of Armadillos, Sloths, and Vermilinguas. Wash-ington, D.C.: Smithsonian Institution Press. Pp. 25–46.

Wheeler, O. C. 1935. Tertiary stratigraphy of the Magdalena valley. Proceedings of the Academy of Natural Sciences, Philadelphia 87: 21–39.

Whitmore, F. C., and R. H. Stewart. 1965. Miocene mammals and Central American seaways. Science 148: 180–85.

Wied-Neuwied, M., P. zu. 1820. *Diclidurus* Klappenschwanz. Ein neues Genus der Chiropteren aus Brasiliens. Isis von Oken 4–5: 1629–30.

Wiel, A. M. vander. 1991. Uplift and volcanism of the S.E. Colombian Andes in relation to Neogene sedimentation in the upper Magdalena valley. Master's thesis, Agricultural University of Wageningem, Netherlands.

Wilkinson, L. 1990. SYSTAT: The System for Statistics. Evanston, Ill.: SYSTAT, Inc.

Willard, B. 1966. The Harvey Bassler Collection of Peruvian Fossils. Bethlehem, Pa.: Lehigh University.

Williams, E. E. 1954. A key and description of the living species of the genus *Podocnemis* (sensu Boulenger). Bulletin of the Museum of Comparative Zoology, Harvard University 111: 279–95.

Williams, E. E., and K. F. Koopman. 1952. West Indian fossil monkeys. American Museum Novitates 1546: 1–16.

Willig, M. R. 1986. Bat community structure in South America: A tenacious chimera. Revista Chilena de Historia Natural 59: 151–68.

Willig, M. R., and M. A. Mares. 1989. A comparison of bat assemblages from phytogeographic zones of Venezuela. In D. W. Morris, Z. Abramsky, B. J. Fox, and M. R. Willig, eds., Patterns in the Structure of Mammalian Communities. Lubbock: Special Publications, Texas Tech University 28: 59–67.

Wilson, D. E. 1973. Bat faunas: A trophic comparison. Systematic Zoology 22: 14–29.

Wilson, D. E., and J. S. Findley. 1977. *Thyroptera tricolor.* Mammalian Species 71: 1–3.

Winge, H. 1895. Jordfunde og nulevende Gumlere (Edentata) fra Lagoa Santa, Minas Gerais, Brasilien. E. Musei Lundi, Copenhagen 3: 1–319.

Wolff, F. G. 1984. New specimens of the primate *Branisella boliviana* from the early Oligocene of Salla, Bolivia. Journal of Vertebrate Paleontology 4: 570–74.

Wood, A. E., and B. Patterson. 1959. The rodents of the Deseadan Oligocene of Patagonia and the beginnings of South American rodent evolution. Bulletin of the Museum of Comparative Zoology, Harvard University 120: 279–428.

Wood, R. C. 1970. A review of the fossil Pelomedusidae (Testudines, Pleurodira) of Asia. Breviora, Museum of Comparative Zoology, Harvard University 357: 1–24.

Wood, R. C. 1975. Redescription of *"Bantuchelys" congolensis,* a fossil pelomedusid turtle from the Paleocene of Africa. Revue Zoologique Africaine 89: 127–44.

Wood, R. C. 1976a. *Stupendemys geographicus,* the world's largest turtle. Breviora, Museum of Comparative Zoology, Harvard University 436: 1–31.

Wood, R. C. 1976b. Two new species of *Chelus* (Testudines, Pleurodira) from the late Tertiary of northern South America. Breviora, Museum of Comparative Zoology, Harvard University 435: 1–26.

Wood, R. C., and M. L. Gamero. 1971. *Podocnemis venezuelensis,* a new fossil pelomedusid (Testudines, Pleurodira) from the Pliocene of Venezuela and a review of the history of *Podocnemis* in South America. Breviora, Museum of Comparative Zoology, Harvard University 376: 1–23.

Wood, R. C., and B. Patterson. 1973. A fossil trionychid turtle from South America. Breviora, Museum of Comparative Zoology, Harvard University 405: 1–10.

Woodburne, M. O. 1977. Definition and characterization in mammalian chronostratigraphy. Journal of Paleontology 51(2): 220–34.

Woodburne, M. O. 1987. Cenozoic Mammals of North America: Geochronology and Biostratigraphy. Berkeley: University of California Press.

Woodruff, F., S. M. Savin, and R. G. Douglas. 1981. Miocene stable isotope record: A detailed deep Pacific Ocean study and its paleoclimatic implications. Science 212: 665–68.

Woods, C. A. 1984. New World porcupines. In D. W. MacDonald, ed., The Encyclopedia of Mammals. New York: Facts on File. Pp. 686–89.

Young, W. G., M. McGowan, and T. J. Daley. 1987. Tooth enamel structure in the koala, *Phascolarctos cinereus:* Some functional interpretations. Scanning Microscopy 1: 1925–34.

Zetti, J. 1972. Los mamíferos fósiles de edad Huayqueriense (Plioceno medio) de la región pampeana. Tésis Doctoral, Universidad Nacional de La Plata, La Plata, Argentina.

Taxonomic Index

Page-number citations in boldface refer to **figures** on those pages.

Abderites, 209
Abderitidae, 207–209
Abramites, 74
Acanthicus, 67, 88
    A. hystrix, 83
    cf. *A.,* **83,** 83–84, 86, 89
Acarechimys, 392, 395, 408, 503, 533
    A. cf. *A. minutissimus,* **394, 395,** 395–
        396, 402, 513
    A. minutissimus, 392, 396
Acarichthys heckeli, 86
Acaronia nassa, 86
Acdestis, 210
Achlysictis, 203, 204, 206. See also *Thylacosmilus atrox*
    A. lelongi, 187, 201, 202, 203, 206
Adinotherium, 338, 339, 340, 343, 345, 346,
    347, 351
Aequidens awani, 86
Ajaia ajaja, **179**
Albertogaudrya, 377, 378
Alces alces, 350
Alitoxodon, 348
Alligator, 136
    A. mississippiensis, **142,** 143, 147
Alligatoridae incertae sedis, 120–121
Allognathosuchus, 121
Alloiomys, 396, 403
Alouatta, 472, 473, 481, 484, 485, 487, 489,
    493, 495
Alzadasaurus colombiensis, 3
Amblydoras, 81
Ameiva, 107

Aminornis excavatus, 180
Anachlysictis, 201–203, 204, 206, 526
    A. gracilis, **ii,** 187, **189,** 201–203, **202,**
        206
Anadasypus, 219, 224, 527, 544
    A. hondanus, 213, 217, **218,** 218–219
Analcimorphus, 241, 247, 258, 259, 262
Analcitherium, 244, 248
Ancistrinae, 83–84
Ancylocoelus, 306, 307, 309, 310, 312, 313,
    314, 316, 317
Anhinga, 171, 181–182
    A. anhinga, 182, 526
    A. cf. *A. grandis,* 181–182
    A. grandis, 181
Anhingidae, 181–182
Aniliidae, 95–96
    cf. Aniliidae, 96
Anisolambda, 297
Anisolornis excavatus, 180
Anostomidae, 73–74
Anoura, 419, 429
Antilope cervicapra, 280
Aotus, 436, 440, 444, 445, 448, 456, 465, 466,
    470, 471, 473, 476, 481, 484, 487, 493,
    494, 495
    A. cf. *A. dindensis,* 468
    A. dindensis, 468–469. See also
        Mohanamico hershkovitzi
    A. trivirgatus, 440, 456
"Aotus" dindensis, 10, 468–469, 484, 487, 494
Aramidae, 179–181
Aramus, 171, 179–181

    A. guarauna, 171, **179,** 180, 182, 526
    A. paludigrus, 171, **179,** 179–181, **180**
Arapaima, 67, 86, 88, 89
    A. gigas, 72
    A. sp., **60,** 72, 88
Arapaimidae, 71
Archaeohyrax, 319, 325, 326, 328, 331, 332
Archaeotherium, 350
Arctodictis, 526
Ardea alba, **179**
Ariidae, **79,** 80
Arius proops, 80
Asmithwoodwardia, 283
Astegotherium, 214, 215, 216
    A. dichotomus, 216
Asterostemma, 213, 214, 222–223, 502, 527
    ?*A. acostae,* 213, 222, 223, 224
    A. barrealense, 222, 223
    A. depressa, 222, 223
    A. gigantea, 213, **215,** 223, 224, 527
    A. venezolensis, 222, 223
Astrapontus, 377, 379
Astrapotheria, 356–379
Astrapothericulus, 360, 363, 366, 374, 377–378,
    379
Astrapotherium, 357, 360, 363, 365, 366, 371,
    374, 378, 379
    A. magnum, 359, 361, 371–372, 377
"Astrapotherium." See *Xenastrapotherium*
Astronotus ocellatus, 86
Ateles, 487, 489, 493, 494
    A. anthropomorphus, 473, 486, 495
Auchenipteridae, 79

Ayllusuchus, 114

Babyrousa babirussa, 280
Badistornis aramus, 180
Baenodon, 316
Balanerodus, 114, 120–121, 153
    B. loginus, 113, 120–121, **121,** 151, 152
Balantiopteryx, 412
Balearica, 180
    B. pavonina, **179**
Bathyopsis, 350
Baurusuchus pachecoi, 114
Blastoceros, 534
Boidae, 96–97
Borhyaenidae, 198–201
Borhyaenoidea, 195–203
    fam. et gen. incertae sedis, 203–204
Bothremys, 156
Brachyplatystoma, 78, 86, 87
    B. cf. *B. vaillanti,* 67, 68, 76–78, **77,** 88,
        89
    B. filamentosum, 78
    B. ("Malacobagrus") sp., **78,** 78–79, 88, 89
    B. rousseauxi, 78
Brachyteles, 487, 493
    B. arachnoides, 459, 486, 491
Brachytherium, 290, 291, 293, 295
Bradypus, 247, 249, 252, 255, 257, 526
Branisella, 473, 494
    B. boliviana, 474, **478**
Brasilosuchus mendesi, 144, 147
Bretesuchus bonapartei, 114
Brosimum, 428
Bucco, 178
Bufo marinus, 525

Cacajao, 483, 487, 489, 536
Caenolestoidea, 207–212
Caiman, 135, 136
    C. cf. *C. jacare,* 152
    C. cf. *C. lutescens,* 113, 151, 152, 153
    C. crocodilus, 135–136
    C. latirostris, 113, 151, 152, 153
    C. neivensis, 4
    C. niteroiensis, 152
Calathea, 430
Callicebus, 436, 440, 444, 445, 448, 453, 456,
    457, 463, 468, 469, 470, 471, 473, 474,
    479, 481, 483, 484, 487, 489, 493, 495
    C. moloch, 440
Callichthyidae, 82–83
Callichthys, 82
Callimico, 435, 436, 438, 440, 442, 444, 445,
    448, 449, 452, 453, 454, 455, 456, 458,
    469, 473, 483, 484, 487, 489, 493, 495, 535
    C. goeldii, 440
Callithrix, 440, 441, 442, 444, 445, 449, 453,
    457, 458, 466, 473, 487, 493, 536
    C. jacchus, 440, 457
Callitrichidae, 438–453
Callopistes, 107, 108, 111
Caluromys, 190, 197

Capra, 326
Carandaisuchus. See also *Mourasuchus*
    C. nativus, 119
Cardiomys, 407, 408
    C. sp., 408
Caririsuchus, 117
Carlocebus, 473, 477, 479, 482, 493, 494, 495
    C. carmenensis, 479, **480**
    C. intermedius, 479
Castor canadiensis, 400
Catonyx, 234
Catoprion, 68, 76, 87
Cavia, 407, 408, 527
    C. aperea, 407
Cayman. See *Purrusaurus*
Cebuella, 454, 457, 466, 473, 487, 493
    C. pygmaea, 459, 491
Cebupithecia, 459, 461–465, **462,** 466, 468,
    469, 470, 471, 474, 487, 489, 491, 493,
    494, 495, 536
    C. sarmientoi, 6, 460–465, **461, 462, 463,**
        464, 465, 470, 483, **484,** 536
Cebus, 435, 436, 445, 448, 453, 457, 458, 463,
    471, 473, 486, 487, 493, 495
    C. apella, 485
Centropus, 174
Ceratosuchus, 121
Ceratotherium, 350
Chaerephon, 428
    C. ansorgei, 425
Chaetobranchus flavescens, 86
Chamops, 107, 112
Chamopsiinae, 112
Characidae, 75
    cf. Tetragonopterinae, **74,** 75
Characiformes, 72
Charactosuchus, 113, 118, 148, 151, 152, 525
    C. fieldsi, 118, 151, 152
    C. kugleri, 118
    C. sansoai, 118, 152
Cheirocerus, 79
Cheiromeles, 428, 430
    C. parvidens, 425
    C. torquatus, 425
Chelus, 168, 525
    C. colombianus, 155, 169
    C. lewisi, 169
Chiropotes, 470, 483, 487, 489, 493, 536
Chiroptera, 411–431
Choloepus, 233, 245, 247, 249, 252, 255, 257,
    526
Chrotopterus, 418, 429
    C. auritus, 417, 418, 429
Cichlasoma, 86
Cichlidae, 85–86
    incertae sedis, **85,** 85–86
Ciconia, 181
Ciconiidae, 181
Ciconiopsis antarctica, 181
Cingulata, 214–226, 228–231
Cladosictis, 203
Clypeotherium, 221, 224–225

Cnemidophorus, 107
Cochilius, 220
Cochlops, 222, 223, 224
Coendou, 406
    C. mexicanus, 392, 394
Coleura, 412
Colombophis, 525
    C. portai, 95–96
Colossoma, 87, 89, 546
    C. macropomum, 67, **75,** 75–76, 87, 88,
        525, 546
Colpodon, 303, 309, 310, 314, 316, 317
Colubroidea, 97
comb. nov. See *Micoureus laventicus; Paradra-*
    caena colombiana
Coraciiformes, 176–179
Cormura, 412
Crassoretitriletes vanraadshooveni, 42
Crenicichla vettata, 86
Crinifer, 173, 174
Crocodilurus, 102, 107, 108, 111, 525
Crocodylia, 114–153
Crocodylidae, 118–119
    cf. Tomistominae, 118
Crocodylus, 136
    C. acutus, **142,** 150
    C. intermedius, 151
    C. niloticus, 150
    C. porosus, 150
Cryptodira, 165–169
Ctenomys haigi, 392, 396
Cuculiformes, 172–176
Cullinia levis, 298
"Cullinia" levis, 299
Cuniculus, 407, 408, 533
    C. paca, 392, 399–400
Cuvierimops parisiensis, 428
Cyclopes, 249, 252, 254, 255, 258, 261, 262
Cynodon gibbus, 73
Cynodontidae, 73

Dabbenea. See *Phoberomys*
Dasypodidae, 214–220
Dasypodinae, 214–219
Dasypodoidea, 214–220, 228–231
Dasyprocta, 392, 406, 407, 408
    D. fuliginosa, 398
    D. punctata, 392, 396, 398
Dasypus, 213, 216, 217, 218, 219, 225, 249,
    255, 258
    D. hybridus, 215
    D. novemcinctus, 217, 249
Desmodus archaeodaptes, 416
Diadiaphorus, 273, 276, 279, 284, 290, 291,
    293, 294, 295, 302
    D. majusculus, 276
    ?*D.* spp., 290
Dianema, 82
Diasemosaurus, 101
Diblosodon, 101
Dicerorhinus sumatrensis, 338, 350
Diceros, 359

D. bicornis, 350, 377
Diclidurus, 411–413, 429, 431, 535
    D. albus, 412
    D. (Depanycteris) isabellus, 412
    D. ingens, 412
    D. scatatus, 412
    D. sp. indet., 411–413, **412**
Dicrodon, 107
Didelphidae, 192–195
Didelphimorpha, 192–195
Didelphinae, gen. et sp. indet., 194–195
Didelphis
    D. albiventris, 197
    D. crucialis, 197
    D. marsupialis, 197, 205
Didelphoidea, 192–195
Didolodus, 283
Diellipsodon, 241
Dilocarcininae, 63
Dinomys, 400, 406
    D. branickii, 400, 406
Dinotoxodon, 335, 336, 340, 348, 352
    D. paranensis, 340, 348
Dinotoxodontinae, 336–352
Diplasiotherium, 291
Disteiromys, 406
Dizzya exsultans, 425
Dolichocebus, 473, 475, 476, 481, 494
    D. gaimanensis, 475, **478**
Dolichotis, 407, 521, 527
    D. patagonum, 407
    D. (Pediolagus) salinicola, 407
    D. salinicola, 392, 400, 402, 408
Dollosuchus, 118
Doradidae, **60,** 67, 79
    incertae sedis no. 1, 80–81, **81**
    incertae sedis no. 2, **79,** 81–82
    incertae sedis no. 3, **81,** 82
Doraops, 81
Dracaena, 101, 104, 107–108, 109, 110, 111,
    525. See also *Paradracaena*
    D. colombiana, 101, 102
    D. guianensis, 100, 102, 109
Dromiciops, 205, 526
Drytomomys, 399
    D. aequatorialis, 398
Dukecynus, 199–201, **200,** 204 '
    D. magnus, 187, **189,** 199–201, **200,** 205,
    206, 526

Echinoprocta, 406
Electron, 177
Emballonura, 412
Emballonuridae, 411–413
Eoastrapostylops, 377, 378
Eocardia, 407
Eogavialis, 113, 149, 150, 152
    E. africanus, 149
    E. gavialoides, **134,** 149
Eomegatherium, 243
Eomicrobiotherium, 190, 191, 205
Epecuneia, 290–291

Ephippiorhynchus, 181
Epipeltephilus, 224
Epitherium, 290, 291, 300
Equus, 272, 283, 286
    E. asinus, 279
    E. caballus, 282
Erethizon, 406
Ernestokokenia, 283
Erythrinidae, 72
Eucholoeops, 241, 247, 257, 259, 260, 262
Eucinepeltus, 223
    E. crassus, 222, 223
    E. petestatus, 223
Eumegamys, 400, 408
Eumomota, 177
Eumops, 424–425, 428, 430, 431
    E. auripendulus, 425
    E. bonariensis, 425
    E. dabbenei, 425
    E. glaucinus, 424, 425
    E. hansae, 425
    E. perotis, 425, 428
    E. trumbulli, 425
    E. underwoodi, 425
    E. sp., 410, 430
    E. sp. indet., **423,** 424–425
Eunectes, 95, 96, 525
    E. stirtoni, 95, 96
    E. sp., 96, **97**
Euphractus, 249, 254, 257
    E. sexcinctus, 249
Eusigmomys, 406
Eusuchia, 118–151
Euthecodon, 148
Eutheria, 304–318

fam. nov.
    Hoazinoididae, 172–176
Felsinotherium ortegense. See *Potamosiren*
    magdalenensis
Ficus, 431
Foro, 176
    F. panarium, 176
Forsteria venezuelensis, 65

Galbula, 171, 176–179, 182, 526, 546
    G. dea, 177, 178, 179
    G. hylochoreutes, 171, 176–179, **177**
    G. ruficauda, 178
Galbulidae, 176–179
Gavialidae, 121–153
Gavialis, 118, 121, 131, 136, 137, 138, 141,
    142, 148, 149
    G. colombianus, 146. See also *Gryposuchus*
        colombianus
    G. gangeticus, 123, 126–140, **130, 134,**
        141–143, **142,** 146, 147, 149, 150,
        152
Gavialosuchus, 118
Gazella granti, 280
gen. nov. See *Anachlysictis; Anadasypus;*
    Dukecynus; Granastrapotherium; Hondathentes;

    Nanoastegotherium; Neoglyptatelus; Paradra-
    caena; Patasola; Pedrolypeutes; Pericotoxodon;
    Potamops; Scirrotherium; Villarroelia
Geochelone, 165–167, 168, 525, 546
    G. carbonaria, 167, 168
    G. denticulata, 167, 168
    G. elephantopus, 166
    G. hesterna, 155, 161, 166, 167
    G. sp. indet., 165–167, **166**
Geograpsus lividus, 65
Geophagus
    G. jurupari, 86
    G. surinamensis, 86
Giraffa camelopardalis, 350
Globorotalia
    G. ciperoensis ciperoensis, 42
    G. fohsi fohsi, 42
Glossophaga, 418, 419, 429
Glossophaginae, 418–427
    gen. et sp. indet., 418–427
Glossotheriopsis, 233, 234, 235, 237–240, 243,
    244, 245, 502, 526
    G. pascuali, 233, 235, 237–240, **238,** 243,
        245, 503, 509, 513
Glossotherium, 234, 235, 237, 240
Glyptatelinae, 221–222
Glyptatelus, 221, 224
    G. malaspinensis, 221
    G. tatusinus, 221
Glyptodontidae, 221–223
Glyptodontoidea, 221–223
Goniopsis cruentata, 65
Goslinia, 78
Goupia, 520, 537
Goupioxylon stutzeri, 39, 537
Granastrapotherium, 355, 366–377, 378, 379,
    518, 534–535, 544
    cf. *G.,* 377–379
    G. snorki, 355, 366–377, **367, 368,** 377,
        496, 509, 512, 514, 515
Grapsidae, 65
Grimsdalea magnaclavata, 42
Gruiformes, 179–181
Gryposuchus, 121–153, 525
    G. colombianus, 113, 122–153, **128–131,**
        134, 138, 139, 140, 141, 142
    G. jessei, 122, 145, **145,** 146, 152
    G. neogaeus, 122, 139, 145, **145,** 146,
        147, 148, 152
Gymnophiona, **98,** 98–99
Gyriabus, 408
    G. royoi, 408
Gyrinodon, 335, 337, 347, 348, 352
    G. quassus, 348

Hapalops, 233, 244, 247, 248, 252, 254, 257,
    258, 259, 262, 264, 526
    cf. *H.,* 242, **496**
    H. longiceps, 247, 252, 257, 260
Hedralophus, 316
Hegetotherium, 319, 326, 327, 330, 331, 332
Heliconia, 430

Hemidoras, 81
Hemigrapsus nudus, 65
Henricofillholia, 316
Henricosbornia, 319, 322, 325, 327, 328, 331
Heros severus, 86
Hesperogavialis, 113, 122, 144, 148, 151, 152
    H. cruxenti, 148
Heterotis niloticus, 72
Hipparion, 501
Hippopotamus amphibius, 377
Hoazinoides, 171, 172–176, 526
    H. magdalenae, 171, 172–176, **174, 175,** 430
Hoazinoididae, 172–176
Holmesina, 230, 232
    H. floridanus, 227, 232
    H. septentrionalis, 232, **232**
Holochilus sciureus, 395
Homo sapiens, 543
Homunculus, 473, 474, 479, 481–482, 491, 493, 494, 495
    H. patagonicus, 481, **481**
"Homunculus." See *Stirtonia tatacoensis*
Hondadelphidae, 195–197
    gen. et sp. indet., 197–198
Hondadelphys, 195–197, 198, 206
    H. fieldsi, 187, **189,** 195–197, 198, 206
    ?*H.* spp., **189,** 197–198, **198,** 204
Hondathentes, 209–211, 526
    H. cazador, 207, 210–211
Hoplias, 67, 73, 86, 88
    H. sp., **71,** 72–73, 88
Hoplophractus, 224
Hoplosternum, 67, 86, 87, 88
    cf. *H.,* **81,** 82–83, 88
    H. magdalenae, 83
Huilatherium, 303, 309–318, 515, 534, 535
    H. pluriplicatum, 303, 304, 305–318, **307, 308, 309,** 509, 513
Hydrochoerus, 408
Hydrolycus, 67, 86, 87, 88, 89
    H. cf. *Myletes* sp., 60
    H. pectoralis, 73
    H. scomberoides, 73
    H. sp., **60,** 73, **74,** 88
Hydromops, 424
Hyemoschus aquaticus, 279, 285–286
Hylochoerus, 348
Hyperdidelphys biforata, 197
Hyperleptus, 247
Hyperoxotodon, 345, 346, 347, 351
Hypostominae, 84
Hypostomus, 88. See also *Acanthicus,* cf. *Hypostomus*
    cf. *H.,* 67, **83,** 84, 88
Hypsiglena, 97

Iguanidae incertae sedis, 101
Iheringichthys, 79
Ikanogavialis, 113, 122, 144, 148, 149, 151
    I. gameroi, **134,** 148
Incamys, 396, 406

Interatherium, 350
Itasuchus, 117

Jabiru, 181
    J. cf. *J. mycteria,* **92**
    J. mycteria, 171, 181, 182, 526
    J. sp., **180**
    J. sp. aff. *J. mycteria,* 181
Jacamerops aurea, 177, 178

Kentropyx, 107
Kiotomops, 411, 428
    K. lopezi, 410, 428
Kondous laventicus, 483
Kraglievichia, 214, 232
    K. cf. *K. paranensis,* 227
    K. paranensis, 227, 229, 232
    K. sp., 227

Laeliichthys, 72
Lagonimico, 474, 489, 491, 494, 495, 535, 536
    L. conclucatus, **432,** 470, 483, **483,** 535
Lagothrix, 487, 493, 536
"Laventatherium hylei," 304. See also *Huilatherium pluriplicatum*
Laventiana, 436, 440, 444, 453, 466, 491. See also *Neosaimiri*
    L. annectens, 436, 453, 466, 482. See also *Neosaimiri annectens*
Leiarius, 80
Lemur macaco, **465**
Leontinia, 306–317
Leontiniidae, 304–318
Leontopithecus, 440, 442, 448, 449, 454, 457, 487
    L. rosalia, 438, 440
Lepidosiren, 86, 88, 89, 546
    L. cf. *L. paradoxa,* 71
    L. megalos, 71
    L. paradoxa, 67, 70–72, **71,** 88
Lepidosirenidae, 70–72
Lepidosireniformes, 70–72
Leporinus, 67, 86
    cf. *L.,* 73–74, 88
    L. cf. *L. fasciatus,* 74
    L. friderici, 74
    L. sp., **74,** 88
Leptochamops, 101, 112
Leptodoras, 81
Leptorrhamphus striatus, 144
Lestodon, 234, 235, 240
Licaphrium, 290, 291, 292, 293, 294, 295, 300
    L. mesopotamiense, 294
Litaletes, 319, 323, 326
    L. disjunctus, 323
Litopterna, 266–276, 290–302
Lonchophylla, 419, 429
Lonchorhina, 418
Loncornis erectus, 180
Loricariidae, 67, 83–85
    incertae sedis 1, **84,** 84–85
    incertae sedis 2, **83,** 85

Loxocoelus, 316
Loxodonta africana, 377
Lumbrerasaurus scaglia, 101
Lutreolina crassicaudata, 197
Lycopsis, 198–199, 201, 204, 503
    L. longirostrus, 187, 198–199, **199,** 200, 206, 526
    L. torresi, 206

Macrauchenia, 299
Macraucheniidae, 297–299
Macrophyllum, 416, 418
Macrotus, 418
"Malacobagrus," 67. See also *Brachyplatystoma* (*"Malacobagrus"*) sp.; *Pimelodus*
Marmosa
    M. laventica, 187, 193. See also *Micoureus laventicus*
    M. sp., 187
"Marmosa," 192
Marmosini, gen. et sp. indet., 194
Megadolodinae, 266–276
Megadolodus, 265–288, 289, 290, 291, 301, 505, 534
    M. molariformis, **ii,** 265, 266–276, **268, 271–272, 274–275,** 280, 282, 283, 287, 291, 300, 301
Megalodoras, 81
Megalonychidae
    gen. et sp. indet.—large species, 241–242
    gen. et sp. indet.—small species, 242
Megalonychotherium, 241
Megathericulus, 243
Megatheriinae, 243
    gen. et sp. indet., 243
Megatherium, 350
Melanosuchus, 152
Meniscognathus, 112
Merycoidodon, 350
Mesocricetus auratus, 392, 396
Mesomys hispidus, 395
Mesonauta festivum, 86
Metaxytherium, 388, 389, 390
    M. calvertense, 388
    M. floridanum, 388, 389
    M. ortegense, 4, 383–391
    M. serresii, 389
Metopotoxus, 223
Metynnis, 68, 76, 87
Micodon, 435, 438, 458, 483
    M. kiotensis, 436, 456, 457, 483
"Micodon kiotensis," **455,** 456
Micoures. See *Micoureus*
Micoureus, 193–194, 205, 503, 505, 526
    M. cinereus, 187, 194
    M. laventicus, 187, **189, 190,** 193–194, 205, **432**
Microbiotheria, 188–192
Microbiotheriidae, 188–192
Microbiotherioidea, 188–192
Microbiotherium, 190
    M. tehuelchum, 190

Microchiroptera, 411–431
Micronycteris, 416, 418
Microscleromys, 392, 396, 402, 406, 533
 M. cribriphilus, **395**, 396
 M. paradoxalis, **395**, 396, 513
Microsteiromys, 527
 M. jacobsi, 392, **394**, 395, **395**, 402, 513
Mimon, 416
 M. crenulatum, 416
Miocochilius, 319, 326, 327, 330, 331, 499, 500,
 509–511, 514, 521, 534, 544
 M. anamopodus, **12**, 509, 510
 M. sp. nov. no. 1, 509, 510–511, 513
 M. sp. nov. no. 2, 511
 ?*M.* spp., 198
Mirandatherium, 191, 205
Mohanamico, 469, 491, 494
 M. hershkovitzi, 9, 468–469, 470, 484,
 485, 536. See also *Aotus dindensis*
Molossidae, 422–427
Molossops, 428
Molossus, 428
 M. ater, 425, 428
Momotus, 178
Monophyllus, 419, 429
Montaneia. See Ateles
Mops, 428
Mormopterus, 422–424, 428, 430
 M. colombiensis, 422–424, **423,** 430, 431
 M. faustoi, 410, 424, 430, 431
 M. helveticus, 424
 M. (Neomops) colombiensis, 410
 M. nonghenesis, 424
 M. stehlini, 424
Morus, 428
Moschus berezovskii, 282
Mourasuchus, 113, 119–120, 151, 152, 525
 M. amazonensis, 119, 120, 152
 M. arendsi, 119, 152
 M. atopus, **60,** 119, 120, 151, 152
 M. nativus, 119, 120, 152
Mycteria, 181
Myletes
 cf. *M.* sp., **60**
Myliobatiformes, 69–70
Mylodon, 234, 240, 526
 M. darwini, 244
Mylodontidae, 234–241
Mylodontinae, 234–240
Myoprocta, 392, 406, 407
 M. acouchy, 392, 395, 397, 402
Myrmecophaga, 247, 249, 252, 254, 255, 257,
 258, 259, 260, 261, 526

Nanoastegotherium, 214–216, 219, 224, 527
 N. prostatum, 213, 214, **215,** 216
Nematherium, 234, 235, 246, 248, 252, 254,
 257, 258, 262, 263
Neoglyptatelus, 221, 224, 509, 527
 N. originalis, 213, **215,** 221–222
Neomops, 424

Neonematherium, 233, 234, 235, 238, 240, 241,
 244, 245, 246, 248, 252, 255, 258, 260,
 262, 263, 500, 502, 526
 N. flabellatum, 233, 240–241, 245, 505
Neoreomys, 392, 393, 396, 403, 513
 N. australis, 393, 396, 399, 407
"*Neoreomys,*" 392, 393, 513, 533
 "*N.*" *huilensis*, 392, **394**, 396, **397,** 402,
 407, 509, 513
Neosaimiri, 435, 438, 440, 444, 445, 448, 449,
 450, 453, 454, 455, 458, 459, 465, 466,
 471, 473, 474, 482, 483, 487, 491, 493,
 494, 495, 536
 N. annectens, 435, 438, 440, 441, 442,
  443, 444, 445, 448, 450, 452, 454,
  455, 456–457, 458, 466–468, 470,
  471
 N. fieldsi, 7, 211, 435, 436, 438, 440,
  441, 442, 443, 444, 445, 448, 450,
  452, 453, 454, 457, 458, **462,** 465–
  466, 470, 471, 482, 536
Neotamandua, 527
 N. borealis, **432,** 513
Neotoma cinerea, 392, 395
Nesodon, 338, 339, 340, 341, 343, 345, 346,
 347, 350, 351
 N. imbricatus, 344, 350
"*Nesodonopsis,*" 347
Nettosuchidae, 119–120
Nettosuchus. See Mourasuchus
Noctilio, 413–415, 535
 N. albiventris, 413–415, **414,** 429, 430,
  431
 N. cf. *N. albiventris*, **60**
 N. leporinus, 414, 415
Nothrotheriinae
 gen. et sp. indet.—large species, 242
 gen. et sp. indet.—small species, 243
Nothrotheriops, 243, 245
Nothrotherium, 243, 244
Notonycteris, 416, 418, 429, 431, 535
 N. magdalenensis, 410, **416,** 417, 418, 428,
  431
Notopithecus, 319, 322, 325, 327, 328, 331
Notostylops, 319, 322, 325, 327, 328, 331, 332
Notoungulata, 336–352
Nyctinomops, 424, 428
 N. femorosacca, 425
 N. macrotis, 425, 428

Ocnerotherium, 347
Octodon degus, 396
Olenopsis, 393, 399, 512
"*Olenopsis,*" 392, 393, 398, 399, 400, 402,
 406, 512, 513, 515, 533
 "*O.*" *aequatorialis*, 512
 "*O.*" sp.—large, 509
 "*O.*" sp.—large adult, **394, 397**
 "*O.*" sp.—large juvenile, **394, 397**
 "*O.*" sp.—small adult, **394, 397**
Opisthocomus, 176, 546
 O. hoazin, 173, 175, 526

Orinocodoras, 81
Orophodon, 238–239, 248
 O. hapaloides, 238
Ortalis, 173, 174, 175, 176
Osteoglossiformes, 71
Otomops, 428
 O. martiensseni, 425, 428
Ovis, 326
Oxydactylus, 350
Oxysdonsaurus, 144

Pachybiotherium, 188–192, 205, 500, 502, 505,
 526
 P. acclinum, 189, 191
 P. minor, 187, 188–192, **189, 190,** 205,
  503
"*Palaeotoxodon*"
 "*P. nazari,*" 348
Paleothentes, 210, 211
 P. boliviensis, 211
 P. minutus, 210
Paleothentidae, 209–211
Palyeidodon, 338, 339, 341, 343, 345, 346, 347,
 351
Pampatheriidae, 228–231
Pampatherium, 227, 230, 231
 P. mexicanum, 231
Paradercetis, 72
Paradracaena, 101, 102, **103,** 107–108, 110,
 112, 525, 546
 P. cf. *P. colombiana*, 106–107
 P. colombiana, 100, 101, 102–106, **104,**
  109, 110, 111
Paralouatta, 473, 486, 495
 P. varonai, 486
Parapimelodus, 79
Parastrapotherium, 357, 360, 363, 365, 366,
 374, 377, 378, 379
 P. sp., 372
Paratrigodon, 338
Paratrygon, 70
Patasola, 435, 438–453, 454–459, 473, 489,
 491, 495, 536
 P. magdalenae, 435, 438–453, **441, 442,**
  443, 444, 455, 458, 470, 482, 536
Paucituberculata, 207–212
Paulicea, 78
Pediolagus, 407
Pedrolypeutes, 219–220, 224, 515, 527
 P. praecursor, 213, **218,** 219, 220, 509, 513
Peirosaurus, 115
Pelecaniformes, 181–182
Pelecyodon, 247, 252, 254, 258, 262
Pelomedusidae, 155–165
Peltocephalus, 163
 P. dumerilianus, 168
Perciformes, 85–86
Pericotoxodon, 319, 326, 327, 330, 337–352,
 534, 544, 546
 P. platignathus, 92, 184, 335, **337,** 337–
  352, **339, 341, 342, 345,** 509, 512–
  513

P. sp., 352
Perrunichthys, 80
Petenia
    P. motaguense, 86
    P. splendida, 86
Phacochoerus, 348
    P. aethiopicus, 286
Philander opossum, 197
Philisis sphingis, 425
Phoberomys, 408
Phoenixauchenia, 298
    ?*P. tehuelcha,* 298
    P. tehuelcha, 298
"Phoenixauchenia" tehuelcha, 299
Phractocephalus, 86, 87, 89
    P. hemiliopterus, 67, **79,** 79–80, 87, 88
Phrynops dahli, 167
Phylloderma, 416, 418
Phyllostomidae, 414–419
Phyllostominae, 414–418
    gen. et sp. indet., 417–418
Phyllostomus, 416
    P. discolor, 416, 418
    P. elongatus, 416, 417, 418
    P. hastatus, 416, 418
Pimelodidae, 67, 76–78
Pimelodinae, 76–78
Pimelodus, 86, 87
    cf. *P.,* 67, 68, **78,** 78–79, **79,** 88
    P. blochi, 79, **79**
    P. clarias, 79
    P. grosskopfii, 79
    P. sp., 79
"Pisanodon," 348. See also *"Palaeotoxodon"*
Pithecia, 459, 461, 466, 468, 469, 470, 471,
    476, 483, 487, 489, 536
    P. monachus, 463, **484**
    P. pithecia, 461, 463, **465**
Pithiculites, 207–209, 503, 526
    P. chenche, 207–209, **208,** 211, 503
    P. minimus, 208, 209
    P. minutus, 211
    P. rothi, 208, 209
Plagioscion, 86
Plaina, 227
Planops, 243, 248, 257
Platydoras, 81
Platygonus, 350
Platymops, 424
Platyrhinni, 438–453
Platysilurus, 78
Plesiotoxodon, 335, 337, 348
    P. amazonensis, 348, 352
Plesiotrygon, 70, **71**
Pleurodira, 155–165
Pleurostylodon, 317, 345
Pliolagostomus, 402
Podocnemis, 155, 156–165, 167, 169, 523, 525
    P. cf. *P. expansa,* 161–162, 163
    P. erythrocephala, 158, **158,** 163, 165
    P. expansa, 155, 159, 160, 162, 163, 164,
        165, 167, 168

P. lewyana, 155, 158, 159, 165, 167, 168
P. medemi, **60,** 155, 159–161, **160,** 162,
    162, 164, 165, 167, 168
P. pritchardi, 155, 156–159, **157,** 160–
    161, 162, **162,** 164, 165, 167, 168
P. sextuberculata, 165
P. sp. indet., 162–165, **163, 165**
P. unifilis, 158, 163, 165, 167, 168
P. vogli, 163, 164, 165, 167, 168
"Podocnemis"
    "P." alabamae, 156. See also *Bothremys*
    "P." antigua, 156. See also *Shweboemys*
    "P." barberi, 156. See also *Bothremys*
    "P." congolensis, 156. See also *Taphrosphys*
    "P." venezuelensis, 156, 169
Pomacea, 180
Poradinotherium leptognathum, 340
Posnanskytherium, 338, 339, 347
Potamops, 424–428, 430, 431
    P. mascahehenes, 410, 424–427, **426,** 428,
        430, 431
Potamosiren, 383–391, 535
    P. cf. *P. magdalenensis,* **386**
    P. magdalenensis, 383–391, **384, 385, 386,**
        535
Potamotrygon, 70
Potamotrygonidae, 67
    cf. Potamotrygonidae gen. et sp. indet.,
        69–70
Potomon, 63
Potomonidae, 63
Prepotherium, 248, 252, 254, 255, 257–258
Priodontes, 527
Pristobrycon, 87
    P. sp., 67, **74,** 76, 88
Proadinotherium, 346, 351, 365
    P. leptognathum, 351
    P. muensteri, 338, 339, 351
Prodolichotis, 400, 402, 503, 505, 521, 527
    P. pridiana, 392, 393, 400, **401,** 407, 505,
        509, 512
Proechimys, 392, 396
Prolagostomus, 402
Prolicaphrium, 289, 290, 291–293, 295, 300,
    503, 534
    P. sanalfonensis, 291–293, **292,** 297, 300,
        301
    P. specillatum, 291, 292, 293
Promegatherium, 243
Promops, 428
Pronothrotherium, 242, 244
Propalaeohoplophorinae, 222–223
Propalaeohoplophorus, 214, 223, 224
    P. australis, 222, 223
    P. minor, 223
Propraopus, 217, 218, 219
Proscelidodon, 234, 241
Prostegotherium, 215
    P. notostylopianum, 215, 216
Proterotheriidae, 266–276, 290–295
    gen. et sp. indet., 295–297, **298**
Proterotherium, 290, 292

P. berroi, 290
    P. cervioides, 290, 291, 295
"Proterotherium," 290, 291, 294, 295
Protheosodon, 285
Prothoatherium, 284, 285, 289, 290, 291, 300,
    500, 502, 503, 505, 534
    P. colombianus, 291, 297, 300, 301
Prothylacyninae, 198–201
Prothylacynus patagonicus, 200, 201
Protomegalonyx, 241–242
Protopithecus, 486, 491
    P. brasiliensis, 473, 485–486, 495
Prototeius, 112
    P. stageri, 112
"Prototrigodon," 352
Pseudodoras, 81
Pseudolycopsis cabrerai, 201
Pseudoplatystoma, 78
Pseudoprepotherium, 234, 235–236, 240, 244,
    245, 246, 248, 252, 254, 255, 257, 258,
    260, 262, 263, 526, 534, 535
    P. confusum, **iii,** 233, 234–235, 245, 503
Pseudostegotherium, 215, 216
    P. glangeaudi, 216
Pseudothylacynus rectus, 200
Psophia, 179
    P. leucoptera, **179**
Pterodoras, 81
Purrusaurus
    P. brasiliensis, 143
    P. neivensis, 525
Pygocentrus, 87
    P. sp., 67, **74,** 76, 88
Pygopristis, 68, 76, 87

Rhamphostoma, 121
Rhamphostomopsis, 121, 123, 145, 147
    cf. *R.,* 145
    R. intermedius, 144–145
    R. neogaeus, 144, 145, **145,** 146. See also
        Gryposuchus neogaeus
"Rhamphostomopsis" neogaeus, 144, 145, **145,**
    146
Rhaphiodon vulpinus, 73
Rhynchippus, 319, 326, 327, 328, 330, 345
Rhynchonycteris, 412
Ribodon, 385, 388, 535
    R. limbatus, 388
    R. magdalenensis. See *Potamosiren*
        magdalenensis
    R. sp., 389
    ?*R.* spp., 390
Ricardomys, 396, 533
    R. longidens, 392, **394, 395,** 396, 402
Rodiotherium, 316. See also *Ancylocoelus*

Saccolaimus, 412
Saguinus, 440, 442, 445, 448, 449, 453, 458,
    465, 473, 487, 536
    S. fuscicollis, 457
    S. oedipus, 440, **441, 442, 443**
Saiga tatarica, 280

Saimiri, 435, 436, 438, 440, 441, 442, 444, 445, 448, 449, 452, 453, 454, 455, 456, 457, 458, 459, 465, 466, 467, 468, 470, 471, 473, 475, 476, 482, 483, 486, 487, 493, 494, 536

    S. fieldsi. See Neosaimiri fieldsi

    S. oerstedii, 467

    S. sciureus, 440, **441, 442, 443,** 466, **470**

"Saimiri" bernensis, 473, 486, 495

Salladolodus deuterotheroides, 285

Sauromys, 424

Scaglia, 377, 379

Scaphops, 316. See also *Leontinia*

    S. grypus, 316

Scarrittia, 306, 308–317

Scelidotheriinae, 240–241

Scelidotherium, 234

Schismotherium, 247, 254, 255, 258, 259, 262

Schizodon, 74

Scincomorpha, 102–111

Scirrotherium, 227, 228–232, 527

    S. hondaensis, 227, **228,** 229, **229,** 231, **232,** 509, 512, 514, 527

Sclerocalyptus ornatus, 223

Scleromys, 5, 393, 396, 511

    S. schurmanni, 4, 5

"Scleromys," 392, 393, 396, 399, 402, 406, 511, 513–514, 515, 533

    "S." cf. *"S." colombianus,* **394,** 398, 511, 513

    "S." cf. *"S." schurmanni,* **394, 397,** 398, 399, 511–512, 513

    "S." colombianus, 392, **397,** 398, 399, 509, 511–512, 513, 514

    "S." schurmanni, 392, 393, 396, 397, **397,** 398, 399, 509, 511, 513, 514

Scolecophidia, 97–98, **98**

Sebecidae, 114–117

Sebecosuchia, 114–117

Sebecosuchus

    S. huilensis, 152

    S. icaeorhinus, 152

Sebecus, 113, 114, 115, 151, 153, 525

    S. huilensis, **92,** 113, 114–117, **115,** 151

    S. icaeorhinus, 113, 114–115, 116

Senodon, 316. See also *Leontinia*

Serpentes, 95–96

Serrasalminae, 75–76

Serrasalmus sp., 67, **74,** 76, 88

Sesarma curacaoense, 65

Shweboemys, 156

Siluriformis, 76–78

Silviocarcinus piriformis, 63–66, **64, 66,** 523

Simplimus, 399, 400, 406

Simpsonotus, 345

Sirenia, 383–391

Soriacebus, 435, 457, 458, 473, 477, 478–479, 482, 493, 494, 495

    cf. *S.* sp., **480**

    S. adrianae, 457, 479

    S. ameghinorum, 458, 479, **480**

Sorubim, 78

sp. nov.

    Anachlysictis gracilis, **ii,** 187, **189,** 201–203, **202,** 206

    Anadasypus hondanus, 213, 217, **218,** 218–219

    Aramus paludigrus, 171, **179,** 179–181, **180**

    Asterostemma gigantea, 213, **215,** 223, 224, 527

    Dukecynus magnus, 187, **189,** 199–201, **200,** 205, 206, 526

    Galbula hylochoreutes, 171, 176–179, **177**

    Granastrapotherium snorki, 355, 366–377, **367, 368,** 377, **496,** 509, 512, 514, 515

    Hondathentes cazador, 207, 210–211

    Mormopterus colombiensis, 422–424, **423,** 430, 431

    Nanoastegotherium prostatum, 213, 214, **215,** 216

    Neoglyptatelus originalis, 213, **215,** 221–222

    Pachybiotherium minor, 187, 188–192, **189, 190,** 205, 503

    Patasola magdalenae, 435, 438–453, **441, 442, 443, 444,** 455, 458, 470, 482, 536

    Pedrolypeutes praecursor, 213, **218,** 219, 220, 509, 513

    Pericotoxodon platignathus, **92, 184,** 335, **337,** 337–352, **339, 341, 342, 345,** 509, 512–513

    Pitheculites chenche, 207–209, **208,** 211, 503

    Podocnemis medemi, **60,** 155, 159–161, **160,** 162, **162,** 164, 165, 167, 168

    Podocnemis pritchardi, 155, 156–159, **157,** 160–161, 162, **162,** 164, 165, 167, 168

    Potamops mascahehenes, 410, 424–427, **426,** 428, 430, 431

    Prolicaphrium sanalfonensis, 291–293, **292,** 297, 300, 301

    Thylamys colombianus, 187, **189,** 193, 205, 513

    Thylamys minutus, 187, **190,** 192–193, 205, 513

    Thyroptera robusta, 410, **419,** 419–420, 422, 431

    Villarroelia totoyoi, 294–295, **295, 297,** 301

    Xenastrapotherium aequatorialis, 358, 361–363, **362,** 379

    Xenastrapotherium chaparralensis, 363, 365, 366, 378, 379

Sparassodonta, 195–204

    fam. et gen. indet., 204

Sphiggurus, 406

Squamata, 102–111

Stehlinia, 420, 422

?*Steiromys,* 393–394, 503

    ?*S.* sp., 394, **395**

?*S.* spp., 393, 527

Stenogenium, 316. See also *Leontinia*

Stereotoxodon, 343, 347

Stirtonia, 472, 473, 474, 481, 485, 487, 491, 509, 513, 518, 544

    S. tatacoensis, 7, 10, **432,** 470, 483, 493, 513, 536

    S. victoriae, 470, 484, **484,** 493, 513, 536

Stupendemys geographicus, 168, 169

Stylocynus, 200

    S. paranensis, 200

subfam. nov. *See* Dinotoxodontinae; Megadolodinae

Sylvilagus, 534

Sylviocarcinus piriformis, 63–66, **64**

Symphysodon discus, 86

Synastrapotherium amazonense, 356, 365, 366. See also *Xenastrapotherium amazonense*

Szalatavus, 473, 494

    S. attricuspis, 474, **478**

Tadarida, 428

    T. faustoi, 422

Tamandua, 249, 252, 254, 255, 258, 260, 262, 263, 526, 527

Taphozous, 412

Taphrosphys, 156

Tapirus, 272, 273, 275, 283, 286, 374, 534

    T. sp., 279, 280

Taubatherium, 303, 309, 316, 317, 318

Tauraco, 174

Tayassu, 267, 282, 287

Teiidae, 100–111

Teius, 107

Testudines, 156–169

Testudinidae, 165–169

Tetragonopterinae. *See* Characidae cf. Tetragonopterinae

Tetragonostylops, 377, 378

Tetramerorhinus, 290

Theosodon, 289, 297–299, 300, 503, 534

    T. aff. *T. gracilis,* 298

    T. fontanae, 298

    ?*T. frenguellii,* 298, 299

    T. garretorum, 298–299

    T. gracilis, 298, 299

    T. hystatus, 298

    T. karaikensis, 298, 299

    T. lallemanti, 298

    T. lydekkeri, 298

    T. patagonicus, 298

    T. pozzi, 298

    T. sp., 298

    T. sp. indet., 297–299, **298**

Thinobadistes, 234

Thoatherium, 279, 290, 291, 292, 293, 295

    T. minisculum, 276, 280, 286

Thylacosmilidae, 201–203

    gen. et sp. indet., **189, 204**

    ?Thylacosmilidae, 203–204

Thylacosmilus atrox, 187, 201. See also *Achlysictis lelongi*

Thylamys, 192–195, 503, 505, 526
    T. colombianus, 187, **189,** 193, 205, 513
    T. elegans, 193
    T. minutus, 187, **190,** 192–193, 205, 513
Thylophorops chapalmalensis, 195
Thyroptera, 419–422, 535
    T. cf. *T. tricolor,* 410, 419, 420–422, **421,**
    431
    T. discifera, 419–420, 422
    T. robusta, 410, **419,** 419–420, 422, 431
    T. tricolor, 419, 420, 422, 430
Thyropteridae, 419–422
Tichodon, 290
Tolypeutes, 213, 219–220, 225, 226, 527
Tolypeutinae, 219–220
Tomistoma, 118, 130, 131, 134, 136, 138, 141,
    142, 143, 149
    T. schlegelii, 123, 126, 127, 131, 132, **134,**
    134–139, 141–143, **142,** 146, 147,
    149, 152
Tomistominae. *See* Crocodylidae, cf.
    Tomistominae
Tonatia, 414–416, 418, 429, 431, 535
    T. bidens, 418, 429
    T. silvicola, 416, 429
    T. sp., 410
    T. sp. indet., 414–416, **416**
Toxodon, 338, 340, 347, 350
Toxodontia, 304–318, 336–352
Toxodontidae, 336–352
Trachops, 415, 418

Trachytherus, 319, 326, 327, 328, 330, 331,
    332–333
Tragulus javanicus, 279, 286
Tremacebus, 473, 475, 476, 477, 481, 495
    T. harringtoni, **479**
Trichechidae, 383–391
Trichechinae, 383–391
Trichechus, 389, 535
Trichodactylidae, 63, 65
Trichodactylus fluviatilis, 65
Trigodon, 338, 339, 343, 347
Trigonostylops, 377, 378
Tupinambinae, 100–112
Tupinambis, 100, 101, 107, 108, 109–110, 525.
    See also *Paradracaena*
    T. huilensis, 100, 101, 102, 110
    T. sp., **109,** 109–110
    T. teguixin, 101
"*Tupinambis huilensis,*" 108
Tyto, 173–174

Uintatherium, 350
Uruguaytheriinae, 356–379
Uruguaytherium, 366, 377, 378
    U. beaulieui, 378

Valdivia
    V. gila, 65
    V. torresi, 65. See also *Silviocarcinus*
    piriformis

V. (Valdivia) piriformis. See *Silviocarcinus*
    piriformis
Vampyrum, 418, 535
    V. spectrum, 429
Vassallia, 214, 227
    V. maxima, 227, 231, 232
    V. minuta, 227, 229, 231, 232
Verrutricolporites rotundiporis, 42
Vicugna, 350
Villarroelia, 289, 291, 293, 295, 300, 302, 534
    V. totoyoi, 294–295, **295, 297,** 301

Wallia scalopidens, 428

Xenarthra, 228–231
Xenastrapotherium, 355, 356–367, 375, 377,
    378, 379, 534, 535, 544
    X. aequatorialis, 358, 361–363, **362,** 379
    X. amazonense, 365–366, 378. See also
    Synastrapotherium amazonense
    X. chaparralensis, 363, 365, 366, 378, 379
    X. christi, 356, 357, 358, 360, 362, 363
    X. kraglievichi, 355, 356, 357–361, **358,**
    362, 363, 365, 366, 369, 370, 372,
    374, 377, 378, 379, 509, 512
Xenothrix, 474, 487, 489, 494, 495
    X. mcgregori, 486–487
Xotodon, 338, 340, 347

Zygodontomys brevicauda, 395
Zygolestes paranensis, 193